Climate Change and Small Pelagic Fish

This book details the effects of climate variability on small pelagic fish and their ecosystems and fisheries. These fish (for example, anchovy, sardine, sprat, and herring) comprise about one-quarter of the global fish catch and are particularly abundant in coastal upwelling regions off the west coasts of the Americas and Africa, off Japan, and in the NE Atlantic. Their stocks fluctuate greatly over the time scale of decades, with large ecological and economic effects. This book describes the nature and cause of these fluctuations, and their effects. It outlines results from paleo-oceanographic studies, showing that fluctuations similar to those at present have also occurred over the past two millennia. The potential effects of future climate change, both natural and anthropogenic, on stocks and fisheries, are considered. It concludes by recommending the continued international study and assessment of small pelagic fish in order to best inform management and policy under a changing climate.

No other book addresses climate change effects on fish in such an extensive manner. The book is also distinctive in being the product of a collaboration of academic and fisheries scientists from each of the regions with major stocks of small pelagic fish. It is written for research scientists, academics, and policy makers in fisheries, oceanography, and climate change.

DAVE CHECKLEY is a biological oceanographer with expertise in the ecology of marine zooplankton and fish and fisheries oceanography. He is a Professor at the Scripps Institution of Oceanography, University of California, San Diego. He has been a NATO Postdoctoral Fellow at DAFS in Aberdeen, Scotland, a member of the faculties of the University of Alaska, the University of Texas, and North Carolina State University, and a Tech Awards Laureate of the Tech Museum, San José, California. He led the development of CUFES, SOLOPC, and REFLICS. He is the co-chair of Small Pelagic Fish and Climate Change (SPACC) and Editor-in-Chief of *Fisheries Oceanography*.

JÜRGEN ALHEIT is a fisheries biologist. His main interest is the impact of climate variability on marine ecosystems. While working at the Intergovernmental Oceanographic Commission at UNESCO in Paris, he was responsible for the Ocean Sciences in Relation to Living Resources Programme of IOC and FAO, which focused on small pelagics. He is the co-founder and former co-chair of SPACC, chair of the GLOBEC Focus 1 Working group on Retrospective Analysis, and chair of the German GLOBEC Project. He serves on the Scientific Steering Committee of the GLOBEC International Programme.

YOSHIOKI OOZEKI is the Chief Scientist of the Fish Ecology Section, National Research Institute of Fisheries Science, Fisheries Research Agency, Japan, and Professor of Marine Life Science at the Tokyo University of Marine Science and Technology. He is a fisheries biologist with expertise in the larval biology and ecology of small pelagic fish, and has led the egg and larval survey project in waters around Japan for 12 years. He has also investigated larva physiology, database management, stock assessment, sampling gear technology, and variation of fish in relation to climate. He received the Uda Award from the Japanese Society of Fisheries Oceanography in 2007.

CLAUDE ROY is a physical oceanographer with expertise in fisheries oceanography and upwelling systems dynamics. As a scientist of the French Research Institute for Development (IRD), he spent extended periods of time in countries bordering upwelling systems during which he contributed to the implementation of several regional "climate and fisheries" research and training projects. He has been involved in SPACC since the late 1990s and served as its co-chair from 2003 to 2008.

Climate Change and Small Pelagic Fish

EDITED BY

DAVID M. CHECKLEY, Jr.
Scripps Institution of Oceanography, University of California, San Diego, USA

JÜRGEN ALHEIT
Baltic Sea Research Institute, Warnemünde, Germany

YOSHIOKI OOZEKI
National Research Institute of Fisheries Science, Kanagawa, Japan

CLAUDE ROY
Laboratoire de Physique des Océans (LPO), Plouzané, France

CAMBRIDGE
UNIVERSITY PRESS

CAMBRIDGE
UNIVERSITY PRESS

University Printing House, Cambridge CB2 8BS, United Kingdom

One Liberty Plaza, 20th Floor, New York, NY 10006, USA

477 Williamstown Road, Port Melbourne, VIC 3207, Australia

4843/24, 2nd Floor, Ansari Road, Daryaganj, Delhi - 110002, India

79 Anson Road, #06-04/06, Singapore 079906

Cambridge University Press is part of the University of Cambridge.

It furthers the University's mission by disseminating knowledge in the pursuit of education, learning and research at the highest international levels of excellence.

www.cambridge.org
Information on this title: www.cambridge.org/9781107434202

© Cambridge University Press, 2009

First published 2009
First paperback edition 2017

A catalogue record for this publication is available from the British Library

Library of Congress Cataloging in Publication data
Checkley, David.
  Climate change and small pelagic fish / David M. Checkley, Jürgen Alheit,
Yoshioki Oozeki.
    p. cm.
  Includes bibliographical references and index.
  ISBN 978-0-521-88482-2 (hardback)
  1. Pelagic fishes–Effect of temperature on.   2. Pelagic fishes–Behavior–Climatic
factors.   3. Fisheries–Climatic factors.   4. Pelagic fishes–Population viability
analysis.   5. Climatic changes.   I. Alheit, J. (Jürgen)   II. Oozeki, Yoshioki.
III. Title.
  QL638.C64C44 2009
  597.1722–dc22   2009007318

ISBN 978-0-521-88482-2 Hardback
ISBN 978-1-107-43420-2 Paperback

To Reuben Lasker

Contents

Acknowledgments

We thank the International Project Office of the Global Ocean Ecosystems Dynamics Program (GLOBEC) for its support during the production of this book and, in particular, Manuel Barange. The Institute of Research and Development (IRD) of France and the University of California, San Diego contributed financial support.

SPACC acknowledges the support it received from GLOBEC and its sponsors, in particular the International Geosphere-Biosphere Programme (IGBP), the International Oceanographic Commission of UNESCO (IOC), and the Scientific Committee on Oceanic Research (SCOR) supported by NSF grant No. OCE-0608600.

Dave Checkley
Jürgen Alheit
Yoshi Oozeki, and
Claude Roy

Abbreviations

AAIW	Antarctic Intermediate Water
ABC	Allowable Biological Catch
AC	Azores Current
ACMRR	Advisory Committee of Experts on Marine Resources Research
AEP	Annual Egg Production
ALP	Aleutian Low Pressure
AMCI	Assessment Model Combining Information
AMO	Atlantic Multidecadal Oscillation
ASAP	Age-Structured Assessment Program
BP	Before Present
BAC	Biological Action Centers
BC	Benguela Current
BITS	Baltic International Trawl Survey
BMSY	Biomass of Maximal Sustainable Yield
CACom	California Commercial Fish Landing Records
CalCOFI	California Cooperative Oceanic Fisheries Investigations
CanCE	Canary Current Ecosystem
CANSAR	Catch-at-Age Analysis for Sardine
CC	California Current
CC	Climate Change (Chapter 14)
CCAMLR	Commission for the Conservation of Antarctic Marine Living Resources
CCE	California Current Ecosystem
CCO	Central California Offshore subpopulation of sardine
CCSR	Center for Climate System Research
CCW	Cold Coastal Water
CEOS	Climate and Eastern Ocean Systems
CHRS	Constant Harvest Rate Strategy
CPR	Continuous Plankton Recorder
CPUE	Catch per Unit Effort
CS	Catalan Sea
CSA	Catch-Survey Analysis
CUFES	Continuous Underway Fish Egg Sampler
CV	Coefficient of Variation
DEPM	Daily Egg Production Method
DFR	Daily Fecundity Reduction
DIN	Dissolved Inorganic Nitrogen
EAF	Ecosystem Approach to Fisheries
EBUS	Eastern Boundary Upwelling Systems
EEZ	Exclusive Economic Zone
ENSO	El Niño Southern Oscillation
EOF	Empirical Orthogonal Function
ERS	European Remote-Sensing Satellite
ESSW	Equatorial Subsurface Water
EU	European Union
EURO-SARP	European Sardine–Anchovy Recruitment Program
EwE	EcoPath with EcoSim
FAO	Food and Agriculture Organization
FiB	Fishing-in-Balance
FL	Fork Length
GAM	Generalized Additive Model
GCM	Global Ocean Circulation Models
GFCM	General Fisheries Council for the Mediterranean
GL	Gulf of Lions
GLM	General Linear Model
GLOBEC	Global Ocean Ecosystem Dynamics
GOOS	Global Ocean Observing System
HAMSOM	Hamburg Shelf Ocean Model
HC	Humboldt Current
HCE	Humboldt Current Ecosystem
HCR	Harvest Control Rule
HSI	Habitat Suitability Index
IBM	Individual-Based Model
IBTS	International Bottom Trawl Survey of the North Sea
ICA	Integrated Catch-at-Age
ICLARM	International Center for Living Aquatic Resources Management
IFOP	Instituto de Fomento Pesquero
IGBP	International Geosphere Biosphere Programme
IGP	Intraguild Predation
IMARPE	Instituto del Mar del Perú
IMECOCAL	Investigaciones Mexicanas de la Corriente de California
IOC	Intergovernmental Oceanographic Commission
IPCC	Intergovernmental Panel on Climate Change
IRD	Institute for Research and Development
IREP	International Recruitment Project
ISPR	Instantaneous Surplus Production Rate
ITQ	Individual Transferable Quota
IUCN	International Union for the Conservation of Nature and Natural Resources
KC	Kuroshio Extension
KCE	Kuroshio Current Ecosystem
KESA	Kuroshio Extension South Area
KO	Kuroshio–Oyashio
KOTZ	Kuroshio–Oyashio Transition Zone
LME	Large Marine Ecosystem
MIN	Minimum Number of Individuals
MLD	Mixed Layer Depth
MLE	Maximum Likelihood Estimator
MSC	Marine Stewardship Council
MSVPA	Multispecies Virtual Population Analysis
MSY	Maximum Sustained Yield
MTI	Mixed Trophic Impact

NA	Northeast Atlantic	SARDYN	Sardine Dynamics and Stock Structure in the Northeast Atlantic
NAC	North Atlantic Current	SARP	Sardine–Anchovy Recruitment Program
NACW	North Atlantic Central Water	SATW	Sub-Antarctic Temperate Water
NAO	North Atlantic Oscillation	SAW	Sub-Antarctic Water
NEA	Northeast Atlantic	SBB	Santa Barbara Basin
NEMURO	North Pacific Ecosystem Model for Understanding Regional Oceanography	SCOR	Scientific Committee on Oceanic Research
NIES	National Institute for Environmental Studies	SDR	Scale Deposition Rate
NISP	Number of Identified Specimens	SECC	Southern Extension of the Cromwell Current
NMFS	National Marine Fisheries Service	SEW	Surface Equatorial Water
NOAA	National Oceanographic and Atmospheric Administration	SL	Standard Length
		SLP	Sea Level Pressure
NPZ	Nutrient–Phytoplankton–Zooplankton	SPACC	Small Pelagic Fish and Climate Change
OMP	Operational Management Procedure	SPF	Small Pelagic Fish
ORSTOM	Office de la Recherche Scientifique et Technique Outre Mer	SSB	Spawning Stock Biomass
		SSB/R	Spawning Stock Biomass/Recruit
OSLR	Ocean Science in Relation to Living Resources Programme	SSS	Sea Surface Salinity
		SST	Sea Surface Temperature
P/B	Production/Biomass	SSW	Subtropical Surface Water
P2P	Prey-to-Predator	STECF	Scientific, Technical and Economic Committee for Fisheries (EU)
PC	Peru Current		
PCC	Peru Coastal Current	STSW	Subtropical Surface Water
PDO	Pacific Decadal Oscillation	SURBA	Survey Based
POM	Princeton Ocean Model	TAB	Total Allowable Bycatch
PROCOPA	Peruvian–German Cooperative Program for Fisheries Investigations	TAC	Total Allowable Catch
		TL	Trophic Level
REX	Recruitment Experiment	TNA	Tropical North Atlantic Index
ROMS	Regional Ocean Model System	VMS	Vessel Monitoring System
RPS	Recruit Per Spawner	VPA	Virtual Population Analysis
RQ	Respiratory Quotient	WA	Western Australia (but also Washington state)
RSHR	Regime-Specific Harvest Rate	WCRP	World Climate Research Program
SACW	South Atlantic Central Water	WSSD	World Summit for Sustainable Development
SAfE	South Africa Experiment	XSA	Extended Survival Analysis

Contributors

Mahfuzuddin Ahmed
The WorldFish Center, Jalan Batu Maung, Batu Maung, 11960 Bayan Lepas, Penang, Malaysia

Jürgen Alheit
Baltic Sea Research Institute, Seestr. 15, 18119 Warnemünde, Germany

Patricia Ayón
Instituto del Mar del Perú, Apartado 22 Callao, Perú

Andrew Bakun
Pew Institute for Ocean Science, Rosenstiel School of Marine and Atmospheric Science, University of Miami, Miami, Florida, 33149–1098, USA

Manuel Barange
Plymouth Marine Laboratory, Prospect Place, Plymouth PL1 3DH, UK

Tim R. Baumgartner
Centro de Investigación Científica y de Educación Superior de Ensenada, Km 107 Carretera Tijuana-Ensenada, Ensenada, Baja California, C.P. 22860, Mexico

Miguel Bernal
Instituto Español de Oceanografía, C/ República Saharaui, s/n, 11510 Puerto Real

Arnaud Bertrand
Institut de Recherche pour le Développement, CRH, Avenue Jean Monnet, Sète, France c/o Instituto del Mar del Perú, Lima, Perú

Antonio Bode
Instituto Español de Oceanografía, Centro Oceanográfico de A Coruña, Spain

Richard Brodeur
Northwest Fisheries Science Center, NOAA, Newport,Oregon, USA

Leonardo R. Castro
Laboratorio de Oceanografía Pesquera y Ecología Larval, Departamento de Oceanografía, Universidad de Concepción, PO Box 160-C, Concepción, Chile

Maria Cristina Cergole
Instituto Brasileiro do Meio Ambiente e dos Recursos Naturais Renováveis (IBAMA), Alameda Tietê 637, CEP 01417–020, São Paulo, SP, Brasil

Francisco P. Chavez
Monterey Bay Aquarium Research Institute, 7700 Sandholdt Road, Moss Landing, California, 95039 USA

David M. Checkley, Jr.
Scripps Institution of Oceanography, University of California, San Diego, La Jolla, California 92093–0218, USA

Kevern Cochrane
FAO, Via delle Terme di Caracalla, 00100 Roma, Italy

Janet C. Coetzee
Marine and Coastal Management, Private Bag X2, Rogge Bay 8012, South Africa

Marta Coll
Dalhousie University, Department of Biology, 1355 Oxford Street, Halifax, Nova Scotia, Canada, B3H4J1 and Institute of Marine Science (CMIMA-CSIC), Passeig Marítim de la Barceloneta, 37–49, 08002, Barcelona, Spain

Luis A. Cubillos
Instituto Español de Oceanografía, Centro Oceanográfico de A Coruña, Spain

Philippe Cury
Centre de Recherche Halieutique Méditerranéenne et Tropicale, IRD – IFREMER & Université Montpellier II Avenue Jean Monnet, BP 171 34203 Sète, Cedex, France

Georgi M. Daskalov
CEFAS, Pakefield Road, Lowestoft, Suffolk NR33 0HT, UK

Carryn L. de Moor (formerly Cunningham)
MARAM (Marine Resource Assessment and Management), Department of Mathematics and Applied Mathematics, University of Cape Town, Rondebosch 7701, South Africa

José A. A. De Oliveira
CEFAS, Pakefield Road, Lowestoft, Suffolk NR33 0HT, UK

Mark Dickey-Collas
IMARES, PO Box 68, 1970 AB Ijmuiden, The Netherlands

Robert Emmett
NOAA Fisheries, 2030 South Marine Science Drive, Newport, Oregon, 97391, USA

Pepe Espinoza
Instituto del Mar del Perú, Lima, Perú

Vicente Ferreira
Centro de Investigación Científica y de Educación Superior de Ensenada, Km 107 Carretera Tijuana-Ensenada, Ensenada, Baja California, C.P. 22860, Mexico

David B. Field
Monterey Bay Aquarium Research Institute, Moss Landing, CA & University of California, Santa Cruz, Santa Cruz, California, USA

Pierre Fréon
ECO-UP Unit, Centre de Recherche Halieutique, IRD, BP 171,
34203 Sète, Cedex, France

Kevin Friedland
National Marine Fisheries Service, Narragansett , Rhode Island,
USA

Susana Garrido
Instituto Nacional de Investigação Agrária e das Pescas, Lisboa,
Portugal

Daniel J. Gaughan
WA Fisheries and Marine Research Laboratory,
Department of Fisheries, PO Box 20, North Beach
WA 6920, Australia

Renato Guevara-Carrasco
Instituto del Mar del Perú, Apartado 22 Callao, Perú

Dimitri Gutierrez
Instituto del Mar del Perú, Callao, Perú

Rognvaldur Hannesson
Norwegian School of Economics and Business Administration,
Helleveien 30, NO-5045, Bergen, Norway

Samuel F. Herrick Jr.
NOAA, NMFS, Southwest Fisheries Science Center, 8604 La
Jolla Shores Drive, La Jolla, California, 92037, USA

Kevin Hill
NOAA-NMFS, Southwest Fisheries Science Center, 8604 La
Jolla Shores Drive, La Jolla, California, 92037, USA

Larry Hutching
Marine and Coastal Management, Private Bag X2, Rogge Bay
8012, South Africa

Leire Ibaibarriaga
AZTI-Tecnalia, Marine Research Unit, Herrera Kaia Portualdea
z/g, 20110 Pasaia, Spain

Xabier Irigoien
Arantza eta Elikaigintzarako Institutu Teknologikoa, Pasaia,
Spain

Larry Jacobson
NOAA-NMFS, Northeast Fisheries Science Center, 166 Water
Street, Woods Hole, Massachusetts, 02543–1026, USA

Astrid C. Jarre
Danish Institute for Fisheries Research, The North Sea Centre,
P.O. Box 101, 9850 Hirtshals, Denmark

Souad Kifani
Institut National de Recherche Halieutique, Casablanca,
Morocco

Fritz W. Köster
Danish Institute for Fisheries Research, Jaegersborgvej 64–66,
Lyngby DK-2800, Denmark

Christophe Lett
Institut de Recherche pour le Développement, UR ECO-UP,
University of Cape Town, Oceanography Department,
Rondebosch, 7701, South Africa

Hector Lozano-Montes
CSIRO Marine and Atmospheric Research, Underwood Avenue,
Floreat, WA, Australia, 6014

Alec D. MacCall
National Marine Fisheries Service, Southwest Fisheries Science
Center, 110 Shaffer Rd., Santa Cruz, California, 95060, USA

Jacques Massé
IFREMER, Rue de l'Ile d'Yeu, B.P. 21105, 44311 Nantes,
France

Bernard A. Megrey
National Oceanic and Atmospheric Administration, National
Marine Fisheries Service, Alaska Fisheries Science Center,
Seattle, Washington, 98115, USA

Todd Miller
Center for Marine Environmental Studies, Ehime University,
Ehime, Japan

Christian Möllmann
Institute for Hydrobiology and Fisheries Science, University of
Hamburg, Hamburg, Germany

Coleen Moloney
Zoology Department and Marine Research Institute, University
of Cape Town, Rondebosch 7701, South Africa

Hideaki Nakata
Faculty of Fisheries, Nagasaki University, 1–14 Bunkyo-cho,
Nagasaki, 852–8521, Japan

Sergio Neira
Marine Biology Research Centre, University of Cape Town,
Rondebosch 7702 Cape Town, South Africa

Miguel Ñiquen
IMARPE, Esquina Gamarra y General Valle S/N, Chucuito,
Callao, Perú

Hiroshi Nishida
National Research Institute of Fisheries Science, Fuku-ura 2–12,
Kanazawa-ku, Yokohama, 236–8648, Japan

Jerrold G. Norton
NOAA, NMFS, Southwest Fisheries Science Center, 1352
Lighthouse Avenue, Pacific Grove, California, 93950, USA

Rosemary E. Ommer
University of Victoria, CEOR, Victoria, British Columbia,
V8W 2Y2, Canada

Yoshioki Oozeki
National Research Institute of Fisheries Science, Fisheries
Research Agency, 2–12–4 Fuku-ura, Kanazawa, Yokohama,
Kanagawa, 236–8648, Japan

Isabel Palomera
Institut de Ciencias del Mar (CSIC), Passeig Marítim de la
Barceloneta 37–49, 08003 Barcelona, Spain

Julio Pena-Torres
Universidad Alberto Hurtado, Erasmo Escala 1835, Santiago, Chile

R. Ian Perry
Fisheries and Oceans Canada, Pacific Biological Station,
Nanaimo, British Columbia, V9T 6N7, Canada

Benjamin Planque
Institute of Marine Research – Tromsø, Postboks 6404, 9294
Tromsø, Norway

Ruben Rodriguez-Sanchez
Centro Interdisciplinario de Ciencias Marinas, La Paz,
Mexico

Kenneth A. Rose
Department of Oceanography and Coastal Sciences, Louisiana
State University, Baton Rouge, Louisiana, 70803, USA

Jean-Paul Roux
Ministry of Fisheries and Marine Resources, Lüderitz, Namibia

Claude Roy
Laboratoire de Physique des Océans (LPO), Plouzané, France

Suzana A. Saccardo
CNR-ISMAR, Largo Fiera della Pesca, 60125 Ancona, Italy

Renato Salvatteci
Centro de Investigación Científica y de Educación Superior de
Ensenada, Km 107 Carretera Tijuana-Ensenada, Ensenada, Baja
California, C.P. 22860, Mexico

Alberto Santojanni
CNR-ISMAR, Largo Fiera della Pesca, 60125 Ancona, Italy

Jake Schweigert
Fisheries & Oceans Canada, Pacific Biological Station, 3190
Hammond Bay Road, Nanaimo, British Columbia, V9T 6N7,
Canada

Rodolfo Serra
Instituto de Fomento Pesquero, Blanco 839, Valparaiso,
Chile

Lynne Shannon
Marine Research (MA-RE) Institute, University of Cape Town,
Private Bag X3, Rondebosch 7702, Cape Town, South Africa

Stylianos Somarakis
University of Patras, Department of Biology, 26500 Patra,
Greece

Andy Soutar
Drumlins Farm, Massena, New York, USA

Yorgos Stratoudakis
Instituto Nacional de Investigação Agrária e das Pescas (INIAP/
IPIMAR), Avenida de Brasilia s/n, 1449–006 Lisboa, Portugal

U. Rashid Sumaila
Fisheries Economics Research Unit, Fisheries Centre,
University of British Columbia, Vancouver, British Columbia,
V6T 1Z4, Canada

Hiroshige Tanaka
Seikai National Research Institute, Fisheries Research Agency,
Nagasaki, Japan

Axel Temming
Institute for Hydrobiology and Fisheries Science, University of
Hamburg, Hamburg, Germany

Andres Uriarte
AZTI, Herrera Kaia Portualde z/g, 20110 Pasaia (Guipuzkoa),
Basque country, Spain

Carl D. van der Lingen
Marine and Coastal Management, Private Bag X2, Rogge Bay
8012, South Africa

Francisco Werner
Marine Sciences Department, University of North Carolina,
Chapel Hill, North Carolina, 27599–3300, USA

Akihiko Yatsu
Hokkaido National Fisheries Research Institute, Katsurakoi 116,
Kushiro 085–0802, Japan

Foreword

The effects of large-scale, environmentally driven changes on the distribution and abundance of fish populations have been a major source of concern for fishery scientists and managers for decades, particularly those dealing with the assessment and management of small pelagic fisheries. While much still needs to be investigated and elucidated, significant progress has been made in describing and understanding the primary aspects of observed large-scale changes in small pelagic fish production and the most likely causal mechanisms, climate–fish abundance interactions, patterns of change, species interactions, and many other important issues. A newer and additional difficulty is that global climate change is altering the structure and functioning of marine ecosystems, which in turn affects availability of ecological resources and benefits, changes the magnitude of some feedbacks between ecosystems and the climate system, and will affect economic systems that depend on marine ecosystems. Newer questions and uncertainties have to be faced. For instance, will there be an increase in variability from season to season and year to year? Predictions of such changes in the future are likely to be less reliable than they may have been in the past, given that the past will become less useful as a guide to the future.

There is, therefore, a pressing need to assess progress made in knowledge development and emerging challenges regarding the variability of and changes in marine ecosystems, including the dynamics of populations of small pelagic fish, which collectively occupy a central role in the food web. This book makes an important and timely contribution to this overall need. It covers the major trends, findings and results of various international research efforts aimed at improving our understanding of climate-induced changes in the production and abundance of small pelagic fish populations, as well as progress made in assessing the predictability of future changes.

The book builds mostly on the results achieved through the collaborative research work undertaken under the auspices of the Small Pelagic Fish and Climate Change regional programme (SPACC) of the Global Ocean Ecosystem Dynamics international project (GLOBEC). Indeed, this publication represents one of the major achievements of SPACC/GLOBEC. Over the subsequent 15 chapters, a well-selected group of international experts has summarized working hypotheses, results and findings in their own fields of expertise. The end result is a successful and comprehensive technical summary and assessment of the current state of knowledge and future research prospects of small pelagics–climate interactions, their fisheries and related ecosystems.

Science is an iterative learning process where current researchers build on challenge as well as modify the findings of their predecessors. To make progress in science, one needs to look at and analyze the past. The historical perspective provided in Chapters 1 and 2 fulfills this purpose by including a brief but thorough review of the main international research initiatives, working hypotheses, mathematical models, research results and findings, and meetings and events where these were publicized and discussed, some of which are now considered major benchmarks in the development of our current understanding of the variability of small pelagic fish populations. These two chapters highlight the progress made in the last three decades towards understanding the effects of climate on the life history and population dynamics of small pelagics and on the use of this knowledge for, first, challenging traditional fisheries management approaches typical of the 1950s, 1960s and 1970s where the aim was maximizing long-term production under steady-state assumptions, and, second, moving towards fully recognizing the importance of climate-induced fluctuations and changes in these fish populations and the various spatial and temporal scales on which they vary. Reference is also made to more recent efforts towards trying to elucidate the role of small pelagics in regulating the ecosystems and to the steps being taken to move towards a fisheries management that takes more into account the ecosystem as a whole, in line with the principles of the Ecosystem Approach to Fisheries (EAF) promoted by the FAO.[1]

The habitat characteristics of the main marine systems hosting particularly large populations of small pelagic fish are described and discussed in Chapter 3. This chapter highlights both the uniqueness and similarities of these habitats, as well as the possible effects of future climate change on them. The natural variability of small pelagics is well covered in Chapter 4, while Chapter 5 describes decadal-scale fluctuations of small pelagics. Analyses of scales in marine sediments, archeological remains and historical records confirm that small pelagics did vary considerably prior to the development of current industrial

[1] Food and Agriculture Organization.

fishing, but also show important differences in the variability patterns of small pelagic populations during the nineteenth and the twentieth centuries, with some extremes in variability occurring on shorter time scales under intensive fishing. This suggests that ecosystem paradigms based on industrial catch records and modern observations may only capture a small part of the range of intrinsic variability of small pelagics, thus limiting our understanding of the possible responses to climate change. Nevertheless, observations of catch records and other usually higher-resolution twentieth and twenty-first century records confirm that populations of small pelagics are characterized by large and relatively long-lasting changes in their abundance and in other population parameters, which are often associated with regime shifts in the dynamics of their marine ecosystems. There is also growing evidence that these ecosystem shifts could be associated with large-scale changes in subsurface processes and basin-scale circulation in the oceans.

Chapters 6, 7 and 8 cover the several biophysical and trophic dynamics models that have been developed and/or are applied to small pelagics to synthesize current understanding, test hypotheses and examine potential consequences of changes in environmental conditions and species interactions. Chapter 6 provides a thorough overview of models that include the physical environment and the population dynamics of either early life stages of fish or juveniles and adults. While referring to the various models, their input requirements, computing challenges and forecasting capabilities, the point is made that much effort is still necessary to get to the stage of having usable "physics-to-fish full life-cycle models" capable of addressing the long-term consequences of climate change on fish and fisheries. Chapter 7 provides an updated review of the trophic dynamics of small pelagics, using seven selected regional case studies and information on feeding-apparatus morphology, diet composition, foraging behavior, and other observations used for the parameterization of bioenergetic and other trophic dynamic models. In discussing the interactions and disparities in the trophic dynamics of coexisting small pelagic species, and between anchovy and sardine in particular, it is suggested that the observed abundance alternations between the two could be trophically mediated. The use of trophic models to examine the trophic dynamic role of small pelagics in five selected marine ecosystems is described in Chapter 8, where it is reported that the models analyzed show how a decrease in small pelagic fish abundance will have detrimental effects on both higher and lower trophic levels of the food web and that, for instance, trophic model simulations consistently suggest that gelatinous zooplankton may increase when small pelagic fish stocks decline. Conversely, a decrease in jellyfish abundance may be expected if pela-

gic fisheries were allowed to rebuild, which is in agreement with circumstantial evidence for some systems.

The assessment and management of small pelagic fisheries, including the social and economic dimensions, are well examined in Chapters 9, 10 and 11. The assessment and management approaches, methods, models, harvesting strategies, and controls and regulatory measures used in selected major small pelagic fisheries are analyzed in Chapter 9. The point is made that the speed of response and the flexibility of management that these highly variable fisheries demand can only be provided though properly tailored scientific assessment and management programs, while noting that improved management of fisheries and related ecosystems is essential in adapting to the impacts of climate change in fisheries. Chapter 10 examines a full range of economic benefits small pelagics can provide, including their direct commercial value as well as their value as prey of predators of higher commercial value and for recreational and non-commercial predators, including international and domestic considerations when looking at different management and conservation options. Social and economic aspects are further dealt with in Chapter 11, where important issues such as social and economic power, institutional frameworks, resource access rights, equity, property rights, differences in temporal and spatial scales, market globalization, ethics, technology, and interactive political agendas between developed and developing countries are examined and discussed.

Future challenges and ways forward are summarized and highlighted in the last four chapters. Chapter 12 examines potential mechanisms for the low-frequency variability in sardine and anchovy populations discussed in other sections. A working theory is proposed in which sardine productivity is linked to low-frequency variability in boundary current flows, with weaker flow periods being favorable for extended sardine reproductive success and stronger flow periods restricting reproduction success to coastal areas. On the other hand, anchovies are always restricted to coastal waters and are more influenced by upwelling and coastal productivity (which tend to be correlated with boundary current fluctuations), giving rise to a tendency of sardine and anchovy alternations. A number of unresolved issues key for improved management and understanding of small pelagics and their related ecosystems are examined in Chapter 13. In particular, this chapter explores further the wasp-waist concept, according to which a dominant highly variable small pelagic fish population, largely responding to its own internal dynamics, may significantly drive the operation of its entire ecosystem. Several types of non-linear feedback mechanisms, breakout thresholds, distributional dynamics, density dependent growth, niche replacement and mechanisms of species alternations are discussed, in support of proposals, worth noting, for research in the years to come. Chapter 14 points out that there is already evidence of sensitivity of small pelagic species and related ecosystems to climate change and of decreased resilience of natural ecosystems caused by overexploitation. Possible scenarios of climate and ecosystem change are then discussed to identify significant gaps in the knowledge of processes and interactions between changes in climate and other ecosystem stressors. Lastly, the book concludes with a thorough synthesis in Chapter 15 of SPACC, its reasons to exist and its major work and findings. Much of it is distilled from the ideas and findings reported in the first fourteen chapters while emphasizing that, for instance, observations from both paleontological and historical records are consistent with a conceptual model in which populations fluctuate due to extrinsic rather than intrinsic factors, which in turn is consistent with the observed out-of-phase variations of sardines and anchovies. Also emphasized is that humans must be considered an important part of the ecosystem and that overexploitation decreases resilience of systems to climate change as well as to more " normal" climate conditions, while reference is made to the wasp-waist concept and the pivotal role small pelagics play in the ecosystems in which they are found. Main gaps in current knowledge and understanding of small pelagic fish dynamics are further discussed, concluding that: the synthetic approach so far focused on small pelagics in highly productive upwelling regions shall be extended to multiple systems at a global scale, including, for example, western boundary currents; future research work be undertaken under the aegis of an international program like SPACC, using the comparative approach and involving scientists from a broad spectrum of disciplines, including climate, fisheries, oceanography, and the social sciences; and that a periodic international assessment of the state of science and climate effects on small pelagic fish be undertaken, as in other fields of science. A key point brought out by the book as a whole is the importance of a more ecosystem-oriented approach to research and fisheries management. Regarding the latter, one may stress that

global climate change requires even more precautionary and adaptive approaches on behalf of all stakeholders.

Jorge Csirke (FAO)
Michael H. Glantz (NCAR)[2]
James Hurrell (NCAR)

Addresses and Affiliations:

Jorge Csirke
Director
Fisheries and Aquaculture Management Division
Fisheries and Aquaculture Department
Food and Agriculture Organization of the United Nations (FAO)
Viale delle Terme di Caracalla
00153 Rome, Italy

Michael H. Glantz[3]
Director
Center for Capacity Building (CCB)
National Center for Atmospheric Research
Box 3000
Boulder, Colorado 80307
USA

James Hurrell
Senior Scientist
National Center for Atmospheric Research
Climate and Global Division
PO Box 3000
Boulder, Colorado 80307-3000, USA

[2] National Center for Atmospheric Research.
[3] Address as of January 1, 2009:
   Michael H. Glantz, Director
   Consortium for Capacity Building
   Institute for Arctic and Alpine Research
   University of Colorado
   Boulder, Colorado 80309–0450
   USA.

Preface

Small pelagic fish include anchovy, sardine, herring, and sprat. They comprise approximately one-quarter of the world's fish catch. The abundance and catch of the small pelagic fish fluctuate greatly on the scale of decades, notably anchovy and sardine off Peru, Japan, Southern Africa, and California. Climate varies on the same scale. How does climate affect small pelagic fish? Can our understanding of this relationship be used to inform management and policy?

Small pelagic fish occupy a key position in marine ecosystems. They respond to change from below (climate, bottom-up) and above (fishing, top-down). In turn, variation of stocks of small pelagic fish affect their prey (plankton) and predators (e.g. fish, marine birds and mammals, and humans). Capture fisheries are now maximal and, as the human population increases, aquaculture will thus continue to grow, exacerbating the demand for small pelagic fish as food for cultured fish. The complex ecosystem and economic roles of small pelagic fish necessitate a holistic view of their dynamics.

The Small Pelagic Fish and Climate Change (SPACC) program is a part of Global Ocean Ecosystem Dynamics (GLOBEC). SPACC was formed to understand and predict climate-induced changes in the production of small pelagic fish. It is unusual in being composed of scientists from both academia and management. This book, a product of SPACC, presents the status of our understanding in 2008. It has 82 authors from 22 countries in Africa, Asia, Australia, Europe, and North and South America. Our hope is that it will form a basis and hence point of departure for future research.

The book consists of 15 chapters. Each, save the first, contains a summary and boxes, the latter connecting the chapters to the overarching theme of climate change and small pelagic fish. Chapters 1 and 2 provide a historical context. Chapter 3 describes habitats of the major stocks. Chapters 4 and 5 concern past variability of small pelagic fish, inferred from sediments and historical records. Chapters 6, 8, and 12 address models, and Chapter 7 trophic dynamics. Chapters 9, 10, and 11 concern the fisheries for small, pelagic fish, stock-by-stock and globally, and the human dimensions of climate change and small pelagic fish. Chapters 13 and 14 address the future. Chapter 15 provides a synthesis and recommendations.

This book would not have been possible without the long-term support of GLOBEC and its contributing members. L'Institut de Recherche pour le Développement (IRD) of France provided partial support for a workshop on and production of the book. Shaw Island provided DC with peace and quiet for writing and editing. Finally, we express our appreciation to John Hunter who, with Jürgen Alheit, founded SPACC and was its former co-chair, leading it with good nature, wisdom, and the insight of a fisheries scientist.

David M. Checkley, Jr.
Jürgen Alheit
Yoshioki Oozeki
Claude Roy

1 History of international co-operation in research

Jürgen Alheit and Andrew Bakun

Over the last 25 years, since about 1980, international co-operation in research on small pelagic schooling fish with pelagic eggs, such as anchovy, sardine, sprat, and sardinella focused, first on processes determining recruitment variability and, then, since the mid 1990s, on the impact of climate variability on ecosystems dominated by small pelagics. Recruitment research was carried out to a large extent under the umbrella of the Sardine–Anchovy–Recruitment Program (SARP) within the Ocean Science in Relation to Living Resources Program (OSLR) run jointly by IOC[1] and FAO[2] and the Climate and Eastern Ocean Systems project (CEOS) conducted by a variety of research institutions.

Lack of scientific understanding of the mechanisms regulating recruitment was widely recognized in the 1980s (and still is) as the key unsolved scientific problem currently hindering effective management of small pelagic fish populations. Their collapses such as the Californian sardine or the Peruvian anchovy have had enormous negative economic and social effects on fishing nations which might have been avoided had there been the opportunity to predict recruitment. Consequently, several international and national initiatives were started in the 1980s to understand the relationship between environmental processes and fish recruitment. At this point, Reuben Lasker's "stable ocean hypothesis" (Lasker, 1975, 1978) had suddenly caught the attention of the fisheries scientific community, and provided a major conceptual basis for motivating and planning the early activity. Simultaneously, two new technologies, the "Daily Egg Production Method" (DEPM) (Lasker, 1985) and a technique for daily age and growth estimates based on measuring and counting daily marks laid down on larval fish otoliths (Methot, 1983), were under development in Lasker's laboratory. By increasing the temporal resolution of demographic studies on fish larvae, these appeared to offer promising new ways to seek improved understanding of fish recruitment variability.

At its 11th Assembly in 1979, the IOC passed resolution XI-17 to promote development of plans for major oceanographic studies of the physical–ecological interactions of importance to fishery resource-related problems, including the formation of a "Group of Four" experts (Bakun *et al.*, 1982) to advise on program formulation. FAO and SCOR[3] were asked in the resolution to develop a comprehensive scientific program for OSLR. SCOR and the ACMRR[4] of FAO responded to the IOC request forming Working Group 67 on "Oceanography, Marine Ecology and Living Resources." This group was formed with the aim to develop a proposal for an international recruitment experiment to investigate the relationships between environmental variability and fluctuations of living resources (Barber *et al.*, 1982). The same year, coincidentally, the "Fish Ecology III Conference" in the USA developed a conceptual framework for REX, a "recruitment experiment" (Rothschild and Rooth, 1982). In the meantime, a "Workshop on the Effects of Environmental Variation on Survival of Larval Pelagic Fishes" (IOC, 1981) was organized in Lima as a contribution to OSLR by FAO and the Peruvian–German technical aid project PROCOPA[5] (Pauly and Tsukayama, 1987) was established at the Peruvian fisheries institute, IMARPE.[6] These efforts finally converged in a joint "Ocean Science in Relation to Living Resources" (OSLR) program, which was co-sponsored by IOC and FAO, the main focus of which was to be the processes governing recruitment to fish populations (IOC, 1983).

The DEPM is a fisheries-independent method to estimate the spawning biomass of small pelagics, including the associated statistical precision of the estimated value. It was developed at the Southwest Fisheries Science Center in La Jolla under the leadership of Reuben Lasker (Lasker, 1985). The breakthrough for its development was the finding of Hunter and Goldberg (1980) that the age of postovulatory follicles can be used to estimate the daily proportion of spawning females. After its first application to the Californian anchovy in 1980, it was successfully carried out for the Peruvian anchovy (Santander *et al.*, 1984) and is now widely used in South America, South Africa and Europe (Alheit, 1993; Stratoudakis *et al.*, 2006). The application of this method for spawning biomass estimates requires extensive knowledge of reproductive parameters such as batch fecundity, spawning frequency and daily egg mortality and, consequently,

Climate Change and Small Pelagic Fish, eds. Dave Checkley, Jürgen Alheit, Yoshioki Oozeki, and Claude Roy. Published by Cambridge University Press. © Cambridge University Press 2009.

as a by-product, has furthered international co-operation in recruitment research.

The Lima Workshop in 1980 (IOC, 1981) brought together an international group of recruitment researchers with scientists working on reproduction and recruitment of the Peruvian anchovy and allowed comparisons of recruitment processes between small pelagics in the Humboldt and California currents. Important outcomes of this meeting were the first quantitative estimate of egg cannibalism (MacCall, 1980) and the first sketch of the "Basin Model," an attempt to explain population dynamics of small pelagics based on an "optimal free distribution" interrelationship between geography, movement, and growth dynamics at the population level (MacCall, 1990). It also produced an early comprehensive review (Bakun and Parrish, 1981) of empirical and conceptual frameworks for applying available environmental data to inferring the primary causative factors in recruitment variability, including the suggestion for the formulation that became known as "Lasker windows" (Peterman and Bradford, 1987; Pauly, 1989).

A most authoritative account on the ecology of marine fish larvae with a focus on anchovy and sardine was published in 1981 by four eminent fish larval researchers from the Southwest Fisheries Center in La Jolla: R. Lasker, J.R. Hunter, H.G. Moser and P.E. Smith (Lasker, 1981). It contains a thorough discussion of the role of larval starvation and predation on fish larvae for recruitment and was a pacemaker for the Sardine–Anchovy Recruitment Program, SARP.

The 12th Assembly of IOC in 1982 adopted the OSLR concept as a long-term program. It promoted coordinated regional research projects to elucidate factors determining recruitment to fish populations with the International Recruitment Project (IREP) as the initial main focus of OSLR. The Assembly established a Guiding Group of Experts for the OSLR Program and initiated the "Workshop on the IREP Component of the IOC Program on Ocean Science in Relation to Living Resources" in Halifax, Canada, in 1983 (IOC, 1983). This workshop, under the chairmanship of R. Beverton, recommended (i) direct investigations of the early life history, particularly SARP, including, *inter alia*, the otolith ring method, the DEPM and relevant oceanographic measurements (IREP Minimum Plan) and (ii) indirect (inferential) approaches by making available the wealth of long-term time series, including relevant information, normally peripheral to the marine field, such as climatic and meteorological data that could help to elucidate the physical and biological coupling controlling recruitment at different scales.

At its first meeting in Paris in 1984, the Guiding Group of Experts for OSLR recommended SARP as the pilot program for IREP. The basic SARP concept involved repeated surveys of larval production during the extended spawning season of small pelagic fish. These surveys were coupled with a comprehensive physical and biological oceanographic sampling program designed to determine variations in conditions related to larval starvation (Lasker, 1981), predation (Lasker, 1981), advection, physiological stress, and other factors leading to mortalities of early life stages. Later in the season, surviving juveniles were to be sampled and their birthdate frequencies determined using daily otolith growth rings (Methot, 1983). These frequencies, when compared with the observed larval production rates corresponding to the various birthdates, provide an index of variation in survival rate of early life stages. This is compared to variations in environmental processes to identify the mechanisms that best explain the observations. The SARP concept therefore, while basically an empirical field approach, offered a major departure from previous empirical approaches to study the recruitment problem in its ability to address higher frequency "within-year" variability. Whereas previous empirical attempts have been defeated by the necessity to combine shorter scale variations, having various causes, into single annual composites, SARP offered the possibility of resolving different causes and effects on the time scales on which they actually act to determine net reproductive success.

Accordingly, a number of regional field-going SARP components (direct investigations) were initiated to test the several recruitment hypotheses (starvation, predation, advection) (Bakun *et al.*, 1991): a US SARP in the California Current on anchovy, an Iberian SARP on anchovy and sardine in a joint Spanish–Portuguese–US project (López-Jamar and Garcia, 1992), a SW Atlantic SARP on anchovy and sardine in a co-operation by Brazil, Uruguay, Argentina, Germany, and Sweden (Alheit *et al.*, 1991) and a EURO-SARP[7] project on anchovy, sardine, and sprat in European waters run by scientists from Germany, Spain, Portugal, and the UK (Alheit and Bakun, 1991; Valenzuela *et al.*, 1991). A major aspect of SARP's scientific rationale was the application of the comparative method of science, whereby the multiple expressions of the problem afforded by various species groups inhabiting different regional ecosystems were considered as "proxy replicates" of similar processes, gaining additional explanatory power (degrees of freedom), to sort out the complex interacting mechanisms involved in recruitment variability. The comparative scientific method (Bakun, 1996) is particularly appropriate to problem areas where experimental controls are unavailable. This is one of the reasons why a coordinated international scientific SARP effort was thought to offer large potential benefits. Whereas all these SARP initiatives provided a wealth of new important information on the life history of small pelagics, it has to be admitted that no breakthrough was made in understanding and predicting recruitment. The main constraint was that no regional initiative had sufficient funding to carry out the complete SARP program. On the other hand, the increase in temporal resolution for the study of larval demography was insufficient by itself to explain differential larval mortality without a comparable increase in resolution in the observation of the potentially causally related oceanographic variables. A serious blow was delivered to the SARP initiative when the most promising US SARP project was stopped suddenly by federal budget restrictions.

Progress was made using inferential methods as suggested by the Guiding Group of Experts. Scientists from the Peruvian–German aid project PROCOPA teamed up in 1981 with an international group to rescue, assemble, and analyze on a monthly basis long-term time series from the period 1953–1982 of all measured variables likely to have affected the Peruvian anchovy and its ecosystem. This resulted in an extremely rich data archive captured in two books (Pauly and Tsukayama, 1987; Pauly *et al.*, 1989), which have served as an important source in later studies on the impact of climate variability on the Humboldt Current and its small pelagic fish resources (e.g. Alheit and Bernal, 1993; Alheit and Niquen, 2004).

A meeting which proved to be a milestone in small pelagics research was the "Expert Consultation to Examine Changes in Abundance and Species Composition of Neritic Fish Resources" organized by G. Sharp and J. Csirke of FAO in 1983 in San José, Costa Rica (FAO, 1983). This meeting gathered most of the information on ecology and fisheries of the large stocks of small pelagics relevant then for SARP and later for Small Pelagic Fish and Climate Change, SPACC, a regional project of the GLOBEC[8] program, particularly from developing countries, and still serves as a rich source of information. A key paper which was an enormous stimulus for climate variability research within SPACC even 20 years thereafter was given by T. Kawasaki (1983) on synchronous large-scale fluctuations of the three sardine stocks in the Pacific.

The CEOS project (Durand *et al.*, 1998) was an international collaborative study of potential effects of global climate change on the living resources of the highly productive eastern ocean upwelling ecosystems and on the ecological and economic issues directly associated with such effects. CEOS involved a variety of research institutions, notably NOAA/NMFS,[9] ORSTOM,[10] and ICLARM,[11] and was devoted to a study of the potential effects of global change on the resources of upwelling systems through identification of global and local effects impacting on these systems. A major focus of the study was the clupeoid fishes (such as anchovy and sardine). The main objectives were: (1) to assemble, summarize, and analyze the data record of the four decades since 1960 regarding the four eastern boundary upwelling ecosystems and other upwelling areas, (2) to apply the comparative method to identify key physical

processes and ecosystem responses, and (3) to resolve underlying global-scale trends that in each individual regional system may be obscured by local interannual and interdecadal variability. Major ideas emanating from the CEOS project are (i) the "Ocean Triad Concept" of Bakun (1996) which suggests that an optimal combination of three physical processes (enrichment, concentration, retention) provides an optimal situation for successful fish recruitment and (ii) the "Optimal Environmental Window" hypothesis (Cury and Roy, 1989), which is a dome-shaped response curve of population growth to increasing intensity of wind stress-associated mixing and transport. The CEOS project paved the way for SPACC, as it started to look not only at recruitment, but also at climatic effects on the dynamics of small pelagics.

SARP had established itself in the international science community so much that it opened access to national funding to carry out recruitment research. Although the large international funding hoped for in the mid 1980s never materialized, several SARP projects were established successfully using national funding. However, due to its complexity the recruitment problem could not be solved by SARP. It became clear in the early 1990s that (i) recruitment studies must be conducted from an ecosystem point of view and that (ii) climatic effects play a major role in population fluctuations of small pelagics leading to so-called regimes (Lluch-Belda *et al.*, 1992).

Consequently, SARP and CEOS researchers widened the scope of their investigations accordingly and jointly created in 1994 the SPACC (Small Pelagic Fish and Climate Change) project which became one of the four core projects of GLOBEC of the IGBP (International Geosphere Biosphere) Program. Science (Hunter and Alheit, 1995) and Implementation (Hunter and Alheit, 1997) plans of SPACC were developed at three international meetings in La Paz, Mexico (1994), Swakopmund, Namibia (1995), and Mexico City (1996). SPACC's objective is to clarify the effect of climate variability on the population dynamics of pelagic fish by comparing the ecosystems that support such populations (Hunter and Alheit, 1995). The goals are (i) to describe the characteristics and variability of the physical environment and of zooplankton population dynamics and their impact on small pelagic fish populations in each key ecosystem and (ii) to improve understanding of the nature and causes of long-term changes in these ecosystems. SPACC uses two general approaches to meet these goals:

(1) Retrospective studies, wherein ecosystem histories are reconstructed by means of fishery data and zooplankton and other time series and paleoecological data. This initiative was started in 1994 when SPACC researchers analyzed long-term data from ecosystems rich in small

pelagics, together with SCOR Working Group 98 on "World-wide Large-Scale Fluctuations of Sardine and Anchovy Populations" (Schwartzlose *et al.*, 1999).

(2) Process studies in which cause-and-effect linkages between zooplankton, fish population dynamics and ocean forcing are inferred from comparisons of standard measurements made in different ecosystems.

The long-range goal is to develop predictive scenarios for the fate of small pelagic fish populations. Results of the SPACC project are presented in the following chapters.

Acknowledgments

The authors wish to thank Patricio Bernal (IOC) and Jorge Csirke (FAO) for reviewing the manuscript and many valuable suggestions.

NOTES

1 Intergovernmental Oceanographic Commission.
2 Food and Agriculture Organization.
3 Scientific Committee on Oceanic Research.
4 Advisory Committee of Experts on Marine Resources Research.
5 Peruvian–German Cooperative Program for Fisheries Investigations.
6 Instituto del Mar del Perú.
7 European Sardine–Anchovy Recruitment Program.
8 Global Ocean Ecosystem Dynamics.
9 US National Oceanic and Atmospheric Administration/ National Marine Fisheries Service.
10 Office de la Recherche Scientifique et Technique Outre Mer.
11 International Center for Living Aquatic Resources Management.

REFERENCES

Alheit, J. (1993). Use of the daily egg production method for estimating biomass of clupeoid fishes: a review and evaluation. *Bull. Mar. Sci.* **53**: 750–767.

Alheit, J. and Bakun, A. (1991). Reproductive success of sprat (*S. sprattus*) in German Bight during 1987. *ICES C.M.* 1991/L:44.

Alheit, J. and Bernal, P. (1993). Effects of physical and biological changes on the biomass yield of the Humboldt Current ecosystem. In Sherman, K., Alexander, L. M., and Gold, B. D. (eds.), *Large Marine Ecosystems – Stress, Mitigation and Sustainability*, pp. 53–68. Washington, DC: American Association for the Advancement of Science, 376 pp.

Alheit, J. and Niquen, M. (2004). Regime shifts in the Humboldt Current ecosystem. *Prog. Oceanogr.* **60**: 201–222.

Alheit, J., Ciechomski, J., Djurfeldt, L. *et al.* (1991). SARP studies on the Southwest Atlantic anchovy, *Engraulis anchoita*, off Argentina, Urugay and Brazil. *ICES C.M./L*: 46, 32 pp.

Bakun, A. (1996). *Patterns in the Ocean: Ocean Processes and Marine Population Dynamics.* University of California, La Jolla: California Sea Grant College System, 323 pp.

Bakun, A. and Parrish, R. H. (1981). Environmental inputs to fishery population models for eastern boundary current regions. In *Workshop on the Effects of Environmental Variation on Survival of Larval Pelagic Fishes, Lima, 20 April – 5 May 1980,* Sharp, G. D., ed. Intergovernmental Oceanographic Commission, UNESCO, Paris, Workshop Rep. 28, p. 67–104.

Bakun, A., Beyer, J., Pauly, D., Pope, J. G., and Sharp, G. D. (1982). Ocean sciences in relation to living resources. *Can. J. Fish. Aquat. Sci.* **39**: 1059–1070.

Bakun, A., Alheit, J., and Kullenberg, G. (1991). The Sardine-Anchovy Recruitment Project (SARP): rationale, design and development. *ICES C.M. 1991/L: 43.*

Barber, R. T., Csirke, J., Jones, R., *et al.* (1982). *Report of SCOR/ ACMRR Working Group 67, Oceanography, Marine Ecology and Living Resources.* SCOR/WG-67, 21pp.

Cury, P. and Roy, C. (1989). Optimal environmental window and pelagic fish recruitment success in upwelling areas. *Can. J. Fish. Aquat. Sci.* **46**: 670–680.

Durand, M.-H., Cury, P., Mendelssohn, R., *et al.* (eds.) 1998. *Global Versus Local Changes in Upwelling Systems.* Paris: ORSTOM editions, 594 pp.

FAO (1983). Proceedings of the Expert Consultation to Examine Changes in Abundance and Species Composition of Neritic Fish Resources. *FAO Fish. Rep.* **291**, 1224 pp.

Hunter, J. R. and Alheit, J. (1995). *International GLOBEC Small Pelagic Fishes and Climate Change Program.* GLOBEC Report No. 8, 72 pp.

Hunter, J. R. and Alheit, J. (1997). *International GLOBEC Small Pelagic Fishes and Climate Change Program. Implementation Plan.* GLOBEC Report No. 11, 36 pp.

Hunter, J. R. and Goldberg, S. R. (1980). Spawning incidence and batch fecundity in northern anchovy, *Engraulis mordax. Fish. Bull. (US)* **77**: 641–652.

IOC (1981). *Workshop on the Effects of Environmental Variation on Survival of Larval Pelagic Fishes, Lima, 20 April– 5 May 1980.* Sharp, G. D. (ed.). Intergovernmental Oceanographic Commission, UNESCO, Paris, Workshop Rep. 28, 323 pp.

IOC (1983). *Workshop on the IREP Component of the IOC Programme on Ocean Science in relation to Living Resources (OSLR).* Intergovernmental Oceanographic Commission, UNESCO, Paris, Workshop Rep. 33, 17 pp.

Kawasaki, T. (1983). Why do some pelagic fishes have wide fluctuations in their numbers? – Biological basis of fluctuations from the viewpoint of evolutionary ecology. In *Proceedings of the Expert Consultation to Examine Changes in Abundance and Species Composition of Neritic Fish Resources,* Sharp, G. and Csirke, J., eds. *FAO Fish. Rep.* **291**: 1065–1080.

Lasker, R. (1975). Field criteria for survival of anchovy larvae: the relation between inshore chlorophyll maximum layers and successful first feeding. *Fish. Bull. (US)* **73**: 453–462.

Lasker, R. (1978). The relation between oceanographic conditions and larval anchovy food in the California Current: identification of factors contributing to recruitment failure. *Rapp. P.-v. Réun. Cons. Int. Explor. Mer.* **173**: 212–230.

Lasker, R. (ed.) (1981). *Marine Fish Larvae: Morphology, Ecology and Relation to Fisheries.* Seattle: University of Washington Press, 131 pp.

Lasker, R. (ed.) (1985). *An egg production method for estimating spawning biomass of pelagic fish: application to the northern anchovy* (Engraulis mordax). NOAA Techn. Rep. NMFS 36, 99 pp.

Lluch-Belda, D., Schwartzlose, R. A., Serra, R. *et al.* (1992). Sardine and anchovy regime fluctuations of abundance in four regions of the world oceans: a workshop report. *Fish. Oceanogr.* **1**: 339–347.

López-Jamar, E. and Garcia, A. (1992). Foreword. *Boletín Instituto Espanol de Oceanografía* **8**: 3–4.

MacCall, A. D. (1980). The consequences of cannibalism in the stock-recruitment relationship of planktivorous pelagic fishes such as *Engraulis*. In *Workshop on the Effects of Environmental Variation on the Survival of Larval Pelagic Fishes*, G. Sharp, ed. Intergovernmental Oceanographic Commission, UNESCO, Paris, Workshop Report 28, pp. 201–220.

MacCall, A.D. (1990). *Dynamic Geography of Marine Fish Populations.* Seattle: University of Washington Press, 153 pp.

Methot, R.D. (1983). Seasonal variation in survival of larval anchovy, *Engraulis mordax*, estimated from the age distribution of juveniles. *Fish. Bull. (US)* **81**: 741–750.

Pauly, D. (1989). An eponym for Reuben Lasker. *Fish. Bull. (US)* **87**: 383–384.

Pauly, D. and Tsukayama, I. (eds.) (1987). *The Peruvian anchoveta and its upwelling ecosystem: Three decades of change.* ICLARM Studies and Reviews **15**, 351 pp.

Pauly, D., Muck, P., Mendo, J., and Tsukayama, I. (eds.) (1989). *The Peruvian upwelling ecosystem: Dynamics and interactions.* ICLARM Conference Proceedings **18**, 438 pp.

Peterman, R. M. and Bradford, M. J. (1987). Wind speed and mortality rate of a marine fish, the northern anchovy (*Engraulis mordax*). *Science* **235**: 354–356.

Rothschild, B. J. and Rooth, C. (eds.) (1982). *Fish Ecology III. A Foundation for REX, A Recruitment Experiment.* University of Miami Tech. Rep. 82008; Miami, Florida, USA, 389 pp.

Santander, H., Alheit, J., and Smith, P. E. (1984). Estimación de la biomasa de la población desovante de anchoveta peruana, *Engraulis ringens*, en 1981 por aplicacion del "Método de Producción de Huevos". *Boletin Instituto del Mar del Perú-Callao* **8**: 208–250.

Schwartzlose, R. A., Alheit, J., Bakun, A. *et al.* (1999). Worldwide large-scale fluctuations of sardine and anchovy populations. *S. Afr. J. Mar. Sci.* **21**: 289–347.

Stratoudakis, Y., Bernal, M., Ganias, K., and Uriarte, A. (2006). The daily egg production method: recent advances, current applications and future challenges. *Fish Fish.* **7**: 35–57.

Valenzuela, G. J., Alheit, J., Coombs, S., and Knust, R. (1991). Spawning patterns of sprat and survival chances of sprat larvae in relation to frontal systems in the German Bight. *ICES C.M. 1991/L:45.*

2 A short scientific history of the fisheries

Alec D. MacCall

CONTENTS

Summary

This chapter briefly summarizes the history of scientific understanding of the fluctuations of small pelagic fishes and fisheries. The classical quantitative models underlying modern fishery analysis and management were developed in the 1950s and 1960s. Although California and Japan had previously experienced collapses of major fisheries for small pelagics in the 1940s and 1950s, it was the collapse of the "scientifically managed" Peruvian anchoveta fishery in the early 1970s that drew worldwide attention to the problem of collapsing small pelagic fisheries. The inability of the anchoveta to regain its former levels of productivity cast doubt on the classical equilibrium fishery models. In the late twentieth century, substantial progress was made toward understanding the environmental influences on these fishes. Some of the major environmental influences (which often may not be specifically identified) fluctuate at interdecadal time scales, giving rise to prolonged periods of high and low fish productivity, abrupt transitions including collapses, global teleconnections and phase relationships. These so-called "regimes" have recently become a major topic of research in fishery science. Despite scientific progress in understanding many facets of these fisheries and their fluctuations, there still is no accepted theory of the fishery–oceanographic dynamics of small pelagic fishes that links their commonly shared properties and that provides the predictive capability needed for ecosystem-based management.

Introduction

This history of small pelagic fisheries focuses specifically on the development of a scientific understanding of their dynamics, especially regarding their problematic fluctuations in abundance. Fréon *et al.* (2005) describe several distinct historical periods in the study of pelagic fish stocks, a system that is adopted for this review. During the pre-1900 "mother nature period", oceanic fish stocks tended to be regarded as inexhaustible, and little attention was given to the patterns and dynamics of their fluctuations. Fréon *et al.* do not give a name to the first half of the twentieth century, but it could be called the "developmental period." They describe the period 1900–1950 as being a time of industrial development, the beginnings of scientific studies, and growing awareness of environmental influences. The third quarter of the twentieth century, which could be called "the classical period" saw the development of the classical models of fish population dynamics, and is where this brief review begins.

The fourth quarter of the twentieth century was aptly termed "the doubt period" by Fréon *et al.* (2005). Whereas the mathematical models developed during the preceding period had engendered a confidence that good science would lead to high and sustainable yields, a growing worldwide list of fishery collapses (Mullon *et al.*, 2005) steadily eroded that confidence. Because ideas and events after 1980 are extensively covered by several other chapters, recent developments are treated more briefly in this chapter.

The classical period

By the mid twentieth century, there was a widespread perception that fisheries for small pelagics tend to be more prone to collapse than are those for other types of marine fishes. Large fisheries for sardines off Japan (*Sardinops melanosticta*) and California (*S. sagax caerulea*) collapsed during the 1940s. After a long decline, the Hokkaido herring (*Clupea harengus pallasi*) fishery finally collapsed in the mid 1950s, shortly followed by a much more sudden

Climate Change and Small Pelagic Fish, eds. Dave Checkley, Jürgen Alheit, Yoshioki Oozeki, and Claude Roy. Published by Cambridge University Press. © Cambridge University Press 2009.

collapse of Norwegian herring (*C. harengus harengus*). In the mid 1960s yet another sardine (*S. sagax ocellata*) fishery collapsed, this time off South Africa. All of these stocks appeared to withstand intense exploitation for an extended length of time, but suddenly failed to exhibit the vigorous productivity that characterized their pre-collapse fisheries. There was intense debate (e.g. Clark and Marr, 1955) whether these collapses were due to the effects of fishing, or whether they were unavoidable consequences of environmental fluctuations.

Development of now-classical quantitative theories of fishery dynamics during the 1950s and 1960s provided convincing evidence that marine fish stocks can be depleted by intense fishing even in the absence of environmental perturbations. The monumental treatise by Beverton and Holt (1957) was especially influential, as it demonstrated the analytical power and insights that could be gained from rigorous mathematical modeling of fish populations. Quantitative fishery scientists developed nearly all of the elements of modern fishery analysis between the mid 1950s and mid 1960s. Some landmark contributions during this period include Ricker's (1954) examination of the stock-recruitment relationship, Schaefer's (1954) development of the stock production model, and development of Virtual Population Analysis (VPA) independently by Murphy (1965) and by John Gulland (1965). One of the first modern stock assessments based on VPA was Murphy's (1966) analysis of the California sardine fishery.

In contrast to the problem of "growth overfishing" seen in the North Sea where individual fish were being harvested at too small a size, other ecosystems faced the problem of "recruitment overfishing" whereby intense fisheries remove individuals faster than they could be generated by the parental reproduction (e.g. Cushing, 1973). The notable small pelagic fishery collapses cited above were generally considered to be examples of recruitment overfishing.

A Peruvian fishery for anchoveta (*Engraulis ringens*) developed rapidly during the late 1950s and early 1960s, accelerated by the transfer of existing equipment and expertise from the collapsed sardine industry in California (Radovich, 1981; Ueber and MacCall, 1992). The Peruvian fishery, which quickly became the largest in the world, was also notable as being managed "scientifically" under advice from the United Nations Food and Agriculture Organization (FAO) and a panel of the world's leading fishery scientists using the newly developed kit of mathematical tools. It came as a shock to both the scientific world and the global economy when the anchoveta fishery collapsed in 1972. As had been prophesied by Paulik (1971), the collapse appeared to be the result of intense fishing on a resource made vulnerable by the El Niño conditions in 1972 that concentrated the fish in the nearshore region. The initial sense of shock slowly evolved into a sense of numbness, as

following years saw no substantial recovery of the anchoveta resource, despite major reductions in fishing pressure. The problem could no longer be associated simply with the El Niño of 1972. Something had mysteriously changed in the ecosystem. In the early 1970s, sardines (*S. sagax sagax*) unexpectedly appeared in abundances sufficient to support an alternative fishery (see below).

Through the 1970s the conventional viewpoint was strongly based on equilibrium fishing assumptions: If fishing pressure could be reduced sufficiently, it should be possible to rehabilitate these collapsed fisheries. In view of a sustained increase in abundance of California's anchovies (*E. mordax*) in the 1950s following the decline of the sardine, simple ecological competition models allowed the equilibrium fishing view to be extended to a multispecies equivalent whereby competition from anchovies was thought to be the reason for low sardine productivity. In California, fishery managers were being urged to "intentionally overfish" anchovies to assist recovery of the sardine resource (Sette, 1969; McEvoy, 1986). A nearly identical rise in anchovies (*E. capensis*) off South Africa during the 1960s generated similar claims of dynamics driven by inter-species competition (Stander and Le Roux, 1968). Immediately following the anchoveta collapse in South America, sardine abundance increased rapidly in the 1970s, leading to widespread acceptance of some form of sardine–anchovy alternation, usually described as "species replacement." Indeed, the South American fishery on sardines was quickly developing toward levels rivaling those of the previous anchoveta fishery. Sardines formerly had been relatively scarce in Peru and Chile, but coincident with the 1970s' increase in apparent abundance, their range also expanded nearly 1000 km southward to Talcahuano, Chile, where they had never before been seen (Serra, 1983).

A decade after the Peruvian anchoveta collapse, Gulland (1983) still viewed the problem in equilibrium terms, although recognition of the increasing abundance of Peruvian sardines extended his concern to interactions in a multispecies system. Gulland (1983, p. 1019) observed that these pelagic species have been especially susceptible to collapse as the result of recruitment overfishing, but also that the collapse of one species frequently coincided with the rise of another. Notably, he says "The collapses have too often followed a period of very heavy fishing to be due solely to chance or environmental effects, and also the rise of a competing species has now occurred too often to be based on chance alone …"

The doubt period

Fréon *et al.* (2005) insightfully called the last quarter of the twentieth century "the doubt period." Ushered in by the catastrophic collapse of the Peruvian anchoveta fishery, which had been the largest fishery in the world, this period saw a growing worldwide list of fishery collapses. It seemed that small pelagic fish were especially prone to collapse. However, it should be noted that in hindsight, Mullon *et al.* (2005) recently found that collapses of small pelagics were no more frequent than in other fisheries. More disturbing than the collapses themselves was the frequent lack of recovery relative to the former productivity of those fisheries. Both "the classical period" and "the doubt period" produced significant advances in understanding many of the underlying patterns and mechanisms of pelagic fish fluctuations, but remarkably little progress was made toward integrating those components into a complete theory of small pelagic fish dynamics that had useful predictive power. By the end of the twentieth century it was apparent that small pelagic fish fluctuations presented a scientific puzzle that was far more difficult than anyone had imagined (e.g. Chavez *et al.*, 2003).

In contrast to the classical equilibrium fishery models, an alternative, non-equilibrium view was beginning to emerge during the 1970s. Soutar and Isaacs (1969, 1974) developed a remarkable time series of prehistoric sardine and anchovy abundances based on fish scales preserved in laminated anerobic sediments found in a nearshore basin in southern California. The 2000-year paleosedimentary record indicated that unfished sardine abundances have always been highly variable off California, with some virtual disappearances (however, at the coarse resolution of ca. 0.5 million tons) even in the absence of fishing. It was not possible to explain those prehistoric fluctuations in terms of equilibrium models, and the new paleosedimentary evidence was eagerly taken to absolve the fishery of responsibility for the disappearance of the resource: "Nor can the virtual absence of the sardine from the waters off Alta California be considered an unnatural circumstance" (Soutar and Isaacs 1974), Another of Soutar and Isaacs' surprising findings was that there was no indication of anchovy–sardine alternations of abundance in the paleosedimentary time series, despite scientific consensus that the two species were competitors (e.g. Sette, 1969). Probably due to the strong circumstantial pattern of anchovy–sardine alternations recently experienced in fisheries off California, South Africa and South America, the lack of paleosedimentary evidence for sardine–anchovy alternation received little attention.

The pattern of sardine fluctuations implied by the paleosedimentary record could not be reconciled with the conventional equilibrium view of an approximately constant "reference" state of the resource corresponding to an unfished condition (i.e. carrying capacity, in ecological terms). Isaacs (1976), expressed this concern clearly, and coined the term "regime" to describe the tendency of ecosystems to fall into prolonged states and patterns that would suddenly change to a new and different pattern (Box 2.1). The terms "regime" and "regime shift" have become common key words in recent climate-related fisheries and oceanographic publications (Fig. 2.1).

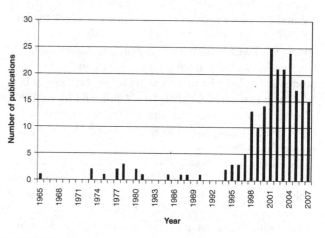

Fig. 2.1. Number of publications containing the keyword "regime" together with "climate" and "fish" or "fisheries" in the *Aquatic Sciences and Fisheries Abstracts* database. Recent years are incomplete.

Box 2.1. The origin of the "regime" concept

At the 1973 Symposium of the California Cooperative Oceanic Fisheries Investigations (CalCOFI), John Isaacs (1976) formalized his concern that sardine–anchovy systems may be far less predictable than our simplistic models would suggest. Not only was this was the first use of the term "regime" in its modern fishery meaning but the explanation of what he meant by the term was extraordinarily clear:

"… there are probably a great number of possible regimes and abrupt discontinuities connecting them, flip–flops from one regime to another… Sardines, for example, are either here or not here… There are internal, interactive episodes locked into persistence, and one is entirely fooled if one takes one of these short intervals of a decade or so and decides there is some sort of simple probability associated with it… organisms must respond to more than just fluctuations around some optimum condition…. Fluctuations of populations must be related to these very large alternations of conditions."

An early development in the field of dynamical systems that gained popularity in the 1970s was "catastrophe theory" (Thom, 1993), which provided a useful explanation of the dynamics of fishery collapses. A number of studies had shown that searching behavior by fishermen, especially for surface-schooling fishes, could result in a tendency for catch-per-unit-effort (CPUE) to be insensitive to declines in underlying fish abundance (Paloheimo and Dickie, 1964; Pope and Garrod, 1975; MacCall,

Box 2.2. Catastrophe theory and fishery collapse

The classical Schaefer model assumes that abundance (as measured by catch-per-unit-effort, CPUE) declines linearly as fishing effort increases (Fig. 2.2, thin lines). However, experience has shown that the fishing mortality rate generated per fishing effort may increase rapidly when abundance is low. Thus at low abundances, fishermen are able to locate and catch the few remaining fish, so that a small amount of fishing effort can catch a very large fraction of the population. Although the underlying relationship between abundance and fishing mortality rate may still be described by a Schaefer model, the relationship between apparent abundance (CPUE) and nominal fishing effort (e.g. vessel–days) is severely distorted (Fig. 2.2, thick lines). There is now a stable upper equilibrium, and an unstable lower equilibrium. As nominal fishing effort increases, the relationship initially behaves like a Schaefer model, but at some intermediate fishing rate sustainability is no longer possible, resulting in sudden collapse. Rebuilding requires severely reducing fishing effort to where it falls on the left side of the equilibrium line (Fig. 2.2, upper), where abundance can increase. The associated rebuilding catch is very low.

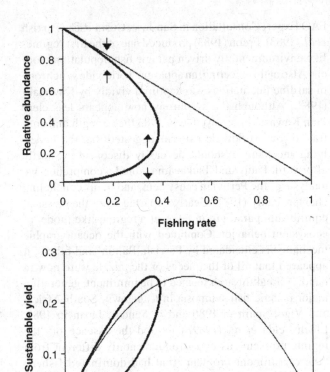

Fig. 2.2. Effect of a population-dependent catchability coefficient on fishery production curves. Thin line: CPUE is proportional to true abundance. Thick line: CPUE varies as the square root of abundance, which is typical in fisheries for small pelagic fish.

1976; Ulltang, 1976). Equivalently, a unit of nominal fishing effort, such as a vessel–day, could generate an ever-increasing fishing mortality rate as stock size becomes small. The role of this mechanism in fishery collapses and its critical importance in management of fisheries for small pelagics was recognized at a 1978 ICES Symposium (Ulltang, 1980). The phenomenon has more recently found to be widespread also in demersal fisheries (Harley, 2002), and is largely due to the effectiveness of modern fish-finding technology. Fox (1974, also reported by Gulland, 1977) incorporated this non-linearity in a production model, and obtained a model akin to a "fold catastrophe" that helped explain fishery collapses and quantified the extreme reductions in fishing pressure needed to rebuild stocks even under equilibrium biological dynamics (Box 2.2).

By the end of the 1970s, fisheries on small pelagics were showing alarming fluctuations around the world. Following the 1960s' collapse of the southern Benguela fishery for sardines, a large fishery developed on the northern Benguela sardine stock off Walvis Bay, but that too collapsed shortly afterward. Anchovies were becoming the dominant small pelagic fish in both subregions of the Benguela system. By the late 1970s, large sardine fisheries had developed in South America and Japan, while at long last there were signs of improvement in sardines off California. The Peruvian anchoveta remained at a low level. In California the anchovy population reached peak abundances earlier in the decade, but decades too late to be explained by competitive release due to lack of sardines (Methot, 1989). A large number of herring stocks in the North Atlantic were in various stages of collapse, with unprecedented fishing bans being imposed on some of them (Jakobsson, 1980; Schumacher, 1980). Fluctuations of small pelagics had become a serious problem in world fisheries. The mechanisms were more puzzling than ever. The good news, especially from the Japanese sardine (Kondo, 1980), was that small pelagic fishery collapses might not be permanent.

The optimism that the puzzle would soon be solved generated a flurry of workshops and symposia in the 1980s. At a workshop in Lima, Peru, Bakun and Parrish (1980) presented an outline of the essential elements of fishery–oceanography associated with small pelagic fishes, and greatly expanded that treatment in a presentation to a 1983

FAO Expert Consultation in San Jose, Costa Rica (Parrish *et al.*, 1983). Fréon (1983) produced one of the first regime-like environmentally driven pelagic fish population models. Also at this meeting, an apparent worldwide synchrony of sardine fluctuations was captured vividly by Kawasaki (1983). Although the synchrony now appears less clear than Kawasaki portrayed it, in 1983 the strength and contrast of the worldwide pattern suggested that the underlying mechanism should be easily discovered. Shortly afterward, Pauly and Tsukayama edited a comprehensive analysis of the Peruvian ecosystem, and in his concluding chapter, Pauly (1987) clearly departed from the classical equilibrium paradigm in favor of a regime-like model of ecosystem behavior. Combined with the oceanographic mechanisms elucidated by Parrish, Bakun, and others, it appeared that all of the pieces of the puzzle were now in hand. A breakthrough seemed to be imminent, generating major pelagic fish symposia in Capetown, South Africa, and Vigo, Spain in 1986 and in Sendai, Japan in 1989. Lluch-Belda *et al.* (1989) captured the essence of "the regime problem" as being fundamentally different from "the recruitment problem" that had dominated fishery science for decades. Yet the answer remained elusive.

There was growing awareness that environmental shifts were associated in some way with changes in pelagic fish productivity in such places as South Africa, Peru and Japan. The "new" sardine fisheries off Japan and Peru–Chile collapsed by the early 1990s, while South Africa experienced a long-awaited resurgence of sardines. An environmentally explicit management policy was adopted for sardines in the US portion of the California Current, based on Jacobson and MacCall's (1995) use of sea surface temperature as a proxy for the admittedly unknown causal mechanism in the stock–recruitment relationship. International research efforts continued, such as the Scientific Committee on Oceanic Resources Working Group 98, titled, "Worldwide Large-Scale Fluctuations of Sardine and Anchovy Populations" (Schwartzlose *et al.*, 1999), and GLOBEC's "Small Pelagics and Climate Change" (reported in the present volume). However, the twentieth century ended without a generally accepted theory of small pelagic fish dynamics capable of explaining the nature and causes of their fluctuations (e.g. Chavez *et al.*, 2003).

A new millennium

Of course, history only becomes clear with the passage of time, and it is perhaps too soon to say whether we have yet emerged from "the doubt period." Fréon *et al.* (2005) optimistically assert that we are now in "the era of ecosystem-based management," and there is certainly widespread interest in moving beyond single-species approaches. Fisheries for small pelagics seem to offer an excellent opportunity for development of an ecosystem approach to management. An open question is whether this can be done without a predictive theory of the dynamics of small pelagic fishes on which to base management decisions. Notwithstanding the repeated disappointments of the past few decades, the science of fisheries–oceanography again seems to be close to achieving this breakthrough.

Acknowledgments

I thank Rob Crawford and Pierre Fréon for their useful reviews of the manuscript. Pierre Fréon suggested reference to the historical periods described in Fréon *et al.* (2005), which I found very useful in structuring this review.

REFERENCES

Bakun, A. and Parrish. R. (1980). Environmental inputs to fishery population models for eastern boundary current regions. In *Workshop on the Effects of Environmental Variation on the Survival of Larval Pelagic Fishes*. IOC Workshop Rep. 28, IOC/UNESCO, Paris, pp. 67–104.

Beverton, R. J. H. and Holt, S. J. (1957). *On the dynamics of exploited fish populations*. HM Stationary Off., London, Fish. Invest., Ser. 2, Vol. **19**, 533 pp.

Chavez, F. P., Ryan, J. Lluch-Cota, S. E., and Ñiquen M. C. (2003). From anchovies to sardines and back: multidecadal change in the Pacific Ocean. *Science* **299**: 217–221.

Clark, F. N. and Marr, J. C. (1955). Population dynamics of the Pacific sardine. *Calif. Coop. Oceanic Fish. Invest. Prog. Rep.*, 1 July 1953 to 31 March 1955: 11–48.

Cushing, D. H. (1973). Dependence of recruitment on parent stock. *J. Fish. Res. Board Can.* **30**: 1965–1976.

Fox, W. W. (1974). An overview of production modeling. pp. 142–156. In *Workshop on Population Dynamics of Tuna. ICCAT Collective Volume of Scientific Papers*, Vol. **3**.

Fréon, P. (1983). Production models as applied to sub-stocks depending on upwelling fluctuations. pp. 1047–1064 In *Proceedings of The Expert Consultation to Examine Changes in Abundance and Species Composition of Neritic Fish Resources*. San Jose, Costa Rica, 18–29 April 1983. FAO Fish. Rep. 291.

Fréon, P., Cury, P., Shannon, L., and Roy, C. (2005). Sustainable exploitation of small pelagic fish stocks challenged by environmental and ecosystem changes. *Bull. Mar. Sci.* **76**: 385–462.

Gulland, J. A. (1965). Estimation of mortality rates. Annex Northeast Arctic Working Group. Council Meeting. G 3. 9pp. Mimeo.

Gulland, J. A. (1977). The stability of fish stocks. *J. Cons. Int. Explor. Mer.* **37**: 199–204.

Gulland, J. A. (1983). World resources of fisheries and their management. pp. 839–1061. In *Marine Ecology*, Vol. **5**, part 2. Chichester, UK: John Wiley and Sons.

Harley, S. J. (2002). *Meta-Analytical Approaches to the Study of Fish Population Dynamics*. PhD dissertation, Dalhousie University, Halifax, Canada. 221pp.

Isaacs, J. D. (1976). Some ideas and frustrations about fishery science. *Calif. Coop. Oceanic Fish. Invest. Rep.* **18**: 34–43.

Jacobson, L. D., and MacCall, A. D. (1995). Stock-recruitment models for Pacific sardine (*Sardinops sagax*). *Can. J. Fish. Aquat. Sci.* **52**: 566–577.

Jakobsson, J. (1980). Exploitation of the Icelandic spring- and summer-spawning herring in relation to fisheries management, 1947–1977. *Cons. Int. Explor. Mer*, Rapp. P.-V. Reuns. **177**: 23–42.

Kawasaki, T. (1983). Why do some pelagic fishes have wide fluctuations in their numbers? – biological basis of fluctuation from the viewpoint of evolutionary ecology. pp. 1065–1080. In *Proceedings of the Expert Consultation to Examine Changes in Abundance and Species Composition of Neritic Fish Resources.* San Jose, Costa Rica, 18–29 April 1983. FAO Fish. Rep. 291.

Kondo, K. (1980). The recovery of the Japanese sardine – The biological basis of stock size fluctuations. *Cons. Int. Explor. Mer*, Rapp. P.-V. Reuns. **177**: 332–354.

Lluch-Belda, D., Crawford, R. J. M., Kawasaki, T. *et al.* (1989). World-wide fluctuations of sardine and anchovy stocks: the regime problem. *S. Afr. J. Mar. Sci.* **8**: 195–205.

MacCall, A. (1976). Density dependence of catchability coefficient in the California Pacific sardine, *Sardinops sagax caerulea*, purse seine fishery. *Calif. Coop. Oceanic Fish. Invest.* **18**: 136–148.

McEvoy, A. F. (1986). *The Fisherman's Problem: Ecology and Law in California Fisheries* 1850–1980. New York: Cambridge University Press, 368pp.

Methot, R. D. (1989). Synthetic estimates of historical abundance and mortality for northern anchovy. *Amer. Fish. Soc. Symp.* **6**: 66–82.

Mullon, C., Fréon, P., and Cury, P. (2005). The dynamics of collapse in world fisheries. *Fish and Fisheries*, **6**: 111–120.

Murphy, G. I. (1965). A solution of the catch equation. *J. Fish. Res. Bd. Canada* **22**: 191–202.

Murphy, G. I. (1966). Population biology of the Pacific sardine (*Sardinops caerulea*). *Proc. Calif. Acad. Sci.* **34**(1): 1–84.

Paloheimo, J. E. and Dickie, L. M. (1964). Abundance and fishing success. *Cons. Int. Explor. Mer*, Rapp. P.-V. Reuns. **155**: 152–163.

Parrish, R. H., Bakun, A. Husby, D. M., and Nelson, C. S. (1983). Comparative climatology of selected environmental processes in relation to eastern boundary current pelagic fish reproduction, pp. 731–777. In *Proceedings of the Expert Consultation to Examine Changes in Abundance and Species Composition of Neritic Fish Resources*, Sharp, G. D. and Csirke, J., eds. San José, Costa Rica, 18–29 April 1983. FAO Fish. Rep. 291.

Paulik, G. J. (1971). Anchovies, birds, and fishermen in the Peru Current. Reprinted in Glantz, M. H., and Thompson, J. D., eds. 1981, pp. 35–79. *Resource Management and Environmental Uncertainty: Lessons From Coastal Upwelling Fisheries.* New York: John Wiley and Sons.

Pauly, D. (1987). Managing the Peruvian upwelling ecosystem: a synthesis. pp. 325–342. In *The Peruvian Anchoveta and its Upwelling Ecosystem: Three Decades of Change*, Pauly, D. and Tsukayama, I., eds. *ICLARM Studies and Reviews* **15**, 351pp.

Pope, J. G. and Garrod, D. J. (1975). Sources of error in catch and effort quota regulations with particular reference to variations in the catchability coefficient. *ICNAF Res. Bull.* **11**: 17–30.

Radovich, J. (1981). The collapse of the California sardine fishery – What have we learned? pp. 107–136. In *Resource Management and Environmental Uncertainty: Lessons from Coastal Upwelling Fisheries*, Glantz, M. H. and Thompson, J. D. , eds. New York: John Wiley and Sons.

Ricker, W. E. (1954). Stock and recruitment. *J. Fish. Res. Board, Can.* **11**: 559–623.

Schaefer, M. B. (1954). Some aspects of the dynamics of populations important to the management of the commercial fish populations. *Bull. Inter-Amer. Trop. Tuna Comm.* **2**: 245–285.

Schumacher, A. (1980). Review of North Atlantic catch statistics. *Cons. Int. Explor. Mer*, Rapp. P.-V. Reuns. **177**: 8–22.

Schwartzlose, R., Alheit, J., Bakun, A. *et al.* (1999). Worldwide large-scale fluctuations of sardine and anchovy populations. *S. Af. J. Mar. Sci.* **21**: 289–347.

Serra, J. R. (1983). Changes in the abundance of pelagic resources along the Chilean coast, pp. 255–284. In *Proceedings of the Expert Consultation to Examine Changes in Abundance and Species Composition of Neritic Fish Resources*, Sharp, G. D. and Csirke, J., eds. San Jose, Costa Rica, 18–29 April 1983. FAO Fish. Rep. (291) Vol. **2**, pp. 1–553.

Sette, O. E. (1969). A perspective of a multi-species fishery. *Calif. Coop. Oceanic Fish. Invest.* Rep. **13**: 81–87.

Soutar, A. and Isaacs, J. D. (1969). History of fish populations inferred from fish scales in anaerobic sediments off California. *Calif. Coop. Oceanic Fish. Invest.* Rep. **13**: 63–70.

Soutar, A. and Isaacs, J.D. (1974). Abundance of pelagic fish during the 19th and 20th centuries as recorded in anaerobic sediment off the Californias. *Fishery Bull.* (*US*) **72**: 257–273.

Stander, G. H. and Le Roux, J.-P. (1968). Notes on fluctuations of the commercial catch of the South African pilchard (*Sardinops ocellata*) 1950–1965. *Invest. Rep. Div. Sea Fish.* **65**: 14pp.

Thom, R. (1993). *Structural Stability and Morphogenesis: An Outline of a General Theory of Models.* Reading, MA: Addison-Wesley.

Ueber, E. and MacCall, A. (1992). The rise and fall of the California sardine empire. pp. 31–48. In Glantz, M. H., ed., *Climate Variability, Climate Change and Fisheries.* Cambridge, UK: Cambridge University Press.

Ulltang, Ø. (1976). Catch per unit effort in the Norwegian purse seine fishery for Atlanto-Scandian (Norwegian spring spawning) herring. *FAO Fish. Tech. Pap.* **155**: 91–101.

Ulltang, Ø. (1980). Factors affecting the reaction of pelagic fish stocks to exploitation and requiring a new approach to assessment and management. *Cons. Int. Explor. Mer*, Rapp. P.-V. Reuns. **177**: 489–504.

3 Habitats

David M. Checkley, Jr., Patricia Ayón, Tim R. Baumgartner, Miguel Bernal, Janet C. Coetzee, Robert Emmett, Renato Guevara-Carrasco, Larry Hutchings, Leire Ibaibarriaga, Hideaki Nakata, Yoshioki Oozeki, Benjamin Planque, Jake Schweigert, Yorgos Stratoudakis, and Carl D. van der Lingen

CONTENTS

Summary

The habitats of populations of small, pelagic fish, especially anchovy and sardine, in the Benguela, California, Humboldt, and Kuroshio-Oyashio current systems, and in the NE Atlantic, are described and discussed in regard to future climate change. These stocks have been the primary concern of the Small Pelagic Fish and Climate Change (SPACC) program of International GLOBEC. Each of these regions and stocks has a unique set of climate and ocean conditions and their variability. However, they also share common characteristics. Spawning and development occurs within broad ranges of temperature (12–26 °C) and salinity (<30–36) and in regions of high plankton production, associated with either upwelling or freshwater. Often, sardine are more oceanic and anchovy more coastal, often associated with wind-driven upwelling and rivers. Sardine tend to make longer migrations between spawning and feeding regions than do anchovy. The habitat of most populations of small, pelagic fish expands when the population size is large and contracts when it is small, often into refugia. Climate change may affect populations of small, pelagic fish by causing poleward shifts in distribution due to warming, some of which have already occurred. Other potential effects are due to changes in winds, hydrology, currents, stratification, acidification, and phenology.

Introduction

Small, pelagic fish, especially anchovy and sardine, abound in many, productive regions of the world ocean. Their habitats include areas with coastal and oceanic upwelling and freshwater influence and can be characterized by both geography (properties of the coast and bottom) and hydrography (properties of the water). The effects of climate change, be it of natural or anthropogenic origin, on populations of small, pelagic fish, are mediated by their habitats. Our objectives in this chapter are to describe the habitats of the major stocks of small, pelagic fish and comment on the possible effects of climate change on these habitats and, in turn, on the populations. Finally, we present future challenges.

Small Pelagic Fish and Climate Change (SPACC) is a program of International Global Ecosystem Dynamics (GLOBEC) that uses the comparative approach (*cf.* Mayr, 1982) to assess the impact of climate variation and change on ecosystems in which small pelagic fish, particularly anchovy and sardine, play an important role. The regions on which SPACC is focused (Fig. 3.1) include, in the Pacific, the California Current (Canada, US, and Mexico), the Humboldt Current (Peru, Chile), and the Kuroshio-Oyashio Region (Japan), and, in the Atlantic, the Benguela Current (South Africa, Namibia), the Canary Current (Morocco, Western Sahara, Mauritania, and Senegal), and the European Atlantic (Portugal, Spain, and France). Other regions, including the Mediterranean and Baltic Seas, the Gulf of California, waters off Australia, Korea, Taiwan, and China, and certain open ocean regions (e.g. the western tropical Pacific), contain significant populations of small, pelagic fish but have not been a focus of SPACC. Stocks of some of these other regions are treated elsewhere in this volume, but not in this chapter.

Anchovy (*Engraulis*) and sardine (or pilchard: *Sardinops, Sardina,* and *Sardinella*) are the genera that have been studied most within SPACC. Other genera receiving attention have been jack, or horse, mackerel (*Trachurus*), true mackerel (*Scomber*), sprat (*Sprattus*), herring (*Clupea*), and round herring (*Etrumeus*). Interactions, or replacements,

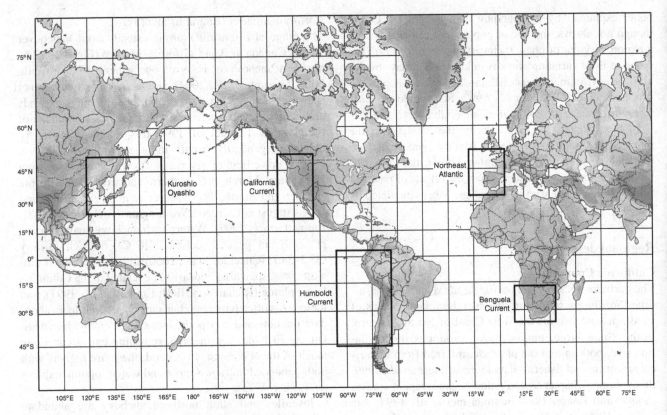

Fig. 3.1. World map showing SPACC regions treated in this chapter.

among these taxa and anchovy and sardine are of particular interest. Genetic studies of anchovy and sardine (Grant and Bowen, 1998; Lecomte *et al.*, 2004) indicate six major regions of speciation of these taxa: the California Current, Humboldt Current, Kuroshio–Oyashio, Australia, Benguela Current, and the European Atlantic. These regions are, in general, synonymous with those of SPACC, with the exception of Australia. These studies show "shallow" evolutionary relationships, indicating recent expansion into current habitats. This, in turn, indicates that habitats of each taxon may share common characteristics. Further, this may allow generalities in regard to the effects of climate change on these taxa.

Anchovy and sardine often differ from one another in regard to their individual characteristics and habitats (Barange *et al.*, 2005; cf. van der Lingen *et al.*, 2005a). In general, anchovy feed on larger particles, have a smaller body (length, weight), and migrate less than sardine. Anchovy are also more often associated with regions of coastal upwelling or freshwater influence, while sardine occur further offshore.

SPACC is unusual among research programs in that all of the populations it studies have been and/or are exploited. In 2004, 25% of world fish landings were small, pelagic fish, including the Peruvian achoveta (*Engraulis ringens*), which supports the world's largest single-species fishery, itself

11% of global landings (http://www.fao.org). In general, these fish inhabit regions of high productivity, often due to nutrients supplied by wind-driven upwelling but also in fresh water, and low diversity, including plankton and fish. This, combined with their planktivory, hence feeding low in the food web, results in productive, single-species populations ideal for exploitation. While these, like all exploited populations, are susceptible to, and have experienced, overexploitation, they may be resilient, due to their fast growth and young age of reproduction and thus high productivity.

Climate change effects on fish populations will be understood only if they are distinguishable from other effects, especially fishing and density dependent processes. To this end, in this chapter we focus on the habitat of fish, as feasible, during all stages of development: egg, larva, juvenile, and adult feeding and spawning. Because our information is from diverse sources worldwide, it is not uniform. Our goal is to assess differences and similarities using this information. We are also particularly interested in comparing anchovy and sardine and the roles of geography and hydrography in their responses to climate change. Here, geography means location in Earth coordinates, e.g. latitude and longitude, and bottom depth. Hydrography refers to currents and water properties, e.g. temperature, salinity, and contained plankton. Variability in the distribution, abundance, and size of populations of pelagic fish may be better explained by hydrography than geography. Often, though not always, the size of populations of these taxa in a region vary out of phase (Schwartzlose *et al.*, 1999). The cause of their variation remains unknown but may involve both the environment, hence habitat-, and density-dependent processes (Lluch-Belda *et al.*, 1992; Schwartzlose *et al.*, 1999; Jacobson *et al.*, 2001; this volume, Chapters 5 and 12).

In this chapter, we first describe the habitats for the major SPACC regions and stocks. We then synthesize these regional descriptions, highlighting similarities and differences. Finally, we briefly address the effects of past and future climate change on small, pelagic fish, as mediated by habitat.

Regional descriptions

California Current (CC)

The northern anchovy (*Engraulis mordax*) and Pacific sardine (*Sardinops sagax*) are the two dominant species of small, pelagic fish off the West Coast of North America. Both exhibit large, multi-decadal variations in population size, both in and out of phase, inferred from historical scientific and fisheries data (Schwartzlose *et al.*, 1999; this volume, Chapter 9) and paleo-oceanographic studies (Soutar and Isaacs, 1974; Baumgartner *et al.*, 1992; this volume, Chapter 4). Data are both fisheries dependent (e.g. landings data during periods of high and moderate population size) and fisheries independent (e.g. ichthyoplankton and adult acoustic-trawl surveys, largely independent of population size).

Physical characteristics

The west coast of North America (Fig. 3.2a), from Vancouver Island, Canada, to Baja California, Mexico, is influenced by the California Current, wind-driven upwelling, and, to a lesser extent, runoff (Hickey, 1979, 1998; Lynn and Simpson, 1987, 1990). The California Current originates off British Columbia and flows south past the US west coast to Baja California. It is of lower salinity (<ca. 33.2) than water of the oceanic N. Pacific to the west and coastal upwelling and the Davidson Current to the east. The latter is a pole-ward-flowing current that is strongest and extends furthest north, beyond Point Conception, in winter; it is subsurface in summer. In spring, NW winds induce upwelling, contributing to the geostrophically balanced California Current. Water upwelled is from the California Undercurrent and typically cool (<15 °C) and high in salinity (> 33.2). Open-ocean upwelling forced by wind-stress curl (Chelton *et al.*, 1982, Pickett and Schwing, 2006), in addition to wind-induced, coastal upwelling, has recently received attention in regard to sardine habitat (Rykaczewski and Checkley, 2008). A nearshore, equator-ward jet can develop in spring, particularly off Washington and Oregon.

Northern anchovy (*Engraulis mordax*)

The range of *Engraulis mordax* extends from Vancouver Island, Canada to Baja California, Mexico (Fig. 3.2b). The northern anchovy consists of one genetic stock (Smith, 2005; Lecomte *et al.*, 2004). However, it has two distinct spawning areas (Richardson, 1981), and its range expands and contracts with increases and decreases in the size of the population (MacCall, 1990; this volume, Chapter 5). Spawning generally occurs in areas with wind-driven, coastal upwelling or river runoff, with peak spawning in January–March off southern and central California (Lasker and Smith, 1977; Moser *et al.*, 1993, 2001) and June–August off Oregon/Washington (Richardson, 1981; Emmett *et al.*, 1997). Waters with anchovy spawning are presently, in general, cold (12–16 °C), of high salinity (33.5–33.7) (Checkley *et al.*, 2000), although off the northwest coast, spawning appears to be related to the Columbia River plume (Richardson, 1981; Emmett *et al.*, 1997), and these areas often have abundant nutrients, resulting in abundant diatoms and copepods, including *Calanus pacificus*. During El Niño, e.g. in 1983, spawning can occur along much of the NW coast. In general, these are regions with both enhanced nutrient supply and water-column stability (cf. Lasker, 1975).

Juveniles and adult northern anchovy are abundant off California, Oregon, and Washington in spring but decline in abundance during the summer, perhaps due to predation. Consistent with variation of its northern limit, the abundance of the northern anchovy off Oregon and Washington exhibits significant interannual fluctuations, and has increased notably in recent years in Puget Sound, Washington. Regular sampling off the Columbia River since 1998 (every 2 weeks at 12 stations at night; Emmett *et al.*, 2006) shows large interannual fluctuations, with peak abundance in April and May and lagging ocean productivity (Peterson and Schwing, 2003) by several years.

Pacific sardine (*Sardinops sagax*)

Pacific sardine is presently more abundant than the northern anchovy and, consistent with past fluctuations, has a large range, extending from Canada to Baja California (Fig. 3.2c) (Parrish *et al.*, 1989). The Pacific sardine consists of three subpopulations (Gulf of California, not further considered; Baja California Sur Inshore, BCSI; and Central California Offshore, CCO; Smith 2005), although these have not been found to be genetically distinct (Smith, 2005, and references therein). During peak population size, the CCO subpopulation is the largest and adults occur from Baja California to Canada. During low population size, the range of the Pacific sardine contracts and the BCSI subpopulation dominates, with the inshore waters off Baja California perhaps being a refuge (Mais, 1974). Spawning and early development of the CCO population occurs primarily in relatively low

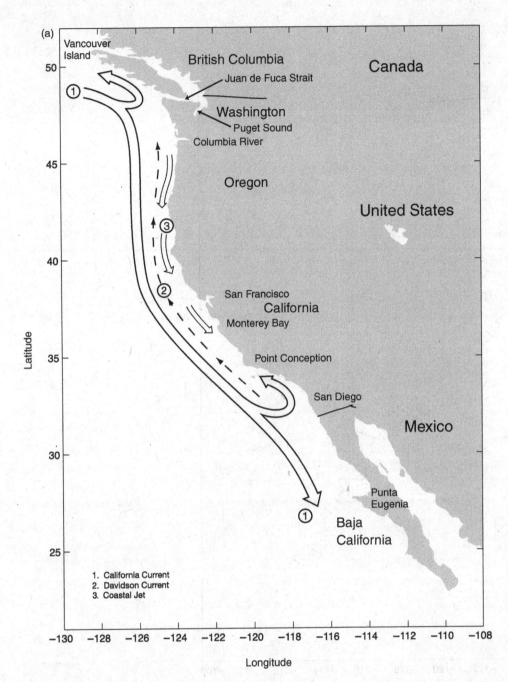

Fig. 3.2. California Current region. (a) Circulation. (b) Distribution of northern anchovy (*Engraulis mordax*). (c) Distribution of Pacific sardine (*Sardinops sagax*). Note that these distributions are the maximal ranges. As discussed in the text, these distributions become smaller at lower population size. (See also this volume, Chapter 5)

salinity water at the inshore edge of the California Current off southern and central California and off Baja California (Parrish *et al.*, 1989; Checkley *et al.*, 2000). Spawning has been noted off Oregon and Washington from 1994–1998 (R. Emmett, unpublished data). Offshore eddies have been hypothesized to be important regions of retention and growth of sardine larvae and juveniles (Logerwell and Smith, 2001). Significant interannual variation occurs in the geographic extent of spawning, extending further offshore during La Niña and being compressed shoreward

during El Niño (e.g. Lynn, 2003). Juvenile and adult Pacific sardine occur off California, Oregon, Washington, and Canada, indicating a northward migration from spawning to juvenile nursery to adult feeding areas. Significant variation in sardine production on the decadal scale is explained by variation in open-ocean upwelling forced by wind-stress curl (Rykaczewski and Checkley, 2008).

Sardine migrate annually from spawning grounds off southern California northward to the Pacific Northwest and Canada and occasionally into southeast Alaska. It is

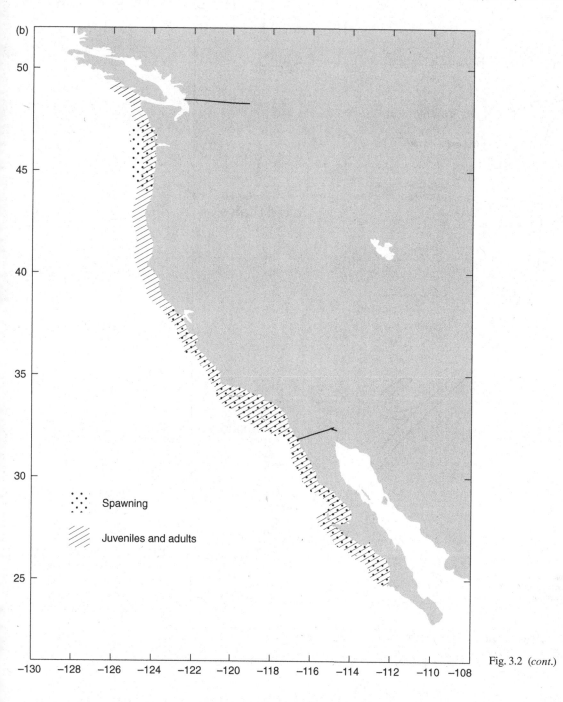

Fig. 3.2 *(cont.)*

generally believed that the northward migration is to spawn and feed, as coastal upwelling in this area results in a ready source of phytoplankton and zooplankton for sardine. Analysis of stomach contents of sardine in Canadian waters indicates an affinity for phytoplankton, primarily diatoms, but there is a definite prevalence of copepods and the eggs and later stages of euphausiids in most years (Emmett *et al.*, 2005; McFarlane *et al.*, 2005).

The mechanism underlying the annual northward movement of sardine is not well understood but the rate of migration may be limited by the rate of northward progression of the local spring and the associated 12 °C isotherm (Emmett *et al.*, 2005; D. Ware, Aquatic Ecosystem Associates, Nanaimo, BC, Canada, unpublished data). During its annual spring and summer northward expansion, the adult sardine population appears to remain relatively offshore but once it reaches the area of the Columbia River and the mouth of Juan de Fuca Strait there is a tendency to move inshore. Sardine generally arrive off the coast of southwest Vancouver Island by late June and are

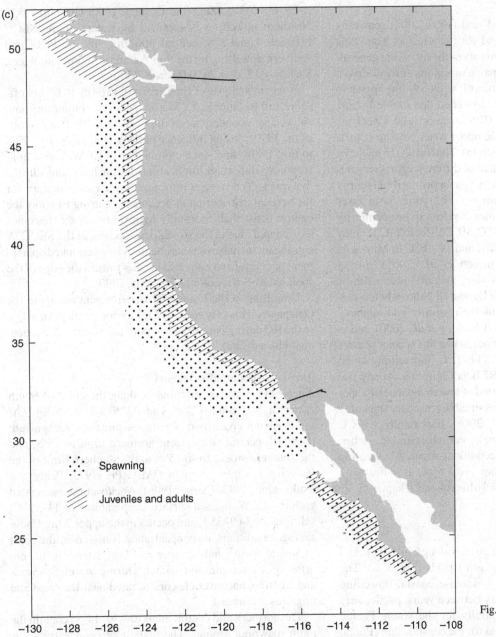

Spawning

Juveniles and adults

Fig. 3.2 *(cont.)*

frequently found in substantial quantities in the inlets of the west coast of Vancouver Island shortly thereafter, possibly due to the elevated sea surface temperatures and high phytoplankton concentrations. McFarlane *et al.* (2005) found that the northern limit of sardines in Canadian waters was broadly related to the sea surface temperature off the west coast of Vancouver Island and northward into Hecate Strait (53–54° N) during the months of June through August. By October, as sea temperatures begin to cool, sardine begin to move southward again, although 0-age sardine do not appear to migrate but stay in coastal waters.

Canadian waters are occasionally conducive to sardine spawning but in general are too cold. During the warm years associated with intense El Niños, such as in 1997 and 1998, limited spawning occurred off Vancouver Island but the fate of the eggs and any resulting progeny is unknown. Although the habitat for sardine becomes unfavorable in the winter and the bulk of the adult sardine population migrates southward again, both juvenile and adult sardine frequently over-winter in inlets of the west coast of British Columbia, at times (e.g. 2005) mixing with schools of Pacific herring (*Clupea pallasi*).

Generalizations

Analyses of both recent and past (1950–2000) spawning of the northern anchovy and Pacific sardine from Baja California to central California allow the following generalizations. Anchovy tends to spawn in regions of wind-driven coastal upwelling and river runoff, while sardine spawn in more oceanic water, perhaps upwelled due to wind-stress curl and Ekman pumping (Rykaczewski and Checkley, 2008). Both taxa spawn in the upper water column (Curtis *et al.*, 2007 and references therein). This results in anchovy and sardine spawning in water of different characteristics and at different times. Anchovy spawn primarily January– March in 12–16 °C, high salinity (> 33.2) water, or in lower salinity water near river plumes. Sardine spawn primarily April–June in water 12–14 °C off California (Checkley *et al.*, 2000; Moser *et al.*, 2001) and 14–16 °C in May–July off Oregon/Washington (Emmett *et al.*, 2005), during periods of large population size, and of lower salinity. Exceptions include spawning during El Niño, when coastally upwelled water is not of high salinity and anchovy spawn in lower salinity water (Checkley *et al.*, 2000), and in periods of low sardine abundance, when this species spawns later in the year (summer) and in 14–21°C, high salinity water, primarily in inshore waters off Baja California. At any one time, particularly during a period of low abundance of a species, not all habitat that appears suitable is used for spawning (cf. Smith, 1990; Reiss *et al.*, 2008). Importantly, the CC system does not exhibit a single, normal state but, rather, variability is the norm, with conditions apparently favoring anchovy, sardine, or both at any given time, as indicated in the paleo-oceanographic record (this volume, Chapter 4).

Humboldt Current (HC)

Physical characteristics

The Humboldt Current system extends along Chile, Peru, and Ecuador, including the Galapagos Islands (Fig. 3.3a). The ecosystem is characterized by intense coastal upwelling and high productivity that vary between years, particularly with ENSO,[1] and multi-decadally for the whole Pacific basin (Chavez *et al.*, 2003) and for the coastal zone (Purca, 2005).

The main surface current flows northward and has two branches: the Peru Current (PC) and the Peruvian Coastal Current (PCC). Between these the Peru–Chile Undercurrent flows southward carrying equatorial waters below 100 m and extending 250 km offshore (Zuta and Guillén, 1970; Morón, 2000). In addition, the Southern Extension of the Cromwell Current (SECC) flows near the coast below 50 m to 300 m, with high oxygen content.

Four types of surface water masses exist: Cold Coastal Water CCW (14–18°C and 34.9–35.0), Surface Subtropical Water SSW (18–27 °C, 35.1–35.7), Surface Equatorial Water SEW (> 20 °C, 33.8–34.8), and Surface Tropical Water

(> 25 °C, < 33.8) (Zuta and Guillén, 1970; Guillén, 1983). Northern upwelling is supplied by waters of the SECC between 4 and 8°S, central upwelling by the PCC, and southern upwelling by the Sub-Antarctic Temperate Water, SATW (< 15 °C, < 34.7) (Morón, 2000).

A recent analysis of time series of salinity to 60 nm offshore and by latitude (O. Morón, personal communication) showed a preponderance of the CCW from 1960 to the start of the 1970s; strong influence of SSW from the early 1970s to mid 1980s; and since, dominance of CCW. These findings have important implications for anchovy and sardine dynamics. The oxygen minimum layer is a boundary for the vertical distribution of organisms. During El Niño, the central coast shelf, normally hypoxic or anoxic, becomes oxygenated, due to the strong development of the SECC. A significant increase of macrobenthos biomass and displacement of coastal and demersal species toward the edge of the shelf have been reported (Tarazona, 1990).

Upwelling of the Equatorial Undercurrent occurs in the Galapagos. This may serve as a refuge for small, pelagic fish of the HC during periods of low population size (M. Niquen, unpublished data).

Anchoveta (Engraulis ringens)

Anchoveta is widely distributed along the coast of South America between 04°30 S and 42°30 S (Fig. 3.3b). Off the Peruvian coast, two population units are recognized: the north-central stock, from northern limit to 15° S, and the southern stock, from 15° S to the southern limit of the Peruvian maritime domain (IMARPE, 1973; Pauly and Tsukayama, 1987; Csirke, 1996). Anchoveta is associated with the CCW and sea surface temperatures of 14–22 °C, salinities of 34.9–35.1, and occurs in the upper 70 m. Under average conditions, its concentration is high near the coast in austral spring and summer and low, extending 100 nm offshore, in autumn and winter. During warming events and El Niño, anchoveta is concentrated near the coast and migrates southward.

The northern and central areas (04°30–14° S) contain the main spawning ground. The highest egg concentration is found in the upper 30 m (Santander and de Castillo, 1973). The spawning area is related to the influence of the CCW with salinities between 34.9 and 35.0.

Three levels of carrying capacity for anchoveta have been proposed (Csirke *et al.*, 1996): the highest in the 1960s until the collapse in 1972, the lowest from 1973 to 1983, and an intermediate level from 1984 to 1994, updated to 2001 by Ayón *et al.* (2004). During the 1990s, anchoveta had five years (1992–96) of high abundance similar to those of the 1960s, but since then has remained at an "intermediate" level.

Different factors affect spawning and recruitment. El Niño is the most striking of these, impacting the

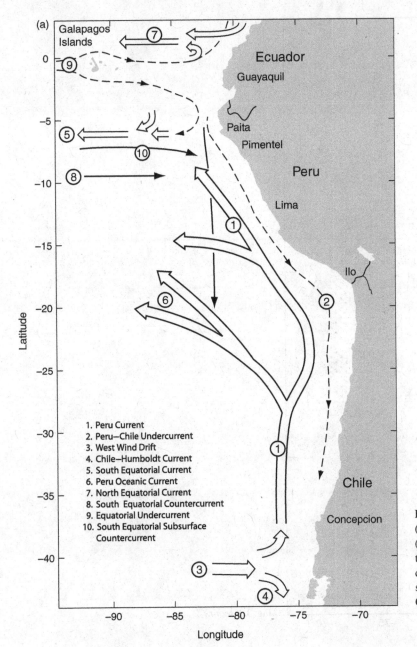

(a)

1. Peru Current
2. Peru–Chile Undercurrent
3. West Wind Drift
4. Chile–Humboldt Current
5. South Equatorial Current
6. Peru Oceanic Current
7. North Equatorial Current
8. South Equatorial Countercurrent
9. Equatorial Undercurrent
10. South Equatorial Subsurface Countercurrent

Fig. 3.3. Humboldt Current region. (a) Circulation. (b) Distribution of anchovy (*Engraulis ringens*). (c) Distribution of sardine (*Sardinops sagax*). Note that these distributions are the maximal ranges. As discussed in the text, these distributions become smaller at lower population size. (See also this volume, Chapter 5).

spawning process and eventually changing the main peak from winter–spring to summer and by lowering its intensity (Santander and Zuzunaga, 1984). Long-term changes have also been observed in the extension of spawning areas with increasing abundance. However, after the 1997–98 El Niño, the areas widened and the center of the distribution shifted northward compared with the 1960s. Another factor is the effect of the fishery on spawning biomass and/or recruits.

The feeding and food environment of the anchoveta has been studied for decades. Anchovy larvae feed mainly on

phytoplankton, with flagellates as the main item of the diet. As larvae grow, their preference changes to zooplankton (Muck, 1989). Studies of stomach contents of postlarval (juvenile and adult) anchoveta collected during 1953–1982 show variation with latitude and distance from the coast. In the north-central region near the coast, it feeds more on phytoplankton but, further offshore, more on zooplankton. In the southern region it feeds more on zooplankton, and this also occurs at a higher temperature (Pauly *et al.*, 1989), probably because of warming associated with a change of water mass. This spatial strategy can help to explain the

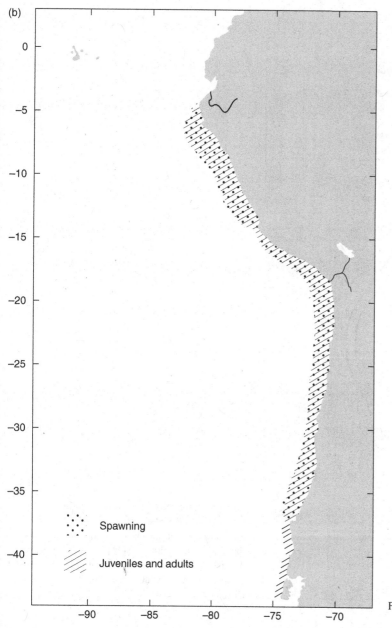

Fig. 3.3 (*cont.*)

observed long-term variability of zooplankton volumes. In the 1960s, zooplankton volumes were high but diminished with decreasing anchoveta abundance. By the end of the 1980s, both started to recover, but this was more marked in anchoveta than in zooplankton (Ayón *et al.*, 2004). Recent analysis suggests that, after the 1980s, anchoveta has been feeding more on zooplankton (P. Espinoza, Instituto del Mar del Perú, personal communication), which could explain why zooplankton abundance has not recovered to the levels observed in the 1960s.

Anchoveta is the prey of higher trophic level species like mackerel and jack mackerel (Muck and Sánchez, 1987;

Pauly *et al.,* 1987); demersal species like hake, though mainly during warm periods (Sánchez *et al.*, 1985; Muck, 1989); seabirds like cormorants, boobies and pelicans (Jahncke and Goya, 1998); and sea mammals (Muck and Fuentes, 1987). When sardine was abundant it preyed on the first life stages of anchoveta. Adult anchoveta cannibalize their own eggs, especially when the population is large (Santander, 1987). Pauly and Soriano (1989) considered that during warm periods, anchoveta eggs and larvae have a higher mortality rate because of the increase of predation by zooplankton and taxa of other trophic levels, including anchoveta.

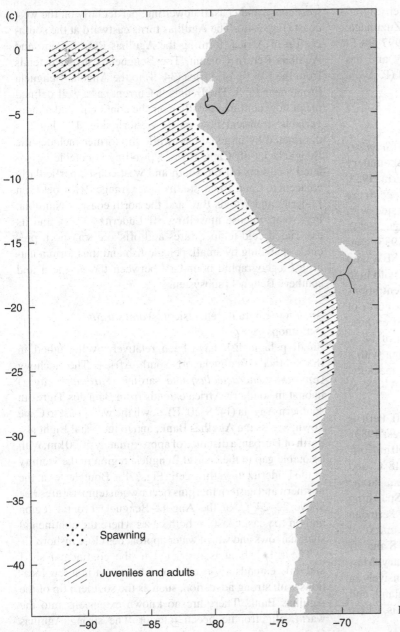

Fig. 3.3 (*cont.*)

Sardine (Sardinops sagax)

Sardine has a wide distribution in the SE Pacific, including off Peru in SSW with temperatures of 19–22 °C and salinities of 35.0–35.3 (Santander and de Castillo, 1981) (Fig. 3.3c). In general, the adult distribution is far offshore of the intense upwelling zone, although juveniles are distributed in the same areas as anchoveta. This species extends its area of distribution during warm periods and El Niño, as in 1982–83, when it occurred near the coast in high densities and in the southern region, because of the southward advection of SSW. Sardine occur as far south as 42° S off Chile (Parrish *et al.*, 1989).

The increase of the sardine population started after the decline of anchoveta in the start of the 1970s. The period of high abundance coincided with the period of major influence of SSW in the first 60 nm offshore, probably favoring good recruitment. By the start of the 1990s, sardine entered into a new period of low abundance (Csirke *et al.*, 1996) that coincided with less SSW near the coast.

The spawning area of sardine varies greatly with distance from the coast and with latitude. Between 1966 and 1971, spawning was poor and restricted to the northern areas. It expanded southward after the anchoveta decline. Major spawning concentrations occurred in SSW

(Santander and Flores, 1983) near the coast after migrating from the equatorial zone (Santander and Zuzunaga, 1984; Arntz and Fahrbach, 1996). After the 1997–98 El Niño, sardine spawning diminished drastically and has been sparse and restricted to areas off Pimentel (P. Ayón, unpublished data).

Jack mackerel *(Trachurus murphyi) and mackerel (Scomber japonicus)*

Jack mackerel (*Trachurus murphyi*) and mackerel (*Scomber japonicus*, also called chub mackerel) are transboundary species widely distributed off Peru and Chile (Serra, 1983; Arcos *et al.*, 2001 and references therein). Both species are frequently caught in the same areas and associated with SSW (Tsukayama, 1982; Santander and Zuzunaga, 1984), although other authors (Santander and Flores, 1983) consider that mackerel is also associated with the SEW and jack mackerel was probably related to the SECC with high oxygen content. Before the 1990s, the major concentrations off Peru were found north of 11°S, and mainly north of 7°S, between depths of 50 and 150 m. However, after those years, their concentrations were restricted south of 14°S. Mackerel has been reported off Chile at 45°S, but with a typical southern boundary of 25°S, while jack mackerel has been observed at 52°S, with a large fishery at 35–38°S, off Talcuhuano (Serra, 1983).

The main spawning area of jack mackerel off Peru is the frontal zone of 14–18°S (Santander and Flores, 1983), where upwelling CCW and SSW meet 100–150 nm offshore, with surface temperatures higher that 18°C and oxygen content higher than 5 ml l⁻¹. In this area, larvae are displaced offshore by the Ekman transport. Spawning has also been observed in other places off the Peruvian coast but with low concentrations of eggs. Jack mackerel occurs off Chile between the Peru border and 25°S and off Talcuhuano (Serra, 1983). Spawning occurs January–May. Jack mackerel is known to migrate to spawn extensively in oceanic waters off Chile, as far west as New Zealand, and return to the coastal nursery and adult feeding areas, where the fishery occurs (Arcos *et al.*, 2001).

Mackerel spawn between August and March with a peak in the austral summer and, under average conditions, off the north-central region of Peru. During El Niño, spawning increases and higher concentrations of eggs and larvae occur (Santander and Flores, 1983), probably associated with SSW. Mackerel spawn off Chile primarily near the border with Peru and, less so, off Talcahuano (Serra, 1983).

Benguela Current (BC)
Physical characteristics

The waters off southern Africa are dominated by the southward-flowing Agulhas Current on the east and south coasts, the northward flowing Benguela Current on the

west coast, and coastal upwelling, particularly on the west coast (Fig. 3.4a). The Agulhas turns eastward at the southern tip of Africa, forming the Agulhas Retroflection and Agulhas Return Current. The Benguela Current extends from the SW tip of Africa (34°S) to the Angola–Benguela Front, near 16°S. The Benguela Current has a well-defined mean flow confined mostly near the continent, and a more variable, transient flow on its western side. The latter is dominated by large eddies, while the former includes the Benguela Jet off Cape Town. Upwelling is notable in isolated locations off the south and west coasts, particularly adjacent to Cape Town, south of the Orange River, between Lüderitz and Walvis Bay, and the north coast of Namibia. Persistent, intense upwelling off Lüderitz (27°S), and its associated cold temperatures and offshore transport, preclude spawning by small, pelagic fish and thus form a natural, biogeographic boundary between the northern and southern Benguela subsystems.

Anchovy *(Engraulis encrasicolus) and sardine (Sardinops sagax)*

Small pelagic fish have been relatively well studied in the southern Benguela off South Africa. The anchovy (*Engraulis encrasicolus*) and sardine (*Sardinops sagax*) habitat in southern Africa extends from Baia dos Tigres in southern Angola (14°S, 10°E), down the west coast to Cape Town, across the Agulhas Bank, and to the Natal Bight just north of Durban, a distance of approximately 5000 km, with a notable gap in the central Benguela region in the vicinity of the Lüderitz upwelling cell (Fig. 3.4b). Boundaries at the northern and eastern margins occur when temperatures rise above 22–24 °C, at the Angola–Benguela Frontal region and on the east coast, in both cases where the continental shelf narrows and warm water approaches close inshore.

The fish habitat is restricted to the continental shelf and only extends a short distance beyond it in a few locations with strong advection, such as the southern tip of the Agulhas Bank. There are no known extensions into the warm, oligotrophic ocean interior. The strong Agulhas Current on the east coast, the jet current on the west coast, and low phytoplankton concentrations offshore on both coasts largely limit the habitat to the continental shelf. Few epipelagic fish occur in the vicinity of the powerful upwelling cell at Lüderitz, where cold water, high turbulence, and offshore transport prevail.

Elsewhere, the cool water coincides with enriched phytoplankton concentrations (Demarcq *et al.*, 2007). This extends far offshore near the Angola–Benguela Front, associated with strong offshore displacement. The width of the zone of enriched plankton decreases at Lüderitz, then widens on the South African west coast as far as 33°S before narrowing again. While surface concentrations of chl *a* decline on the Agulhas Bank, there are extensions offshore

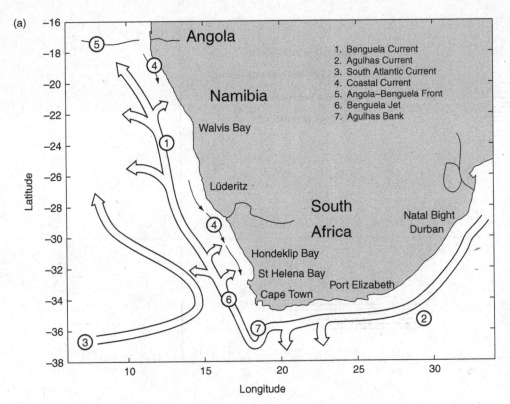

(a)

1. Benguela Current
2. Agulhas Current
3. South Atlantic Current
4. Coastal Current
5. Angola–Benguela Front
6. Benguela Jet
7. Agulhas Bank

Fig. 3.4. Benguela Current region. (a) Circulation. (b) Distribution of anchovy (*Engraulis encrasicolus*). (c) Distribution of sardine (*Sardinops sagax*).

associated with a subsurface ridge at 22°E and at 26°E, where the Agulhas Current diverges from the coastline. A narrow belt of cool, phytoplankton-rich water extends up the east coast in winter, with a spawning microhabitat, consisting of a wider shelf and upwelling in the Natal Bight (Lutjeharms *et al.*, 2000) in the extreme east. The east coast is the location of the famous "Sardine Run," which occurs in June–July each year and comprises about 3%–13% of the total biomass in the southern Benguela region.

Mesozooplankton concentrations are moderate to high throughout the Benguela upwelling system and peak on the west coast near St Helena Bay (33° S) and just downstream (NW) from the Lüderitz upwelling cell at 23–25° S (Hutchings *et al.*, 1991) with declining levels in warmer water of the boundary zones. Pelagic fish therefore migrate against food gradients to spawn in locations upstream of the main nursery grounds. These costly migrations may account for the poor fish yield relative to primary productivity in the Benguela relative to the Humboldt Current system (Hutchings, 1992; Ware, 1992).

Anchovy eggs are found in spring and early summer predominantly over the Agulhas Bank between Cape Point and Port Elizabeth (Fig. 3.4b), although they are reported from the east coast as far north as Durban and also off the west coast during years of anomalously warm water (van der Lingen and Huggett, 2003). An eastward shift in anchovy egg (and spawner) distributions from being predominantly

to the west of Cape Agulhas to predominantly to the east occurred in 1996 and has persisted since (van der Lingen *et al.*, 2002; Roy *et al.*, 2007). Sardine eggs occur around the South African coast from Hondeklip Bay (32° S) to Durban, principally during summer off the west and/or south coasts but also in winter off the east coast during the annual migration known as the sardine run (Connell, 2001) (Fig. 3.4c). Both the west and the south coasts have been the dominant site of summer spawning by sardine during different periods, with a recent shift from the west coast over the period 1994–2000 to the south coast (to the east of Cape Agulhas) from 2001 onwards (van der Lingen *et al.*, 2005b). Winter spawning off the central Agulhas Bank may be linked to a phytoplankton surface maximum in the austral autumn (March to June) (Demarcq *et al.*, 2007).

Spawning habitat for both species has been characterized in terms of a variety of environmental variables, both physical (e.g. SST, salinity, water depth, mixed layer depth and current speed) and biological (e.g. phytoplankton and zooplankton biomass and production) using co-inertia analysis, single parameter quotient (SPQ) analysis, and temperature–salinity plots (Twatwa *et al.*, 2005; van der Lingen, 2005). Those studies indicate that anchovy egg habitat is characterized by a narrower range for many environmental variables than is sardine egg habitat, and that anchovy eggs are found in waters of 17–21 °C and sardine at 15–21 °C, with other variables being roughly similar for both species.

(b)

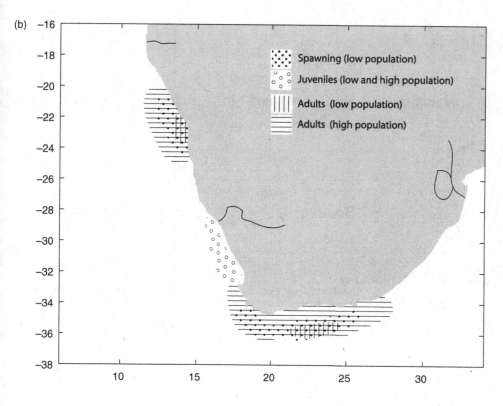

Fig. 3.4 (*cont.*)

(c)

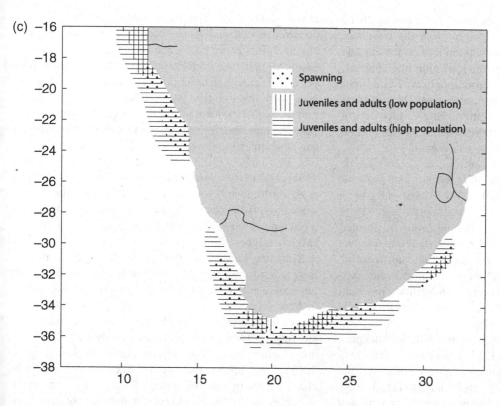

Fig. 3.4 (*cont.*)

Anchovy spawning was significantly related to temperature, salinity, phytoplankton and zooplankton biomass, but not to current speed, wind speed, mixed-layer depth, and zooplankton production. Sardine spawning was significantly related to water depth and zooplankton biomass, but not to other variables. The recent shift eastwards in spawning habitat of sardine was accompanied by an upward shift in spawning temperature to 22–24 °C in November (summer). Minimum surface temperature thresholds for spawning of 13–14 °C characterize the west coast between 22° S and 33° S, which, combined with strong offshore advection of buoyant eggs, preclude this area as a spawning habitat. However, the broad shelf on the west coast of South Africa is a suitable nursery area for juveniles of both species, indicating tolerance to both warm and cool waters at different life stages.

Eggs and resultant larvae of both species spawned on the South African south coast are transported to the west coast nursery area by a shelf-edge jet current (van der Lingen and Huggett, 2003). Anchovy larva size increases progressively both alongshore (equatorward) and cross-shelf, indicating transport at the shelf edge and shoreward movement of larvae as they grow. The shoreward movement of pre-recruits to the nearshore habitat against the predominant offshore Ekman drift appears to be via active transport through swimming and/or passive transport in surface waters during a relaxation of upwelling (Parada *et al.*, 2008), as both active and passive transport are required to move young fish shorewards in sufficient numbers to account for observed recruitment. Active directed swimming by a 20 mm larva at one body length s^{-1} for 12 hours day^{-1} would require 105 days to swim the 180 km from offshore of the continental shelf to the nearshore region (van der Lingen *et al.*, 2006). A decline in upwelling-favorable wind stress in the austral autumn (March–May) may facilitate this movement, and whilst vertical migration of larvae to below the Ekman layer should minimize offshore transport during strong southerly winds in summer months, the modelling studies of Parada *et al.* (2008) did not support this hypothesis. Single-parameter quotient analysis for pre-recruits and early juveniles of 20–40 mm SL (van der Lingen, unpublished data) indicates that sardine are found in warmer waters than are anchovy, offshore on the west coast and on the Agulhas Bank.

The juvenile habitat was most recently described by Barange *et al.* (1999). Recruits of both species are found primarily off the west and southwest coasts, although some south coast recruitment of sardine has been observed, but generally much less than that observed off the west coast (Coetzee *et al.*, 2008). Juveniles migrate, against a food gradient, to the south coast, where they metamorphose and, at the end of their first (anchovy) or second (sardine) year, spawn. Low oxygen occurs in a bottom mixed layer close

inshore along the entire west coast, reaching a maximum from February to May each year, but it does not appear to geographically restrict the habitat of pelagic fish, as they occupy the well-oxygenated and productive upper mixed layer and the pycnocline.

The geographic characterization of adult distribution in the southern Benguela was most recently performed also by Barange *et al.* (1999). Anchovy are found over the Agulhas Bank off the south coast whereas adult sardine have been found off both the south and west coasts during different periods. There has been a noticeable eastward shift in distributions of anchovy and particularly sardine over the past decade (van der Lingen *et al.*, 2005b; Fairweather *et al.*, 2006; Roy *et al.*, 2007; Coetzee *et al.*, 2008).

Much less is documented on the life history stages of pelagic fish off Namibia. There have, until recently, been few studies of the early life history stages of pelagic fish in the northern region (e.g. Stenevik *et al.*, 2001). Despite the massive decline in biomass of both species compared to historical population sizes (Griffiths *et al.*, 2004), both sardine and anchovy still spawn in localized areas in similar habitats, although within a much smaller range and with much lower densities than previously observed. Spawning of both species occurs in summer and autumn towards the Angola–Benguela frontal area at 15–20° S at 19–21 °C, and older sardine further southwards on the central Namibian shelf at 20–23° S and during austral spring in cooler upwelled water of 14–17 °C. Juveniles of both species occur close inshore in the central–northern Namibian shelf (18–22° S). Low dissolved oxygen concentration influences the distributions of anchovy and sardine eggs and larvae (Kreiner *et al.*, 2009), and may impact their recruitment success (Ekau and Verheye, 2005). A combination of a decaying algal blooms and accumulated oxygen debt can result in hydrogen sulfide eruptions which result in mass mortalities, although pelagic fish are less affected than rock lobster or inshore species such as mullet (Bakun and Weeks, 2004). The collapse of the Namibian sardine population, together with the eastward shift of southern Benguela sardine to the Agulhas Bank, has meant that there is a dearth of sardine in the entire cool-water upwelling regime of the Benguela system from Cape Point to Cape Frio, despite high phytoplankton and zooplankton levels and productivity.

Population changes are dramatic in the Benguela system, with contrasting trends in the northern and southern populations (this volume, Chapter 9). In the southern Benguela, peak annual catches of sardine (400 000 tonnes) occurred in 1961–2, with a VPA^{2}-estimated biomass of 2 million tonnes (Butterworth, 1983). As sardine declined, emphasis shifted to anchovy and catch levels were maintained at 300 000–400 000 tonnes for the next 30 years, but dominated by anchovy. Sardine recovered in the period 1984–2003, peaking at approximately 4 million tonnes biomass. Anchovy fluctuated through the 1990s, but good recruitment in 2000 and 2001 led to record high population levels of both species. In the northern Benguela, sardine dominated the catches, peaking at 1.3 million tonnes, from a VPA-estimated population biomass of 6–7 million tonnes (Butterworth, 1983). A steep decline saw a shift to target anchovy and juvenile horse mackerel and a midwater fishery for horse mackerel developed. Fishing effort was directed at anchovy, perceived to be a direct competitor of sardine, but after a few years anchovy biomass and catches decreased to low levels and horse mackerel became the dominant planktivore in the northern Benguela. Sardine started improving in the early 1990s, but a combination of a huge low oxygen event in 1994, a Benguela Niño in 1995, and sustained warming in the northern Benguela has continued to suppress sardine biomass in the northern Benguela, in stark contrast to the southern Benguela (Boyer and Hampton, 2001; Boyer *et al.*, 2001). There was some speculation that sardine moved into the southern Benguela following poor environmental conditions in the north, but no fish were detected at intermediate localities during the past decade. Ironically, some fish were caught in southern Angola in Namibian purse seines in 1994 and 1995, indicating a portion of the stock was still present in the northern portion of the Angola–Benguela frontal region, despite the warm water. Most of the fish were shifted southwards several degrees of latitude, making them more available to the purse-seine fleet based at Walvis Bay (Boyer *et al.*, 2001). Speculation about the role of large carnivorous jellyfish in predation of fish eggs and larvae continues, but no historical time series of jellyfish abundance exists. Alternatively, predation by horse mackerel on sardine eggs and larvae and enhanced mortality of adult sardine from other predators, such as seals, may be keeping sardine repressed, when it is limited in its range, in addition to the warm water effects of diminished plankton on the nursery grounds. Fishing pressure has been severely limited in Namibia, yet despite these low levels of effort, fishing mortality may still be a significant part of the total mortality of sardine, which has still not recovered.

Northeast Atlantic (NA)

Physical characteristics

The Northeast Atlantic (European Atlantic and Canary Current) is a more complex oceanographic region than other major eastern boundary systems, mainly due to the greater irregularity of the European and northwest African coastline and bathymetry (Fig. 3.5a). The largely meridional orientation of the coast is interrupted by significant zonal stretches in the Bay of Biscay and the Gulf of Cadiz. Also, the Straits of Gibraltar form a salient topographic feature where dense Mediterranean water enters from a shallow sill (ca. 300 m) in the Gulf of Cadiz and rapidly sinks

to below 1000 m with a profound impact to slope dynamics and regional circulation. The dominant offshore sources of water in the NA are two eastward flowing basin-scale currents, both transporting North Atlantic Central Water (NACW, 8–18 °C, 35.2–36.7): the North Atlantic Current (NAC) at the north of Iberia (48–53° N), with a branch flowing into the Bay of Biscay, and the Azores Current (AC) at the south of Iberia (34–35° N), with its northerly edge affecting coastal circulation off western Iberia and its southerly component feeding the Canary Current. Some exchange between the two (NAC and AC) is enabled by the broad, slow, generally southward-flowing Portugal Current. Cape Blanc (21° N), on northwest Africa, has NACW to the north and South Atlantic Central Water (8–18 °C, 34.7–36.3) to the south (Sverdrup *et al.*, 1960). These features are major contributors to a complex and variable circulation system onto which are superimposed multi-scale seasonal variations in atmospheric forcing, heating, and input of buoyancy through river discharges (e.g. Hernandez-Leon *et al.*, 2007 and references therein; Relvas *et al.*, 2007).

Given that sardine and anchovy in the NA are almost exclusively constrained within the continental shelf (with the bulk of both species in the Iberian Peninsula being found at the inner shelf and only juvenile anchovy in the Bay of Biscay having a clearly oceanic phase), shelf area and coastal circulation are important physical characteristics of their habitat. Shelf width is generally narrow off the Iberian Peninsula but becomes progressively wider at the French coast. Off western Iberia, the Iberian Poleward Current (IPC), subsurface and density driven, is a dominant feature of mesoscale circulation outside the summer upwelling season, flowing northwards along the shelf edge with warmer, less saline and oligotrophic water (Peliz *et al.*, 2005). Outside the season of upwelling, the IPC has been shown to interact with the Ekman layer, enriched by river runoff off northern Portugal, creating a barrier to offshore advection of eggs and larvae (Santos *et al.*, 2004). During the upwelling season (April–September), the surface signature of the IPC is lost but there are recent indications that it may persist as a subsurface poleward current (Relvas *et al.*, 2007). Off northern Spain, the shelf is narrow and circulation controlled by local features, such as small river plumes, local wind-induced upwelling, and capes, bays, and deep canyons. The main hydrographic feature on the shelf is the variable extension of the IPC off northern Spain, creating fronts parallel to the coast, reducing cross-shelf fluxes, and fronts perpendicular to the coast as the IPC advances eastwards off the north Iberian coast (González-Quirós *et al.*, 2004; Llope *et al.*, 2006). Off France, the shelf hydrography is predominantly under the influence of river runoff from the Loire and Gironde but also from the Adour, which flows in front of the Cap Breton Canyon, an important area for anchovy and sardine

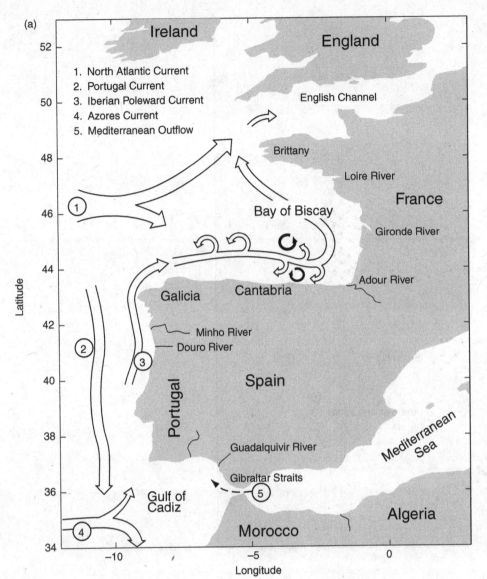

(a)

1. North Atlantic Current
2. Portugal Current
3. Iberian Poleward Current
4. Azores Current
5. Mediterranean Outflow

Fig. 3.5. Northeast Atlantic region. (a) Circulation. (b) Distribution of anchovy (*Engraulis encrasicolus*). (c) Distribution of sardine (*Sardina pilchardus*). Note that these distributions are the maximal ranges. As discussed in the text, these distributions become smaller at lower population size.

spawning. Wind-driven upwelling partially controls primary production and offshore transport along the coast north of the Adour river in spring. The IPC, which flows in winter along the northern coast of Spain, turns counterclockwise in the south-western part of the Bay of Biscay and flows northward (Pingree and Le Cann, 1990). At the shelf break, large amplitude internal waves can result in the vertical pumping of sub-pycnocline water, which can reach the surface. These colder and productive waters can be directly observed via satellite, for they generate frontal structures at the surface (Bardey *et al.*, 1999). Off the shelf, hydrography is characterized by intense mesoscale activity, with cyclonic and anticyclonic eddies (Pingree and Le Cann, 1992; van Aken, 2002). This ensemble of oceanic features results in a mosaic of potential habitats for

anchovy and sardine in the Bay of Biscay (Koutsikopoulos and Le Cann, 1996).

Bay of Biscay anchovy (Engraulis encrasicolus)

The main spawning areas of the Bay of Biscay anchovy are the Gironde and Adour river plumes, the smaller but abundant Cantabrian river plumes, shelf edge fronts, and oceanic eddies (Fig. 3.5b) (Motos *et al.*, 1996, 2004; ICES 2004; Sagarminaga *et al.*, 2004; Bellier *et al.*, 2007; Ibaibarriaga *et al.*, 2007). All these are recurrent features where the potential for high biological production exists (Valencia *et al.*, 2004). Spawning starts at low rates in the southeast corner of the Bay of Biscay and extends to most of the southeast Bay of Biscay, with two consistent centers: the coastal region in front of the Gironde estuary and the shelf,

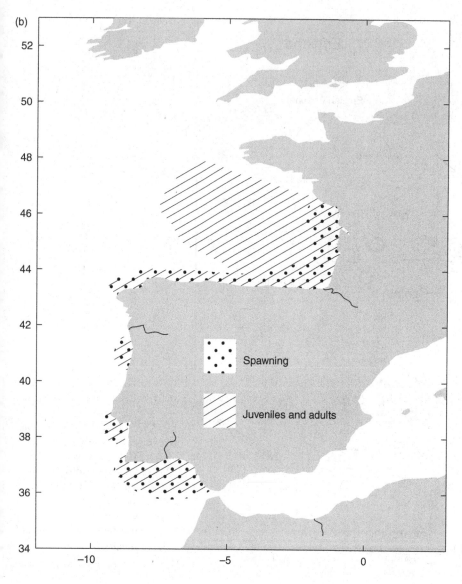

Fig. 3.5 (*cont.*)

shelf break and oceanic regions in the southernmost Bay. In June, the coastal and oceanic spawning regions appear well separated. During the summer, spawning is scattered all along the shelf and close to the shelf break in northern areas. Moreover, some authors have suggested that coastal and oceanic regions are used by different components of the anchovy population. One-year-old anchovy spawn in coastal areas, whereas two- or more year old individuals are in oceanic areas close to or beyond the shelf (Massé, 1996; Motos *et al.*, 1996; Uriarte *et al.*, 1996; Vaz and Petitgas, 2002). Macroscopic maturity studies conducted by Lucio and Uriarte (1990) showed that smallest anchovy reach maturity slightly later than bigger individuals, consistent with younger fish spawning in coastal areas and slightly later than older fish at the shelf break and oceanic areas.

The spawning season of the Bay of Biscay anchovy population begins in mid March, triggered by the warming of surface waters, and extends to August with a peak in May and June, when the maximum rate of warming and the onset of stratification occur (Motos *et al.*, 1996).

Irigoien *et al.* (2005) characterized the environmental and hydrographic conditions of the realized spawning habitat of anchovy using single quotient analysis and bivariate plots. The main variables used for those analyses were sea surface temperature (SST) and sea surface salinity (SSS). Anchovy eggs are present in warm waters ($15–19\,°C$) with low salinity (<35). Instead of considering the absolute values of SST and SSS, the anomalies of SST and SSS indicate a preference for a combination of both variables with respect to the average situation: either warmer waters with lower salinities or colder water with higher salinities. These two situations may represent the two main spawning sites: the Gironde river plume in the north and the shelf break and oceanic regions in the south, respectively. Ibaibarriaga

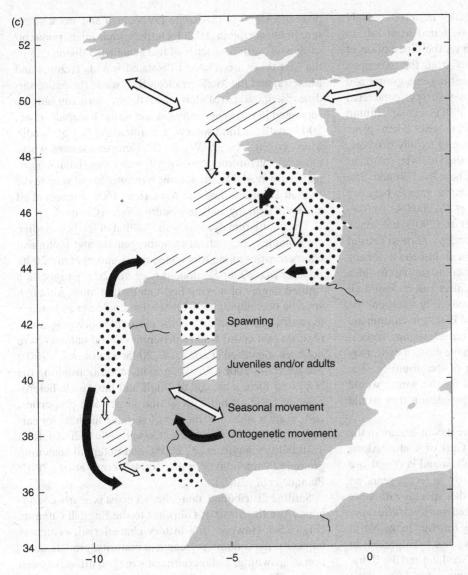

(c)

Spawning

Juveniles and/or adults

Seasonal movement

Ontogenetic movement

Fig. 3.5 (*cont.*)

et al. (2005) use generalized additive models (GAMs), including a bivariate smoothing of SST and SSS, to model the realized egg production in space. Planque *et al.* (2007) go a step further and uses GAMs on presence/absence of CUFES (Continuous Underway Fish Egg Sampler; Checkley *et al.*, 1997) egg abundance to predict the anchovy potential spawning habitat as a function of environmental variables derived from hydrological model simulations. The best predictor was bottom temperature, followed by surface temperature and mixed-layer depth. Surface and bottom salinity were not significant. Even if potential habitat can be characterized in terms of environmental and hydrographic variables, its relationship to realized and successful habitats will depend on additional factors, such as the population abundance, and the age structure and the environmental and hydrographic conditions after spawning. For instance, it is well known that the anchovy spatial

distribution changes depending on the overall level of the population. In years with high adult abundances, eggs cover most of the sampling area, going well beyond the habitat predicted from hydrological conditions, whereas in years with low adult abundance eggs are found in discrete aggregations restricted to the more favorable sites. Uriarte *et al.* (1996) and Somarakis *et al.* (2004) show a linear relationship between the positive area and the total spawning biomass for anchovy. Similar relationships have been observed for anchovy and sardine in the CC (Reiss *et al.*, 2008).

After spawning, anchovy larvae first appear in April and are generally present until August. In June, young larvae are found in the middle French shelf, whereas older larvae are found in the southeastern area, on the Cantabrian shelf, and in oceanic waters. This age gradient could be due to advection, due to NE winds, preferential areas, or their combination (Cotano *et al.*, 2008; Irigoien *et al.*, 2008).

Larvae metamorphose into juveniles at approximately 30–40 days of age and these can be found from July to November. Until 1998, information on the distribution of juvenile anchovy was obtained only from the fishermen that use them as live bait for tuna fishing in summer and early autumn. However, since the pilot surveys conducted in 1998 and 1999 (Uriarte *et al.*, 2001), regular autumn surveys targeting anchovy juveniles have taken place (ICES, 2005). Early juveniles are found mainly offshore, beyond the Cantabrian and French shelves, whereas older juveniles are found on the shelf and near the Gironde. The lack of a significant difference in larva growth between areas (Cotano *et al.*, 2008; Irigoien *et al.*, 2008) is consistent with the hypothesis of Uriarte *et al.* (2001) that larvae are advected off the shelf and juveniles return to coastal zones to overwinter. Data from the commercial fishery (Uriarte *et al.*, 1996 and references therein) support the idea that in October and November juveniles can be found all along the Cantabrian and French coasts. In addition, the hydrological conditions of the Bay of Biscay in autumn are characterized by downwelling, high turbulence and vertical mixing which, when combined with the eastern drift, may facilitate homing during migration of the juveniles. The juveniles moving towards the coast for the winter would comprise the new recruits to the population that would spawn the next spring.

Anchovy occur elsewhere than the Bay of Biscay in the NA. There are populations in the Gulf of Cádiz. Along the Spanish Cantabrian coast, to Galicia and Portugal, are areas where anchovy appear irregularly. In recent years, an increase of anchovy, together with other species with more southern affinities, like sardine and red mullet (*Mullus surmeletus*), has been detected in survey trawling in the North Sea (Beare *et al.*, 2004a,b). Anchovy has been present in the North Sea up to the east coast of Scotland and the Baltic in the past (Calderwood, 1892; Cunningham, 1890, 1895, 1896), with spawning and supporting fisheries on the Dutch coast (Cunningham, 1896). The recently detected increase (Beare *et al.*, 2004a,b) is associated with the regional increase in sea temperatures hypothesized to be associated with climate change.

*European Atlantic sardine (*Sardina pilchardus*)*
Sardine is widely and continuously distributed along the eastern Atlantic shelf from Mauritania to the English Channel (Parrish *et al.*, 1989), with occasional occurrences along Senegal in the south (Fréon *et al.*, 1979) and off the coast of Scotland in the north (Beare *et al.*, 2004a) (Fig. 3.5c). Sardine is also found in all oceanic eastern Atlantic islands, including the Azores, Madeira and the Canary Islands, as well as within most of the Mediterranean coastal habitat, with the exception of the southwest corner. Within this broad distribution, recent results from the European Union

project SARDYN (SARdine DYNamics and stock structure in the northeast Atlantic, http://ipimar-iniap.ipimar.pt/sardyn) show the existence of five genetically distinct units: the first three are related to isolated islands (Azores and Madeira) and the Mediterranean Sea, while the entire sardine distribution area along the Atlantic coast can only be separated in two genetically distinct units (Kasapidis *et al.*, 2004), with the limit somewhere around the Bay of Agadir, Morocco (30.3° N, 9.6° W). Within European waters, where more data on sardine population dynamics and habitat occupation are available, the species is mainly found around the Iberian Peninsula and the Armorican (NW France) shelf and up to Brittany and the western English Channel.

The population dynamics and habitat of the NA sardine are complex. A gradient in some genetic and biological characteristics along this wide area (Laurent *et al.*, 2006; Silva *et al.*, 2006; Stratoudakis *et al.*, 2007) suggests a limited degree of mixing between distant units, although specific oceanographic features that could act as barriers to sardine gene flow or specific permanent oceanographic regimes that could lead to different coastal habitats have not been identified (Peliz *et al.*, 2005; Santos *et al.*, 2005; Mason *et al.*, 2006). Oceanographic conditions along the NA coast form a mosaic of small and mesoscale heterogeneous habitats, but with similar large-scale properties, and create a wide distribution of suitable habitats for sardine in terms of temperature (Coombs *et al.*, 2006), food availability (Smyth *et al.*, 2005), and potential spawning grounds (Stratoudakis *et al.*, 2004; Bernal *et al.*, 2007; Planque *et al.*, 2007).

Sardine distribution along the NA coast is nearly continuous from the Strait of Gibraltar to the English Channel (Fig. 3.5c). However, life history characteristics (such as longevity, maximum age, length at maturity, growth, duration of spawning, and recruitment strength) differ between regions. The spatial distribution of young and old individuals is different, with old fish being observed more offshore over the continental shelf and young fish more inshore. Spatial fidelity of young fish to recruitment hotspots has been observed off northern Portugal and in the Gulf of Cadiz at depths < 30 m during the summer months, while, for adult populations, pronounced local variations have been reported. In both cases, the physical and/or biological conditions which could promote recruitment hotspots or variability in the adult population distribution are still unclear (ICES, 2006). Spawning habitats also show some permanent features, with spawning confined to the shelf and to temperatures of 12–17 °C, but with variation along the time series that is not clearly related to environmental variables or to the stock size (Bernal *et al.*, 2007). Bathymetric constraints and some unknown environmental variables allow the separation of the spawning habitat into four geographic nuclei, located in south Iberia, west Iberia, the Cantabrian coast, and the Armorican shelf (Bernal *et al.*, 2007; Planque *et al.*, 2007). Recruitment areas, i.e. areas dominated by juveniles, are located near the main spawning areas, but further restricted to inshore grounds in the Armorican shelf, and west and south Iberia. Increased retention in those areas (Santos *et al.*, 2004) has been generally suggested as the main reason for the spatial persistence of these recruitment spots. Adult habitat spreads beyond the offshore edge of these recruitment areas or else to more oceanic areas (near the shelf edge in the Armorican shelf), as suggested by the adult distribution found in acoustic surveys.

The general picture of sardine habitat along the European Atlantic coast is of some spatial persistence in the distribution of distinct life stages that seem to be associated with some known (temperature, retention areas, bathymetry) and unknown physical properties of the ocean. Nevertheless, this average picture can be perturbed, sometimes greatly, by variations of the mean distribution patterns of the adult population, which may be caused by oceanographic or other environmental forcing. The interannual variability in spatial patterns, associated with spatial gradients in life history characteristics, makes it difficult to establish the limits of the potential habitat for the entire sardine population. In addition, recent observations off the northwest North Sea suggest that the habitat of European sardine may be expanding northward as a result of increasing sea temperatures (Beare *et al.*, 2004 a, b). If this hypothesis is confirmed, it is expected that the spatial extent of sardine habitat in European waters will expand to northern waters, as already seen for a number of species (Perry *et al.*, 2005).

The Canary Current ecosystem (not shown) can be divided into two areas that overlap at the Cape Barbas (22° 30 N)–Cape Blanc (20° 20 N) transition region. The temperate *Sardina pilchardis* is confined to the NACW and extends from off Iberia to the Cape Juby–Cape Barbas (29° 20 N–22° 30 N) region. The more tropical *Sardinella* sp. is present south of Cape Blanc to Senegal (14° N). The intermediate area between Cape Barbas and Cape Blanc is a transition zone in which either species exists depending on the prevailing hydrological conditions (S. Kifani, Institut National de Recherche Halieutique, Morocco, pers. commun.; Hernandez-Leon *et al.*, 2007).

Kuroshio-Oyashio (KO)
Physical characteristics
The near-surface currents of the region of spawning of anchovy and sardine are dominated by the Kuroshio from the south, Oyashio from the north, their convergence (KO Transition Zone) off Honshu, and the Kuroshio Extension to the east (Fig. 3.6a) (Ichikawa and Beardley, 2002; Yasuda, 2003). A fraction of the Kuroshio also flows, as the Tsushima (Warm) Current, into the Japan/East Sea. These northward-flowing currents are warm and of high salinity,

while the southward-flowing currents (e.g. Oyashio) are cold and of lower salinity. Fronts, meanders, and eddies are characteristic of this region. There is significant interannual to interdecadal variation in these features, particularly the path of the Kuroshio and the trajectory of the Kuroshio Extension. These have important implications for the dynamics of anchovy and sardine (Nakata *et al.*, 2000; Yasuda, 2003).

*Japanese anchovy (*Engraulis japonicus*)*
Three stocks of Japanese anchovy are thought to exist, based on fishing and migration patterns: the Pacific stock, Tsushima Current stock, and the Seto Inland Sea stock (Fig. 3.6b) (Fisheries Agency and Fisheries Research Agency of Japan, 2005). Distributions of the two offshore stocks depend on their size. The Seto Inland Sea stock is confined to that sea. The Pacific stock is distributed along the Pacific coast of Japan, with an eastern boundary of ca. 155° E during low abundance and beyond 180° E during high abundance. The western and southern boundaries are 29° N 129° E in years of small population size. The northern limit of its migration area corresponds to the northern coast of Honshu during low abundance and north Kuril Island (47° N 150° E) during high abundance.

Spawning grounds of the Pacific stock are located in coastal areas from 32° N 131° E to 38° N 142° E during low abundance (Mori *et al.*, 1988; Kikuchi and Watanabe, 1990; Ishida and Kikuchi, 1992; Zenitani *et al.*, 1995; Kubota *et al.*, 1999). During high abundance, spawning is observed not only in the inshore areas off the Pacific coast but also in the offshore areas beyond the Kuroshio axis, from 29° N 130° E to 43° N 155° E, and the spawning grounds of this stock connect to those of the Tsushima Current stock. Spawning of the Seto Inland Sea stock is throughout that sea, and the offspring of the Pacific stock also migrates into this area.

Spawning takes place from February to October, and juveniles are transported by the Kuroshio and Kuroshio Extension currents beyond 170° E longitude in spring. Feeding grounds are typically located in the Oyashio and Kuroshio–Oyashio Transition regions, from 42° N 144° E to 45° N 155° E, in summer and autumn, although the northern limit of migration areas is 41° N 142° E during low abundance. Major fishing grounds of the purse-seine fishery are located along the Pacific coast of northern Japan. Adult fish migrate southward to their spawning grounds in late autumn; however, young of the year tend to stay at the middle of their spawning migration path (35–37° N 141–143° E).

Distribution area of the Tsushima Current stock is limited to the coastal areas from 30° N 129° E to 41° N 140° E during low abundance (Fisheries Agency and Fisheries Research Agency of Japan, 2005). However, the same

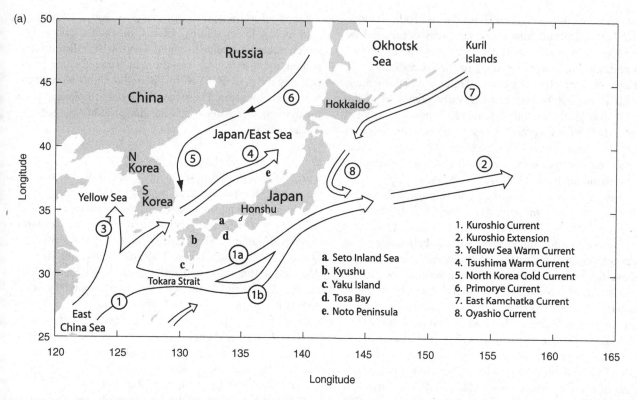

Fig. 3.6. Kuroshio–Oyashio region. (a) Circulation. 1a and 1b refer to straight and meandering patterns, respectively, of Kuroshio flow, with implications for anchovy and sardine dynamics (Nakata *et al.*, 2000; Yasuda, 2003). (b) Distribution of anchovy (*Engraulis japonicus*). (c) Distribution of sardine (*Sardinops melanostictus*).

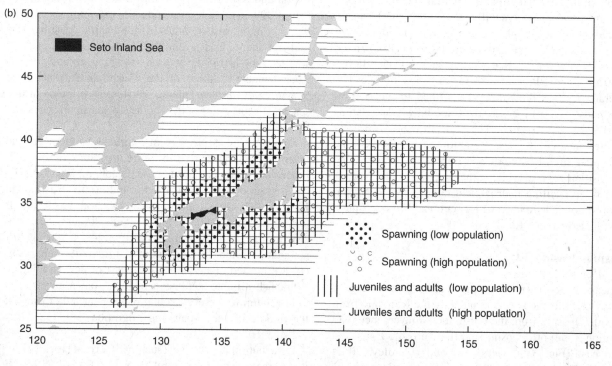

Fig. 3.6 (*cont.*)

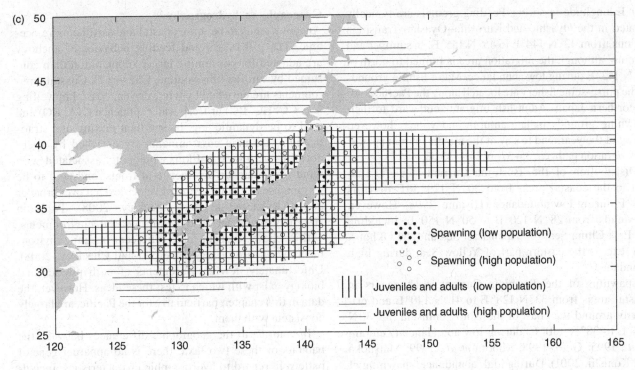

Fig. 3.6 (*cont.*)

stock is distributed from 25° N 120° E to 50° N 150° E, including the East China Sea, the Japan/East Sea and the Okhotsk Sea, except the eastern part of Yellow Sea, during high abundance.

Spawning grounds of the Tsushima Current stock are located in the coastal areas from 30° N 130° E to 41° N 140° E during low abundance (Zenitani *et al.*, 1995; Kubota *et al.*, 1999). Spawning activities of the Tsushima Current stock are observed not only in the coastal areas of Japan/East Sea but also in the offshore areas of Japan/East Sea and the East China Sea (26° N 125° E to 41° N 140° E), and major spawning grounds are formed in the western area of the Japan/East Sea and the eastern area of East China Sea (30° N to 36° N and from 128° E 133° E, respectively). Spawning takes place mainly from February to June, although details are still unknown because survey cruises did not cover the whole spawning areas and spawning seasons. Migration behaviors of adult fish have not been reported in detail (Fisheries Agency and Fisheries Research Agency of Japan, 2005).

Japanese sardine (Sardinops melanostictus)
Two stocks of Japanese sardine are thought to be distributed around Japan (Fig. 3.6c) (Fisheries Agency and Fisheries Research Agency of Japan, 2005): the Pacific stock and Tsushima Current stock, based on distribution and migration patterns, although mitochondrial

DNA analysis indicate no difference between these two stocks (Okazaki *et al.*, 1996). Distributions of these two stocks depend on their population size. The Pacific stock is distributed off the Pacific coast of Japan and its eastern boundary extends to ca. 155° E longitude during low abundance. The northern limit of its migration is the northern coast of Honshu during low abundance; however, the same stock migrates to the coast of north Kuril Island (46° N 150° E) and the eastern boundary extends beyond 180° E during high abundance. The southern and western boundaries are 29° N and 129° 30 E, the former coinciding with that of the spawning grounds in the years of small population size.

Spawning grounds of the Pacific stock are located in the coastal areas from 31° N 131° E to 41° N 142° E and mainly around the coast of the Tosa Bay from 32° N 132° E to 33° N 135° E during low abundance (Mori *et al.*, 1988; Kikuchi and Watanabe, 1990; Ishida and Kikuchi, 1992; Zenitani *et al.*, 1995; Kubota *et al.*, 1999). During high abundance, spawning is observed not only in the coastal areas but also offshore and beyond the Kuroshio axis, from 29° N 130° E to 41° N 144° E. Major spawning grounds are also around Yaku Island and the spawning grounds of both stocks merge in these areas during high abundance.

Spawning takes place from October to May, and juveniles are transported by and move in (migrate to) the Kuroshio and Kuroshio Extension currents to beyond

170° E longitude in spring. Feeding grounds are typically located in the Oyashio and Kuroshio–Oyashio Transition regions, from 42° N 144° E to 45° N 155° E, in summer and autumn, although the migration area is limited to north of 41° N 142° E during low abundance. Major fishing grounds of the purse-seine fishery are located along the Pacific coast of northern Japan. Adult fish migrate southward to their spawning grounds in late autumn and winter; however, young of the year tend to stay at the middle of their spawning migration path, ca. 35–37° N 141–143° E.

Distribution of the Tsushima Current stock is limited to the coastal areas from 30° N 129° 30 E to 41° N 140° E during low abundance (Hiyama, 1998). However, it extends from 28° N 120° E to 50° N 150° E, including the East China Sea, the Japan/East Sea and the Okhotsk Sea, except the eastern part of Yellow Sea, during high abundance.

Spawning of the Tsushima Current stock occurs in coastal areas from 33° N 128° E to 41° N 140° E and primarily around the coast of the Noto Peninsula (35° 30 N 134° E to 38° N 140° E) during low abundance (Zenitani *et al.*, 1995; Goto, 1998; Kubota *et al.*, 1999; Matsuoka and Konishi, 2001). During high abundance, spawning of the Tsushima Current stock occurs not only in the coastal areas but also in the offshore areas of the Japan Sea and the East China Sea, from 29° N 131° E to 41° N 140° E, and major spawning occurs around the Yaku Island, connecting with the spawning ground of the Pacific stock (Matsuoka and Konishi, 2001). Spawning takes place mainly from March to May, and juveniles move in the Tsushima Current northward in spring. Adult fish migrate northward along the coast of Japan from 37° N to 40° N in March, from 37° N to 42° N in April, and from 39° N to 43° N in May (Hiyama, 1998).

Synthesis

Anchovy and sardine populations worldwide appear to have been established relatively recently (ca. 10^4–10^6 years) (Grant and Bowen, 1998). Speciation of anchovy and sardine, conversely, occurred far earlier (ca. 10^7 y) (Grant and Bowen, 1998). One predicts, therefore, significant differences between these two groups and substantial consistency within the groups in regard to inherited traits. Populations of anchovy are expected to share characteristics among regions, and similarly for sardine, and the two groups are expected to differ within regions. Are the habitats of these fish consistent with these expectations? If so, what are the similarities within species and between regions, and differences between species across regions? Can this information be used to predict the effects of climate change on populations of these fish, their fisheries, and their role in the ecosystem? We address these issues below.

Geography vs. hydrography

Anchovy tends to be more coastal and sardine more oceanic. The gill rakers and feeding behavior of anchovy are adapted for consuming larger zooplankton than consumed by sardine (this volume, Chapter 7). Coastal environments are enriched by wind-driven, coastal upwelling (e.g. CC, HC, BC) or fresh water influence (NA, KO) and tend to be dynamic (e.g. intermittent mixing and stratification). Oceanic environments are enriched by open-ocean upwelling, e.g. Ekman pumping, associated with wind-stress curl or divergence at fronts, and tend to be less intense (e.g. lower vertical velocities) but over broader areas (Rykaczewski and Checkley, 2008). Plankton size may be larger in dynamic, coastal environments, with anchovy, than in more quiescent oceanic environments with sardine (Rykaczewski and Checkley, 2008). Unfortunately, few measurements of both physics and biology exist with which to test these ideas. However, the data in this chapter, particularly for the Pacific, are largely consistent with them.

In contrast to the geographic differences between the habitats of these two taxa, there is no apparent general pattern in regard to hydrographic characteristics, including temperature and salinity (Table 3.1). Thus, in some cases (HC, KO, NA), anchovy tend to spawn in less saline water than do sardine. Conversely, in the CC, anchovy tend to spawn in more saline and warmer water than do sardine. Spawning seasons of anchovy and sardine are slightly (e.g. Bay of Biscay, CC, and Tsushima Current) or completely (e.g. Gulf of Cadiz and Kuroshio Current) out of phase. Anchovy and sardine in the BC appear to not differ from one another in this regard. However, these patterns are consistent with the relative source of nutrients for each of the stocks in question. Thus, for the KO and NA, anchovy spawns in waters with nutrients associated with freshwater (e.g. Oyashio (Fig. 3.6a) and Adour, Gironde and Guadalquivir Rivers (Fig. 3.5a), while sardine spawning occurs further offshore throughout the shelf in more saline water. Conversely, anchovy of the CC and HC inhabit waters enriched by upwelling. For the CC, this water is of higher salinity than at the surface, while the reverse is so for the HC. In addition, some anchovy in the CC spawn near river plumes (Fig. 3.2b), low in salinity and rich in nutrients. Thus, in each case, the hydrographic characteristics of the habitats of these fish are consistent with the water masses containing their source nutrients. Exceptions to these generalizations occur, e.g. under extraordinary conditions such as during an intense El Niño. Thus, in the CC, water upwelled during a strong El Niño off California is of lower salinity than during other years, and salinity of the spawning habitats of anchovy and sardine can be quite similar during an El Niño (Checkley *et al.*, 2000).

Table 3.1. *Temperature and salinity of habitats for systems with populations of small, pelagic fish considered in this chapter*

System	Anchovy Temperature (°C)	Salinity	Sardine Temperature (°C)	Salinity	Comments
California Current (CC)	12–25	>33.2	12–24	<33.2	Spawning; lower salinity for anchovy in Columbia River plume
Humboldt Current (HC)	14–22	34.9–35.1	<19–22	35.0–35.3	Spawning
Benguela Current (BC)	12–24	34.8–36.0	12–24	34.8–36.0	Juvenile and adult distribution
Northeast Atlantic (NA)	15–19 (18)	<30–35.5 (30.1)	12–17 (13)	<30.0–>36.0 (35.3)	Spawning (range and mode)
Kuroshio-Oyashio (KO)	21	32.4	18	34.4	Spawning, Pacific (weighted mean)
	17	34.0	15	33.9	Spawning, Tsushima Current (weighted mean)
	22	31.6	n.a.	n.a.	Spawning, Seto Inland Sea (mean weighted by egg abundance)

These data are approximate, as significant variation occurs for each stock in regard to location, date, and developmental stage (n.a.: not available). References for these data are in the sections for the respective systems.

Transport and migration

All stocks, to a degree, inhabit different areas during different life stages. This results from the passive transport of eggs and larvae and the active movement (migration) of juveniles and adults. Sardine are larger and live longer, on average, than anchovy and, in some regions (CC, HC, BC, NA), migrate greater distances. For example, Pacific sardine that spawn off central and southern California are believed to migrate to feed, and perhaps spawn, as far north as off Vancouver Island, British Columbia, whereas the northern anchovy does not appear to make such a migration. Similarly, anchovy in the Bay of Biscay reside in a mosaic of locations influenced by runoff and local upwelling, while sardine spawn in similar waters but migrate longer distances, both coastally and offshore, to feed. However, both anchovy and sardine in the BC and KO undergo both significant transport downstream, as plankton, and migration upstream, as juveniles and adults. Such migrations also have implications for interpretation of the paleo-oceanographic record of fish scale deposition and inference therefrom about population size (this volume, Chapter 4).

Expansion, contraction, and refugia

Stocks of small, pelagic fish expand their geographic distribution when more abundant and contract when less abundant. Because their populations typically vary in abundance on the scale of decades (this volume, Chapter 9), their geographic distribution also varies. MacCall (1990; this volume, Chapter 12) proposed the basin hypothesis to explain this phenomenon. A related issue is habitat use. Planque *et al.* (2007) classified spawning habitat (SH) as potential SH, where conditions are suitable for fish to spawn, realized SH, where spawning occurs, and successful SH, where spawning results in successful recruitment. Recent work has demonstrated that potential SH may be much greater than realized or successful SH for stocks at low population size (Planque *et al.*, 2007; Reiss *et al.*, 2008).

Refugia at small population size may exist for most stocks. However, data are sparse due to the lack and/or difficulty of both fishing and scientific surveys of stocks at low abundance. Anchovy in the CC contract into coastal upwelling areas, estuaries, and the southern California Bight. Sardine in the CC appear to contract into the inshore waters off Baja California, Mexico. Anchoveta off Peru are believed to move close to shore in isolated pockets of upwelling during low abundance. Sardine of the HC may contract to waters around the Galapagos Islands (M. Niquen, personal communication). The Japanese anchovy and sardine spawning grounds contract to waters around Honshu and Kyushu during low abundance, and expand offshore during high abundance. Hence, a general feature appears to be expansion during high abundance and contraction at low abundance to predictable locations which may be refugia.

Mortality

A process that has not been quantified spatially is mortality of all stages due to predation, starvation, or both.

> **Box 3.1. Habitats of small, pelagic fish under future climate change**
>
> The habitat of small, pelagic fish is a three-dimensional environment comprised of water and dissolved and particulate and living and dead matter. In contrast to demersal and benthic fish, which relate to geography, e.g. the sea bottom, pelagic fish relate to hydrography, perhaps especially temperature, salinity, and planktonic food. A key challenge to scientists is to understand these relations to be able to predict the response of populations of small, pelagic fish to future climate change.
>
> Future climate change, associated with human-caused enhancement of CO_2 concentration and global mean temperature, may affect the environment of small, pelagic fish, and thus their populations, in several ways. Average warming of 1.1–6.4 °C is predicted globally by 2100 (IPCC, 2007), but change will vary regionally, and thus have different effects in the different SPACC regions. Latitudinal shifts in plankton (Beaugrand *et al.*, 2002) and anchovy and sardine (Beare *et al.*, 2004a,b) are already apparent in the northeast Atlantic. Temporal variation may occur abruptly, as in the past (Alley *et al.*, 2003; IPCC, 2007), and become more episodic (IPCC, 2007). The timing, or phenology, of ecological events, including, for example, the spring bloom, may affect fish feeding, survival, and recruitment. Interactions of changed phenologies, may lead to mismatches in food supply and demand, *sensu* Cushing (1974). The hydrological cycle is predicted to change, with potential effects on the habitats of some species and stocks, e.g. anchovy that spawn in regions affected by freshwater. Effects on small pelagic fish of changing ocean chemistry, including enhanced concentration of CO_2 and acidification, are largely unknown but of potential significance. Simultaneous monitoring of populations of small, pelagic fish and their habitats is needed, including continued and new observing programs and the development of new technologies to this end. Perhaps the greatest challenge is to achieve a predictive understanding of the cumulative effects of future climate change on the habitats, and thus stocks, of small, pelagic fish. This is particularly difficult given the unprecedented nature of that change.

eggs and early larvae made when applying the Daily Egg Production Method (Stratoudakis *et al.*, 2006; this volume, Chapter 15, and references therein). One might hypothesize that the oceanic habitat of sardine may have lower concentrations of both food and predators and, conversely, anchovy may inhabit waters with both more food and predators. Conversely, greater predation offshore is invoked in the school-mix feedback hypothesis (this volume, Chapter 13). Cannibalism can be a significant source of mortality for anchovy in the CC (Hunter and Kimbrell, 1980), HC (Santander, 1987), and other regions (this volume, Chapter 7 and references therein).

Future climate change

At the time of writing, we know the Earth's climate is changing due to both natural and anthropogenic causes (http://www.ipcc.ch). We are beginning to achieve skill in predicting the effects of present and future climate change on ocean physics, including its state (e.g. temperature, salinity, and stratification) and dynamics (e.g. currents, upwelling, and fluxes). Collectively, these physical changes will alter the habitat of small, pelagic fish in decades and centuries to come. Fréon *et al.* (this volume, Chapter 14) discusses these issues and their implications. The past behavior of systems with small, pelagic fish is the subject of several chapters in this book. However, knowledge of the past behavior of these systems may not be sufficient to predict their behavior under future climate change. Rather, an understanding of their dynamics is needed. Below, we present topics that merit consideration for the study of habitat under future climate change.

Ocean temperature

Globally, the Earth's surface is predicted to warm 1.1–6.4 °C this century (http://www.ipcc.ch). Temperature, as a habitat variable, may induce shifts in population distributions and rates (e.g. feeding, respiration, growth, birth, and death). Warming will vary regionally and is, in general, predicted to be greater at high latitudes. This may result in more rapid poleward shifts of the high-latitude (cool) than low-latitude (warm) range limits. Continued high-resolution mapping of the distribution and abundance of fish, at all stages of development, and environmental variables is needed to resolve such shifts. To this end, equipping vessels of opportunity, e.g. fishing vessels, with devices to measure and log environmental data and survey vessels with continuous egg samplers and acoustics is desirable. An increased contrast of land and sea temperatures may lead to greater upwelling, with consequences, albeit they uncertain, for small pelagic fish (Bakun, 1990; Bakun and Weeks, 2004).

Species fitness depends on the balance of birth and death. Species evolve to maximize the difference between these two processes. Thus, both food and predators must be considered as habitat characteristics. Unfortunately, little is known of natural predators and mortality for any stage of development. Exceptions are estimates of the mortality of

Winds

Currents, upwelling, and turbulence are key habitat characteristics likely to be affected by changes in wind speed and direction. Currents affect transport from spawning to nursery areas. Coastal, wind-driven upwelling is particularly important for anchovy. Both winds themselves and their effectiveness in causing upwelling are related to geography, including coastal headlands and mountains. Offshore winds, and their curl and associated Ekman pumping (oceanic upwelling), are of particular importance to some populations of sardine. Turbulence affects the encounter rate and capture success of fish larvae, which, in turn, affects feeding and mortality. Expanded measurement of wind by instruments deployed at sea and in space is desirable.

Ocean stratification

Increased stratification, due to enhanced heat content of the upper layer of the ocean, will result in a more intense thermocline. This, in turn, will reduce vertical nutrient flux due to both diffusion and mixing at the thermocline and wind-driven upwelling. Changed winds may also affect retention and dispersal of planktonic eggs and larvae. Expanded measurement of hydrography from ships and autonomous devices (e.g. buoys, floats, and gliders) is desirable, particularly in conjuction with measurements of fish distribution and abundance of all stages.

Hydrology

A primary result of global climate change will be altered hydrology (IPCC, 2007). Anchovy habitat includes river plumes (CC, NA). Anchovy and sardine off Japan inhabit currents influenced by freshwater (i.e. Oyashio). The pattern of precipitation may have important effects on the habitat of anchovy, particularly off Peru (e.g. El Niño), the NA (e.g. Ardour and Gironde Rivers), and the CC (e.g. Columbia River). Expanded monitoring of coastal ocean hydrography is needed.

Phenology

The temporal occurrences of related processes, especially the production of planktonic food and the spawning, hence first feeding, of fish is likely to be affected by climate change. Such effects are best known on land (Visser and Both, 2005). The timing of the biomass peak of *Neocalanus* in the N. Pacific is known to vary naturally on the scale of decades (Mackas *et al.*, 1998). The existence and consequence of such changes in phenology on the habitat of small, pelagic fish is largely unknown but of potential significance. However, many fish stocks considered here have protracted (months) spawning seasons and, thus, may be less susceptible to phenological changes than are stocks, generally at higher latitude, that require a closer (weeks) match of food production and spawning (Edwards and Richardson, 2004). In fact, recent data for some stocks of anchovy and sardine (CC, KO) are consistent with variation of survival being more important in the juvenile than the larval stage (M. Takahashi, Scripps Institution of Oceanography, La Jolla, California, USA, personal communication). Continuous sampling from towed (e.g. the Continuous Plankton Recorder, CPR, Reid *et al.*, 2003) and autonomous (see above) devices is needed to resolve important changes in phenology.

Events

The frequency and magnitude of both single and repeating events must be considered. While predictions of change, e.g. of temperature, are gradual, past change has at times been abrupt and episodic (Alley *et al.*, 2003). The consequences of this to marine populations, including small, pelagic fish, is unknown. Storms and longer-term events, e.g. El Niño, may change in frequency and intensity, also with unknown consequence to small, pelagic fish. Continued and new time series of observations of both the environment and fish are needed. Examples are the CPR survey (Reid *et al.*, 2003) and the California Cooperative Oceanic Fisheries Investigations (Bograd *et al.*, 2003).

Acidification

The habitat of small, pelagic fish includes the ambient seawater. Carbon dioxide produced by human activities is acidifying the ocean (Orr *et al.*, 2005). The solubility of biogenic minerals, particularly aragonite, will increase, particularly at high latitudes. This has known consequences for organisms with external shells, e.g. pteropods. The effects on internal structures, including otoliths of fish, is unknown. Experimental and observational studies of the effects of acidification on fish are warranted.

These effects, combined with those from other causes, particularly fishing, will have cumulative effects on populations of small, pelagic fish in ways as yet unknown. A challenge is to gain sufficient understanding of these processes and relationships to enable prediction with acceptable confidence.

A related question is how refugia may be affected by climate change. Will geographic regions which have in the past served as refugia change, e.g. due to winds and hydrography, to no longer be acceptable as refugia? Similarly, are there "hot spots," or loci of enhanced activity (e.g. spawning, feeding), that may change with climate change and, if so, how will this affect the dynamics of the populations? The recently observed eastward movement of both anchovy and particularly sardine stocks in the southern BC may be an example of this. Finally, we note (e.g. Table 3.1) that anchovy and sardine occur in a broad range of temperatures and salinities. Are these taxa able to adapt to new conditions and, if so, what are the mechanisms, rates, and limits, particularly in regard to regional changes? Genetic analyses of populations of anchovy and sardine indicate shallow population structure and low genetic diversity, implying susceptibility of these populations to change (Grant and Bowen, 1998).

Two general, overarching questions exist in regard to the anticipated rise in global and ocean temperatures in the twenty-first century. First, will the net effect of climate change on a population of small, pelagic fish be a latitudinal shift in distribution (e.g. Beare *et al.*, 2004a,b; Perry *et al.*, 2005), a change in overall abundance, or both? We need mechanistic models in which we have appropriate confidence. Second, what will the effects of climate change be on the ecosystem containing populations of small, pelagic fish? Such ecosystems have been named wasp-waist (this volume, Chapters 8 and 13), implying strong interaction of small pelagic fish with both lower and higher trophic levels. How climate change will affect such systems is unknown.

Conclusions

Work of SPACC has shown consistent differences between taxa in habitat characteristics but similarities within each major taxon, particularly *Engraulis* spp. and *Sardinops sagax*. While variation occurs between regions (CC, HC, BC, NA, and KO), generalities exist. Anchovy tends to be coastal, derive nutrients from water influenced by coastal upwelling or freshwater, consume larger food and migrate less, while sardine tends to be more oceanic, derive nutrients from oceanic processes, eat smaller food, and migrate more, with exceptions. The occupied habitat of small, pelagic fish expands and contracts with increasing and decreasing population size, and refugia may exist for most populations.

Key questions are how the distribution, abundance, and production of small, pelagic fish will be affected by future climate change, due to natural and anthropogenic causes. Greater skill has been achieved in forecasting physical aspects of climate change. The challenge will be to achieve similar success in predicting biological characteristics, including those of small, pelagic fish.

Critical actions will be to adequately observe essential features of the habitat and the populations themselves, to develop predictive models, and to test the skill of forecasts from those models. The results of these actions can then be used to inform those who manage populations of small, pelagic fish and make policy affecting those fish, the ecosystems, and climate.

Acknowledgments

We thank Salvador Lluch-Cota, Alec MacCall, and Jürgen Alheit for constructive comments on the manuscript. Souad Kifani contributed information on the Canary Current system and its sardine populations. Jo Griffith assisted in drafting figures. GLOBEC International and the Institute for Research and Development (IRD) of France provided financial support for meetings, workshop, and other costs leading to publication of this chapter and the book.

NOTES
1   El Niño-Southern Oscillation.
2   Virtual Population Analysis.

REFERENCES

Alley, R. B., Marotzke, J., Nordhaus, W. D. *et al.* (2003). Abrupt climate change. *Science* **299**: 2005–2010.

Arcos, D. F., Cubillos, L. A., and Núñez, S. P. (2001). The jack mackerel fishery and El Niño 1997–98 effects off Chile. *Prog. Oceanogr,* **49**: 597–617.

Arntz W. and Fahrbach, E. (1996). *El Niño: experimento climático de la naturaleza.* 1st edn. Fondo de Cultura Económica México.

Ayón, P., Purca, S., and Guevara-Carrasco, R. (2004). Zooplankton volume trends off Peru between 1964 and 2001. *ICES J. Mar. Sci.* **61**: 478–484.

Bakun, A. (1990). Global climate change and intensification of coastal ocean upwelling. *Science* **247**: 198–201.

Bakun, A. and Weeks, S. J. (2004). Greenhouse gas buildup, sardines, submarine eruptions and the possibility of abrupt degradation of intense marine upwelling ecosystems. *Ecol. Lett.* **7**: 1015–1023.

Barange, M., Hampton, I., and Roel, B. A. (1999). Trends in the abundance and distribution of anchovy and sardine on the South African continental shelf in the 1990s, deduced from acoustic surveys. *S. Afr. J. Mar. Sci.* **21**: 367–391.

Barange, M., Coetzee, J. C., and Twatwa, N. M. (2005). Strategies of space occupation by anchovy and sardine in the southern Benguela: the role of stock size and intra-specific competition. *ICES J. Mar. Sci.* **62**: 645–654.

Bardey, P., Garnesson, P., Moussu, P., and Wald, L. (1999). Joint analysis of temperature and ocean colour satellite images for mesoscale activities in the Gulf of Biscay. *Int. J. Remote Sens.* **20**: 1329–1341.

Baumgartner, T., Soutar, A., and Ferreira-Bartrina, V. (1992). Reconstruction of the history of Pacific sardine and northern anchovy populations over the past two millenia from sediments of the Santa Barbara basin. *CalCOFI Rep.* **33**: 24–40.

Beare, D. J., Burns, F., Peach, K. (2004a). An increase in the abundance of anchovies and sardines in the northwestern North Sea since 1995. *Glob. Change Biol.* **10**: 1209–1213.

Beare, D. J., Burns, F., Greig, A. *et al.* (2004b). Long-term increases in prevalence of North Sea fishes having southern biogeographic affinities. *Mar. Ecol. Prog. Ser.* **284**: 269–278.

Beaugrand, G., Reid, P. C, Ibanez, F., *et al.* (2002). Reorganization of North Atlantic marine copepod biodiversity and climate. *Science* **296**: 1692–1694.

Bellier, E., Planque, B., and Petitgas, P. (2007). Historical fluctuations in spawning location of anchovy (*Engraulis encrasicolus*) and sardine (*Sardina pilchardus*) in the Bay of Biscay during 1967–73 and 2000–2004. *Fish. Oceanogr.* **16**: 1–15.

Bernal, M., Stratoudakis, Y. , Coombs, S. *et al.* (2007). Sardine spawning off the European Atlantic coast: characterization of and spatio-temporal variability in spawning habitat. *Prog. Oceanogr.* **74**: 210–227.

Bograd, S.J., Checkley, D.M., and Wooster, W. S., ed. (2003). CalCOFI: A half century of physical, chemical, and biological research in the California Current System. *Deep-Sea Res. II* **50**: 2349–2594.

Boyer, D. C. and Hampton, I. (2001). An overview of living marine resources of Namibia. In *A Decade of Namibian Fisheries Science. S. Afr. J. Mar. Sci.* **23**: 5–35.

Boyer, D. C, Boyer, H. J., Fossen, I., and Kreiner, A. (2001). Changes in abundance of the Northern Benguela sardine stock during the decade 1990–2000, with comments on the relative importance of fishing and the environment. *S. Afr. J. Mar. Sci.* **23**: 67–84.

Butterworth, D. S. (1983). Assessment and management of pelagic stocks in the southern Benguela region. In Sharp, G. D. and Csirke J., eds. *Proceedings of the expert consultation to examine changes in abundance and species composition of neritic fish resources, San José, Costa Rica, 18–29 April 1983. FAO Fish. Rep.* **291**: 329–406.

Calderwood, W.L. (1892). Experiments on the relative abundance of anchovies off the South Coast of England. *J. Mar. Biol. Assoc. UK* **2**: 268–271.

Chavez, F. P., Ryan, J., Lluch-Cota, S. E., and Niquen, M. (2003). From anchovies to sardines and back: multidecadal change in the Pacific Ocean. *Science* **299**: 217–221.

Checkley, D. M. Jr., Ortner, B., Settle, L. R., and Cummings, S. R. (1997). A continuous, underway fish egg sampler. *Fish. Oceanogr.* **6**: 58–73.

Checkley, D. M., Jr., Dotson, R. C., and Griffith, D. A. (2000). Continuous, underway sampling of eggs of Pacific sardine (*Sardinops sagax*) and northern anchovy (*Engraulis mordax*) in spring 1996 and 1997 off southern and central California. *Deep-Sea Res. II* **47**: 1139–1155.

Chelton, D. B., Bernal, P. A., and McGowan, J. A. (1982). Large-scale interannual physical and biological interaction in the California Current. *J. Mar. Res.*, **40**: 1095–1125.

Coetzee, J. C., van der Lingen, C. D., Hutchings, L., and Fairweather, T. P. (2008). Has the fishery contributed to a major shift in the distribution of South African sardine? *ICES J. Mar. Sci.* **65**: 1676–1688.

Connell, A. D. (2001). Pelagic eggs of marine fishes from Park Rynie, KwaZulu-Natal, South Africa: seasonal spawning patterns of the three most common species. *Afr. Zool.* **36**: 197–204.

Coombs, S. H., Smyth, T. J., Conway, D. V. P., *et al.* (2006). Spawning season and temperature relationships for sardine (*Sardina pilchardus*) in the eastern North Atlantic. *J. Mar. Biol. Assoc. UK* **86**: 1245–1252.

Cotano, U., Etxebeste, E., Irigoien, X., *et al.* (2008). Distribution and survival of early life anchovy (*Engraulis encrasicolus* L.) in relation to hydrodynamic and trophic environment in the Bay of Biscay. *J. Plankt. Res.* **30**: 467–481.

Csirke J. (1996). Situación y perspectivas de la ordenación de la pesquería en el Perú y propuesta de un plan de ordenación de las pesquerías de anchoveta y sardina. Documento de campo No. 7, *Proyecto FAO TCP/ PER/4451*, pp. 36.

Csirke, J., Guevara-Carrasco, R., Cárdenas, G., *et al.* (1996). Situación de los recursos anchoveta (*Engraulis ringens*) y Sardina (*Sardinops sagax*) a principios de 1994 y perspectivas para la pesca en el Perú, con particular referencia a las regiones norte y centro de la costa peruana. *Bol. Inst. Mar Perú* **15**: 3–23.

Cunningham, J. T. (1890). Anchovies in the English Channel. *J. Mar. Biol. Assoc. UK* **1**: 328–339.

Cunningham, J. T. (1895). The migration of the anchovy. *J. Mar. Biol. Assoc. UK* **3**: 300–303.

Cunningham, J. T. (1896). Physical and biological conditions in the North Sea. *J. Mar. Biol. Assoc. UK* **4**: 233–263.

Curtis, K. A, Checkley, D. M., Jr., and Pepin, P. (2007). Predicting the vertical profiles of anchovy (*Engraulis mordax*) and sardine (*Sardinops sagax*) eggs in the California Current System. *Fish. Oceanogr.* **16**: 68–84.

Cushing, D. H. (1974). The natural regulation of fish populations. In *Sea Fisheries Research*, F. R. Harden Jones, F.R., ed. London: Elsk, pp. 399–412.

Demarcq, H., Barlow, R., and Hutchings, L. (2007). Application of a chlorophyll index derived from satellite data to investigate the variability of phytoplankton in the Benguela ecosystem. *Afr. J. Mar. Sci.* **29**: 271–282.

Edwards, M. and Richardson, A. J. (2004). Impact of climate change on marine pelagic phenology and trophic mismatch. *Nature* **430**: 881–884.

Ekau, W. and Verheye, H. M. V. (2005). Influence of oceanographic fronts and low oxygen on the distribution of ichthyoplankton in the Benguela and southern Angola currents. *Afr. J. Mar. Sci.* **27**: 629–639.

Emmett, R. L., Brodeur, R. D., Miller, T. W., *et al.* (2005). Pacific sardine (*Sardinops sagax*) abundance, distribution and ecological relationships in the Pacific Northwest. *CalCOFI Rep.* **46**: 122–143.

Emmett, R. L., Bentley, P. J., and Schiewe, H. M. (1997). Abundance and distribution of northern anchovy eggs and larvae (*Engraulis mordax*) off the Oregon coast, mid-1970s and 1994 and 1995., In *Forage Fishes in Marine Ecosystems, Proceedings International Symposium on the Role of Forage Fishes in Marine Ecosystems. U. Alaska Sea Grant College Prog. Rep. 97–01*, Fairbanks: U. Alaska, pp. 505–508.

Emmett, R. L., Krutzikowsky, G. K., and Bentley, P. (2006). Abundance and distribution of pelagic piscivorous fishes in the Columbia River plume during spring/early summer 1998–2003, Relationship to oceanographic conditions, forage fishes, and juvenile salmonids. *Prog. Oceanogr.* **68**: 1–26.

Fairweather, T. P., van der Lingen, C. D., Booth, A. J., *et al.* (2006). Indicators of sustainable fishing for South African sardine (*Sardinops sagax*) and anchovy (*Engraulis encrasicolus*). *Afr. J. Mar. Sci.* **28**: 661–680.

Fisheries Agency and Fisheries Research Agency of Japan. (2005). Marine fisheries stock assessment and evaluation for Japanese waters (fiscal year 2005/2006). Japanese Fisheries Agency, pp. 1049.

Fréon P. B. and Stequert, B. (1979). Note sur la présence de *Sardina pilchardus* (Walb.) au Sénegal: étude de la biométrie et interprétation. *Cybium* **6**: 65–90.

González-Quirós, R., Pascual, A., Gomis, D., and Anadón, R. (2004). Influence of mesoscale physical forcing on trophic pathways and fish larvae retention in the central Cantabrian Sea. *Fish. Oceanogr.* **13**: 351–364.

Goto, T. (1998). Abundance and distribution on the eggs of the sardine, *Sardinops melanostictus*, in the Japan Sea during spring, 1979–1994. *Bull. Jap. Sea Nat. Fish. Res. Inst.* **48**: 51–60.

Grant, W. S. and Bowen, B. W. (1998). Shallow population histories in deep evolutionary lineages of marine fishes: Insights from sardines and anchovies and lessons for conservation. *J. Hered.* **89**: 415–426.

Griffiths, C. L., van Sittert, L., Best, P. B. *et al.* (2004). Impacts of human activities on marine animal life in the Benguela: a historical overview. *Oceanogr. Mar. Biol. Ann. Rev.* **42**: 303–392.

Guillén O. (1983). Condiciones oceanográficas y sus fluctuaciones en el Pacífico Sur Oriental. In *Proceedings of the Expert Consultation to Examine Changes in abundance and species of neritic fish resources, 18–19 April 1983, San José, Costa Rica*, Sharp G. D., and Csirke J. J., eds. *FAO Fish. Rep.*, **291** (3), Rome.

Hernández-León, S., Gómez, M., and Arístegui, J. (2007). Mesozooplankton in the Canary Current System: The coastal-ocean transition zone. *Progr. Oceanogr.* **74**: 397–421.

Hickey, B. M. (1979). The California Current System – hypotheses and facts. *Prog. Oceanogr.* **8**: 191–279.

Hickey, B. (1998). Coastal oceanography of western North America from the tip of Baja California to Vancouver island. In *The Sea. Vol. 11. The Global Coastal Ocean*, Robinson, A. R., and Brink, K. H., eds. New York: John Wiley & Sons, pp. 345–393.

Hiyama, Y. (1998). Migration range and growth rate in the Tsushima Current area. In *Stock Fluctuations and Ecological Changes of the Japanese Sardine*, Watanabe, Y. and Wada, T., eds. Tokyo, Japan: Koseisha-Koseikaku, pp. 35–44. (in Japanese)

Hunter, J. R. and Kimbrell, C. A. (1980). Egg cannibalism in the northern anchovy, *Engraulis mordax*. *Fish. Bull. US* **78**: 811–816.

Hutchings, L. (1992). Fish harvesting in a variable, productive environment: searching for rules or searching for exceptions? *S. Afr. J. Mar. Sci.* **12**: 297–318.

Hutchings, L., Pillar, S. C., and Verheye, H. M. (1991). Estimates of standing stock, production and consumption of meso- and macro-zooplankton in the Benguela ecosystem. *S. Afr. J. Mar. Sci.* **11**: 499–512.

Ibaibarriaga, L., Santos, M., Uriarte, A. *et al.* (2005). Application of generalized additive models to the DEPM for the Bay of Biscay anchovy (*Engraulis encrasicolus*). In *Report of the SPACC meeting on small pelagic fish spawning habitat dynamics and the Daily Egg Production Method (DEPM)*, ed. Castro, L., Fréon, P., van der Lingen, C. D., and Uriarte A. *GLOBEC Rep.* **22**, 107 pp.

Ibaibarriaga, L., Irigoien, X., Santos, M. *et al.* (2007). Egg and larvae distribution of seven fish species in the north-east Atlantic waters. *Fish. Oceanogr.* **16**: 284–293.

ICES (2004). Report of the study group on regional scale ecology of small pelagics (SGRESP). *ICES CM 2004/**G:06***.

ICES (2005). Report of the working group on acoustic and egg surveys for sardine and anchovy in ICES areas VIII and IX (WGACEGG). *ICES CM 2006/**LRC: 01.***

ICES (2006). Report of the working group on the assessment of mackerel, horse mackerel, sardine and anchovy. *ICES CM 2006/**ACFM: 08.***

Ichikawa, H. and Beardsley, R. C. (2002). The current system in the Yellow and East China Seas. *J. Oceanogr.* **58**: 77–92.

IMARPE (1973). Report of the third session of the panel of experts on the population dynamics of Peruvian anchoveta. *Bol. Inst. Mar Perú Callao* **2** (7): 373–345.

IPCC (2007). *Fourth Assessment Report, Climate Change 2007*, http://www.mnp.nl/pages_media/AR4-chapters.html, 4 volumes.

Irigoien, X., Cotano, U., Boyra, G., *et al.* (2008). From egg to juvenile in the Bay of Biscay: spatial patterns of anchovy (*Engraulis encrasicolus*) recruitment in a non-upwelling region. *Fish. Oceanog.* **17**: 446–462.

Irigoien, X., Ibaibarriaga, L., Santos, M., *et al.* (2005). Anchovy and sardine spawning habitat in the Bay of Biscay. *GLOBEC Rep.* **21**, 16–17.

Ishida, M. and Kikuchi, H. (1992). *Monthly egg productions of the Japanese sardine, anchovy, and mackerels of the Southern Coast of Japan by egg censuses: January 1989 through December, 1990.* Japan Fisheries Agency, p. 86.

Jacobson, L. D., de Oliveira, J. A. A., Barange, M. *et al.* (2001). Surplus production, variability, and climate change in the great sardine and anchovy fisheries. *Can. J. Fish. Aquat. Sci.* **58**: 1891–1903.

Jahncke, J. and Goya, E. (1998). Las dietas del guanay y del piquero Peruano como indicadoras de la abundancia de anchoveta. *Bol. Inst. Mar Perú* **17**: 15–23.

Kasapidis, P., Planes, S., Laurent, V. *et al.* (2004). Stock discrimination and temporal and spatial genetic variation of sardine (*Sardina pilchardus*) in northeastern Atlantic, with a combined analysis of nuclear (microsatellites and allozymes) and mitochondrial DNA markers. *ICES CM 2004/**Theme Session Q:21.***

Kikuchi, H. and Watanabe, Y. (1990). *Monthly egg productions of the Japanese sardine, anchovy, and mackerels of the southern coast of Japan by egg censuses: January 1987 through December 1988.* Japan Fisheries Agency, p. 72.

Koutsikopoulos, C. and Le Cann, B. (1996). Physical processes and hydrological structures related to the Bay of Biscay anchovy. *Sci. Mar.* **60** (Suppl. 1): 9–19.

Kreiner, A., Stenevik, E. K., and Ekan, W. (2009). Sardine *Sardinops sagax* and anchovy *Engraulis encrasicolus* larvae avoid regions with low dissolved oxygen concentration in the northern Benguela Current system. *J. Fish. Biol.* **74**: 270–277.

Kubota, H., Oozeki, Y., Ishida, M. *et al.* (1999). *Distributions of eggs and larvae of Japanese sardine, Japanese anchovy, mackerels, round herring, jack mackerel and Japanese common squid in the waters around Japan, 1994 through 1996.* Japanese Fisheries Agency, p. 352.

Lasker, R. (1975). Field criteria for survival of anchovy larvae: The relation between inshore chlorophyll maximum layers and successful first feeding. *Fish. Bull. US,* **73**: 953–1163.

Lasker, R. and Smith, P. E. (1977). Estimation of the effects of environmental variations of the eggs and larvae of the northern anchovy. *CalCOFI Rep.* **19**: 128–137.

Laurent, V., Voisin, M., and Planes, S. (2006). Genetic clines in the Bay of Biscay provide estimates of migration for *Sardina pilchardus. J. Hered.* **97**: 81–88.

Lecomte, F., Grant, W. S., Dodson, J. J., *et al.* (2004). Living with uncertainty: Genetic imprints of climate shifts in east Pacific anchovy (*Engraulis mordax*) and sardine (*Sardinops sagax*). *Mol. Ecol.* **13**: 2169–2182.

Llope, M., Anadón, R., Viesca, L., *et al.* (2006). Hydrography of the southern Bay of Biscay shelf-break region: integrating the multiscale physical variability over the period 1993–2003. *J. Geophys. Res.* **111** (C9): C09021.

Lluch-Belda, D., Schwartzlose, R. A., Serra, R., *et al.* (1992). Sardine and anchovy regime fluctuations of abundance in four regions of the world oceans: a workshop report. *Fish. Oceanogr.* **1**: 339–347.

Logerwell, E. A. and Smith, P. E. (2001). Mesoscale eddies and survival of late stage Pacific sardine (*Sardinops sagax*) larvae. *Fish. Oceanogr.* **10**: 13–25.

Lucio, P. and Uriarte, A. (1990). Aspects of the reproductive biology of the anchovy (*Engraulis encrasicolus* L.) during 1987 and 1988 in the Bay of Biscay. *ICES CM* **1990/H:27**.

Lutjeharms, J. R. E., Valentine, H. R., and van Ballegooyen, R. C. (2000). The hydrography and water masses of the Natal Bight, South Africa. *Cont. Shelf Res.* **20**: 1907–1939.

Lynn, R. J. (2003). Variability in the spawning habitat of Pacific sardine (*Sardinops sagax*) off southern and central California. *Fish. Oceanogr.* **12**: 541–553.

Lynn, R. J. and Simpson, J. J. (1987). The California Current System: The seasonal variability of its physical characteristics. *J. Geophys. Res.* **92**: 12 947–12 966.

Lynn, R. J. and Simpson, J. J. (1990). The flow of the undercurrent over the continental borderland off southern California. *J. Geophys. Res.* **95**: 12 995–13 008.

MacCall, A. D. (1990). *Dynamic Geography of Marine Fish Populations.* Seattle: Washington Sea Grant Program.

Mackas, D. L., Goldblatt, R., and Lewis, A. G. (1998). Interdecadal variation in developmental timing of *Neocalanus plumchrus* populations at Ocean Station P in the subarctic North Pacific. *Can. J. Fish. Aquat. Sci.* **55**: 1878–1893.

Mais, K. F. (1974). Pelagic fish surveys in the California Current. *Fish Bull. US* **162**, 79 pp.

Mason, E., Coombs, S., and Oliveira, P. B. (2006). An overview of the literature concerning the oceanography of the eastern North Atlantic region. *Relat. Cient. Téc. IPIMAR: Digital Series* **33**, 58 pp.

Massé, J. (1996). Acoustic observations in the Bay of Biscay: schooling, vertical distribution, species assemblages and behaviour. *Sci. Mar.* **60** (Suppl. 2): 227–234.

Matsuoka, M. and Konishi, Y. (2001). Abundance and distributional changes of Japanese sardine eggs around Kyusyu, Japan, from 1979 to 1995. *Bull. Japan. Soc. Fish. Oceanogr.* **65**: 67–73. (in Japanese).

Mayr, E. (1982). *The Growth of Biological Thought.* Cambridge, MA: Harvard University Press.

McFarlane, G. A., Schweigert, J., MacDougall, L., and Hrabok, C. (2005). Distribution and biology of Pacific sardines (*Sardinops sagax*) off British Columbia, Canada. *CalCOFI Rep.* **46**: 144–160.

Mori, K., Kuroda, K., and Konishi, Y. (1988). *Monthly Egg Productions of the Japanese Sardine, Anchovy, and Mackerels of the Southern Coast of Japan by Egg Censuses: January, 1978 through December, 1986.* Japanese Fisheries Agency, p. 321.

Morón, O. (2000). Características del ambiente marino frente à la costa peruana. *Bol. Inst. Mar. Perú* **19** (1–2): 179–204.

Moser, H. G., Charter, R. L., Smith, P. E. (1993). Distributional atlas of fish larvae and eggs in the California Current region: taxa with 1000 or more total larvae, 1951 through 1984. *CalCOFI Atlas* **31**: 1–233.

Moser, H. G., Charter, R. L., Smith, P. E. *et al.* (2001). Distributional Atlas of fish larvae and eggs in the Southern California Bight region: 1951–1998. *CalCOFI Atlas* **34**: 1–166.

Motos, L., Uriarte, A., and Valencia, V. (1996). The spawning environment of the Bay of Biscay anchovy (*Engraulis encrasicolus* L.). *Sci. Mar.* **60** (Suppl. 2): 117–140.

Motos, L., Cotano, U., Coombs, S. H., *et al.* (2004). Ichthyoplankton assemblages. In *Oceanography and Marine Environment of the Basque Country*, Borja, A. and Collins, M., eds. Elsevier Oceanography Series 70. Amsterdam: Elsevier, pp. 425–454.

Muck, P. (1989). Major trends in the pelagic ecosystem off Peru and their implications for management. In *The Peruvian Upwelling Ecosystem: Dynamics and Interactions*, Pauly, D., Muck, P., Mendo, J. and Tsukayama, I., eds. Callao: IMARPE, pp. 386–403.

Muck, P. and Fuentes, H. (1987). Sea lion and fur seal predation on the Peruvian anchoveta, 1953–1982. In *The Peruvian Anchoveta and Its Upwelling Ecosystem: Three Decades of Change*, Pauly, D. and Tsukayama, I., eds. Manilla, Philippines: International Center for Living Aquatic Resources Management, pp. 234–247.

Muck, P. and Sanchez, G. (1987). The importance of mackerel and horse mackerel predation for the Peruvian anchoveta stock (a population and feeding model). In *The Peruvian Anchoveta and Its Upwelling Ecosystem: Three Decades of Change*, Pauly, D. and Tsukayama, I., eds. Manilla, Philippines: International Center for Living Aquatic Resources Management, pp. 276–293.

Nakata, H., Funakoshi, S., and Nakamura, M. (2000). Alternating dominance of postlarval sardine and anchovy caught by coastal fishery in relation to the Kuroshio meander in the Enshu-nada Sea. *Fish. Oceanogr.* **9**: 248–258.

Okazaki, T., Kobayashi, T., and Uozumi, Y. (1996). Genetic relationships of pilchards (genus: *Sardinops*) with anti-tropical distributions. *Mar. Biol.* **126**: 585–590.

Orr, J.C., Fabry, V.J., Aumont, O., *et al.* (2005). Anthropogenic ocean acidification over the twenty-first century and its impact on calcifying organisms. *Nature* **437**: 681–686.

Parada, C., Mullon, C., Fréon, P., *et al.* (2008). Does vertical migratory behavior retain fish larvae onshore in upwelling ecosystems? A modelling study of anchovy in the southern Benguela. *Afr. J. Mar. Sci.* **30**: 437–452.

Parrish, R. H., Serra, R., and Grant, W. S. (1989). The monotypic sardines, *Sardina* and *Sardinops*: their taxonomy, distribution, stock structure and zoogeography. *Can. J. Fish. Aquat. Sci.* **41**: 414–422.

Pauly, D. and Soriano, M. (1989). Production and mortality of anchoveta (*Engraulis ringens*) eggs off Peru. In *The Peruvian Upwelling Ecosystem: Dynamics and Interactions*, Pauly, D., Muck, P., Mendo, J. and Tsukayama, I., eds. Manilla, Philippines: International Center for Living Aquatic Resources Management Conference Proceedings **18**, pp. 155–167.

Pauly, D. and Tsukayama, I., ed. (1987). *The Peruvian Anchoveta and its Upwelling Ecosystem: Three Decades of Change*. Manilla, Philippines: International Center for Living Aquatic Resources Management Contribution, p. 351.

Pauly, D., Chirinos de Vildoso, A., Mejía, J., *et al.* (1987). Population dynamics and estimated anchoveta consumption of bonito (*Sarda chiliensis*) off Peru, 1953 to 1982. In *The Peruvian Anchoveta and Its Upwelling Ecosystem: Three Decades of Change*, Pauly, D., and Tsukayama I., eds. Manilla, Philippines: International Center for Living Aquatic Resources Management, pp. 248–267.

Pauly D., Jarre A., Luna S. and Alamo A. (1989). On the quantity and types of food ingested by Peruvian anchoveta 1953–1982. In *Proceedings of the Expert Consultation to Examine Changes in Abundance and Species Composition of Neritic Fish Resources, San José, Costa Rica, 18–29 April 1983*, Sharp, G.D., and Csirke, J., eds. *FAO Fish. Rep.* **291** (3): 109–124.

Peliz, A., Dubert, J., Santos, A. M. P., *et al.* (2005). Winter upper ocean circulation in the Western Iberian Basin – fronts, eddies and poleward flows: an overview. *Deep-Sea Res. I* **52**: 621–646.

Perry, A. L., Low, P. J., Ellis, J. R., and Reynolds, J. D. (2005). Climate change and distribution shifts in marine species. *Science* **308**: 1912–1915.

Peterson, W. T. and Schwing, F. B. (2003). A new climate regime in Northeast Pacific ecosystems. *Geophys. Res. Lett.* **30**: 1896, doi:10.1029/2003GLO15.

Pickett, M. H. and Schwing, F. B. (2006). Evaluating upwelling estimates off the west coasts of North and South America. *Fish. Oceanogr.* **15**: 256–269.

Pingree, R. D. and Le Cann, B. (1990). Structure, strength and seasonality of the slope currents in the Bay of Biscay region. *J. Mar. Biol. Assoc. UK* **70**: 857–885.

Pingree, R. D. and Le Cann, B. (1992). Three anticyclonic Slope Water eddies (SWODDIES) in the southern Bay of Biscay in 1990. *Deep-Sea Res. A* **39**: 1147–1175.

Planque, B., Bellier, E., and Lazure, P. (2007). Modelling potential spawning habitat for sardine (*Sardina pilchardus*) and anchovy (*Engraulis encrasicolus*) in the Bay of Biscay. *Fish. Oceanogr.* **16**: 16–30.

Purca, S. (2005). Variabilidad temporal de baja frecuencia en el ecosistema de la Corriente Humboldt frente a Peru. Thesis, Universidad de Concepción, pp. 79.

Reid, P. C., Colebrook, J. M., Matthews, J. B. L., and Aiken, J. (2003). The continuous plankton recorder: concepts and history, from plankton indicator to undulating recorders. *Prog. Oceanogr.* **58**: 117–173.

Reiss, C. S., Checkley, D. M., Jr., and Bograd, S. J. (2008). Remotely sensed spawning habitat of Pacific sardine (*Sardinops sagax*) and Northern anchovy (*Engraulis mordax*) within the California Current. *Fish. Oceanogr.* **17**: 126–136.

Relvas, P., Burton, E. D., Dubert, J., *et al.* (2007). Physical oceanography of the western Iberia ecosystem: latest views and challenges. *Prog. Oceanogr.* **74**: 149–173.

Richardson, S. L. 1981. Spawning biomass and early life of northern anchovy, *Engraulis mordax*, in the northern subpopulation off Oregon and Washington. *Fish. Bull. US*, **78**: 855–876.

Roy, C., van der Lingen, C. D., Coetzee, J. C., and Lutjeharms, J. R. E. (2007). Abrupt environmental shift links with changes in the distribution of Cape anchovy (*Engraulis encrasicolus*) spawners in the Southern Benguela. *Afr. J. Mar. Sci.* **29**: 309–319.

Rykaczewski, R. R., and Checkley, D. M., Jr. (2008). Influence of ocean winds on the pelagic ecosystem in upwelling regions. *Proc. Nat. Acad. Sci.* **105**: 1965–1970.

Sagarminaga, Y., Irigoien, X., Uriarte, A., *et al.* (2004). Characterization of the anchovy (*Engraulis encrasicolus*) and sardine (*Sardina pilchardus*) spawning habitats in the Bay of Biscay from the routine application of the annual DEPM surveys in the Southeast Bay of Biscay. *ICES CM 2004/Q: 06.*

Sánchez, G., Alamo, A., and Fuentes, H. (1985). Alteraciones en la dieta alimentaria de algunos peces comerciales por efecto del fenómeno El Niño. *Bol. Inst. Mar Perú (Vol. Extraordinario)*, pp. 135–142.

Santander, H. (1987). Relationship of anchoveta egg standing stock and parent biomass off Peru, 4–14° S. In *The Peruvian Anchoveta and its Upwelling Ecosystem: Three Decades of Change*, Pauly, D. and Tsukayama, I., eds. Callao, Perú: IMARPE, pp. 179–207.

Santander, H. and de Castillo, O. S. (1973). Estudio sobre las primeras etapas de vida de la anchoveta. *Inf. Inst. Mar Perú*, **41**: 30.

Santander, H. and de Castillo, O. S. (1981). Algunos indicadores biológicos del ictioplancton. In *Memorias del Seminario Sobre Indicadores del Plancton, Seminario Realizado en el Instituto del Mar del Perú 8–11 de Setiembre de 1980*. *UNESCO Rep. Mar. Sci.*, **11**, Paris.

Santander, H. and Flores, R. (1983). Los desoves y distribución larval de cuatro especies pelágicas y sus relaciones con las variaciones del ambiente marino frente al Perú. In *Proceedings of the Expert Consultation to Examine Changes in Abundance and Species Composition of Neritic Fish Resources, San José, Costa Rica, 18–29 April 1983*, ed. G. D. Sharp and J. Csirke. *FAO Fish. Rep.* **291**: 835–867.

Santander, H. and Zuzunaga, J. (1984). Cambios en algunos componentes del ecosistema marino frente al Perú durante el fenómeno El Niño 1982–1983. *Rev. Com. Perm. Pacífico Sur* **15**: 311–331.

Santos, A. M. P., Peliz A., Dubert J., *et al.* (2004). Impact of a winter upwelling event on the distribution and transport of sardine eggs and larvae off western Iberia: A retention mechanism. *Cont. Shelf Res.* **24**: 149–165.

Santos, A. M. P., Kazmin, A. S., and Peliz, A. (2005). Decadal changes in the Canary upwelling system as revealed by satellite observations: Their impact on productivity. *J. Mar. Res.* **63**: 359–379.

Schwartzlose, R. A., Alheit, J., Bakun, A. *et al.* (1999). Worldwide large-scale fluctuations of sardine and anchovy populations. *S. Afr. J. Mar. Sci.* **21**: 289–347.

Serra, J.R. (1983). Changes in the abundance of pelagic resources along the Chilean coast. In *Proceedings of the Expert Consultation to examine changes in abundance and species composition of neritic fish resources. San José, Costa Rica, 18–29 April 1983*, Sharp, G. D. and Csirke, J., ed. *FAO Rep.* **291**: 255–284.

Silva, A., Santos, M. B., Caneco, B. *et al.* (2006). Temporal and geographic variability of sardine maturity at length in the northeastern Atlantic and the western Mediterranean. *ICES J. Mar. Sci.* **63**: 663–676.

Smith, P. E. (1990). Monitoring interannual changes in spawning area of Pacific Sardine (*Sardinops sagax*). *CalCOFI Rep.* **31**: 145–151.

Smith, P. E. (2005). A history of proposals for subpopulation structure in the pacific sardine (*Sardinops sagax*) population off western North America. *CalCOFI Rep.* **46**: 75–82.

Smyth, T. J., Coombs, S. H., Kloppmann, M. H. F. (2005). Use of satellite data for modelling food availability and survival of marine fish larvae. In *Proceedings of RSPSoc 2005: Measuring, Mapping and Managing a Hazardous World, 6–9 September 2005, Portsmouth UK*, Teeuw, R., Whitworth, M. and Laughton, K., eds. The Remote Sensing and Photogrammetry Society (RSPSoc), pp. 1–13.

Soutar, A. and Isaacs, J. D. (1974). Abundance of pelagic fish during the 19th and 20th centuries as recorded in anaerobic sediments off California. *Fish. Bull.* **72**: 259–275.

Somarakis, S., Palomera, I., García, A. *et al.* (2004). Daily egg production of anchovy in European waters. *ICES J. Mar. Sci.* **61**: 944–958.

Stenevik, E. K., Sundby, S., and Cloete, R. (2001). Influence of buoyancy and vertical distribution of sardine *Sardinops sagax* eggs and larvae on their transport in the northern Benguela system. *S. Afr. J. Mar. Sci.* **23**: 85–97.

Stratoudakis, Y., Coombs, S., Halliday, N. *et al.* (2004). Sardine (*Sardina pilchardus*) spawning season in the North East Atlantic and relationships with sea surface temperature. *ICES CM 2004/Q:19*.

Stratoudakis, Y., Bernal, M., Ganias, K., and Uriarte, A. (2006). The daily egg production method: Recent advances, current applications and future challenges. *Fish and Fisheries* **7**: 35–57.

Stratoudakis, Y., Coombs, S., Lago de Lanzós, A. *et al.* (2007). Sardine (*Sardina pilchardus*) spawning seasonality in European waters of the northeast Atlantic. *Mar. Biol.* **152**: 201–212.

Sverdrup, H. U., Johnson, M. W., and Fleming, R. H. (1960). *The Oceans, Their Physics, Chemistry, and General Biology*. New York: Prentice-Hall.

Tarazona, J. (1990). Disturbance and stress associated to El Niño and their significance for the macrobenthos of shallow areas of the Peruvian Upwelling Ecosystem. Dissertation, U. Bremen, Germany, pp. 82.

Tsukayama, I. (1982). Recursos pelágicos y sus pesquerías en Perú. *Rev. Com. Perm. Pacífico Sur.* **13**: 25–63.

Twatwa, N. M., van der Lingen, C. D., Drapeau, L., *et al.* (2005). Characterizing and comparing the spawning habitats of anchovy (*Engraulis encrasicolus*) and sardine (*Sardinops sagax*) in the southern Benguela upwelling ecosystem. *Afr. J. Mar. Sci.* **27**, 487–499.

Uriarte, A., Prouzet, P., and Villamor, B. (1996). Bay of Biscay and Ibero-Atlantic anchovy populations and their fisheries. *Sci. Mar.* **60**: 237–255.

Uriarte, A., Sagarminaga, Y., Scalabrin, C., and Valenand Jiménez, M. (2001). Ecology of anchovy juveniles in the Bay of Biscay four months after peak spawning: do they form part of the plankton? *ICES CM 2001/W:20*.

Valencia, V., Franco, J., Borja, Á., and A. Fontán. (2004). Hydrography of the southeastern Bay of Biscay. In *Oceanography and Marine Environment of the Basque Country*, Borja A. and Collins M., eds. *Elsevier Oceanogr. Ser.*, 70: 159–94. Amsterdam: Elsevier.

van Aken, H. M. (2002). Surface currents in the Bay of Biscay as observed with drifters between 1995 and 1999. *Deep-Sea Res. I*: **49**, 1071–1086.

van der Lingen, C. D. (2005). Characterizing spawning habitat of anchovy (*Engraulis encrasicolus*), redeye round herring (*Etrumeus whiteheadi*) and sardine (*Sardinops sagax*) from CUFES sampling in the southern Benguela. *GLOBEC Rep.* **21**: 29–30.

van der Lingen, C. D., Coetzee, J. C., and Hutchings, L. (2002). Temporal shifts in the distribution of anchovy spawners and their eggs in the southern Benguela: implications for recruitment. *GLOBEC Rep.* **16**: 46–48.

van der Lingen, C. D. and Huggett, J. A. (2003). The role of ichthyoplankton surveys in recruitment research and management of South African anchovy and sardine. In *The Big Fish Bang: Proceedings of the 26th Annual Larval Fish*

Conference, Browman, H. I. and Skiftesvik, A. B., Bergen, Norway: Inst. Mar. Res., pp. 303–343.

van der Lingen, C. D., Castro, L., Drapeau, L., and Checkley, D., Jr., eds. (2005a). Report of a GLOBEC-SPACC Workshop on characterizing and comparing the spawning habitats of small pelagic fish. *GLOBEC Rep.* **21**: 33 pp.

van der Lingen, C. D., Coetzee, J. C., Demarcq, H., *et al.* (2005b). An eastward shift in the distribution of southern Benguela sardine. *GLOBEC International Newsletter* **11**: 17–22.

van der Lingen, C. D., Fréon, P., Hutchings, L., *et al.* (2006). Forecasting shelf processes of relevance to living marine resources in the BCLME. In Shannon, V., Hempel, G., Malanotte-Rizzoli, P., *et al. Predicting a Large Marine Ecosystem.* Large Marine Ecosystem Series **14**: 309–347.

Vaz, S. and Petitgas, P. (2002). Study of the Bay of Biscay anchovy population dynamics using spatialised age-specific matrix models. *ICES CM 2002/O: 07.*

Visser, M.E. and Both, C. (2005). Shifts in phenology due to global climate change: the need for a yardstick. *Proc. Roy. Soc. B – Biol. Sci.* **272**: 2561–2569.

Ware, D. M. (1992). Production characteristics of upwelling systems and the trophodynamic role of hake. *S. Afr. J. Mar. Sci.* **12**: 501–513.

Yasuda, I. (2003). Hydrographic structure and variability in the Kuroshio–Oyashio Transition area. *J. Oceanogr.* **59**: 389–402.

Zenitani, H., Ishida, M., Konishi, Y., *et al.* (1995). *Distributions of Eggs and Larvae of Japanese Sardine, Japanese Anchovy, Mackerels, Round Herring, Japanese Horse Mackerel and Japanese Common Squid in the Waters around Japan, 1991 through 1993.* Japan Fisheries Agency, 368 pp.

Zuta, S. and Guillén, O. (1970). Oceanografía de las aguas costeras del Perú. *Bol. Inst. Mar Perú* **2**: 157–324.

4 Variability from scales in marine sediments and other historical records

David B. Field, Tim R. Baumgartner, Vicente Ferreira, Dimitri Gutierrez, Hector Lozano-Montes, Renato Salvatteci, and Andy Soutar

CONTENTS

Summary

Records of variability in populations of small pelagic fishes exist from a variety of historical sources that precede industrial fishing catch records. We review the historical records of artisanal fisheries, archeological remains, and fish remains from marine sediments. Fish scale deposition rates from ocean sediments offer the most quantitative records with little bias from anthropogenic factors. As quantitative estimates from fish scale deposition rates and their comparison with other records depend on chronostratigraphies, we discuss chronological development in detail, as well as the preservation and significance of fish scale flux. The different historical records indicate considerable variability in small pelagics prior to industrial fishing. However, the historical records provide little support for paradigms of ecosystem variability based on industrial catch records, such as synchronous worldwide fluctuations in abundance of small pelagic from different boundary currents or alternations of sardines and anchovies within a given boundary current. Rather, a variety of different modes of variability in small pelagics is consistent with paleoceanographic evidence for many different climate states and modes of variability.

Introduction

Some of the best evidence of long-term variability in marine populations comes from different records of pelagic fishes. There is evidence from industrial catch records of many decades in length, artisanal catch records, historical observations, archeological remains, and fossil remains in marine sediments. Small pelagics reflect many aspects of climate change effects on fisheries, since their recruitment and population size are sensitive to environment conditions. However, understanding variability in population size of small pelagics is intertwined with understanding their migrations (presumably in search of ideal environmental conditions). This chapter reviews the collective evidence for variability in small pelagic populations observed prior to industrial catch records and what these different records reveal about past variability in small pelagic fishes.

The use of fish scales as indicators of variability in pelagic fish populations began with the innovative work of Soutar (1967) and Soutar and Isaacs (1969). By sifting different layers of Santa Barbara Basin (SBB) sediments, Soutar and Isaacs (1969) quantified changes in the numbers of scales of sardines and anchovies deposited to the sediments during different decades in time. Finding a large amount of natural variability in scale deposition rates (SDR) to the sediments they inferred natural fluctuations, independent of industrial fishing, in the populations of *Sardinops sagax* (Pacific sardine), *Engraulis mordax* (northern anchovy), and *Merluccius productus* (Pacific hake) off California throughout the last two millennia. Their results implied that the collapse of the California sardine fishery may have been largely due to natural variability.

Interannual ENSO variability was just becoming recognized as having an impact on ocean circulation, biogeochemistry, and ecosystems that could be discernable over meso-scale variability when fish scale deposition rates to the Santa Barbara Basin were reported (Soutar, 1971). Nonetheless, the Holocene was generally considered a stable time period with little consideration of decadal-scale variability in marine populations. Skepticism remained over the significance of changes in the number of scales from a single core taken from a single point in the ocean. Baumgartner *et al.* (1992) showed that the results of several cores from the SBB were consistent with one another, and could be used to define multiple collapses and expansions of sardines over interdecadal timescales. The variability in small pelagics documented by Baumgartner *et al.* (1992) was persuasive in illustrating that a shift in climate and ecosystems that occurred in the mid 1970s was not unlike prior variability. While it is clear that there is some influence of the warming trend on the mid 1970s climate shift in the North Pacific (Field *et al.*, 2006), it is now accepted that there is substantial variability in marine populations on decadal to millennial timescales, which we refer to here as long-term variability.

Understanding the dynamics of long-term variability requires multiple historical records of small pelagic fish populations across the habitat range (see Figs. 4.1 and 4.2). We first discuss records from historical observations and archeological remains that primarily reflect the presence of populations in coastal environments. We then focus on sedimentary sequences, as these offer continuous records of quantifiable variations in time. However, the inferred temporal variations are also highly susceptible to issues associated with chronological development and degradation of remains. Therefore, we discuss the assumptions and problems associated with developing chronostratigraphies, beginning with the biogeochemical conditions necessary for the preservation of fish scales in marine sediments. We then discuss the assumptions and significance of fish scale deposition records followed by the major lessons learned from all of the different historical records.

Historical observations

A compilation of historical records of sardine and herring fisheries off northern Europe and comparison of these records with large-scale climate change associated with the North Atlantic Oscillation (NAO) was done by Alheit and Hagen (1997). They reported on multiple records of fisheries on small pelagics from France, England, and Sweden over the last 500 years, with records of herring fishing off the Swedish coast of Bohuslän extending back nearly a thousand years.

Bohuslän herring events represent periods where mass abundances of herring were frequently available to Swedish fisherman in skerries and fjords during fall and winter. Decades of Bohuslän herring catch using beach seines were followed by decades of absence. Herring presence (absence)

along the Bohuslän coastline occurred during periods of absence (presence) of spring-spawning herring off northern and northwestern Sweden. The variations in catches of different stocks of Norwegian spring-spawning herring and Bohuslän herring on different parts of the coast of Sweden appear to be related to changes in the NAO, which affected both total stock size and the migration of stocks to nearshore environments (Alheit and Hagen, 1997).

Records of coastal fishery catches in northern France and England also reflect changes in availability of different species and stocks to coastal fisheries. The English Channel is the northern biogeographic edge of sardine and southern edge of herring. There was an apparent alternation between sardine and herring fisheries around the English channel, with sardines occurring during warmer periods and herring during cooler episodes. As fishing was local with little capacity to search for fish offshore across the range of their distribution, the records reflect some combination of fish stock migrations and abundance (Alheit and Hagen, 1997). Remarkably, the periods of presence or absence of different stocks in the English Channel generally coincided with the variations off the Norwegian coast, reiterating the role of large-scale climate forcing associated with the NAO (Alheit and Hagen, 1997).

Several qualitative descriptions of good and poor sardine fishing years in the Japan Sea have been kept since the sixteenth century while records from multiple areas off Japan have been documented since the eighteenth century (Hiramoto, 1991 and references therein). These records generally reflect fishing success in coastal areas by qualifying each year as "good" or "poor." The records probably reflect an increase in population size associated with an expansion of its range since sardine distribution and commercial fishing effort off Japan extend well offshore during years of an expanded population. Fishing success records indicate persistence of good periods lasting from 7–45 years and poor periods lasting from 14–35 years, which may be related to broader North Pacific climate (Yasuda *et al.*, 1999).

Observations from early explorers, fur-traders, naturalists, and surveys also offer some glimpses of past population states. While such records are not consistently maintained in time, they can reflect presence/absence across biogeographic ranges. For example, the presence of sardine off Oregon, Washington, and British Columbia in the Northeast Pacific is believed to reflect population range expansion associated with warmer conditions. Historical observations of sardines in the Northeast Pacific have been reported by Field *et al.* (2001). Sardines were observed by multiple sources in the late 1700s, including by trained naturalists. However, multiple naturalists, observations, and surveys done in the late nineteenth century, including investigations of the US Fish Commission in 1880–81, failed to locate sardines in the northern California Current. Sardines were

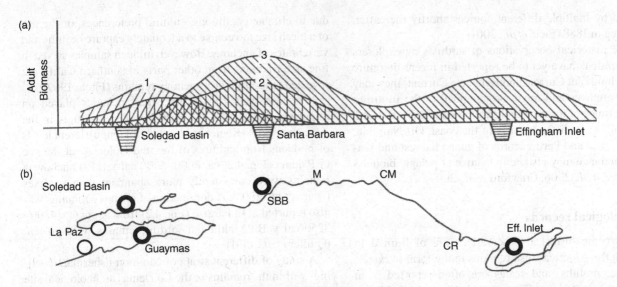

Fig. 4.1. (a) Model of how multiple sedimentary environments from the California Current region (British Columbia to the right and Baja California to the left) could reflect different scenarios for population abundance and distribution of the California sardine: (1) A diminished population contracted to Baja California and the Gulf of California; (2) a moderate population distributed throughout the coast, with highest densities off central California; and (3) an expanded population with a distribution centered off central California and the northern regions of the California Current. (b) Location of known sedimentary sites capable of resolving SDR (bold circles) and potential sites (thin circles). Monterey (M), Cape Mendocino (CM), and the Columbia River (CR) are indicated for reference. Orientation as in (a).

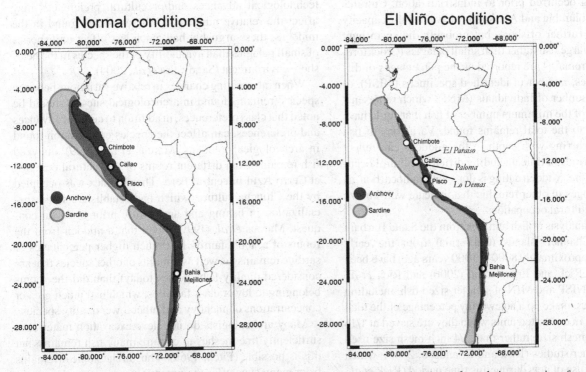

Fig. 4.2. Theoretical distribution of anchovy and sardine off Peru and northern Chile during times of population expansion of anchoveta during normal conditions and contraction of anchoveta towards the coast and south during El Niño conditions. Sardine are found further offshore than anchoveta (normal conditions) but move south and closer to shore during population expansions (El Niño conditions). Location of sedimentary sites (circles) where SDR records are being developed and may resolve shifts in population density and distribution. Also shown are archeological sites where fish remains have been recovered.

reported by multiple different sources shortly thereafter, beginning in 1888 (Field *et al.*, 2001).

While historical observations of sardines by explorers and naturalists have yet to be reported in recent literature for the Humboldt Current and Benguela Current, they may exist in archives and might be compiled in the future to indicate times of a substantial sardine population size and its expansion to different parts of the coast. Off Namibia, South Africa, and Peru, records of guano harvest and seabird abundance may also be indicative of pelagic biomass (Gutierrez *et al.*, 2006; Crawford *et al.*, 2007).

Archeological records

While written historical observations are of limited in length in the Americas, fish remains in the form of bones, vertebrae, otoliths, and scales are often reported from archeological middens. These records have revealed information on the range of marine diets of Native Americans, and presumably the fishes available to them (Fitch, 1969; Rick and Erlandson, 2000; Sandweiss *et al.*, 2004), including terrestrial communities that utilized salmonids (Butler and O'Conner, 2004).

Fish remains from middens may reflect the availability of different populations in the nearshore environment where most fishing occurred prior to industrialization. Cultures of British Columbia and Alaska may have had the capacity to navigate further offshore, but probably did so to hunt whales and large fish rather than small pelagics (Crockford, 1997). Fish remains are reported as the percentages of different species, number of identified specimens (NISP), or minimum number of individuals (MNI), which represents an estimate of the minimum number of fish that could have contributed to the total remains found. Variations in fish remains occurring with depth in the middens can reflect different periods in time, including different cultural occupations of a site. Midden dating is done with radiocarbon, as well as changes in other remains that indicate well-known changes in cultural occupations.

Detailed analysis of fish remains from the Santa Barbara area and Channel Islands (California) from the early Holocene (approximately 8000–10000 years ago) have been reported by Rick and Erlanderson (2000) and Rick *et al.* (2001). The NISP and MIN of smaller-sized fish, including small pelagics, make up a noteworthy percentage of the total fish remains in excavated units when they are sieved at 1/16 or 1/8 inch mesh size, rather than 1/4 inch mesh size used by many other studies (Rick and Erlandson, 2000), which suggests the use of nets during this time period (Rick *et al.*, 2001). There is evidence for varying amounts of clupeoids (sardine and herring) relative to other fish species in different strata, but neither of these studies reported the presence of anchovy remains. The absence of anchovy could be due to climate conditions, cultural preferences, or the use of a mesh size too coarse to adequately capture otoliths and vertebrates of anchovy. However, midden samples sieved in fine size fractions from other parts of southern California revealed predominantly anchovy remains (Fitch, 1969).

It is also well known that small pelagics played an important dietary role for early Holocene cultures in the Atacama desert (Keefer *et al.*, 1998). In four different levels of middens from cultures of the mid-Holocene at the site of Paloma (Fig. 4.2; ca. 8600–5400 cal y B.P.[1]) anchoveta remains were consistently more abundant than sardines (Reitz *et al.*, 2003). A domination of anchovy remains was also reported at El Paraíso (Fig. 4.2) from about (ca. 4200–3250 cal y B.P.), although sardine remains were present (Quilter *et al.*, 1991).

A study of different strata of common fisherman dwellings within the remains of the Lo Demás archeological site in the Chincha Valley, Peru (Fig. 4.2) were chronologically constrained by changes in ceramics and introduction of foreign seeds (Sandweiss *et al.*, 2004). The Chincha culture was conquered by the Inca in 1479, who in turn were invaded by the Spanish in 1532. Sandweiss *et al.* (2004) found a change from an anchovy dominant MNI during the Chincha period to a more equal split between sardine and anchovy MNI during the Inca period at Lo Demás. While technological advances and/or cultural preferences may affect the relative numbers of these species found in the middens, these arguably had much less effect on fisheries of small pelagics than availability of the species in the nearshore environment (Sandweiss *et al.*, 2004).

When interpreting changes in relative variations between species' remains found in archeological sites, it should be noted that class preferences, in addition to cultural advances and preferences, can affect the species composition found in archeological middens. Marcus *et al.* (1999) analyzed fish remains from different rooms of a structural complex at Cerro Azul in central Peru. The complex was occupied by the Chincha culture, which had established some specialization in fishing and agriculture prior to Incan conquest. Marcus *et al.* (1999) found that a midden from the rooms of a noble family had much higher percentages of sardine remains (as well as remains of other species that are considered quality fish species today) than did the rooms belonging to lower class families, which had much greater concentrations of anchovy and other lower quality species.

Many archeologists do not sieve excavation material at sufficiently fine meshes to capture many fish remains, but this is possible (Fitch, 1969). Thus archeological remains from many sites offer the potential to examine changes in species available in nearshore environments, particularly if independent records, such as those from sedimentary environments, suggest a change in abundances of small pelagics or climate that could be further corroborated.

Sedimentary records

Oceanographic settings for sedimentary records

The presence of pelagic fishes over a particular sedimentary environment at some point in time is, of course, a prerequisite for using fish scales to reconstruct population variability. However, records from both the fringes of historical ranges, or possible historical ranges, as well as those from the center of the population's distribution, are all important to understanding past variation in population dynamics (see Figs. 4.1 and 4.2).

The most limiting factor is finding sedimentary environments with sufficiently low oxygen and high sedimentation rates to result in a continuous, high-resolution sedimentary sequence. Disoxia helps to both preserve fish remains and eliminates the presence of benthic fauna that bioturbate sedimentary sequences. Fortuitously, low oxygen conditions that are necessary for preserving stratigraphic sequences and fossil remains often coincide with regions of high productivity, including pelagic fishes. Sufficiently low oxygen concentrations occur due to some combination of the following processes: (1) intermediate waters that are relatively old and depleted of oxygen through remineralization of organic matter during water mass passage through the ocean conveyor circulation, (2) active upwelling of nutrient rich (and oxygen deficient) water resulting in high primary productivity and increased vertical transport of organic matter to the subsurface, which further reduces local oxygen concentrations, and (3) reduced circulation due to a basin or sill, which prevents water mass renewal and results in depleted oxygen concentrations.

The type of sedimentary environment confines the degree to which taking cores across lateral distances can be done for chronostratigraphic development and to span different points of a population's range. The Santa Barbara Basin (SBB) and Soledad Basin (San Lazaro) have all three of the aforementioned characteristics (Fig. 4.1). Fjords such as Effingham Inlet or Saanich Inlet have shallow sills that result in very long residence times and hence disoxia, even though waters entering the basin are not initially as depleted in oxygen as other regions. Although oxygen levels are much higher in the western Pacific, a bay in Japan apparently has adequate productivity to result in high sedimentation rates and preservation of fish scales (Kuwae *et al.*, 2007). Margins that have large ranges of laminated sediments throughout a very depleted oxygen minimum zone include the Peru shelf, Guaymas in the Gulf of California, and the Benguela Current. In these regions, cores can be taken across large lateral distances of 10s to 100s of kilometers to develop chronostratigraphies and comparisons of records between cores (Fig. 4.3).

Chronostratigraphies and fish scale deposition rates (SDRs)

There are a multitude of sedimentary factors to consider in the development and interpretation of records of fish scale deposition rates (SDRs). Of these, the chronostratigraphy is of fundamental importance to calculating relative dates and sedimentation rates, which give the temporal period of any given sampling interval for estimating SDR. Other important factors are core type, sediment area and volume provided, sampling resolution, and sampling replication.

Coring and sampling techniques

Recovery of surface sediments is important for calibration and development of the chronostratigraphy and can be achieved with a multi-core or Soutar box core. A Soutar box core provides more sediment area (20 × 20 cm) for quantification, rectangular slabs that facilitate X-radiography, and development of other proxy records of climate change. The recovery of deeper sedimentary histories can be done with Kasten cores and piston cores. Kasten cores provide much greater sampling area (15 × 15 cm) and can capture up to ~3 m of sediments. Piston cores generally provide insufficient material (~4–12 cm diameter) for reliable presence/absence estimates of many fish species or decadal-scale resolution of abundant species, but can capture much deeper sedimentary histories (Tunicliffe *et al.*, 2001).

Fish remains can also be recovered by sieving sediments of outcropped sediments (Fitch, 1969). There are also numerous fossil imprints of scales and even whole fish throughout the Monterey formation, a Miocene analog of the SBB sedimentary environment and other uplifted sedimentary sequences. Similar sequences off Peru also contain abundant scales and other remains (DeVries, personal communication).

Chronostratigraphy

Chronological development is a critical factor for records of temporal variability. The ideal scenario is that of the SBB or Saanich Inlet where counts of annual laminae or "varves" facilitates chronostratigraphy development. The uncertainty in estimating varves (and hence years and SDR) may stem from erasure of laminae, presence of seasonal laminae within an annual varve pair, or the presence of slumps that appear to be varves or bioturbated sequences. While error in absolute dates increases downcore in all records, the confidence in the number of years in a given sampling interval remains high in varved records because they are identifiable despite changes in sedimentation rates, hiatuses, or instantaneous deposits. The error in estimating SDR and absolute dates increases in sediments that are laminated, but not varved, and increases even more in sediments that are bioturbated.

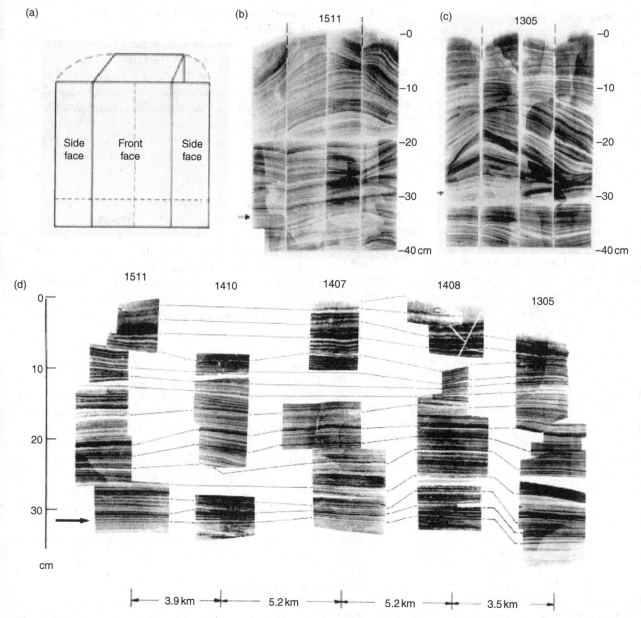

Fig. 4.3. Comparison of laminae structure both within and between cores from the Guaymas slope (Gulf of California) illustrating the horizontal variability in laminae structure that can confound chronostratigraphies. (a) Diagram of the slabs of core faces used for X-radiograph composites of the upper 40 cms of box core. (b) Radiographs of four sides of core GCBC-7807–1511. (c) Radiographs of four sides of core GCBC-7807–1511. (d) Reconstructed lamina sequences from five box cores from the Guaymas slope. Vertical continuity within each core is maximized by fitting selected sections from different regions of the cores. Correlation lines mark prominent laminae which are clearly identifiable among the five cores. Blank areas within the lamina sequences indicate discontinuities which occur across an entire core. Distances between core sites are shown at the bottom of the figure. Arrows indicate a sediment gravity flow forming the base of each reconstruction. The gravity flow is dated at ~1907 by ²¹⁰Pb, which is consistent with a known 7.5 seismic event in the Gulf of California in 1907 (from Baumgartner *et al.*, 1991).

Chronostratigraphies are usually most accurate in the last 100–200 years where there is good control with radioisotopes and less accumulation of errors. The half-life of ^{210}Pb is 22.3 years, meaning that it can be used to estimate sediment ages up to ~100–150 years ago. ^{241}Am in the atmosphere resulted from bomb testing in the 1950s and 1960s, and thus serves as a good tracer for these time periods. Beyond the last 100 years, ^{14}C is considered the most reliable constraint on absolute dates and sedimentation rates where varves are not present. However, ^{14}C values must be corrected for the oceanic reservoir age, which can vary as intermediate water masses change. Tephra layers (volcanic ash) or seismicly induced slumps attributable to historically known dates may provide additional chronological constraints (e.g. Baumgartner *et al.*, 1991).

Chronostratigraphic obstacles

It is well known that varve thicknesses, and hence sedimentation rates, vary considerably on interannual to centennial timescales due to changes in productivity and terrestrial run-off (Soutar, 1967; Soutar and Crill, 1977; Lange *et al.*, 1996). Sediment winnowing and erosion by bottom currents can also occur on continental platforms. Changes in sedimentation rate are occasionally taken into consideration at coarse resolution in a core, but rarely considered at the sampling resolution. However, such changes can have major effects on estimated SDR and chronologies.

Physical disturbances, such as bioturbation and slumps, complicate the development of a chronostratigraphy in otherwise continuous sedimentary sequences. Slumps can be associated with both instantaneous deposits of sediments from upslope and/or erasures of material to the downslope. Bioturbation disturbs the continuity of a record but generally does not result in a hiatus. Physical disturbances also include variations, sometimes dramatic, on the scales of centimeters (Fig. 4.3b, 4.3c).

Multiple cores, separated by 100s to 1000s of meters, can reveal the differences in structure of physical disturbances to indicate the mechanism and significance of the disturbance. Figure 4.3d shows how cores taken from kilometers apart in the Gulf of California vary widely in their preservation of particular sedimentary sequences and laminae structure (Baumgartner *et al.*, 1991). There are both discontinuities and changes in laminae structure that would otherwise be missed in a single core, or even a single slab of a particular core (Fig. 4.3). X-radiographs of Kasten cores from different locations throughout the SBB also reveal slumps of varying thickness (Fig. 4.4). The relative different thickness of each slump not only indicates that they are instantaneous deposits from upslope (rather than bioturbation events) but is also indicative of the area of greatest intensity of the slide (Fig. 4.4). Additionally, some cores from different areas of the SBB preserve particular laminae or varve sequences that are not observed in all cores.

The most well-studied sites show that there is considerable horizontal variability in laminae sequences with many slumps that are not resolvable with single cores. Consequently, records having weak chronological constraints should only be compared with other records (sediment cores, tree rings, corals, ice cores, historical data, etc.) within the limits of the chronology. Due to possible uncertainties in chronologies that increase downcore, sufficient confidence regarding decadal-scale variability is probably only resolvable up to 100–150 years ago. Beyond 200 years ago, reliable comparisons between records with

independent chronologies may be efficient only on the interdecadal timescale while centennial-scale variability can probably be effective up to several thousand years ago.

Variability in scale deposition rate (SDR) between sites

Multiple cores within a region can also verify the extent to which records of SDR for one area of sediment reflect fluxes to the sediment environment within a broader region. Differences may arise from random events, sedimentary differences, and/or in association with movements of fish populations. Variability in SDR across centimeters has been quantified in different slabs of sediments from the same core (Table 4.1). The correlations are low in part because the number of scales in each slab generally ranges from 0–10 and are insufficient to result in a reliable estimate of SDR (Lozano-Montes, 1997). Estimates of sardine SDR between a Kasten core and two piston cores in the SBB show that the principal downcore signals are quite coherent and can be reproduced despite being separated by kilometers (Fig. 4.4; Table 4.1; Baumgartner *et al.*, 1992). While the records in Fig. 4.3 were developed with independent chronologies, the decadal and centennial-scale patterns of SDR are clearly coherent between cores.

O'Connell and Tunnicliffe (2001) also showed clear coherence in herring SDR between cores from Saanich Inlet. On a larger scale, there is some coherence in anchovy SDR over multidecadal to centennial timescales from Peruvian sediments taken 100s of kilometers. apart (Gutierrez *et al.*, 2006). Lack of coherence at decadal time scales could be from chronological differences, erosive or more oxic sedimentary environments, or from differences in local populations.

Interpreting scale deposition rates

SDRs calculated from a given sampling interval are generally representative of the actual fluxes to a sedimentary environment when sufficient areas of sediment and number of scales are examined. We here discuss the basis for interpreting SDR to the sediments. Multiple lines of evidence show that fish SDR reflect local abundances of fish, primarily from shedding of scales from live fish; the deposition of a piece of a fish from a mortality event is very rare. We discuss the population dynamics that can affect relationships between the local abundance of small pelagics and their regional variability. Finally, we review how geochemical composition of scales may contain environmental histories of fish and SDR calibrations to biomass.

Scales and scale shedding

Clupeoids and many other fish are caducous by nature (they shed their scales), which may enhance escape from predators via visual confusion by a flurry of scales. New

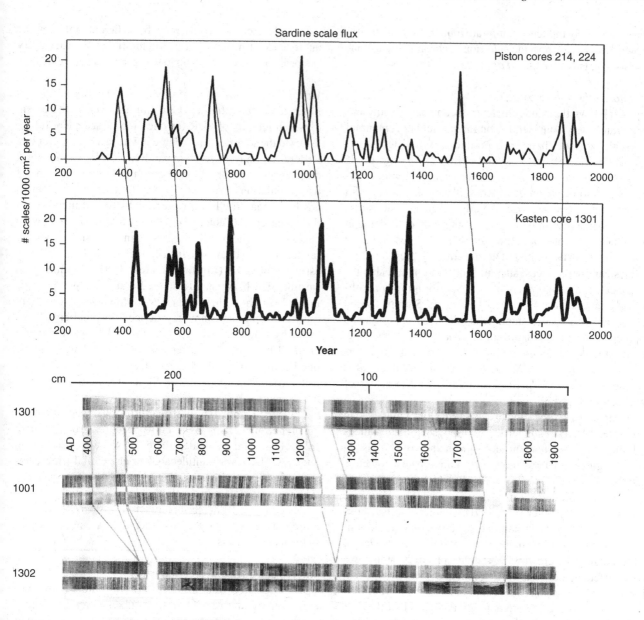

Fig. 4.4. Comparison of sardine scale deposition rates (SDRs) from the composite counts of two piston cores (from Soutar and Isaacs, 1969) with Kasten core SBKC 9110–1301. Lines indicate correlations between SDR records, each of which is dated with independent chronologies. Note the offset of nearly 50 years in the chronology from independent varve counts from two piston cores relative to three Kasten cores. X-radiographs from three Kasten cores were used to anchor the master chronology used for SBKC 9110–1301. Lines between each two sets of X-radiographs from a Kasten core show slumps of varying thickness between cores.

scales that grow where previously scales were lost (which are generally thinner and without clear circuli) are common on living fish (Shackleton, 1986). Fish scales are made of hydroxyapatite $Ca_{10}(PO_4)_6(OH)_2$ embedded within an organic matrix of fibrous protein collagen. As fish grow, scales grow laterally through deposition of both hydroxyapatite and the organic matrix. Thickening of the organic matrix continues during fish growth (with no addition of hydroxyapatite).

The types of scales found in the sediments indicate that they are primarily derived from scale shedding. Scale morphology varies over the body of clupeoids and can be characterized into five different categories (Fig. 4.5; Shackleton, 1986; Lozano-Montes, 1997). Typical scales ("type T" in sardines; e.g. Fig. 4.5) are found along the lateral line and make up ~35% of the total scales on a sardine. In anchovy, Typical scales have been found to make up from 42%–49% of the total number of scales and have been

Table 4.1. *Shared variability between different time series of scale deposition rates (SDRs) from Santa Barbara Basin sediments*

Time series	R value
Sardine between slabs	0.30 (0.38)
Anchovy between slabs	0.37 (0.43)
Sardine between cores	0.53
Anchovy between cores	0.54
Sardine with anchovy	0.32

Variability between slabs is the average coefficient of correlation between four different slabs of sediment from Kasten core 1301 (and 1302) after Lozano-Montes (1997). Correlations between cores come from the comparison of two piston cores with a low amount of sediment area quantified. The correlation between sardine and anchovy is the average of the two piston cores (after Baumgartner *et al.*, 1992).

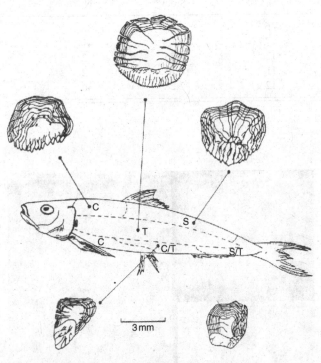

Fig. 4.5. Five representative scale morphologies and their locations on the body of a Pacific sardine (*Sardinops sagax*) using the nomenclature proposed by Shackleton (1986) for *Sardinops ocellata*. Typical scales ("type T") are those along the lateral line.

classified as "type X or Y" (Shackleton, 1988; Table 4.2). Collections of scales shed from fish in controlled aquarium environments have shown that Typical scales are highly disproportionately shed from living fish (Table 4.2). Since the majority of sardine and anchovy scales in the sedimentary records are Typical as well (Table 4.2), most of these scales originate from shedding.

Scales are shed during schooling behavior, but scale loss is much higher during pursuit, predation, and/or mortality events (Lozano-Montes, 1997; O'Connell and Tunnicliffe, 2001). A predation event would most likely result in the loss of scales during pursuit or impact. Scales do not pass through the guts of most predators, while bones, vertebrae, and otoliths have been documented to pass through the guts of some predators, but not others (DeVries and Pearcy, 1982; O'Connell and Tunnicliffe, 2001). Other scale types are as durable as Typical scales, but are not observed in as high proportions in the sediments or aquariums as observed on fish (Lozano-Montes, 1997). Thus only scale shedding, rather than mortality and consequential passage through guts, can account for the disproportionate representation of typical scales in marine sediments.

Several independent observations show that, on very rare occasions, deposition of many scales (and bones) can be instantaneous, likely derived from a part of a single fish. Observations include many hake scales and bones found in part of a core from the SBB with rapidly diminishing numbers in slabs several centimeters further away (Soutar, 1967; Field, personal observation). There is also an observation of four vertebrae lined up with each other within a piece of sediment from Peru with scales and bones lining up on either side, which is clearly the consequence of the preservation of a part of an anchovy. In contrast, one frequently encounters individual scales while cutting sediment, with no nearby scales, bones, or vertebrae associated (Ferriera, Field, and Salvateci, personal observation). Additionally, O'Connell and Tunnicliffe (2001) noted that one of four box cores analyzed had a large spike in herring scales in one sampling interval that was an order of magnitude higher than the background levels and not present in other cores.

Replicate sampling can clearly identify rare events of instantaneous deposition of many scales from a portion of a fish. After elimination of one outlier interval, O'Connell and Tunnicliffe (2001) found consistent downcore variability in SDR between four box cores spanning 130 years. Their result is consistent with the signal to noise ratios found between different slabs of the same core (Table 4.1) and different cores from the SBB (Baumgartner *et al.*, 1992; Table 4.1; Fig. 4.4). The coherent signals across horizontal sampling intervals indicate that SDR stem primarily from a rain of scales from shedding and predation that varies with fish density.

Scale degradation

The oceans are undersaturated in apatite, which could result in dissolution of scales and bones (Suess, 1981). Bacterially mediated degradation may be common within the organic matrix of the scale in oxygenated environments. In scales from laminated sediments, there is a loss of organic matter in the fossil scales relative to scales from living fish (Table 4.3).

Table 4.2. *Percent of "Typical" scales (those derived from the lateral line characterized by having a symmetric form) found on living fish (see Fig. 4.5) shed during aquarium experiments and found in ocean sediments*

Source of scales	Shackleton (sardine)	Lozano (sardine)	Shackleton (anchovy)	Lozano (anchovy)	Salvatteci (anchovy)
Living fish	34%	36%	47%	49%	42%
Aquarium	65%	85%	n.a.	n.a.	55%
Sediments	67%	66%	n.a.	n.a.	85%

Typical scales from sardine are 'type T' (Fig. 4.5) while those from anchovy are 'type X and Y' (after Shackleton, 1986). Observations from Shackleton (1986) are for pilchard (*Sardinops ocellata*) and anchovy (*Engraulis japonicus*) off Namibia, Lozano-Montes (1997) examined California sardine (*Sardinops caeruls sagax*) and northern anchovy (*Engraulis mordax*) from the California Current, and R. Salvatteci (unpublished data) examined Peruvian anchoveta (*Engraulis ringens*) from the Humboldt Current (following Shackleton, 1986).

Table 4.3. *Percent of total scale weight that is composed of nitrogen (N) and carbon (C) and their standard deviations for living California sardine (*Sardinops sagax*) and northern anchovy (*Engraulis mordax*) as well as fossil scales from these same species and Peruvian anchoveta (*Engraulis ringens*)*

Source of scales	% N	Stdev	% C	Stdev	N
Living Northern anchovy	14.2	0.32	46.8	1.17	38
Living Pacific sardine	13.8	0.37	47.7	1.13	41
Fossil sardine scales (La Paz)	3.9	0.17	12.5	0.35	2
Fossil anchovy scales (La Paz)	3.0	0.17	11.2	0.10	2
Fossil anchoveta scales (Peru)	3.6	0.76	12.9	2.28	36

Each measurement is done on 4–15 scales (D. Field, unpublished data).

Scales from the Peruvian margin have been observed to have fissures forming throughout the scale (Fig. 4.6a, 4.6c with 4.6b and 4.6d). Scales with large quantities of fissures become brittle and more likely to break during handling (D. Field and R. Salvatteci, unpublished data). Some scales vary in color (D. Field and R. Salvatteci, unpublished data; Patterson *et al.*, 2004). O'Connell and Tunnicliffe (2001) found that sections of piston cores sampled 19 months after initial sampling had vertebrae, but no scales, indicating post-coring degradation of scales. Furthermore, initial sampling showed a large decrease in the ratio of scales to vertebrae downcore, suggesting degradation of scales prior to coring (O'Connell and Tunnicliffe, 2001). There is a greatly diminished number of scales from the slope of the SBB (Soutar, 1971), but no observation of degradation of scales or development of fissures from the deepest part of the Santa Barbara Basin, despite the fact that many cores have been quantified years after coring (D. Field, unpublished data). In fact, there is a trend towards greater scale abundances in the SBB further back in time (Baumgartner *et al.*, 1992). While degradation is not currently understood, it undoubtedly varies with sedimentary chemistry and could bias SDR estimates through time.

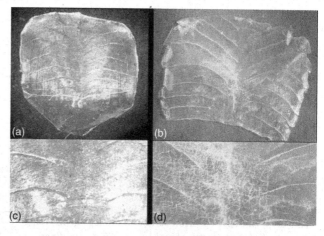

Fig. 4.6. Photos of typical sardine scales ('type T') from (a) living sardine (*Sardinops sagax sagax*) and (b) fossil sardine from sediments off Peru. (c), (d) Close-up of the foci of each scale. (d) Note the formation of fissures, which are neither the circuli around the scale or radii extending to the focus, in the center of the fossil scale from the Peruvian shelf.

One way to address the degree to which degradation may be affecting downcore records is by examining the ratio of scales to otoliths, vertebrae, and/or bones. Otoliths are aragonite and quite rare, while bones and vertebrae are generally present and composed of hydroxyapatite, but with lower surface area to volume ratios and less organic matter than scales. Scales can be identified to species more accurately. Otoliths often have dissolution affecting key characteristics, and bones and spines are not usually distinguishable among species. DeVries (1979) documented that oxygenated areas of the Peruvian shelf have considerably lower scale to vertebrae ratios as well as lower total concentrations of both scales and vertebrae. However, changes in scale to bone ratios could also be derived from changes in community assemblage towards species with much different SDRs.

Another method for assessing degradation is by noting the different stages of development of fissures on scales (Fig. 4.6). Other paleo proxies of oxygen concentration, such as Mo concentrations, can also be useful in assessing the potential degree of degradation affecting scale abundances in the sediments.

SDR and regional variations in fish abundance

SDR to a sedimentary environment should vary directly with the integrated fish abundance (as schools move across the area in time). But how well do local fish abundances over a given area reflect fish abundance over a broad region of a boundary current? Pelagic fish populations generally expand both their habitat range and regional densities during population expansions (this volume, Chapter 12). Figure 4.1 shows three different scenarios for the relative abundance and distribution of sardines in the California Current System, and how the population state could be reflected in SDR records from different anoxic areas. For example, an expanded population distributed in the northern range of the California Current would be captured by SDR to Effingham Inlet and SBB (Fig. 4.1a, ex. 3), while a contracted and southern population would be limited to sites around Baja California (Fig. 4.1a, ex. 1).

While migrations of sardine adult biomass could complicate SDR to the SBB as an indicator of total biomass, sardine SDR in this region probably reflects recent recruits, rather than adult biomass (Baumgartner *et al.*, 1992). Figure 4.7 shows very clear differences in sardine scale widths in the SBB as compared with scale widths from Effingham Inlet. Although different scale types vary in width from one another (e.g. Fig. 4.5), there is a strong correlation between scale width and fish length within each scale type (Lozano-Montes, 1997). Scale widths from the SBB, which are primarily "'type T," indicate that the record is composed primarily of 0–2-year-old fish (Fig. 4.7; Lozano-Montes, 1997). Thus the SDR to SBB sediments

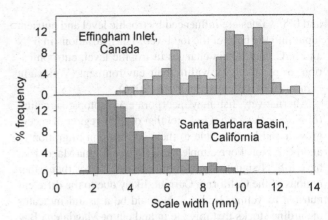

Fig. 4.7. Histogram of sardine scale widths from the Santa Barbara Basin (SBB) and Effingham Inlet, British Columbia.

can be considered an integrated estimate of sardine recruitment over multi-annual timescales. In contrast, scales from Effingham Inlet are from much older fish (Fig. 4.7), suggesting that presence of sardines in the northern region occurs when an expanded population of sardines has reached sufficient age to migrate and persist to its northern range (e.g. Fig. 4.1a, ex. 3).

Anchovy in the California Current (and other regions) tend to be found in the more coastal upwelling regions, are generally less migratory (this volume, Chapters 3 and 12), and local densities may more closely reflect regional biomass. Soutar and Isaacs (1969) found that anchovy scale widths in the SBB generally correspond to 1–3-year-old fish, which is the age distribution of the population in that region.

Off Peru, it is well known that anchovy modify their spatial distribution both alongshore and cross-shore on seasonal and interannual timescales. There are tendencies towards more northern and offshore distributions during winter months of greater upwelling. In contrast, anchovy are found closer to the coast, further south, and deeper during summer and El Niño events when cool, productive waters are limited to the coastal upwelling regions (Fig. 4.2). The existence of laminated sedimentary records off Chimbote, Callao, Pisco, and Bahia Mejillones (a coastal bay off northern Chile) provides a means to examine spatial variability in small pelagic populations along different latitudinal (and cross-shore) environments of the coast (Fig. 4.2).

Scale geochemistry

Additional information on environmental conditions and/or population movements may be inferred from both the hydroxyapatite and organic matrix components of fish scales. There is generally sufficient organic matter within a fossilized scale to measure the $\delta^{15}N$ and $\delta^{13}C$ of the organic matter within the scale (Struck *et al.*, 2004). Fish scale $\delta^{15}N$

and $\delta^{13}C$ values are influenced by trophic level and nutrient input into the base of the food web. Thus variations in $\delta^{15}N$ and $\delta^{13}C$ may reflect changes in trophic level, eutrophication, or productivity within their environment (Wainright *et al.*, 1993; Gerdeaux and Perga, 2006).

Alternatively, fish may incorporate an isotopic signature ($\delta^{15}N$ of scales or $\delta^{18}O$ of otoliths) of their region of recent growth and be traceable to their origin upon migration to another region. For example, sardines from Bahia Magdalena have a $\delta^{15}N$ signature of several per mil higher than other regions of the California Current, likely due to the different nutrient recycling. Hence $\delta^{15}N$ could be a useful indicator of sardine stocks that migrate in and out of Magdalena Bay (Field and Vetter, 2005). Trace elements in scales and otoliths also offer potential to resolve life histories of individuals with respect to environmental conditions, particularly within different rivers and bays, but the trace elements are not always stable in fish scales (Wells *et al.*, 2003).

Calibrations to biomass variability

Scale shedding rates (and thus SDRs) vary with species and require independent calibrations with each species and ocean region (Shackleton, 1987; Baumgartner *et al.*, 1992; Lozano-Montes, 1997). Calibrations of sardine SDR to estimated biomass have been done for averaged five-year intervals from the SBB up to 1969 (Soutar and Isaacs, 1974; Baumgartner *et al.*, 1992). This coarse calibration indicates that, at a population size of approximately 400 000 (200 000) tons of sardines (anchovies) off California, zero scale counts become common and the calibration loses definition. Recent sampling of several box cores in two-year intervals shows the increase in northern anchovy during the 1960s and 1970s and the presence of sardines scales again in the SBB since the early 1990s. While more material is needed for an accurate high-resolution comparison, a reliable calibration is difficult due to number of scales counted per limited temporal interval. Yet the calibration is useful in indicating that past times of high SDR were likely associated with very high sardine recruitment off central California relative to the early twentieth century (Baumgartner *et al.*, 1992).

Variations in combined SDR of Peruvian anchoveta from cores taken at Callao and Pisco reflect nearly 50% of the variance in catch records during the developed fishery, indicating that SDR is fairly consistent with catch records (Salvatteci *et al.*, 2006). Both SDR and catch records can be considered biased. SDR is sometimes based on a small area of sediment and/or low number of scales and fish can migrate to other regions. Changes in sedimentation rates and sampling interval may also affect calibrations. Fish catch records and biomass estimates are affected by sampling effort as well.

O'Connell and Tunnicliffe (2001) reported the most rigorous examination of SDR to date by examining herring SDR from five box cores taken in the Saanich Inlet. Visual comparison of the combined record from these cores shows that it reflects the principal variations in estimated herring biomass of the area.

Significance of the historical records

The historical records of SDRs reveal many aspects of long-term variability. Many ideas of interdecadal-scale ecosystem change in the North Pacific relate to inferred changes between basin-wide, bimodal states or "regime shifts" associated with the Pacific Decadal Oscillation (PDO) over 50–60-year time scales. Some of these ecosystem paradigms are that expansions and contractions of the Aleutian Low on 50–60 year time scales, as reflected by the PDO, result in out-of-phase alternations in abundance of sardines and anchovies (Lluch-Belda *et al.*, 1992; Chavez *et al.*, 2003), out-of-phase fluctuation in salmon between the California Current and the Gulf of Alaska (Mantua *et al.*, 1997), and in-phase fluctuations between populations of small pelagics in different boundary currents of the Pacific and South Atlantic (Kawasaki, 1991; Lluch-Belda *et al.*, 1992). There are many indications that both the modes and time scales of ocean variability and the response of pelagic fish populations far exceed the range of variations observed in the twentieth century.

Presence/absence

The presence or absence of scales (or other remains) can be an important indicator of variability when sufficient sediment is quantified. Anchovies appeared in the Gulf of California sardine fishery for the first time in the late 1980s, raising speculations that anchovy were occupying an ecological niche recently opened from the fishing of sardines (Holmgren-Urba and Baumgartner, 1993). However, the presence of anchovy scales in Guaymas sediments decades and centuries prior reiterated the importance of natural variability not just in the California Current System, but in the surrounding seas as well (Holmgren-Urba and Baumgartner, 1993; Fig. 4.8).

The occurrence of many species in Saanich Inlet by about 6000 years ago indicates that colonization of the fjord region since deglaciation was complete by this time, if not earlier (Tunnicliffe *et al.*, 2001). The presence of bluefin tuna remains in British Columbia middens suggests that bluefin tuna, generally found further south and offshore, have reached those latitudes in the past (Crockford, 1997).

Remains of many species of small pelagics in the California Current today have been found in Native American middens from the Holocene and in outcropped Pleistocene deposits around southern California spanning both glacial and interglacial periods (Fitch, 1969). Noteworthy observations by Fitch (1969) are that otoliths of

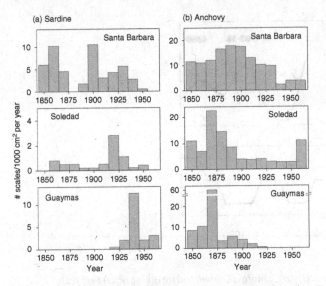

Fig. 4.8. SDRs for (a) Pacific sardine and (b) northern anchovy from central California (Santa Barbara Basin), Baja California (Soledad), and the Gulf of California (Guaymas slope). Locations shown in Fig. 4.1 (from Holmgren-Urba and Baumgartner, 1993).

northern anchovy were found continuously throughout the last 12 million years, including during different glacial and interglacial periods as well as warmer Pliocene periods. In contrast, herring remains were only found during cool glacial periods. Neither sardines nor Pacific Mackerel otoliths were found in glacial sediments and sardine otoliths were not found from Pliocene sediments, despite the fact that the Pliocene was a relatively warm period. Absence of sardine may reflect its shallow evolutionary history in the California Current (see MacCall, this volume, Chapter 12).

Relationships between sites and records

The case of sardine in the California Current offers the potential to compare different historical records. Changes in the distribution of sardines have been documented by Holmgren-Urba and Baumgartner (1993). Variations in sardine SDRs in cores from the SBB, Soledad Basin, and Guaymas shelf shown in Fig. 4.8 show how sardines from the central California Current contracted to a reduced population with a more southern distribution during the collapse of the California fishery, as proposed in Fig. 4.1.

Following the model of population expansions and contractions in Fig. 4.1, high abundances of sardines off central California would correspond with an expansion of sardine populations into the northeast Pacific. However, Holmgren-Urba (2001) reported the presence of sardine scales in Effingham Inlet, British Columbia to be out of phase with sardine scale deposition in the Santa Barbara Basin. Yet chronological uncertainties associated with a homogeneous layer question this result and reiterate the importance of accurate chronostratigraphies. Considering the homogeneous section as a slump rather than bioturbated event would result in simultaneous sardine scale deposition in Effingham Inlet and the SBB (T. Baumgartner, unpublished data).

Chronologies from historical observations and annually laminated sediments are more reliable. Figure 4.9a compares fish scale abundance in the SBB with historical observations in the Pacific Northwest. The increase in scale fluxes to the SBB around 1885 corresponds with observations of sardine presence in the Pacific Northwest in 1888. Although surveys in the Pacific Northwest did not detect sardines in the early 1880s when scale flux to the SBB was low, there were no positive observations in the mid nineteenth century either, when scale flux indicates high sardine abundances off Central California (Fig. 4.9). While observational data may not be exhaustive, the combination of historical records does not support the hypothesis that all sardine population expansions are similar.

The combination of historical records of artisanal sardine catch off Japan and sardine scale deposition records in the SBB do not support the hypothesis of simultaneous basin-scale variations in small pelagics. The basin-scale hypothesis asserts that sardine and anchovy populations in different regions of the Pacific (and South Atlantic) vary coherently. While sardines were apparently abundant in the California Current from 1890 through the early twentieth century, this was a period of poor catch throughout different regions of Japan (Fig. 4.9; Hiramoto, 1991). The other most conspicuous period of poor catch of sardine throughout Japan began around 1730 and lasted until around 1775, which corresponds with a large peak in sardine scale deposition in the SBB (Fig. 4.9). Furthermore, good fishing years occurred throughout Japan from 1690–1725 and again from 1790–1830, but these two periods are characterized by moderate, not high sardine scale fluxes to the Santa Barbara Basin (Fig. 4.9). Although these observations do not capture the whole distributional ranges of sardine populations, the compilation of historical records of sardine do not support the basin-scale hypothesis of coherent variations in sardine population abundance.

Schwartzlose *et al.* (1999) reported a general correspondence between anchovy and sardine SDR in the Santa Barbara Basin and Bohuslän herring events off Sweden for several centuries, but their proposed relationship is not present throughout the record. An analysis of instrumental records indicates that the atmospheric circulation in the N. Pacific and N. Atlantic has varied both in phase and out of phase during the twentieth century (Schwing *et al.*, 2003). Just as large-scale coherency between regions observed in the twentieth century may have been weaker in the past, there may be relationships between oceanic regions in the past that were not predominant in the twentieth century.

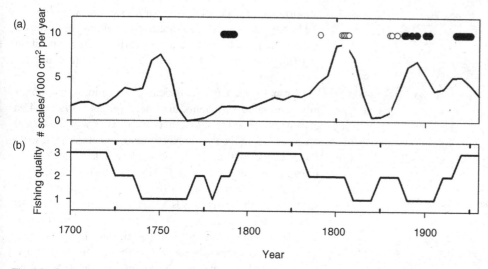

Fig. 4.9. Comparison of different historical records of sardine abundance in (a) the California Current and (b) off Japan. (a) Fish scale deposition rate of sardines in the Santa Barbara Basin (central California) are shown as a three-term smoothing of 5-year sampling intervals. Closed (open) symbols are historical observations of sardine presence (absence) in the northern California Current region, around Puget Sound and British Columbia (from Field *et al.*, 2001). (b) Quality of sardine fishing around Japan (after Hiramoto, 1991) corresponding to good sardine fishing in all areas (3), a mix between good and poor fishing in different regions (2), and poor fishing throughout Japanese waters (1).

However, there was no reported relationship of Bohuslän herring events with sardine SDR in the SBB, which may be a better indicator of the PDO. Their observation reiterates the importance of chronological uncertainties when comparing records, both with respect to finding relationships or absence of relationships, since even the annually laminated SBB sediments have considerable dating uncertainty beyond several centuries.

Off Peru, cores from several sites that have been examined to date and lie within the population range of anchovies shown in Fig. 4.2 (Pisco and Callao, Peru) indicate diminished abundance of anchovy scales prior to ~1820 (Gutierrez *et al.*, 2006). Part of this diminished abundance may be due to scale degradation. This period of diminished anchovy scale deposition was characterized by a reduction of scales of all species, rather than an increase in scale deposition rates of sardine (which are more resistant to degradation). However, a relatively greater increase in sardine remains has been reported from Peruvian middens during the time period coinciding with the decrease in total scales (Sandweiss *et al.*, 2004). While sediment chronologies this far back in time are associated with considerable uncertainties, the increase in sardine remains in middens might reflect a shift of the sardine population towards the nearshore rather than an increase in total biomass.

Relationships between species
The records of SDR from the SBB are the most high-resolution, continuous, and reliable records to date. Scale

deposition records of sardine and anchovy show a weak positive relationship with each other throughout the SBB record (Fig. 4.10; Table 4.1; Baumgartner *et al.*, 1992) rather than a negative relationship, which characterizes the twentieth century (Chavez *et al.*, 2003). Although there are some periods of abundance of one species or another, the sediment records do not support the idea of negative fluctuations in population recruitment and abundances of these species as a predominant mode of variability in prior centuries. There are many periods when both species have relatively high or low SDR. However, Lasker and MacCall (1983) found that anchovy scale widths, and thus lengths, were smaller when sardine SDRs were moderately high, suggesting some reduction in anchovy growth during periods of high sardine recruitment and abundance.

Scale deposition rates of pilchards and anchovy off Namibia show no clear relationship with one another (Shackleton, 1987). In two different cores examined at 4-year sampling intervals, periods of both high or low scale flux of pilchards occur during periods of high and low flux of anchovy scales (Shackleton, 1987).

Holmgren-Urba (2001) found no consistent relationships between SDR of different species in recent sediments from Effingham Inlet, although one period dominated by anchovy SDR has fewer remains of other species. There is no clear relationship between herring and anchovy SDR for Effingham Inlet on longer timescales, they have weak positive covariability during the past five thousand years (Patterson *et al.*, 2004). Patterson *et al.* (2004) proposed that herring and anchovy varied out of phase within the

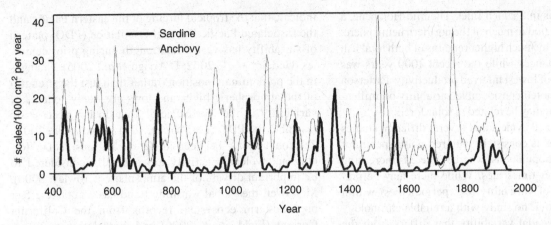

Fig. 4.10. Sardine and anchovy scale deposition rates in the Santa Barbara Basin from Kasten core SBKC 9110–1301. Lines are three-term smoothing of continuous 5-year sampling intervals (updated from Baumgartner *et al.*, 1992). The positive relationship between SDR of each species is very weak, but significant ($r^2=0.03$, $P<0.005$).

centennial-scale time domain, although their chronology was only constrained by three ^{14}C dates and homogeneous layers were not defined as bioturbations or slumps.

While there appears to be a positive relationship between anchovy and hake SDR in sediments off Peru on decadal to millennial timescales, there is no clear relationship between sardine SDR with either of these species (DeVries and Pearcy, 1982; Salvatteci *et al.*, 2006).

Only the records from historical fisheries in the N. Atlantic show clear alternations of species prior to the twentieth century (Alheit and Hagen, 1997). The alternations between species in the N. Atlantic are probably due primarily to changes in biogeographic distributions.

The lack of clear relationships between SDR of different species of small pelagics in cores from different oceanic regions indicates that abundances of each species are determined by different combinations of environmental changes that could include circulation, productivity, and the abundances of top-down predators, which affect mortality and recruitment. The majority of the evidence points to climatically driven changes modifying recruitment, with interactions between species playing a less significant role.

Timescales of variability

Inferred changes in fish populations varies across a large range of timescales. While the average period of "good" or "poor" sardine fishing off Japan is around 20–30 years and coincides with the timescale of variability observed in the twentieth century, Yasuda *et al.* (1999) noted that a given time period of "good" or "poor" fishing could last from 7–45 years and 14–35 years respectively. Baumgartner *et al.* (1992) emphasized that sardine SDR could be characterized by multiple periods of recoveries and collapses that last from 20–30 years in duration (Fig. 4.10). However, periods of sustained sardine SDR persisted for nearly a century

in different parts of the record, indicating nearly continuous sardine recruitment off central California beyond the multidecadal timescale (Fig. 4.10). After removing the low frequency variability, there are peaks in spectral power of both anchovy and sardine SDR from 50–70 years, consistent with observations from the twentieth century (Baumgartner, 1992). While this is an important characterization of the variability, these time series have not been shown to have a preferred periodicity that differs statistically from the null hypothesis of an oceanic red-noise spectrum, which results from the oceanic integration of atmospheric white noise (Pierce, 2001).

Off Namibia there are large variations in the composition of scales from pilchard, anchovy, hake, and mackerel on both decadal and millennial timescales (Shackleton, 1987; Baumgartner *et al.*, 2004). Although records from the last three thousand years are not currently continuous, the differences in species composition between different centennial-scale portions of the core examined may be associated with large variations in the alkenone unsaturation index, a coccolithopohorid-based proxy of near-surface temperatures (Baumgartner *et al.*, 2004). Taken together, the records are suggestive of large centennial to millennial-scale variations that exceed the magnitude of fluctuations observed in the twentieth century.

The strongest downcore signal from records of anchovy SDR off Peru show a centennial-scale period of low abundance that is associated with changes in numerous other proxies of ocean variability (Gutierrez *et al.*, 2006). The only time period in which sardine SDR was higher than that of anchovy off Peru was in the early Holocene, a time period when varied insolation resulted in a different hemispheric heat balance (DeVries and Pearcy, 1982).

Tunnicliffe *et al.* (2001) documented large millennial-scale variability in total fish remains that varies with diatom abundances in Saanich Inlet. The mid-Holocene, a relatively warm period in much of the northern hemisphere, was characterized by much higher remains of both total fish and diatom abundances while the recent 1000 years was one of the periods of lowest inferred productivity. Patterson *et al.* (2004) also noted considerable variability on millennial timescales that may be related to solar forcing.

Thus the combined evidence of several different studies from different sites is consistent with red-noise spectra of variability in fish populations, whereby there is increasing variability at longer timescales. While there may be preferred timescales of variability (e.g. periodicities) within a red-noise spectrum, no study with a reliable chronology has shown interdecadal variability that differs from the null hypothesis of a red-noise spectrum (Pierce, 2001). The historical records of abundance do not show interdecadal oscillations of sufficient strength for predicting future changes. Furthermore, modelling studies have shown that oceanic red-noise is a more dominant source of variability than periodic variations that may arise from gyre-scale circulation (Yeh and Kirtman, 2006).

Implications for ocean variability

The historical records clearly suggest many different patterns of ocean and ecosystem variability that have not been observed in the twentieth century. The relationship between SDR of sardines and anchovies in the California Current with proxy records of salmon abundance in the Gulf of Alaska is opposite to the relationship observed in the twentieth century (Finney *et al.*, 2002). The paleo records are consistent with the speculations of Isaacs (1976) that "there are probably a great number of possible regimes; multifarious regimes involving biology or climate, or oceanography, or migrations, temperature, or weather, or combinations of these."

In the Santa Barbara Basin, no clear relationship has been observed between sardine or anchovy SDR and other proxy records of ocean variability prior to the twentieth century (D. Field and T. Baumgartner, unpublished data). While this does not mean a relationship doesn't exist, it does imply that it may not be sufficiently strong to be clearly observed between different types of paleo records. In contrast, sediment records from Peru do show consistent relationships between anchovy scale deposition, organic carbon flux, and other proxy records of productivity and oxygen, suggesting a stronger coupling between primary productivity and anchovy production in this system (Gutierrez *et al.*, 2006).

Paleoecological patterns that are considerably more complex than the twentieth century are consistent with other studies of past climate changes in the Pacific. The relationships between coral records and tree rings

indicate that the tropical forcing of the eastern Pacific and the associated Pacific Decadal Oscillation (PDO) pattern of variability has varied in strength during prior centuries (Gedalof *et al.*, 2002; D'Arrigo *et al.*, 2005). Changes in the predominant position (rather than just the strength) of the atmospheric highs and lows are likely sources of variability in the North Pacific that do not necessarily resemble the PDO (La Marche, 1974; Bond *et al.*, 2003). Teleconnections between different regions, solar forcing and volcanic activity are generally considered to be important drivers of the atmosphere (Crowley, 2000). Moreover, the global warming trend has a notable effect on long-term ecosystem records from the California Current (Field *et al.*, 2006) and may dominate future changes. Climate and ecosystem variability may shift from familiar modes of variability to an entirely different climate regime.

What could be learned in the future?

There remain many issues concerning variability in small pelagics that can be answered with fish scales. Most hypotheses concerning decadal-scale variability in small pelagics are based on observations in the twentieth century that span one to two fluctuations. The apparent synchronization of fluctuations in abundance of small pelagics, both within and among boundary currents, has attracted considerable attention, yet effectively only two decadal-scale fluctuations have been documented. To date, the historical evidence shows little support for this hypothesis. However, cores from additional regions with very good dating may enable testing global synchrony of pelagic fish stocks as well as the different scenarios shown in Fig. 4.1 within a boundary current. In particular, sufficient chronological controls to test the hypothesis might be obtained for the last 150–200 years.

Perhaps one of the most insightful lessons for the future of small pelagics will be learned from different periods of longer-term climate forcing, such as the mid-Holocene maximum or different periods of insolation. Having a time period far different from the recent Holocene, as well as a relatively warm time period with many differences in global circulation, may be instructive to mechanisms of change. The future of ocean climate and marine populations will undoubtedly differ from this more recent time period of the Holocene and even from the last 2000 years. Although preservation issues may be prevalent at some sites, there are many clues to be found deeper in the sedimentary records indicating the response of small pelagics to variable climatic conditions not seen in recent centuries or millennia.

Box 4.1. What if paradigms of small pelagic population variability were based on nineteenth-century observations?

Are paradigms of decadal-scale changes in pelagic ecosystems too strongly linked with the changes observed in the twentieth century? It is well known that each El Niño event evolves differently and is associated with different regional climate and ecosystem responses. Our understanding of the effects of El Niño on climate and ecosystems would be greatly skewed if it came from only two El Niño events. Thus it should come as no surprise that historical records indicate a wide range of long-term climate and ecosystem variations relative to twentieth century observations. Moreover, the twentieth-century observations have been affected by increasing anthropogenic alterations of ecosystems, biogeochemical cycles, and climate.

The twentieth century fluctuations in fish catch have been characterized as oscillating variations between sardines and anchovies in different upwelling environments with approximately 50–60-year periodicity (Kawasaki, 1991; Lluch-Belda *et al.*, 1992; Mantua *et al.*, 1997; Chavez *et al.*, 2003). These paradigms are based on observations that are, at most, a few decades longer than the inferred periodicity and patterns. Would our scientific perspectives on long-term ocean and ecosystem variability be much different if they were formed from observations in the nineteenth century?

If we take the implications of some of the historical records at face value, then we can imagine the following scenarios for N. Pacific variability during the nineteenth century. Historical records of fishing off Japan and sardine scale deposition off California are suggestive of a negative relationship between sardine abundance off Japan and California (Fig. 4.9), rather than a positive relationship as seen in the twentieth century. It is not clear that all sardine population expansions off central California were associated with sardine presence off British Columbia (Fig. 4.9). In addition, scale fluxes from Santa Barbara Basin sediments indicate that both anchovy and sardines were abundant off southern California throughout much of the nineteenth century, rather than having alternated in abundance (Fig. 4.10). While it is debatable whether historical records are sufficient to reject hypothesized paradigms, they do not support them as dominant modes of variability. The different historical records suggest that the ecosystem paradigms of the twentieth century are only a small part of the range of ecosystem response to climate change and intrinsic variability. Likewise, studies of tree rings and coral records indicate that PDO type variability was not a dominant form of variability during the nineteenth century (Gedalof *et al.*, 2002; D'Arrigo *et al.*, 2005).

Does the paradigm of 50–60-year fluctuations, based on twentieth century observations, offer much hope of long-term predictions? Some tree ring evidence suggest that variability on 8–16-year timescales was more dominant than variability in 50–70-year timescales in the nineteenth century and beyond (Biondi *et al.*, 2001), while other evidence supports a continuous 50–70-year periodicity (Minobe *et al.*, 1997). Persistent sardine scale deposition to the SBB has persisted for periods of 70–100 years in the past (Fig. 4.10). With the strong La Niña of 1999, some were expecting a persistence of negative PDO conditions for several decades. Looking back, there were three negative years followed by four positive years – hardly a persistent negative PDO phase. Finally, the advent of anthropogenic global change will likely result in little practicality of using past periodicities for predicting future changes.

Undoubtedly, the twenty-first century ecosystem and climate changes will be dramatically different from those of the twentieth century and will result in paradigm shifts. Useful paradigms of ecosystem variability will be those that are consistent with past variations, can be successfully modeled, and are strongly linked with well-founded mechanisms of change, rather than with broad correlations. Like fossils in ocean sediments, paradigms of climate and ecosystem regimes based on just twentieth-century observations and unfounded correlations will be buried in the archives as we continue into the twenty-first century.

Acknowledgments

We gratefully thank Alec MacCall, John Field, Jürgen Alheit, and an anonymous reviewer for providing comments and ideas that greatly improved the manuscript.

NOTE
1 Before Present.

REFERENCES

Alheit, J. and Hagen, E. (1997). Long-term climate forcing of European herring and sardine populations. *Fish. Oceanogr.* **6**: 130–9.

Baumgartner, T. R., Ferreira-Bartrina, V., Cowen, J., and Soutar, A. (1991). Reconstruction of a 20th century varve chronology from the central Gulf of California. In *The Gulf and Peninsular Province of the Californias*, Dauphin, J. R. and Simoneit, B. R. T., eds. pp. 603–616.

Baumgartner, T. R., Soutar, A., and Ferreira-Bartrina, V. (1992). Reconstruction of the history of Pacific sardine and northern anchovy populations over the last two millennia from sediments of the Santa Barbara Basin, California. *CalCOFI Rep.* **33**: 24–40.

Baumgartner, T. R., Struck, U., and Alheit, J. (2004). GLOBEC investigation of interdecadal to multi-centennial variability in marine fish populations. *PAGES News* **12** (1): 19–21.

Biondi, F., Gershunov, A., and Cayan, D. R. (2001). Winter PDO reconstructed from Southern California tree-ring records. *J. Climate* **14**: 5–10.

Bond, N. A., Overland, J. E., Spillane, M., and Stabeno, P. (2003). Recent shifts in the state of the North Pacific. *Geophys. Res. Lett.* **30** (2183): doi:10.1029/2003GL018597.

Butler, V. L. and O'Conner, J. E. (2004). 9000 years of salmon fishing on the Columbia River, North America. *Quat. Res.* **62**: 1–8.

Chavez F. P., Ryan, J., Lluch-Cota, S. E., and Niquen, M. (2003). From anchovies to sardines and back: multidecadal change in the Pacific Ocean. *Science* **299**: 217–221.

Crawford, R. J. M., Dundee, B. L., Dyer, B. M., *et al.* (2007). Trends in numbers of Cape gannets (*Morus capensis*), 1956/57–2005/06, with a consideration of the influence of food and other factors. *ICES J. Mar. Sci.* **64**: 169–177.

Crockford, S. J. (1997). Archaeological evidence of large northern bluefin tuna, *Thunnus thynnus*, in coastal waters of British Columbia and northern Washington. *Fish. Bull.* **95**: 11–24.

Crowley, T. J. (2000). Causes of climate change over the past 1000 years. *Science* **289**: 270–276.

D'Arrigo, R., Wilson, R., Deser, C., *et al.* (2005). Tropical-North Pacific climate linkages over the past four centuries. *J. Climate* **18** (24) : 5253–5265.

DeVries, T. J. (1979). Nekton remains, diatoms, and Holocene upwelling off Peru. MS thesis, Oregon State University, Corvallis, Oregon. 85 pp.

DeVries, T. J. and Pearcy, W. G. (1982). Fish debris in sediments of the upwelling zone off central Peru: a late Quaternary record. *Deep-Sea Res.* **28**: 87–109.

Field, D. B. and Vetter, R. (2005). Can carbon and nitrogen isotopes be used to infer trophic level and movement in Pacific sardine?, In *Trinational Sardine Forum*, November 14–15, Ensenada, Mexico.

Field, D. B., Baumgartner, T. R., Charles, C., Ferriera-Bartrina, V., and Ohman, M. D. (2006). Planktonic foraminifera of the California Current reflect twentieth century warming, *Science* **311**: 63–66.

Field, J. C., Francis, R. C., and Strom, A. (2001). Toward a fisheries ecosystem plan for the Northern California Current. *CalCOFI Rep.*, **42**, 74–87.

Finney, B. P., Gregory-Eaves, I., Douglas, M. S. V., and Smol, J. P. (2002). Fisheries productivity in the northeastern Pacific Ocean over the past 2,200 years. *Nature* **416**: 729–733.

Fitch, J. E. (1969). Fossil records of certain schooling fishes of the California Current System. *CalCOFI Rep.* **13**: 71–80.

Gedalof, Z., Mantua, N. J., and Peterson, D. L. (2002). A multi-century perspective of variability in the Pacific Decadal Oscillation: new insights from tree rings and coral. *Geophys. Res. Lett.* **29**: 54–57.

Gerdeaux, D. and Perga, M. E. (2006). Changes in whitefish scales d¹³C during eutrophication and reoligotrophication of subalpine lakes. *Limnol. Oceanogr.* **51**: 772–780.

Gutiérrez, D., Field, D., Salvatteci, R., *et al.* (2006). Decadal to centennial variability of the Peruvian upwelling ecosystem during the last centuries as inferred from fish scale deposition rates of anchovy and other marine sediment records. In *International Conference on the Humboldt Current System*, November 2006.

Hiramoto, K. (1991). The sardine fishery and ecology in the Joban and Boso waters of central Japan. In *Long-Term Variability of Pelagic Fish Populations and their Environment*, Kawasaki, T., Tanaka, S., Toba Y., and Taniguchi, A., eds., Oxford, UK: Pergamon Press, pp. 117–128.

Holmgren-Urba, D. (2001). Decadal-centennial variability in marine ecosystems of the northeast Pacific Ocean: the use of fish scale deposition in sediments, Ph.D. dissertation, University of Washington.

Holmgren-Urba, D. and T. R. Baumgartner. (1993). A 250-year history of pelagic fish abundance from anaerobic sediments of the central Gulf of California. *CalCOFI Rep.* **34**: 60–68.

Isaacs, J. D. (1976). Fishery science: Fact, fiction, and dogma. *CalCOFI Rep.* **18**: 34–43.

Kawasaki, T. (1991). Long term variability in the pelagic fish populations. In *Long-Term Variability of Pelagic Fish Populations and their Environment*, Kawasaki, T., Tanaka, S., Toba Y., and Taniguchi, A., eds. Oxford, UK: Pergamon Press, pp. 47–60.

Keefer, D. K., deFrance, S., Moseley, M., *et al.* (1998). Early maritime economy and El Niño Events at Quebrada Tacahuay, Peru. *Science* **281**: 1833–1835.

Kuwae, M., Okuda, N., Miyasaka, H., *et al.* (2007). Decadal- to centennial-scale variability of sedimentary biogeochemical parameters in Kagoshima Bay, Japan, associated with climate and watershed changes. *Estuar. Coast. Shelf Sci.* **73**: 279–289.

LaMarche, V. C. (1974). Paleoclimatic inferences from long tree-ring records. *Science*, **183**, 1043–1048.

Lange, C. B., Schimmelmann, A., Yasuda, M. and Berger, W. H. (1996). Marine varves off Southern California. *Scripps Inst. Oceanogr. Ref. Ser.* **96**: 122.

Lasker, R. and MacCall, A. D. (1983). New ideas on the fluctuations of the clupeoid stocks off California. In *C.N.C./SCOR Proceedings of the Joint Oceanographic Assembly 1982 – General Symposia, Ottawa, Canada*, pp. 110–120.

Lluch-Belda D., Schwartzlose, R. A., Serra, R., *et al.* (1992). Sardine and anchovy regime fluctuations of abundance in four regions of the world oceans: a workshop report. *Fish. Oceanogr.* **1**: 339–347.

Lozano-Montes, H. 1997. Reconstruction of marine fish populations using fossil fish scales deposited in Santa Barbara Basin (USA). MS thesis. Centro de Investigación Científica y de Educación Superior de Ensenada, Mexico. 97 pp.

Mantua, N. J., Hare, S. R., Zhang, Y., *et al.* (1997). A Pacific interdecadal climate oscillation with impacts on salmon production. *Bull. Amer. Meteorolog. Soc.* **78**: 1069–1079.

Marcus, J., Sommer, J. D., and Glew, C. P. (1999). Fish and mammals in the economy of an ancient Peruvian kingdom. *Proc. Natl. Acad. Sci.* **96**: 6564–6570.

Minobe, S. (1997). A 50–70 year climatic oscillation over the North Pacific and North America. *Geophys. Res. Lett.* **24**: 683–686.

O'Connell, J. M. and Tunnicliffe, V. (2001). The use of sedimentary fish remains for interpretation of long-term fish population fluctuations. *Mar. Geol.* **174**: 177–195.

Patterson, R. T., Prokoph, A., Wright, C., *et al.* (2004). Holocene solar variability and pelagic fish productivity in the NE Pacific. *Palaeontol. Electron.* **7**, 17 pp.

Pierce D. W. (2001). Distinguishing coupled ocean-atmosphere interactions from background noise in the North Pacific. *Progress in Oceanography*, **49**, 331–352.

Quilter, J., Ojeda, E. B., Pearsall, D. M., *et al.* (1991). Subsistence economy of El Paraíso, an early Peruvian site. *Science*, **251**, 277–283.

Reitz, E. J. (2003). Resource use through time at Paloma, Peru. *Bull. Florida Mus. Nat. Hist.* **44**: 65–80.

Rick, T. C. and Erlandson, J. M. (2000). Early Holocene fishing strategies on the California coast: Evidence from CA-SBA-2057. *J. Archaeolog. Sci.* **27**: 621–33, doi:10.1006/jasc.1999.0493.

Rick, T. C., Erlandson, J. M., and Vellanoweth, R. L. (2001). Paleocoastal marine fishing on the Pacific coast of the Americas: Perspectives from Daisy Cave, California. *Am. Antiquity* **66**: 595–613, doi:10.2307/2694175.

Salvatteci, R., Gutierrez, D., Field, D., *et al.* (2006). Fluxes of scales and other fish remains en marine sediments in front of Callao, Perú in the last 550 years. *XIII Congress of Geology, Society of Geologists of Peru, Lima, Perú*.

Sandweiss, D. H., Maasch, K. A., Chai, F., *et al.* (2004). Geoarchaeological evidence for multidecadal natural climatic variability and ancient Peruvian fisheries. *Quat. Res.* **61**: 330–334.

Schwartzlose, R. A., Alheit, J., Bakun, A., *et al.* (1999). Worldwide large-scale fluctuations of sardine and anchovy populations. *S. Afr. J. Mar. Sci.* **21**: 289–347.

Schwing, F. B., Jiang, J., and Mendelssohn, R. (2003). Coherency of multi-scale abrupt changes between the NAO, NPI, and PDO. *Geophys. Res. Lett.* **30** (1406): doi:10.1029/2002GL016535.

Shackleton, L. Y. (1986). An assessment of the reliability of fossil pilchard and anchovy scales as fish population indicators off Namibia, MSc thesis, Univ. of Cape Town, South Africa, 141 pp.

Shackleton, L. Y. (1987). A comparative study of fossil fish scale from three upwelling regions. *S. Afr. J. Mar. Sci.* **5**: 79–84.

Shackleton, L. Y. (1988). Scale shedding: an important factor in fossil fish studies. *J. Cons. Int. Explor. Mer.* **44**: 259–263.

Soutar, A. (1967). The accumulation of fish debris in certain California coastal sediments. *CalCOFI Rep.* **11**: 136–139.

Soutar, A. (1971). Micropalaeontology of anaerobic sediment and the California Current. In *The Micropalaeontology of the Oceans*, Funnel, B. M. and Riedel, W. R., eds. Cambridge, UK: Cambridge University Press, pp. 223–230.

Soutar, A. and Crill, p. A. (1977). Sedimentation and climatic patterns in the Santa Barbara Basin during the 19th and 20th centuries. *Geol. Soc. Am. Bull.* **88**: 1161–1172.

Soutar, A. and Isaacs, J. D. (1969). History of fish populations inferred from fish scales in anaerobic sediments off California. *CalCOFI Rep.* **13**: 63–70.

Soutar, A. and Isaacs, J. D. (1974). Abundance of pelagic fish during the 19th and 20th centuries as recorded in anaerobic sediment of the Californias. *Fish. Bull.* **72**: 257–273.

Struck, U., Heyn, T., Altenbach, A., *et al.* (2004). Distribution and nitrogen isotope ratios of fish scales in surface sediments from the upwelling area off Namibia. *Zitteliana* **A44**: 125–132.

Suess, E. (1981). Phosphate regeneration from sediments of the Peru continental margin by dissolution of fish debris. *Geochim. et. Cosmochim. Acta* **45**: 577–588.

Tunicliffe, V., O'Connell, J. M., and McQuoid, M. R. A. (2001). Holocene record of marine fish remains from the Northeastern Pacific. *Mar. Geol.* **174**: 197–210.

Wainright, S. C., Fogarty, M. J., Greenfield, R. C., and Fry, B. (1993). Long-term changes in the Georges Bank food web: trends in stable isotope composition of fish scales. *Mar. Biol.* **115**: 481–493.

Wells, B. K., Thorrold, S. R., and Jones, C. M. (2003). Stability of elemental signatures in the scales of spawning weakfish (*Cynoscion regalis*). *Can. J. Fish. Aquat. Sci.* **60**: 3619.

Yasuda, I., Sugisaki, H., Watanabe, Y., and Minobe, S.-S. (1999). Inter-decadal variations in Japanese sardine and Ocean/Climate. *Fish. Oceanogr.* **8**: 18–24.

Yeh S.-W. and Kirtman B. P. (2006). Origin of decadal El Niño–Southern Oscillation-like variability in a coupled general circulation model. *J. Geophys. Res.* **111**: C01009, doi:10.1029/2005JC002985.

5   Decadal-scale variability in populations

Jürgen Alheit, Claude Roy, and Souad Kifani

CONTENTS

Summary

Decadal-scale dynamics of small pelagic fish populations from five large marine boundary currents (Kuroshio, California, Humboldt, Benguela, and Canary Currents) and their possible links to climate variability are described and compared. Small pelagic clupeiform fish species such as anchovies, sardines, sardinellas, herring and sprat are characterized by decadal drastic fluctuations of biomass, which are often associated with regime shifts in large marine ecosystems and their dynamics are governed by long-term, decadal-scale physical processes. Consequently, small pelagics are excellent indicators of regime shifts. When occurring in the same system, anchovies and sardines usually fluctuate out of phase. These shifts between sardine-dominated and anchovy-dominated states seem to restructure the entire ecosystem, as concomitant qualitative and quantitative changes in ecosystem components other than sardines and anchovy populations have been observed. The small pelagic fish seem thereby to respond to a bottom-up forcing of the ecosystem which itself is driven by changing ocean conditions. Evidence is emerging that these ecosystem shifts are associated with large-scale changes in subsurface processes and basin-scale circulation. In the Humboldt Current ecosystem, the shifts seem to be linked to lasting
periods of warm or cold water anomalies related to the approach or retreat of warm oceanic subtropical surface water (SSW) of high salinity to the coast of Peru and Chile. The famous collapse of the Peruvian anchovy around 1970 was the result of such a regime shift, not the consequence of the El Niño 1972/73, which happened after the anchovy decline was already initiated. Dynamics of Japanese sardines and anchovies and their Kuroshio Current ecosystem exhibit a surprising synchrony with processes in the Humboldt system. Although changes in basin-scale circulation must be tightly interwoven with climate variability, the direct association of shifts in the dynamics of small pelagic fish and their ecosystems and climate dynamics and mechanisms linking them, at least in the Pacific, are still largely obscure. The climate regime shift observed in 1976/77 in the North Pacific did not cause clear reactions in the dynamics of small pelagic fish and zooplankton populations of the Pacific boundary currents.

Introduction

Sufficient evidence has been accumulated to show that marine ecosystems undergo decadal-scale fluctuations which seem to be driven by climate variability (e.g. Beamish, 1995; Bakun, 1996). Climate variability can reorganize marine communities and trophodynamic relationships and can induce regime shifts where the dominating species are replacing each other on decadal time scales. One way to predict how marine ecosystems will react to future climate variability or to climate change is to search for causal relationships of past patterns of natural variability and to draw conclusions based on retrospective studies. Long-term biological time series are essential for retrospective analysis of climate impact on marine ecosystems; however, they are scarce. Fish populations usually provide longer records than other biological components of marine ecosystems because of their economical importance. The dynamics of exploited fish populations are affected by natural environmental variability and man-made activities (fishing, habitat alteration) and retrospective studies will help to distinguish between the two. Although the potential impact of climate

Climate Change and Small Pelagic Fish, eds. Dave Checkley, Jürgen Alheit, Yoshioki Oozeki, and Claude Roy. Published by Cambridge University Press. © Cambridge University Press 2009.

Box 5.1. Small pelagic fish

Small pelagic fish such as sardine, anchovy, sardinella, sprat and others represent about 20%–25% of the total annual world fisheries catch. They are widespread and occur in all oceans. They support important fisheries all over the world and the economies of many countries depend on those fisheries. They respond dramatically and quickly to changes in ocean climate. Most are highly mobile, have short, plankton-based food chains, and some even feed directly on phytoplankton. They are short-lived, highly fecund and some can spawn all year-round. These biological characteristics make them highly sensitive to environmental forcing and extremely variable in their abundance (Hunter and Alheit, 1995). Thousandfold changes in abundance over a few decades are characteristic for small pelagics and well-known examples include the Japanese sardine, sardines in the California Current, anchovies in the Humboldt Current, sardines in the Benguela Current or herring in European waters. Their drastic stock fluctuations often caused dramatic consequences for fishing communities, entire regions and even whole countries. Their dynamics have important economic consequences as well as ecological ones. They are the forage for larger fish, seabirds and marine mammals. The collapse of small pelagic fish populations is often accompanied by sharp declines in marine bird and mammal populations that depend on them for food (Hunter and Alheit, 1995). Major changes in abundance of small pelagic fishes may be accompanied by marked changes in ecosystem structure. The great plasticity in growth, survival and other life history characteristics of small pelagic fishes is the key to their dynamics and makes them ideal targets for testing the impact of climate variability on marine ecosystems.

small pelagics which fluctuate antagonistically with each other (Lluch-Belda *et al.*, 1989; Schwartzlose *et al.*, 1999). This chapter will investigate these earlier suggestions and present an update on this debate, now that more than 20 years have passed since the classical paper of Kawasaki (1983) and almost 10 years since Schwartzlose *et al.* (1999). This chapter is also a contribution to the regime shift debate which has been, and still is, the focus of GLOBEC[1] retrospective studies. We present new views on the existence and causes of regime shifts in ecosystems where small pelagic fish such as anchovies and sardines play an important role, particularly making use of recent insights about regional oceanographic processes.

Small pelagic fish such as anchovies and sardines are ideal targets for testing the impact of climate variability on marine ecosystems (Box 5.1). This chapter describes interdecadal variability of the large anchovy and sardine populations in the four eastern boundary currents (California, Humboldt, Benguela and Canary Currents) and the Kuroshio Current, a western boundary current. The analysis of each ecosystem is structured according to (i) changes in the biota, (ii) regime shifts, and (iii) mechanisms linking climate to decadal-scale population dynamics. The final discussion focuses on tropho-dynamic aspects, climate relationships, synchronous ecosystem dynamics and possible teleconnection patterns.

Decadal-scale regime shifts

Huge populations of sardines and anchovies are dwelling in the upwelling ecosystems of the eastern boundary currents (California, Humboldt, Canary and Benguela Currents) and in the waters around Japan. They support important fisheries, mainly for fish meal, and the well-being of the economy of the riparian countries of upwelling systems depends heavily on these fisheries. The dynamics of these anchovy and sardine populations are characterized by their inverse relationships. When one species supports a large biomass and high production, the other species usually sustains a rather low biomass (Fig. 5.1). The changes in biomass are accompanied by enormous expansions and contractions of the areas of distribution (Fig. 5.2). These shifts between sardine-dominated and anchovy-dominated states seem to restructure the entire ecosystem, as concomitant qualitative and quantitative changes in ecosystem components other than sardines and anchovy populations have been observed. The small pelagic fish species seem thereby to respond to a bottom-up restructuring of the ecosystem, which itself is forced by changing ocean conditions as explained below for several ecosystems. Because of their dramatic and long-lasting nature, these switches have been termed "regime shifts" (Lluch-Belda *et al.*, 1989, 1992). The first use of the term regime was by Isaacs

variability on marine ecosystems and their fisheries has been described in a number of cases (e.g. Cushing, 1982; Laevastu, 1993), rigorous studies on these relationships were started only in the 1990s. This was certainly stimulated by the world-wide public awareness of global changes and the predicted greenhouse effect. The initiation of global international research programmes, such as the World Climate Research Programme (WCRP) and the International Geosphere Biosphere Programme (IGBP), vastly improved co-operation across disciplinary boundaries accumulating knowledge on the impact of climate variability on marine ecosystems, particularly on the decadal scale.

It has been suggested that there exist teleconnections among the low-frequency fluctuations of anchovies and sardines in the Pacific which swing in synchrony (Kawasaki, 1983; Chavez *et al.*, 2003) and among Pacific and Atlantic

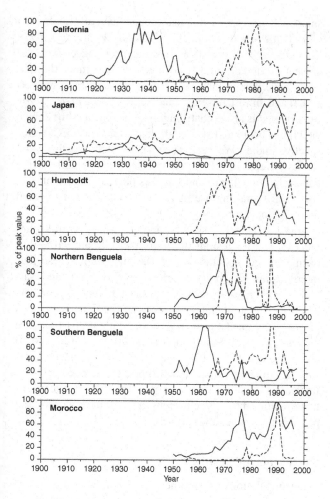

Fig. 5.1. Normalized catches of anchovies (stippled line) and sardines (black line) from California Current, Japan, Humboldt Current, Northern and Southern Benguela Current and from NW Africa. Respective peak catches were set at 100%. California Current: anchovy peak, 316 × 10³ mt; sardine peak, 718 × 10³ mt. Japan: anchovy peak, 430 × 10³ mt; sardine peak, 4488 × 10³ mt. Humboldt Current: anchovy peak, 12 866×10³ mt; sardine peak, 5621 × 10³ mt. Northern Benguela: anchovy peak, 376 × 10³ mt; sardine peak, 1400 × 10³ mt. Southern Benguela: anchovy peak, 597 × 10³ mt; sardine peak, 410 × 10³ mt. Morocco: sardine peak, 1087 × 10³ mt; anchovy peak, 264 × 10³ mt.

short period, an ecological regime shift cannot necessarily be pinpointed to one or two single years. The reason behind this is that marine populations often react with time lags to physical forcing due to complex recruitment processes. A difference in the timing of the shift of different populations can also be expected as different species react differently to climatic forcing depending on their particular physiological threshold values and their life history traits (Beaugrand and Reid, 2003; Beaugrand, 2004). The start and end point (turning points) of regime shifts are defined here as those brief periods where marked changes in ecosystems have been observed. For example, a shift from an anchovy to a sardine period starts when the anchovy stock begins to decrease and/or the sardine stock starts to increase. At this time, at the early stages of a sardine period, the anchovy stock might still have a much larger biomass than the sardine stock; however, it is assumed that the ecosystem has begun a change from a state favorable for anchovy recruitment to a state favorable for sardine recruitment (Jacobson *et al.*, 2001). In other words, the new "sardine" regime begins when the sardine stock shows the first signs of a longer-lasting increase. Often, what has been considered to be the beginning of a sardine regime is the time when the biomass of sardine surpasses that of anchovies or when negative biomass anomalies change to positive anomalies. However, at those times the "sardine" regime has usually already been going on for several years. Consequently, in the following, the focus will be on the "turning points," i.e. those brief periods when relevant changes in the physics and several trophic levels of an ecosystem have been recorded approximately at the same time (within a period of 1–3 years).

The bulk of the data on fish population dynamics stems from the fisheries. Catch data are a somewhat crude measure of fish abundance; however, they usually give an acceptable signal of the trends in population dynamics. Additional data arise from research surveys conducted by fisheries management institutions. Figures of long-term catches and biomass of anchovies and sardines from the ecosystems discussed in this chapter are in Chapter 9, this volume and are, consequently, not displayed here. Habitats are described in Chapter 3, this volume.

California Current ecosystem (CCE)

Changes in the biota

The California sardine fishery peaked at over 700 000 mt (metric tons) in 1936 and then decreased rapidly from the mid 1940s to the early 1950s to catches of 13 000 mt in 1952 (Fig. 5.1; Fig. 9.3a). A fisheries moratorium was established from 1966–1985 which was lifted in 1986 because biomass exceeded the minimum of 20 000 mt. The stock recovered in the early 1980s and annual catches have exceeded 100 000 mt since 1997 (Lluch-Belda *et al.*, 1989; Schwartzlose

(1976) to describe distinct environmental or climatic states and regime shifts are transitions between different regimes (Lluch-Belda *et al.*, 1989, 1992; MacCall, 1996). During a Workshop on Regime Shifts in Villefranche-sur-Mer in April 2003, regime shifts in the marine realm were defined in a pragmatic way as "changes in marine system functioning that are relatively abrupt, persistent, occurring at a large spatial scale, observed at different trophic levels and related to climate forcing" (DeYoung *et al.*, 2004). In contrast to a climate regime shift, which might happen within a very

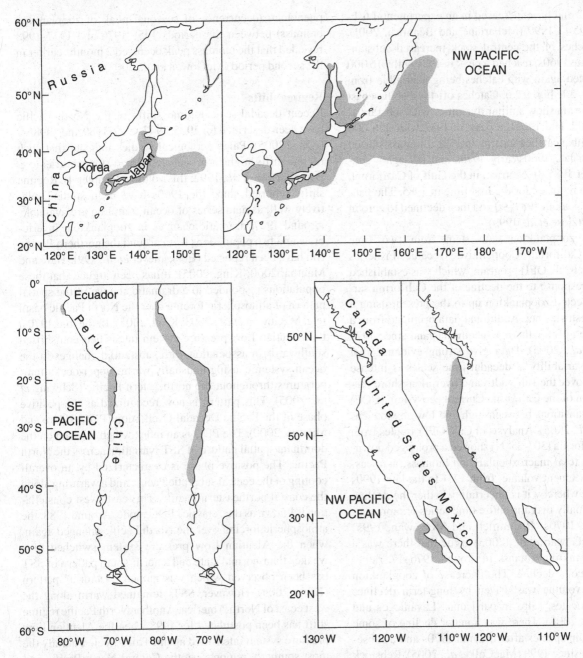

Fig. 5.2. Minimum (left) and maximum (right) areas of distribution of Japanese, Humboldt Current and California Current sardines (redrawn after Hunter and Alheit, 1995).

et al., 1999; McFarlane *et al.*, 2002). Biomass peaked in 1934 with over 4 million mt and then declined to values of under 3000 mt in the 1970s. In 1984, biomass increased from unmeasurably low values, rose steadily to 1.5 million mt in 1996, and subsequently declined to about 1 million mt (Fig. 9.3a) (Hill *et al.*, 2006). In the Gulf of California, a sardine fishery started in the late 1960s and caught a peak of 294 000 mt in 1988–89 when the fishery had fully expanded. It declined rapidly to 7000 mt in 1991–92 and increased again to 203 000 mt in 1996–97. Biomass increased from

1976 to a peak in 1985–86 of 1.2 million mt and decreased subsequently. Recruitment of the Gulf of California population increased after 1975 until 1984–85 and then fell drastically (Schwartzlose *et al.*, 1999). In British Columbia, sardines were the largest fishery from the mid 1920s to the mid 1940s when catches ranged between 5000 and 80 000 mt with an annual average of 40 000 mt. This fishery collapsed in 1947 and sardines disappeared completely from waters of British Columbia (Fig. 5.2). After 40 years of total absence, sardines were caught again off Vancouver

Island in 1992 and have increased in an experimental fishery from 1995 to 1999 (McFarlane and Beamish, 2001). Anchovy catches of the central stock increased substantially in the late 1960s, reached a peak in 1981 with 316 000 mt and declined again with catches being negligible from 1990 on (Fig. 5.1; Fig. 9.2a). Catches off the Pacific coast of Baja California show similar dynamics with substantial increases in the late 1960s, a peak in 1982 with 256 000 mt and negligible catches starting in 1991. Biomass started to increase in late 1960s/early 1970s and fell again substantially after 1985. In contrast, in the Gulf of California, anchovy were reported for the first time in 1985. The fishery reached a peak in 1989–90 and then declined to zero in 1997 (Schwartzlose *et al.*, 1999).

Long-term zooplankton data stem from two time series. The California Cooperative Oceanic Fisheries Investigations (CalCOFI) program, which was established in 1951 as a response to the decline of the California sardine, has collected zooplankton up to the present using a 0.505 mm-mesh size net. Additional data originate from a series collected off southern Vancouver Island since 1979 (MacCall *et al.*, 2005). Data give strong evidence that zooplankton variability at decadal time scales is intense and coherent over the full width and over alongshore distance >400 km of the California Current ecosystem (CCE) with abrupt transitions between high and low abundances (MacCall *et al.*, 2005). Analysis of CalCOFI samples from southern California (30°–35° N) showed a prolonged downward trend in total macrozooplankton biomass, as measured by displacement volume, from 1951 to the late 1990s by about 80% whereby it is uncertain whether the decline occurred gradually over the entire time series or more rapidly since the 1970s (Roemmich and McGowan, 1995) (Fig. 5.3). McGowan *et al.* (2003) argue that there was a shift in zooplankton biomass in around 1976–77, rather than a continuous decline. The decrease of zooplankton displacement volume was driven by long-term declines of pelagic tunicates, salps in particular (Lavaniegos and Ohman, 2003, 2007). There was a major decline of some (but not all) salp species after the mid 1970s and a subsequent increase since 1999 (MacCall *et al.*, 2005). Rebstock (2001, 2002) did not find long-term trends in copepod abundance. In contrast, MacCall *et al.* (2005) state, based on Rebstock data, that abundance of calanoid copepods increased after 1977 and declined again around 1990 and seems to have increased again in 1998–99. This statement is confirmed by Lavaniegos and Ohman (2007) (using carbon biomass of copepods). However, they also state that they did not find long-term trends of copepods. The solution for understanding these somewhat confusing, contradicting statements is probably that over the long period of 56 years, there was no overall change, but short-term fluctuations lasting several years. A comparison of zooplankton

phenology (occurrence of seasonal peak of zooplankton biomass) between the periods 1951–1976 and 1977–1998 revealed that the biomass peak occurred 2 months earlier in the second period (McGowan *et al.*, 2003).

Regime shifts

Recent decadal-scale regime shifts in the North Pacific have been described for 1925, 1947, 1977, 1989 and 1998/99 (King, 2005). Paleo-ecological studies indicate that such regime shifts might have occurred already for centuries (e.g. Baumgartner *et al.*, 1992; this volume, Chapter 4). Regime shifts reported since the 1960s have been studied relatively well, as data series of ocean sampling are available. Around 1977, dramatic changes in zooplankton, invertebrate and fish populations from around the northern Pacific rim have been reported (Ebbesmeyer *et al.*, 1991; Hare and Mantua, 2000; King, 2005). It has been argued that these populations responded to a dramatic change of the spatial pattern of atmospheric forcing over the North Pacific basin (e.g. McGowan *et al.*, 2003; King, 2005). In the mid 1970s, the Aleutian Low pressure system intensified and shifted southwards in association with substantial changes in the ocean system causing unusually warm, upper-ocean temperatures throughout the northeastern Pacific (Schwing *et al.*, 2005). This pattern is now recognized as the positive phase of the Pacific Decadal Oscillation (PDO) (Hare and Mantua, 2000). The PDO is an index which is based on the dominant spatial pattern of SST[2] variation across the North Pacific. The positive phase is characterized by an overall cooling of the central subarctic Pacific and a warming along the coastal northeastern Pacific. This east–west constellation is known as the "classic" PDO mode. Around 1988, the atmospheric forcing over the North Pacific changed again, when the Aleutian Low pressure system switched to a weaker-than-normal state and a north-south pattern of SST has been observed which was named "Victoria" pattern (King, 2005). However, SSTs remained warm along the west coast of North America. Another North Pacific regime shift has been postulated for 1998 when the Aleutian Low pressure system intensified again. This affected mainly the most southerly regions, i.e. the Central North Pacific and the California Current (CC). In the CCE, cooling of coastal waters and enhanced southward flow of water and organisms were recorded (King, 2005). The associated deepening of the thermocline and decreased stratification resulted in increasing phytoplankton biomass, both in amount and seaward extent, and higher zooplankton biomass throughout the CCE, whereby the species composition returned to patterns similar to those during the mid 1980s.

Recently, a debate was initiated about the exact beginning of the 1970s regime shift in the North Pacific, particularly in the CCE. Schwing *et al.* (2004) challenged the dogma of a Pacific-wide regime shift in 1976–77 and provided

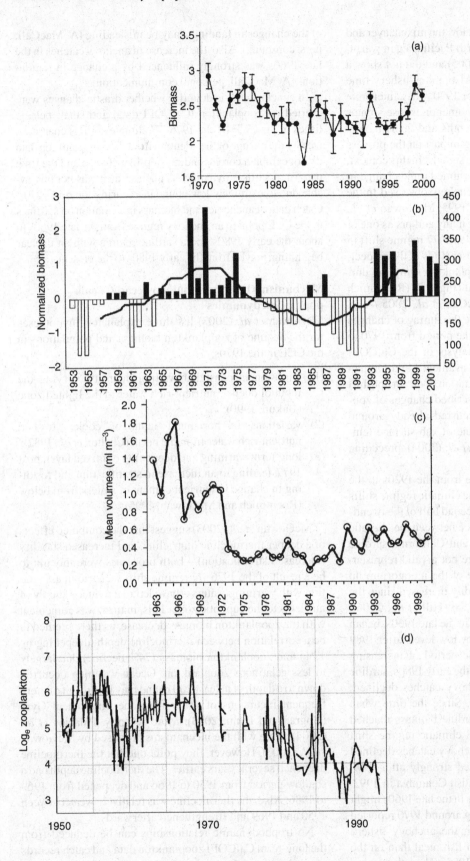

Fig. 5.3. Zooplankton time series. (a) Copepods > 1 mm in Kuroshio in winter (reprinted from Nakata and Hidaka, 2003, with permission from Wiley–Blackwell). (b) Total zooplankton wet weight (mg m⁻³; May–July mean) in Oyashio region (reprinted from Chiba *et al.*, 2006, with permission from Wiley–Blackwell). (c) Meso-zooplankton volumes, excluding jellyfish, from Peru (reprinted from Ayón *et al.*, 2004, with permission from Oxford University Press). (d) Zooplankton volumes from California Current (reprinted from Roemmich and McGowan, 1995, with permission from AAAS).

evidence that ocean temperatures below the mixed layer and in the northern extremes of the North Pacific began warming around 1970 (Schwing *et al.*, 2005) and did not show a clear shift at the surface in 1976. Also, many fishery time series suggest population shifts near 1970. They interpret a regime shift as an "evolving phenomenon whose signals propagate into different regions, depths and fields having different response times, depending in part on the process that is directly supplying the climate signal." In this context, it is interesting to re-evaluate the assumed rapid change in CCE zooplankton biomass and phenology observed in the 1970s (Roemmich and McGowan, 1995; McGowan *et al.*, 2003), which has been presented by many authors as one of the most convincing proofs for the 1976–77 regime shift in the CCE (e.g. Hayward, 1997). A closer study of the respective data reveals that CalCOFI zooplankton sampling during the 1970s was only intermittent (Fig. 5.3) (Roemmich and McGowan, 1995 – Fig. 2; MacCall *et al.*, 2005 – Fig. A2.5), and was too sparse to resolve the timing of changes in zooplankton abundance during the period from 1970 to 1976. Also, the more objective analysis of the CalCOFI zooplankton time series done by Lavaniegos and Ohman (2007) does not support the interpretation of an abrupt shift in the mid 1970s. All the above described changes of zooplankton might very well have occurred already around 1970, in association with the decline of sub-surface temperatures as recorded by Schwing *et al.* (2004), preceding the 1976–77 climatic shift.

There was a clear sardine regime from the 1920s to the 1940s, very likely associated with the climatic regime shifts in 1925 and 1947. However, for the second half of the twentieth century, not much correspondence between the climatic regime shifts in the North Pacific and California sardine dynamics can be found. Catches are not a valid representation of sardine dynamics because of the moratorium up to 1985. However, biomass has steadily increased since the early to mid 1980s to 1996 (Fig. 9.3a) (Hill *et al.*, 2006). Anchovy catches started to increase in the late 1960s, began to decrease in 1982 and reached very low levels after 1989 (Fig. 9.2a). Interestingly, during the period of increasing anchovy catches from about 1965 to the early 1980s, sardine biomass was very low. When anchovy catches declined, sardine biomass began to increase. Since the time when anchovy catches were very low, sardine biomass reached rather high levels. Maybe the 1988 climatic regime shift had an impact on both species as anchovy catches declined thereafter, sardine biomass increased strongly after 1990 and sardines occurred again off British Columbia in 1992. Also, the increase in anchovy catches in the late 1960s might be related to the subsurface warming around 1970 reported by Schwing *et al.*, 2004. However, the anchovy fishery was more dependent on the price of fish meal than on the availability of fish (Thomson *et al.*, 1985). Thus, the timing

of the changes in landings may be misleading (A. MacCall, pers. commun.). Also, the increase in anchovy catches in the late 1960s was strongly influenced by a change in regulations (A. MacCall, personal communication).

In conclusion, it is doubtful whether drastic changes were exerted on zooplankton (CalCOFI data) and small pelagic fishes in the CCE by the 1976–77 climate shift as characterized by the change of the Aleutian Low. The zooplankton data do not exhibit a corresponding step-like change and the time pattern of sardine dynamics is unclear, as it was not observable due to extremely low abundance during the mid 1970s. Catch data of anchovies and biomass information on sardines in the CCE point to an anchovy regime from the late 1960s to about the early 1980s and a sardine regime with an unclear beginning, maybe from the early 1980s to the present.

Mechanisms linking climate to decadal-scale population dynamics

McGowan *et al.* (2003) list three explanatory hypotheses for the decline of zooplankton biomass and populations in the CCE in the 1970s:

(1) variation in coastal upwelling intensity, varying the input of deeper, nutrient-rich water to the lighted zone (Bakun, 1990);
(2) variations in horizontal input of cooler, fresher, nutrient-rich water from the north (Chelton *et al.*, 1982);
(3) long-term warming and deepening of mixed layer, post 1977, leading to an increase in stratification and resulting in change in nutrient content of waters from below (Roemmich and McGowan, 1995).

McGowan *et al.* (2003) suggest that the combined effects of a deeper thermocline (nutricline) and increased stability (increased stratification) – both processes were thought to be a result of the 1976–77 regime shift scenario in association with warming waters – have led to a reduced supply of nutrients to the euphotic zone. This, in turn, was coincident with the zooplankton biomass decrease, as there seemed to be a correlation between thermocline depth (deepening by 17%) and zooplankton biomass (74% decline) due largely to less gelatinous zooplankton. Ocean warming occurred down to a depth of at least 200 m and may be a result of more frequent incursions of water from the Subtropical Gyre (Bograd and Lynn, 2003). More recently, Palacios *et al.* (2004) agree with the mechanisms proposed by McGowan *et al.* (2003). However, they point out that the thermocline deepened several years earlier. The thermocline depth had a shallow period from 1950 to 1966 and deepened from 1969 to 1986. Also, the thermocline was relatively weak between 1950 and 1969 and strengthened afterwards.

No trophodynamic relationships can be deduced from the long-term CalCOFI zooplankton data and catch records of sardines and anchovies. Both anchovies and sardines increased during the period of plankton sampling, whereas zooplankton biomass declined steadily. However, the lack of any trophodynamic relation is not surprising considering that (i) the plankton net mesh size was 0.505 mm (which is too large for major food items of sardines and anchovies) and (ii) the major contribution to the zooplankton decline was due to salps which do not seem to be a relevant food item of sardines and anchovies. Because of their low energy content, the long-term decline in salp biomass is not likely to be of much consequence to planktivorous fish; however, in the Benguela Current, sardine stomachs were found which were packed with salps (H. Verheye, personal communication). Unfortunately, there is no long-term information on zooplankton smaller than 0.505 mm.

Kuroshio Current ecosystem (KCE)

Changes in the biota

The sardine population of Japan is distributed in the Sea of Japan and in the waters east of Japan, namely coastal waters, the Kuroshio Current Ecosystem (KCE) and the Kuroshio Current Extension (KC Extension) and the Oyashio Current. Catch records for the Japanese sardine have been kept since at least 1894. Yields were between 0.1 and 0.6 million mt until 1927 (Schwartzlose *et al.*, 1999). Then, mainly based on the Sea of Japan subpopulation, they increased sharply until 1942 with a peak of 1.6 million mt in 1936 (Fig. 5.1). Thereafter, they decreased again to former levels until the late 1950s when they dropped dramatically to 0.02 million mt and less during the 1960s. Sardine catches, mainly from the Pacific subpopulation, started to increase in 1971 and those from the Sea of Japan in 1972. Catches reached levels well above 1 million mt from 1976 to 1994 with a peak of 4.5 million mt in 1988 when they started to decrease steadily again. The transition from an anchovy to a sardine regime in the KCE ecosystem in 1969–1971 was further indicated by (i) the instantaneous surplus production rate (ISPR) turning positive (Fig. 5A in Jacobson *et al.*, 2001), (ii) the decrease of the economically important post-larval anchovy (shirasu) fishery around 1969 (Fig. 5.4) (Nakata *et al.*, 2000), (iii) the sardine mass spawning in Tosa Bay in 1970 (Kawasaki, 1993), (iv) the first appearance of sardine eggs in the Enshu-nada Sea in 1971 (Nakata *et al.*, 2000), and (v) the decrease of summer zooplankton biomass in the Oyashio region, which serves as an important feeding ground for sardines and started to decline around 1972 (Chiba *et al.*, 2006) (Fig. 5.3). Also, new results of fish scale accumulation rates in anaerobic sediments demonstrate that the Japanese anchovy population decreased around 1970 (Poster, M. Kuwae, ASLO Summer Meeting 2006, Victoria). The commercially important squid, *Todarodes pacificus*, also started to decrease around 1970 (Sakurai *et al.*, 2000). After a decline since the late 1950s, catches of

the Pacific saury, *Cololabis saira*, started to increase again in 1971 (Watanabe *et al.*, 2003). In the mid 1980s, Japanese waters switched back to anchovy dominance. ISPR turned negative (Jacobson *et al.*, 2001). In 1986, anchovy shirasu (post-larvae) catches started to increase again (Fig. 5.4), whereas sardine recruitment began to decline and a dramatic decrease of sardine juveniles was observed. In 1987, anchovy catches began their increase and, in 1988, sardine catches reached their peak and began then to decrease (Fig. 5.1). At the same time, sardine shirasu catches started to decline (Fig. 5.4), whereas anchovy shirasu appeared in 1988 in Suruga bay and Enshu-nada to Ise and Mikawa bays (Kondo, 1991). Anchovy increase and sardine decrease in the late 1980s are reflected well by catches from the post-larval fish fishery (Nakata *et al.*, 2000). Also, squid increased again since the late 1980s (Sakurai *et al.*, 2000). Zooplankton biomass in the Oyashio as collected with a 0.333-mm mesh net increased from the early 1950s to 1970, with a short decline in the early 1960s (Fig. 5.3) (Odate, 1994; Chiba *et al.*, 2006). From around 1970 on, it decreased again up to the mid 1980s and increased again until 2000. Biomass of KC Extension copepods > 1mm decreased from 1970, when sampling began, to the early 1990s and then increased again (Nakata and Hidaka, 2003) (Fig. 5.3). All these changes are summarized in Table 5.1.

Regime shifts

A number of physical processes indicate that substantial changes occurred in the waters east of Japan in the late 1960s which reversed around the mid to late 1980s, for example changes in the winter mixed-layer depth (MLD) in the KC Extension and in the movement of the Oyashio Current (Yasuda, 2003). The MLD was deeper than the long-term mean between 1967 and 1985 (Yasuda *et al.*, 2000; Yatsu and Kaeriyama, 2005) (Fig. 5.5). The dynamics of SST in the southern recirculation area of the KC Extension followed the MLD by a lag of 2–3 years, as they decreased in 1969 and increased again in 1988 (Noto and Yasuda, 1999; Yasuda *et al.*, 2000; Yatsu and Kaeriyama, 2005). Data on dynamics of sardine and anchovy (biomass, catches, mass appearances of spawn and post-larvae, anchovy scale accumulation rates), squid and zooplankton show turning points at the same time. In addition, sardine recruitment was negative during most years of the 1960s and in most years after 1988 (Yatsu and Kaeriyama, 2005). Noto and Yasuda (1999, 2003) report a significant positive correlation between the mortality coefficient of sardine post-larvae to age 1 and the SST of the KC Extension and its southern recirculation area (Fig. 5.6). The high SSTs in this area from 1950–1969 and the abrupt increase in SSTs since 1988 (Fig. 5.6) were accompanied by low sardine production. Wada and Jacobson (1998) and Yatsu *et al.* (2005) also considered conditions in the KCE during the period from 1969–1987

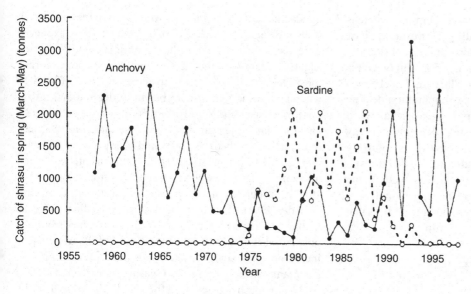

Fig. 5.4. Annual catches of post-larval anchovy and sardine (shirasu fishery) in western Enshu-nada Sea (reprinted from Nakata *et al.*, 2000, with permission from Wiley–Blackwell).

as a regime favorable for sardines, whereas it was unfavorable during 1951–67 and after 1988. The Oyashio Front near the island of Honshu has shifted north- and southwards on decadal time scales (Oyashio intrusions) (Yasuda, 2003). In the late 1960s, there was a change from northward shifts to southward shifts until the mid 1980s, after which northward shifts dominated again (Yasuda, 2003). The Pacific saury fishing grounds are known to be related to this meridional shift of the Oyashio Front (Yasuda and Watanabe, 1994). The turning points of Oyashio zooplankton biomass exhibit similar dynamics with a high around 1970 and a low in the late 1980s (Odate, 1994; Chiba *et al.*, 2006) (Fig. 5.3). All these data indicate that the ecosystem of the Japanese anchovies and sardines east of Japan experienced a regime shift around 1969 to 1971, when the system changed to sardine dominance, and in the mid to late 1980s, when it reversed to a state more favorable for anchovies. Clearly, the dynamics of major pelagic populations are associated with these shifts, not with the 1977 climatic shift of the Northeastern Pacific (Yasunaka and Hanawa, 2002; Yatsu and Kaeriyama, 2005).

Mechanisms linking climate to decadal-scale population dynamics

Japanese sardines and anchovies spawn in inshore waters of central and southern Japan and in the KCE. The KCE and the KC Extension are important for recruitment success as they transport larvae to nursery areas in the northeast of Japan into the KC Extension area (Nakata *et al.*, 1994; Noto and Yasuda, 1999; Yatsu *et al.*, 2005). The Kuroshio–Oyashio Transition Zone is a nursery and feeding ground for juveniles (Yatsu *et al.*, 2005). Consequently, hydrographical processes in all these areas are potentially of importance to sardines and anchovies.

Changes in the MLD, which affect the depth and rate of vertical mixing, appear to be an important mechanism through which climatic variations can induce biological changes (Limsakul *et al.*, 2001) as shown for several locations in the central and western North Pacific. Polovina *et al.* (1995), working with data from the northwest Hawaiian islands, argued that, when the mixed layer deepens, nutrients from subthermocline depths are brought into the mixed layer, thereby increasing phytoplankton production. Yasuda *et al.* (2000) demonstrated that the winter (Jan–Mar) mixed layer depth (MLD) in the Kuroshio Extension area was particularly deep from the mid 1960s to the mid 1980s and that there was a drastic change from a deep to a shallow phase between 1984 and 1988 (Fig. 5.5). Limsakul *et al.* (2001) showed that MLD in the south of the Kuroshio Current, about 450 km south of Shikoku Island, Japan, increased from around 1970 to early/mid 1980s and then decreased again. In the KC Extension, the decadal-scale variation pattern of the January and February MLD is quite similar to winter SST dynamics (Yasuda, 2003), both of which started to change in the mid to late 1960s and reversed in the mid to late 1980s (Noto and Yasuda 1999; Yasuda *et al.*, 2000), slightly preceding the shifts from anchovy to sardine (1969–71) and back to anchovy (1986–88) in the KC Extension. Spring and summer MLD (April–June) in the Oyashio also deepened in 1970 and started shoaling again in 1985 (Chiba *et al.*, 2006). Shortly thereafter, zooplankton biomass began to decline and started to increase again in the late 1980s, a short time after the shallowing of the MLD (Odate, 1994; Chiba *et al.*, 2006) (Fig. 5.3). Chiba *et al.* (2006) argue that, in contrast to the impact of a deepening MLD on productivity as reported by Polovina *et al.* (1995), a deepening MLD in the more northern latitudes of the Oyashio might decrease productivity as this region is not nutrient, but light limited.

Table 5.1. *Timing of events in Humboldt and Kuroshio Current ecosystems*

Year	Biological changes HCE	Biological changes KCE	Physical changes HCE	Physical changes KCE
1969	Peru zooplankt.[1] Chil. zooplankt.[2] shift rel. abund. mesopel. larval fish[3] anch. surplus prod.[4]	anch. shirasu[13]	salinity[21] subtrop. surface water approaches coast[10]	Kuroshio Ext. SST[25]
1970	rel. abund. Chil. Sard.[5] ; rel. abund. Chil. horse mackerel[5] ; Chil. hake catches[6] ; bonito catches[7] sard. spawn.[8] ;	sard. spawn. Tosa Bay[14] ; anch. scale accumulation in anoxic sediments[15] squid catches[16]	thermocline (model)[22] SST anomalies[23] ; turbulence mixing index[24] ;	
1971	anch. biomass[9] anch. catches[10] anch. recruitment[11]	sard. catches in Pac.[17] ; sard. eggs Enshu-nada Sea[13] ; saury catches[18] ; Oyashio summ. zoopl.[19]	PDO ;	
1972			El Niño	
1973			El Niño	
1982			El Niño	
1983	lowest anch. biomass ever[10]		El Niño thermocline (model)[22] ;	
1984	anch. recruitment[9] ; anch. biomass[9] ;		SST anomalies[23] salinity[21]	
1985	sard. catches[10]			
1986	offshore phyto- and micro-zooplankton[12] ;	juvenile sard.[20] anch. shirasu catches[13] ; sard. recruitment[20] squid catches[16] ;		
1987	coastal phytoplankton[12] ;	anch. catches[17] ;		PDO Kuroshio Ext. SST[25] ;
1988		sard. shirasu[13] sard. catches[17]		

[1] Carrasco and Lozano, 1989; [2] Bernal *et al.*, 1983; [3] Loeb and Rojas, 1988; [4] Jacobson *et al.*, 2001; [8] Zuta *et al.*, 1983; [9] Csirke *et al.*, 1996; [10] Alheit and Niquen, 2004; [11] Mendelsohn and Mendo, 1987; [12] Sanchez *et al.*, 2000; [13] Nakata *et al.*, 2000; [14] Kawasaki, 1993; [15] M. Kuwae, poster at ASLO Summer Meeting 2006, Victoria, Canada; [16] Sakurai *et al.*, 2000; [17] Schwartzlose *et al.*, 1999; [18] Watanabe *et al.*, 2003; [19] Chiba *et al.*, 2006; [20] Yatsu and Kaeriyama, 2005; [21] Checkley *et al.*, 2007; [22] Pizarro and Montecinos, 2004; [23] Montecinos *et al.*, 2003; [24] Bakun, 1987; [25] Noto and Yasuda, 2003.

Interestingly, in the KC Extension, the shoaling of the MLD preceded the abrupt increase in SSTs in 1987–88 which, in turn, preceded the 1989–90 climatic shift (Yasuda *et al.*, 2000; Yasuda, 2003). According to Yasuda (2003), variation in the KC could induce the climate regime shift. The increase in the heat transport by the KC Extension may be related to the spin-up of the subtropical gyre (Yasuda and Hanawa, 1997; Yasuda *et al.*, 2000). Consequently, Yasuda *et al.* (2000) suggest that variations in the KC system are a major driving force in the control of the long-term climate variability in the North Pacific. Further, they propose to use the change in MLD to predict dynamics of the Japanese sardine stock.

Anchovies and sardines in the waters around Japan have been alternating over the last 100 years on a decadal-scale pattern (Fig. 5.1). The mechanisms causing this alternation are largely unknown, however, the timing of the changes (turning points) from sardine to anchovy periods and back to sardines can now be determined rather exactly. At the time of these ecosystem shifts substantial changes in physical and biological variables in the waters off the east coast of Japan have been observed. Evidence is emerging that these ecosystem shifts are associated with large-scale changes in sub-surface processes and basin-scale circulation. Sardines seem to thrive at periods of reduced biomass of meso-zooplankton in the KC

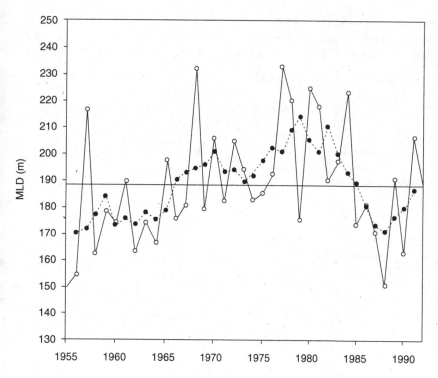

Fig. 5.5. Dynamics of the mixed layer depth (MLD) in the Kuroshio Current Extension (reprinted from Yasuda *et al.*, 2000, with permission from Elsevier).

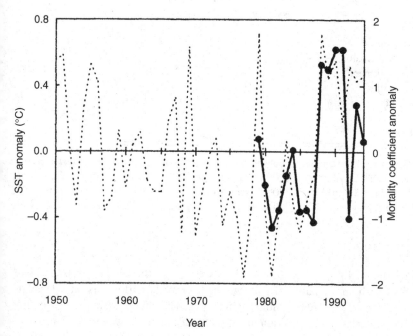

Fig. 5.6. February SST anomalies in Kuroshio Extension Southern Area (broken line) and natural mortality coefficient anomalies of sardines, from post-larva to age 1 (solid line). (Reprinted from: Noto and Yasuda, 1999, with permission from NRC Research Press.)

(Nakata and Hidaka, 2003) and in the Oyashio Current ecosystems (Odate, 1994; Chiba *et al.*, 2006) as collected with 0.300-mm mesh nets which consists mainly of larger copepods, whereas anchovies appear to suffer during these periods.

Humboldt Current ecosystem (HCE)

Changes in the biota

Peruvian anchovy catches of the northern/central stock (Fig. 5.1; Fig. 9.2a) peaked in 1970 at 11 million mt, fell dramatically from 1970 to 1972, remained between 0.5 and

3 million mt until 1982 and decreased to an extremely low level during the early 1980s, also in response to the El Niño of 1982–83. In 1984, the stock recovered and catches rose to 3 million mt in 1986. Catches increased steadily thereafter to a peak of 9.8 million mt in 1994, dropped to 1 million mt in 1998 because of another strong El Niño and have maintained at 6–9 million mt since (Alheit and Niquen, 2004). Sardine spawning (Zuta *et al.*, 1983) and catches (Serra, 1983) were insignificant during the 1950s and 1960s. From 1964 to 1971, the only distinct spawning areas were in northern Peru and northern Chile (Bernal *et al.*, 1983). After 1971, the sardine expanded to the northern and southern extremities of both refuge areas. Sardine spawning off Peru from 1966 to 1968 was poor and limited to the region between 6 and 10°S (Zuta *et al.*, 1983). After 1969, an increase of spawning was observed and, after the El Niño event of 1972–73, sardine spawning increased strongly and the spawning area expanded considerably (Fig. 5.2). From 1973 on, the area of distribution and the abundance of sardines increased notably in Ecuador, Peru and Chile (Zuzunaga, 1985). Between 1976 and 1980, spawning increased further (Zuta *et al.*, 1983). From 1964 to 1973, no eggs or larvae were observed in Chile south of 25°S. However, a new spawning area off Talcahuano was established subsequently (Fig. 5.2) (Bernal *et al.*, 1983; Serra, 1983). Sardine spawning increased and the geographic distribution of spawning expanded during warm years. Sardine catches of Peru and Chile increased steadily from less than 15 000 mt in 1970 to more than 3 million mt in 1979, peaked 1985 with 5.5 million mt and decreased thereafter to almost zero at present (Fig. 5.1; Fig. 9.3a). Zooplankton volumes declined in 1969 and preceded the anchovy crash in 1970 (Ayón *et al.*, 2004) (Fig. 5.3), whereas the recovering anchovy stock in 1984 preceded the increase of zooplankton biomass by about two years. However, up to 2001, zooplankton biomass had by far not reached the high values of the 1960s. As for zooplankton volumes, phytoplankton volumes started to increase again around 1987, after the recovery of the anchovy (Sánchez, 2000).

Regime shifts

All seven HCE anchovy and sardine stocks (Alheit and Niquen, 2004) show clear decadal variability in abundance and, in spite of the wide geographic distances between their habitats, they seem to swing in synchrony. When *E. ringens* supports high biomasses, *S. sagax* exhibits low population levels and *vice versa*. Consequently, the HCE has passed through alternating anchovy and sardine periods on the decadal time scale. Critical periods of transition, turning points, were 1969–1971 when the famous Peruvian anchovy stock started to collapse, and 1985–88, when the HCE switched back from a sardine to an anchovy system. The fact that all seven anchovy and sardine populations

showed dramatic changes around these periods indicates the regime shift character of these processes. Major changes in zooplankton and fish populations were observed between 1969 and 1974, well before 1976–77.

Over the last 40 years, two regime shifts of the HCE have been recorded (Alheit and Niquen, 2004). 1969–71 was the turning point when the HCE changed from an anchovy- to a sardine-dominated system and dramatic changes were observed during this short period. Anchovy recruitment collapsed in 1971 (Cushing, 1996; Gulland, 1982; Mendelssohn and Mendo, 1987). Zooplankton biomass decreased drastically along the entire Peruvian coast (Alheit and Niquen, 2004; Ayón *et al.*, 2004; Carrasco and Lozano, 1989). Off northern Chile zooplankton biomass also diminished (Bernal *et al.*, 1983) and the composition of the ichthyoplankton community changed markedly (Loeb and Rojas, 1988). Subsequently, the once largest fish population in the world, the Peruvian anchovy, collapsed and increasing sardine spawning was recorded. Whereas the anchovy went down to very low biomass levels, the sardine biomass increased steadily and surpassed that of the anchovy in the mid 1970s. Zooplankton biomass stayed at very low levels, just as the anchovy. In the mid 1980s, the HCE shifted back to an anchovy system, but zooplankton biomass increased only several years later. However, phytoplankton biomass also started to increase in the mid 1980s (Sanchez, 2000). Thus, when the HCE shifted in the late 1960s, changes were recorded in zooplankton and fish, whereas, when it changed back, signals were observed in phytoplankton and fish (Alheit and Niquen, 2004). All these changes are summarized in Table 5.1.

Mechanisms linking climate to decadal-scale population dynamics

In the HCE, the regime shifts seem to be linked to lasting periods of warm or cold water anomalies related to the approach or retreat of warm subtropical oceanic waters to the coast of Peru and Chile (Santander and Flores, 1983; Tsukayama, 1983; Alheit and Bernal, 1993; Alheit and Niquen, 2004). When the SST anomalies of the El Niño 4 region are subtracted from those of the El Niño 1 and 2 region (Trenberth and Stepaniak, 2001) and the results plotted cumulatively, the resulting cumsums indicate changes in 1970 and 1986 (Fig. 5.7). This is interpreted as oceanic warm water masses moving to the coast between 1970 and 1986 (E. Hagen, personal communication). Phases during the descending part of the curve parallel anchovy regimes (1950s to about 1970; 1985 up to the present) and the ascending phase from about 1970–1985 was characterized by sardine dominance (Alheit and Niquen, 2004). These results were recently confirmed by analysis of long-term salinity data from waters off Peru, up to 60 nm offshore (Poster, O. Morón, Humboldt Current Symposium, Lima,

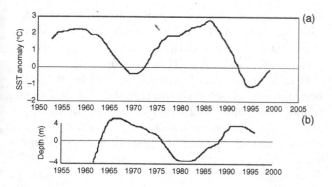

Fig. 5.7. (a) Cumulative plot of 7-year running mean of standardized and annually averaged SST anomalies. The SST anomaly values of El Niño region 4 were subtracted from those of El Niño regions 1 and 2 and plotted cumulatively (E. Hagen, unpublished data). (b) Decadal variability of modeled thermocline anomaly off coast of Chile (modified after Pizarro and Montecinos, 2004).

Peru, December 2006; this volume, Chapter 3). These data demonstrate the approach of oceanic subtropical high salinity water masses toward the Peruvian coast. From 1960 to the late 1960s, coastal waters off Peru were dominated by cold coastal water (CCW; 14–18 °C; 34.9–35.0 salinity). Thereafter, subtropical surface water (SSW; 18–27 °C; 35.1–35.7 salinity) approached the coastal realm from the late 1960s to the mid 1980s. When the SSW retreated again offshore, it was replaced by CCW. Decadal SST variability along the western coast of South America has been studied by Montecinos *et al.* (2003). The first principal component (PC1) of the normalized SST time series (at 23.5° S) shows a temperature increase from about 1970 up to 1983 when SSTs decrease again. Associated with this decadal-scale variability of SSTs is a decadal-scale oscillation of the (modeled) thermocline (Pizarro and Montecinos, 2004). During the late 1960s, the thermocline was very shallow. It then deepened continuously up to the early 1980s, when it became shallower again (Fig. 5.7). At the end of the 1960s, the thermocline was, on average, 10 m shallower than during the beginning of the 1990s (Pizarro and Montecinos, 2004). Interestingly, the turning points of the thermocline are both a few years before the turning points of the cumsums curve of the temperature anomalies of the El Niño regions (Fig. 5.7). These processes described for the HCE are very similar to the suggestions by Chavez *et al.* (2003) that the California Current weakened and moved shoreward during the sardine period, whereas a stronger and broader California Current during the anchovy period was associated with a shallower coastal thermocline from California to British Columbia, leading to enhanced primary production.

From the timing of all these physical processes, it emerges that the waters off Peru, and probably Chile, were dominated by CCW from the beginning of measurements

in the early 1960s up to the late 1960s, a period of cooler SSTs and a shallow thermocline. This environment seems to have favoured a high biomass of meso-zooplankton > 0.300 mm and provided excellent recruitment conditions for the anchovy (Alheit and Niquen, 2004). The intrusion of SSW beginning in the late 1960s resulted in warmer SSTs and a deeper thermocline. This new "subtropical" environment led to the decline of meso-zooplankton biomass and the crash of the Peruvian anchovy stock in 1970–71 which provided about 20% of the world fisheries yield at that time. Clearly, the anchovy stock collapsed before the 1972–73 El Niño as a combined result of a changing environment and, probably, also of heavy fishing pressure. The advance of SSW caused an entire suite of adverse conditions for the anchovy (Alheit and Niquen, 2004). Predation on all life stages increased whereas the abundance of their most important food source, large calanoid copepods, decreased enhancing recruitment failure of the anchovy population. The new environment seems to have provided feeding conditions favourable for sardine and less favourable for anchovy which need larger particles than sardines (Alheit and Niquen, 2004; this volume, Chapter 7). As zooplankton was collected with 0.300-mm mesh nets, there is no information on long-term dynamics of zooplankton which passes through the meshes of these nets. In the mid 1980s, the SSW moved again offshore. Consequently, anchovy biomass increased again and the sardine populations started to decline.

Obviously, dynamics of anchovies and sardines in the HCE are governed by long-term, decadal-scale physical processes. Whereas single El Niño events cause short-term perturbations of the environment which are unfavorable for the anchovy, the anchovy has the ability to recover very fast during periods when CCW is preponderant as observed, for example, after the strong El Niño events in 1982–83 and 1997–98 (Alheit and Niquen, 2004).

Benguela Current ecosystem (BCE)

Changes in the biota

The Benguela Current ecosystem consists of two subsystems: the northern part, off Namibia, from the Angola–Benguela front (14° S–16° S) to the permanent upwelling cell off Lüderitz (26° S), and the southern part, off South Africa, from the Lüderitz cell to East London (28° E). The southern part includes the upwelling region along the southwestern coast of southern Africa and extends over the Agulhas Bank along the south coast (Cury and Shannon, 2004). Each of the two sub-systems has an independent sardine and anchovy stock. Spawning and fishing areas of sardines expand and contract with periods of high and low abundance (Lluch-Belda *et al.*, 1989). Decadal-scale changes in the BCE have been described and discussed recently by Crawford *et al.*

(2001), Cury and Shannon (2004) and van der Lingen *et al.* (2006b).

Northern Benguela

Catches of northern Benguela sardines increased throughout the 1950s and 1960s up to a maximum of 1.4 million mt in 1968 (Fig. 5.1; Fig. 9.3b) (Schwartzlose *et al.*, 1999). They collapsed in 1969–70, were between 0.3 and 0.7 million mt from 1970–77 and finally fell to very low levels since. There might have been a slight recovery in the early 1990s. Virtual population analysis of spawner biomass demonstrates similar dynamics with a peak in 1965. Anchovy fisheries started in 1964, but they never reached the high yields as for sardine. In 1978, when catches peaked at 0.4 million mt, anchovy catch for the first time surpassed that of sardine. However, in 1984 catches fell to very low levels and have stayed there since, with the exception of 1987. At present, both species are at extremely low levels. It seems that anchovy replaced the sardine after the collapse in the late 1960s. Now that both species are very low they have apparently been replaced by a suite of different pelagic species such as horse mackerel (*Trachurus capensis*) and bearded goby (*Sufflogobius bibarbatus*), which were recorded in the purse-seine fishery for the first time in 1971 (Crawford *et al.*, 2001). Anomalies in the condition factor of sardine were negative from 1953–67 and then positive from 1968–84, probably because of a density-dependent response to high and low biomass, respectively (Crawford *et al.*, 2001). Also, after 1967, the proportion of spawners to non-spawners in the sardine population switched from 10%–50% to 70%–95%, indicating a density-dependent response whereby the population put more effort into reproduction as biomass declined (Crawford *et al.*, 2001). Sardine egg concentrations have decreased and egg distribution has been contracted to the north since the collapse. Also, anchovy egg distribution has shifted northwards (van der Lingen *et al.*, 2006b). At higher population levels, anchovy and sardine used to migrate between the northern spawning and the southern feeding grounds in the Walvis Bay area. Intense fishing may have interrupted this migration pattern (Boyer *et al.*, 2001). The substantial decrease of the Walvis Bay spawning is suggested to be the result of removal of older, migratory specimens (Daskalov *et al.*, 2003) or of a selective change in migratory behavior as a reaction to heavy fishing activities in the Walvis Bay area (Bakun, 2001). The change in pelagic fish community structure was reflected in predator diets. The proportion of sardines in the diet of Cape gannets, Cape cormorants and African penguins declined whereas the pelagic goby gained importance (Cury and Shannon, 2004). Also, sardine lost and the pelagic goby gained importance in seal diets after the sardine collapse.

There appears to be a decade-scale decline in zooplankton biomass and copepod abundance off Walvis Bay during the 1960s and 1970s until about the mid 1980s when the trend reversed and a subsequent increasing trend over the next two decades became evident. This cyclic pattern, which closely follows that of sea surface temperature in the same area, declining since the mid 1960s to a minimum in 1984 and rising subsequently until the present, is thought to be related to advective loss and retention processes associated with coastal upwelling (H.M. Verheye, personal communication).

Southern Benguela

Catches of southern Benguela sardine increased from the 1950s to a peak of 0.4 million mt in 1962 and then collapsed from 1963 throughout the mid 1960s to 0.07 million mt in 1967 (Fig. 5.1; Fig. 9.3b). Sardine increased again from 1985 onwards and has surpassed now the levels of the 1960s (van der Lingen *et al.*, 2006b). Southern Benguela anchovies were caught in larger quantities in the 1960s, particularly as a reaction to the sardine collapse. For 30 years, from 1966 to 1996, anchovy catches were larger than sardine catches with a peak of 0.6 million mt in 1987 and 1988 (Fig. 5.1; Fig. 9.2b). Catches decreased thereafter, but increased again since 2000. As a density-dependent reaction, conditon factor of sardine started to increase from 1969 on and stayed high until 1985, when it declined steadily (Crawford *et al.*, 2001). At present, both species are at rather high levels. Considerable shifts in spawning grounds of anchovies and sardines have been reported (van der Lingen *et al.*, 2006b). Anchovy shifted its major spawning activities since 1996 from the western to the central and eastern Agulhas Bank. The sardine has two major spawning grounds, the west coast from north of Cape Columbine to Cape Point and the south coast, the Agulhas Bank. Relative importance of the two areas kept on changing periodically since the 1960s. As with the anchovy, most sardine spawning has been recorded at the south coast since 2000.

The dynamics of anchovies and sardines as inferred from catch data are confirmed by other sources. The seabirds producing guano eat mainly small pelagics. Fluctuations in the quantity of guano produced are based on the availability of these fish to birds. Thus, the sardine collapse in the mid 1960s resulted in substantially reduced guano yields (Crawford and Shelton, 1978). The long-term dynamics of anchovies and sardines are reflected in their predators' diets (Crawford *et al.*, 2001). The frequency of occurrence of sardine in snoek (*Thyrsites atun*) stomachs fell from more than 50% in 1962 to less than 10% in 1964, just at the time when the sardine stock collapsed (Fig. 7a in van der Lingen *et al.*, 2006b). At the same time, the percentage of anchovy increased to over 70%. Studies on the diet of Cape gannets show that there was a good sardine year class in 1983 which triggered the subsequent increase in sardine biomass (Crawford *et al.*, 2001). Sardine increased steadily in the diet of Cape gannets from 1983–90. Biomass of sardine increased considerably after 1985 and at the same time the condition factor began to decrease (Crawford *et al.*, 2001). Around 1962–63, there was a change in the relative abundance of larvae of non-harvested mesopelagic fish species (Loeb, 1988). Total copepod abundance in a single location in the southern Benguela increased by two orders of magnitude from 1951–96 (Verheye *et al.*, 1998). During the period of the preferentially filter-feeding sardine dominance, smaller-sized cyclopoid copepods made up 41% of the crustacean zooplankton. In contrast, when the preferentially particle-feeding anchovy was dominant, the proportion of cyclopoid crustacean zooplankton was significantly higher (56%) (Fig. 7.16) (Verheye and Richardson, 1998; Verheye, 2000; this volume, Chapter 7). This differential predation impact of sardine and anchovy focusing on different size spectra of crustacean zooplankton might be a mechanism leading to the alternating dominance of sardines and anchovies (Alheit and Niquen, 2004; van der Lingen *et al.*, 2006a) (see below).

Regime shifts

The northern and southern Benguela ecosystems, which are separated by the strong Lüderitz upwelling cell, provide quite distinct habitats for anchovies and sardines. In the northern Benguela, two kinds of environmental anomalies have been recorded periodically, which lead to shifts in distribution and decline in recruitment and catches of pelagic resources, as was observed particularly during the 1990s. The Benguela Niño events, which occur about once every decade, are associated with warm waters penetrating onto the Namibian shelf. Frequently occurring low-oxygen or anoxic events (Weeks *et al.*, 2002) also lead to high mortalities in the pelagic realm of the northern Benguela. The low-oxygen event in 1993–94 and the immediately following Benguela Niño in early 1995 probably inhibited the potential recovery of northern Benguela sardines in the mid 1990s, beside the heavy fishing pressure. Another substantial difference between the two Benguela subsystems is that anchovies and sardines in the southern Benguela make use of the non-upwelling Agulhas Bank environment for completing their life cycle. Consequently, their dynamics are not directly comparable with those from pure upwelling ecosystems. This might also explain why the alternation between anchovy and sardine periods, which is typical for upwelling systems, has not been observed in the southern Benguela recently. Whereas there seems to have been a shift in the southern Benguela from a sardine to an anchovy regime in the early to mid 1960s with concomitant changes in the zooplankton (Verheye *et al.*, 1998) and predator communities (Cury and Shannon, 2004) and large changes in the

structure and functioning of the pelagic component of the southern Benguela ecosystem (Crawford *et al.*, 2001), both species exhibited recently very high population sizes at the same time since the late 1990s in the southern Benguela. Dynamics of small pelagics in the northern Benguela took a different turn. After the collapse of the sardine in the late 1960s, catches of anchovies increased. However, the increase in biomass of other pelagic species such as horse mackerel, pelagic goby and even jellyfish was much more conspicuous and led Cury and Shannon (2004) to the assumption that there was a regime shift from sardines to other non-anchovy pelagic species. There was probably a substantial transfer of production to the mid water, which implies that the trophic functioning of the system changed considerably. Changes in the trophic regime of the northern Benguela ecosystem continued into the 1970s, whereas the regime shift in the southern Benguela was completed in the 1960s (Crawford *et al.*, 2001). In addition, both anchovies and sardines in the northern Benguela have been at extremely low biomass levels since the early 1990s, partly caused by the devastating warm water and low-oxygen events of 1993 to 1995. The major changes in the entire pelagic Benguela system took place in the 1960s and the early 1980s, when, respectively, the two sardine populations decreased and that of the southern system recovered again (Crawford *et al.*, 2001). Apparently, these changes in the trophic structure and functioning of the northern Benguela ecosystem were associated with intrusions of warm, saline surface waters onto the Namibian shelf, the Benguela Niños (Crawford *et al.*, 2001). Also, the major changes in the southern Benguela were approximately coincident with Benguela Niños (Crawford *et al.*, 2001). The causes of the regime shifts in both Benguela subsystems remain largely obscure.

Canary Current ecosystem (CanCE)

Changes in the biota

In contrast to the other ecosystems analyzed in this article, the CanCE is not dominated by a *Sardinops/Engraulis* species pair. Instead, it is characterized by *Sardina pilchardus* and *Sardinella aurita* and *S. maderensis*. Anchovies (*E. encrasicolus*), although present in this ecosystem, do not play such an important role as in the other systems. Biogeographically, the CanCE can be divided into two regions with a transition between Cape Barbas (23° N) and Cape Blanc (21° N), the northern one of which is characterized by North Atlantic Central Water (NACW) whereas the southern one is dominated by South Atlantic Central Water (SACW) (Mittelstaedt, 1983). *S. pilchardus* and *E. encrasicolus* are predominant in the NACW area, off the Iberian and Moroccan coasts, whereas *Sardinella aurita* and *S. maderensis* are dwelling in the SACW area, off Mauritania and Senegal (Fréon, 1988; Marchal, 1991).

Off NW Africa, between 20° N and 36° N, there are three important fishing areas for *S. pilchardus:* in the north (36° N–33° N), the centre (32°30'N–27° N), and the south (26° N–20° N) (Belvèze and Erzini, 1983). It is assumed that they correspond to separate stocks, but some seasonally occurring partial mixing is likely. The biomass in the central and southern areas has fluctuated widely during the last century. Sardine has largely dominated the total landings of pelagic fish north of Cape Bojador (26° N) since the beginning of fishing activities in the 1920s. The northern and central stocks provided the bulk of the catches until the 1970s. After 1966, an increase of sardine spawning was observed south of Cape Juby (28° N) and, from 1970 onward, the area of distribution and the abundance of sardine increased notably toward the south after several years of good spawning (Holzlöhner, 1975; Barkova and Domanevsky, 1976). In contrast, in the historical Moroccan sardine fishery grounds (the northern part of the central area), a constant decrease of the catches, spreading from north to south, began around 1968 and in the early 1990s the fishery that operated between 32° N and 30° N collapsed (Belvèze and Erzini, 1984; Kifani, 1998) (Fig. 5.1). During the 1970s, south of Cape Bojador (26° N) to Cape Blanc (20° N), the rather tropical *Sardinella* and *Scomber japonicus* were replaced in the catch composition by *Sardina*, which had expanded southwards. During this period of high abundance of sardine off the Saharan area (south of 25° N), significant catches have been recorded as far south as Cape Verde (14° N) (Fréon, 1988).

Other remarkable events that were recorded during the period of sardine outburst off the Sahara are the dramatic increase in the early 1970s of snipefish (*Macrorhamphosus scolopax* and *M. gracilis*), both in Moroccan and Iberian waters (Brêthes, 1979), as well as the sudden explosion of the triggerfish (*Balistes carolinensis*) which spread geographically from Ghana to Mauritania, occupying the pelagial during the first two years of its life cycle (Caverivière, 1991; Fréon and Misund, 1999). Triggerfish and snipefish collapsed simultaneously in NW African waters after they had expanded their biomass respectively to more than 1 million mt in the late 1970s to the early 1980s (Sætersdal *et al.*, 1999 Belvèze, 1984). Outburst and collapse of snipefish remain largely unexplained, while the triggerfish expansion was linked to South Sahelian rivers runoff deficit during the 1970s and 1980s (Gulland and Garcia, 1984; Caverivière, 1991).

In the mid 1990s; the biomass of the Saharan sardine stock between 20° N and 26° N crashed drastically from more than 5 million mt in 1995 to about 1 million mt in 1997. This collapse does not seem to be linked to fishing pressure, as the USSR fishing fleet had left the area already in the early 1990s. Thereafter, the sardine recovered again gradually. In the early 1980s, most of the total biomass of

sardinellas was located between Cape Verde (14°40'N) and Cape Roxo (12°20'N) (Sætersdal *et al,*, 1999), while, after the mid 1990s, more than 50% of the total biomass of these species was found between 20° N and 25° N, indicating a clear northward shift in the distribution of sardinellas towards the Saharan region. Recently, over the last 2 years, some scattered concentrations of sardinella have been caught further north, off Cape Juby. Zooplankton data collected over the NW African continental shelf and slope between 16° N and 32°30'N revealed over the Mauritanian shelf a pronounced increase up to 1998 of zooplankton species, which are usually prevailing in the waters off Senegal and Guinea (Sirota *et al.*, 2004).

Changes in geographic distribution and potential regime shifts

Although a large amount of decadal-scale variability has been described for small pelagic fish populations off NW Africa, neither regime shifts nor alternating population fluctuations between anchovies and sardines have been observed, in contrast to other eastern boundary systems. Possible reasons, as stated above, are that a different sardine species is dwelling in the CanCE and that anchovy biomass is rather negligible (Fig. 5.1). In addition, there are no decadal-scale long-term time series for other biological components than fish available from the waters off NW Africa, which would indicate regime shifts. Instead of regime shifts, this region seems to be governed by large-scale shifts of distributional boundaries of small pelagic fish populations moving the centers of gravities of the populations and the transition zones between *S. pilchardus* and *Sardinella* spp. northwards and southwards along the NW African coast (Kifani, 1998; Binet *et al.*, 1998).

Over the period in which there have been major fisheries operating, two transformations of habitat geography of *S. pilchardus* off NW Africa have been observed and, over a multi-decadal time scale, gradual 2000 km southward expansions along the coast have occurred (Binet *et al.*, 1998; Kifani, 1998; Bakun, 2005). In the 1920s, the southward limit of sardines has been reported to be off northern Morocco. By the 1950s, the species had become abundant as far south as Mauritania (Belvéze, 1984), and, around the mid 1970s, sardines were fairly common even off Senegal (Fréon, 1988; Bakun, 2005). Thereafter, the distribution area contracted again northwards and sardine abundance was drastically reduced south of 25° N in 1982–84 (Bakun, 2005). Subsequently, sardines expanded their range far southward and were caught again in Senegalese waters during the 1990s (Binet *et al.*, 1998; Bakun, 2005). The southward migration in the late 1960s was followed by the emergence of a new sardine fishery off the Sahara, which increased quickly from 80 000 to 650 000 mt (Binet *et al.*, 1998). At the same time, in the early 1970s, the small pelagic community in Saharan waters switched from a community dominated by horse mackerel, mackerel and sardinellas to a sardine-dominated one (Gulland and Garcia, 1984). When the sardine range expanded to the south, the distribution area of sardinella shifted equatorward (Kifani, 1998). In 1974, some sardines were fished off Senegal. A few years later the southern boundary of the sardine population started moving back to the north again and by 1982–83 sardine had almost disappeared from Mauritanian waters (Binet *et al.*, 1998). A second southward extension of the sardine population occurred in the late 1980s all the way to Senegal leading again to high catches off the Sahara. Thus, twice in 20 years a southward extension of the geographic range of the sardine was recorded, followed by high catches. Again, in the mid 1990s, when the abundance of sardine off the Saharan region declined, sardinellas extended northwards (Fréon *et al.*, 2006). It is not clear whether these alternate shifts in biogeographic boundaries of sardines and sardinellas might indicate regime shifts.

Mechanisms linking climate to decadal-scale population dynamics

The reason for the collapse in the early 1990s of the traditional Moroccan sardine fisheries in the area surrounding Cap Sim and Agadir Bay (32° N–30° N) is still an unresolved issue. Several hypotheses have been proposed (Belvèze and Erzini, 1984; Do-Chi and Kiefer, 1996; Kifani, 1998; Bakun, 2005). It seems that this collapse could have resulted from a reduced sardine migration from the primary reproductive zone in the south (near Tantan, between 28° N and 29° N) to the adult feeding grounds located further north in the region of active upwelling between Cap Sim and Agadir Bay where the major fishing ports are located. A decrease of the wind-induced upwelling and the associated increasing trend in sea surface temperature has been proposed as an explanation for the cessation of the migration to the feeding grounds (Kifani, 1998; Bakun, 2005).

The increase in sardine biomass that occurred during the 1970s in the southern part of the Canary current (26° N to 20° N) and the related southward extension of the sardine population was associated with the strengthening of upwelling off the Sahara region and further south off Mauritania and Senegal (Holzlöhner, 1975; Sedykh, 1978; Binet, 1988; Binet *et al.*, 1998). There is also indication that the second southward extension of the sardine population that occurred in the late 1980s was related to colder than usual temperatures and increased upwelling (Binet *et al.*, 1998; Roy and Reason, 2001). The mechanisms involved are still unclear. Binet *et al.* (1998) hypothesized that intensification of the trade winds along the NW African coast enhanced upwelling activity and southward transport whereby phytoplankton production was most probably boosted by enhanced upwelling, but not matched by

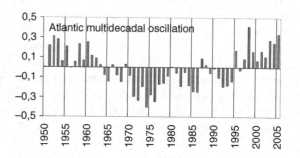

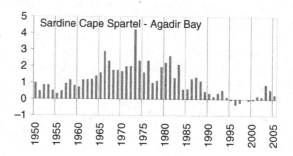

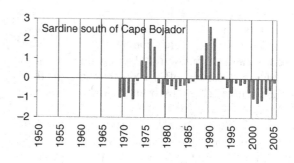

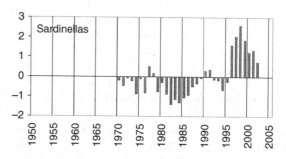

Fig. 5.8. Atlantic multidecadal oscillation and normalized catch anomalies of sardine and sardinella off NW Africa.

zooplankton grazing, due to the brevity of the water residence time over the shelf. This led to new distribution patterns of primary and secondary production that are in favor of sardines, which can feed on phytoplankton in contrast to the zooplankton-feeding sardinellas (Binet *et al.*, 1998). A recent study of the impact of the continental shelf geometry on the structure of a wind-driven upwelling explains the mechanism, which separates the upwelling area from the coast, that is observed off Sahara (Estrade, 2006). It has been shown that enhanced upwelling-favorable wind over this wide and shallow shelf tends to move the core of the upwelling from the nearshore domain to the mid shelf area. This creates an inner front that allows retention of biological material such as larvae in the highly productive nearshore environment. This separation of the upwelling core from the near-coastal area that is observed and modeled under enhanced wind forcing is another potential mechanism to explain the postive link between upwelling intensity and sardine abundance off the Sahara region.

The increased occurrences of tropical species north of Cap Blanc after the mid 1990s seem to be associated with a northward shift of the boundary between the NACW and the SACW toward 23°–25° N latitudes and warmer SSTs off NW Africa and changes in the main current pattern (Ostrowski and Strømme, 2004; Sirota *et al.*, 2004). The Atlantic Multidecadal Oscillation, AMO (Kerr, 2000), a distinct signal of multidecadal variability of North Atlantic SST was in a positive phase between 1925 and 1965 (Fig. 5.8). Another positive period has been in progress since the mid 1990s. In contrast, cold periods were recorded from the end of the nineteenth century until 1925 and from the mid 1960s to the mid 1990s. Interestingly, the timing of the turning points of AMO anomalies seems to be consistent with the timing of the shifts recorded in the CanCE, such as trade wind intensity, current and species distribution and catches (Fig. 5.8). Also, Moses *et al.* (2006), using a century length proxy record from corals (*Siderastrea radians*) at Pedra de Lume (16°45.6'N, 22° 53.3'W) on the island of Sal, Cape Verde Islands, in the eastern tropical North Atlantic, show that long-term salinity correlates positively with the two major North Atlantic SST indices, the Tropical North Atlantic Index (TNA) and Atlantic Multidecadal Oscillation (AMO).

Discussion

Temperature

For many authors (e.g. Chavez *et al.*, 2003), it has been a puzzle that sardine biomass increased in the eastern boundary currents when SSTs increased, whereas the Japanese sardine increased when temperatures decreased. However, the argument that Japanese sardines increase in abundance when temperatures decrease is not valid, as a closer look at their ambient temperatures and the respective literature reveals. Already Tomosada (1988) stated that sardine catch increased in the 1930s and the 1970s when the temperature in the coastal sardine spawning grounds increased. He also reported that, at the same time, temperature in the fishing grounds decreased. Also, when the KC adopts its large meandering phase (Kasai *et al.*, 1993, 1996), the frequency

of warm-water Kuroshio intrusions into the coastal zone, such as the Enshu-nada Sea, which is one of the most important larval nursery areas, is increased. Consequently, during the high abundance phase of the Pacific sardine population off Japan, shelf temperatures were mainly elevated (Nakata *et al.*, 2000; Yasuda *et al.*, 1999). After 1988, shelf temperatures started to decrease again. In addition, it has to be considered that the Pacific sardine of Japan expands during its high abundance phase from cooler coastal waters into a warm water current, the KC. Consequently, the effect is the same in eastern boundary currents (CC, HC[3]) as well as in the western boundary current (KC): sardine populations in their expansion phase extend their area of distribution into warmer waters.

Trophodynamic aspects

The alternation between anchovy and sardine periods may be trophodynamically mediated (van der Lingen *et al.*, 2006a; this volume, Chapter 7). Intermittent mixing such as upwelling leads to relatively cool temperatures in the upper layers and favors food chains dominated by diatoms and large calanoid copepods. This is the favorite feeding environment for the preferentially particulate-feeding anchovy as demonstrated for the Humboldt (Alheit and Niquen, 2004) and the Benguela Current ecosystems (van der Lingen *et al.*, 2006a). More stable water column situations caused by, e.g. relaxed upwelling and/or El Niño, leading to warmer temperatures in the upper layers, lead to flagellate-dominated food chains and a shift in the size spectrum of the crustacean zooplankton towards small-sized copepods such as cyclopoid copepods (Alheit and Niquen, 2004; van der Lingen *et al.*, 2006a). For example, abundances of small cyclopoid copepods (*Oncaea* spp. and *Oithona* spp.) in the Humboldt Current ecosystem increased from 3-fold to 1 order of magnitude (*Oithona*) as the El Niño event developed from January 1997 to January 1998 (González *et al.*, 2000). This is the feeding environment more favorable for the non-selective filter-feeding sardine (Alheit and Niquen, 2004; van der Lingen *et al.*, 2006a). Physical forcing leading to different size spectra of phyto- and zooplankton provides, by trophodynamical mediation, feeding environments favorable for anchovies or for sardines, respectively (van der Lingen *et al.*, 2006) (Fig. 7.16; this volume, Chapter 7).

Regime shifts and mechanisms linking climate to decadal-scale population dynamics

Major hypotheses about potential mechanisms for the observed low frequency population swings of anchovies and sardines and their alternations are discussed by MacCall (this volume, Chapter 12). A climatic regime shift has been reported for the North Pacific region for 1976–77 (e.g. Hare and Mantua, 2000; King *et al.*, 2005) and numerous publications from Pacific rim ecosystems in the western and eastern North Pacific and the eastern South Pacific have described ecosystem regime shifts occurring in reaction to this climate shift. However, more exact investigation of biological data series reveals that in several cases dramatic changes in the abundance of key populations did not occur in the mid 1970s, but some years earlier, between the late 1960s and the early 1970s. Often, the occurrence of a regime shift was concluded because biomass anomaly values of time series were crossing the zero line. However, the crossing of the zero line has no biological meaning. Consequently, the concept of a North Pacific or even a pan-Pacific ecosystem regime shift in the mid 1970s has to be seriously questioned. The scrutiny of zooplankton and pelagic fish time series of three major Pacific ecosystems, such as the waters east of Japan, the CCE and the HCE, reveals that many of these populations did not exhibit abrupt changes in response to the 1976–77 climate shift of the North Pacific. However, a critical period for anchovies, sardines and zooplankton in waters east of Japan and the HCE was around 1969–71 when anchovy populations (Fig. 5.1) and meso-zooplankton in the Oyashio, off Peru and off Chile declined (Fig. 5.3) and sardine populations started to increase. For the same period, a number of major changes in physical variables such as SST and subsurface processes (depth changes in MLD and thermocline) have been described. Around the mid to late 1980s, all these changes reversed. Obviously, the waters east of Japan and the HCE experienced regime shifts in 1969–71 and from the mid to late 1980s. Studies of major changes in physical subsurface processes, which were observed at the same time as the changes in major fish and zooplankton populations, give rise to speculations that they were caused by changes in basin-scale gyre circulation, particularly by "more frequent incursions of water from the subtropical gyre in the CCE" (Bograd and Lynn, 2003; McGowan et al., 2003) or "a spin-up of the subtropical gyre in the Kuroshio Extension" (Yasuda and Hanawa, 1997; Yasuda et al., 2000). Also, the approach and retreat of the warm SSW to and from the Peruvian coast indicates that changes in basin-scale circulation might be involved in decadal-scale dynamics of anchovy and sardine populations and respective ecosystem regime shifts. The same argument with respect to circulation is brought forward by MacCall (this volume, Chapter 12) who suggests that flow patterns in the Japanese system and eastern boundary currents are strongly associated not only with characteristic shifts in ocean temperature, but also with shifts in the physical and biological structure of entire ecosystems. In conclusion, alternations between anchovy and sardine populations and regime shifts in respective ecosystems seem to be brought about by changes in basin-scale circulations which alter the feeding (and maybe the

reproductive) environment such that they are either favorable to anchovies or to sardines.

The possible association between Pacific ecosystem regime shifts and basin-scale circulation changes seems to be compatible with the flow hypothesis of MacCall (this volume, Chapter 12). Although changes in basin-scale circulation are tightly interwoven with climate variability, the direct association of shifts in the dynamics of small pelagic fish and their ecosystems and climate dynamics, at least in the Pacific, is still obscure. The major change in the PDO was observed 1976–77 when the anomalies jumped from negative to positive values, thereby crossing the zero line. However, at the time of the observed regime shifts in 1969–71 and the mid to late 1980s, there are also clear signals in the PDO which exhibits turning points during these brief periods. From the late 1950s to 1970, PDO anomalies became more and more negative. Then, the trend turned around and anomalies became more and more positive until the mid/late 1980s. Thus, the timing of the turning points of PDO anomalies corresponds to the shifts recorded in the ecosystems. Also, there is an interesting link between historic sardine fishing periods of Japanese sardines and the dynamics of the Aleutian Low (Yasuda et al., 1999). Spring temperatures over the NW coast of N. America are correlated with the dynamics of the Aleutian Low and Kuroshio temperatures. Using tree rings, a NW coast temperature curve was reconstructed for several centuries indicating that high sardine yields are in synchrony with positive temperature anomalies over the NW coast. Consequently, it is concluded that the dynamics of the Japanese sardine are governed by Aleutian Low dynamics.

Synchronies and teleconnections

The long-term dynamics of the HCE are characterized by alternating sardine and anchovy regimes and an associated restructuring of the entire ecosystem from phytoplankton to the top predators (Alheit and Niquen, 2004). These regime shifts seem to be linked to lasting periods of warm or cold water anomalies related to the approach or retreat of warm subtropical oceanic waters (SSW) to the coast of Peru and Chile. Phases with mainly negative temperature anomalies parallel anchovy regimes (1950 to about 1970; 1985 up to now) and the rather warm period from about 1970–1985 characterized by sardine dominance. The transition periods (turning points) from one regime to the other were 1969–1971 and 1985–88 which are much earlier than those suggested by Chavez et al. (2003) who suggested mid 1970s and mid to late 1990s, respectively. The KCE is similarly characterized by alternating periods of sardines and anchovies. The most recent transition periods between the two Kuroshio species were strikingly synchronous to those of their Humboldt congeners (Fig. 5.9). The striking synchronies between the HCE and the KCE extend to the

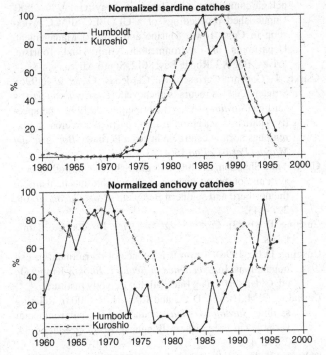

Fig. 5.9. Normalized anchovy and sardine catches from Humboldt and Kuroshio Currents.

timing of (i) changes between temperature regimes, (ii) subsurface processes (changes in thermocline depth, MLD position), (iii) anchovy and sardine periods, (iv) dynamics of zooplankton, and (v) other nektonic populations. The question remains: how were the changes in temperature regimes and subsurface processes in both ecosystems synchronized? A very likely mechanism is synchronized flow patterns of basin-scale circulation of the gyres in the North and South Pacific. Coastal sea level is a well-established indicator of integrated boundary current flow (Chelton *et al.*, 1982; this volume, Chapter 12). Using sea level data, MacCall (this volume, Chapter 12) demonstrates a Pacific-wide tendency toward synchrony in fluctuations in sea levels and associated boundary current strengths. The claim of Kawasaki (1983) and Chavez *et al.* (2003) that Pacific anchovy and sardine populations fluctuate in synchrony cannot be confirmed. Whereas fluctuations of sardines off Japan and California might have been in synchrony during the 1930s and 1940s, they are clearly out of phase since the 1970s (Fig. 5.1).

Acknowledgments

A. MacCall, D. Checkley, H. Verheye, and an anonymous reviewer provided very valuable comments, which improved the manuscript.

NOTES

1 Global Ocean Ecosystem Dynamics.
2 Sea Surface Temperature.
3 Humboldt Current.

REFERENCES

Alheit, J. and Bakun, A. (2009). Population synchronies within and between ocean basins: apparent teleconnections and implications as to physical–biological linkage mechanisms. *J. Mar. Syst.* **4**, doi:10.1016/j.jmarsys.2008.11.029

Alheit, J. and Bernal, P. (1993). Effects of physical and biological changes on the biomass yield of the Humboldt Current ecosystem. In *Large Marine Ecosystems – Stress, Mitigation, and Sustainability*, Sherman, K. Alexander, L.M., and Gold, B., eds. American Association for the Advancement of Science, pp. 53–68.

Alheit, J. and Niquen, M. (2004). Regime shifts in the Humboldt Current ecosystem. *Prog. Oceanogr.* **60**: 201–222.

Ayón, P., Purca, S., and Guevara-Carrasco, R. (2004). Zooplankton volume trends off Peru between 1964 and 2001. *ICES J. Mar. Sci.*, **61**, 478–484.

Bakun, A. (1987). Monthly variability in the ocean habitat off Peru as deduced from maritime observations, 1953–84. In *The Peruvian Anchoveta and Its Upwelling Ecosystem: Three Decades of Change*, Pauly, D., and Tsukayama, I., eds. Manila: International Center for Living Aquatic Resources Management (ICLARM), pp. 46–74.

Bakun, A. (1990). Global climate change and intensification of coastal ocean upwelling. *Science* **247**: 198–201.

Bakun, A. (1996). *Patterns in the Ocean: Ocean Processes and Marine Population Dynamics*, San Diego: University of California Sea Grant.

Bakun, A. (2001). "School-mix feedback": a different way to think about low frequency variability in large mobile fish populations. *Prog. Oceanogr.* **49**: 485–511.

Bakun, A. (2005). Regime shifts. In *The Sea*, vol.13, Robinson, A.R. and Brink, K, eds. Cambridge, MA: Harvard University Press, pp. 971–1026.

Barkova N.A. and Domanevsky L.N. (1976). Some peculiarities of sardine (*Sardina pilchardus*) distribution and spawning along Northwest Africa. *ICES CM 1976/J*:6, 15 pp.

Baumgartner, T.R., Soutar, A., and Ferreira-Bartrina, V. (1992). Reconstruction of the history of Pacific sardine and northern anchovy populations over the past two millenia from sediments of the Santa Barbara Basin, California. *CalCOFI Rep.*, **33**, 24–40.

Beamish, R.J., ed. (1995). Climate Change and Northern Fish Populations. *Can. Spec. Publ. Fish. Aquat. Sci.* **121**: 1–793.

Beaugrand, G. (2004). The North Sea regime shift: evidence, causes, mechanisms and consequences. *Prog. Oceanogr.* **60**: 245–262.

Beaugrand, G. and Reid, P.C. (2003). Long-term changes in phytoplankton, zooplankton and salmon related to climate. *Glob. Change Biol.* **9**: 1–17.

Belvèze, H. (1984). Biologie et dynamique des populations de sardine peuplant les côtes atlantiques marocaines et propositions pour un aménagement des pêcheries. Unpublished Ph.D. thesis, Université de Bretagne Occidentale, Brest.

Belvèze, H. and Erzini, K. (1983). The influence of hydroclimatic factors on the availability of the sardine (*Sardina pilchardus*, Walbaum) in the Moroccan Atlantic fishery. *FAO Fish. Rep.* 291: 285–327.

Bernal, P., Robles, F.L., and Rojas, O. (1983). Variabilidad fisica y biologica en la region meridional del sistema de corrientes Chile-Peru. *FAO Fish. Rep.* 291: 683–711.

Binet, D. (1988). Rôle possible d'une intensification des alizés sur le changement de répartition des sardines et sardinelles le long de la côte ouest africaine. *Aquat. Living Resour.* 1: 115–32.

Binet, D., Samb, B., Sidi, M.T., et al. (1998). Sardine and other pelagic fisheries changes associated with multi-year trade wind increases in the southern Canary current. In *Global versus Local Changes in Upwelling Systems*, Durand, M.H., Cury, P., Mendelssohn, R., et al. eds. Paris: ORSTOM Editions, pp. 211–233.

Bograd, S.J. and Lynn, R.J. (2003). Long-term variability in the Southern California Current System. *Deep-Sea Res. II*, 50, 2355–2370.

Boyer, D.C., Boyer, H.J., Fossen, I., and Kreiner, A. (2001). Changes in abundance of the northern Benguela sardine stock during the decade 1990 to 2000 with comments on the relative importance of fishing and the environment. *S. Afr. J. Mar. Sci.* 23: 67–84.

Brêthes, J.C. (1979). Contribution à l'étude des populations de *Macrorhamphosus scolopax* (L. 1759) et *Macrorhamphosus gracilis* (Lowe, 1839) des côtes atlantiques marocaines. *Bull. de l'Inst. Pêches Marit.* 24 (juillet 1979).

Carrasco, S. and Lozano, O. (1989). Seasonal and long-term variations of zooplankton volumes in the Peruvian sea, 1964–1987. In *The Peruvian Upwelling Ecosystem: Dynamics and Interactions*, Pauly, D., Muck, P., Mendo J., and Tsukayama, I., eds. Manila: International Center for Living Aquatic Resources Management (ICLARM), pp. 82–85.

Caverivière, A. (1991). Quelques considérations méthodologiques sur les campagnes scientifiques de chalutage de fond en Afrique Occidentale. *ICES CM 1991/D*: 10.

Chavez, F.P., Ryan, J., Lluch-Cota, S.E., and Niquen, M. (2002). From anchovies to sardines and back: multidecadal change in the Pacific Ocean. *Science* 299: 217–221.

Chelton, D.B., Bernal, P.A., and McGowan, J.A. (1982). Large-scale interannual physical and biological interaction in the California Current. *J. Mar. Res.* 40: 1095–1125.

Chiba, S., Tadokoro, K., Sugisaki, H., and Saino, T. (2006). Effects of decadal climate change on zooplankton over the last 50 years in the western subarctic North Pacific. *Glob. Change Biol.* 12: 907–920.

Crawford, R.J.M. and Shelton, P.A. (1978). Pelagic fish and sea-bird interrelationships off the coasts of South West and South Africa. *Biol. Conserv.* 14: 85–109.

Crawford, R.J.M. and 23 co-authors. (2001). Periods of major change in the structure and functioning of the pelagic component of the Benguela ecosystem, 1950–2000. Unpublished proceedings of a GLOBEC/SPACC workshop in Cape Town, Marine and Coastal Management, Department of Environmental Affairs and Tourism, Private Bag X2, Rogge Bay, 8012, South Africa.

Csirke, J., Guevara-Carrasco, R., Cardenas, G., et al. (1996). Situación de los recursos anchoveta (*Engraulis ringens*) y sardina (*Sardinops sagax*) a principios de 1994 y perspectivas para la pesca en el Perú, con particular referencia a las regiones norte y centro de la costa Peruana. *Bol. Inst. del Mar del Perú-Callao* 15: 1–23.

Cury, P. and Shannon, L. (2004). Regime shifts in upwelling ecosystems: observed changes and possible mechanisms in the northern and southern Benguela. *Prog. Oceanogr.* 60: 223–243.

Cushing, D. (1982). *Climate and Fisheries*. London: Academic Press.

Cushing, D.H. (1996). Towards a science of recruitment in fish populations. In *Excellence in Ecology. Book 7*, Kinne, O., ed. Oldendorf/Luhe: International Ecology Institute.

Daskalov, G.M., Boyer, D.C., and Roux, J.-P. (2003). Relating sardine *Sardinops sagax* abundance to environmental variables in the northern Benguela. *Prog. Oceanogr.* 59: 257–274.

DeYoung, B., Harris, R., Alheit, J., et al. (2004). Detecting regime shifts in the ocean: data considerations. *Prog. Oceanogr.* 60: 143–164.

Do Chi, T. and Kiefer, D.A., eds. (1996). *Report of the Workshop on the Coastal Pelagic Resources of the Upwelling System off Northwest Africa: Research and Predictions. FI:TCP/MOR/4556(A) Field Document 1*. Rome: FAO.

Ebbesmeyer, C., Cayan, D.R., McLain, D.R., et al. (1991). 1976 step in the Pacific climate: forty environmental changes between (1968–1975 and 1977–1984). In *Proc. Seventh Ann. Pacific Climate (PACLIM) Wkshp, April (1990)*, Betancourt, J.L., and Tharp, V.L., eds. California Department of Water Resources Interagency Ecological Studies Program Tech. Rep. 26, pp. 129–141.

Estrade, P. (2006). Mécanisme de décollement de l'upwelling sur les plateaux larges et peu profond d'Afrique du Nord-Ouest. Thèse de doctorat, Université de Bretagne Occidentale, Brest, France, 136 pp.

Fréon, P. (1988). Réponses et adaptations des stocks de clupéidés d'Afrique de l'ouest à la variabilité du milieu et de l'explotation: Analyse et réflexion à partir de l'exemple du Sénégal. Etudes et Thèses, ORSTOM.

Fréon, P. and Misund, O.A. (1999). *Dynamics of Pelagic Fish Distribution and Behavior: Effect on Fisheries and Stock Assessment*. London: Blackwell Fishing News Books.

Fréon, P., Alheit, J., Barton, E.D., et al. (2006). Modelling, forecasting and scenarios in comparable upwelling ecosystems: California, Canary and Humboldt. In *Benguela: Predicting a Large Marine Ecosystem*, Shannon, V., Hempel, G., Melanotte-Rizzoli, P., et al., eds. Amsterdam: Elsevier, pp. 185–220.

González, H.E., Sobarzo, M., Figueroa, D., and Nöthig, E.-M. (2000). Composition, biomass and potential grazing impact of the crustacean and pelagic tunicates in the northern Humboldt Current area off Chile: differences between El Niño and non-El Niño years. *Mar. Ecol. Progr. Ser.* **195**: 201–220.

Gulland, J. (1982). Management policies and models of pelagic stocks: The Peruvian experience 1960–1972. *Faculdad de Ciencias Biológicas, Pontifica Universidad Católica de Chile, Monografías Biológicas* **2**: 65–74.

Gulland, J. A. and Garcia, S. (1984). Observed patterns in multispecies fisheries. In *Exploitation of Marine Communities*, May, R. M., ed. Berlin: Springer-Verlag, pp. 155–190.

Hare, S. R. and Mantua, N. J. (2000). Empirical evidence for North Pacific regime shifts in 1977 and 1989. *Prog. Oceanogr.* **47**: 103–145.

Hayward, T. L. (1997). Pacific ocean climate change: atmospheric forcing, ocean circulation and ecosystem response. *Trends Ecol. Evol.* **12**: 150–154.

Hill, K. T., Lo, N. C. H., Macewicz, B. J., and Felix-Uraga, R. (2006). Assessment of the Pacific sardine (*Sardinops sagax caerulea*) population for U.S. management in 2006. *NOAA Tech. Memo. NMFS-SWFSC*, **386**, pp. 1–103.

Holzlöhner, S. (1975). On the recent stock developement of *Sardina pilchardus* Walb. off Spanish Sahara. *ICES CM 1975/J*: **13**.

Hunter, J. R. and Alheit, J. (1995). International GLOBEC Small Pelagic Fishes and Climate Change program. Report of the First Planning Meeting, La Paz, Mexico, June 20–24, 1994.. *GLOBEC Rep.* **8**, 72 pp.

Isaacs, J. D. (1976). Some ideas and frustrations about fishery science. *Cal. Coop. Oceanic Fish. Invest. Rep.* **18**: 34–43.

Jacobson, L. D., de Oliveira, J. A. A., Barange, M., *et al.* (2001). Surplus production, variability, and climate change in the great sardine and anchovy fisheries. *Can. J. Fish. Aquat. Sci.* **58**: 1891–1903.

Kasai, A., Kimura, S., and Sugimoto, T. (1993). Warm water intrusion from the Kuroshio into the coastal areas south of Japan. *J. Oceanogr.* **49**: 607–624.

Kasai, A., Koizumi, N. and Sugimoto, T. (1996). Transport and survival of Japanese sardine larvae (*Sardinops melanostictus*) in the Kuroshio–Oyashio region: a numerical approach. In *Survival Strategies in Early Life Stages of Marine Resources*, Watanabe, Y., Yamashita, Y., and Oozeki, Y., eds. Rotterdam: A. A. Balkema, pp. 219–226.

Kawasaki, T. (1983). Why do some pelagic fishes have wide fluctuations in their numbers? – Biological basis of fluctuation from the viewpoint of evolutionary ecology. *FAO Fish. Rep.*, **291**, 1065–1080.

Kawasaki, T. (1993). Recovery and collapse of the Far Eastern sardine. *Fish. Oceanogr.* **2** 244–53.

Kerr, R. A. (2000). A North Atlantic climate pacemaker for the centuries. *Science* **288**: 1984–1986.

Kifani, S. (1998). Climate dependant fluctuations of the Moroccan sardine and their impact on fisheries. In *Global Versus Local Changes in Upwelling Systems*, Durand, M. H., Cury, P., Mendelssohn, R., *et al.*, eds. Paris: Editions ORSTOM, pp. 235–248.

King, J. R. (2005). Report of the Study Group on Fisheries and Ecosystem Responses to Recent Regime Shifts. In *Report of the Study Group on Fisheries and Ecosystem responses to Recent Regime Shifts*, King, J. R., ed. *PICES Scientific Rep.* **28**.

Kondo, K. (1991). Interspecific relation between Japanese sardine and anchovy populations that reflects the essential mutual relation between fluctuation mechanisms of the two species based on 'organism-environment' coupling. In *Long-term Variability of Pelagic Fish Populations and their Environment*, Kawasaki, T., Tanaka, S., Toba, Y., and Taniguchi, A., eds. Pergamon Press, pp. 129–134.

Laevastu, T. (1993). *Marine Climate, Weather and Fisheries*. Oxford: Blackwell Scientific Publications.

Lavaniegos, B. E. and Ohman, M .D. (2003). Long term changes in pelagic tunicates of the California Current. *Deep-Sea Res. II*, **50**, 2473–2498.

Lavaniegos, B. E. and Ohman, M. D. (2007). Coherence of long-term variations of zooplankton in two sectors of the California Current system. *Prog. Oceanogr.* doi: 10.1016/j.pocean.2007.07.002.

Limsakul, A., Saino, T., Midorikawa, T., and Goes, J. (2001). Temporal variations in lower trophic level biological environments in the northwestern North Pacific Subtropical Gyre from 1950 to 1997. *Prog. Oceanogr.* **49**: 129–149.

Lluch-Belda, D., Crawford, R. J. M., Kawasaki, T., *et al.* (1989). World-wide fluctuations of sardine and anchovy stocks: the regime problem. *S. Afr. J. Mar. Sci.* **8**: 195–205.

Lluch-Belda, D., Schwartzlose, R. A., Serra, R., *et al.* (1992). Sardine and anchovy regime fluctuations of abundance in four regions of the world oceans: a workshop report. *Fish. Oceanogr.* **1**: 339–347.

Loeb, V. J. (1988). *Report on the analysis of the Cape area historical ichthyoplankton data base*. Unpubl. Rep., Sea Fisheries Research Institute, Cape Town.

Loeb, V. J. and Rojas, O. (1988). Interannual variation of ichthyoplankton composition and abundance relations off northern Chile, 1964–83. *Fish. Bull.* **86**: 1–24.

MacCall, A. D. (1996). Patterns of low-frequency variability in fish populations of the California Current. *CalCOFI Rep.* **37**: 100–110.

MacCall, A. D., Batchelder, H., King, J., *et al.* (2005). Appendix 2: Recent ecosystem changes in the California Current system. In *Report of the Study Group on Fisheries and Ecosystem responses to Recent Regime Shifts*, King, J. R., ed. *PICES Scientific Rep.* **28**, pp. 65–86.

Marchal, E. (1991). Location of the main West African pelagic stocks. In *Variabilité, instabilité et changement dans les pêcheries ouest africaines*, Cury, P., and Roy, C. Paris: ORSTOM, pp. 187–191.

McFarlane, G. A. and Beamish, R. J. (2001). The re-occurrence of sardines off British Columbia characterises the dynamic nature of regimes. *Prog. Oceanogr.* **49**: 151–165.

McFarlane, G. A., Smith, P. E., Baumgartner, T. R., and Hunter, J. R. (2002). Climate variability and Pacific sardine populations and fisheries. *Am. Fish. Soc. Symp.* **32**: 195–214.

McGowan, J., Bograd, S. J., Lynn, R. J., and Miller, A. J. (2003). The biological response to the 1977 regime shift in the California Current. *Deep-Sea Res. II* **50**: 2567–2582.

Mendelssohn, R. and Mendo, J. (1987). Exploratory analysis of anchoveta recruitment off Peru and related environmental series. In *The Peruvian Anchoveta and its Upwelling Ecosystem: Three Decades of Change*, Pauly, D., and Tsukayama, I., eds. *ICLARM Studies and Reviews* **15**: 294–306.

Mittelstaedt, E. (1983). The upwelling area off Northwest Africa. A description of phenomena related to coastal upwelling. *Prog. Oceanogr.* **12**: 307–331.

Montecinos, A., Purca, S., and Pizarro, O. (2003). Interannual-to-interdecadal sea surface temperature variability along the western coast of South America. *Geophys. Res. Lett.* **30** (11): 1570, (24–1 – 24–4).

Moses C. S., Swart, P. K., and Rosenheim, B. E. (2006). Evidence of multidecadal salinity variability in the eastern tropical North Atlantic. *Paleoceanography* **21** PA3010, doi: 10.1029/2005PA001257.

Nakata, H., Funakoshi, S., and Nakamura, M. (2000). Alternating dominance of postlarval sardine and anchovy caught by coastal fishery in relation to the Kuroshio meander in the Enshu-nada Sea. *Fish. Oceanogr.* **9**: 248–258.

Nakata, K. and Hidaka, K. (2003). Decadal-scale variability in the Kuroshio marine ecosystem in winter. *Fish. Oceanogr.* **12**: 234–244.

Nakata, K., Hada, A., and Matsukawa, Y. (1994). Variations in food abundance for Japanese sardine larvae related to the Kuroshio meander. *Fish. Oceanogr.* **3**: 39–49.

Noto, M. and Yasuda, I. (1999). Population decline of the Japanese sardine, *Sardinops melanostictus*, in relation to sea surface temperature in the Kuroshio Extension. *Can. J. Fish. Aquat. Sci.* **56**: 973–983.

Noto, M. and Yasuda, I. (2003). Empirical biomass model for the Japanese sardine, *Sardinops melanostictus*, with sea surface temperature in the Kuroshio Extension. *Fish. Oceanogr.* **12**: 1–9.

Odate, K. (1994). Zooplankton biomass and its long-term variation in the western North Pacific Ocean, Tohoku Sea area. *Japan. Bull. Tohoku Natl. Fish. Res. Inst.* **56**: 115–173 (in Japanese, Engl. Abstr.).

Ostrowski, M. and Strømme, T. (2004). Evolution of coastal SST in Northeast subtropical Atlantic 1990–2001 and distribution patterns of small pelagic fish from Mauritania to Morocco. *ICES Symposium on The Influence of Climate Change on North Atlantic Fish Stocks. 11–14 May 2004, Bergen* (Poster B4).

Palacios, D. M., Bograd, S. J., Mendelssohn, R., and Schwing, F. B. (2004). Long-term and seasonal trends in stratification in the California Current. *J. Geophys. Res.* **109**: C10016 doi: 10.1029/2004JC002380.

Pizarro, O. and Montecinos, A. (2004). Interdecadal variability of the thermocline along the west coast of South America. *Geophys. Res. Lett.* **31**: L20307 (1–5).

Polovina, J. J., Mitchum, G. T., and Evans, G. T., (1995). Decadal and basin-scale variation in mixed layer depth and the impact on biological production in the central and North Pacific, 1960–88. *Deep-Sea Res. I* **42**: 1701–1716.

Rebstock, G. A. (2001). Long-term stability of species composition in calanoid copepods off southern California. *Mar. Ecol. Prog. Ser.* **215**: 213–224.

Rebstock, G. A. (2002). Climatic regime shifts and decadal-scale variability in calanoid copepod populations off southern California. *Glob. Change Biol.* **8**: 71–89.

Roemmich, D. and McGowan, J. (1995). Climatic warming and the decline of zooplankton in the California Current. *Science*, **267**: 1324–1326.

Roy, C. and Reason, C. (2001). ENSO related modulation of coastal upwelling in the eastern Atlantic. *Prog. Oceanogr.* **49**: 245–255.

Sætersdal, G., Bianchi, G., Strømme, T., and Venema, S. C. (1999). The DR. FRIDTJOF NANSEN Programme 1975–1993. Investigations of fishery resources in developing countries. History of the programme and review of results. *FAO Fish. Techn. Pap.* **391**: 1–434.

Sakurai, Y., Kiyofuji, H., Saitoh, S., *et al.* (2000). Changes in inferred spawning areas of *Todarodes pacificus* (Cephalopoda: Ommastrephidae) due to changing environmental conditions. *ICES J. Mar. Sci.* **57**: 24–30.

Sánchez, S. (2000). Variación estacional e interanual de la biomasa fitoplanktonica y concentraciones de chlorofila a, frente a la costa Peruana durante 1976–2000. *Bol. Inst. Mar Perú-Callao* **19**: 29–43.

Santander, H. and Flores, R. (1983). Los desoves y distribucion larval de quatro especies pelagicas y sus relaciones con las variaciones del ambiente marino frente al Peru. *FAO Fish. Rep.* **291**: 835–887.

Schwartzlose, R. A., Alheit, J., Bakun, A., *et al.* (1999). Worldwide large-scale fluctuations of sardine and anchovy populations. *S. Afr. J. Mar. Sci.* **21**: 289–347.

Schwing, F. B., Mendelssohn, R., and Bograd, S. J. (2004). When did the 1976 regime shift occur? *North Pacific Marine Science Organization, 13th Annual Meeting*, Honolulu, Hawaii, USA, abstract.

Schwing, F., Batchelder, H., Crawford, W., *et al.* (2005). Appendix 1. Decadal-scale climate events. In *Report of the Study Group on Fisheries and Ecosystem Responses to Recent Regime Shifts*, King. J. R. ed. *PICES Scientific Rep.* **28**: pp. 9–36.

Sedykh, K. A. (1978). The coastal upwelling off Northwest Africa. *ICES. CM1978/C*:12.

Serra, J. R. (1983). Changes in the abundance of pelagic resources along the Chilean coast. *FAO Fish. Rep.* **291**: 255–284.

Sirota, A., Chernyshkov, P., and Zhigalova, N. (2004). Water masses distribution, currents intensity and zooplankton assemblage off Northwest African coast. *ICES CM 2004/*N:2.

Thomson, C. A., Grover, A., and Craig, W. L. (1985). Status of the California coastal pelagic fisheries. *NMFS Southwest Region Admin. Rep.* **SWR 85–1**, 27 pp.

Tomosada, A. (1988). Long term variation of sardine catch and temperature. *Bull. Tokai Reg. Fish. Res. Lab.* **126**: 1–9 (in Japanese, Engl. abstr.).

Trenberth, K. E. and Stepaniak, D. P. (2001). Indices of El Niño evolution. *J. Clim.* **14**: 1697–1701.

Tsukayama, I. (1983). Recursos pelagicos y sus pesquerias en el Peru. *Rev. Com. Perm. Pac. Sur* **13**: 25–63.

van der Lingen, C. D., Hutchings, L., and Field, J. G. (2006a). Comparative trophodynamics of anchovy *Engraulis encrasicolus* and sardine *Sardinops sagax* in the southern Benguela: are species alternations between anchovy and sardine in the southern Benguela trophodynamically mediated? *Afr. J. Mar. Sci.* **28**: 465–477.

van der Lingen, C.D., Shannon, L.J., Cury, P., *et al.* (2006b). Resource and ecosystem variability, including regime shifts, in the Benguela Current ecosystem. In *Benguela: Predicting a Large Marine Ecosystem*, ed. Shannon, V., Hempel, G., Malanotte-Rizzoli, P., *et al.* Amsterdam: Elsevier, pp. 147–185.

Verheye, H.M. (2000). Decadal-scale trends across several marine trophic levels in the southern Benguela upwelling system off South Africa. *Ambio* **29** (1): 30–34.

Verheye, H.M. and Richardson, A.J. (1998). Long-term increase in crustacean zooplankton abundance in the southern Benguela upwelling region (1951–1996): bottom-up or top-down control? *ICES J. Mar. Sci.* **55**: 803–807.

Verheye, H.M., Richardson, A.J., Hutchings, L., *et al.* (1998). Long-term trends in the abundance and community structure of coastal zooplankton in the southern Benguela system, 1951–1996. *S. Afr. J. Mar. Sci.* **19**, 317–332.

Wada, T. and Jacobson, L.D. (1998). Regimes and stock recruitment relationships in Japanese sardine (*Sardinops melanostictus*), 1951–1995. *Can. J. Fish. Aquat. Sci.* **55**: 2455–2463.

Watanabe, Y., Kurita, Y., Noto, M., *et al.* (2003). Growth and survival of Pacific saury (*Cololabis saira*) in the Kuroshio–Oyashio transitional waters. *J. Oceanogr.*, **59**: 403–414.

Weeks, S.J., Currie, B., and Bakun, A. (2002). Massive emissions of toxic gas in the Atlantic. *Nature* **415**: 493–494.

Yasuda, I. (2003). Hydrographic structure and variability in the Kuroshio–Oyashio Transition Area. *J. Oceanogr.* **59**: 389–402.

Yasuda, I. and Watanabe, Y. (1994). On the relationship between the Oyashio front and saury fishing grounds in the northwestern Pacific: a forecasting method for fishing ground locations. *Fish. Oceanogr.* **3**: 172–181.

Yasuda, I., Sugisaki, H., Watanabe, Y., *et al.* (1999). Interdecadal variations in Japanese sardine and ocean/climate. *Fish. Oceanogr.* **8**, 18–24.

Yasuda, I., Tozuka, T., Noto, M., and Kouketsu, S. (2000). Heat balance and regime shifts of the mixed layer in the Kuroshio Extension. *Prog. Oceanogr.* **47**: 257–78.

Yasuda, T. and Hanawa, K. (1997). Decadal changes in the mode waters in the midlatitude North Pacific. *J. Phys. Oceanogr.* **27**: 858–870.

Yasunaka, S. and Hanawa, K. (2002). Regime shifts found in the northern hemisphere SST field. *J. Meteor. Soc. Japan* **80**: 119–135.

Yatsu, A. and Kaeriyama, M. (2005). Linkages between coastal and open-ocean habitats and dynamics of Japanese stocks of chum salmon and Japanese sardine. *Deep-Sea Res. II* **52**: 727–737.

Yatsu, A., Watanabe, T., Ishida, M., *et al.* (2005). Environmental effects on recruitment and productivity of Japanese sardine *Sardinops melanostictus* and chub mackerel *Scomber japonicus* with recommendations for management. *Fish. Oceanogr.* **14**: 263–278.

Zuta, S., Tsukayama, I., and Villanueva, R. (1983). El ambiente marino y las fluctuaciones de las principales poblaciones pelagicas de la costa peruana. *FAO Fish. Rep.* **291**: 179–253.

Zuzunaga, J. (1985). Cambios del equilibrio poblacional entre la anchoveta (*Engraulis ringens*) y la sardina (*Sardinops sagax*), en el sistema de afloramiento frente al Perú. In *El Niño, Su Impacto en la Fauna Marina*, Arntz, W., Landa, A., and Tarazona, J., eds. *Boletin Instituto del Mar del Perú-Callao, Volumen Extraordinario*, pp. 107–111.

6 Biophysical models

*Christophe Lett, Kenneth A. Rose, and
Bernard A. Megrey*

CONTENTS

Summary

The objective of this chapter is to provide an overview of biophysical models of marine fish populations, with a particular focus on those applied, or potentially applicable, for examining the consequences of climate change on small pelagic fish species. We focus on models that include physics and are therefore spatially explicit, and review models under the categories of those that include lower trophic level dynamics (NPZ[1]), early life stages of fish (eggs and larvae), and juvenile and adult stages. We first give an overview of the methods that are used to represent transport, growth, mortality, and behavior in biophysical models of early life stages. Second, we detail several case studies of such models, focusing on those applied to anchovy and sardine in SPACC regions and those involving small pelagic fish in "non-SPACC" regions. Some questions related to climate change require models that include juveniles and adults. Models that include juveniles and adults differ from the early life stage models in the important role played by behavior in fish movement. We briefly discuss several approaches used for modeling behavioral movement of fish, and then summarize several case studies of biophysical models that include adults that are relevant, or potentially relevant, to small pelagic species. Finally, we conclude with a discussion of the potential use of biophysical models of early life stages and adults for investigating some of the issues associated with forecasting the effects of climate change on small pelagic fish species.

Introduction

Understanding and forecasting the effects of climate change on small pelagic fish involves coupling the physics and lower trophic level dynamics to the growth, mortality, reproduction, and movement processes of key life stages that govern fish recruitment and population dynamics. Fish exhibit large changes in body weight, and often dramatic changes in body form, habitat usage, and diet, with ontogenetic development through the egg, larval, juvenile, and adult life stages. Early life stages (eggs and larvae) are heavily influenced by advective processes, which determine where they go in the system and to what environmental conditions they are exposed. There are many models that have coupled physics with eggs and larvae dynamics, and examined how physical transport under different conditions can affect the growth and mortality rate experienced by individuals during these early life stages (Werner *et al.*, 2001; Runge *et al.*, 2005; Miller, 2007). The "Workshop on advancements in modeling physical–biological interactions in fish early-life history: recommended practices and future directions" held on 3–5 April 2006 in Nantes, France (co-chairs: A. Gallego, E. North, and P. Petitgas), attracted more than 50 participants from 14 countries, an indication of the international interest in the topic. There are now modeling tools that allow the design of simulations coupling physics with ichthyoplankton dynamics (e.g. Ichthyop *et al.*, 2008; http://www.eco-up.ird.fr/projects/ichthyop/).

Many questions posed about climate effects on fish and fisheries can be only partially answered by models that predict growth and survivorship to the larval or early juvenile life stage. Recruitment for some species may not be determined until later in the life cycle than is simulated with these early life stage models. Furthermore, some questions require simulating the effects of climate on adults (e.g. how will climate affect migration, spatial distributions and spawning success), and other questions require multi-generational simulations that include all life stages in order to forecast the long-term consequences of climate effects (e.g. climate effects on sustainable harvest

Climate Change and Small Pelagic Fish, eds. Dave Checkley, Jürgen Alheit, Yoshioki Oozeki, and Claude Roy. Published by Cambridge University Press. © Cambridge University Press 2009.

levels). The coupling of physics and the lower trophic levels to models that include the adult life stages is much less common than models restricted to early life stages. Biophysical models that include adult life stages are being increasingly considered and their use, especially with the need to investigate the effects of climate and fishing in marine ecosystems, will accelerate over the next decade (Travers *et al.*, 2007).

The objective of this chapter is to provide an overview of biophysical models of fish, with a particular focus on those applied, or potentially applicable, for examining the consequences of climate change on small pelagic species. We consider models of early life stages and models that include adults. Our review is not intended to be comprehensive, and others have previously reviewed biophysical models of marine populations (Werner *et al.*, 2001; Runge *et al.*, 2005; Miller, 2007). We focus our review on models of fish that include physics and possibly lower trophic level dynamics; inclusion of physics implies the models must be spatially explicit. We divided these models into two general categories: single-species individual-based models (IBMs) of fish early life history and models of adult stages (either adults only, or full life cycle that include all life stages). In the early life stage IBMs, the abiotic environment is described with outputs from a hydrodynamic model, or from a hydrodynamic model coupled to a biogeochemical (e.g. NPZ) model. Werner *et al.* (2001) classified IBMs in which the physics is used for transport but the biotic (NPZ) environment is absent or represented by static prey fields derived from field data as "hydrodynamics and simple behaviors" or "hydrodynamics and static prey." Building on the classification proposed by Werner *et al.*, we also include in our review IBMs that use outputs of biogeochemical models as a dynamic representation of the prey for the fish, and these we term "hydrodynamics and dynamic prey." The adult models are much less common and, while we focus on models that are spatially explicit and coupled to hydrodynamic and biogeochemical models, we also include models that do not fulfill all of the criteria. We included some adult models because they could be adapted to examine questions about climate change effects that require the inclusion of adult life stages.

First, we give an overview of the methods used to represent transport, growth, mortality, and behavior of individuals in biophysical models of early life stages. Second, we detail several case studies of such models, organized by models of anchovy and sardine in SPACC regions (first in the Benguela upwelling system, then in other major upwelling systems), and for small pelagic fish in general in "non-SPACC" regions. We then turn to models that include adults. Because fish behavior comes much more into play when adult stages are considered, advective transport is no longer sufficient and we discuss methods used to represent movement. We then detail several case studies of biophysical models that include adults that are relevant, or potentially relevant, to small pelagic species and climate change. Finally, we conclude with a discussion of the potential use of these biophysical models to investigate some of the issues raised in the context of climate change.

Biophysical models of early life stages

Modeling the fish egg and larva environment

The abiotic environment

The abiotic environment for the fish eggs and larvae is usually provided by simulations of hydrodynamic models. In some studies, field data were directly used to generate the needed physics for model simulations (e.g. Heath *et al.*, 1998; Rodríguez, 2001; Santos *et al.*, 2004). The hydrodynamic models were generally used to provide temporally dynamic, three-dimensional fields of physical variables such as current velocities, temperature, and salinity.

The biotic environment

The use of biogeochemical models or field data to generate the biotic environment for the fish eggs and larvae has received less consideration. Most applications focus on the prey field for influencing the feeding and growth rates of the fish larvae. Prey fields have been generated as input to the early life stage models using biogeochemical models (e.g. Hinckley, 1999; Koné, 2006), interpolated from field data (e.g. Hinrichsen *et al.*, 2002; Lough *et al.*, 2006), or assessed using remotely sensed data (e.g. Bartsch and Coombs, 2004). On the other hand, use of models or data to specify spatially and temporally varying predation on fish eggs and larvae is rare (but see Suda and Kishida, 2003; Suda *et al.*, 2005; Vikebø *et al.*, 2007a). Predation is usually considered part of the total mortality rate, with the changing vulnerability of early life stages represented by mortality rates being stage specific or size dependent.

Modeling fish egg and larva dynamics

Most biophysical models of the early life stages are individual-based models (Lagrangian), although a few models used an Eulerian approach (e.g. Zakardjian *et al.*, 2003, who applied their model to zooplankton dynamics). There are a variety of different methods for coupling a hydrodynamic model to an individual-based model of eggs and larvae (Hermann *et al.*, 2001). Most common is to run the hydrodynamic model, store the outputs, and then use the outputs as inputs to the fish model (Fig. 6.1).

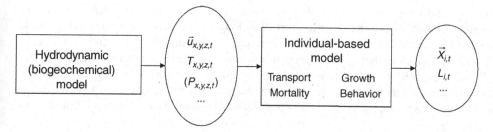

Fig. 6.1. A schematic view of the approach generally used for implementing biophysical models of the early life history of marine species (modified from Hermann *et al.*, 2001). A hydrodynamic model (or a hydrodynamic–biogeochemical coupled model) provides three-dimensional dynamic fields of current velocities $\vec{u}$, temperature T, and other variables (e.g. phytoplankton biomass P), to an individual-based model that tracks location $\vec{x}$, length L and other variables of interest for a collection of individuals i over time t.

Transport

Transport of individuals in biophysical models consists in updating the individuals' position $\vec{x}$ using the following equation:

$$d\vec{x}/dt = \vec{u} + \vec{u}' \qquad (6.1)$$

The deterministic advection term $\vec{u}$ is typically derived from spatial and temporal interpolations of the flow fields provided by the hydrodynamic model. A stochastic term $\vec{u}'$ is often added to take into account the small-scale fluctuations in the currents that are lost due to the grid resolution of the hydrodynamic model and temporal averaging of the hydrodynamic model outputs. The stochastic term is often represented using an ad hoc diffusion term (i.e. a random walk process) or using a random displacement process if a spatially non-uniform diffusivity is used (North *et al.*, 2006; Huret *et al.*, 2007). Recent theoretical results using the dispersal characteristics of drifters show promise for objectively specifying the stochastic term (Haza *et al.*, 2007). Equation (1) is numerically integrated using schemes such as a forward Euler or a Runge-Kutta. Ådlandsvik's (abstract[2]) has recently proposed several standard test cases to evaluate the advection scheme used in biophysical models. Some studies have specifically attempted to "validate" the flow fields and the transport scheme used by comparing trajectories of virtual and real drifters (Gutiérrez *et al.*, 2004; Thorpe *et al.*, 2004; Edwards *et al.*, 2006; Fach and Klinck, 2006).

Growth

A general relationship between the attributes of individuals (e.g. length L and age) and environmental variables (e.g. temperature T and food biomass F) was proposed by Heath and Gallego (1997) to simulate growth:

$$L_{age} = \int_{t=0}^{t=age} f_1(age)f_2(T)f_3(F)dt \qquad (6.2)$$

In most biophysical models, the growth algorithm used various simplified versions of equation (6.2) in which stage durations (e.g. Miller *et al.*, 1998), length (e.g. Fox *et al.*, 2006), or weight (e.g. Vikebø *et al.*, 2005) were functions of temperature only. When the effect of prey fields on growth was considered, relatively more complex bioenergetics submodels were developed (e.g. Hinckley, 1999; Megrey and Hinckley, 2001; Hinrichsen *et al.*, 2002; Lough *et al.*, 2005). These bioenergetics models allow for weight and temperature effects on consumption and the loss terms (respiration, egestion, excretion), and use the prey field information to determine the actual realized consumption rate. There is an extensive set of bioenergetics models for fish, many of which use the same notation and formulations in what is termed the Wisconsin model that allows for easy inter-specific and inter-life stage comparisons (Ney, 1993; Hanson *et al.*, 1997).

Mortality

Mortality has been represented in early life stage models in a variety of ways, including constant rates (e.g. Brown *et al.*, 2005; Tilburg *et al.*, 2005), and depending on life stage (e.g. Miller *et al.*, 1998), length (e.g. Hinckley, 1999), weight (e.g. Brickman *et al.*, 2001), temperature (e.g. Mullon *et al.*, 2003), and growth rate (e.g. Hinrichsen *et al.*, 2002; Bartsch and Coombs, 2004). Suda and colleagues (Suda and Kishida, 2003; Suda *et al.*, 2005) is one of the few examples that included inter-specific effects and density-dependent effects on mortality. Vikebø *et al.* (2007a) considered predation rates on larvae as functions of their size and light intensity. The high mortality rates of early life stage IBMs can result in numerical problems because many individuals need to be followed in order to obtain enough survivors to allow analysis of the results. One method for dealing with the high mortality problem is to use the concept of super-individuals (Scheffer *et al.*, 1995), in which each simulated individual is assumed to represent some number of identical population individuals (e.g. Hinckley, 1999; Megrey and Hinckley, 2001; Allain, 2004; Bartsch and Coombs, 2004). Rather than mortality resulting in the removal of model individuals, the worth of each super-individual is decreased based on the mortality rate. Models that use super-individuals can therefore use a constant, and an *a priori* determined number of model individuals throughout their simulations. Caution is needed in using super-individuals to ensure numerical accuracy of model simulations. When processes are density dependent or the introduction of new individuals occurs over an extended period of time or space, proper balancing of how new super-individuals are introduced (number, initial population number they represent, timing and spatial location of their introduction) is needed to ensure accurate model results.

Behavior

Egg buoyancy schemes have been included in determining the vertical position of individuals. Typically, a vertical velocity term is used that is assumed to be proportional to the difference between egg density and water density. Egg densities have been assumed to be constant (e.g. Parada *et al.*, 2003; North *et al.*, 2005), stage dependent (e.g. Hinckley, 1999), and time dependent (e.g. Brickman *et al.*, 2001). Because growth, mortality, and horizontal advection are all functions of depth, a key behavioral trait for drifting larvae is vertical positioning (Fiksen *et al.*, 2007). It is therefore no surprise that vertical migration of larvae has received a lot of attention, even in early works (Bartsch *et al.*, 1989). Most studies use a diel vertical migration scheme (e.g. Rice *et al.*, 1999; Peliz *et al.*, 2007), but other vertical migration approaches include the use of a depth-by-age curve (Ådlandsvik *et al.*, 2004), stage-specific vertical velocities (Pedersen *et al.*, 2006), and length-specific depth distributions (Bartsch and Coombs, 2004). Vikebø *et al.* (2007a) have used an approach where larvae are assumed to know the conditions within the upper 100 m and to migrate where they have an optimal blend of growth and survival. Horizontal swimming of larvae has also been considered in a few studies (Rodríguez *et al.*, 2001; Yeung and Lee, 2002; Fox *et al.*, 2006; Fiksen *et al.*, 2007). Specific experiments have been conducted to assess swimming abilities of larvae, and described as an integral part of the modeling process (e.g. Guizien *et al.*, 2006). To our knowledge, schooling behavior, which commonly appears in small pelagic fish larvae (Hunter and Coyne, 1982), has never been taken into account. However, there are many models of schooling that focus on how schools form and persist (Lett and Mirabet, 2008).

Case studies

Models of anchovy and sardine in the Benguela upwelling system

Most biophysical models of early life stages developed for the Benguela Current upwelling system have focused on anchovy (*Engraulis encrasicolus*) in the southern Benguela (Huggett *et al.*, 2003; Mullon *et al.*, 2003; Parada, 2003; Parada *et al.*, 2003; Skogen *et al.*, 2003; Koné, 2006). Early life stage models have also been developed for sardine (*Sardinops sagax*) in the southern Benguela (Miller, 2006; Miller *et al.*, 2006) and in the northern Benguela (Stenevik *et al.*, 2003). All these models, except Koné (2006), fall into the "hydrodynamics and simple behaviors" category; Koné's (2006) model is an example of an "hydrodynamics and dynamic prey" IBM.

Most of these Benguela Current models used the same regional PLUME configuration (Penven, 2000; Penven *et al.*, 2001) of the ROMS[3] hydrodynamic model (Shchepetkin and McWilliams, 2005). The PLUME configuration covers the southern Benguela from 28 to 40° S and from 10 to 24° E, at a horizontal resolution ranging from 9 km at the coast to 16 km offshore and with 20 terrain-following vertical levels. Koné (2006) used the same PLUME configuration of ROMS but coupled with a biogeochemical model (Koné *et al.*, 2005). Skogen *et al.* (2003) and Stenevik *et al.* (2003) used a different configuration and the NORWECOM[4] hydrodynamic model (Skogen, 1999). This configuration covered the whole Benguela area, from 12 to 46° S and from 4 to 30° E, at a horizontal resolution of 20 km and with 18 terrain-following vertical levels.

Recently, the ROMS model has been applied to the Benguela region using an embedding procedure that places a high-resolution small-scale (child) grid nested into a low-resolution large-scale (parent) grid (Penven *et al.*, 2006a). The parent grid covers the whole of southern Africa from 5 to 46° S and from 2° W to 54° E, at a horizontal resolution of 0.25° ranging from 19 km in the south to 27 km in the north (Southern Africa Experiment or SAfE, Penven *et al.*, 2006b). The first child grid (SAfE south coast) was designed to study the interactions between the Agulhas and Benguela systems and covers the area from 28 to 39° S and 12 to 27° E at a horizontal resolution of about 8 km (N. Chang, pers. commun.). The second child grid (SAfE west coast) was designed to encompass most of the Benguela from 18 to 35° S and 10 to 20° E, and also has a horizontal resolution of about 8 km (J. Veitch, pers. commun.). The parent and child grids have 32 terrain-following vertical levels. Hydrodynamics provided on the SAfE west coast child grid was recently used in a Lagrangian model (Lett *et al.*, 2007b).

The development of anchovy early life stage models for the southern Benguela that used the PLUME configuration of ROMS has followed a step-by-step progression from simple to more complex models. The first model developed by Huggett *et al.* (2003) only tracked eggs and larvae with transport completely determined passively by currents. Parada *et al.* (2003) introduced a buoyancy scheme for the eggs, and Parada (2003) added temperature-dependent and stage-dependent growth and mortality, and vertical swimming behavior of larvae. A synthesis of these simulation experiments can be found in Mullon *et al.* (2003). Koné (2006) then added food dependency to the temperature-dependent larval growth.

The progression of anchovy model development provides an opportunity for determining how increasing biological complexity affected model results. All of these models used the same configuration of the same hydrodynamic model (PLUME implementation of ROMS), and all relied on the same method for designing simulations and analyzing the results. Simulations were performed using all combinations of pre-defined parameter values, and using comparable graphical and statistical analysis to determine the effects of the parameters and their interactions on the output variables. The main output variable used was the percentage of larvae transported from the anchovy spawning grounds near the South African south coast to nursery grounds off the west coast. This percentage was referred to as the simulated transport success. The major results from the progression of models and analyses were:

(1) Simulated transport success has a strong seasonal pattern, peaking in the austral spring and summer time periods (October to March).
(2) There is little chance for virtual eggs released on the eastern side of the spawning grounds to be transported to the west coast.
(3) Simulated transport success is highest for an egg density of 1.025 g cm^{-3}, and use of this density generated results similar to those obtained using purely passive transport.
(4) Simulated transport success increases with spawning depth within 0 and 75 m.

These modeling results corresponded reasonably well to the field observations of anchovy in the southern Benguela. The main spawning season of anchovy is austral spring and summer, with major spawning areas located in the western and central parts of the spawning grounds. Measured egg density is about 1.025 g cm^{-3}. However, anchovy eggs are mainly found in the upper 20 m and rarely deeper than 60 m (Dopolo *et al.*, 2005). Preliminary results obtained from an anchovy biophysical model using new hydrodynamics (SAfE south coast implementation of ROMS) suggest a slight decrease of simulated transport success with spawning depth, which is more in accordance with the observations.

While the early life stage models of the Benguela described above used hydrodynamics based on seasonal forcing, other models (Skogen *et al.*, 2003; Stenevik *et al.*, 2003; Miller, 2006; Miller *et al.*, 2006) used hydrodynamics based on interannual forcing. Interannual indices of retention or transport success derived from these models did not correlate well with time series of sardine

recruitment (Stenevik *et al.*, 2003; Miller, 2006; Miller *et al.*, 2006). However, the main objective of these analyses was to investigate the factors that affected retention and transport of sardine eggs and larvae, and not to derive indicators of recruitment success. Skogen *et al.* (2003) showed that different indices of transport to the South African west coast were correlated with anchovy recruitment. The contrasting results of model-derived indices being correlated or not correlated with recruitment data was also found in other studies. Parada *et al.* (in press) found that pre-recruitment indices and retention indices derived from their walleye pollock (*Theragra chalcogramma*) model were not correlated with recruitment estimates in the Gulf of Alaska, while Baumann *et al.* (2006) found significant correlations between a drift index and sprat (*Sprattus sprattus*) recruitment success in the Baltic Sea. The recent availability of interannual simulations covering the period 1957–2001 for the SAfE south coast and SAfE west coast implementations (N. Chang and J. Veitch, personal communication), presents an opportunity for attempting to correlate model-generated indices to recruitment data for the Benguela region.

The numerical experiments performed by Mullon *et al.* (2002), Lett *et al.* (2006) and Lett *et al.* (2007b) provide some basic building blocks for analyses of anchovy and sardine early life stage dynamics. Mullon *et al.* (2002) developed an evolutionary model that explored how environmental constraints (e.g. avoiding offshore advection and cold water) affect the spatial and temporal patterns in spawning. These relatively simple constraints imposed for multiple generations led to the selection of spawning patterns that were in surprisingly good agreement with those observed for anchovy and sardine in the southern Benguela. This approach of allowing the temporal and spatial patterns in spawning to emerge from a selective process and environmental conditions offers an alternative to the usual approach of specifying the location and timing of egg deposition that is used in most biophysical models.

Lett *et al.* (2006) used a Lagrangian approach to simulate and quantify enrichment and retention processes in the southern Benguela. These two processes, along with the concentration process, form a triad of processes that are fundamental for the survival and recruitment of early life stages of pelagic fish (Bakun, 1996). The results of Lett *et al.* (2006) reinforce the view of Cape Agulhas as a "dividing line" in the southern Benguela pelagic fish recruitment system, with a transport-based subsystem to the west and a retention-based subsystem to the east. These two subsystems were considered by Miller (2006) and Miller *et al.* (2006) in their sardine biophysical model. Lett *et al.* (2007b) also used a Lagrangian approach to investigate the processes responsible for the absence of anchovy and sardine spawning in the central Benguela off the Lüderitz region. They examined the flow field and temperature conditions that particles would experience in the Lüderitz region, and concluded that the combination of a surface hydrodynamic and a subsurface thermal barrier could limit the possibility for anchovy and sardine ichthyoplankton to be transported from the southern to the northern Benguela. Recent remote sensing data also suggested a poor trophic environment in the Lüderitz region (C.H. Bartholomae and J. Basson, personal communication; Demarcq *et al.*, 2007).

Models of anchovy and sardine in other upwelling systems

We are not aware of biophysical models of anchovy early life stages developed in Eastern Boundary Current upwelling systems outside the Benguela. A preliminary biophysical model of sardine (*Sardina pilchardus*) has been developed in the Iberian system (Santos *et al.*, 2005). Other models of sardine, in the Kuroshio Current (*Sardinops melanostictus*, Heath *et al.*, 1998) and in the Iberian (Santos *et al.*, 2004) upwelling systems, used hydrodynamics derived from field data. However, three anchovy (*Engraulis ringens*) biophysical models using hydrodynamics provided by ROMS are under development for the Humboldt Current upwelling system (Brochier *et al.*, 2008; Brochier *et al.*, abstract[5]; Soto-Mendoza *et al.*, abstract[6]; Chai *et al.*, abstract[7]), and there are anchovy models in the vicinity of major upwelling systems. Models exist for European anchovy (*Engraulis encrasicolus*) in the Bay of Biscay (Allain *et al.*, 2003; Allain, 2004; Allain *et al.*, 2007a,b; A. Urtizberea, personal communication) and for Japanese anchovy (*Engraulis japonicus*) in the Yellow Sea (Hao *et al.*, 2003). Models of the early life stages of taxa in upwelling systems besides small pelagic fish include crab (*Carcinus maenas*) in the Iberian upwelling system (Marta-Almeida *et al.*, 2006; Peliz *et al.*, 2007), zooplankton in the California Current upwelling system (Carr *et al.*, 2008), and blue shrimp (*Litopenaeus stylirostris*) and brown shrimp (*Farfantepenaeus californiensis*) in the Gulf of California (Marinone *et al.*, 2004).

Santos *et al.* (2004) used a combination of measured and modeled velocities to estimate the surface flow field experienced by particles during an upwelling event off Portugal, and obtained qualitatively similar patterns of retention at the shelf-break for modeled particles and for observed sardine eggs and larvae. In the same region, Marta-Almeida *et al.* (2006) used the ROMS model on a domain extending from 37.5 to 44.3° N and 12.8 to 8.4° W and simulated different vertical migration behaviors of crab larvae. They showed that migrating larvae were retained within the inner shelf under a wider range of upwelling and downwelling conditions than non-migrating larvae. This was further studied by Peliz *et al.* (2007). They used a three-level embedding procedure in ROMS, with a high-resolution (~2 km) small-scale domain nested into a medium-resolution (~6 km) medium-scale domain, both nested into a low-resolution

(~16 km) large-scale domain. They showed that a model including both diel vertical migration for larvae and river outflow allowed simulating patterns of crab larvae concentrations that were closer to the observed patterns than when only one of the two processes was included (Fig. 6.2). This is a good example of a modeling study using a pattern-oriented approach (Grimm *et al.*, 2005). We can anticipate a similar study on sardine (Santos *et al.*, 2007) to complement the preliminary one of Santos *et al.* (2005), as data on the vertical distribution and behavior of sardine larvae have recently been collected off Portugal (Santos *et al.*, 2006).

Allain and colleagues (Allain *et al.*, 2003; Allain 2004; Allain *et al.*, 2007a,b) used hydrodynamics derived from the MARS[8] three-dimensional circulation model (Lazure and Dumas, 2008) that covered the Bay of Biscay south of 49° N and east of 8° W at a 5 km horizontal resolution with 30 terrain-following vertical levels. Allain *et al.* (2003) used virtual passive buoys released in the simulated flow fields to reconstruct the putative origins in time and space of collected anchovy larvae and juveniles. Allain (2004) and Allain *et al.* (2007a,b) used a biophysical model of anchovy that included transport, growth, and mortality. They used the super-individual approach (Scheffer *et al.*, 1995), with each simulated buoy representing a large number of eggs, larvae, and juveniles. Every time step, each buoy was characterized by a distribution of growth rates (function of age and the physical environment it experienced) and by a survival probability (the probability that growth rates in the distribution were above a pre-defined age-specific and stage-specific threshold). They used the simulations to form an index of recruitment (survival rate multiplied by egg production, summed over the year and spatial domain) that correlated remarkably well to a (short) time series of anchovy recruitment data.

Rodríguez *et al.* (2001) released particles in a static two-dimensional flow field calculated from observations to assess the potential for retention of organisms around the Canary Island of Gran Canaria. They showed that particles released north of Gran Canaria would need limited "swimming ability" (speeds of ~0.5 cm s^{-1}) in order for them to be retained, whereas those particles released east or west of Gran Canaria would require a much stronger swimming ability (>5 cm s^{-1}) in order to be retained. Hunter (1977) reported mean swimming speeds of 0.3–0.45 cm s^{-1} for early (5 mm) northern anchovy (*Engraulis ringens*) larvae at 17 °C to 18 °C. A configuration of ROMS for the northern Canary is under development (E. Machu, pers. commun.). This configuration uses an embedding procedure with a parent grid covering the area from 10 to 40° N and 5.5° E to 30° W at a horizontal resolution of 1/4°, and a child grid from 21 to 32.5° N and 9 to 20.5° W at a horizontal resolution of about 8 km. A biophysical model will be developed in the near future to study sardine early life history in this region (T. Brochier and A. Ramzi, pers. commun.).

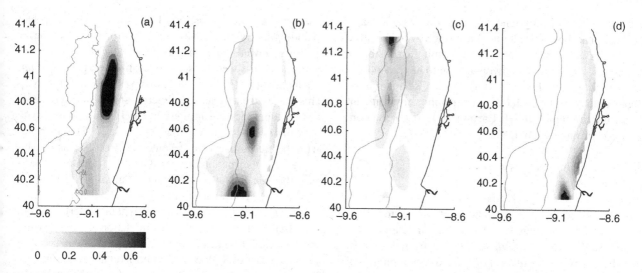

Fig. 6.2. Normalized concentrations of crab larvae (a) observed and (b)–(d) obtained from a biophysical model that included (b) diel vertical migration for larvae and river outflow (c) only diel vertical migration for larvae (d) only river outflow. (Reproduced from Peliz *et al.* (2007) with permission from Elsevier.)

Lett *et al.* (2007a) used a similar approach for the northern Humboldt upwelling system as Lett *et al.* (2006) did in the southern Benguela. They derived indices of enrichment, concentration, and retention based on Lagrangian simulations using hydrodynamics from a regional ROMS configuration. This configuration covered the domain off Peru from 20° S to 3° N and 90 to 70° W at a horizontal resolution of 1/9° (~10 km) and with 32 terrain-following vertical levels (Penven *et al.*, 2005). Lett *et al.* (2007a) analyzed the spatial distribution and seasonal variability in the simulated indices, and discussed their results in relation to the distributions of anchovy (*Engraulis ringens*) eggs and larvae off Peru. A coastal area of enhanced enrichment was identified between Punta Falsa and Pisco (6° –14° S), which corresponds to the zone where most anchovy eggs and early larvae are found. Preliminary modeling results suggest that the early life stages of anchovy are found mainly in the northern part of this zone (6° –9° S) because this area provides high retention and concentration. Another striking characteristic of anchovy spawning in the region is the bimodal seasonal pattern, with peaks in January–March and August–October. A biophysical model of anchovy early life stages is currently being developed to better understand the bimodal spawning pattern in this region (Brochier *et al.*, 2008). Other biophysical models of the early life stages of anchovy are under development for the northern Humboldt (Y. Xu, personal communication) and southern Humboldt (S. Soto-Mendoza, personal communication) upwelling systems.

Heath *et al.* (1998) used static two- and three-dimensional flow fields calculated from observations for two years to assess the dispersal of sardine (*Sardinops melanostictus*) eggs and larvae in the Kuroshio Current upwelling system.

They obtained very different results for the two years, with most egg production being transported towards the Kuroshio Current extension in the first year, while most of it was retained in the coastal areas in the second year. In contrast, different scenarios for transport (without and with diffusion) and for behavior (without and with vertical redistribution of individuals according to observed depth distributions) had little effect on model simulations. They argued that, at least during the time of their study, hydrodynamic models could not easily simulate the instabilities in the Kuroshio path that were responsible for the contrasting flow fields they observed.

Hao *et al.* (2003) conducted Lagrangian simulations in the nearby Yellow Sea using a regional configuration of the HANSOM[9] hydrodynamic model. This configuration covered the Yellow Sea and the East China Sea from about 25 to 40° N and 120 to 130° W at a horizontal resolution of 1/12° and with 12 vertical layers. They showed that including a tidal component in the hydrodynamic model affected the predicted transport pattern of simulated particles, and discussed their results in relation to the high concentrations of anchovy (*Engraulis japonicus*) eggs and larvae that are observed in tidal fronts.

Despite the preliminary Lagrangian experiments conducted in the California Current upwelling system to investigate the effects of vertical migration on larval drift trajectories (Botsford *et al.*, 1994), there has not been much follow-up with more biologically based models of the early life stages of small pelagic species. Lagrangian experiments have been designed to study the transport patterns of other taxa, including zooplankton (Carr *et al.*, 2008) and shrimp larvae (Marinone *et al.*, 2004). Carr *et al.* (2008) used a nested approach in ROMS with a parent grid covering the entire California upwelling system (7.5 km resolution) and a child grid focusing on the Monterey Bay region (2.5 km resolution), and with 32 terrain-following vertical levels. They investigated the effects of diel vertical migration of planktonic organisms, and obtained results suggesting that migration into subsurface onshore currents during the day would not compensate for surface offshore transport during the night and thus retention would be low within the Monterey Bay. Marinone *et al.* (2004) explored different scenarios for when advection of particles occurred (only during the day, the night, or when the current flows northwards), and discussed their results in relation to transport patterns of shrimp larvae to their nursery areas. They used a configuration of HAMSOM, developed by Marinone (2003), that covered the Gulf of California at a horizontal resolution of 1/24° (~4 km) and with 12 vertical layers.

Models of small pelagic fish in "non-SPACC" regions
In this section, we briefly mention several early life stage models for small pelagic species and regions outside the main focus of SPACC. We mention these examples because the methods and results may be of interest, and because they provide a broad view of early life stage biophysical models.

A. Vaz (unpublished data) used outputs from a western South Atlantic configuration of the POM[10] as input to a biophysical model of anchovy (*Engraulis anchoita*) eggs and early larvae. They investigated the spatial and seasonal patterns in the levels of retention and transport along the coast off northern Argentina, Uruguay, and southern Brazil.

Bartsch and colleagues (Bartsch *et al.*, 1989; Bartsch, 1993; Bartsch and Knust, 1994a,b) used hydrodynamics derived from early versions of the HAMSOM model as input to biophysical models of herring (*Clupea harengus*) and sprat (*Sprattus sprattus*) early life stages for the North Sea. Their seminal studies incorporated transport and size-dependent vertical migration of larvae, and focused on qualitative comparisons between observed and simulated distributions of larvae. Sætre *et al.* (2002) used the NORWECOM model to assess the transport routes and retention areas for herring along the Norwegian coast. They showed that herring tended to spawn in areas where retention was enhanced by the presence of topographically induced quasi-stationary eddies. Hinrichsen *et al.* (2005) developed a model of sprat for the Baltic Sea and showed that the degree of overlap between observed and simulated distributions of juveniles was higher when individuals were constrained to remain close to the surface than when they were allowed to migrate to deeper waters during the day. Baumann *et al.* (2006) proposed different empirical models fitted to observed values of sprat recruitment, and obtained the best result when they incorporated a drift index derived from the biophysical model simulations.

Bartsch and Coombs (2001) developed a mackerel (*Scomber scombrus*) early life stage biophysical model using an eastern North Atlantic configuration of the HAMSOM model. They used initial conditions for egg distribution and abundance that were based on field data, and simulated the drift of eggs and larvae and their growth using a function dependent on temperature and age. Bartsch and Coombs (2004) later introduced additional biological processes into their model, including vertical migration, feeding, and mortality. They used size-dependent diel vertical migration, and dynamic prey fields calculated from satellite-derived sea-surface temperature and chlorophyll-*a* concentration. Like Allain (2004), Bartsch and Coombs (2004) also used super-individuals, where every simulated entity initially represented 10^6 individuals who experienced the same environment and who died according to growth-dependent and length-dependent mortality rates. Using this model, Bartsch *et al.* (2004) derived indices of mackerel early post-larvae survival that they then compared to juvenile catch data for different sub-areas during 1998 to 2000. These simulations, however, all used the same initial conditions of egg distribution and abundance (those observed for 1998). Bartsch (2005) showed that simulating survival for 2001 using 1998 egg data as the initial conditions led to a "wrong" 60% increase in survival compared with using actual 2001 egg data.

Biophysical models that include adults

There are two major categories of models that include adult fish and physics. The first category is models in which the key processes of growth, mortality, reproduction, or movement depend on the physics or variables greatly influenced by the physics. The second category is models for which the questions to be addressed require full life cycle simulations (i.e. include all life stages) so that the long-term (multigenerational) effects of impacts or environmental changes can be examined. These full life cycle models thus must deal with eggs, larvae, and adults in a single model. Such models are relatively rare now but model development is headed in that direction. We expect biophysical full life cycle models to be the focus of much effort during the next decade, and anticipate that full life cycle biophysical models will be especially important for addressing issues related to the effects of climate change on fish populations and fisheries.

In this section, we review biophysical models that include adult stages of fish. We focus our review on those models that are spatially explicit, and that use physics, or variables derived from the physics, as inputs. We also include the NEMURO[11] family of models because some versions of the NEMURO models fit our criteria for inclusion, but also because this effort is ongoing and provides an example of a spatially explicit full life cycle approach. We do not include models that use environmental variables as inputs (e.g. temperature) but only simulate a single spatial box (e.g. Robinson and Ware, 1999; Clark *et al.*, 2003), and also do not include other population modeling approaches (e.g. spawner-recruit models, Fiksen and Slotte, 2002; surplus production models, Jacobson *et al.*, 2005). These other modeling approaches are likely useful for addressing certain questions related to climate change effects on fish. Our focus in this chapter is on biophysical spatially explicit models that include adult fish.

Abiotic and biotic environment, growth, and mortality

Models that include adults generally use similar information from the physics and lower trophic models as used by the egg and larval models. These outputs include: current velocities (advection), salinity, temperature, and prey fields. The growth and mortality formulations in adult models are also similar to those by the early life stage models. Growth in weight is generally either based on empirical relationships (e.g. von Bertalanffy equation, Shin and Cury, 2004) or bioenergetics (e.g. Megrey *et al.*, 2007), and natural mortality rate is often treated as a constant. There are also some predator–prey biophysical models that include adults wherein mortality depends on fish encounters with other fish (Ault *et al.*, 1999; Shin and Cury, 2004). With the inclusion of adult life stages, representing harvest rates and fisheries can also become important.

Movement and behavior

The process of movement is where models that include adults differ from early life stage models. Unlike the egg and larval stages, movement of juvenile and adult fish no longer is necessarily dominated by the physics. Movement of adults can be influenced by the physics, neutral with respect to the physics, or even opposite to the physics. In many situations, the physics (or variables directly derived from the physics, e.g. salinity, temperature, food) influences some portion of the fish's movement, while another component is due to other factors dependent on behavioral decisions. An extreme example of movement of adult fish opposite to the physics is migration upstream or counter to the prevailing currents.

How to represent the movement of adult fish in spatially explicit biophysical models is unclear and there are several approaches that have been proposed. Conceptually, movement of adults adds terms to equation (6.1) attributable to behavior, and then there must be some weighting of the relative contribution of the physics-based term and the behavior-based term. We highlight here two of the approaches for modeling movement to illustrate some of the variety in how one can represent movement of adults in biophysical models.

Railsback *et al.* (1999) proposed an approach based on fitness considerations in which individuals search neighboring cells, and move towards cells that provide an optimal blend of growth and survival. In our context, the physics would provide the temperature and prey fields, and possibly other variables, that are needed to compute growth and mortality in the neighboring cells. The growth and mortality in each neighboring cell is then combined into a fitness measure that is the projection of the likelihood of an individual surviving and obtaining some target size at some time in the future (e.g. size at maturity next year), assuming conditions in the cell remained constant into the future. Individuals move towards the cell in their neighborhood that has the highest fitness.

Humston *et al.* (2004) implemented a different approach to modeling movement of fish based on the concept of kinesis that does not require that the fish has knowledge of the conditions in neighboring cells. Rather, individuals evaluate the conditions in their present cell against some specified optimal value. Movement has inertial and random components; inertial movement involves smaller angles and shorter distances than the random component. They used a weighting scheme to combine the inertial and random components into distances moved and the angle of movement. The closer the conditions in the present cell were to optimal, the more the inertial movement dominated and fish would make smaller moves going in about the same direction and therefore tended to stay in the good areas. Poor conditions resulted in the random movement being weighted more heavily and the appearance of individuals searching over relatively large areas and in all directions.

The Railsback and Humston approaches illustrate a fundamental schism in movement modeling approaches: whether organisms can sense the conditions in neighboring cells sufficiently to project how conditions there would affect their growth or mortality (Tyler and Rose, 1994). With an appropriate weighting scheme, one can also add terms for the contribution of advective transport to the mix in both approaches. Other approaches to modeling movement include multi-step approaches that first determine the type of behavior from a list of discrete options and then implement the specifics of movement associated with that behavior (Blackwell, 1997; Anderson, 2002; Goodwin *et al.*, 2006), and the use of genetic algorithms to train neural networks (Huse and Giske, 1998).

Given the uncertainty in how to represent movement, it is important to note that some biophysical models bypassed the issue. Heath *et al.* (1997) did not try to model movement but rather simply assumed movement would occur and would result in the spatio-temporal distributions derived from field data. Even more extreme is the growth potential approach (Luo *et al.*, 2001). They ignored movement and spatial distributions, and simply predicted what would happen if fish were present in all locations.

Case studies

We have organized this section differently from the early life stage case studies section. There are far fewer examples that include physics and adult fish, and not enough examples to divide by species and geographic area as was done with the early life stage examples. Rather, we have grouped the adult models together according to two general categories: models of adults that use physics and full life cycle models. We emphasize the methods used because of the much wider diversity of approaches used with adult models than with the more standardized approaches generally used with early life stage models.

Models of adults that use physics or physics-related variables

Luo *et al.* (2001) used the output of a three-dimensional hydrodynamic model coupled to a water quality model as input to a fish feeding and bioenergetics model to determine the carrying capacity of Chesapeake Bay for young-of-the-year juvenile menhaden (*Brevoortia tyrannus*). The water quality model had over 4000 cells, arranged as a surface grid of 729 cells, with vertical cells every 2 m. They averaged the 2-hour simulated values of temperature, dissolved oxygen, and chlorophyll-*a* to obtain daily values for each cell for June through December in an average freshwater inflow year. Temperature was used directly in the bioenergetics model, and affected maximum consumption rate and respiration rate. Menhaden are filter feeders and so chlorophyll-*a* concentrations were multiplied by the area of the mouth, swimming speed, and an efficiency term to obtain realized consumption rate. Consumption, with the rest of the bioenergetics model, enabled prediction of daily growth rate in each cell. Carrying capacity for each cell was then computed based on the predicted consumption rate, prey production rates, and prey biomass, and further adjusted for low dissolved oxygen, to obtain the biomass of menhaden that could be supported in that cell on each day.

Luo *et al.* (2001) showed spatial maps and reported the percent of the bay volume able to support different growth rates and carrying capacities. Their results showed that there was large spatial and temporal variation in growth rate potential and biomass supportable due to the non-linear functional forms in the feeding and bioenergetics models and as a result of combining the effects of the multiple factors of temperature, chlorophyll-*a*, and dissolved oxygen. Menhaden must occupy cells with growth potential greater than 0.005 to 0.01 g g^{-1} day^{-1} during the June through December growing season in order to achieve their observed weights in December, and the daily total volume of such good habitat in the bay varied between practically zero and 80%.

There are many examples of the use of habitat suitability indices (HSI) that use spatially explicit environmental variables estimated from field data or outputted from hydrodynamic models. Most of these examples involve how changes in stream and river flows would affect the habitat for specific species downstream because the HSI approach was initially developed for evaluating how water releases from dams would affect fish downstream (Acreman and Dunbar, 2004). Use of HSI avoids the issues of representing how adult fish move because an endpoint can be simply the changes in the quantity and quality of the habitat. The HSI approach has also been used as an intermediate variable that is then used to affect movement of adult fish (e.g. Lehodey *et al.*, 2003).

Rubec *et al.* (2001) offers an example of HSI applied in an estuarine environment. We summarize an estuarine example here even though it uses field data to derive the environmental variables because the same analysis would be used with model-predicted environmental variables. Others have computed HSI values from the output of hydrodynamic and water quality models (e.g. Guay *et al.*, 2000). Rubec *et al.* interpolated temperature and salinity values to obtain seasonal values for 18.5 m^2 cells in the surface and bottom layers for Tampa Bay, Florida. Additional information on depth, bottom substrate type, and species abundances were also used. Relationships between suitability (0 to 1) and each variable were formulated from abundance data, and the product of the suitabilities in each cell (geometric mean) was computed as the overall suitability of that cell for that life stage. They then checked whether overall suitability was correlated to species abundances over the four seasons, and swapped suitability functions with those developed for Charlotte Harbour to see if suitability functions were transferable among locations.

Karim *et al.* (2003) coupled a hydrodynamic-NPZ model to a model of fish movement and dissolved oxygen-related mortality. Their simulations were based on the Marbled Sale (*Pleuronectes yokohamae*), a demersal species, in Hakata Bay. The spatial domain was a horizontal grid of 300 m × 300 m cells with five vertical layers. The grid was selected to correspond to the horizontal distance typically traveled by an individual fish in the 30-minute time step. Movement could then be based on individuals moving to neighboring cells each time step. The hydrodynamic-NPZ model was solved using its own time step, and temperature and dissolved oxygen values were obtained for each grid cell for use with the fish movement and mortality models. Karim *et al.* used a series of laboratory experiments and field tracking data of individual fish to develop and calibrate the movement model. Movement in the model was based on computing the preference of each neighboring cell in three dimensions from functions that related temperature to a preference level and dissolved oxygen concentration to a preference level, and then a function that combined the preference levels into a single overall preference value for the cell. Individuals moved to the cells with the highest overall preference. This approach to modeling movement could be considered a simplified version of the more general Railsback *et al.* (1999) fitness-based approach. Horizontal position was updated every 30 minutes, while vertical position was evaluated every 10 seconds. Exposure to low dissolved oxygen caused increased mortality.

After performing simulations that satisfactorily mimicked the laboratory and field tracking results, Karim *et al.* (2003) performed 3-day simulations with individuals released in different vertical layers (surface or bottom) and horizontal locations (whole grid or inner bay). They computed the mortality rate of the cohort from exposure to low dissolved oxygen as mortality accumulated over the 3 days. They concluded that hypoxic conditions in the inner portion of the bay during the summer can cause significant mortality of demersal fish, and that future model developments should include the simulation of hypoxia effects on pelagic species.

Ault *et al.* (1999) developed a predator–prey model of spotted seatrout (*Cynoscion nebulosus*) and pink shrimp (*Penaeus duorarum*) that was embedded in a two-dimensional hydrodynamic model of Biscayne Bay, Florida. The hydrodynamic model consisted of 6346 triangular elements and 3407 nodes, with grid spacing between nodes of about 500 m. The model was driven with tide, winds, and freshwater discharge data for 1995. A 1-minute time step was used to solve the hydrodynamic model, and currents and salinities were output at 10-minute intervals for use with the fish model. Shrimp and seatrout were followed as super-individuals (what Ault *et al.* termed "patches"). Patches were introduced monthly for shrimp, and at night on incoming tides for seatrout. Patches were tracked using continuous x and y positions; each time step the patch experienced the conditions of the cell it inhabited. Circulation was used to move around the patches of egg and yolk-sac larval seatrout, circulation and behavior together were used with pre-settlement shrimp, and active behavior only was used for moving settled shrimp and juvenile and adult seatrout. Salinity affected the angle and distance moved (based on swimming speed) of pre-settled shrimp, which was added to the circulation-based movement. Horizontal movement only occurred at night; shrimp went to bottom during the day. Once shrimp settled to the bottom, they were moved around based on their body length and habitat quality. Seatrout movement was based on the growth rate in cells within a specified distance (detection range) of their present location, and patches moved towards cells with the highest growth potential. Movement within the Ault *et al.* model of the different life stages of shrimp and seatrout illustrates the diversity of movement algorithms and approaches, including purely hydrodynamics driven, an approach rooted in cellular automation, an approach that

was later generalized into the Humston *et al.* (2004) kinesis model, and a hybrid mix of the kinesis and Railsback's fitness-based approach. Shrimp grew based on the cumulative temperature they were exposed to; seatrout grew based on bioenergetics with consumption based on ingested shrimp co-located in their present cell.

Ault *et al.* (1999) reported the results of one year simulations that illustrated model behavior and the usefulness of the model for examining how environmental and biological factors affect predator–prey dynamics. They concluded that June-spawned seatrout were transported more into the bay, settled over a wider range of habitats, and grew faster than August-spawned cohorts. They also showed that spatial variation in habitat quality can affect growth rates of fish, even causing difference among individuals from within the same spawning cohort.

Full life cycle models
The EcoPath with EcoSim (EwE) family of models (Walters *et al.*, 1997) are biomass-based compartment models and consist of a steady-state version (EcoPath), which is also used as initial conditions for a single spatial box time-dynamic version (EcoSim). Walters *et al.* (1999) extended the EwE models to be spatially-explicit (EcoSpace) by imbedding an EcoSim model on each cell of a two-dimensional grid of cells. Differential equations are constructed for each fish compartment (species or functional group) in an analogous manner to the equations for phytoplankton and zooplankton in NPZ models (net effect of gain and loss terms corresponding to growth and mortality processes). In early versions of EwE, fish were represented as biomass, which was recognized as inadequate because of the large ontogenetic changes in growth and mortality in many fish species and because a single biomass compartment did not allow for explicit representation of recruitment dynamics which is critical to understanding and forecasting fish population dynamics. Thus, EcoSim was extended to allow for each fish compartment to be subdivided into smaller compartments (e.g. juveniles versus adults) as a crude way to allow for ontogenetic differences and to make recruitment dynamics semi-distinct from total biomass (Walters *et al.*, 2000). Recent versions of EcoSpace went further and allowed for monthly cohorts (age structure) of key species or groups. EcoSpace is presently undergoing modification to allow for even more detailed representation (e.g. following individuals or packets) for selected taxa, and for easy coupling to hydrodynamic and NPZ models. While there have been many applications of EcoPath and quite a few applications of EcoSim (e.g. Shannon *et al.*, 2004; Field *et al.*, 2006), the spatially explicit EcoSpace version has not yet received much attention. We mention EwE here because we anticipate increasing attention being paid to the EcoSpace approach in the future.

One example of the use of the EwE models is that of Shannon *et al.* (2004), who applied EcoSim to the Southern Benguela ecosystem. The model was composed of about 30 species or groups, and they adjusted previous parameter values and assumptions (e.g. Shannon *et al.*, 2003) based on the model's ability to replay the historical biomasses of selected species for the 1978 to 2002 period. Using the EwE fitting algorithm, Shannon *et al.* (2004) explored how changes to historical fishing pressure patterns, prey vulner-abilities to predation of key prey–predator combinations, and environmental forcing of primary production affected the model fit to the observed biomass time series. The fit of modeled biomasses to the observed time series was insensi-tive to alterations in the fishing pressure time series, and moderately sensitive to interannual variation in environ-mental forcing of primary production. The model's rela-tively high sensitivity to prey predator vulnerabilities was consistent with the idea of wasp-waist control (Cury *et al.*, 2000), in which small pelagic fish control their zooplank-ton prey while simultaneously exerting bottom-up control on their predators.

The IGBEM and BM2 models (Fulton *et al.*, 2004a,b) are also biomass-based but they are truly spatially explicit, cou-pled to an elaborate water quality model, and separate each fish species or group into age classes and follow the aver-age weight and numbers of individuals in each age class. BM2 was developed in an attempt to reduce the complexity and parameter demands of its more complicated predeces-sor IGBEM (Fulton *et al.*, 2004b). The application to Port Phillip Bay, Australia, followed 29 living compartments, plus compartments related to detritus, nutrients, sediment, and some physical variables in a three-layer grid with about 59 cells in each layer. Transport among cells was derived from the output of a hydrodynamic model. A daily time step was used, although if rates were too fast, then a finer time step was adopted. Four fish groups were followed, with each represented by means of an age-structured cohort approach. Spawning occurred outside of the model domain and recruits were injected into the grid on a specific day as the initial number of individuals in the first age class and then the oldest age class was removed. The average body weight of an individual in each age class was followed using two separate weight variables (structural and reserve) based on simulated growth. Growth depended on the summed prey biomass, whose vulnerability was sized-based, in the same cell as the predator and adjusted for feeding efficiency and crowding. Consumed prey was imposed as mortality on the appropriate prey compartments in the cell. Movement of fish age classes was simulated using fluxes among cells based on specified quarterly target densities in cells and how many days were left in the quarter. Fulton *et al.* (2004b) acknowledged that the recruitment and movement formula-tions of the model were likely the weakest aspects of their model, and they have investigated using spawner–recruit relationships and forage-based and density-based move-ment approaches. They stated that they used the constant recruitment and simple movement because the results did not differ much between the simple and more complicated alternatives.

Fulton *et al.* (2004b) performed a variety of simula-tions of BM2 and compared the outputs to field data for Prince Philip Bay, to field data for other estuaries, and to predictions from the more complicated IGBEM version. Simulations repeated a 4-year time series of forcings as input and they determined that 30-year simulations were sufficient because the model state after 30 years was simi-lar to that predicted after 100 years. Examples of model outputs compared to field data or to IGBEM predictions included: averaged biomasses of key compartments, pre-dicted community composition, relationship between DIN[12] and chlorophyll-*a*, predicted and expected size-spectra features, system-wide indices such as P/B[13] ratios and cyc-ling indices, and the temporal and spatial dynamics of key compartments.

Adamack (2007) and Adamack *et al.* (abstract[14]) coupled a three-dimensional hydrodynamic-water quality model to an individual-based population model of bay anchovy (*Anchoa mitchilli*) in Chesapeake Bay. The water quality model is a three-dimensional model that simulated 24 state variables, including dissolved oxygen, four forms of nitro-gen, four forms of phosphorus, two phytoplankton groups, and two zooplankton groups in a 4073 (729 surface layer cells) cell grid. Anchovy were introduced weekly during the summer spawning season as individual recruiting juve-niles and followed until they reached their end of third year when they were removed from the model. Growth and mor-tality rates of individual bay anchovy within a water qual-ity model cell were calculated every 15 minutes. Growth depended on a bioenergetics model with the predicted zooplankton densities from the water quality model pro-viding the prey for bay anchovy consumption. Anchovy consumption was summed by cell and, with the predicted diet, imposed as an additional mortality term back onto the zooplankton. Anchovy excretion and egestion also contrib-uted to the nutrient recycling dynamics of the water quality model. Mortality rate of anchovy was assumed to decrease with their length. Movement of individual anchovy was simulated both vertically (hourly) and horizontally (daily) based upon temperature, salinity, and prey densities using the kinesis approach of Humston *et al.* (2004).

Adamack (2007) reported the results of simulations that used the dynamically coupled models to predict the effects of changes in nitrogen and phosphorous loadings on bay anchovy growth rates and survival. Ten-year simula-tions using historical sequences of low, average, and high freshwater inflow years were performed under baseline, increased, and reduced nutrient loadings. Results showed that the anchovy response to changes in nutrient loadings was a complex function of changes in high-quality habitat, prey densities, assumptions about movement, and the magnitude and temporal pattern of the introduction of young-of-the-year recruits. This analysis provides an example of a biophysical model in which the physics is not used directly because the egg and larval stages were bypassed and the adults did not need circulation information, but the physics was needed to properly simulate the NPZ portion.

Lehodey et al. (2003) used the output of a three-dimensional hydrodynamic and NPZ model of the Pacific Ocean (45–65° N, 100° E-70° W) as input to an Eulerian-based tuna population dynamics model (SEPODYM). The NPZ model used cells that were 2° in longitude by 2° in latitude at the extreme north and south boundaries of the model domain and 0.5° square near the equator. Forty vertical layers were represented with a layer every 10 m within euphotic zone and thicker below. Predicted currents were averaged over the 0–30 m surface layer, primary production was integrated over the euphotic zone (1 to 120 m) and with SST, were interpolated on a two-dimensional grid of 1° resolution. Lehodey et al. added separate population models for tuna and for what they termed "tuna forage." The forage model simulated the biomass of tuna prey in each cell using advection and diffusion, and assuming continuous recruitment based on new primary production predicted by the NPZ model and a time lag to account for the delay until the primary production would show up as new forage biomass. The mortality rate and time lag of the forage were related to SST. Tuna population dynamics were modeled by following the numbers in age classes. Length-at-age was determined from a von Bertalanffy relationship; length was then converted to weight. Two habitat indices (adult and spawning) were computed for each cell based on SST in the cell. The spawning index was used to divide up the annual recruitment among the individual cells. For larvae and juveniles (i.e. until about 4 months of age), tuna movement was purely advection–diffusion. Movement of adult tuna also used the advection terms but these were adjusted by tuna length and by the adult habitat index. They assumed movement was proportional to the length of the age class and increased with poor habitat quality in the local cell. Natural and fishing mortality was applied to each age class, with natural mortality rate increased when the habitat was poor (spawning index used for first age class; adult index used for the rest of the ages). Multiple (six) fisheries with specific gear types and with effort varying by month and by cell or subregion were simulated to drive the fishing mortality rate. The SEPODYM application shared information and used outputs from a traditional fisheries stock assessment model called MULTIFAN-CL (Fournier et al., 1998). Annual recruitment in SEPODYM

was calibrated to match the overall recruitment estimated by MULTIFAN-CL.

Lehodey et al. (2003) reported the results of a NPZ and SEPODYM simulation that spanned 1960 to 1999. General features of the simulation were described, such as the effects of ENSO events and the 1976–77 regime shift, to check the realism of the NPZ simulation. Tuna recruitment and biomass simulated by SEPODYM qualitatively agreed with the estimates from the MULTIFAN-CL model, and predicted catches by grid cell and month were well correlated with observed catches. Most recruitment was predicted to occur in the western and central Pacific region, with large variability caused by El Niño versus La Niña years. The out-of-phase dynamics of simulated primary production between the western and central Pacific predicted by the NPZ model was also seen in the SEPODYM-simulated tuna recruitment. They emphasized the importance of a carefully constructed and evaluated NPZ model. The SEPODYM model is now being applied to small pelagic species in the Humboldt system (Gaspar and Lehodey, abstract[15]).

Shin and Cury (2004) described a general simulator of fish communities (OSMOSE) that uses an individual-based size-based approach, and Shin et al. (2004) applied the model to the Benguela Current system. They used the super-individual approach, with each super-individual assumed to represent a school of identical fish. A distinction was made in the model between non-piscivorous and piscivorous behavior, with species assigned a behavior based on stage or age. In the Benguela application, 12 fish species were simulated on a 40 cell by 40 cell horizontal grid using a 6-month time step. OSMOSE does not explicitly use the output of hydrodynamic or NPZ models, but rather represents the prey field effects by specifying a system-wide carrying capacity for non-piscivorous fish biomass. When total biomass of non-piscivorous species biomass exceeded the carrying capacity, then mortality was imposed on all non-piscivorous individuals on the grid, disproportionately on age 0 versus older individuals, until the total biomass fell below the carrying capacity. Piscivorous stages consumed prey species if they co-occurred together in the same spatial cell and if the prey were vulnerable based on predator to prey size ratios. All fish species grew in length according to von Bertalanffy equations, with the growth of piscivorous species predicted by the von Bertalanffy equation affected by a predation efficiency computed for each super-individual from its present consumption rate. Piscivorous super-individuals moved to their neighboring cell that had the highest prey biomass that was vulnerable to them. In addition to predation, there were terms for starvation and harvesting mortality. Reproduction closed the life cycle by initiating new super-individuals based on total egg production computed annually from the mature female spawners; larval and juvenile survival determined subsequent recruitment.

Shin *et al.* (2004) performed a series of 200-year simulations with increased fishing mortality rates on selected species and compared predicted responses to those from a comparably constructed EcoSim model. Predicted biomass of each species in OSMOSE was averaged over the last 100 years of each simulation, and compared to the average biomass under the reference or baseline simulation. Increased fishing morality on sardine (*Sardinops sagax*), anchovy (*Engraulis encrasicolus*), and round herring (*Etrumeus whiteheadi*) showed that sardine would be the first to collapse, anchovy would collapse second, and round herring were highly resistant because of the small initial value of their fishing mortality rate. The biomass of some species, such as chub mackerel (*Scomber japonicus*), increased due to relaxed competition. Increased fishing on Cape hake (*Merluccius capensis*) also showed the expected decrease in hake biomass and a general increase in the biomasses of hake's competitors. Both sets of increased fishing mortality simulations were compared to similar simulations preformed with an EcoSim version of the Benguela Current system and, at the qualitative level (increase or decrease), both models generated similar responses. While OSMOSE does not explicitly use the output of a NPZ model, with some creativity, one could perhaps assume how climate change would affect prey and one could adjust the carrying capacity appropriately. An ongoing effort is attempting to link an NPZ model to OSMOSE so that the growth of non-piscivorous individuals can be modelled dynamically (Y. Shin, personal communication).

The NEMURO family of models was a milestone in an ongoing large international collaboration focused on the development of standard NPZ model for application to the North Pacific, and the coupling of fish growth and population dynamics models to this standard NPZ model. The biological food web represented in NEMURO was fairly detailed with two groups of phytoplankton and three groups of zooplankton, plus the usual nitrogen, silicate, and detrital recycling dynamics. Using this common formulation for the NPZ, several models and applications were developed (Werner *et al.*, 2007). Of particular interest here are the spin-offs in which NEMURO was imbedded in a three-dimensional hydrodynamic model configured for the North Pacific, a version that coupled an age-structured fish model to the NPZ (termed NEMURO.FISH), and the latest incarnation (NEMURO.SAN) which is an individual-based, full life cycle, spatially explicit model of sardine and anchovy interactions.

NEMURO.FISH dynamically couples an adult fish bioenergetics-based population dynamics model to the NEMURO NPZ model. The coupled models have been configured for Pacific herring (*Clupea harengus pallasii*) on the west coast of Vancouver Island (Megrey *et al.*, 2007) and for Pacific saury (*Cololabis saira*) off Japan (Ito *et al.*,

2007). The herring application uses an age-structured approach, with bioenergetics used to describe the changes in the average body weight of individuals in each age class and mortality rates used to describe the changes in the numbers in each age class. Recruitment was knife-edge and computed from spawning biomass and environmental variables (SST, air temperature, and North Pacific Pressure Index) using a spawner–recruit relationship. New recruits become the new youngest age class. The dynamics of the three zooplankton groups in the NEMURO model determine the consumption rate in the fish bioenergetics model through a multispecies functional response formulation. Herring consumption affects the zooplankton through predation mortality, and fish egestion and excretion contribute to the nitrogen dynamics.

Using the NEMURO.FISH model of herring, Megrey *et al.* (2007) presented baseline simulations of herring weight-at-age and population dynamics for Vancouver Island, Canada, and Rose *et al.* (2008) used the model to examine how climate change could affect herring growth and population dynamics. Rose *et al.* performed simulations that mimicked the conditions in each of the four documented climate regimes in the North Pacific (1962–1976; 1977–1988; 1989–1997; 1998–2002). Climate regimes differed in the values assumed for environmental variables used in the spawner–recruit relationship, and in the water temperature, mixed layer depth, and nutrient influx rate used by the NPZ model. In agreement with general opinion and with the herring data from West Coast Vancouver Island, model-predicted estimates of weight-at-age, recruitment, and spawning stock biomass were highest in regime 1 (1962–1976), intermediate in regime 2 (1977–1988), and lowest in regime 3 (1989–1999). The regime effect on weight-at-age was a mix of recruitment effects and lower trophic level effects that varied in direction and magnitude among the four regimes.

Isolating the growth component of NEMURO.FISH, Rose *et al.* (2007) used the output of the three-dimensional NEMURO for the North Pacific and simulated weight-at-age (not population dynamics) for the west coast Vancouver Island, Prince William Sound, and Bering Sea regions for 1948–2002. The NEMURO application was a 3-D implementation for the Northern Pacific (Aita *et al.*, 2007). The NEMURO-3D simulation represented the NPZ dynamics for 1948 to 2002 using, as much possible, observed data for driving variables. The output of the NEMURO-3D simulation at the three locations, averaged over cells in the top 50 m, was used as input to the bioenergetics model and daily growth of herring was simulated for 1948 to 2000. Rose *et al.* applied the sequential t-test analysis to detect regime shifts (STARS) algorithm (Rodionov and Overland, 2005) to the simulated temperatures, zooplankton, and herring growth rate (annual change in weight between ages 3 and 4) to identify statistical shifts in their average values. All three locations showed shifts in simulated herring growth rate around the 1977 regime shift. While the NEMURO-3D output showed warming temperatures at all three locations beginning in the late 1970s, herring growth was predicted to decrease in the west coast of Vancouver Island and Prince William Sound, and to increase in the Bering Sea. Interannual variation in zooplankton densities caused the time series response in herring growth for west coast Vancouver Island. Temperature and zooplankton densities both played roles in Prince William Sound and the Bering Sea herring growth responses, with zooplankton dominating the response for Prince William Sound and temperature dominating the response for Bering Sea.

NEMURO.SAN is under development and extends the NEMURO approach to simulating sardine and anchovy as individuals on a two-dimensional spatial grid of cells (Rose et al., abstract[16]). The vertical dimension is represented as the volume of water in each cell based on the volume of the water above the mixed layer depth. Each year, recruits are computed from spawning biomass at a point in time and the recruits, after a suitable time delay, are slowly introduced over the next year as new model individuals on the grid. Growth, mortality, and movement of individuals are evaluated daily. Positions of individuals are tracked in continuous x and y space, and each day their cell is determined and they experience the conditions in that cell. Alternative approaches to movement, including the use of Railsback and Humston approaches, are being investigated. How the spatial (among cells) and temporal (daily or monthly) variation in the mixed layer depth, nutrient influx, and other inputs to NPZ portion are specified allows for flexibility in configuring NEMURO.SAN to different locations. To date, 100-year simulations have been performed in an exploratory mode for a version configured to roughly resemble the California Current system. Alternative hypotheses about climate conditions can be specified via changing the inputs to NEMURO, and then using the coupled models to predict the long-term responses of the anchovy and sardines in terms of their growth, survival, and spatial distributions.

Biophysical models and climate change

In this section we discuss the use of biophysical models of small pelagic fish in the context of predicting the effects of climate change. Harley et al. (2006) recently reviewed the potential impacts of climate change on coastal marine ecosystems. They discussed ecological responses to climate change at the individual, population, and community levels. We envision most analyses using biophysical models being single-species analyses. The early life stage models focus on a single species, and while some of the full life cycle models include multiple species, there is still great

uncertainty in how to represent the food web dynamics of the upper trophic level species (Rose and Sable, 2009). Despite the recognition of the importance of food web interactions and the push towards ecosystem-based fisheries management (Rose and Sable, 2009), community-level responses to climate change that require multi-species simulations will likely remain in the demonstration mode, rather than for management decision making, for the foreseeable future. We envision most analyses in the near term will focus mainly on the early life stages of key species, with some analysis using single-species full life cycle models for well-studied locations and theoretical-oriented analyses using the multispecies full life cycle models.

Among the factors reviewed by Harley et al. (2006), the ecological responses to changes in circulation, temperature, and productivity are relatively easy to investigate using biophysical models. Indeed, changes in circulation would induce changes in advective and dispersive transport, a process used directly in the early life stage models, and used directly or indirectly in the adult-based models. Some of the reviewed models used the direct effects of transport to at least influence the movement of adult fish, and many of the reviewed models used the effects of transport indirectly via transport's effects on temperature, salinity, and prey fields that affected the growth, mortality, reproduction, and movement rates of adult fish. Harley et al. also discussed the potential shifts (vertical and horizontal) in species distributions that could occur under climate change. Initial conditions in early life stage models typically use data based on observed egg distributions. Shifts in egg distributions induced by climate change could be hypothesized and simulations performed to explore potential consequences on early life stage transport, growth, and survival.

To our knowledge, the only example of a biophysical model used in the context of climate change comes from the recent study of Vikebø et al. (2007). In the Barents Sea off Norway, they used the ROMS hydrodynamic model forced by a global climate model in which the river runoff was increased by a factor three over the current value, causing the thermohaline circulation to slow down. Then they used a biophysical model of the early life stages of Arcto-Norwegian cod (Gadus morhua) to study the impact of the anomalous circulation and ocean temperature on transport and growth of cod larvae and juveniles. They showed clear differences between the drift and growth patterns obtained under the current and the reduced thermohaline circulation (Fig. 6.3).

We have more confidence in the early life stage models for simulating climate change scenarios because they are more tightly coupled to the physics and NPZ, and because early life stages of fish tend to be the focus of field and modeling studies in the marine ecosystems. Early life stages can be measured and they are often the focus as part of the

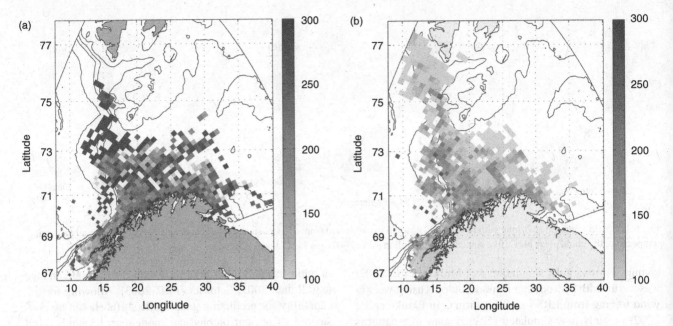

Fig. 6.3. Simulated distribution and wet weight (grayscale, in mg) of 2–4-month-old juvenile cod, obtained with a biophysical model using hydrodynamics under (a) current conditions and (b) reduced thermohaline circulation. (Reproduced from Vikebø *et al.* (2007b), with permission from Blackwell Publishing.)

search for recruitment indices (Kendall and Duker, 1998). Imposing climate change scenarios would seem to be possible almost from first principles because the early life stage models are tightly coupled to the hydrodynamic and NPZ models. Simulating climate change scenarios becomes more complicated for the adult models because many of the adult models used surrogates for the physics and prey outputs of the hydrodynamic and NPZ models as their inputs (e.g. carrying capacity). This implies another step is required to convert the hydrodynamics and NPZ outputs into the variables needed by the adult models.

Predicting responses of early life stages is necessary, but not sufficient, to address many but not all of the climate change issues. Many issues require that the predictions of early life stage models be related to population dynamics and health, and some questions require the simulation of the responses of the adult life stages. One challenge is to balance the use of early life stage, adult only, and full life cycle modeling approaches to ensure predicted responses are most relevant to the scientific questions and to management issues. There are also practical computing challenges. For example, the numerical considerations that arose with using super-individuals with early life stage models become more complicated with full life cycle models, as these models often include density-dependent effects and require new sets of individuals be introduced each year of their multi-year simulations. Also, full life cycle models typically simulate decades in order for the full effects of environmental and climate changes to be manifested in the population

response. Yet, hydrodynamic and NPZ models typically simulate a few years. How one creates realistic long-term scenarios for use as input to the full life cycle model from the relatively limited number of hydrodynamic and NPZ results can affect predicted responses to climate change.

Perhaps the greatest challenge relates to the assumption that the predicted responses of the hydrodynamic and NPZ models under climate change scenarios are sufficiently accurate and precise, and on the correct spatial and temporal scales to be used as inputs to the fish models. There is still disagreement about how some important driving variables to the hydrodynamic and NPZ model (e.g. boundary conditions, precipitation, wind) will change under a given climate change scenario. Furthermore, climate change scenarios, even once agreed upon, can push the hydrodynamic and NPZ models beyond their calibrated and validated domain. Also, most climate change simulation experiments have been conducted at a large spatial scale (e.g. the entire Southern Ocean, Russell *et al.*, 2006), whereas the early life stage and adult models tend to operate at smaller scales. Regional-level climate change simulation experiments should become increasingly available in the near future (e.g. for the California Current system, Auad *et al.*, 2006). Meanwhile, creative use of long-term time series of historical hydrodynamic interannual simulations should be melded with climate change scenarios to ensure realistic outputs of the hydrodynamic and NPZ models that act as inputs to the fish models.

Models such as the Mullon *et al.* (2002) model could be used to derive expected egg distributions under climate

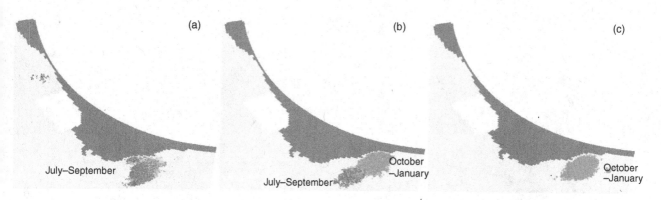

Fig. 6.4. Spatio-temporal spawning pattern after 200 generations, obtained from a biophysical model that used 100 000 particles and a lethal temperature threshold of (a) 14 °C (like simulation 4 in Mullon *et al.*, 2002), (b) 15 °C, and (c) 16 °C.

change. For example, simulation 4 of Mullon *et al.* (2002) was re-run with hydrodynamic model outputs using weekly wind forcing from ERS[17] satellites (run C in Blanke *et al.*, 2002) to study how simulated selected spawning patterns would change according to evolutionary constraints that involve a lethal temperature threshold for the transported particles. Two main spatio-temporal spawning patterns emerged depending on the value of the temperature threshold. A cooler threshold selected spawning in the Central Agulhas Bank in July–September (Fig. 6.4a), while a warmer threshold resulted in the selection of spawning in the Eastern Agulhas Bank in October–January (Fig. 6.4c); an intermediate temperature threshold resulted in a mix of both patterns (Fig. 6.4b). Assuming that the spawning pattern of Fig. 6.4b corresponds to the current situation, and that climatic change would only result in homogeneous increase of sea surface temperature, one could infer that spawning would shift more towards the pattern shown in Fig. 6.2a (i.e. cooler threshold mimics warmer water). These results can then be used as inputs to early life stage biophysical models.

Simulating geographic shifts in adult life stages will be more problematic with biophysical adult models. The use of adult models for simulating geographic shifts is limited because the distances that could be moved by adult fish in response to climate change would likely exceed the spatial domain of most of the models. A few of the adult-based fish models used very large spatial domains (e.g. Lehodey *et al.*, 2003), but whether the biological aspects of these models are sufficiently general to allow for prediction of geographic shifts needs to be tested. This was partially addressed in the SEPODYM model because they had to deal with the highly migratory tuna. Whether the current models of small pelagic fish are sufficiently robust to simulate large-scale shifts in distribution remains to be determined. To date, more statistically oriented empirical approaches have been used to predict geographic shifts in the distributions

of adult fish and other taxa in response to climate change (e.g. Rahel, 2002; Schmitz *et al.*, 2003). Allowing for the capability for predicting geographic shifts should be considered as present biophysical models are expanded and new models are developed.

Biophysical models of fish early life stages could likely be improved by using spawning and nursery habitats defined by environmental variables rather than simply being geo-referenced (P. Fréon, personal communication). Characterizations of spawning habitats based upon environmental variables such as temperature and salinity have been performed for anchovy and sardine in California (Checkley *et al.*, 2000; Lynn, 2003), the southern Benguela (Twatwa *et al.*, 2005), the Bay of Biscay (Planque *et al.*, 2007), and elsewhere (Castro *et al.*, 2005; van der Lingen *et al.*, 2005). Relating spawning habitat to environmental conditions would allow for the simulation of an "environmental homing" reproduction strategy as opposed to a "natal homing" strategy (Cury, 1994), and allow investigation of climate-driven changes in the habitat and their consequences on early life stages dynamics. Some caution is appropriate for such a "climate envelope" approach (Davis *et al.*, 1998) because the characterizations by environmental variables may themselves change in response to climate change due to adaptation, acclimation, and changes in the food web.

Early life stage models are ready for species-specific and site-specific analysis of climate change effects of small pelagic fish in upwelling systems. Some of the details of such applications need careful attention, especially the matching of the spatial and temporal scales of the hydrodynamic and NPZ models with the fish models, and how growth, mortality, behavior-based movement and trophic feedbacks are represented (Runge *et al.*, 2005). We see an accelerating trend from models of "hydrodynamics and simple behaviors" to models of "hydrodynamics and dynamic prey." It may be possible to include invertebrate predators in the models, but it will remain difficult to include fish

Box 6.1. Biophysical models

Biophysical models of fish early life stages are used to simulate the dynamics of eggs and larvae in virtual marine environments. Environments are characterized by temporally dynamic three-dimensional fields of physical (e.g. current velocities, temperature) and biogeochemical (e.g. phytoplankton and zooplankton concentrations) variables provided by hydrodynamic or hydrodynamic–biogeochemical coupled models. These fields are used as inputs to fish models that simulate the dynamics of eggs and larvae including their transport, growth, mortality and behavioral processes (Fig. 6.1). Biophysical models of early life stages are ready for investigating the effects of climate change on egg and larval growth, mortality, and spatial distributions. The quality of the predictions of the biophysical models will depend on sufficient accuracy in the physical and biogeochemical models under climate change scenarios, whether the spatial and temporal scales of the physical and biogeochemical model predictions are appropriate to be used as inputs to biophysical fish models, and the degree to which the growth (including feeding), mortality (including predation), and behavior-based movement processes represented in the biophysical model have adequate species-specific and site-specific details.

Many questions related to the effects of climate change on fish population dynamics (e.g. sustainable harvest) require the inclusion of juvenile and adult life stages in the biophysical models. Unlike for eggs and larvae where physics plays a dominant role in their movement, behavior plays an important role in the movement of juvenile and adult fish. Biophysical models that include the adult life stages of fish are of high research interest but will likely be used for general analysis of climate effects or application to a few, extremely well-studied species and locations. The uncertainty in how to model movement will limit the development of a generally agreed upon modeling approach that has helped advance the early life stage models. Continued effort on adult models is necessary in order to get to "end-to-end" (physics to fish), full life cycle models capable of addressing the long-term consequences of climate change on fish and fisheries.

Biophysical models that include the adult life stages of fish are of high research interest but will likely be used for general analysis or application to a few, extremely well-studied species and locations. The uncertainty in how to model the behavioral aspects of movement will limit the development of a generally agreed upon modeling approach (i.e. physics-based) that has helped advance the early life stage models. New measurement methods are becoming available that should provide the empirical basis for evaluating the alternative movement options (Cooke *et al.*, 2004). Sugden and Pennisi (2006) recently introduced a special section in Science on "movement ecology." Continued effort on adult models are necessary in order to get to "end-to-end" (physics to fish), full life cycle models capable of addressing the long-term consequences of climate change on fish and fisheries.

NOTES
1 Nutrient–Phytoplankton–Zooplankton.
2 Ådlandsvik, B. The particle-tracking method for transport modeling. Workshop on advancements in modeling physical–biological interactions in fish early-life history: recommended practices and future directions. Nantes, France, April 3–5, 2006.
3 Regional Ocean Modeling System.
4 Norwegian Ecological Model.
5 Brochier, T., Tam, J., and Ayón, P. IBM for the anchovy in the northern Humboldt Current ecosystem: identification of processes affecting survival of early life stages. International conference on the Humboldt Current System: climate, ocean dynamics, ecosystem processes, and fisheries. Lima, Peru, November 27–December 1, 2006.
6 Soto-Mendoza, S., Castro, L., Parada, C., *et al.* Modeling the egg and early larval anchoveta (*Engraulis ringens*) transport/retention in the southern spawning area of the Humboldt Current. International conference on the Humboldt Current System: climate, ocean dynamics, ecosystem processes, and fisheries. Lima, Peru, November 27–December 1, 2006.
7 Chai, F., Shi, L., Xu, Y., *et al.* Modeling Peru upwelling ecosystem dynamics: from physics to anchovy. International conference on the Humboldt Current System: climate, ocean dynamics, ecosystem processes, and fisheries. Lima, Peru, November 27–December 1, 2006.
8 Model for Applications at Regional Scale.
9 Hamburg Shelf Ocean Model.
10 Princeton Ocean Model.
11 North Pacific Ecosystem Model for Understanding Regional Oceanography.
12 Dissolved Inorganic Nitrogen.
13 Production/Biomass.
14 Adamack, A.T., Rose, K.A., and Cerco, C.F. Simulating the effects of nutrient loadings on bay anchovy population dynamics in Chesapeake Bay using coupled water quality and individual-based models. ECSA 41st International

predators of eggs and larvae. A large uncertainty may be getting agreement about how climate change will affect hydrodynamic outputs that are then used as inputs to the fish models, and obtaining these outputs on the spatial and temporal scales needed by the biophysical fish models.

Conference – Measuring and Managing Changes in Estuaries and Lagoons, Venice, Italy, October 15–20, 2006.

15 Gaspar, P. and Lehodey, P. Application of a spatial Eulerian ecosystem and population dynamic model (SEPODYM) to small pelagic fish: Modelling approach and preliminary tests. International conference on the Humboldt Current System: climate, ocean dynamics, ecosystem processes, and fisheries. Lima, Peru, November 27–December 1, 2006.

16 Rose, K. A., Agostini, V. N., Jacobson, L., *et al.* Towards coupling sardine and anchovy to the NEMURO lower trophic level model. North Pacific Marine Science Organization (PICES) 15th Annual Meeting, Yokohama, Japan, October 13–22, 2006.

17 European Remote-Sensing Satellite.

REFERENCES

Acreman, M. and Dunbar, M. J. (2004). Defining environmental river flow requirements – a review. *Hydrol. Earth Syst. Sci.* **8**: 861–876.

Adamack, A. (2007). Predicting water quality effects on bay anchovy (*Anchoa mitchilli*) growth and production in Chesapeake Bay: linking water quality and individual-based fish models. Ph.D. thesis, Louisiana State University, Baton Rouge, LA, USA.

Ådlandsvik, B., Gundersen, A. C., Nedreaas, K. H., *et al.* (2004). Modeling the advection and diffusion of eggs and larvae of Greenland halibut (*Reinhardtius hippoglossoides*) in the north-east Arctic. *Fish. Oceanogr.* **13**: 403–415.

Aita, M. N., Yamanaka, Y., and Kishi, M. J. (2007). Interdecadal variation of the lower trophic ecosystem in the northern Pacific between 1948 and 2002, in a 3-D implementation of the NEMURO model. *Ecol. Mod.* **202**: 81–94.

Allain, G. (2004). Modélisation biophysique pour la prévision du recrutement: couplage stochastique d'un modèle individu-centré de croissance larvaire avec un modèle hydrodynamique 3D pour développer un indice de recrutement de l'anchois dans le golfe de Gascogne. Ph.D. thesis, Ecole Nationale Supérieure Agronomique de Rennes, France.

Allain, G., Petitgas, P., Grellier, P., and Lazure, P. (2003). The selection process from larval to juvenile stages of anchovy (*Engraulis encrasicolus*) in the Bay of Biscay investigated by Lagrangian simulations and comparative otolith growth. *Fish. Oceanogr.* **12**: 407–418.

Allain, G., Petitgas, P., and Lazure, P. (2007a). Biophysical modeling of larval drift, growth and survival for the prediction of anchovy (*Engraulis encrasicolus*) recruitment in the Bay of Biscay (NE Atlantic). *Fish. Oceanogr.* **16** (6): 489–505.

Allain, G., Petitgas, P., and Lazure, P. (2007b). The influence of environment and spawning distribution on the survival of anchovy (*Engraulis encrasicolus*) larvae in the Bay of Biscay (NE Atlantic) investigated by biophysical simulations. *Fish. Oceanogr.* **16**: 506–514.

Anderson, J. J. (2002). An agent-based event driven foraging model. *Nat. Resour. Mod.* **15**: 55–82.

Auad, G., Miller, A., and Di Lorenzo, E. (2006). Long-term forecast of oceanic conditions off California and their biological implications. *J. Geophys. Res.* **111**: C09008, doi:09010.01029/02005JC003219.

Ault, J. S., Luo, J., Smith, S. G., *et al.* (1999). A spatial dynamic multistock production model. *Can. J. Fish. Aquat. Sci.* **56** (Suppl. 1): 4–25.

Bakun, A. (1996). *Patterns in the ocean. Ocean processes and marine population dynamics.* University of California Sea Grant, California, USA, in cooperation with Centro de Investigaciones Biologicas de Noroeste, La Paz, Baja California Sur, Mexico.

Bartsch, J. (1993). Application of a circulation and transport model system to the dispersal of herring larvae in the North Sea. *Cont. Shelf Res.* **13**: 1335–1361.

Bartsch, J. (2005). The influence of spatio-temporal egg production variability on the modelled survival of the early life history stages of mackerel (*Scomber scombrus*) in the eastern North Atlantic. *ICES J. Mar. Sci.* **62**: 1049–1060.

Bartsch, J. and Coombs, S. H. (2001). An individual-based growth and transport model of the early life-history stages of mackerel (*Scomber scombrus*) in the eastern North Atlantic. *Ecol. Mod.* **138**: 127–141.

Bartsch, J. and Coombs, S. H. (2004). An individual-based model of the early life history of mackerel (*Scomber scombrus*) in the eastern North Atlantic, simulating transport, growth and mortality. *Fish. Oceanogr.* **13**: 365–379.

Bartsch, J. and Knust, R. (1994a). Simulating the dispersion of vertically migrating sprat larvae (*Sprattus sprattus* (L.)) in the German Bight with a circulation and transport model system. *Fish. Oceanogr.* **3**: 92–105.

Bartsch, J., Brander, K., Heath, M., *et al.* (1989). Modeling the advection of herring larvae in the North Sea. *Nature* **340**: 632–636.

Bartsch, J. and Knust, R. (1994b). Predicting the dispersion of sprat larvae (*Sprattus sprattus* (L.)) in the German Bight. *Fish. Oceanogr.* **3**: 293–296.

Bartsch, J., Reid, D., and Coombs, S. H. (2004). Simulation of mackerel (*Scomber scombrus*) recruitment with an individual-based model and comparison with field data. *Fish. Oceanogr.* **13**: 380–391.

Baumann, H., Hinrichsen, H.-H., Möllmann, C., *et al.* (2006). Recruitment variability in Baltic Sea sprat (*Sprattus sprattus*) is tightly couplesd to temperature and transport patterns affecting the larval and early juvenile stages. *Can. J. Fish. Aquat. Sci.* **63**: 2191–2201.

Blackwell, P. G. (1997) Random diffusion models for animal movement. *Ecol. Mod.* **100**: 87–102.

Blanke, B., Roy, C., Penven, P., *et al.* (2002). Linking wind and interannual upwelling variability in a regional model of the southern Benguela. *Geophys. Res. Lett.* **29**: 2188, doi:2110.1029/2002GL015718.

Botsford, L. W., Moloney, C. L., Hastings, A., *et al.* (1994). The influence of spatially and temporally varying oceanographic conditions on meroplanktonic metapopulations. *Deep-Sea Res. II* **41**: 107–145.

Brickman, D., Shackell, N. L., and Frank, K. T. (2001). Modeling the retention and survival of Browns Bank haddock larvae using an early life stage model. *Fish. Oceanogr.* **10**: 284–296.

Brochier, T., Lett, C., Tam, J., *et al.* (2008). An individual-based model study of anchovy early life history in the northern Humboldt Current system. *Prog. Oceanogr.* **79**: 313–325.

Brown, C. A., Jackson, G. A., Holt, S. A., and Holt, G. J. (2005). Spatial and temporal patterns in modeled particle transport to estuarine habitat with comparisons to larval fish settlement patterns. *Estuar. Coast. Shelf Sci.* **64**: 33–46.

Carr, S. D., Capet, X. J., McWilliams, J. C., *et al.* (2008). The influence of diel vertical migration on zooplankton transport and recruitment in an upwelling region: estimates from a coupled behavioral–physical model. *Fish. Oceanogr.* **17**: 1–15.

Castro, L. R., Fréon, P., van der Lingen, C. D., and Uriarte, A. (2005). Report of the SPACC meeting on small pelagic fish spawning habitat dynamics and the daily egg production method (DEPM). *GLOBEC Rep.* **22**: xiv, 107 pp.

Checkley, D. M., Jr., Dotson, R. C., and Griffith, D. A. (2000). Continuous, underway sampling of eggs of Pacific sardine (*Sardinops sagax*) and northern anchovy (*Engraulis mordax*) in spring 1996 and 1997 off southern and central California. *Deep-Sea Res. II* **47**: 1139–1155.

Clark, R. A., Fox, C. J., Viner, D., and Livermore, M. (2003). North Sea cod and climate change – modelling the effects of temperature on population dynamics. *Glob. Change Biol.* **9**: 1669–1680.

Cooke, S. J., Hinch, S. G., Wikelski, M., *et al.* (2004). Biotelemetry: a mechanistic approach to ecology. *Trends Ecol. Evol.* **19**: 334–343.

Cury, P. (1994). Obstinate nature: an ecology of individuals. Thoughts on reproductive behavior and biodiversity. *Can. J. Fish. Aquat. Sci.* **51**: 1664–1673.

Cury, P., Bakun, A., Crawford, R. J. M., *et al.* (2000). Small pelagics in upwelling systems: patterns of interaction and structural changes in "wasp-waist" ecosystems. *ICES J. Mar. Sci.* **57**: 603–618.

Davis, A. J., Jenkinson, L. S., Lawton, J. H., *et al.* (1998). Making mistakes when predicting shifts in species range in response to global warming. *Nature* **391**: 783–786.

Demarcq, H., Barlow, R. G., and Hutchings, L. (2007). Application of a chlorophyll index derived from satellite data to investigate the variability of phytoplankton in the Benguela ecosystem. *Afr. J. Mar. Sci.* **29** (2): 271–282.

Dopolo, M. T., van der Lingen, C. D., and Moloney, C. L. (2005). Stage-dependent vertical distribution of pelagic fish eggs on the western Agulhas Bank, South Africa. *Afr. J. Mar. Sci.* **27**: 249–256.

Edwards, K. P., Hare, J. A., Werner, F. E., and Blanton, B. O. (2006). Lagrangian circulation on the southeast US continental shelf: implications for larval dispersal and retention. *Cont. Shelf Res.* **26**: 1375–1394.

Fach, B. A. and Klinck, J. M. (2006). Transport of Antarctic krill (*Euphausia superba*) across the Scotia Sea. Part I: Circulation and particle tracking simulations. *Deep-Sea Res. I* **53**: 987–1010.

Field, J. C., Francis, R. C., and Aydin, K. (2006). Top-down modeling and bottom-up dynamics: linking a fisheries-based ecosystem model with climate hypotheses in the northern California Current. *Prog. Oceanog.* **68**: 238–270.

Fiksen, O. and Slotte, A. (2002). Stock-environment recruitment models for Norwegian spring spawning herring (*Clupea harengus*). *Can. J. Fish. Aquat. Sci.* **59**: 211–217.

Fiksen, O., Jørgensen, C., Kristiansen, T., *et al.* (2007). Linking behavioral ecology and oceanography: larval behavior determines growth, mortality and dispersal. *Mar. Ecol. Prog. Ser.* **347**: 195–205.

Fournier, D. A., Hampton, J., and Sibert, J. R. (1998). MULTIFAN-CL : a length-based, aged-structured model for fisheries stock assessment, with application to South Pacific albacore, *Thunnus alalunga*. *Can. J. Fish. Aquat. Sci.* **55**: 2105–2116.

Fox, C. J., McCloghrie, P., Young, E. F., and Nash, R. D. M. (2006). The importance of individual behavior for successful settlement of juvenile plaice (*Pleuronectes platessa* L.): a modeling and field study in the eastern Irish Sea. *Fish. Oceanogr.* **15**: 301–313.

Fulton, E. A., Smith, A. D. M., and Johnson, C. R. (2004a). Biogeochemical marine ecosystem models I: IGBEM – a model of marine bay ecosystems. *Ecol. Model.* **174**: 267–307.

Fulton, E. A., Parslow, J. S., Smith, A. D. M., and Johnson, C. R. (2004b). Biogeochemical marine ecosystem models II: the effect of physiological detail on model performance. *Ecol. Model.* **173**: 371–406.

Goodwin, R. A., Nestler, J. M., Anderson, J. J., *et al.* (2006). Forecasting 3-D fish movement behavior using a Eulerian–Lagrangian-agent method (ELAM). *Ecol. Model.* **192**: 197–223.

Grimm, V., Revilla, E., Berger, U., *et al.* (2005). Pattern-oriented modeling of agent-based complex systems: lessons from ecology. *Science* **310**: 987–991.

Guay, J. C., Boisclair, D., Leclerc, M., *et al.* (2000). Development and validation of numerical habitat models for juveniles of Atlantic salmon (*Salmo salar*). *Can. J. Fish. Aquat. Sci.* **57**: 2065–2075.

Guizien, K., Brochier, T., Duchene, J. C., *et al.* (2006). Dispersal of *Owenia fusiformis* larvae by wind-driven currents: turbulence, swimming behavior and mortality in a three-dimensional stochastic model. *Mar. Ecol. Prog. Ser.* **311**: 47–66.

Gutiérrez, O. Q., Marinone, S. G., and Parés-Sierra, A. (2004). Lagrangian surface circulation in the Gulf of California from a 3D numerical model. *Deep-Sea Res. II* **51**: 659–672.

Hanson, P. C., Johnson, T. B., Schindler, D. E., and Kitchell, J. F. (1997). *Fish bioenergetics 3.0.* University of Wisconsin Sea Grant Institute, WISCU-T-97-001, Madison, Wisconsin.

Hao, W., Jian, S., Ruijing, W., *et al.* (2003). Tidal front and the convergence of anchovy (*Engraulis japonicus*) eggs in the Yellow Sea. *Fish. Oceanogr.* **12**: 434–442.

Harley, C. D. G., Hughes, A. R., Hultgren, K. M., *et al.* (2006). The impacts of climate change in coastal marine systems. *Ecol. Lett.* **9**: 228–241.

Haza, A. C., Piterbarg, L. I., Martin, P., *et al.* (2007). A Lagrangian subgridscale model for particle transport improvement and application in the Adriatic Sea using the Navy Coastal Ocean Model. *Ocean Model.* **17** (1): 68–91.

Heath, M. and Gallego, A. (1997). From the biology of the individual to the dynamics of the population: bridging the gap in fish early life studies. *J. Fish Biol.* **51**:1–29.

Heath, M., Scott, B., and Bryant, A.D. (1997). Modeling the growth of herring from four different stocks in the North Sea. *J. Sea Res.* **38**: 413–436.

Heath, M., Zenitani, H., Watanabe, Y., *et al.* (1998). Modeling the dispersal of larval Japanese sardine, *Sardinops melanostictus*, by the Kuroshio Current in 1993 and 1994. *Fish. Oceanogr.* **7**: 335–346.

Hermann, A.J., Hinckley, S., Megrey, B.A., and Napp, J.M. (2001). Applied and theoretical considerations for constructing spatially explicit individual-based models of marine larval fish that include multiple trophic levels. *ICES J. Mar. Sci.* **58** 1030–1041.

Hinckley, S. (1999). Biophysical mechanisms underlying the recruitment process in walleye pollock *(Theragra chalcogramma)*. Ph.D. thesis, University of Washington, USA.

Hinrichsen, H.-H., Möllmann, C., Voss, R., *et al.* (2002). Biophysical modeling of larval Baltic cod *(Gadus morhua)* growth and survival. *Can. J. Fish. Aquat. Sci.* **59**: 1858–1873.

Hinrichsen, H.-H., Kraus, G., Voss, R., *et al.* (2005). The general distribution and mixing probability of Baltic sprat juvenile populations. *J. Mar. Syst.* **58**: 52–66.

Huggett, J., Fréon, P., Mullon, C., and Penven, P. (2003). Modeling the transport success of anchovy *Engraulis encrasicolus* eggs and larvae in the southern Benguela: the effect of spatio-temporal spawning patterns. *Mar. Ecol. Prog. Ser.* **250**: 247–262.

Humston, R., Olson, D.B., and Ault, J.S. (2004). Behavioral assumptions in models of fish movement and their influence on population dynamics. *T. Am. Fish. Soc.* **133**: 1304–1328.

Hunter, J.R. (1977). Behavior and survival of northern anchovy *Engraulis mordax* larvae. *CalCOFI Rep.* **19**: 138–146.

Hunter, J.R. and Coyne, K.M. (1982). The onset of schooling in northern anchovy larvae, *Engraulis mordax. CalCOFI Rep.* **23**: 246–251.

Huret, M., Runge, J.A., Chen, C., *et al.* (2007). Dispersal modeling of fish early life stages: sensitivity with application to Atlantic cod in the western Gulf of Maine. *Mar. Ecol. Prog. Ser.* **347**: 261–274.

Huse, G. and Giske, J. (1998). Ecology in the Mare Pentium: an individual-based spatio-temporal model for fish with adapted behavior. *Fish. Res.* **37**: 163–178.

Ito, S., Megrey, B.A., Kishi, M.J., *et al.* (2007). On the interannual variability of the growth of Pacific saury (*Cololabis saira*): a simple 3-box model using NEMURO.FISH. *Ecol. Model.* **202**: 174–183.

Jacobson, L.D., Bograd, S.J., Parrish, R.H., *et al.* (2005). An ecosystem-based hypothesis for climatic effects on surplus production in California sardine (*Sardinops sagax*) and environmentally dependent surplus production models. *Can. J. Fish. Aquat. Sci.* **62**: 1782–1796.

Karim, M.R., Sekine, M., Higuchi, T., *et al.* (2003). Simulation of fish behavior and mortality in hypoxic water in an enclosed bay. *Ecol. Model.* **159**: 27–42.

Kendall, A.W. and Duker, G.J. (1998). The development of recruitment fisheries oceanography in the United States. *Fish. Oceanogr.* **7**: 69–88.

Koné, V. (2006). Modélisation de la production primaire et secondaire de l'écosystème du Benguela sud. Influence des conditions trophiques sur le recrutement des larves d'anchois. Ph.D. thesis, Université Pierre & Marie Curie (Paris VI), France.

Koné, V., Machu, E., Penven, P., *et al.* (2005). Modeling the primary and secondary productions of the southern Benguela upwelling system: A comparative study through two biogeochemical models. *Global Biogeochem. Cy.* **19**: GB4021, doi:4010.1029/2004GB002427.

Lazure, P. and Dumas, F. (2008). A 3D hydrodynamics model for applications at the regional scale (MARS-3D): application to the Bay of Biscay. *Adv. Wat. Res.* **31**: 233–250.

Lehodey, P., Chai, F., and Hampton, J. (2003). Modeling climate-related variability of tuna populations from a coupled ocean-biogeochemical-population dynamics model. *Fish. Oceanogr.* **12**: 483–494.

Lett, C., Roy, C., Levasseur, A., *et al.* (2006). Simulation and quantification of enrichment and retention processes in the southern Benguela upwelling ecosystem. *Fish. Oceanogr.* **15**: 363–372.

Lett, C., Penven, P., Ayón, P., and Fréon, P. (2007a). Enrichment, concentration and retention processes in relation to anchovy (*Engraulis ringens*) eggs and larvae distributions in the northern Humboldt upwelling ecosystem. *J. Mar. Syst.* **64**: 189–200.

Lett, C., Veitch, J., van der Lingen, C.D. and Hutchings, L. (2007b). Assessment of an environmental barrier to transport of ichthyoplankton from the southern to the northern Benguela ecosystems. *Mar. Ecol. Prog. Ser.* **347**: 247–259.

Lett, C., Verley, P., Mullon, C., *et al.* (2008). A lagrangian tool for modelling ichthyoplankton dynamics. *Environ. Model. Softw.* **23**: 1210–1214.

Lett, C., and Mirabet, V. (2008). Modelling the dynamics of animal groups in motion. *S. Afr. J. Sci.* **104**: 192–198.

Lough, R.G., Buckley, L.J., Werner, F.E., *et al.* (2005). A general biophysical model of larval cod (*Gadus morhua*) growth applied to populations on Georges Bank. *Fish. Oceanogr.* **14**: 241–262.

Lough, R.G., Broughton, E.A., Buckley, L.J., *et al.* (2006). Modeling growth of Atlantic cod larvae on the southern flank of Georges Bank in the tidal-front circulation during May 1999. *Deep-Sea Res.* **53**: 2771–2788.

Luo, J., Hartman, K.J., Brandt, S.B., *et al.* (2001). A spatially-explicit approach for estimating carrying capacity: an application for the Atlantic menhaden (*Brevoortia tyrannus*) in Chesapeake Bay. *Estuaries* **24**: 545–556.

Lynn, R.J. (2003). Variability in the spawning habitat of Pacific sardine (*Sardinops sagax*) off southern and central California. *Fish. Oceanogr.* **12**: 541–553.

Marinone, S.G. (2003). A three-dimensional model of the mean and seasonal circulation of the Gulf of California. *J. Geophys. Res.* **108**: C10, 3325, doi:3310.1029/2002JC001720.

Marinone, S.G., Gutierrez, O.Q., and Pares-Sierra, A. (2004). Numerical simulation of larval shrimp dispersion in the Northern Region of the Gulf of California. *Estuar. Coast. Shelf Sci.* **60**: 611–617.

Marta-Almeida, M., Dubert, J., Peliz, A., and Queiroga, H. (2006). Influence of vertical migration pattern on retention of crab larvae in a seasonal upwelling system. *Mar. Ecol. Prog. Ser.* **307**: 1–19.

Megrey, B.A. and Hinckley, S. (2001). Effect of turbulence on feeding of larval fishes: a sensitivity analysis using an individual-based model. *ICES J. Mar. Sci.* **58**: 1015–1029.

Megrey B.A., Rose, K.A., Klumb, R.A., *et al.* (2007). A bioenergetics-based population dynamics model of Pacific herring (*Clupea harengus pallasi*) coupled to a lower trophic level nutrient-phytoplankton-zooplankton model: Description, calibration, and sensitivity analysis. *Ecol. Model.* **202**: 144–164.

Miller, C.B., Lynch, D.R., Carlotti, F., *et al.* (1998). Coupling of an individual-based population dynamic model of *Calanus finmarchicus* to a circulation model for the Georges Bank region. *Fish. Oceanogr.* **7**: 219–234.

Miller, D.C.M. (2006). An individual-based modelling approach to examine life history strategies of sardine (*Sardinops sagax*) in the southern Benguela ecosystem. Ph.D. thesis, University of Cape Town, South Africa.

Miller, D.C.M., Moloney, C.L., van der Lingen, C.D., *et al.* (2006). Modelling the effects of physical–biological interactions and spatial variability in spawning and nursery areas on transport and retention of sardine eggs and larvae in the southern Benguela ecosystem. *J. Mar. Syst.* **61**: 212–229.

Miller, T.J. (2007). Contribution of individual-based coupled physical biological models to understanding recruitment in marine fish populations. *Mar. Ecol. Prog. Ser.* **347**: 127–138.

Mullon, C., Cury, P., and Penven, P. (2002). Evolutionary individual-based model for the recruitment of anchovy (*Engraulis capensis*) in the southern Benguela. *Can. J. Fish. Aquat. Sci.* **59**: 910–922.

Mullon, C., Fréon, P., Parada, C., *et al.* (2003). From particles to individuals: modeling the early stages of anchovy (*Engraulis capensis/encrasicolus*) in the southern Benguela. *Fish. Oceanogr.* **12**: 396–406.

Ney, J.J. (1993). Bioenergetics modeling today: growing pains on the cutting edge. *T. Am. Fish. Soc.* **122**: 736–748.

North, E.W., Hood, R.R., Chao, S.-Y., and Sanford, L.P. (2005). The influence of episodic events on transport of striped bass eggs to the estuarine turbidity maximum nursery area. *Estuaries* **28**: 108–123.

North, E.W., Hood, R.R., Chao, S.Y. and Sanford, L.P. (2006). Using a random displacement model to simulate turbulent particle motion in a baroclinic frontal zone: a new implementation scheme and model performance tests. *J. Mar. Syst.* **60**: 365–380.

Parada, C. (2003). Modeling the effects of environmental and ecological processes on the transport, mortality, growth and distribution of early life stages of Cape anchovy (*Engraulis capensis*) in the Bengula system. Ph.D. thesis, University of Cape Town, South Africa.

Parada, C., van der Lingen, C.D., Mullon, C., and Penven, P. (2003). Modeling the effect of buoyancy on the transport of anchovy (*Engraulis capensis*) eggs from spawning to nursery grounds in the southern Benguela: an IBM approach. *Fish. Oceanogr.* **12**: 170–184.

Parada, C., Hinckley, S., Horne, J., *et al.* (in press). Comparing simulated walleye pollock recruitment indices to data and stock assessment models from the Gulf of Alaska. *Mar. Ecol. Prog. Ser.*

Pedersen, O.P., Nilssen, E.M., Jorgensen, L.L., and Slagstad, D. (2006). Advection of the red king crab larvae on the coast of North Norway – a Lagrangian model study. *Fish. Res.* **79**: 325–336.

Peliz, A., Marchesiello, P., Dubert, J., *et al.* (2007). A study of crab larvae dispersal on the Western Iberian Shelf: physical processes. *J. Mar. Syst.* **68** (1–2): 215–236.

Penven, P. (2000). A numerical study of the Southern Benguela circulation with an application to fish recruitment. Ph.D. thesis, Université de Bretagne Occidentale, France.

Penven, P., Roy, C., Brundrit, G.B., *et al.* (2001). A regional hydrodynamic model of upwelling in the Southern Benguela. *S. Afr. J. Sci.* **97**: 472–475.

Penven, P., Echevin, V., Pasapera, J., *et al.* (2005). Average circulation, seasonal cycle, and mesoscale dynamics of the Peru Current System: A modeling approach. *J. Geophys. Res.* **110**, C10021: doi:10010.11029/12005JC002945.

Penven, P., Debreu, L., Marchesiello, P., and McWilliams, J.C. (2006a). Evaluation and application of the ROMS 1-way embedding procedure to the central California upwelling system. *Ocean Model.* **12**: 157–187.

Penven, P., Lutjeharms, J.R.E., and Florenchie, P. (2006b). Madagascar: a pacemaker for the Agulhas Current system? *Geophys. Res. Lett.* **33**: L17609: doi:17610.11029/12006GL026854.

Planque, B., Bellier, E., and Lazure, P. (2007). Modeling potential spawning habitat of sardine (*Sardina pilchardus*) and anchovy (*Engraulis encrasicolus*) in the Bay of Biscay. *Fish. Oceanogr.* **16**: 16–30.

Rahel, F.J. (2002). Using current biogeographic limits to predict fish distributions following climate change. In *Fisheries in a Changing Climate*, McGinn, N.A., ed. American Fisheries Society Symposium 32, Bethesda, Maryland, pp. 99–110.

Railsback, S.F., Lamberson, R.H., Harvey, B.C., and Duffy, W.E. (1999). Movement rules for individual-based models of stream fish. *Ecol. Model.* **123**: 73–89.

Rice, J.A., Quinlan, J.A., Nixon, S.W., *et al.* (1999). Spawning and transport dynamics of Atlantic menhaden: inferences from characteristics of immigrating larvae and predictions of a hydrodynamic model. *Fish. Oceanogr.* **8**: 93–110.

Robinson, C.L.K. and Ware, D.M. (1999). Simulated and observed response of the southwest Vancouver Island pelagic ecosystem to oceanic conditions in the 1990's. *Can. J. Fish. Aquat. Sci.* **56**: 2433–2443.

Rodionov, S. and Overland, J. E. (2005). Application of a sequential regime shift detection method to the Bering Sea. *ICES J. Mar. Sci.* **62**: 328–332.

Rodríguez, J. M., Barton, E. D., Eve, L., and Hernández-León, S. (2001). Mesozooplankton and ichthyoplankton distribution around Gran Canaria, an oceanic island in the NE Atlantic. *Deep-Sea Res.* **48**: 2161–2183.

Rose, K. A. and Sable, S. E. (2009). Multispecies modeling of fish populations. In *Computers in Fisheries Research*, Vol. 2, Moksness, E. and Megrey, B. A., eds. New York: Chapman and Hall (ISBN 978-1-4020-8635-9): 373–397.

Rose, K. A., Werner, F. E., Megrey, B. A., *et al.* (2007). Simulated herring growth responses in the Northeastern Pacific to historic temperature and zooplankton conditions generated by the 3-dimensional NEMURO nutrient–phytoplankton–zooplankton model. *Ecol. Model.* **202**: 184–195.

Rose, K. A., Megrey, B. A., Hay, D. E., *et al.* (2008). Climate regime effects on Pacific herring growth using coupled nutrient–phytoplankton–zooplankton and bioenergetics models. *T. Am. Fish. Soc.* **137**: 278–297.

Rubec, P. J., Smith, S. G., Coyne, M. S., *et al.* (2001). Spatial modeling of fish habitat suitability in Florida estuaries. In *Spatial Processes and Management of Marine Populations*, Kruse, G. H., Bez, N., Booth, A., *et al.* eds. University of Alaska Sea Grant, AK-SG-01–02, Fairbanks, pp. 1–18.

Runge, J. A., Franks, P. J. S., Gentleman, W. C. *et al.* (2005). Diagnosis and prediction of variability in secondary production and fish recruitment processes: developments in physical–biological modelling. In *The Sea, Vol. 13, The Global Coastal Ocean: Multiscale Interdisciplinary Processes*, Robinson, A. R. and Brink, K. H., eds. Harvard University Press, pp. 413–473.

Russell, J. L., Stouffer, R. J., and Dixon, K. W. (2006). Intercomparison of the Southern Ocean circulations in IPCC coupled model control simulations. *J. Climate* **19**: 4560–4575.

Sætre, R., Toresen, R., Søiland, H., and Fossum, P. (2002). The Norwegian spring-spawning herring – spawning, larval drift and larval retention. *Sarsia* **87**: 167–178.

Santos, A. M. P., Peliz, A., Dubert, J., *et al.* (2004). Impact of a winter upwelling event on the distribution and transport of sardine (*Sardina pilchardus*) eggs and larvae off western Iberia: a retention mechanism. *Cont. Shelf Res.* **24**: 149–165.

Santos, A. J. P., Nogueira, J., and Martins, H. (2005). Survival of sardine larvae off the Atlantic Portuguese coast: a preliminary numerical study. *ICES J. Mar. Sci.* **62**: 634–644.

Santos, A. M. P., Ré, P., Dos Santos, A., and Peliz, A. (2006). Vertical distribution of the European sardine (*Sardina pilchardus*) larvae and its implications for their survival. *J. Plankton Res.* **28**: 523–532.

Santos, A. M. P., Chícharo, A., Dos Santos, A., *et al.* (2007). Physical–biological interactions in the life history of small pelagic fish in the Western Iberia Upwelling Ecosystem. *Prog. Oceanogr.* **74**: 192–209.

Scheffer, M., Baveco, J. M., DeAngelis, D. L., *et al.* (1995). Super-individuals a simple solution for modeling large populations on an individual basis. *Ecol. Model.* **80**: 161–170.

Schmitz, O. J., Post, E., Burns, C. E., and Johnston, K. M. (2003). Ecosystem responses to global climate change: moving beyond color mapping. *BioScience* **53**: 1199–1205.

Shannon, L. J., Moloney, C. L., Jarre, A., and Field, J. G. (2003). Trophic flows in the southern Benguela during the 1980s and 1990s. *J. Mar. Syst.* **39**: 83–116.

Shannon, L. J., Christensen, V., and Walters, C. J. (2004). Modeling stock dynamics in the southern Benguela ecosystem for the period 1978–2002. *Afr. J. Mar. Sci.* **26**: 179–196.

Shchepetkin, A. F. and McWilliams, J. C. (2005). The regional oceanic modeling system (ROMS): a split-explicit, free-surface, topography-following-coordinate oceanic model. *Ocean Model.* **9**: 347–404.

Shin, Y. J. and Cury P. (2004). Using an individual-based model of fish assemblages to study the response of size spectra to changes in fishing. *Can. J. Fish. Aquat. Sci.* **61**: 414–431.

Shin Y. J., Shannon, L. J., and Cury, P. M. (2004). Simulations of fishing effects on the southern Benguela fish community using an individual-based model: learning from a comparison with ECOSIM. *Afr. J. Mar. Sci.* **26**: 95–114.

Skogen, M. (1999). A biophysical model applied to the Benguela upwelling system. *S. Afr. J. Mar. Sci.* **21**: 235–249.

Skogen, M. D., Shannon, L. J., and Stiansen, J. E. (2003). Drift patterns of anchovy *Engraulis capensis* larvae in the southern Benguela, and their possible importance for recruitment. *Afr. J. Mar. Sci.* **25**: 37–47.

Stenevik, E. K., Skogen, M., Sundby, S., and Boyer, D. (2003). The effect of vertical and horizontal distribution on retention of sardine (*Sardinops sagax*) larvae in the Northern Benguela – observations and modeling. *Fish. Oceanogr.* **12**: 185–200.

Suda, M. and Kishida, T. (2003). A spatial model of population dynamics of the early life stages of Japanese sardine, *Sardinops melanostictus*, off the Pacific coast of Japan. *Fish. Oceanogr.* **12**: 85–99.

Suda, M., Akamine, T., and Kishida, T. (2005). Influence of environment factors, interspecific-relationships and fishing mortality on the stock fluctuation of the Japanese sardine, *Sardinops melanostictus*, off the Pacific coast of Japan. *Fish. Res.* **76**: 368–378.

Sugden, A. and Pennisi, E. (2006). When to go, where to stop. *Science* **313**: 775.

Thorpe, S. E., Heywood, K. J., Stevens, D. P., and Brandon, M. A. (2004). Tracking passive drifters in a high resolution ocean model: implications for interannual variability of larval krill transport to South Georgia. *Deep-Sea Res. I* **51**: 909–920.

Tilburg, C. E., Reager, J. T., and Whitney, M. M. (2005). The physics of blue crab larval recruitment in Delaware Bay: A model study. *J. Mar. Res.* **63**: 471–495.

Travers, M., Shin, Y. J., Jennings, S., and Cury, P. (2007). Towards end-to-end models for investigating the effects of climate and fishing in marine ecosystems. *Progr. Oceanogr.* **75**: 751–770.

Twatwa, N. M., van der Lingen, C. D., Drapeau, L., *et al.* (2005). Characterising and comparing the spawning habitats of anchovy *Engraulis encrasicolus* and sardine *Sardinops sagax* in the southern Benguela upwelling ecosystem. *Afr. J. Mar. Sci.* **27**: 487–500.

Tyler, J. A. and Rose, K. A. (1994). Individual variability and spatial heterogeneity in fish population models. *Rev. Fish Biol. Fisher.* **4**: 91–123.

van der Lingen, C. D., Castro, L., Drapeau, L., and Checkley, D. (2005). Report of a GLOBEC-SPACC workshop on characterizing and comparing the spawning habitats of small pelagic fish. *GLOBEC Rep.* **21**: xii, 33 pp.

Vikebø, F. B., Jørgensen, C., Kristiansen, T., and Fisken, Ø. (2007a). Drift, growth, and survival of larvae Northeast Arctic cod with simple rules of behavior. *Mor. Ecol. Prog. Ser.* **347**: 207–219.

Vikebø, F. B., Sundby, S., Ådlandsvik, B., and Fiksen, Ø. (2005). The combined effect of transport and temperature on distribution and growth of larvae and pelagic juveniles of Arcto-Norwegian cod. *ICES J. Mar. Sci.* **62**: 1375–1386.

Vikebø, F. B., Sundby, S., Ådlandsvik, B., and Otterå, O. H. (2007b). Impacts of a reduced thermohaline circulation on transport and growth of larvae and pelagic juveniles of Arcto-Norwegian cod (*Gadus morhua*). *Fish. Oceanogr.* **16**: 216–228.

Walters C., Christensen V., and Pauly, D. (1997). Structuring dynamic models of exploited ecosystems from trophic mass-balance assessments. *Rev. Fish Biol. Fisher.* **7**: 139–172.

Walters, C., Pauly D., and Christensen, V. (1999). Ecospace: prediction of mesoscale spatial patterns in trophic relationships of exploited ecosystems, with emphasis on the impacts of marine protected areas. *Ecosystems* **2**: 539–554.

Walters, C., Pauly D., Christensen, V., and Kitchell, J. F. (2000). Representing density dependent consequences of life history strategies in aquatic ecosystems: EcoSim II. *Ecosystems* **3**: 70–83.

Werner, F. E., Quinlan, J. A., Lough, R. G., and Lynch, D. R. (2001). Spatially-explicit individual based modeling of marine populations: a review of the advances in the 1990s. *Sarsia* **86**: 411–421.

Werner, F. E., Ito, S., Megrey, B. A., and Kishi, M. J. (2007). Synthesis of the NEMURO model studies and future directions of marine ecosystem modeling. *Ecol. Model.* **202**: 211–223.

Yeung, C. and Lee, T. N. (2002). Larval transport and retention of the spiny lobster, *Panulirus argus*, in the coastal zone of the Florida Keys, USA. *Fish. Oceanogr.* **11**: 286–309.

Zakardjian, B. A., Sheng, J. Y., Runge, J. A., *et al.* (2003). Effects of temperature and circulation on the population dynamics of *Calanus finmarchicus* in the Gulf of St. Lawrence and Scotian Shelf: study with a coupled, three-dimensional hydrodynamic, stage-based life history model. *J. Geophys. Res.* **108**: 8016, doi:8010.1029/2002JC001410.

7 Trophic dynamics

Carl D. van der Lingen, Arnaud Bertrand, Antonio Bode, Richard Brodeur, Luis A. Cubillos, Pepe Espinoza, Kevin Friedland, Susana Garrido, Xabier Irigoien, Todd Miller, Christian Möllmann, Ruben Rodriguez-Sanchez, Hiroshige Tanaka, and Axel Temming

CONTENTS

Summary

Literature on the trophic ecology of small pelagic fish (primarly anchovy *Engraulis* spp. and sardine *Sardinops* spp. but including the genera *Brevoortia*, *Clupea*, *Sardina*, *Sprattus*, and *Strangomera*) and their interactions with plankton are reviewed using case studies describing research on some economically and ecologically important small pelagic fish from upwelling and temperate non-upwelling ecosystems. Information from morphological studies of the feeding apparatus, field studies on dietary composition and foraging behaviour, and laboratory studies that have provided data for the parameterization of bioenergetic and other models of these small pelagic fish are presented, where available. Two or more small pelagic fish species are described in each case study, and disparities in trophic dynamics between co-occurring anchovy and sardine are consistently seen, supporting the hypothesis that alternations between the two species could be trophically mediated. Linkages between climate and fish are described for many of the systems, and possible impacts of climate change on some of the species are described.

Introduction

Small pelagic fish are, in general, microphagous planktivores, and their high abundance levels in upwelling systems, in particular, was attributed to their ability to feed directly on phytoplankton and hence benefit from a short and efficient food chain (Ryther, 1969; Walsh, 1981). This two-step food chain hypothesis, with small pelagic fish being regarded as essentially phytophagus and feeding on large, chain forming diatoms such as *Chaetoceros* and *Fragilaria* (Yoneda and Yoshida, 1955; Bensam, 1964; Loukashkin, 1970; King and Macleod, 1976) was initially well supported (Longhurst, 1971; Durbin, 1979; Walsh, 1981). However, subsequent studies challenged this hypothesis and suggested that clupeoids consume both phytoplankton and zooplankton (Cushing, 1978), with significant feeding on phytoplankton considered likely where strong upwelling is a persistent feature of the environment (e.g. off Namibia and Peru), whereas feeding on phytoplankton would be less common and zooplankton would become the dominant food source of species living where upwelling is weaker and less persistent (e.g. the southern California coast; Blaxter and Hunter, 1982). A comprehensive review of the diets of commercially important clupeids concluded that few true phytophagist species exist, and that most clupeids are omnivorous microphagists that derive the bulk of their energy from zooplankton (James, 1988). Additionally, most microphagous clupeoids possess two feeding modes and switch between the two when conditions dictate, generally filter feeding on smaller food particles and particulate feeding on larger food particles (Blaxter and Hunter, 1982). The ability to switch between these feeding modes makes these species highly opportunistic and flexible foragers which are able to maximize their energy intake through employing the feeding mode most appropriate to a particular food environment. The high abundance and success of small pelagic fish in upwelling areas in particular was attributed to this flexibility in feeding behavior, which

Climate Change and Small Pelagic Fish, eds. Dave Checkley, Jürgen Alheit, Yoshioki Oozeki, and Claude Roy. Published by Cambridge University Press. © Cambridge University Press 2009.

Table 7.1 *System type, region, and species used in case studies in which the trophic dynamics of small pelagic fishes are described*

System Type	Region	Species (and common name)
Upwelling	Benguela Current system (SE Atlantic)	*Engraulis encrasicolus* (Cape anchovy) *Sardinops sagax* (sardine)
	California Current system (NE Pacific)	*Engraulis mordax* (Northern anchovy) *Sardinops sagax* (Pacific sardine)
	Humboldt Current system (SE Pacific)	*Engraulis ringens* (anchovy) *Sardinops sagax* (sardine) *Strangomera bentincki* (common sardine)
Temperate non-upwelling	Kuroshio Current system (NW Pacific)	*Engraulis japonicus* (anchovy) *Sardinops melanosticus* (sardine)
	Iberian Peninsula, Bay of Biscay, and Mediterranean Sea (NE Atlantic)	*Engraulis encrasicolus* (anchovy) *Sardina pilchardus* (Atlantic Iberian sardine)
	US East Coast and Gulf of Mexico (W Atlantic)	*Brevoortia gunteri* (Finescale menhaden) *Brevoortia patronus* (Gulf menhaden) *Brevoortia tyrannus* (Atlantic menhaden)
	Baltic Sea	*Sprattus sprattus* (sprat) *Clupea harengus* (herring)

enables them to efficiently utilize a wide range of particle sizes and hence take advantage of their dynamic trophic environment (James, 1988).

The objective of this chapter is to provide an updated synthesis on the trophic dynamics of small pelagic fishes by using seven case studies selected to describe trophic aspects of a wide range of economically and ecologically important small pelagic fishes, primarily anchovy and sardine, from upwelling and temperate non-upwelling ecosystems (Table 7.1). Published literature describing morphological, experimental, and field studies is reviewed, and some unpublished data, where available, is incorporated into each case study. Two or more species of small pelagic fish co-occur and often have shown alternating species dominance in the systems reviewed (see this volume, Chapter 9), and whilst Blaxter and Hunter (1982) considered that different species of clupeoids existing in the same habitat would tend to show strong overlap in food habits, Schwartzlose *et al.* (1999) hypothesized that species alternations between anchovy and sardine might be linked to changes in habitat (such as food or temperature fields) that resulted in one species being favored over the other. However, for the two species to show different responses to the same environmental forcing (e.g. a changed trophic environment) requires that they show some ecological differences, and trophic disparities between co-occurring species from each region are highlighted where they occur. The case studies describe trophic aspects of small pelagic fish in each system and

provide the bulk of this chapter, and consistent patterns among and between co-existing species are identified and implications of these discussed. Whether anchovy and sardine are sufficiently trophically distinct such that changed food environments could impact their population variability is assessed, climate–fish interactions for many of the systems are also detailed, and possible impacts of climate change on some of the species are described.

The Benguela Current upwelling system

A substantial amount of research has been conducted to examine the trophic dynamics of anchovy (*Engraulis encrasicolus*) and sardine (*Sardinops sagax*) in the Benguela Current system. This research has consisted of morphological studies that described and compared the structure and development of the feeding apparatus; field studies that investigated diet and assessed feeding periodicity; and laboratory studies that examined aspects of feeding behavior and enabled the parameterization of a variety of processes such as ingestion, respiration, excretion, and evacuation rates. Empirical results from laboratory experiments were used to construct carbon and nitrogen budget models for each species, which were used to quantify the effect of different food environments upon fish growth. Together, these studies have provided a wealth of information that has permitted detailed comparison of the trophic ecologies of anchovy and sardine from this system. Unfortunately, however, almost all of the work has been conducted on juveniles and adults, and knowledge concerning the trophic ecology of larvae of these small pelagic fishes is limited to descriptions of the diet of only three individuals (King and Macleod, 1976). Whereas preliminary studies on the trophic ecology of a third small pelagic species from the Benguela (the redeye round herring *Etrumeus whiteheadi*) have been performed (Wallace-Fincham, 1987), only results from the research on anchovy and sardine are described in more detail, below.

The gape diameter of juvenile anchovy appears larger than that of sardine of a similar size (Fig. 7.1a) and indicates that anchovy are capable of ingesting larger prey than are sardine (Booi, 2000). The branchial basket of both species show many similarities in structure and development (King and Macleod, 1976); gill arch length and gill raker length increase with increasing fish length for both species, and juveniles of both species and adult sardine show an increase in the number of gill rakers with increasing fish length although the number of gill rakers remains constant for anchovy of >80 mm standard length (SL) (Fig. 7.1b). Gill raker gap also increases with increasing fish length and for fish up to ± 40 mm SL the gill raker gap is about the same for both species; however, the rate of increase in gill raker gap with fish length is higher in anchovy than in sardine such that anchovy >40 mm have a larger gill raker gap than do sardine, and an adult anchovy has a gill raker gap about twice as large as does an adult sardine (Fig. 7.1c). Additionally, there is a marked dissimilarity in the structure and alignment of gill raker denticles between the two species, which in sardine are elaborate spine-like projections terminating in a serrated nodule that are unidirectionally aligned along the entire gill raker length, whereas in anchovy gill raker denticles are simple, spine-like projections in structure that are randomly arranged along the gill raker axis (King and Macleod, 1976).

Early studies of the diet of anchovy and sardine in the Benguela reported that sardine stomach contents were dominated by phytoplankton (predominantly diatoms) with a mean annual ratio of 2:1 by volume of phytoplankton to zooplankton (Davies, 1957), and that anchovy displayed a "preference" for phytoplankton (Robinson, 1966). Similarly, King and Macleod (1976) considered adults of both species to be phytoplanktophagus and juveniles-zooplanktophagous, the dietary switch being attributed to a decrease in porosity of the filtering mechanism with increasing fish size. Additionally, stomach contents of adults showed good correlation with the ambient plankton, leading those authors to conclude that both species were essentially non-selective feeders feeding primarily on phytoplankton. However, those studies assessed relative dietary importance using frequency-of-occurrence data and estimation of the volume of different food types, both

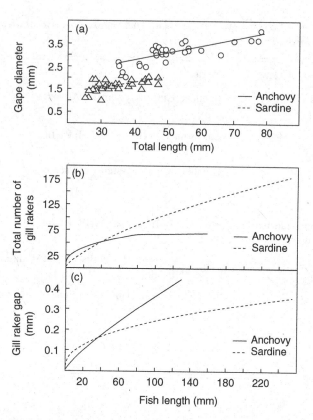

Fig. 7.1. Morphometrics of some of the feeding apparatus of anchovy and sardine from the Benguela Current system, showing relationships between (a) gape size and total length for juvenile anchovy and sardine from the southern Benguela (redrawn from Booi, 2000), (b) the total number of gill rakers and fish length, and (c) gill raker gap and fish length, for anchovy and sardine from the northern Benguela (redrawn from King and Macleod, 1976).

of which are likely to overestimate the contribution made by phytoplankton (James, 1987); the volumetric method particularly so because of the low carbon:volume ratio of phytoplankton compared to zooplankton (which is even lower for nitrogen; one unit volume of copepod has almost eight times as much nitrogen as that contained within one unit volume of phytoplankton; van der Lingen, 2002).

More recent dietary studies of sardine and anchovy in the southern Benguela ecosystem that assessed relative dietary contribution according to the carbon content of ingested prey have shown that zooplankton contributes a far greater amount to dietary carbon than does phytoplankton for both species, although phytoplankton can be an important dietary contributor in localized regions or at particular times of the year for both species (James, 1987; van der Lingen, 2002). Results obtained from measurements of the stable isotope ratios of carbon and nitrogen in plankton and anchovy, which integrate the relative contributions of isotopically distinct dietary components over a period of time as opposed to the "snapshot" provided by stomach content

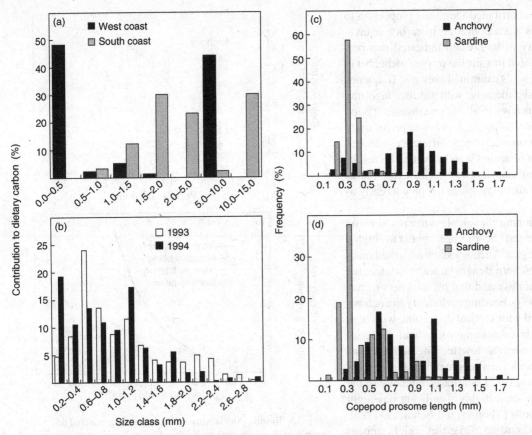

Fig. 7.2. Dietary composition of anchovy and sardine in the southern Benguela Current system, showing (a) the contribution to dietary carbon by prey size class for adult anchovy of 100–109 mm standard length on the South African west and south coasts (redrawn from James, 1987), (b) the contribution to dietary carbon by prey size class for adult sardine sampled off the South African southwest and south coasts in 1993 and 1994 (redrawn from van der Lingen, 2002), (c, d) frequency distributions of copepod prosome length in stomach contents of juvenile anchovy and sardine from presumed mixed shoals sampled off the South African west coast (each plot represents a separate shoal; redrawn from Louw *et al.*, 1998).

analysis, did not support the suggestion that adult anchovy are predominantly phytophagous (Sholto-Douglas *et al.*, 1991). However, whereas zooplankton is the dominant food source for both species, anchovy and sardine consume different fractions of the zooplankton and appear to partition this resource on the basis of food particle size. Anchovy derive the bulk of their carbon from larger (>1.5 mm; Fig. 7.2a) zooplankton, typically calanoid copepods and euphausiids captured through size-selective particulate feeding, although smaller zooplankton and phytoplankton captured via filter feeding can make a substantial contribution to dietary carbon, for example off the west coast (James, 1987). In contrast, sardine derive the bulk of their dietary carbon from smaller (<1.2 mm; Fig. 7.2b) zooplankton, typically calanoid and cyclopoid copepods and crustacean eggs and nauplii captured through non-selective filter feeding (van der Lingen, 2002).

Size-based partitioning of zooplankton also appears to apply to younger life history stages of the two species, which shoal together as juveniles in the southern Benguela ecosystem and aggregate inshore along the South African west

coast during austral winter (Mar–Aug). Studies on the diet of juveniles caught in the same midwater trawl (and hence from presumably mixed-species shoals or at least from a similar, if not the same, ambient food environment) found that, whereas both species consumed similar types of food (namely crustacean eggs and nauplii, and copepods, with very little phytoplankton being found), sardine ingested significantly smaller copepods than did anchovy (Fig. 7.2c, d; Louw *et al.*, 1998), this difference being attributed to the product of differences in gill raker morphology, feeding behavior, and within-shoal position of the two species. Similar work comparing the stomach contents of anchovy and sardine pre-recruits (i.e. post-larvae and juveniles of 25–75 mm total length) caught in the same Methot net haul (and hence from presumably mixed-species shoals) also indicated that sardine consumed smaller prey than did anchovy (Booi, 2000). Pre-recruits of both species fed on phytoplankton (principally the diatom *Coscinodiscus gigas* and the dinoflagellate *Peridinium* spp.) and zooplankton (copepod eggs and nauplii, and cyclopoid copepods), and although phytoplankton was numerically dominant zooplankton contributed the major proportion to the diet of both species when converted to carbon equivalents. Overlap in dietary niche breadth indicated that pre-recruits of both species feed in a similar trophic niche, but in eight of the ten mixed shoals examined prey size frequency distributions differed significantly, with sardine ingesting smaller prey than did anchovy in all but one instance (Booi, 2000). Despite the limited sampling and hence preliminary nature of the research examining trophic differences between early life history stages of anchovy and sardine, the results obtained to date suggest that the two species exhibit size-based partitioning of available zooplankton prey as early as the pre-recruit stage.

In addition to partitioning the zooplankton resource on the basis of size, sardine and anchovy also appear to employ different foraging strategies. Anchovy show a marked feeding periodicity in the southern Benguela, with peak feeding by recruits occurring at dusk and that by spawners occurring at dawn (James, 1987). Feeding periodicity in anchovy appears to be associated with vertical migration, with high feeding activity at night coinciding with shoal dispersal in the surface waters, whereas low feeding activity during the day coincides with shoal aggregation and descent into deeper water (James, 1987). Similarly, anchovy in the northern Benguela also show highly significant day/night differences in shoal depth (Thomas and Schülein, 1988). In contrast to anchovy, sardine >25 g wet weight appear to feed continuously and show no peaks in feeding activity throughout the diel cycle, although fish <25 g do show a peak in feeding activity at or around sunset (van der Lingen, 1998a). In the southern Benguela, vertical migration by sardine appears to be highly variable, with fish being observed throughout the water column during the day (in many cases being on or near the bottom) but forming a scattering layer close to the surface at night (Coetzee, 1997). Sardine in the northern Benguela show no significant difference in shoal depth between night and day, but tend to form small, scattered schools by night and dense schools by day, whilst generally remaining in the upper 20 m of the water column (Hampton *et al.*, 1979; Kruger and Cruikshank, 1982; Thomas and Schülein, 1988).

Laboratory experiments have shown that both anchovy and sardine filter-feed on small particles and particulate-feed on large particles (James and Findlay, 1989; van der Lingen, 1994) but have different threshold sizes at which they change from filter to particulate feeding (Fig. 7.3a). Filter feeding is the principal feeding mode of sardine, with food particles of <1.2 mm maximum dimension eliciting a filter feeding response whilst larger particles elicit particulate feeding at low concentrations but filter feeding at high concentrations (van der Lingen, 1994). Sardine are less efficient at retaining particles <0.4 mm but are able to entrap particles down to 0.01 mm in size, and also display

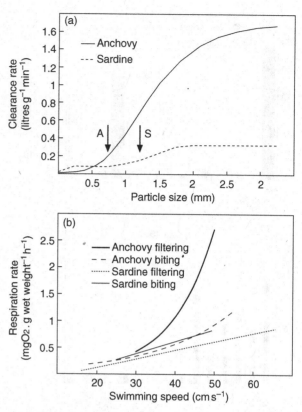

Fig. 7.3. Results from laboratory experiments conducted on southern Benguela anchovy and sardine, showing relationships between (a) weight-standardized clearance rate [l (g min)⁻¹] and prey size (mm) for each species (arrows labeled A and S indicate the approximate prey size at which the transition from filter feeding to particulate feeding occurs for anchovy and sardine, respectively; redrawn from van der Lingen, 1994; permission granted by Inter-Research); and (b) respiration rate [mg O_2 (g wet weight h)⁻¹] and swimming speed (cm s⁻¹) for each species engaged in filter feeding ("filtering") and particulate feeding ("biting"); redrawn from van der Lingen, 1995, with permission from Inter-Research.

size-selectivity during particulate feeding. In contrast to sardine, particulate feeding is the primary feeding mode for anchovy, which switch from filter to particulate feeding at a threshold prey size of 0.7 mm (James and Findlay, 1989). The minimum particle size that can be entrapped by anchovy during filtering is 0.20–0.25 mm, hence a large portion of phytoplankton is unavailable to anchovy. Anchovy are highly size-selective, selecting for the largest particle available. A comparison of predictive equations of weight-standardized clearance rate as a function of particle size for both species demonstrates that sardine are more efficient at removing particles <0.6 mm in size, anchovy having distinctly higher clearance rates on particles larger than this size (Fig. 7.3a; van der Lingen, 1994). Sardine are therefore more efficient removers of small particles, whilst anchovy remove large particles more effectively. The smaller gill raker gap and more elaborate denticle structure of adult sardine compared to adult anchovy are likely to be responsible for the higher retention of small prey such as phytoplankton by sardine compared to anchovy.

Energetic costs of feeding for both sardine and anchovy were determined through laboratory experiments that measured respiration rates during each of the feeding modes (James and Probyn, 1989; van der Lingen, 1995). For both species, respiration rate increased with swimming speed for both modes, although the shape of the functional relationship was linear for sardine but log-linear for anchovy. At a given swimming speed, filter feeding by sardine is energetically cheaper than particulate feeding, which is in contrast to anchovy where particulate feeding is the energetically cheapest feeding mode (Fig. 7.3b; van der Lingen, 1995). James and Probyn (1989) argued that the change in body shape and increased drag associated with flared operculae during filter feeding was responsible for the increased metabolic costs of filter feeding by anchovy. Whilst opercular flaring by sardine during filter feeding does occur and must increase drag and therefore energetic costs, the higher cost of particulate feeding in this species has been ascribed to the higher relative importance of inertial forces compared to viscous (= drag) forces for sardine relative to anchovy (van der Lingen, 1995). Additionally, the formation of a compact shoal during filter feeding by sardine in particular may confer a hydrodynamic advantage and hence reduce energy expenditure during filter feeding. The experimental procedures used in determining the energetic costs of feeding also allowed estimation of the respiratory quotient (RQ, an index of the type of physiological fuel involved in metabolism) for each species; anchovy had an RQ of 0.92 ± 0.18 (James and Probyn, 1989), whereas that for sardine was 0.96 ± 0.10 (van der Lingen, 1995). Although this difference is not statistically significant, it suggests that sardine is able to utilize carbohydrate to a greater degree than is anchovy, and implies a more herbivorous diet of the former compared to the latter (van der Lingen, 1995).

Experiments conducted to determine excretion and absorption efficiencies of anchovy and sardine, and assess the effect of food type on these, showed that both species excrete the majority of their nitrogen in the form of ammonia, both show greater absorption efficiencies for nitrogen than for carbon, and both absorb these elements more efficiently from zooplankton than from phytoplankton (James *et al.*, 1989b; van der Lingen, 1998b). Where the two species differ, however, is in their retention of nitrogen; sardine excrete more than half of their ingested and absorbed nitrogen rations (62 and 66%, respectively), whereas anchovy excrete less than half of their ingested and absorbed nitrogen rations (42 and 48%, respectively). Hence anchovy appear to be more efficient retainers of nitrogen than are sardine.

Carbon and nitrogen budget models that used the same basic growth equations and examined how growth varied under different food environments (phytoplankton, microzooplankton and mesozooplankton) as a function of foraging time and prey size and concentration have been developed for anchovy (James *et al.*, 1989a) and sardine (van der Lingen, 1999). Despite the fact that budget models were not identical in terms of their input parameters and hence cannot be directly compared, they differ notably in two respects. Firstly, the two species regulate their swimming speed during feeding (and hence their energetic output) according to different properties of the food environment. When feeding on zooplankton, sardine regulate their swimming speed according to prey concentration and swim faster at higher concentrations (van der Lingen, 1999), whilst anchovy regulate their swimming speed according to prey size, with larger zooplankton eliciting faster swimming speeds (James *et al.*, 1989a). The second major difference in the budget models derived for sardine and anchovy pertains to the effect of the food environment upon potential growth. Sardine showed positive growth over most of the input (particle size, prey concentration, and foraging time) ranges in each of the three food type-feeding behavior scenarios (filter feeding on phytoplankton, filter feeding on microzooplankton, and particulate feeding on mesozooplankton; van der Lingen, 1999), with the maximum scope for growth occurring for sardine filter feeding on dense concentrations of microzooplankton. Anchovy, on the other hand, only showed positive growth over a limited portion of the input range when filter feeding on either phytoplankton or microzooplankton, but showed positive growth over most of the input ranges, and maximum scope for growth, when particulate feeding on mesoplankton (James *et al.*, 1989a). Additionally, modeled maximum scope for growth values for anchovy were almost three times as high as sardine maximum values.

The studies described above have been summarized in van der Lingen *et al.* (2006a), and provide strong evidence that anchovy and sardine in the Benguela Current upwelling system are trophically distinct, and minimize their dietary overlap by resource partitioning according to prey size, this partitioning arising from the different morphological features of the branchial apparatus and different feeding behaviors and associated metabolic attributes (e.g. relative energetic costs of the different feeding modes) characteristic of sardine and anchovy in the Benguela. Van der Lingen *et al.* (2006a) considered sardine to have evolved a "steady-state" trophic strategy in which they feed primarily on that fraction of the zooplankton (small calanoid and cyclopoid copepods) characterized by a relatively high degree of population stability, and show positive growth over a wide range of prey types, sizes, and concentrations. However, the sardine's relatively slow metabolic rate and inefficient (compared to anchovy) retention of nitrogen suggests that the transfer of primary production into sardine tissue would occur at a moderate rate. In contrast, because anchovy feed primarily on zooplankton (large calanoids) that appear to show a greater degree of population variability, they were considered to have evolved a "feast-or-famine" trophic strategy – under suitable (i.e. a food environment dominated by large zooplankton) conditions they would show rapid growth as a result of their fast metabolic rate and efficient retention of ingested nitrogen, but under unsuitable conditions their growth would·be minimal or even negative. Those authors hypothesized that the trophic difference between anchovy and sardine implies that the size spectrum of the planktonic food environment will affect the feeding success of each species, with food environments dominated by small particles favoring sardine over anchovy since sardine are more efficient removers of small particles and can collect such food particles through employing relatively cheap filter feeding. Hence sardine should have a greater net energetic gain in small particle-dominated environments than anchovy, as was shown by the respective carbon and nitrogen budget models constructed for these species. In contrast, anchovy are inefficient removers of small particles, and filter feeding is the energetically most expensive feeding mode for this species. Food environments comprising mainly large particles will favor anchovy over sardine, due to the anchovy's greater efficiency at removing large particles through relatively cheap particulate feeding.

The California Current upwelling system

The trophic ecologies of Pacific sardine *Sardinops sagax* and northern anchovy *Engraulis mordax* are described in this case study, which encompasses research conducted on these small pelagic fish species from the Gulf of California and Baja California as well as recent studies from the northern California Current system as far north as British Columbia in Canada. Additional studies exist on the diets of other clupeoid fishes such as Pacific herring (*Clupea pallasi*) and American shad (*Alosa sapidissima*) but are described elsewhere (Brodeur *et al.*, 1987; Miller, 2006; Miller and Brodeur, 2007).

The first gill rakers of Pacific sardine have been observed on the lower limb of the first gill arch in the form of five small protusions in a larva of 13.4 mm SL (Villalobos and Rodriguez-Sanchez, 2002). Fairly long (0.5–1.2 mm) gill rakers occur up to a fish size of about 35 mm SL (Villalobos-Ortiz, 1998), this size being the same as that recorded by Matarese *et al.* (1989), based on external morphological characters, as the size at which the transformation from the larval to the juvenile stage occurs. Because of the relatively poor development of gill rakers until the fish is about this size, larval feeding is selective. Arthur (1956) concluded that the crustacean food of sardine larvae is procured by visual detection and active attack. Northern

anchovy larvae adopt an S-posture, approach their prey by pectoral fin sculling and finfold undulation, then open the mouth, straighten the body and engulf the prey (Hunter, 1972), and this is almost certainly true for *Sardinops* larvae as well as for many other clupeoids (Hunter, 1981; Blaxter and Hunter, 1982).

At 50 mm SL the gill rakers of Pacific sardine are longer (1.4–2.6 mm) and number around 42 on the lower part of the first gill arch (Villalobos-Ortiz, 1998), the denticles are just beginning to develop (Scofield, 1934), and the fish can perhaps filter-feed. At about 70 mm SL the gill rakers have reached their adult proportion and filter-feeding appears likely, and at 100 mm SL, the denticles are considered to be fully developed (Scofield, 1934). The pattern of increase in gill raker number through the ontogeny of Pacific sardine can be described using an asymptotic equation developed by Villalobos and Rodriguez-Sanchez (2002). Detailed information on ontogenic changes in gill raker morphology of northern anchovy are not available, but the number of gill rakers remains constant over the size range 80–140 mm SL (James and Chiappa-Carrara, 1990; Chiappa-Carrara and Gallardo-Cabello, 1993) and is substantially less than sardine of the same size range. In Pacific sardine, the gap between the gill rakers on the first gill arch increases with fish size, from about 64 µm at 35 mm SL to 250 µm at 290 mm SL (Villalobos-Ortiz, 1998), this increase also being recorded by CICIMAR (1983). Comparative analysis of Pacific sardine sampled from commercial catches at Magdalena Bay, Baja California Sur, has shown that the gap between the gill rakers on the first arch is relatively wide, being 127.6–132.0 µm (CICIMAR, 1983), whereas that for sardine sampled from the Gulf of California was 90.2–96.5 µm (Molina-Ocampo, 1993). Whereas some of the co-occurring small pelagic fish species in these areas have similar or narrower gill raker gaps than shown by Pacific sardine (round herring *Etrumeus teres* has a gill raker gap of 97.5–107.0 µm; thread herring *Opisthonema libertate* of 60.6–64.4 µm; and anchoveta *Cetengraulis mysticetus* of 59.5–61.3 µm; CICIMAR, 1983; Molina-Ocampo, 1993), these species lack the highly specialized gill raker denticles characteristic of sardine. The gill raker gap of northern anchovy also increases with fish size, from around 360 µm at 80 mm to 590 µm for fish of 145 mm SL (Chiappa-Carrara and Gallardo-Cabello, 1993), substantially wider than that of the other species.

The most comprehensive study of the feeding of larval sardine and anchovy is the thesis by Arthur (1956) who reported qualitative results from the examination of prey taxa of larvae sampled in 1951 and 1952. Both species fed only during the day, and eggs, nauplii, and juvenile stages of copepods composed almost all the identifiable food of sardine (Fig. 7.4a). The diet of anchovy larvae (Fig. 7.4b) was found to be very similar, except that this species is more euryphagous than that of sardine with about 40% of their diet (by number) consisting of noncrustacean food particles. Laboratory studies have shown that northern anchovy larvae are able to subsist for up to 20 days on a diet of the dinoflagellate *Gymnodinium splendens*, albeit at a depressed growth rate (Hunter, 1981). Copepod nauplii become increasingly important as anchovy larvae increase in length and compose the bulk of particulate food when all sizes of larvae are considered. The increase of prey size with larval growth is reported for both Pacific sardine and northern anchovy larvae (Fig. 7.4c). Arthur (1976) reported that, in one of his samples, most of the larger (>10 mm) specimens of both species were "literally crammed" with the pteropod *Limacina bulmunoides*, with up to 50 pteropods in the intestine (compared with the normal one or two food items in the gut).

Berner (1959) also reported that copepod eggs and nauplii were found to be the most important element in the diet of larvae of *Engraulis mordax* sampled in 1954, although he pointed out that the size intervals of food (minimum and maximum) ingested by anchovy larvae of various sizes was wider than previously suggested. Berner (1959) showed monthly maps of distribution of feeding and non-feeding anchovy larvae captured in 1954 between Point Conception and the southern end of the Baja California peninsula and observed that stations occupied by feeding anchovy larvae were concentrated mostly north of Bahia Magdalena, Punta Eugenia and the Southern California Bight. The first two areas have been defined as Biological Action Centers (BAC; Lluch-Belda *et al.*, 2000), where productivity is enhanced by topography when seasonal upwelling conditions are diminished or lacking. In theory, both processes develop favorable mixing conditions resulting in enhanced biological production. Thus, the Southern California Bight and BACs are areas of higher biological productivity than the rest of the California Current system and the occurrence of feeding larvae at these locations suggests that they take advantage of the favorable conditions encountered in these areas.

Lewis (1929) found a significant similarity between the species and numbers of diatoms and dinoflagellates in the digestive tract of adult Pacific sardine and those in surface plankton hauls, although crustaceans (particularly copepods) were at times also common in stomach contents. From this, Parr (1930) concluded that zooplankton was the principal food of sardine: "both diatoms and dinoflagellates being merely incidental." Radovich (1952) also showed that the bulk of the food was crustaceans. Sardine dietary data are less complete from the northern California Current as sardines were absent from this region for almost 50 years during their period of low abundance (see this volume, Chapters 5 and 9). In earlier samples from Vancouver Island, Hart and Wailes (1932) found sardine diet to be "mainly diatoms supplemented with copepods

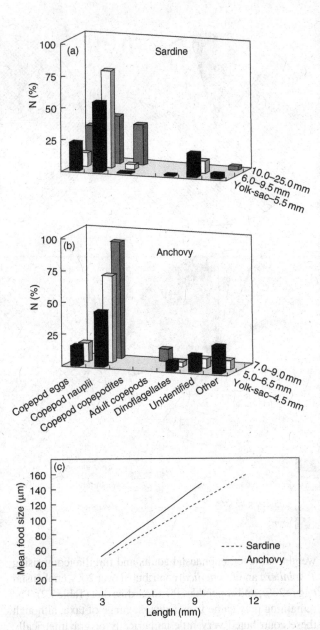

Fig. 7.4. Changes in dietary composition (percentage number) with fish size for (a) Pacific sardine and (b) northern anchovy larvae from the California Current system (drawn using data in Arthur, 1976; note that the length classes used by that author were based on size distributions in samples and not on any definite changes in larvae with respect to age), and (c) increase in average prey width (μm) with size for sardine and anchovy larvae (drawn using data in Arthur, 1976).

and other animals and plants". Sardines again became available in the northern California Current in the 1990s (Emmett *et al.*, 2005; McFarlane *et al.*, 2005), and copepods (both adults and copepodites) and euphausiid eggs were the dominant prey by number in this region, with all other taxa together contributing only about 20% to the diet of sardine off Oregon and Washington. In terms of wet

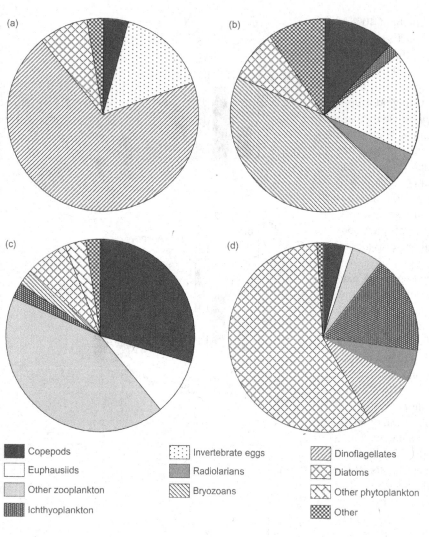

Fig. 7.5. Dietary composition (percentage number) of Pacific sardine off (a) southern California and (b) central Baja California (drawn using data in Radovich, 1952), and of northern anchovy from (c) northern Baja California to central California (drawn using data in Loukashkin, 1970) and (d) northern Baja California (drawn using data in Chiappa-Carrara and Gallardo-Cabello, 1993).

weight, however, euphausiid adults and furcilia (comprising *T. spinifera* and *E. pacifica*) contributed over 20%, euphausiid eggs ~20%, and copepods 30% to sardine diet (Table 7.2). The remaining prey came from a diverse range of taxa, although these contributed very little numerically or gravimetrically to the total diet. There was substantial spatial variability in sardine diet in the northern California Current, with fish off Washington feeding predominantly on phytoplankton (84% by weight) whereas those from Oregon, particularly those collected off the continental shelf, fed mainly on crustaceans (Emmett *et al.*, 2005). There was also substantial between-cruise variability in the composition and amount of food consumed; however, there were no differences in feeding intensity among different times of day. Off British Columbia, McFarlane *et al.* (2005) found that five prey groups (phytoplankton, euphausiids, euphusiid eggs, copepods, and larvaceans) made up the majority of prey consumed, but there was substantial interannual variability in their diets.

Fewer studies have been made of the food and feeding habits of *Sardinops* in Mexican waters. However, Radovich

(1952) recorded gut contents in 24 fish (80–184 mm SL) from four Pacific localities off central Baja California in late September and early October and concluded that sardine was omnivorous and probably a non-selective filter-feeder on principally diatoms and dinoflagellates (Fig. 7.5a, b), and that the diet should therefore closely match what is available. Ramírez-Granados (1957) also reported on sardine off the Pacific coast and confirmed a general opinion that juvenile sardine fed mainly on crustaceans, but became almost exclusively phytoplankton feeders (on diatoms) when larger (CICIMAR, 1983). Diatoms such as *Rhizosolenia, Bacteriastrum, Thalassiothrix, Melosira, Coscinodiscus, Amphora* and *Navicula* dominated ingested phytoplankton (almost 70% by number of total plant species), and adult and juvenile copepods the zooplankton component (nearly 90% of total animal groups); zooplankton predominated in fish of 84–100 mm SL in September, and in larger fish of 168–190 m SL in November and December. Hence, in addition to feeding on diatoms, sardine are also able feed on zooplankton, especially copepods and invertebrate eggs.

In the Gulf of California, a detailed study was made by López-Martínez (1991), based on stomach contents from 90 sardine (95–184 mm SL) caught in February 1990 just north of Isla Tiburon. She showed that zooplankton (especially calanoid copepods and larval brachyurans) was important in the smaller specimens (16% by volume), but that in the large size groups the diet became increasingly phytoplanktonic, with the diatom *Coscinodiscus* and dinoflagellate *Protoperidinium* predominating (zooplankton comprising only 2% volume); in general, sardine apparently ate what was available.

On an examination of stomachs of northern anchovy collected in northern Baja California, southern, and central California, Loukashkin (1970) suggested that the northern anchovy is an omnivorous species subsisting either on phytoplankton or zooplankton, or on both at the same time, feeding indiscriminately on planktonic organisms available within the area occupied. He found that zooplankters seem to be preferred in the anchovy diet (comprising 86.5% by number), of which the most frequently and abundantly dominant prey in the diets were crustaceans (50.8%), mainly copepods (29.3%) and euphausiids (9.5%) in all stages from eggs to adults, with the remainder of the diet consisting of other zooplankters (35.8%), unidentified material (2.5%), and phytoplankton (11.0%; mainly diatoms (7.2%) belonging to the genera *Chaetoceros*, *Coscinodiscus*, and *Thalassiosira*). However, despite the fact that, at times, phytoplankton contributed up to 100% of the stomach contents (mainly in anchovy sampled in May and June, the season of peak phytoplankton blooms) their role in anchovy diet on the whole seems negligible. Baxter (1967) suggested that anchovy are indiscriminate filter feeders, and observed them to be predatory on small fish at times, even their own kind. Loukashkin (1970) agreed that northern anchovy is primarily a filter feeder, feeding mostly during the day, but he pointed out that it may also be a particulate or selective feeder, depending on the size of the available food. Loukashkin (1970) also pointed out that the same dual tendency in feeding patterns was observed in Pacific sardine, and that other planktophagous pelagic fish occurring with anchovy schools or in the general vicinity feed on the same plankton available to the anchovy.

Working in the Southern California Bight, Koslow (1981) made direct field measurements of northern anchovy feeding by collecting plankton samples from the wake of feeding schools and comparing zooplankton species composition and size structure with control samples taken in front of or to the side of the school. Results indicated consistent, size-selective feeding on the dominant zooplankton taxa, which ranged from chaetognaths to small copepods such as *Clausocalanus*, *Ctenocalanus* and *Paracalanus*, other copepods, and larvaceans. Koslow (1981) observed a significant positive relationship between the proportion of a taxa consumed and its size, and considered northern anchovy to be an effective size-selective planktivore over a 1000-fold range in prey size. Preferential feeding on the basis of prey type was only observed for the large calanoid copepod *Calanus pacificus*.

Analysis of stomach contents of northern anchovy caught off the west coast of Baja California showed that, whilst diatoms, anchovy eggs, crustaceans, dinoflagellates and protozoans numerically dominated anchovy diet (Fig. 7.5c, d), zooplankton (principally anchovy eggs and crustaceans) provided the bulk of dietary carbon (Chiappa-Carrara and Gallardo-Cabello, 1993). Peak feeding occurred shortly after dawn, and whereas northern anchovy filter-feed on prey 0.05–1.5 mm and particulate-feed on larger prey, size-selective, particulate feeding on zooplankton was considered to be the dominant feeding mode (James and Chiappa-Carrara, 1990).

The numerically dominant prey taxa for anchovy in 1981 in the northern California Current were calanoid copepods and the pteropod, *Limacina helicina*, which together accounted for around 89% of the total number of prey consumed (Table 7.2). However, during the last few years, copepods were still important by number but the euphausiid, *Thysanoessa spinifera*, was the dominant prey by wet weight (~97%), with other taxa contributing less than 1% (Miller, 2006; Miller and Brodeur, 2007). There was no evidence of phytophagy in either time period but, since the sample size was relatively limited, we cannot say that anchovy in this region do not subsist partially on phytoplankton, but various types of microzooplankton and mesozooplankton appear to be the dominant prey overall (see also Brodeur *et al.*, 1987).

Like many other clupeoid fishes with rather fine and numerous gill rakers, northern anchovy and Pacific sardine are considered both filter feeders and particulate feeders, although the relative importance of these two feeding modes is still not clear. Laboratory studies by Leong and O'Connell (1969) showed that northern anchovy capture *Artemia* nauplii by filter feeding, but captured the much larger *Artemia* adults individually by direct biting (particulate feeding) when each food was presented separately, and that food biomass was accumulated more rapidly by biting than by filtering. The authors hypothesized that the anchovy could not sustain its daily food requirements by filtering alone off southern California, except in limited areas of high plankton concentration. Later, O'Connell (1972) analyzed the role of biting activity as response of the fish to mixed size assemblages of food organisms. He concluded that the ratio of biting to filtering activity in small schools varies with the relative concentration of *Artemia* adults and nauplii in the water, and suggested that the anchovy is a selective feeder and that food-organism size is a major determinant of feeding mode choice.

Table 7.2. *Summary of Pacific sardine (n=151) diets collected from the northern California Current system during June and August 2000 and 2002 GLOBEC surveys, and summary of northern anchovy diets collected from the northern California Current system during the August 1981 Oregon State University survey (n=18) and the June and August 2000 and 2002 GLOBEC surveys (n=45)*

Prey taxa	Sardine (2000 and 2002) %N	Sardine (2000 and 2002) %W	Anchovy (1981) %N	Anchovy (1981) %W	Anchovy (2000 and 2002) %N	Anchovy (2000 and 2002) %W
Phytoplankton	19.62	–				
Mollusca						
Pteropoda	<0.01	0.1				
Limacina helicina			35.52	18.56	0.03	<0.01
Cephalopoda			0.09	0.18		
Chaetognatha	<0.01	<0.01				
Arthropoda						
Crustacea						
Copepoda (eggs)	6.16	0.3				
(nauplii)	5.51	0.03				
(copepodites–adults)	66.91	30.4	53.84	5.43	46.83	1.59
Cirripedia (nauplii–cyprids)	0.04	<0.01				
(cyprids)					0.59	0.01
Cladocera	<0.01	<0.01				
Decapoda (larvae)	<0.01	0.01				
Cancer magister (megalopae)			5.96	2.75		
Caridea (zoea)					0.03	<0.01
Amphipoda						
Hyperiidea	0.02	0.06				
Themisto pacifica			0.28	0.27		
Euphausiidae (eggs)	11.61	0.16			1.66	<0.01
(furcilia–adults)	1.54	0.40				
Euphausia pacifica	0.34	19.59	1.07	7.42	0.15	0.64
Thysanoessa spinifera	<0.01	23.98	1.28	1.24	48.89	97.62
Unidentified			1.61	4.97		
Crustacea (material)	<0.01	<0.01				
Chordata						
Osteichthyes (eggs)	<0.01	0.1	0.43	0.39	0.30	0.04
(larvae–juveniles)	<0.01	0.01			0.03	0.09
Appendicularia	1.16	0.52				
Other						
Fish scales	<0.01	<0.01				
Invertebrate eggs	1.48	0.10			1.48	0.01
Gelatinous material	<0.01	<0.01				
Unidentified remains	4.63	5.21	–	58.61		

Anchovy and sardine showed some overlap in the prey items consumed in the northern California Current, although there were some major differences in the relative amounts of prey taken. Diet overlap (Percent Similarity Index; PSI) was only about 3% by number but was close to 25% by weight, due to the common use of *T. spinifera* by both species. Anchovy generally showed a higher overlap (mean PSI = 32.2%) with other nekton collected in the same sampling than did sardine (7.9%), indicating a greater potential for trophic interactions (Miller, 2006; Miller and Brodeur, 2007). Stable isotope analyses revealed differences in diets of the two species with anchovies having higher $\delta^{13}C$ values, and more dependent on recently upwelled production, than sardine. Anchovy also showed consistently higher $\delta^{15}N$ values than sardine when collected during the same sampling period, indicating that they fed at a slightly higher trophic level than sardine (Miller, 2006). Examination of the prey sizes showed further differentiation between the feeding of these species (Fig. 7.6). Although the northern anchovy examined were substantially smaller than the Pacific sardine (mean length = 149 mm vs. 237 mm), they actually consumed significantly

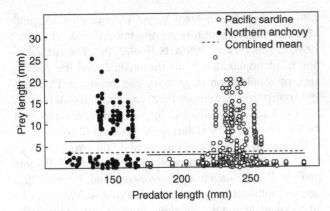

Fig. 7.6. Relationships between predator length and prey length for northern anchovy (closed circles) and Pacific sardine (open circles) from the northern California Current system. The dashed line indicates the overall mean and the solid lines indicate the mean for each species (from T. Miller and R. Brodeur, unpublished data).

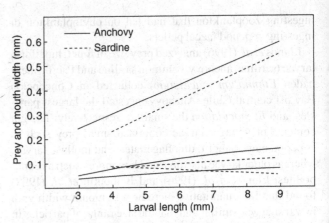

Fig. 7.7. Relationships between mouth width and larval length (upper lines), and prey width and larval length (lower lines show the maximum and minimum prey widths) for anchovy and sardine larvae from the Humboldt Current system (redrawn from Muck *et al.*, 1989).

larger (*t*-test on log-transformed data; $P < 0.001$) prey items (mean ± S.E. = 6.12 ± 0.29 mm) than did sardines (2.98 ± 0.09 mm), despite a similar range of prey sizes taken.

The Humboldt Current upwelling system

This case study reviews the trophic dynamics of small pelagic fish in the Humboldt Current system with the emphasis on anchovy *Engraulis ringens* and sardine *Sardinops sagax*, and notes on some trophic aspects for common sardine (*Strangomera bentincki*) off Chile.

In the first published work on feeding of larval anchovy off Peru, Rojas de Mendiola (1974) showed that feeding was mainly diurnal and that the main prey items were copepod eggs and nauplii, centric diatoms, and dinoflagellates, with selection by size appearing to occur. As larvae increased in size they showed an increase in the zooplankton fraction in the total diet, and by 10 mm larvae were feeding exclusively on zooplankton. Rojas de Mendiola and Gomez (1981), and Ware *et al.* (1981), conducted *ex situ* experiments that showed that the prolarval stage lasted 2.75 days at 18 °C and that feeding started 40 h after the disappearance of the yolk, when the eyes were completely pigmented and the mouth open. The minimum concentration of food necessary to initiate first feeding was about 80 particles ml^{-1} for prey of 14–20 µm, and larvae did not feed on chain-forming phytoplankton but on round diatoms and dinoflagellates. The percentage of larvae feeding and gut fullness increased with age, and ingested prey width was also proportional to larval size and averaged 1%–2% of body length.

In a comparative study on feeding by larval anchovy and sardine off Peru, Muck *et al.* (1989) showed that these species presented significant differences in the relationships between body length and mouth width, and between body length and gut length, with anchovy having a larger mouth than a sardine of the same size (Fig. 7.7), but with a

shorter gut. In addition, the relationship between the proportion of larvae with mouth and eye functionality and SL showed that first feeding in anchovy starts at around 2.7 mm and is completed at a length of ca. 4 mm, whereas sardine start to first feed at ca. 4 mm and are all feeding by 5 mm. The minimum size of anchovy with food in the gut was 3.02 mm and for sardine was 4.07 mm. Additionally, the species and size composition of prey in the guts of first feeding larvae (anchovy of 3–3.5 mm SL and sardine of 4–4.5 mm SL) were significantly different, with anchovy feeding almost exclusively on a pure phytoplankton diet (comprising 77% by number of phytoflagellates and dinoflagellates, and with a mean width of 29.6 ± 25.3 µm) and sardine feeding almost exclusively on zooplankton (copepod eggs and nauplii, with a mean width of 72.1 µm; Muck *et al.*, 1989). For older anchovy and sardine larvae (>5.5 mm) there was a wide overlap in diet for all stages of calanoid copepods, and phytoplankton was a less important prey item.

Valenzuela *et al.* (1995) studied the dietary composition and prey size of sardine and anchovy larvae off central Chile (32° S–33° S) and found that the diet of both species was composed mostly of copepod eggs (frequency of occurrence > 60%) but also some phytoplankton, with the diatoms *Skeletonema* and *Chaetoceros* contributing 89% by number of ingested phytoplankton cells. Anchovy did not show a significant relationship between prey width and mouth width whereas sardine did, indicating that the latter tends to feed on larger prey items with increasing mouth width and also with increasing body length. Anchovy ingested small particles independently of body size, with around 70% of the diet composed by prey having a width <10% of mouth width. In sardine, almost 60% of the diet consisted of prey of a width 20%–30% of the mouth width. According to Valenzuela *et al.* (1995), the presence of phytoplankton in gut contents could be explained by fish ingesting zooplankton that had fed on phytoplankton or ingesting copepod faecal pellets.

Llanos *et al.* (1996) analyzed prey size of four Clupeiform larvae (sardine, anchovy, common sardine and Pacific menhaden *Ethmidium maculatum*) collected in Concepción Bay off central Chile. Anchovy ingested the largest particles, and *E. maculatum* the smallest, respectively. The gut contents of younger larvae contained small prey such as copepod eggs, nauplii, dinoflagellates, and mollusc larvae, whereas older larvae also ingested larger prey such as copepodites. Llanos *et al.* (1996) and Balbontín *et al.* (1997) found that the quantitative increment in mouth width with larval size was similar and that the size range of particles in the diet overlapped widely in all four species, indicating a similar use of food resources.

According to Rojas de Mendiola (1989), the first reports on the food and feeding habits of adult anchovy off the Peruvian coast were published by Vogt (1940) and Sears (1941). However the first dedicated study of anchovy diet was performed by Rojas (1953), who observed a clear numerical dominance of diatoms in the diet with the most important species being *Coscinodiscus centralis*, *C. perforatus*, *Gyrosigma* spp. and *Thalassiothrix frauenfeldii*. Rojas de Mendiola (1989) compiled a database based on the frequency of occurrence of prey in anchovy stomach contents and synthesized information from studies conducted over the period 1953–1974. Key points that emerged were that the food and feeding habits of anchovy change with length, with larvae ingesting round diatoms, postlarvae feeding on zooplankton (primarily copepod eggs and nauplii), juveniles being largely zooplanktivorous and adults being mainly phytoplanktivorous. Pauly *et al.* (1989) summarized the database on anchovy dietary information from Alamo (1989) and Rojas de Mendiola (1989) for the period 1953–1982, and listed a total of 259 different food items that had been identified in anchovy stomach contents. Using results based on numerical abundance, Pauly *et al.* (1989) suggested that in the north-central part of Peru anchovy rely less on zooplankton than in southern Peru; that the mean zooplankton fraction of anchovy diet increased with distance offshore; and that more zooplankton was consumed when sea surface temperature was high.

Research conducted during the 1990s, also based on numerical abundance, showed that chain-forming (*Skeletonema*, *Chaetoceros* and *Thalassiosira*), group-forming (*Asterionellopsis*), and solitary (*Coscinodiscus*), diatoms, dinoflagellates (*Protoperidinium*), and copepods (*Centropages*, *Corycaeus*, and *Oncaea*), were important contributors to anchovy diet, with anchovy eggs making a less important contribution (Alamo *et al.*, 1996a, b; Alamo *et al.*, 1997a, b; Alamo and Espinoza, 1998; Espinoza *et al.*, 1998a, b, 1999, 2000). However, during the 1997/98 El Niño, the main prey items were copepods and euphausiids

with amphipods, anchovy eggs, *Vinciguerria* sp., and Myctophidae being of minor importance (Blaskovic *et al.*, 1998; Espinoza *et al.*, 1998a, b, 1999, 2000). The consumption of phytoplankton items diminished, and the dominance of zooplankton in anchovy diet persisted, from fall 1997 until the end of winter 1998, covering the shift from El Niño to La Niña conditions. Similar results were reported by Sánchez *et al.* (1985) during the 1982–83 El Niño.

During the main El Niño period, the trophic spectrum of sardine was represented by 23 genera of diatoms (mainly *Pseudo-nitzchia*, *Chaetoceros*, and *Asterionella*), seven dinoflagellate genera (mainly *Protoperidinium*), one silicoflagellate genus, copepods, euphausiids, amphipods, decapod zoea and megalope larvae, molluscs, and anchovy eggs (Blaskovic *et al.*, 1998; Espinoza *et al.*, 1998a). The numerical proportion of phytoplankton items was higher in sardine stomach contents than seen in anchovy (Fig. 7.8). Prey common to both fish species included the copepods *Calanus*, *Clausocalanus*, and *Corycaeus*, and the diatoms *Chaetoceros*, *Coscinodiscus*, and *Rhizosolenia*.

The large majority of published work describing the diet of anchovy and sardine off Peru considered phytoplankton to be as or more important than zooplankton. The first work that questioned this paradigm was by Konchina (1991), who showed that assessing the diet with qualitative, numerical ratios or frequencies of phyto- to zooplankton produced

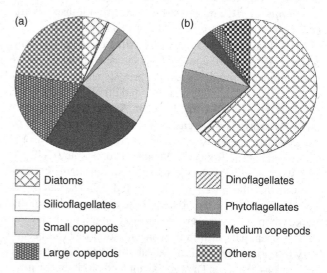

Fig. 7.8. Dietary composition (percentage number) of (a) anchovy (*n* = 86) and (b) sardine (*n* = 56) collected off Peru during 1998 (drawn using data in Blaskovic *et al.*, 1998); small copepods were comprised of *Acartia*, *Calocalanus*, *Clausocalanus*, *Clymnestra*, *Corycaeus*, *Euterpina*, *Oithona*, *Oncaea*, *Paracalanus*, Harpactacoida, copepodites and copepod remains; medium copepods were comprised of *Centropages*, *Lucicutia*, *Macrosetella*, *Mecynocera* and *Microsetella*; and large copepods were comprised of *Calanus*, *Candacia*, *Euaetideus*, *Eucalanus*, *Euchaeta*, *Euchirella*, *Pontellina* and *Rhincalanus*.

Table 7.3. *Estimates of ingestion rate, gastric evacuation rate, and daily ration of Peruvian anchovy and sardine*

Season	Year	Ingestion rate (g h^{-1})	Evacuation rate (h^{-1})	Daily ration (g d^{-1})	Feeding period	Source
Anchovy						
Pooled data	1954–1982	0.034	0.1518	0.4480	11:55–23:45	Pauly *et al.* (1989)
	1996–2003	–	–	–	07:00–18:00	Espinoza and Bertrand (2008)
Summer	1997	0.0479	0.0902	0.4391	10:09–19:20	Alamo *et al.* (1997b)
Winter–spring	1997	0.0453	0.0811	0.4428	09:23–19:00	Alamo and Espinoza (1998)
Fall	1998	0.0157	0.1646	0.2657	09:58–02:48	Espinoza *et al.* (1998a)
End of winter	1998	0.0192	0.1332	0.3252	11:00–04:00	Espinoza *et al.* (1998b)
Spring–summer	1999	–	–	0.445	09:00–18:50	Espinoza *et al.* (2000)
Sardine						
Fall	1998	0.1165	0.2345	0.3243	12:00–24:00	Espinoza *et al.* (1998a)

erroneous results. When prey weight was considered, the zooplankton fraction comprised 90% of anchovy stomach content mass and 50% for sardine, leading Konchina (1991) to conclude that sardine is situated closer to the base of the food web than anchovy. Additionally, anchovy is able to eat macroplankton including migrating mesopelagic fish 5 cm long, and this ability to seek large, heavy, and high calorie food items over the continental slope provides this species with the opportunity to reach high levels of prey biomass, far surpassing that available to sardine.

Konchina's (1991) conclusion regarding the trophic position of anchovy is supported by a recent extensive study performed by Espinoza and Bertrand (2008) who studied more than 21 000 anchovy stomachs sampled during 23 acoustic surveys conducted over the period 1996–2003. Numerically, phytoplankton dominated anchovy diet and comprised 99.52% of prey items, whilst copepods only accounted for 0.07% by number. However, this view dramatically changes when the carbon content of prey items is considered; zooplankton was by far the most important dietary component (comprising 98.0% of dietary carbon), with a strong dominance of euphausiids (67.5%) followed by copepods (26.3%) (Espinoza and Bertrand, 2008). Previous studies that concluded that the diet of Peruvian anchovy was mainly based on phytoplankton, or that phytoplankton and zooplankton played a similar role, were based on qualitative descriptions of anchovy diet such as frequency of occurrence and numerical percentage, rather than carbon content. Espinoza and Bertrand (2008) also showed that stomach fullness varied with latitude, with higher values encountered off northern (<6° S) and southern Peru (>13° S), and that stomach fullness also increased with distance from the coast and often reached maximum values at stations more than 120 km from the coast. However, these mean patterns were not a good representation of reality at

any one moment, and those authors emphasized the marked variability in stomach content fullness data between surveys, illustrating anchovy's plasticity: "inside its range of overall viable conditions anchovy is therefore able to forage efficiently at any time, any place, and any temperature" (Espinoza and Bertrand, 2008).

Estimates of ingestion rate, gastric evacuation rate, and daily ration of Peruvian anchovy have been estimated through analysis of diel dynamics in stomach contents, and are summarized in Table 7.3. Pauly *et al.* (1989) reported that feeding starts near noon and is continuous until near midnight, and estimated a daily ration of 0.45 g d^{-1} (mixed phyto- and zooplankton) for anchovy with a mean live weight of 20.4 g. During the El Niño of 1997–98, estimated daily ration showed a declining trend from March 1997 until end of winter 1998 (Table 7.3), but by November–December 1999 it was similar to that computed by Pauly *et al.* (1989), suggesting the return to pre-El Niño values. All daily ration estimates assume anchovy to have a marked diurnal feeding cycle. Espinoza and Bertrand (2008) studied variability in stomach fullness with respect to the time of the day using generalized additive models (GAMs), and their results confirmed that the main feeding activity occurred during daylight hours, between 07h00 and 18h00 (Fig. 7.9). Whilst this general pattern was observed for most surveys, nighttime feeding was important in some cases.

Off Chile, the first descriptive studies on the feeding of *E. ringens* were carried out by Mann (1954) and De Buen (1958). Trophic studies including analysis of the trophic spectrum and seasonal variations in the diet of common sardine (*Strangomera bentinckii*) and anchovy from the central-south area of Chile were carried out by Arrizaga and Inostroza (1979), Arrizaga (1983), and Arrizaga *et al.* (1993). Those authors concluded that the two species exhibited the same feeding strategy and used the same

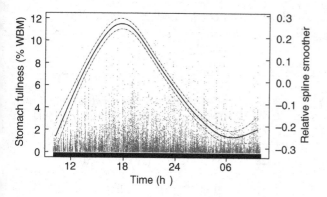

Fig. 7.9. Scatter plot (gray dots) and cubic spline smoother fit (black solid line) of GAM models based on anchovy stomach fullness according to time of day for the complete set of 21 203 stomach contents collected off Peru over the period 1996–2003. The black dotted lines show the 95% confidence limits of GAM models. The right *y*-axis is a relative scale, and shows the spline smoother that was fitted to the data, so that a *y*-value of zero is the mean effect of the variable on the response. (From Espinoza and Bertrand, 2008.)

prey resources, both species feeding almost exclusively on a pure phytoplankton diet, particularly on diatoms of the genus *Skeletonema*. Zooplankton was scarcely represented, accounting for <1% by number and consisting mainly of copepod remains. Nevertheless, a recent study carried out by Rebolledo and Cubillos (2003) points out that, on the basis of frequency of occurrence data, phytoplankton and zooplankton are of comparable importance in the diet of both species. Trophic studies conducted off Chile indicate that the diatoms *Skeletonema*, *Chaetoceros*, *Nitzschia*, and *Coscinodiscus* are the most important phytoplankton prey for both common sardine and anchovy, and both species appear to show a preference for *Skeletonema*, although this may be an artifact of the tendency of this diatom to form chains and its high level of abundance. Of the zooplankton, copepods and other crustaceans showed the highest frequency of occurrence for both species, followed by invertebrate eggs in the case of common sardine and cirripede larvae in the case of anchovy (Rebolledo and Cubillos, 2003). Both species exhibit a high degree of trophic overlap for phyto- and zooplankton prey, and hence are assumed to be positioned at the same trophic level (Arrizaga *et al.*, 1993). Finally, using stable isotope analysis Hückstädt *et al.* (2007) estimated the trophic level of anchovy to be 3.63, indicating a strong preference for zooplankton and supporting the results obtained by Espinoza and Bertrand (2008) for anchovy in the northern Humboldt.

Balbontín *et al.* (1979) carried out laboratory experiments by supplying pieces of fish liver in order to determine the preference for prey size for *E. ringens* and *S. bentincki*, and showed that a particle having a volume <6 mm³ is preferred

by anchovy, while common sardine preferred particles <4 mm³. Filter feeding was the most usual feeding mode in common sardine whereas anchovy tended to particulate-feed (bite) on food particles. Those authors hypothesized that the larger particles ingested by anchovy could be a comparative advantage in terms of energy intake compared to common sardine.

The Northwestern Pacific (Kuroshio Current) system

This case study reviews the trophic ecologies of Japanese sardine *Sardinops melanostictus* and Japanese anchovy *Engraulis japonicus*, two dominant and alternating small pelagic fish distributed in the northwestern Pacific, including the warm Kuroshio Current system. Some studies on the diets of other small pelagic fish from this region have been reported, including Pacific round herring (*Etrumeus teres*) and Pacific herring (*Clupea pallasii*) (see Nakai *et al.*, 1955; Yokota *et al.*, 1961; Hirota *et al.*, 2003; and Tanaka *et al.*, 2006 for round herring; and Irie *et al.*, 1979 for herring), but are not described here.

Japanese sardine has been recognized as planktivorous since the early study by Kishinoue (1908), although the size and type of plankton that comprises the food of this species changes with development. Copepod eggs and nauplii are the dominant food items during the post-larval stage (<10 mm total length [*TL*]; Yokota *et al.*, 1961; Nakai *et al.*, 1962; Kidachi, 1968; Nakata, 1988; Nakata, 1995; Hirakawa and Goto, 1996). Yokota *et al.* (1961) reported that copepod nauplii of 150–200 μm in length are fed on by larvae of several pelagic fishes including Japanese sardine, and Nakai *et al.* (1962) revealed that the size of nauplii ingested is related to fish size; sardine <5.8 mm *TL* fed mainly on nauplii 60–80 μm wide whereas sardine of 11–14 mm *TL* fed on nauplii of 200–300 μm wide. Hirakawa and Goto (1996) identified ingested nauplii to genus, finding that nauplii of *Oithona* spp. and *Paracalanus* spp. dominated the alimentary tract of sardine larvae of 4.1–8.0 mm SL in Toyama Bay in the southern Japan Sea. Nakata (1995) also noted the importance of nauplii of *Paracalanus* spp. and Cyclopoida (including *Oithona* spp.) for larvae of 5–7 mm *TL* in the region of the Kuroshio Current.

Juvenile sardine tend to feed on larger prey items such as copepodite or adult stages of copepods, other crustacean zooplankton, appendiculata, and planktonic larvae of various animals (Yamashita, 1955; Yokota *et al.*, 1961; Nakai, 1962; Watanabe and Saito, 1998; Tanaka, 2006). Copepods are generally dominant, with calanoid (e.g. *Paracalanus*, *Calanus*) and poecilostomatoid (e.g. *Oncaea*, *Corycaeus*, and *Microsetella*) copepods frequently found in stomach contents. Phytoplankton such as diatoms are ingested by sardine from a size of approximately 40–50 mm body length, arising from the development of the gill rakers (Yamashita, 1955; Nakai, 1962), although phytoplankton is more often found in the stomachs of larger fish.

Sub-adult and adult sardine feed omnivorously on plankton, and whilst copepods are important prey items, phytoplankton such as diatoms have often been reported as "dominating" sardine stomach contents (Kishinoue, 1908; Nakai, 1938; Nakai *et al.*, 1955; Yamashita, 1955; Li *et al.*, 1992), sometimes by volume (Kawasaki and Kumagai, 1984) or weight (Hiramoto, 1985). Genera of diatoms frequently found in sardine stomachs include *Chaetoceros, Skeletonema, Rhizosolenia, Coscinodiscus, Thalassiosira, Thalassionema, Nitzschia*, and *Eucampia* that are captured via filter feeding (Azuma, 1994). Differences in the specific composition of stomach contents (e.g. dominance by copepods or diatoms) depend upon locality and season (Nakai, 1962) and sardine migration patterns (Hiramoto, 1985). However, Kishinoue (1908) and Nakai (1938) both noted that fish with a high degree of stomach fullness always had zooplankton-filled stomachs, whereas those having a low degree of fullness had stomach contents dominated by phytoplankton. In some cases, adult sardine feed on fish eggs and larvae; Kidachi (1968) noted that the eggs of Japanese anchovy are sometimes found to dominate sardine stomach contents, and Azuma (1994) reported egg cannibalism by sardine during their spawning season. Feeding usually occurs in the daytime for all sardine life history stages from larvae to adult (Yoneda and Yoshida, 1955; Nakai *et al.*, 1962).

The feeding habits and trophic level of Japanese anchovy are generally considered to be similar to those of Japanese sardine. Anchovy post-larvae of <10 mm *TL* feed mainly on copepod eggs and nauplii (e.g. Nakai *et al.*, 1962; Kuwahara and Suzuki, 1984; Hirakawa and Ogawa, 1996; Hirakawa *et al.*, 1997), and both size and the taxonomic groups of copepods ingested (e.g. *Paracalanus* spp. and *Oithona* spp.) are similar to those fed on by sardine. Copepodite and adult stages of copepods are important in the diet of anchovy of 10–40 mm *TL*, among them *Paracalanus* spp. and *Oithona* spp. that were frequently found in anchovy from Pacific coastal waters (Uotani *et al.*, 1978; Uotani, 1985; Mitani, 1988). Poecilostomatoid copepods (e.g. *Oncaea* spp. and *Corycaeus* spp. of approximately 200–400 μm in width) often dominate stomach contents of juvenile and mature anchovy of >40 mm *TL* (Takasuka, 2003; Tanaka, 2006; Tanaka *et al.*, 2006). Diatoms have also been found in stomach contents of juvenile and mature anchovy (Nishikawa, 1901; Nakai *et al.*, 1955; Yamashita, 1957; Shen, 1969; Li *et al.*, 1992), following gill raker development and therefore the ability to filter-feed. Large organisms such as euphausiids, larvae and eggs of fish and other animals are also ingested by anchovy, and the larvae of benthic animals

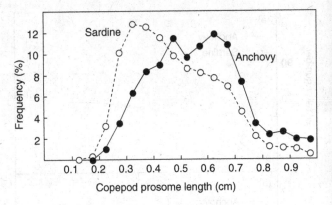

Fig. 7.10. Frequency distributions of copepod prosome length in stomachs of Japanese anchovy and sardine from the northwestern Pacific (redrawn from Li *et al.*, 1992).

such as bivalves and decapods are found in anchovy stomachs from fish collected from inshore habitats compared to those collected from offshore habitats (Tanaka, 2006). Cannibalism is sometimes found in sub-adult and mature fish (Takasuka *et al.*, 2004). In general, however, copepods of small size are considered to be the dominant dietary component of Japanese anchovy juveniles and adults. Feeding by anchovy occurs usually in the daytime (Nakai *et al.*, 1962; Uotani, 1985).

Whereas the feeding habits of Japanese sardine and anchovy are generally similar for all developmental stages, comparative studies suggest some notable differences. Nakai *et al.* (1955) reported that although sardine and anchovy juveniles and adults both feed mainly on copepods and diatoms, the frequency of occurrence of phytoplankton is higher in sardine stomach contents than in anchovy stomach contents. Li *et al.* (1992) compared the prosome length of copepods ingested by the two species and showed that the mode for sardine occurs at a smaller size than for anchovy (Fig. 7.10). These differences may be due to differences in gill raker structure and hence differing retention abilities, since sardine have a higher number of gill rakers, and a smaller gap between gill rakers, than do anchovy of the same size (Fig. 7.11; Nakai, 1938; Tanaka, 2006). However, Li *et al.* (1992) concluded that competition for food might play an important role in anchovy and sardine dynamics because of the overlap in dietary spectrum between the two species. Recently, Tanaka (2006) compared stomach contents of the two species collected at the same time and found that in the case of juveniles, stomach contents were very similar for anchovy and sardine of a similar size. However, significant differences were observed in the diets of sub-adults of similar size, with anchovy stomach contents being mainly composed of small copepods whereas sardine stomachs were filled with thousands of euphausiid eggs. Interestingly, the width of the euphausiid eggs fed on

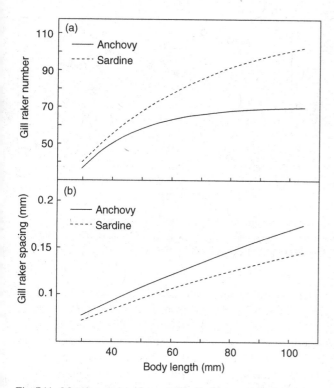

Fig. 7.11. Morphometrics of some of the feeding apparatus of anchovy and sardine from the northwestern Pacific, showing relationships between (a) the total number of gill rakers and fish length, and (b) gill raker gap and fish body length (redrawn from Tanaka, 2006).

by sardine was similar to, or even larger than, that of the copepods fed on by anchovy. This phenomenon may be explained by a differential ability to retain euphausiid eggs (e.g. due to differences between anchovy and sardine of the mouth or other feeding structures such as the epibranchial gland) rather than by differences in gill raker spacing. The fact that sardine shows a feeding behavior known as "gulping", which is intermediate behavior between particulate- and filter-feeding and during which fish ingest a few particles in one mouth-opening act (Azuma, 1994), may possibly also be involved. In summary, substantial dietary overlap between anchovy and sardine occurs from the larval to juvenile stages, which implies potential competition between the two species at these life history stages. In the sub-adult and adult stages, however, sardine shows a higher degree of herbivory, and consumes smaller zooplankton, than does anchovy. Additionally, sardine also appear to be better able to retain what may be called "elusive" prey organisms (such as euphausiid eggs) that have a smooth membrane and are not efficiently retained by anchovy, most likely due to differences in morphology (of the gill raker but possibly also other feeding structures such as the jaw) and in feeding behavior.

The Northeastern Atlantic and Mediterranean systems

This case study is an updated review of published and unpublished information on feeding habits and trophic ecology of sardine (*Sardina pilchardus*) and anchovy (*Engraulis encrasicolus*) in waters around the Iberian Peninsula, including the Atlantic and Mediterranean shelves. The southern and particularly the western Iberian Peninsula are at the northern limit of the Canary Current upwelling system and hence are not temperate non-upwelling systems whilst the Bay of Biscay and the Mediterranean Sea are, but research on small pelagic fish trophic dynamics conducted off the western and southern Iberian peninsula has been included in this case study because of the geographical proximity of the three regions. The original information was provided by studies of morphology (Andreu, 1953, 1960, 1969), abundance (ICES, 2005) and trophic ecology (see Tables 7.4 and 7.5) of the selected species. Trophic studies included either analysis of stomach contents (generally by prey abundance but also by biovolume; most recently by estimation of prey carbon content) or stable isotope composition (carbon and nitrogen isotopes in muscle).

Studies on the morphology of the feeding apparatus in the Iberian region are available mostly for sardine, although some measurements were also made on anchovy (Andreu, 1953, 1960, 1969). The larvae of both species lack the mechanism for concentrating planktonic particles and must capture individual prey. The filtering apparatus of juvenile and adult *S. pilchardus* was described in detail by Andreu (1969), with some data for *E. encrasicolus* for comparison. The main components of this apparatus are the five pairs of branchial arches, each supporting one series of gill rakers covered with denticles on the anterior side and branchiae on the posterior (the latter are only found on the first four branchial arches). The gill rakers develop along the supporting basihyal bones and those of the first arch over the sides of the tongue, thus forming an efficient filter for food particles. Gill rakers appear when sardines reach 15 mm in length and increase in number and size with growth (Fig. 7.12a). In contrast, the number of gill rakers does not vary significantly with body length in anchovy. The separation between gill rakers, as well as the number and separation between denticles, increases as a power or logarithmic function of size (Fig. 7.12b). The power coefficients are <1, as the rate of increase in both number and separation of gill rakers and denticles decreases with size. Considering the sardines sampled at Vigo (Galicia, NW Spain), the morphology of the gill rakers changes little after they reach 15–20 cm in length, suggesting that the filtering apparatus is fully developed in sardines after they reach the first year of life (ICES, 2005). The mesh formed by gill rakers and denticles is about 40 μm on average, although sardines can capture particles smaller than this size (Garrido et al., 2007a). Gill rakers are also present in anchovy, but

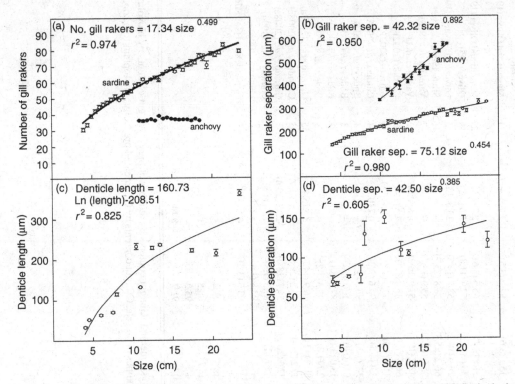

Fig. 7.12. Morphometrics of some of the feeding apparatus of anchovy and sardine from Vigo, NW Spain, in the northeastern Atlantic, showing relationships between (a) mean number of gill rakers and fish length for both species, (b) gill raker separation and fish length for both species, (c) denticle length and fish length for sardine only, and (d) denticle separation and fish length for sardine only. (Redrawn using data in Andreu, 1953, 1960, 1969). Measurements were taken from post-larval and adult individuals and regressed against size (total length in 0.5 cm size classes), and standard errors were computed using individual measurements from 4 to >50 individuals per size class.

are present in lower numbers and are more widely separated than in sardine (Fig. 7.12a, b). In addition, the gill raker gap increases at a faster rate with fish size in anchovy than in sardine (Fig. 7.12b), indicating that anchovy is less suited to filter feeding than is sardine. Other co-occurring planktivorous species such as *Trachurus trachurus* and *Sprattus sprattus* display gill raker gaps similar to those of anchovy and clearly larger than those of sardine (Andreu, 1969).

The morphological studies conducted by Andreu (1969) also detected substantial differences in the number of gill rakers and the separation between gill rakers of sardine from different regions; those from the Mediterranean Sea have fewer gill rakers that are more widely separated than sardine from Atlantic waters. This was explained as an adaptation to the higher plankton abundance in the Atlantic, which would benefit filter feeding, while Mediterranean sardines would preferentially capture individual prey (Andreu, 1969).

The variability of prey items found in the stomachs of *S. pilchardus* has puzzled scientists for a long time; earlier reports indicated the presence of items as varied as mud (Lebour, 1920, 1921) or even pollen from terrestrial plants (Oliver, 1951; Oliver and Navarro, 1952), but mostly cited the numerical dominance of phytoplankton as preferred prey. The composition of diet varied largely among the first quantitative studies, ranging from reports of almost purely herbivorous

to zoophagous diets (Table 7.4). As a result, *S. pilchardus* became considered as an opportunistic feeder, the composition of stomach contents reflecting prey availability in the water. In this way, phytoplankton cells were less abundant in the stomachs of sardines from the Mediterranean compared to those from the Atlantic (see review in Andreu, 1969).

When estimates of sardine diet composition were made taking account of prey biovolume or weight, the dominance of zooplankton became apparent. The same occurred when the carbon content of the prey was estimated (Garrido *et al.*, 2008a). The dominance of zooplankton biomass in sardine stomach contents (on average ca. 70%) is reflected by the stable isotope composition of muscle, which represents the diet integrated over a longer period than can be observed when analyzing stomach contents. Isotopic studies suggested that most of the nitrogen-forming structural muscle proteins in adult sardines originated from the assimilation of zooplankton, while only a small portion of carbon could be derived from phytoplankton (Bode *et al.*, 2004). Copepods, generally the group dominating total zooplankton biomass such as *Acartia clausi*, *Temora* spp. and other calanoids, or cyclopoids such as *Oithona nana* and *Oncaea* spp. (Massuti and Oliver, 1948; Varela *et al.*, 1990; Garrido, 2002; Cunha *et al.*, 2005; Garrido *et al.*, 2008a), were also dominant in sardine stomach contents. In

Table 7.4. *Summary of studies on the quantitative composition of* S. pilchardus *diet in Iberian waters indicating average percent composition of prey items*

Year	Season	Zone	Stage (size; cm)	n1	Anal.	% phy. x (S.E.)	% zoo. x (S.E.)	% other x (S.E.)	n2	Source
1944	Su	Mar Cantábrico	Adults (13.0–17.5)	6	Gn	75.0 (11.2)	10.0 (2.6)	–	6	Navarro and Navaz (1946)
1946	Sp-Su	Mediterranean	Juveniles? (3.7–9.9)	90	Gn	15.1 (3.6)	82.2 (3.5)	2.7 (1.3)	11	Massuti and Oliver (1948)
1946	All	Mediterranean	Adults (10.0–12.5)	125	Gn	31.8 (3.7)	65.5 (3.6)	2.7 (1.1)	27	Massuti and Oliver (1948)
1948–49	Sp-Su	Galicia	Adults	9	Gn	36.6 (11.3)	57.9 (14.8)	5.6 (3.5)	9	Oliver (1951)
1950	Sp	Galicia	Adults	?	Gn	28.0 (18.2)	58.7 (14.3)	13.3 (13.3)	3	Oliver and Navarro (1952)
1952	All	Portugal W	Adults	300	Gn	66.2 (25.7)	33.8 (25.7)	–	15	Silva (1954)
1987	Wi-Sp	Galicia and Mar Cantábrico	Adults (>20.0)	70	Gn	35.0 (13.7)	55.0 (13.7)		7	Varela et al. (1988)
1987	Wi-Sp	Galicia and Mar Cantábrico	Adults (>20.0)	150	Gn	42.2 (16.1)	57.0 (16.2)	0.8 (0.3)	9	Varela et al. (1990)
2001–02	Su, Sp	Portugal W and S	Adults (20.0)	110	Gw	66.5 (7.3)	29.0 (6.4)	4.5 (1.4)	6	Garrido (2002)
2001–02	Su, Sp	Portugal W and S	Adults (20.0)	110	Gw	11.3 (1.9)	77.3 (7.8)	11.5 (7.5)	6	Garrido (2002)
2002–03	Wi-Sp, Au	Portugal W and S	Adults	30	Gw	5.4 (4.5)	94.6 (2.2)	0.7 (0.2)	4	Cunha et al. (2005)
1998–01	Sp	Galicia and Mar Cantábrico	Adults (11.0–17.9)	108	SIC	3.7 (1.5)	96.3 (1.5)	–	3	Bode et al. (2004)
1998–01	Sp	Galicia and Mar Cantábrico	Adults (11.0–17.9)	108	SIN	0.0 (0.0)	100.0 (0.0)	–	3	Bode et al. (2004)
1998–01	Sp	Galicia and Mar Cantábrico	Adults (18.0–23.0)	142	SIC	13.0 (3.1)	87.0 (3.1)	–	3	Bode et al. (2004)
1998–01	Sp	Galicia and Mar Cantábrico	Adults (18.0–23.0)	142	SIN	2.7 (1.2)	97.3 (1.2)	–	3	Bode et al. (2004)
2003–04	All	Portugal W and S	Adults (18.0)	491	Gn	80.3 (17.0)	16.9 (14.7)	2.8 (5.4)	55	Garrido et al. (2008a)
2003–04	All	Portugal W and S	Adults (18.0)	491	Gw	17.2 (21.0)	55.2 (29.8)	27.5 (31.9)	55	Garrido et al. (2008a)
2003–04	All	Portugal W and S	Adults (18.0)	491	CC	17.6 (20.2)	59.7 (28.4)	22.7 (30.4)	55	Garrido et al. (2008a)
1991–92	Sp	Galicia and Mar Cantábrico	Larvae (0.4–2.4)	1429	Gn	0.1 (0.1)	87.5 (1.8)	12.4 (1.8)	3	Conway et al. (1994)
2003–04	All	Mar Cantábrico	Larvae (0.3–2.2)	97	Gn	–	94.8 (1.4)	5.2 (1.4)	7	Munuera (2006)
2000	Sp	Mar Cantábrico	Larvae (0.4–2.2)	618	Gn	–	93.3 (3.8)	6.6 (3.8)	3	Fernández and González-Quirós (2006)
Mean			Adults		Gn	48.3	45.4	5.3		
			Adults		Gw	18.7	68.8	12.6		
			Adults		SIC/SIN	4.8	95.2	–		
			Larvae		Gn	0.1	91.9	8.14		

Su: summer; Au: autumn; Wi: winter; Sp: spring; $n1$: number of individuals analyzed; Anal.: analysis method; Gn: gut contents by number of prey; Gw: gut contents by biovolume or weight of prey; SIC: estimations from stable C isotopes in muscle; SIN: estimations from stable N isotopes in muscle; CC: gut contents by C content of prey; phy.: phytoplankton; zoo.: zooplankton; x: mean; S.E.: standard error; $n2$: number of estimations averaged some cases, reports focus on species well recognizable in stomach contents such as *Euterpina acutifrons* and species of *Microsetella, Corycaeus* and *Centropages*, noting that many others, particularly small individuals, are difficult to recognize because of partial digestion (e.g. Massuti and Oliver, 1948). Other zooplankton groups such as tintinnids or cladocerans, were also among the most abundant prey both in the Mediterranean (Massuti and Oliver, 1948) and the Atlantic (Navarro and Navaz, 1946; Varela *et al.*, 1988, 1990; Garrido, 2002; Cunha *et al.*, 2005; Garrido *et al.*, 2008a). Fish eggs in sardine stomach contents have also been noted (Silva, 1954; Varela *et al.*, 1988; Garrido, 2002), and one study has considered fish eggs as a major contributor to sardine dietary carbon off the Iberian Peninsula, especially during the winter months (Garrido *et al.*, 2008a).

There are no detailed data on diel variability in sardine feeding but some information from the Mediterranean suggests that feeding activity occurs mainly during daylight (Andreu, 1969). Seasonal variability in sardine feeding intensity has been recently documented for fish sampled off western and southern Portugal (Garrido *et al.*, 2008a), with the relative contribution by phytoplankton (including dinoflagellates and diatoms) being highest in spring and summer.

Phytoplankton appears to increasingly contribute to the food of sardine as they become larger, as shown by both stomach content (Massuti and Oliver, 1948) and isotope studies (Bode *et al.*, 2003). Such ontogenic change in diet can be related to changes in the morphology of the feeding apparatus, which is fully developed at sizes above 15 cm in length (Andreu, 1969). Large sardine are able to take full advantage of filter feeding because they can swim at the required speed for capturing diluted prey (Cushing, 1978). In contrast, the filtering apparatus of young individuals is insufficiently fine, hence small fish rely on the capture of individual prey via particulate feeding. Recent experimental work has shown that filter feeding appears to be the dominant feeding mode of adult *S. pilchardus* and that the switch to particulate feeding occurs at a particle size of around 750 μm, that this species is able to entrap phytoplankton cells as small as 4 μm during filter feeding, and that sardine show selective feeding, preferentially ingesting fish eggs compared to other prey types when fed cultured, mixed prey assemblages and selecting copepods and decapods over other zooplankton prey when fed wild-collected, mixed prey assemblages (Garrido *et al.*, 2007a). Whereas phytoplankton may be captured incidentally by sardines when using filter feeding to feed on copepods, as suggested by Cushing (1978), the large amount of phytoplankton biomass that is occasionally found in sardine stomach contents (e.g. Cunha *et al.*, 2005) calls for more studies. Recent estimates indicate that phytoplankton makes a significant contribution to the diet of *S. pilchardus*, with average annual values of phytoplankton carbon ranging between 14% and 19.4% of total stomach content carbon for fish sampled off the

south and west costs of Portugal, respectively (Garrido *et al.*, 2008a). The higher contribution made by phytoplankton to the diet of west coast sardine resulted in higher levels of polyunsaturated fatty acids (PUFAs) in their stomach contents, as well as in their muscles and oocytes, compared to south coast fish (Garrido *et al.*, 2007b, 2008b). Dinoflagellates were generally dominant over diatoms in stomach contents in both Mediterranean (Massuti and Oliver, 1948) and Atlantic (Garrido *et al.*, 2008a) waters, although the presence of diatoms may increase in spring in the latter (e.g. Varela *et al.*, 1990; Garrido, 2002; Garrido *et al.*, 2008a). An additional interest regarding feeding on phytoplankton by sardines is their role as vectors of phytoplankton toxins to upper trophic levels (Costa and Garrido, 2004).

In contrast to results found for adults, the food of sardine larvae is almost exclusively composed of zooplankton. Among these, eggs and early developmental stages of copepods constitute the bulk of contents, either by numerical abundance (Lebour, 1920, 1921; Conway *et al.*, 1994; Fernández and González-Quirós, 2006; Munuera, 2006) or by their contribution to total ingested carbon (Munuera, 2006). The maximum size of prey consumed increases with larval size, although larger larvae continue to feed on a high proportion of small particles. Copepod eggs up to 183 μm in diameter were consumed mostly by larvae of intermediate sizes (10–15 mm), and the largest larvae (13–22 mm) fed almost exclusively on nauplii and copepodites of calanoids such as *Paracalanus, Psedocalanus* and *Clausocalanus* (Conway *et al.*, 1994; Fernández and González-Quirós, 2006; Munuera, 2006). Whilst phytoplankton remains were generally rare in larval sardine stomach contents, larvae have been reported as feeding exclusively on phytoplankton in some instances (Rasoanarivo *et al.*, 1991), and phytoplankton may also constitute unidentified residuals in stomach contents (Conway *et al.*, 1994). Seasonal variations in stomach contents of larvae are small, suggesting that they are largely specialized in their food preferences, although their stable isotopes content reveals that larvae are able to track seasonal changes in the isotopic signature of plankton (Munuera, 2006). Significant changes, however, were found in larval stomach contents during daily cycles (Conway *et al.*, 1994; Munuera, 2006), with maximum feeding following dawn (Fernández and González-Quirós, 2006) but also occurring in the afternoon. Diurnal feeding patterns varied according to location, particularly in coastal areas (Munuera, 2006).

There is surprisingly little information on the diet and feeding rates of *E. encrasicolus* in European waters (Table 7.5). To our knowledge, there are only three papers dealing with the diet of adults in the Mediterranean (Tudela and Palomera, 1995, 1997; Plounevez and Champalbert, 2000), one for adults in the Bay of Biscay (Plounevez and Champalbert, 1999), two on larvae in the Mediterranean (Conway *et al.*, 1998, Tudela *et al.*, 2002), and a Ph.D. thesis with some information on larval anchovy diet in the

Table 7.5. *Summary of studies on the quantitative composition of E. encrasicolus diet in Europoean waters indicating average percent composition of prey items*

Year	Season	Zone	Stage (size; cm)	n	Anal.	% phy.	% copep.	% cladoc.	% other crust.	% fish eggs + larvae	% other	Source
1995–96	Su	Mediterranean, Gulf of Lions	Adults (10.2–15.0)	857	PI	–	85.0	2.0	8.9	1.9	3.3	Plounevez and Champalbert (2000)
1997	Sp	Bay of Biscay, Gironde plume	Adults (14.0)	195	PI	–	96.3	0.0	3.4	0.0	0.1	Plounevez and Champalbert (1999)
1997	Sp	Bay of Biscay, shelf-break	Adults (12.5)	198	PI	–	99.5	–	0.5	–	–	Plounevez and Champalbert (1999)
1991	Sp	Portugal W	Larvae (0.3–7.0)	545	Gn	2.1	8.4	–	87.5*	–	4.3	Ferreira and Ré (1993)
1996	Sp–Su	Adriatic Sea	Larvae (0.2–1.9)	889	Gn	0	95	–	–	–	5.0	Conway et al. (1998)
1992	Su	Mediterranean NW	Larvae (0.3–1.6)	936	Gn	3.7	79.5	1.4	0.4	–	14.1	Tudela et al. (2002)
2004	Su	SE Bay of Biscay	Larvae (0.3–1.3)	66	Gn	–	94.0	–	–	–	6.0	Munuera (2006)
Mean			Adults			0	93.6	2.0	4.3	1.0	1.7	
			Larvae			1.5	69.2	0.4	22	–	29.4	

Su: summer; Sp: spring; n: number of individuals analyzed; Anal.: analysis method; PI: preponderance index (biovolume); Gn: gut contents by number of prey; phy.: phytoplankton; copep.: copepods; cladoc.: cladocerans; other crust.: other crustaceans.

Bay of Biscay (Munuera, 2006). In all cases, however, the bulk of the diet determined from stomach content analysis is composed of zooplankton. Whereas the scarcity of data precludes the extraction of general patterns for this species, there are a number of observations worth mentioning.

Copepods dominate the diet composition of adult anchovy, with the genera *Centropages* (*C. typicus* and *C. chierchiae*) and *Temora* (*T. stylifera* and *T. longicornis*) appearing as important food items both in the Mediterranean and the Bay of Biscay. Small cyclopoids of the genera *Oncaea, Oithona* and *Corycaeus*, and small calanoids such as *Clausocalanus, Pseudocalanus,* and *Paracalanus* also occur regularly in stomach contents of anchovy from in both areas (Plounevez and Champalbert, 1999, 2000). Two copepod species, *Candacia armata* and the harpaticoid *Microsetella rosae*, appear as an important food source in the Mediterranean but not in the Bay of Biscay. Large items such as fish larvae, euphausiids and larger copepods (*Candacia, Temora, Centropages*) often dominate in terms of prey biovolume in adult anchovy stomach contents (Plounevez and Champalbert, 2000).

Tudela and Palomera (1997) report that, although *E. encrasicolus* is mainly a diurnal feeder, there is a difference between the size of prey consumed during the day and during the night, with copepods and other small prey ingested during the day and the occasional presence of decapod larvae and mysids observed in the stomach contents of fish sampled at night. Plounevez and Champalbert (1999) also observed that gut fullness in the plume of the Gironde River was about half that in offshore waters, although the concentration of zooplankton in the river plume was twice as high that in offshore waters. Both observations can perhaps be related to visual acuity (Fiksen *et al.*, 2005), since anchovy are only able to detect large prey at night, and will also experience reduced visibility in the turbid waters of the Gironde river plume.

Studies on anchovy in the Mediterranean do not report cannibalism (Tudela and Palomera, 1997; Plounevez and Champalbert, 2000), whereas in the Bay of Biscay anchovy eggs were systematically observed in stomach contents of adult anchovies (Plounevez and Champalbert, 1999). It is difficult to ascertain whether this difference reflects a difference in behavior or methodologic differences (season and time between capture and spawning allowing for digestion of eggs).

A detailed study by Conway *et al.* (1998) on anchovy larvae in the Adriatic Sea indicated that copepod eggs and nauplii, in particular cyclopoid nauplii, dominated the diet of smaller (< 6 mm) larvae; from this size onwards the copepod *Oithona* spp. became the dominant item in the diet, comprising 25%–75% (by number) of the items in the gut. Conway *et al.* (1998) also reported observing the dinoflagellate *Peridinium* and pollen in the guts of larvae. Two other studies also report a diet dominated by copepod eggs, nauplii and copepodites (Tudela *et al.*, 2002; Munuera, 2006), with larger larvae showing an increase in the size range of ingested prey in parallel with an increase in the incidence of copepodites in the diet. However, biochemical measurements such as lipid composition (Rossi *et al.*, 2006) suggest that microzooplankton might be an important component of the diet of early anchovy larvae. Furthermore, Munuera (2006) did not find differences in the isotopic signature of sardine and anchovy larvae from the Bay of Biscay, indicating that both are at the same trophic level. Bergeron (2000) analyzed digestive enzymatic activity of anchovy larvae in the Bay of Biscay, and found a clear spatial pattern in nutritional condition which increased from the middle shelf towards the shelf edge, perhaps associated with the increased planktonic productivity of the shelf edge (Albaina and Irigoien, 2004). Bergeron (2000) also observed that, after a strong wind event, the nutritional condition of the larvae decayed, suggesting a negative effect of strong turbulence on ingestion.

Information on the diets of anchovy and sardine indicates that both species share a common trophic resource, as the identified plankton species are roughly the same and the two species are distributed over basically the same geographic area. Thus trophic level estimates using either average dietary data (Froese and Pauly, 2006) or stable isotope content of muscle (Bode *et al.*, 2006; Munuera, 2006) are remarkably similar for both species. Such estimates, however, are only valid for large spatial and temporal scales, where individual and instantaneous feedings are largely integrated. The ample variability of actual diets with spatial locations and the mobility of these species would likely compensate for their trophic overlap. In addition, the morphology of the feeding apparatus of sardine is more adapted to filter feeding than that of anchovy, thus creating a potential mechanism to avoid food competition.

The Northwestern Atlantic system

Menhaden (the genus *Brevoortia*) species occupy a unique ecological niche in the food webs of estuarine and coastal ecosystems of North and South America. Like other small pelagic fish, menhaden are a schooling species that feed on plankton; what is atypical about menhaden is the size spectra of the food items they consume – menhaden feed directly on nanoplankton. Coupled with the fact that some of the species of menhaden occur in great numbers, they take on an ecological role in the trophic structure of these ecosystems. The menhaden species of the US East Coast and Gulf of Mexico, *Brevoortia tyrannus*, *B. gunteri*, and *B. patronus*, have been the most studied species of the genus, thus most of this discussion is conditioned on data from experimental and field studies on these species.

For a relatively large animal, Atlantic menhaden are able to filter extremely small plankton particles. The small and fragile nature of their food items has made it difficult to characterize their diet, a problem compounded by the fact that Atlantic menhaden diet changes ontogenetically with changes in the morphology of their mouth parts and branchial basket used to feed. First feeding Atlantic menhaden larvae feed on individual plankton particles such as large dinoflagellates and zooplankton (Chipman, 1959; June and Carlson, 1971), and continue to feed in this manner until they metamorphose into juveniles, at which time their branchial baskets and gill rakers adapt to filter feeding.

As juveniles and adults, Atlantic menhaden are omnivorous, feeding on both phytoplankton and zooplankton. Chipman (1959) and June and Carlson (1971) examined the stomachs of post-metamorphic juveniles and observed a variety of phytoplankton and zooplankton prey in the diet. They also noticed that juveniles consumed smaller food items than younger larvae. Adults, like the juveniles, also feed on phytoplankton and zooplankton (Peck, 1893). These observations led Peck (1893) to suggest that Atlantic menhaden diet was simply a reflection of the composition of the plankton where the fish were feeding; this hypothesis holds true today and is further supported by field studies and laboratory experiments. Jeffries (1975) characterized the contribution of zooplankton to the diet of juvenile Atlantic menhaden and found the fish adaptable to local plankton availability. Lewis and Peters (1994) showed that Atlantic menhaden can derive nutritive value from the detrital cellulose they consume along with phytoplankton and zooplankton. These studies reinforce the idea that Atlantic menhaden diet can be qualitatively described by the composition of the plankton. Owing to difficulties in measuring menhaden stomach contents, we must rely on feeding studies to predict what portion of the plankton community menhaden consume.

Ontogenetic change in the structure of the gill rakers results in an ontogenetic change in the filtering ability of menhaden over the size spectra of juvenile to adult. Atlantic menhaden filtering efficiency has been measured with two experiments that characterized particle retention over an ecologically relevant size spectrum of plankton. The study done with large migratory adult Atlantic menhaden (ca. 260 mm fork length, FL) showed that older menhaden had a minimum particle threshold size of 13 μm diameter (Durbin and Durbin, 1975); however, the clearance rate data for these fish also suggests that significant retention (10% efficiency) does not occur for particles <30 μm in diameter. Adult fish were highly efficient when filtering zooplankton. The second study was done with Atlantic menhaden of a transitional size between juvenile and adult (≈14 cm FL), and showed that the menhaden feeding apparatus changes ontogenetically since these fish were able to retain particles

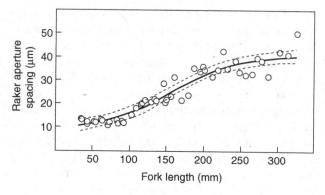

Fig. 7.13. Relationship between gill raker gap and fish length for Atlantic menhaden (redrawn from Friedland *et al.*, 2006.)

as small as 5–7 μm and showed significant retention efficiencies for particles of 7–9 μm (Friedland *et al.*, 1984). However, these fish filtered some zooplankton at a lower efficiency than phytoplankton, suggesting they have maximum filtration efficiency at a size intermediate between phytoplankton and zooplankton.

The gill rakers of Atlantic menhaden form a physical sieve that retains plankton particles (Monod, 1961; Friedland, 1985). Menhaden, like other filter-feeding species (MacNeill and Brandt, 1990; Matsumoto and Kohda, 2001; Tanaka *et al.*, 2006), show an ontogenetic shift in the physical dimensions of the apertures of the sieve formed by their gill rakers (Fig. 7.13). In Atlantic menhaden, this shift is allometric, adapting various sized menhaden to the habitats they utilize. Menhaden juveniles retain the ability to filter the smallest size fractions of plankton while they reside in estuaries. The transition to larger gill raker apertures coincides with their migration to coastal ocean habitats characterized by plankton communities with lower concentrations of phytoplankton biomass and a general shift to larger sized particles, including higher concentrations of zooplankton (Hulbert, 1963; Malone and Chervin, 1979; Muylaert and Sabbe, 1999; Sin *et al.*, 2000; Kimmel *et al.*, 2006). Therefore, as migratory fish they filter more water to concentrate food, which would be hydrodynamic and energetic burdens if they retained the raker spacing they had as juveniles. The allometry in gill raker apertures also affects the foraging of the size range of adult Atlantic menhaden, which find themselves competing with other clupeid fish such as American shad (*Alosa sapidissima*; Munroe, 2002), and Atlantic herring (*Clupea harengus*; Gibson and Ezzi, 1992). These species are able to particulate-feed in addition to filter feeding, and are thus less likely to make much use of phytoplankton by direct filtration. The ontogenetic changes in Atlantic menhaden appear adapted to allow this species to filter smaller particles than other clupeid fishes.

Resource partitioning among menhaden species has also been observed based on functional morphology and feeding

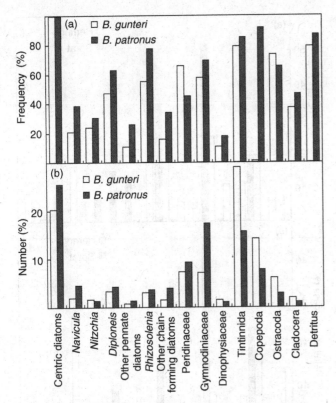

Fig. 7.14. Percentage frequency (a) and percentage number (b) of food items in the diets of the menhadens *Brevoortia gunteri* and *B. patronus* (drawn using data in Castillo-Rivera *et al.*, 1996).

studies of fish from the Gulf of Mexico. Castillo-Rivera *et al.* (1996) examined the trophic morphology of finescale menhaden (*Brevoortia gunteri*) and gulf menhaden (*B. patronus*), and reported that *B. patronus* has more denticles per mm of gill raker and significantly longer gill rakers, than does *B. gunteri*. Hence *B. patronus* has a finer-meshed branchial apparatus, and should be able to retain smaller food items, than *B. gunteri*. Analysis of the dietary components of these two species showed a significant difference in their diets, with zooplankton (tintinids, ostracods and calanoids) predominating in *B. gunteri* stomachs whereas phytoplankton comprised the majority of stomach contents of *B. patronus* (Fig. 7.14). In addition, *B. patronus* has longer epibranchial organs (responsible for the concentration of food particles by microphagous fish, with development of this organ appearing to be related to the degree of microphagy; Blaxter and Hunter, 1982) and a significantly longer intestine, than does *B. gunteri*, providing further support for the hypothesis that the former exhibits a higher degree of phytophagy than the latter.

The relationship between menhaden and plankton is also distributional in nature. Of the suite of physical and biological parameters that control menhaden migration and local movement, the distribution of food resources appears to be of primary importance. The distribution of Atlantic menhaden juveniles is correlated with gradients of plankton biomass reflecting gradient search behaviour of the fish (Friedland *et al.*, 1989; Friedland *et al.*, 1996). Though not as clearly defined, the meso-scale distribution of adult gulf menhaden in the Gulf of Mexico also appears to be conditioned by similar mechanisms (Kemmerer *et al.*, 1974; Kemmerer, 1980).

The Baltic Sea system

The clupeids herring (*Clupea harengus* L.) and sprat (*Sprattus sprattus* L.) are the commercially important small pelagic fish species in the Central Baltic Sea, and the main planktivores in the area (Möllmann *et al.*, 2004). Herring spawn benthic eggs in coastal areas and use the deep basins of the area as feeding grounds. There are both autumn and spring spawners with the latter presently dominating for unknown reasons (Parmanne *et al.*, 1994). Beside the abundant Main Basin Herring (stocks assessed in ICES Sub-division 25–29+32, excluding the Gulf of Riga; ICES, 2006), a number of coastal and gulf stocks exist in the Baltic Sea (Aro, 1989). Even more uncertainty exists about the structure of the sprat stock(s). Sprat are generally spring-spawners and show the opposite behaviour to herring, i.e. they spawn in the deep basins and have their main feeding areas in shallower waters (Aro, 1989). The biomass of the main basin herring stock declined continuously during the recent three decades, while the sprat stock increased to unusually high levels since the 1990s (see this volume, Chapter 9). In this case study we describe the trophic ecology of the two species and summarize the processes which potentially led to these opposite trends in stock size.

Numerous studies on the feeding ecology of herring and sprat have been performed in different areas of the Baltic Sea (e.g. Popiel, 1951; van Khan *et al.*, 1972; Zalachowski *et al.*, 1975, 1976; Aro *et al.*, 1986; Hansson *et al.*, 1990; Davidyuk *et al.*, 1992; Flinkman *et al.*, 1992, 1998; Ostrowski and Mackiewicz, 1992; Rudstam *et al.*, 1992; Starodub *et al.*, 1992; Fetter and Davidyuk, 1993; Raid and Lankov, 1995; Davidyuk, 1996; Szypula *et al.*, 1997; Möllmann and Köster, 1999; Möllmann *et al.*, 2004). Möllmann *et al.* (2004) reviewed interannual variability and differences in the diets of both herring and sprat, based on long-term data. Both species preyed mainly upon calanoid copepods, with *Pseudocalanus acuspes* (formerly called *P. elongatus*; Renz and Hirche, 2006) dominating the diet of herring whereas sprat generally preferred *Temora longicornis*. The highest feeding activity of both fish species occurred in spring and summer, the main reproductive periods of calanoid copepods (Hansen *et al.*, 2006; Renz and Hirche, 2006). The most important food item for both predators in spring was *P. acuspes*, whereas in summer sprat switched to *T. longicornis* and to a lesser degree to *Acartia* spp. Since the late 1970s the total stomach fullness has decreased and the fraction of empty stomachs increased in both species, and this has occurred in parallel with a decrease in the amount of *P. acuspes* in the diets of both herring and sprat. The decrease in *P. acuspes* in the diet of both fish species can be related to the long-term decrease in this major Baltic mesozooplankton species during the 1980s and 1990s, caused by a climate-driven reduction in salinity (Möllmann *et al.*, 2000, 2003b). In contrast, the populations of *T. longicornis* and *Acartia* spp. have increased, profiting from warming since the late 1980s (Möllmann *et al.*, 2000, 2003b).

Generally, the main differences in the feeding ecology of herring compared to sprat are the inclusion of larger food items, e.g. macrozooplankton and 0-group fish in larger herring specimens (Hardy, 1924; Last, 1989), and the ability to switch between filter feeding and particulate feeding modes (Gibson and Ezzi, 1992). In the Baltic Sea, mysids (larger than copepods) provide an additional food for herring in winter and autumn. Generally, older copepodite stages and adult copepods dominate the diets of herring and sprat (Fig. 7.15), indicating particulate feeding by both fish species (Möllmann *et al.*, 2004). Baltic herring are occasionally found feeding on younger copepodite stages, indicating a low availability of older stages and forcing the fish to filter-feed on smaller prey (Möllmann *et al.*, 2004). Both planktivores are known to selectively feed on certain prey species and stages. Flinkman *et al.* (1992) showed that herring select reproducing female copepods, because of their larger body size and better visibility due to their attached egg sacs. Casini *et al.* (2004) compared autumn and winter herring and sprat diets with *in-situ* prey availability and found a preference for *T. longicornis* and *P. acuspes* but avoidance of *Acartia* spp. Recent investigations within the GLOBEC-GERMANY project (http://www.globec-germany.de) support this preference for spring–summer, and also indicate that both fish show a strong selection for C4–5 and adult female copepods (M. Bernreuther, Hamburg University, Hamburg, Germany, unpublished data).

Considerable dietary overlap between these species is typical, being highest among large sprat and small herring (Möllmann *et al.*, 2004). Consequently, competition may play an additional role in determining the diets of both species, and in the Baltic appears to be most pronounced in spring when herring return from their coastal spawning areas into the deep basins, where sprat are reproducing (Möllmann *et al.*, 2003a). Both fish species feed at this time of the year on reproducing *P. acuspes* stages which dwell in the halocline of the deep basins (Hansen *et al.*, 2006; Renz and Hirche, 2006). Later in summer sprat move into their coastal feeding areas (Aro, 1989; Parmanne *et al.*, 1994), which reduces competition (Möllmann *et al.*, 2004).

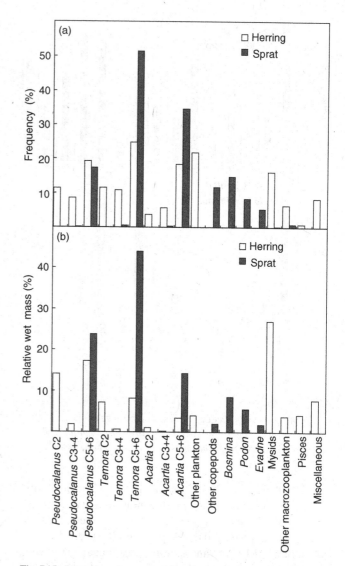

Fig. 7.15. Percentage frequency (a), and percentage relative wet mass to the total stomach content wet mass (b), of the major prey items of Baltic Sea herring and sprat summarized from samples collected over the period 1977–1999 (drawn using data in Möllmann *et al.*, 2004.)

Strong inter- and intraspecific competition is usually manifested in pronounced changes in fish growth and condition. These have been observed for herring and sprat in the Baltic, with individual weight and condition of herring decreasing since the 1980s (e.g. Cardinale and Arrhenius, 2000; Möllmann *et al.*, 2003a; Möllmann *et al.*, 2005), and of sprat later in the early 1990s (Cardinale *et al.*, 2002; Möllmann *et al.*, 2005). Recent studies indicate the importance of the composition of the zooplankton community for fish growth and condition; Rönkkönen *et al.* (2004) showed growth rates of herring in the northern Baltic to be explicitly dependent on the abundance of *P. acuspes*. The condition of herring in the Central Baltic Sea was shown to depend on the population size of this copepod (Möllmann *et al.*, 2003a; Möllmann *et al.*, 2005). Food availability, especially *P. acuspes* population size, has been further hypothesized to be crucial for the growth of sprat as well (Cardinale *et al.*, 2002; Möllmann *et al.*, 2004). However, while the decreasing availability of *P. acuspes* and the resulting competition between herring and sprat may affect both fish species, sprat growth and condition during the 1990s in particular seemed to be mainly determined by intra-specific competition within the unusually large sprat stock (Möllmann *et al.*, 2005; Casini *et al.*, 2006).

Providing evidence of food limitation requires sound estimates of individual food intake, and two approaches have been applied for deriving consumption estimates for Baltic clupeids: (i) gastric evacuation modeling (Köster and Schnack, 1994; Temming, 1996), and bioenergetic modeling (Rudstam, 1989; Arrhenius and Hansson, 1993). Using the latter approach, Arrhenius and Hansson (1999) demonstrated young-of-the-year herring to be food-limited. Estimates of individual daily ingestion rates using the gastric evacuation model approach indicate that juvenile sprat (1–2 g body weight) have evacuation rates that are 1.5 times higher than those of similar sized herring, which indicates that small sprat realise higher consumption rates even if their stomach content mass and dietary composition are very similar to those of herring. Recent investigations indicate that sprat stomach contents, at least in the deep Baltic basins, are too low to support the observed annual growth, supporting the hypothesis of food limitation in these regions (A. Temming, unpublished data).

The difference in individual weight trajectories over time leads to the question of whether one of the species is able to out-compete the other when co-occurring in an identical environment. Monthly data on mean stomach contents of sprat and herring of the same length-class do not show significant differences (M. Bernreuther, unpublished data), a result confirmed by a similar study performed off the Scottish West coast (De Silva, 1973).

In summary, Baltic herring and sprat are ecologically quite distinct from each other, with herring growing larger, having shallow spawning areas and spawning earlier in the year, and being able to filter-feed but also consuming larger prey items at larger fish sizes. Sprat spawn deeper and later, and do not appear to be able to filter-feed. Despite these differences, when observed together in the deep central basins of the Baltic Sea the diet of herring and sprat is remarkably similar, consisting almost exclusively of three dominant copepod species. Filter feeding does not play a role in these regions as generally only older and larger copepodite stages occur in the diet. When stomach contents are compared between individual herring and sprat of the same size no consistent difference is observable. However, differences in gastric evacuation indicate that equal stomach contents translate into higher consumption rates of sprat compared to herring, pointing to a possible competitive advantage for sprat. Another advantage might emerge from the potential of sprat to exploit the most preferred (and energy rich) copepod species earlier in the season, with the consequence of a severe depletion of this resource prior to the arrival of herring in the basin. These two effects might serve as an explanation for the earlier decline (in the 1980s) in herring weight-at-age when compared to a similar decline in sprat (in the 1990s).

Discussion

The detailed descriptions of aspects of the trophic dynamics of small pelagic species provided in the case studies above may be synthesized into generalizations that appear to be broadly applicable between different systems.

First, small pelagic fish are omnivorous and most species described above derive the bulk of their nutrition from zooplankton (see Table 7.6), as suggested by James (1988). Previous studies indicating that phytoplankton was the dominant dietary component were erroneous and were biased by the methodology used which assessed prey importance on the basis of numerical occurrence or frequency of occurrence instead of biovolume or contribution by mass or carbon content. Larvae and juveniles of most small pelagic species feed predominantly on zooplankton, although the almost exclusive feeding on phytoplankton reported for first-feeding Humboldt anchovy larvae makes this species the exception to this generalization. Although zooplankton is generally the dominant dietary component of small pelagic fish, phytoplankton increases in dietary importance (to differing degrees for different species) as fish grow, arising from the ontogenic development of the branchial basket which allows the retention of small particles.

Second, in all but one of the case studies described above, co-occurring species pairs show size-based partitioning of the zooplankton resource, particularly at the adult stage. For the anchovy/sardine species pair, sardine generally feed on smaller zooplankton (often cyclopoid, poecilostomatoid and small calanoid copepods) than do anchovy (larger calanoid copepods and euphausiids), and phytoplankton is generally more important for sardine than for anchovy. Hence sardine consistently feed at a lower trophic level than do anchovy. These results are clear from analyses of stomach contents of anchovy and sardine conducted in the southern Benguela Current system, the northern California Current system, the northern Humboldt Current system, and the northwestern Pacific. Stable isotope analyses of both species in the northern California Current system showed anchovy to have higher isotopic values for both $\delta^{13}C$ and $\delta^{15}N$ than sardine, which adds support to this generalization. However, whereas mean isotopic values for anchovy were also slightly higher than those for sardine from the Iberian system the results were not statistically significant, indicating that, on average, both species occupied equivalent trophic positions. Clear resource partitioning by prey size also occurs between finescale menhaden and Gulf menhaden in the Gulf of Mexico, but is not obvious for herring and sprat in the Baltic Sea, indicating that competition between members of this species pair is higher than in the other systems and possibly arising from the fact that the Baltic system is a relatively oligotrophic system compared to the other case studies described. Size-based partitioning of zooplankton prey may also occur for early life history stages since anchovy larvae tend to have a larger mouth width (gape), and to ingest slightly larger prey, than do sardine larvae of similar size in some systems. However, these differences are small and not obvious in all case studies, and there is generally substantial overlap in the type and size of prey ingested by larvae of co-occurring species. Pre-recruit and recruit anchovy and sardine in the southern Benguela Current system do appear to show resource partitioning by prey size.

Differences in diet are considered to arise primarily from differences in morphology of the feeding apparatus, with one of the species pair having a finer branchial basket with more gill rakers and a smaller gill raker gap than the other species (Table 7.6). Gulf menhaden have a finer branchial basket than do finescale menhaden, and sardine have a finer branchial basket than do anchovy in the Benguela Current system, the California Current system, the northwestern Pacific, and the northeastern Atlantic; unfortunately, morphological studies on the branchial basket of Humboldt Current species have not been conducted to date. The ability to retain very small particles by sardine has been demonstrated by experimental studies, with southern Benguela sardine able to retain particles of 13–17 μm and Iberian sardine particles as small as 4 μm; southern Benguela anchovy can only retain particles >200 μm.

Resource partitioning is also likely to arise from different trophic behaviors (Table 7.6). Anchovy and sardine are capable of both filter feeding and particulate feeding, and whilst feeding mode choice will depend on factors such as prey size and concentration and ambient light levels, sardine appear to be primarily filter feeders and are generally considered as being non-selective in most systems although they have been observed to exhibit prey size selectivity or prey type selectivity in laboratory studies, particularly when particulate feeding. Additionally, sardine, particularly adults, generally do not show feeding periodicity and have been reported as feeding throughout the diel cycle in the Benguela Current, Humboldt Current, and northeastern Atlantic systems. These characteristics are in marked contrast to anchovy which are strongly size selective, appear to feed primarily by particulate feeding, and show strong feeding periodicity, primarily feeding during daylight, in most systems.

Anchovy and sardine are therefore trophically distinct, and this trophic distinction appears in some but not all other pairs of small pelagic fish species. The differences between anchovy and sardine described in the case studies above may be at least partly responsible for the alternations or regime shifts observed between these species as suggested by Schwartzlose *et al.* (1999). For the southern Benguela Current system, van der Lingen *et al.* (2006a) suggested how different physical scenarios can lead to food environments being dominated by either small or large particles that might favor one species over the other (Fig. 7.16). Phytoplankton biomass and size structure in that system are closely related to physico-chemical conditions, being dominated by large, chain-forming diatoms under intermittent mixing conditions (such as occur during upwelling) but dominated by small, nanoflagellates during stable, warm conditions (Mitchell-Innes and Pitcher, 1992). In turn, zooplankton community structure is affected by phytoplankton community and size structure, with large copepods such as *Calanoides* and *Calanus* exhibiting higher rates of ingestion of large phytoplankton cells compared with small cells (Peterson, 1989), and displaying enhanced growth rates under diatom-dominated conditions compared to flagellate-dominated conditions (Walker and Peterson, 1991). In contrast, small copepod species such as *Oithona* appear to be favoured when small cells predominate (H.M. Verheye, Marine and Coastal Management, Cape Town, South Africa, unpublished data). Based on this, van der Lingen *et al.* (2006a) suggested that different physical scenarios can lead to food environments being dominated by either small or large particles, which would tend to favor sardine or anchovy, respectively.

Support for the hypothesis that changes in zooplankton community size structure can drive alternations between sardine and anchovy in the southern Benguela was provided by Shannon *et al.* (2004), who examined the effect of changing the relative availability of phytoplankton and zooplankton (comprising meso- and macrozooplankton) to anchovy and sardine in trophic models that simulated alternations between the two species. Shannon *et al.* (2004) concluded that modeled shifts between an anchovy-dominated system and a sardine-dominated system were "... likely to have been caused by changes in the availability of mesozooplankton to anchovy and sardine," and that those changes occurred in opposite directions for the two species and were assumed to be mediated through changes in environmental conditions.

Recent observations of simultaneously high abundance levels of both anchovy and sardine in the southern Benguela (van der Lingen *et al.*, 2006b) appear contrary to the hypothesis that trophic differences between these two species might be partly responsible for alternating periods of species dominance. However, during species alternations the subdominant genus may initiate a recovery, while the other is still abundant (Schwartzlose *et al.*, 1999), which may have been the case in the southern Benguela. Additionally, zooplankton abundance in that system during austral autumn/winter increased 100-fold between the 1950s and the mid 1990s (Verheye *et al.*, 1998), apparently due to intensified upwelling and increased primary and secondary productivity (Verheye, 2000). This substantially increased production by the lower trophic levels may also account for the high combined biomass of anchovy and sardine in the southern Benguela.

Regime shifts between anchovy and sardine in the northern Humboldt Current system off Peru have been linked to differences in their trophic position and a restructuring of the ecosystem, with changes in trophic relationships occurring during warm and cool phases (Alheit and Ñiquen, 2004; this volume, Chapter 5). Schwartzlose *et al.* (1999) proposed that food and temperature may drive shifts in abundance between sardine and anchovy, and indeed during warm conditions (El Niño–El Viejo; see Chavez *et al.*, 2003) smaller species of phytoplankton and zooplankton occur, favoring sardines (Ulloa *et al.*, 2001; Chavez *et al.*, 2002; Fiedler, 2002; Alheit and Ñiquen, 2004). Furthermore, sardine recruitment is probably enhanced by warm conditions (Schwartzlose *et al.*, 1999; McFarlane *et al.*, 2002). On the other hand, during cold conditions (La Niña – La Vieja; see Chavez *et al.*, 2003) larger plankton species that should favour anchovy are present and abundant, and anchovy is also adapted to cooler water than sardine in the northern Humboldt system. An El Niño event should thus favor sardine more than anchovy. However at the end of the 1997–98 El Niño, the anchovy population was high while the sardine population was depleted. Consequently, considering only temperature and plankton composition is insufficient to explain anchovy and sardine alternations, and Bertrand *et al.* (2004) used an integrated approach to propose a hypothesis related simply to the range of habitat available to explain species alternations. Indeed, as these species do not share the same ecological niche in terms of water masses, tolerance to abiotic factors, migrating and feeding capacities, climatic oscillations at different temporal scales (El Viejo–La Vieja, El Niño–La Niña) lead to oscillations of the range of their respective habitat (*sensu lato*), and then to alternation of anchovy and sardine populations. In addition, the hypothesis of Bertrand *et al.* (2004) explains a paradox in sardine dynamics which arises from the observation that sardines are abundant during the same periods in the eastern and western Pacific, i.e. during "warm" periods in the eastern Pacific that correspond to "cold" periods in the western Pacific (Chavez *et al.*, 2003), even if the Kuroshio Current is much warmer than the Humboldt Current (see this volume, Chapter 5 for further discussion on temperature and

sardine in the Kuroshio Current). Off Peru, production is associated with coastal upwelling, which is stronger during cold conditions. When strong, coastal upwelling allows extension of the habitat (mainly cold coastal waters) favorable to anchovy, anchovy attain higher absolute biomass levels than do sardine, which take advantage of the periods when weakening of the equatorial trade winds disrupts upwelling. On the contrary, off Japan, high production is related to offshore upwelling present along the frontal zone of the Kuroshio (Kimura *et al.*, 1997, 2000a), and in this system the dominant small pelagic fish is sardine, which is strongly associated with the Kuroshio Current. The frontal zone provides better feeding conditions for first-feeding sardine larvae than does coastal waters (Kimura *et al.*, 2000b), and during "eastern" El Niño–El Viejo conditions, the Kuroshio Current is strongest which leads to higher productivity in the frontal zone, with benefits to sardine recruitment (Sugimoto *et al.*, 2001). Thus, the western Pacific sardine would profit from the increase in frontal productivity related to "eastern" El Niño–El Viejo periods that also favor sardine in the eastern Pacific.

One climate descriptor for the northwestern Pacific that has been related to population dynamics of Japanese sardine is the Aleutian Low Pressure (ALP) (Yasuda *et al.*, 1999; Yatsu and Kaeriyama, 2005). Yasuda *et al.* (1999) hypothesized that an amplified ALP intensifies westerly winds during winter and early spring, which results in the low spring SST east of Japan including the Kuroshio extension and Kuroshio–Oyashio mixed water region where sardine migrate and feed. The intensified wind and low SST imply a higher biological productivity owing to the nutrient supply due to wintertime deepening of the surface mixed layer (see this volume, Chapter 5), and this higher productivity might sustain the recruitment of the sardine in the fragile metamorphosis stage. In fact, significant positive correlations were found between the natural mortality coefficient during the period from the postlarval stage to age 1 and winter–spring SST in the Kuroshio extension and its southern recirculation area (Noto and Yasuda, 1999). In this context, however, it remains to be discussed why anchovy population level has been low and/or decreasing during periods of high sardine population levels, and *vice versa*, even under the same level of plankton biomass. Species alternations between Japanese anchovy and sardine in the northwestern Pacific may be a consequence of the difference in preferred prey size between sardine and anchovy in the subadult and adult stages. For example, it is possible that fluctuations in zooplankton biomass caused by changing climate regimes differ for small and large sized zooplankton, possibly even in opposite phases, which would result in food environments better suited to sardine and anchovy, respectively, as described above. This may cause differences in prey availability for the two species during

Table 7.6. *Comparative summary of aspects of the trophic dynamics of small pelagic fish species pairs described in the case studies*

System	Species 1	Species 2
Benguela Current	Anchovy (*Engraulis encrasicolus*) – coarse branchial basket; particulate feeding dominant and energetically cheapest feeding mode; efficient remover of large (>0.6 mm) particles; respiratory quotient = 0.92; assimilation of N most efficient from zooplankton; excretes <50% of ingested nitrogen; regulates swimming speed according to prey size; phytoplankton generally unimportant in juvenile and adult diet and majority of adult dietary carbon from larger (>1.0 mm) zooplankton (calanoid copepods and euphausiids); strong size selectivity; exhibit feeding periodicity with peak feeding at dawn and dusk.	Sardine (*Sardinops sagax*) – fine branchial basket; filter feeding dominant and energetically cheapest feeding mode; efficient remover of small (<0.6 mm) particles; respiratory quotient = 0.96; assimilation of N most efficient from zooplankton; excretes >50% of ingested nitrogen; regulates swimming speed according to prey concentration; phytoplankton generally unimportant in juvenile diet but occasionally important in adult diet and majority of adult dietary carbon from small (<1.2 mm) zooplankton (calanoid and cyclopoid copepods); weak size selectivity; feed throughout diel cycle.
California Current	Anchovy (*Engraulis mordax*) – coarse branchial basket; larvae feed on zooplankton (copepod eggs, nauplii and copepodites); phytoplankton generally unimportant in adult diet and majority of adult dietary carbon from zooplankton (copepods, anchovy eggs, euphausiids, pteropods); strong size selectivity over a 1000-fold range in prey size; exhibit feeding periodicity with peak feeding at dawn.	Sardine (*Sardinops sagax*) – fine branchial basket; larvae feed on zooplankton (copepod eggs, nauplii and copepodites); phytoplankton occasionally important in adult diet but majority of adult dietary carbon from zooplankton (crustaceans, euphausiids), although substantial spatial and temporal variability in sardine diet; feed throughout diel cycle.
Humboldt Current	Anchovy (*Engraulis ringens*) – particulate feeding dominant feeding mode; larvae feed on phytoplankton (centric diaoms) and zooplankton; phytoplankton occasionally important in adult diet but majority of adult dietary carbon from zooplankton (copepods and euphausiids); latitudinal differences in adult diet (phytoplankton vs zooplankton) related to differences in trophic morphology; generally feed during daytime with some night-time feeding.	Sardine (*Sardinops sagax*) – larvae feed on zooplankton (copepod eggs and nauplii) and some phytoplankton (diatoms); phytoplankton and zooplankton (small copepods and tunicates) make equivalent contributions to adult dietary mass. Common sardine (*Strangomera bentincki*) – filter feeding dominant feeding mode; larvae feed on zooplankton (copepod eggs and nauplii) and some phytoplankton (dinoflagellates); phytoplankton and zooplankton of comparable importance to adult diet.
Northwestern Pacific	Anchovy (*Engraulis japonicus*) – coarse branchial basket; larvae feed on zooplankton (copepod eggs and nauplii); phytoplankton occasionally important in adult diet but small copepods dominate zooplankton dietary component; larvae of benthic animals (bivalves, decapods) consumed when fish inshore; feed during daytime.	Sardine (*Sardinops melanosticus*) – fine branchial basket; larvae feed on zooplankton (copepod eggs and nauplii); phytoplankton frequently important in adult diet but small copepods dominate zooplankton dietary component; seasonal and spatial differences in sub-adult and adult diet (phytoplankton vs zooplankton); feed during daytime.
Northeastern Atlantic and Mediterranean	Anchovy (*Engraulis encrasicolus*) – coarse branchial basket; larvae feed on zooplankton (copepod eggs and nauplii); phytoplankton unimportant in adult diet and zooplankton (large copepods, euphausiids and fish larvae) dominate prey biovolume; generally feed during daytime with some night-time feeding; diel variability in dominant dietary components.	Sardine (*Sardina pilchardus*) – fine branchial basket but substantial spatial variability in number and spacing of gill rakers; filter feeding dominant feeding mode; larvae feed on zooplankton (copepod eggs, nauplii and copepodites); phytoplankton occasionally important in adult diet but majority of adult dietary carbon from zooplankton (crustacean eggs, copepods, decapod zoea, and fish eggs); feed during daytime.

Table 7.6. *(cont.)*

System	Species 1	Species 2
Northwestern Atlantic		Atlantic menhaden (*Brevoortia tyrannus*) – fine branchial basket and able to filter extremely small plankton particles; larvae feed on large phytoplankton and zooplankton whilst juveniles and adults are omnivorous; ontogenic changes in branchial basket linked to movement from estuarine (juvenile) to coastal (adult) habitats.
	Fine-scale menhaden (*Brevoortia gunteri*) – coarser branchial basket; zooplankton (tintinnids, ostracods and calanoid copepods) dominate adult diet.	Gulf menhaden (*Brevoortia patronus*) – finer branchial basket; phytoplankton (diatoms) dominates adult diet.
Baltic Sea	Sprat (*Sprattus sprattus*) – appear unable to filter-feed; diet dominated by calanoid copepods; exhibit selective feeding.	Herring (*Clupea harengus*) – can filter-feed and particulate-feed; diet dominated by calanoid copepods; exhibit selective feeding; larger fish consume larger (macrozooplankton, 0-group fish) prey.

the same regime, and may finally influence population dynamics through biological aspects such as egg production. There are some data that show inter-annual and inter-decadal changes in plankton biomass (Odate, 1994; Nakata and Hidaka, 2003), and Nakata and Hidaka (2003) in particular showed inter-annual fluctuation in large (prosome length >1 mm) and small (<1 mm) copepods separately, although the two patterns were apparently not consistent with patterns in fluctuations of anchovy and sardine. This suggests that plankton dynamics should be further studied in order to account for their stock dynamics.

Alternatively, it may be that physical environmental factors directly impact on the survival of early life stages. For example, Takasuka *et al.* (2007) revealed a difference in optimal temperature for growth between anchovy and sardine larvae, and hypothesized that temperature shifts caused by climate regimes may account for the difference in survival through growth, and therefore the observed species alternations. If so, feeding might accessorily accelerate species alternation, since competition between early life stages arising from their dietary overlap could mean that the species with higher survival rates would consume the bulk of available food. For either the trophic or optimal temperature hypothesis, however, there has not yet been clear biological evidence and further work on these are required. Futhermore, it should be considered that species alternations in the western Pacific have occurred not only in the Kuroshio system, but also in waters to the west of Japan (the Tsushima Current system; see this volume, Chapter 5).

Co-existing populations of sardine and anchovy in areas around the Iberian Peninsula and in the Mediterranean (e.g. Andreu, 1969) have shown multidecadal oscillations in population size in both Atlantic (Anadón, 1950) and Mediterranean fisheries (Muzinic, 1958), with periods of 25–35 years in which one species clearly dominated. Although early studies did not find clear relationships between the alternating periods of sardine and anchovy in Galicia (Anadón, 1950), longer time series of fishery data integrated over the NW Iberian Atlantic coastline (ICES, 2005, 2007) have identified multidecadal periods of relative dominance of sardine and anchovy (Fig. 7.17). Quasi-decadal scales are characteristic of climatic, oceanographic and fish abundance indices, as found in other upwelling regions (e.g. Chavez *et al.*, 2003; this volume, Chapter 5). In the Iberian case, however, sardine and anchovy showed synchrony in positive and negative phases up to 1978, as their populations increased and decreased simultaneously. This pattern was broken and moved to asynchrony thereafter, with sardine and anchovy showing opposite phases. Direct effects of climate on plankton and sardine populations have been shown in subzones of the study area: for instance, north wind intensity in winter and early spring is related with poor sardine recruitment and landings off Portugal and Galicia, likely by increasing larval dispersion (Guisande *et al.*, 2004; ICES, 2007), whilst moderate upwelling enhanced anchovy recruitment in the eastern Bay of Biscay (Borja *et al.*, 1998; ICES, 2007). Indirect effects of climate and oceanographic conditions may operate

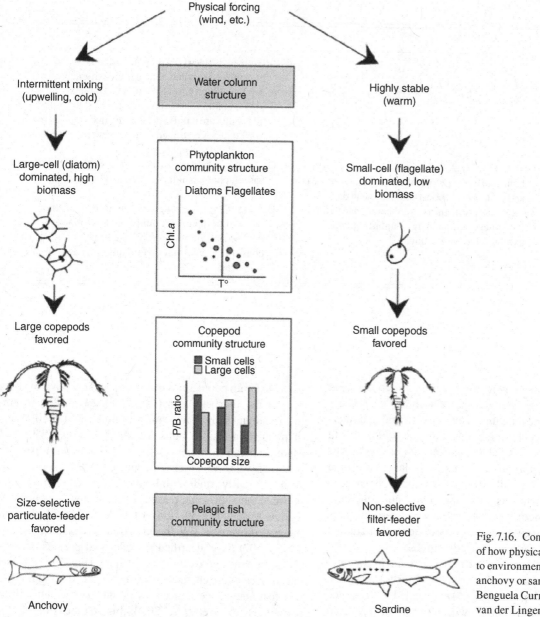

Fig. 7.16. Conceptual schematic of how physical forcing may lead to environments that favor either anchovy or sardine in the southern Benguela Current system (from van der Lingen *et al.*, 2006a.)

through changes in productivity and structure of plankton communities (Chavez *et al.*, 2003, van der Lingen *et al.*, 2006a) while the availability of appropriate prey would favor the dominance of one or another species of planktivorous fish (Fig. 7.16). Adult sardines are better suited to filter feeding on small copepods than anchovies, but the latter are efficient predators of large copepods (van der Lingen *et al.*, 2006a). In this way, sardine dominated landings in the Iberian Atlantic when small copepods (e.g. *Acartia*) were abundant, while periods of anchovy dominance were related to relatively high abundances of large copepods (e.g. *Calanus*, Fig. 7.17). Variations in the intensity and periodicity of upwelling, caused by climatic factors, were identified as the drivers for changes in the size-structure and composition of plankton (van der Lingen *et al.*, 2006a; Fig. 7.16), although the association between upwelling intensity and a particular fish species seems to vary among geographic locations (Chavez *et al.*, 2003; Bertrand *et al.*, 2004). This suggests that other effects not directly related to differences in food preference, such as Bakun's loophole hypothesis

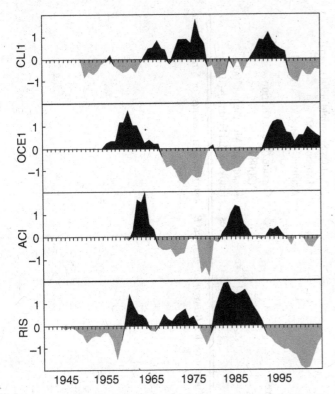

Fig. 7.17. Comparison of anomalies of climatic (CLI1), oceanographic (OCE1), plankton (ACI) and sardine dominance indices (RIS) in the northeastern Iberian Atlantic. CLI1 and OCE1 were extracted from principal component analysis of climate indices (ftp://ftp.cpc.ncep. noaa.gov) and oceanographic variables of the northern Atlantic, respectively. Climatic variability was mainly related to latitudinal anomalies in north winds (positive values) but also in the influence of subtropical winds (negative values). Oceanographic variability was reflected by periods of relative stability and high temperature of the upper water column (positive values) alternating with others of high turbulence and low temperature (negative values). ACI indicated the relative dominance of small (e.g. *Acartia*, positive values) or large copepods (e.g. *Calanus*, negative values) and was computed from data from the Continuous Plankton Recorder survey (http://www.sahfos. org). RIS expressed the relative dominance of sardine (positive values) or anchovies (negative values) in commercial landings (data from ICES, 2005). Further details on the data sources and analysis can be found in ICES (2007).

(Bakun and Broad, 2003) or intraguild predation (Valdes Szeinfeld, 1991), might have to be considered to explain observed species shifts.

Intraguild predation (IGP) seems to be a common interaction between small pelagic fishes (Table 7.7), as is the case in most systems (Arim and Marquet, 2004). Intraguild predation is defined as competitors that eat each other (Polis *et al.*, 1989), in this case adults that compete for food and at the same time prey on the early stages (eggs and larvae) of the competitor. In that sense, cannibalism, also widespread in small pelagics (Table 7.7), can be considered as an extreme case of intraguild predation, where adults of the same species compete for food and at the same time eat their own offspring. Cannibalism makes sense from the evolutionary point of view of the gene or individual, as it increases the reproductive condition of the adult whilst at the same time reducing the number of competitors for the offspring. It would be a problem if the adult was to eat more of its own offspring than that of the other adults, but this is unlikely in the frame of a school spawning in a mixed environment.

Although recognized, little attention is usually paid to the role of IGP. This effect can be divided in three components: i) the individual, (ii) population dynamics of the species, and (iii) consequences for the ecosystem. At the individual level there is little information about the role of IGP in the energetic budget of small pelagic fishes and IGP is often considered to play a minor role. However, this may not necessarily be the case; in a detailed seasonal study, Garrido *et al.* (2008a) found cannibalism to play an important role in the diet of Atlantic–Iberian sardine. At the population dynamics level, predation on the early stages may play quite an important role in recruitment (Valdes-Szeinfeld and Cochrane, 1992). For example, cannibalism accounted for 22% of anchovy egg mortality off Peru (Alheit, 1987); up to 56% of anchovy egg mortality could be due to predation by sardine off South Africa (Valdes Szeinfeld, 1991), and Baltic sprat at times consume >60% of their own egg production (Köster and Möllmann, 2000). Finally, IGP has important consequences on the ecosystem structure because predation between competitors releases pressure on their common prey and may increase the dominance of one species over the other when food is abundant (Mylius *et al.*, 2001). This final property could influence the observed cycles of alternating dominance shown by small pelagic fish and described above, as well as amplify climatic signals (Hsieh *et al.*, 2005).

In addition to its hypothesized effect in impacting on population variability and species alternations, the trophic difference between anchovy and sardine is also likely to have habitat implications. Both anchovy and sardine are adapted to the coastal upwelling domain, but anchovy are likely to be favored in such an environment because the blooms of large phytoplankton (diatoms) there promote the development of communities of large zooplankton, as described above. However, because of their ability to efficiently feed on small prey, sardine should be better adapted to survive in the small particle-dominated oceanic environment than anchovy, and this, together with their better-developed swimming capabilities arising from their larger body size, may enable sardine to exploit the oceanic domain during certain periods as hypothesized by MacCall (this volume, Chapter 12).

Table 7.7. *Incidence of cannibalism (Can) and/or intraguild predation (IGP) shown by small pelagic fish from some of the case studies described above*

	Benguela		California		Humboldt		NW Pacific		NE Atl. / Med.		NW Atlantic		Baltic		References
	Can	IGP	Can	IGP	Can	IGP	Can	IGP	Can	IGP	Can	IGP	Can	IGP	
Engraulis encrasicolus	Yes	Yes							Yes	Yes					Plounevez and Champalbert (1999, 2000); Valdes *et al.* (1987)
Engraulis mordax			Yes	Yes											Hunter and Kimbrell (1980); Loukashkin (1970)
Engraulis ringens					Yes	Yes									Alamo and Espinoza (1998); Alheit (1987); Santander *et al.* (1983)
Engraulis japonicus							Yes								Takasuka *et al.* (2004)
Sardinops sagax		Yes				Yes									Alamo and Espinoza (1998); Valdes Szeinfeld (1991); van der Lingen (2002)
Sardinops melanostictus							Yes	Yes							Azuma (1994); Kidachi (1968)
Sardina pilchardus									Yes	Yes					Garrido (2002); Garrido *et al.* (2008a)
Sprattus sprattus													Yes		Köster and Möllmann (2000)
Clupea harengus														Yes	Patokina and Feldman (1998)

Note that the authors are not aware of any publications describing cannibalism or intraguild predation for *Brevoortia gunteri, B. patronus, B. tyrannus,* or *Strangomera bentincki.*

Herring and sprat in the Baltic Sea have similar diets and are not markedly trophically distinct, yet they show opposite trends in stock size, emphasizing the fact that mechanisms other than trophodynamics impact on species alternations. Baltic sprat reproductive success has been shown to be sensitive to climate-driven temperature conditions; MacKenzie and Köster (2004) found a positive correlation between sprat recruitment and the NAO index, as well as average water column temperature in May, and Alheit *et al.* (2005) also reported an association between the NAO index and sprat biomass. In an exploratory analysis relating recruitment to monthly depth-specific temperatures, Baumann *et al.* (2006) observed significant temperature–recruitment correlations between March and July in mid-water depths, confirming the results of MacKenzie and Köster (2004). However, in July to August, correlations shifted to surface waters, with August temperatures in surface waters explaining most of the variability in sprat recruitment (Baumann *et al.*, 2006). These results indicate the importance of direct and indirect effects of temperature on Baltic sprat recruitment. After spawning in spring, sprat eggs occur at depths where the water temperature is affected by winter cooling, and egg and larval development is influenced by extremely low water temperatures. Consequently, weak year classes of Baltic sprat have been associated with severe winters when temperatures of <4 °C were observed in the intermediate water layer (Köster *et al.*, 2003; Nissling, 2004). This direct effect of low temperature on eggs and larvae may explain the observed correlations in spring (MacKenzie and Köster, 2004; Baumann *et al.*, 2006).

The stronger correlations of surface temperature and recruitment in summer (Baumann *et al.*, 2006), may be due to food availability and match–mismatch with the food production (Voss *et al.*, 2006). Sprat larvae prey mainly on the copepod *Acartia* spp. (Voss *et al.*, 2003), and higher water temperatures during the 1990s have resulted in a drastic increase in the standing stock of this copepod (Möllmann *et al.*, 2003b). Voss *et al.* (2006) computed an index of larval mortality which suggests a higher survival of summer-over spring-born sprat larvae, with pronounced differences in survival for older larvae. Independent indices of larval growth (from analyses of the RNA:DNA ratio) support this survival pattern and could be linked to the temporal variability in prey abundance. Growth was faster but less variable in spring- than in summer-born larvae, indicating a strong selection for fast growth in April–May but a less selective environment in June–July (Voss *et al.*, 2006). The advantage for summer-born larvae may be due to the demographic status of the *Acartia* spp. population; while nauplii are abundant and only very low concentrations of larger copepodites can be found in spring, the situation is reversed in summer. These results suggest that the increased availability of *Acartia* spp. during the critical late-larvae stage

has contributed to the high reproductive success and eventually to the unusually high sprat stock during the 1990s. The state of knowledge on herring recruitment in the Baltic is generally lower than for sprat, but the observation that the decreasing growth and condition of herring observed since the 1980s occurred in parallel with a negative trend in recruitment suggests an effect of condition and growth on recruitment. The Gulf populations (e.g. Gulf of Riga) of herring seemed to have profited from the warmer regime, especially through enhanced copepod production assuring larval survival (G. Kornilovs, Latvian Fish Resources Agency, Riga, Latvia, unpublished data). The populations of coastal copepods (e.g. *Acartia* spp.) have increased in the recent warmer period (Möllmann *et al.*, 2000). Unfortunately, almost nothing is known about the larval ecology of the open-sea herring stocks, but the fact that warmer temperatures and enhanced food production did not positively influence recruitment of the Baltic main basin herring stocks points to the importance of the link between growth and condition.

Our present understanding of the effects of variability in climate on herring and sprat populations in the Baltic Sea is summarized in Fig. 7.18. Climate variability affects salinity and oxygen (S/O$_2$) through runoff and inflows of North Sea water, and temperature (T) through direct air–sea interaction. Changes in S/O$_2$ influence *P. acuspes* availability for small pelagic species and hence their growth and condition. Increased inter- and intra-specific competition with the enlarged sprat stock further contributed to herring and sprat growth changes. The large sprat stock during the 1990s is to a large degree a result of high reproductive success, caused by the direct effect of temperature on egg and larval survival, and the indirect effect of *Acartia* spp. availability.

Because menhaden distributions in the northwestern Atlantic are correlated with plankton distributions, there is reasonable evidence to predict where menhaden will distribute and what components of the plankton community they will consume, thus we are equipped to consider what the impact of future changes to climate and ecosystem condition may have on menhaden, and *vice versa*. Climate change impacts and anthropogenic effects on estuaries may be inexorably intertwined since many aspects of climate change could affect the nutrient dynamics of estuaries. The thermal requirement of an organism will be challenged by the anticipated increases in global temperature, but in addition, climate change will produce changes in precipitation, solar radiation, humidity, wind, and potential evaporation over a drainage system that will affect the physical and biological suitability of menhaden habitats (Hulme, 2005). Estuarine habitat for menhaden is already in a fragile state because of chronic nutrient loading and hypoxic conditions (Hagy *et al.*, 2004). Eutrophication is creating hypoxic conditions that are lethal to menhaden populations, in particular

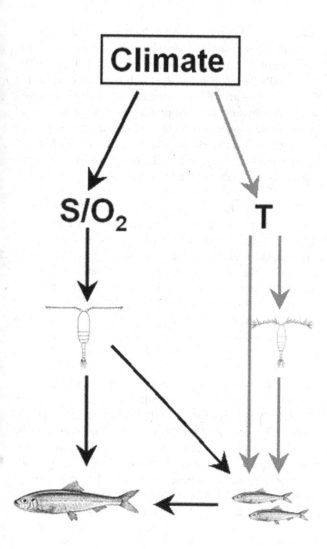

Fig. 7.18. Schematic representing the effects of variability in climate on Baltic herring (left side) and sprat (right side) growth (black arrows), and sprat recruitment (gray arrows), copepod to the left is *Pseudocalanus acuspes* and to the right *Acartia* spp.; *T* – temperature, S – salinity, O_2 – oxygen. The black arrow between sprat and herring indicates competition.

and detrital material, which menhaden have been shown to digest at low levels (Lewis and Peters, 1994). Menhaden may accelerate the deposition of nutrients to the benthos via the formation of their fecal pellets thus affecting the distribution of oxygen-consuming material in the water column and benthos (Lane *et al.*, 1994). Cyanobacteria appear viable after gut passage through menhaden (Friedland *et al.*, 2005), which may contribute to the development of summer bloom patterns in estuaries (Davis *et al.*, 1997). The diminishment of menhaden populations caused by eutrophication and climate change effects will alter these filtering functions provided by menhaden, thus contributing to the instability of estuarine and coastal ecosystems.

Predicted effects of climate change on upwelling systems have been described in this volume, Chapter 14, and changes in productivity are one area of focus of that chapter. Given the trophic difference between sardine and anchovy, and between members of some other small pelagic species pairs, a change in productivity that results in a change in zooplankton composition, particularly zooplankton size composition, seems likely to have an impact on the population variability of and species alternations between these species. Altered wind stress that would enhance coastal upwelling (Bakun, 1990, 1992; Shannon *et al.*, 1996) and result in a trophic environment dominated by large particles would tend to favor anchovy over sardine, whereas a reduction in coastal upwelling could act to favor sardine. Conversely, increases in sea surface temperature (e.g. Roux, 2003) or an increased frequency and intensity of the advection of warm tropical water into upwelling systems (e.g. Siegfried *et al.*, 1990) that result in a trophic environment dominated by small particles may also tend to favor sardine over anchovy. Whilst trophic differentiation represents one possible mechanism that may impact on population variability and species alternations of small pelagic fish, it should not be considered alone, and the effects of other mechanisms on population variability, including those mentioned above, should be considered.

juvenile populations, which are estuarine dependent and more sensitive to oxygen levels than adults (Shimps *et al.*, 2005). Climate change is likely to affect each estuarine system differently due to difference in the physical nature of each estuary and the amount of development and nutrient loading associated with the watershed.

Plankton feeding organisms provide a service to the function of estuaries and coastal ecosystems by removing particles from the water column and facilitating nutrient cycling (Breitburg, 2002; Jung and Houde, 2003, 2005). Menhaden populations are capable of filtering significant quantities of water (Oviatt *et al.*, 1972), which results in the removal of phytoplankton and zooplankton from the water column,

Conclusions

General conclusions regarding the trophic dynamics of small pelagic fish species, in particular the so-called SPACC species (i.e. anchovy and sardine), are listed below:

1. Anchovy (*Engraulis* spp.) and sardine (*Sardinops* spp. and *Sardina pilchardus*) are omnivorous but derive the bulk of their dietary carbon from zooplankton, and previous descriptions of exclusive or near-total phytophagy by these species are erroneous and were based on inappropriate methodology.
2. The two species show ontogenetic changes in diet: larvae and juveniles are typically almost exclusively

---

Box 7.1. Trophically mediated impacts of climate change on anchovy and sardine

An hypothesized outcome of climate change is alterations in physical forcing that are likely to affect the productivity of marine ecosystems and the composition of their lower trophic levels, particularly the phytoplankton (see this volume, Chapter 14 and references therein). If such changes in phytoplankton composition result in changes in the zooplankton community, in particular zooplankton size structure, then this could favor anchovy over sardine or *vice versa*, depending on how zooplankton size has changed.

For example, altered wind stress that results in an increase in the frequency and/or intensity of coastal upwelling could result in a food environment dominated by large particles (i.e. diatoms and larger zooplankton) as suggested in Fig. 7.16, which would provide a more favorable trophic environment for anchovy than for sardine. This enhanced food environment could have a positive impact on reproductive output (e.g. through an increase in the quantity and/or quality of eggs produced during the spawning season) and hence on population size. However, increased upwelling will also result in increased Ekman transport of surface layers that, should it occur over the reproductive season, would increase the dispersal of eggs and larvae to unfavorable offshore habitat. Hence, whereas increased upwelling would have a positive impact on anchovy adults this would be balanced by the negative impact on early life history stages, of both anchovy and sardine. Additionally, increased upwelling would also reduce water temperatures, which could result in different larval survival of the two species under the optimal growth hypothesis. Similarly, the increased stratification of surface waters predicted to arise from global warming will decrease the vertical supply of nutrients to the euphotic zone. Because small dinoflagellates will be less negatively impacted by this reduction in nutrients than the larger diatoms, this could also act to change phytoplankton composition, with subsequent impacts on zooplankton community structure. Such a scenario would provide a more favorable trophic environment for sardine than for anchovy, but the likely overall reduction in productivity could negatively impact both species.

From a trophic perspective the predicted impacts of climate change on anchovy and sardine will be indirect, since such impacts appear more likely to affect the prey of these fish than the fish themselves. These indirect effects may act in concert with, or in opposition to, the direct impacts of climate change on the fish, such as changes in distribution arising from altered habitat or changes in survival through growth. These interactions between direct and indirect impacts of climate change on anchovy and sardine substantially reduce our ability to make predictions regarding the response of small pelagic fish to climate change.

---

zooplanktivorous (an exception being anchovy larvae in the northern Humboldt Current system), and the relative dietary importance of phytoplankton increases with fish size due to elaboration of the branchial basket;

3. Size-based partitioning of the zooplankton resource between anchovy and sardine is typically observed, with sardine feeding on smaller zooplankton than do anchovy;

4. This trophic partitioning arises from different trophic morphologies, with anchovy having a coarser branchial basket and sardine a finer branchial basket, and different feeding behaviors, anchovy being predominantly size-selective particulate feeders and typically showing feeding periodicity whereas sardine appear to be predominantly filter feeders and typically do not show feeding periodicity;

5. Phytoplankton tends to be more important in the diet of sardine than anchovy, and this, together with the trophic partitioning referred to above, means that sardine feed at a lower trophic level than do anchovy;

6. Because of this trophic partitioning, changes in the size and/or species composition of the zooplankton community may be implicated in species alternations between anchovy and sardine, and periods of anchovy or sardine dominance have coincided with periods of changed zooplankton community structure in some of the case studies described above; and

7. From a trophic perspective, climate change appears more likely to have an indirect (i.e. impacting on the food) rather than a direct impact on anchovy and sardine (see Box 7.1).

Acknowledgments

The authors wish to express their thanks and appreciation to Juergen Alheit (Baltic Sea Research Institute, Warnemünde, Germany) and to an anonymous reviewer for constructive comments on an earlier version of the chapter; to Cathy Boucher (Marine and Coastal Management) for redrawing and standardizing the figures; and to the editors of this book for their vision, encouragement and support.

REFERENCES

Alamo, A. (1989). Stomach contents of anchoveta (*Engraulis ringens*), 1974–1982. In *The Peruvian Upwelling*

Ecosystem: Dynamics and Interactions, Pauly, D., Muck, P., Mendo, J., and Tsukayama, I., eds. *ICLARM Conf. Proc.* **18**: 105–108.

Alamo, A. and Espinoza, P. (1998). Variaciones alimentarias en *Engraulis ringens* y otros recursos pelágicos durante invierno-primavera de 1997. *Inf. Inst. Mar Perú* **130**: 45–52 (in Spanish with English abstract).

Alamo, A., Navarro, I., Espinoza, P., and Zubiate, P. (1996a). Espectro alimentario y ración de alimentación de *Engraulis ringens* y de *Sardinops sagax sagax*, y mortalidad de huevos de la anchoveta peruana por predación. *Inf. Inst. Mar Perú* **119**: 34–42 (in Spanish with English abstract).

Alamo, A., Navarro, I., Espinoza, P., and Zubiate, P. (1996b). Relaciones tróficas, espectro alimentario y ración de alimentación de las principales especies pelágicas en el verano 1996. *Inf. Inst. Mar Perú* **122**: 36–46 (in Spanish with English abstract).

Alamo, A., Espinoza, P., Zubiate, P., and Navarro, I. (1997a). Comportamiento alimentario de la anchoveta peruana *Engraulis ringens*, durante el invierno de 1996. Crucero BIC Humboldt 9608–09. *Inf. Inst. Mar Perú*, **123**: 38–46 (in Spanish with English abstract).

Alamo, A., Espinoza, P., Zubiate, P., and Navarro, I. (1997b). Comportamiento alimentario de los principales recursos pelágicos peruanos en verano y comienzos de otoño 1997. *Inf. Inst. Mar Perú*, **127**: 82–89 (in Spanish with English abstract).

Albaina, A. and Irigoien, X. (2004). Relationships between frontal structures and zooplankton communities along a cross-shelf transect in the Bay of Biscay (1995–2003). *Mar. Ecol. Prog. Ser.* **284**: 65–75.

Alheit, J. (1987). Egg cannibalism versus egg predation: their significance in anchovies. In *The Benguela and Comparable Ecosystems*, Payne, A.I.L., Gulland, J.A., and Brink, K.H., eds. *S. Afr. J. Mar. Sci.* **5**: 467–470.

Alheit, J. and Ñiquen, M. (2004). Regime shifts in the Humboldt Current ecosystem. *Prog. Oceanogr.* **60**: 201–222.

Alheit, J., Möllmann, C., Dutz, J., *et al.* (2005). Synchronous ecological regime shifts in the central Baltic and the North Sea in the late 1980s. *ICES J. Mar. Sci.* **62** (7): 1205–1215.

Anadón, E. (1950). Sobre la sustitución alternativa en el litoral gallego de los llamados peces emigrantes (sardina, espadín, anchoa y jurel). *Bol. Inst. Esp. Oceanogr.* **24**: 1–20 (in Spanish).

Andreu, B. (1953). Sobre la relación entre el número de branquispinas y la talla de la sardina (*Sardina pilchardus*, Walb.) española. *Bol. Inst. Esp. Oceanogr.* **62**: 3–28 (in Spanish).

Andreu, B. (1960). Sobre la aparición de las branquispinas en las formas juveniles de sardina (*Sardina pilchardus* Walb.). *Bol. R. Soc. Esp. Hist. Nat., Sec. Biol.* **58**: 199–216 (in Spanish).

Andreu, B. (1969). Las branquispinas en la caracterización de las poblaciones de *Sardina pilchardus* (Walb.). *Inv. Pesq.* **33**: 425–607 (in Spanish).

Arim, M. and Marquet, P. A. (2004). Intraguild predation: a widespread interaction related to species biology. *Ecol. Lett.* **7**: 557–564.

Aro, E. (1989). A review of fish migration patterns in the Baltic. *Rapp. P.-v. Réu. Cons. Int. l'Exp. Mer.* **190**: 72–96.

Aro, E., Uitto, A., Vuorinen, I., and Flinkman, J. (1986). The food selection of Baltic herring in late summer in the northern Baltic. *ICES CM:1986/J:* **26**: 1–19.

Arrhenius, F. and Hansson, S. (1993). Food consumption of larval, young and adult herring and sprat in the Baltic Sea. *Mar. Ecol. Prog. Ser.* **96**: 125–137.

Arrhenius, F. and Hansson, S. (1999). Growth of Baltic Sea young-of-the-year herring *Clupea harengus* is resource limited. *Mar. Ecol. Prog. Ser.* **191**: 295–299.

Arrizaga, A. (1983). Variación estacional en la alimentación de la sardina común *Clupea* (*Strangomera*) *bentincki* Norman 1936 (Pisces, Clupeidae) in the Bio-Bio Región, Chile. *Bol. Soc. Biol. Concepción, Chile* **54**: 7–26 (in Spanish).

Arrizaga, A. and Inostroza, I. (1979). Estudio preliminar del contenido estomacal de la sardina común, *Clupea* (*Strangomera*) *bentincki*, Norman 1936, Pices, Clupeidae, en la VIII Región-Chile. *Acta Zool. Lilloana* **35**: 509–515 (in Spanish).

Arrizaga, A., Fuentealba, M., Espinoza, C., *et al.* (1993). Hábitos tróficos de dos especies de peces pelágicos: *Strangomera bentincki* (Norman, 1936) y *Engraulis ringens* (Jenyns, 1842) en el litoral de la región del Biobío, Chile. *Bol. Soc. Biol. Concepción Chile* **64**: 27–35 (in Spanish).

Arthur, D. K. (1956). The particulate food and the food resources of the larvae of three pelagic fishes, especially the Pacific sardine *Sardinops caerulea* (Girard). Ph.D. thesis, University of California, Scripps Institution of Oceanography, La Jolla, California, USA, 231 pp.

Arthur, D. K. (1976). Food and feeding of larvae of three fishes ocurring in the California Current, *Sardinops sagax*, *Engraulis mordax* and *Trachurus symmetricus*. *Fish. Bull.*, **74** (3): 517–529.

Azuma, N. (1994). Behavioral ecology of feeding in Japanese sardine *Sardinops melanostictus*. Ph.D. thesis, University of Tokyo, Tokyo, Japan, 283 pp. (in Japanese).

Bakun, A. (1990). Global climate change and intensification of coastal upwelling. *Science* **247**: 198–201.

Bakun, A. (1992). Global greenhouse effects, multidecadal wind trends, and potential impacts on coastal pelagic fish populations. *ICES Mar. Sci. Symp.* **195**: 316–325.

Bakun, A. and Broad, K. (2003). Environmental 'loopholes' and fish population dynamics: comparative pattern recognition with focus on El Niño effects in the Pacific. *Fish. Oceanogr.* **12**: 458–473.

Balbontín, F., Carretó N.M., and Maureira, J. (1979). Estudio experimental sobre selección de alimento y comportamiento alimentario en anchoveta y sardina de Chile (Pises, Clupeiformes). *Rev. Biol. Mar., Valparaíso* **16** (3): 211–220 (in Spanish).

Balbontín, F., Llanos, A., and Valenzuela, V. (1997). Sobreposición trófica e incidencia alimentaria en larvas de peces de Chile central. *Revis. Chilena Hist. Nat.* **70**: 381–390 (in Spanish).

Baumann, H., Hinrichsen, H. -H., Möllmann, C., *et al.* (2006). Recruitment variability in Baltic Sea sprat (*Sprattus sprattus*) is tightly coupled to temperature and transport patterns affecting the larval and early juvenile stages. *Can. J. Fish. Aquat. Sci.* **60** (10): 2191–2201.

Baxter, J.L. (1967). Summary of biological information on the northern anchovy *Engraulis mordax* Girard. *CalCOFI Rep.* **11**: 110–116.

Bensam, P. (1964). Differences in the food and feeding adaptations between juveniles and adults of the Indian oil sardine, *Sardinella longiceps* Valenciennes. *Indian J. Fish. Ser. A* **11**: 377–390.

Bergeron, J.P. (2000). Effect of strong winds on the nutritional condition of anchovy (*Engraulis encrasicolus* L.) larvae in the Bay of Biscay, Northeast Atlantic, as inferred from an early field application of the DNA/C index. *ICES J. Mar. Sci.* **57**: 249–255.

Berner, L., Jr. (1959). The food of the larvae of the northern anchovy *Engraulis mordax*. *Inter-Amer. Trop. Tuna Comm.* **4** (1): 1–22.

Bertrand, A., Segura, M., Gutiérrez, M., and Vásquez, L. (2004). From small-scale habitat loopholes to decadal cycles: a habitat-based hypothesis explaining fluctuation in pelagic fish populations off Peru. *Fish Fisher.* **5**: 296–316.

Blaskovic, V., Torriani, F., and Navarro, I. (1998). Características tróficas de las principales especies pelágicas durante el otoño 1998. Crucero BIC José Olaya Balandra 9805–06 de Tacna a Máncora. *Inf. Inst. Mar Perú* **137**: 72–79 (in Spanish with English abstract).

Blaxter, J.H.S. and Hunter, J.R. (1982). The biology of the clupeoid fishes. *Adv. Mar. Biol.* **20**: 1–223.

Bode, A., Carrera, P., and Lens, S. (2003). The pelagic foodweb in the upwelling ecosystem of Galicia (NW Spain) during spring: natural abundance of stable carbon and nitrogen isotopes. *ICES J. Mar. Sci.* **60**: 11–22.

Bode, A., Alvarez-Ossorio, M.T., Carrera, P., and Lorenzo, J. (2004). Reconstruction of trophic pathways between plankton and the North Iberian sardine (*Sardina pilchardus*) using stable isotopes. *Sci. Mar.* **68**: 165–178.

Bode, A., Carrera, P., and Porteiro, C. (2006). Stable nitrogen isotopes reveal weak dependence of trophic position of planktivorous fishes from individual size: a consequence of omnivorism and mobility. In *International Conference on Isotopes and Environmental Studies*, Povinec, P. and Sánchez-Cabeza, J.A., eds. *Isotopes in the Environment* Vol. 8, Amsterdam, the Netherlands: Elsevier, pp. 281–293.

Booi, K. (2000). Characterization and comparison of diets of anchovy (*Engraulis capensis*) and sardine (*Sardinops sagax*) pre-recruits from mixed shoals in the southern Benguela ecosystem. M. Phil. thesis, University of Bergen, Bergen, Norway, v, 74 pp.

Borja, A., Uriarte, A., Egaña, J., *et al.* (1998). Relationship between anchovy (*Engraulis encrasicolus* L.) recruitment and environment in the Bay of Biscay. *Fish. Oceanogr.* **7**: 375–380.

Breitburg, D. (2002). Effects of hypoxia, and the balance between hypoxia and enrichment, on coastal fishes and fisheries. *Estuaries* **25**: 767–781.

Brodeur, R.D., Lorz, H.V., and Pearcy, W.G. (1987). Food habits and dietary variability of pelagic nekton off Oregon and Washington, 1979–1984. *NOAA Tech. Rep. NMFS* **57**: 32 pp.

Cardinale, M. and Arrhenius, F. (2000). Decreasing weight-at-age of Atlantic herring (*Clupea harengus*) from the Baltic Sea between 1986 and 1996: a statistical analysis. *ICES J. Mar. Sci.* **57**: 882–893.

Cardinale, M., Casini, M., and Arrhenius, F. (2002). The influence of biotic and abiotic factors on the growth of sprat (*Sprattus sprattus*) in the Baltic Sea. *Aquat. Living Res.* **15**: 273–282.

Casini, M., Cardinale, M., and Arrhenius, F. (2004). Feeding preferences of herring (*Clupea harengus*) and sprat (*Sprattus sprattus*) in the southern Baltic Sea. *ICES J. Mar. Sci.* **61**: 1267–1277.

Casini, M., Cardinale, M., and Hjelm, J. (2006). Inter-annual variation in herring (*Clupea harengus*) and sprat (*Sprattus sprattus*) condition in the central Baltic Sea: what gives the tune? *Oikos*, **112**, 638–650.

Castillo-Rivera, M., Kobelkowsky, A., and Zamayoa, V. (1996). Food resource partitioning and trophic morphology of *Brevoortia gunteri* and *B. patronus*. *J. Fish. Biol.* **49**: 1102–1111.

Chavez, F.P., Pennington, J.T., Castro, C.G., *et al.* (2002). Biological and chemical consequences of the 1997–1998 El Niño in central California waters. *Prog. Oceanogr.* **54**: 205–232.

Chavez, F.P., Ryan, J.P., Lluch-Cota, S.E., and Ñiquen, M. (2003). From anchovies to sardines and back: multidecadal change in the Pacific Ocean. *Science* **299**: 217–221.

Chiappa-Carrara, X. and Gallardo-Cabello, M. (1993). Feeding behavior and dietary composition of the northern anchovy, *Engraulis mordax* Girard (Pisces: Engraulidae), off Baja California, Mexico. *Cienc. Mar.* **19** (3): 285–305.

Chipman, W.A. (1959). Use of radioisotopes in studies of the foods and feeding activities of marine animals. *Publ. Sta. Zool. Napoli* **31** (suppl.): 154–175.

CICIMAR. (1983). Pesquería de Sardina en Baja California Sur. *Informe Final a la Secretaria de Pesca. Tech. Rep. September* **20**, 1983, 279 pp. (in Spanish).

Coetzee, J.C. (1997). Acoustic investigation of the shoaling dynamics of sardine *Sardinops sagax* populations: implications for acoustic surveys. M.Sc. thesis, University of Cape Town, Cape Town, South Africa, 117 pp.

Conway, D.V.P., Coombs, S.H., Fernández de Puelles, M.L., and Tranter, P.R.G. (1994). Feeding of larval sardine, *Sardina pilchardus* (Walbaum), off the north coast of Spain. *Bol. Inst. Esp. Oceanogr.* **10** (2): 165–175.

Conway, D.V.P., Coombs, S.H., and Smith, C. (1998). Feeding of anchovy *Engraulis encrasicolus* larvae in the northwestern Adriatic Sea in response to changing hydrobiological conditions. *Mar. Ecol. Prog. Ser.* **175**: 35–49.

Costa, P.R. and Garrido, S. (2004). Domoic acid accumulation in the sardine *Sardina pilchardus* and its relationship to *Pseudo-nitzschia* diatom ingestion. *Mar. Ecol. Prog. Ser.* **284**: 261–268.

Cunha, M.E., Garrido, S., and Pissarra, J. (2005). The use of stomach fullness and colour indices to assess *Sardina pilchardus* feeding. *J. Mar. Biol. Ass. UK* **85**: 425–431.

Cushing, D. H. (1978). Upper trophic levels in upwelling areas. In *Upwelling Ecosystems*, Boje, R. and Tomczak, M., eds. New York: Springer-Verlag, pp. 101–110.

Davidyuk, A. (1996). Herring and sprat feeding in 1994 and 1995 in the eastern Baltic. *ICES CM:1996/J:***24**: 1–7.

Davidyuk, A., Fetter, M., and Hoziosky, S. (1992). Feeding and growth of Baltic herring (*Clupea harengus m. membras* L.). *ICES CM:1992/J:***27**: 1–8.

Davies, D. H. (1957). The South African pilchard (*Sardinops ocellata*). Preliminary report on feeding off the west coast, 1953–56. *Investl. Rep. Div. Fish. S. Afr.* **30**: 1–40.

Davis, L. N., Phillips, K. A., and Marshall, H. G. (1997). Seasonal abundance of autotrophic picoplankton in the Pagan River, a nutrient enriched subestuary of the James River, Virginia. *Virginia J. Sci.* **48**: 211–218.

De Buen, F. (1958). Peces de la superfamilia Clupeoidae, en aguas de Chile. *Rev. Biol. Mar.* **8** (1–3): 83–110 (in Spanish).

De Silva, S. S. (1973). Food and feeding habits of the herring *Clupea harengus* and the sprat *C. sprattus* in inshore waters of the west coast of Scotland. *Mar. Biol.* **20**: 282–290.

Durbin, A. G. (1979). Food selection by plankton feeding fishes. In *Predator–Prey Systems in Fisheries Management*, Clepper, H., ed. Washington, DC: Sport Fishing Institute, pp. 203–218.

Durbin, A. G. and Durbin, E. G. (1975). Grazing rates of the Atlantic menhaden, *Brevoortia tyrannus*, as a function of particle size and concentration. *Mar. Biol.* **33**: 265–277.

Emmett, R. L., Brodeur, R. D., Miller, T. W., *et al.* (2005). Pacific sardines (*Sardinops sagax*) abundance, distribution, and ecological relationships in the Pacific Northwest. *CalCOFI Rep.* **46**: 122–143.

Espinoza, P. and Bertrand, A. (2008). Revisiting Peruvian anchovy (*Engraulis ringens*) trophodynamics provides a new vision of the Humboldt Current system. In *The Northern Humboldt Current System: Ocean Dynamics, Ecosystem Processes, and Fisheries*, Bertrand, A., Guevara-Carrasco, R., Soler, P., *et al.*, eds. *Prog. Oceanog.* **79**: 215–227.

Espinoza, P., Navarro, I., and Torriani, F. (1998a). Variaciones en el espectro alimentario de los principales recursos pelágicos durante otoño 1998. Crucero BIC Humboldt 9803–05 de Tumbes a Tacna. *Inf. Inst. Mar Perú* **135**: 134–142 (in Spanish with English abstract).

Espinoza, P., Blaskovic, V., and Navarro, I. (1998b). Comportamiento alimentario de *Engraulis ringens*, a finales del invierno 1998. Crucero de evaluación hidroacústica de recursos pelágicos 9808–09. *Inf. Inst. Mar Perú* **141**: 67–71 (in Spanish with English abstract).

Espinoza, P., Blaskovic, V., Torriani, F., and Navarro, I. (1999). Dieta de la anchoveta *Engraulis ringens* según intervalos de talla. Crucero BIC José Olaya Balandra y BIC Humboldt 9906. *Inf. Inst. Mar Perú* **149**: 41–48 (in Spanish with English abstract).

Espinoza, P., Navarro, I., and Torriani, F. (2000). Variaciones espaciales en la dieta de la anchoveta a finales de la primavera 1999, Crucero BICs José Olaya Balandra y SNP. 2 9911–12. *Inf. Inst. Mar Perú* **157**: 72–76 (in Spanish with English abstract).

Fernández, I. M. and González-Quirós, R. (2006). Analysis of feeding of *Sardina pilchardus* (Walbaum, 1792) larval stages in the central Cantabrian Sea. *Sci. Mar.* **70** (S1): 131–139.

Ferreira, P. L. and Ré, P. (1993). Feeding of larval anchovy, *Engraulis encrasicolus* (L.), in the Mira Estuary (Portugal). *Port. Zool.* **2** (3): 1–37.

Fetter, M. and Davidyuk, A. (1993). Herring feeding and growth in the eastern Baltic Sea during 1977–1990. *ICES CM:1993/J:***27**: 1–6.

Fiedler, P. C. (2002). Environmental change in the eastern tropical Pacific Ocean: review of ENSO and decadal variability. *Mar. Ecol. Prog. Ser.* **244**: 265–283.

Fiksen, O., Eliassen, S., and Titelman, J. (2005). Multiple predators in the pelagic: modelling behavioural cascades. *J. Anim. Ecol.* **74**: 423–429.

Flinkman, J., Vuorinen, I., and Aro, E. (1992). Planktivorous Baltic herring (*Clupea harengus*) prey selectively on reproducing copepods and cladocerans. *Can. J. Fish. Aquat. Sci.* **49**: 73–77.

Flinkman, J., Aro, E., Vuorinen, I., and Viitasaalo, M. (1998). Changes in northern Baltic zooplankton and herring nutrition from 1980s to 1990s: top-down and bottom-up processes at work. *Mar. Ecol. Prog. Ser.* **165**: 127–136.

Friedland, K. D. (1985). Functional morphology of the branchial basket structures associated with feeding in the Atlantic menhaden, *Brevoortia tyrannus* (Pisces: Clupeidae). *Copeia* **10**: 1018–1027.

Friedland, K. D., Haas, L. W., and Merriner, J. V. (1984). Filtering rates of the juvenile Atlantic menhaden *Brevoortia tyrannus* (Pisces: Clupeidae), with consideration of the effects of detritus and swimming speed. *Mar. Biol.* **84**: 109–117.

Friedland, K. D., Ahrenholz, D. W., and Guthrie, J. F. (1989). Influence of plankton on distribution patterns of the filter-feeder *Brevoortia tyrannus* (Pisces: Clupeidae). *Mar. Ecol. Prog. Ser.* **54**: 1–11.

Friedland, K. D., Ahrenholz, D. W., and Guthrie, J. F. (1996). Formation and seasonal evolution of Atlantic menhaden juvenile nurseries in coastal estuaries. *Estuaries* **19**: 105–114.

Friedland, K. D., Ahrenholz, D. W., and Haas, L. W. (2005). Viable gut passage of cyanobacteria through the filter-feeding fish Atlantic menhaden, *Brevoortia tyrannus*. *J. Plank. Res.* **27**: 715–718.

Friedland, K. D., Ahrenholz, D. W., Smith, J. W., *et al.* (2006). Sieving functional morphology of the gill raker feeding apparatus of Atlantic menhaden. *J. Exp. Zoo. Part A – Comp. Exp. Biol.* **305A** (12): 974–985.

Froese, R. and Pauly, D. (2006). FishBase. Version 03/2006. World Wide Web electronic publication (http://www.fishbase.org).

Garrido, S. (2002). Alimentaçao de *Sardina pilchardus* (Walbaum, 1792) ao largo da costa continental portuguesa e implicaçoes da condiçao nitricional das fêmeas na qualidade dos oócitos. M.Sc. thesis, Universidade do Porto, Porto, Portugal, 85 pp. (in Portuguese)

Garrido, S., Ben-Hamadou, R., Oliveira, P. B., *et al.* (2008a). Diet and feeding intensity of sardine *Sardina pilchardus*:

correlation with satellite-derived chlorophyll data. *Mar. Ecol. Prog. Ser.* **354**: 245–256.

Garrido, S., Marçalo, A., Zwolinski, J., and van der Lingen, C.D. (2007a). Laboratory investigations on the effect of prey size and concentration on the feeding behaviour of *Sardina pilchardus*. *Mar. Ecol. Prog. Ser.* **330**: 189–199.

Garrido, S., Rosa, R., Ben-Hamadou, R., *et al.* (2007b). Effect of maternal fat reserves on the fatty acid composition of sardine (*Sardina pilchardus*) oocytes. *Comp. Biochem. Physiol. B.* **148** (4): 398–409.

Garrido, S., Rosa, R., Ben-Hamadou, R., *et al.* (2008b). Spatio-temporal variability in fatty acid trophic biomarkers in stomach contents and muscle of Iberian sardine (*Sardina pilchardus*) and its relationship with spawning *Mar. Biol.* **154**: 1053–1065.

Gibson, R.N. and Ezzi, I.A. (1992). The relative probability of particulate-feeding and filter-feeding in the herring, *Clupea harengus* L. *J. Fish. Biol.* **40**: 577–590.

Guisande, C., Vergara, A.R., Riveiro, I., and Cabanas, J.M. (2004). Climate change and abundance of the Atlantic–Iberian sardine (*Sardina pilchardus*). *Fish. Oceanogr.* **13**: 91–101.

Hagy, J.D., Boynton, W.R., Keefe, C.W., and Wood, K.V. (2004). Hypoxia in Chesapeake Bay, 1950–2001: long-term change in relation to nutrient loading and river flow. *Estuaries* **27**: 634–658.

Hampton, I., Agenbag, J.J., and Cram, D.L. (1979). Feasibility of assessing the size of the South West African pilchard stock by combined aerial and acoustic measurements. *Fish. Bull. S. Afr.* **11**: 10–22.

Hansen, F.C., Möllmann, C., Schütz, U., and Neumann, T. (2006). Spatio-temporal distribution and production of calanoid copepods in the Central Baltic Sea. *J. Plank. Res.* **28**: 39–54.

Hansson, S., Larsson, U., and Johannson, S. (1990). Selective predation by herring and mysids, and zooplankton community structure in a Baltic Sea coastal area. *J. Plank. Res.* **12**: 1099–1116.

Hardy, A.C. (1924). The herring in relation to its animate environment. *Fish. Invest. Ser. II.* **VII** Nr. (3): 1–53.

Hart, J.L. and Wailes, C.H. (1932). The food of pilchard, *Sardinops caerulea* (Girard), off the coast of British Columbia. *Contrib. Can. Biol. Fish.* **7** (19): 247–254.

Hirakawa, K. and Goto, G. (1996). Diet of larval sardine, *Sardinops melanostictus* in Toyama Bay, southern Japan Sea. *Bull. Jap. Sea Nat. Fish. Res. Inst.* **46**: 65–75 (in Japanese with English abstract).

Hirakawa, K. and Ogawa, Y. (1996). Characteristics of the copepod assemblage in the southwestern Japan Sea and its implication for anchovy population dynamics. *Bull. Jap. Sea Natl. Fish. Res. Inst.* **46**: 45–64.

Hirakawa, K., Goto, T., and Hirai, M. (1997). Diet composition and prey size of larval anchovy, *Engraulis japonicus*, in Toyama Bay, southern Japan Sea. *Bull. Jap. Sea Natl. Fish. Res. Inst.* **47**: 67–78.

Hiramoto, K. (1985). Life pattern of the Japanese sardine, *Sardinops melanosticta*, and the fluctuations on its stock size. *Aquabiol.* **7** (3): 170–182 (in Japanese with English abstract).

Hirota, Y., Honda, H., Ichikawa, T., and Mitani, T. (2003). Stomach contents of round herring *Etrumeus teres* in Tosa Bay. *Fish. Biol. Oceanog. Kuroshio* **4**: 35–44 (in Japanese).

Hsieh, C.-H., Glaser, S.M., Lucas, A.J., and Sugihara, G. (2005). Distinguishing random environmental fluctuations from ecological catastrophes for the North Pacific Ocean. *Nature* **435** (7040): 336–340.

Hückstädt, L.A., Rojas, C.P., and Antezana, T. (2007). Stable isotope analysis reveals pelagic foraging by the Southern sea lion in central Chile. *J. Exp. Mar. Biol. Ecol.* **347**: 123–133.

Hulbert, E.M. (1963). The diversity of phytoplanktonic populations in oceanic, coastal, and estuarine regions. *J. Mar. Res.* **21**: 81–93.

Hulme, P.E. (2005). Adapting to climate change: is there scope for ecological management in the face of a global threat? *J. Appl. Ecol.* **42**: 784–794.

Hunter, J.R. (1972). Swimming and feeding behavior of larval anchovy *Engraulis mordax*. *Fish. Bull. US* **70**(3): 821–838.

Hunter, J.R. (1981). Feeding ecology and predation of marine fish larvae. In *Marine Fish Larvae. Morphology, Ecology and Relation to Fisheries*, Lasker. R., ed. Seattle: Washington Sea Grant Program, pp. 33–77.

Hunter, J.R. and Kimbrell, C.A. (1980). Egg cannibalism in the northern anchovy *Engraulis mordax*. *Fish. Bull.* **78**: 811–816.

ICES (2005). Report of the ICES Advisory Committee on Fishery Management, Advisory Committee on the Marine Environment and Advisory Committee on Ecosystems, 2005, 7. International Council for the Exploration of the Sea, Copenhagen, 117 pp.

ICES (2006). Report of the Baltic Fisheries Assessment Working Group. *ICES CM:2006/ACFM24*, 672 pp.

ICES (2007) Report of the ICES/GLOBEC Workshop on Long-term Variability in SW Europe (WKLTVSWE). *ICES CM2007/LRC02*, 111 pp.

Irie, T., Kobayashi, T., and Inomata, A. (1979). Studies on herring in the northern Okhotsk Sea. I. Feeding habits. *Bull. Hokk. Reg. Fish. Res. Lab.* **44**: 25–37 (in Japanese with English abstract).

James, A.G. (1987). Feeding ecology, diet and field-based studies on feeding selectivity of the Cape anchovy *Engraulis capensis* Gilchrist. In *The Benguela and Comparable Ecosystems*, Payne, A.I.L., Gulland, J.A., and Brink, K.H., eds. *S. Afr. J. Mar. Sci.* **5**: 597–611.

James, A.G. (1988). Are clupeoid microphagists herbivorous or omnivorous? A review of the diets of some commercially important clupeids. *S. Afr. J. Mar. Sci.* **7**: 161–177.

James, A.G. and Chiappa-Carrara, X. (1990). A comparison of field based studies on the trophic ecology of *Engraulis capensis* and *E. mordax*. In *Trophic Relationships in the Marine Environment*, Barnes, M., and Gibson, R.N., eds. Aberdeen, UK: Aberdeen University Press, pp. 208–221.

James, A. G. and Findlay, K. P. (1989). Effect of particle size and concentration on feeding behaviour, selectivity and rates of food ingestion by the Cape anchovy *Engraulis capensis*. *Mar. Ecol. Prog. Ser.* **50**: 275–294.

James, A. G. and Probyn, T. (1989). The relationship between respiration rate, swimming speed and feeding behaviour in Cape anchovy *Engraulis capensis* Gilchrist. *J. Exp. Mar. Biol. Ecol.* **131**: 81–100.

James, A. G., Probyn, T., and Hutchings, L. (1989a). Laboratory-derived carbon and nitrogen budgets for the omnivorous planktivore *Engraulis capensis* Gilchrist. *J. Exp. Mar. Biol. Ecol.* **131**: 124–145.

James, A. G., Probyn, T., and Seiderer, L. J. (1989b). Nitrogen excretion and absorption efficiencies of the Cape anchovy *Engraulis capensis* Gilchrist fed upon a variety of plankton diets. *J. Exp. Mar. Biol. Ecol.* **131**: 101–124.

Jeffries, H. P. (1975). Diets of juvenile Atlantic menhaden (*Brevoortia tyrannus*) in three estuarine habitats as determined from fatty acid composition of gut contents. *J. Fish. Res. Bd. Can.* **32**: 587–592.

June, F. C. and Carlson, F. T. (1971). Food of young Atlantic menhaden, *Brevoortia tyrannus*, in relation to metamorphosis. *Fish. Bull.* **68**: 493–512.

Jung, S. and Houde, E. D. (2003). Spatial and temporal variabilities of pelagic fish community structure and distribution in Chesapeake Bay, USA. *Estuar. Coast. Shelf. Sci.* **58**: 335–351.

Jung, S. and Houde, E. D. (2005). Fish biomass size spectra in Chesapeake Bay. *Estuaries* **28**: 226–240.

Kawasaki, T. and Kumagai, A. (1984). Food habits of the far eastern sardine and their implication in the fluctuation pattern of the sardine stocks. *Bull. Jap. Soc. Sci. Fish.* **50** (10): 1657–1663.

Kemmerer, A. J. (1980). Environmental preferences and behavior patterns of gulf menhaden (*Brevoortia patronus*) inferred from fishing and remotely sensed data. In *Fish Behavior and its uses in the Capture and Culture of Fishes*, Bardach, J. E., Magnuson, J. J., May, R. C. and Reinhart, J. M., eds. *ICLARM Conf. Proc.* **5**: 345–370.

Kemmerer, A. J., Benigno, J. A., Reese, G. B., and Minkler, F. C. (1974). Summary of selected early results from the ERTS-1 menhaden experiment. *Fish. Bull.* **72**: 375–389.

Kidachi, T. (1968). Maiwashi no shokusei ni tsuite (Feeding habit of sardine). *Choki Gyokaikyo Yohou, Tokaiku* **15**: 14–15 (in Japanese).

Kimmel, D. G., Roman, M. R., and Zhang, X. (2006). Spatial and temporal variability in factors affecting mesozooplankton dynamics in Chesapeake Bay: evidence from biomass size spectra. *Limnol. Oceanogr.* **51**: 131–141.

Kimura, S., Kasai, A., Nakata, H., *et al.* (1997). Biological productivity of meso-scale eddies caused by frontal disturbances in the Kuroshio. *ICES J. Mar. Sci.* **54**: 179–192.

Kimura, S., Nakata, H., and Okazaki, Y. (2000a). Biological production in meso-scale eddies caused by frontal disturbances of the Kuroshio Extension. *ICES J. Mar. Sci.* **57**: 133–142.

Kimura, R., Watanabe, Y. and Zenitani, H. (2000b). Nutritional condition of first-feeding larvae of Japanese sardine in the coastal and oceanic waters along the Kuroshio Current. *ICES J. Mar. Sci.* **57**: 240–248.

King, D. P. F. and Macleod, P. R. (1976). Comparison of the food and filtering mechanism of pilchard *Sardinops ocellata* and anchovy *Engraulis capensis* off South West Africa, 1971–1972. *Investl. Rep. Sea Fish. Brch. S. Afr.* **111**: 1–29.

Kishinoue, K. (1908). Iwashi Gyogyou Chousa (Notes on the natural history of the sardine (*Clupea melanosticta* Schlegel). *Suisan Chousa Houkoku (J. Imp. Fish. Bur. Tokyo)* **14** (3): 71–110 (in Japanese).

Konchina, Y. V. (1991). Trophic status of the Peruvian anchovy and sardine. *J. Ichthyol.* **31**: 59–72.

Koslow, J. A. (1981). Feeding selectivity of schools of northern anchovy, *Engraulis mordax*, in the southern California Bight. *Fish. Bull. Wash.* **79** (1): 131–142.

Köster, F. W. and Möllmann, C. (2000). Egg cannibalism in Baltic sprat *Sprattus sprattus*. *Mar. Ecol. Prog. Ser.* **196**: 269–277.

Köster, F. W. and Schnack, D. (1994). The role of predation on early life stages of cod in the Baltic. *Dana* **10**: 179–201.

Köster, F. W., Möllmann, C., Neuenfeldt, S., *et al.* (2003). Fish stock development in the Central Baltic Sea (1976–2000) in relation to variability in the physical environment. *ICES Mar. Sci. Symp.* **219**: 294–306.

Kruger, I. and Cruickshank, R. A. (1982). Environmental aspects of a few pelagic fish shoals off South West Africa. *Fish. Bull. S. Afr.* **16**: 99–114.

Kuwahara, A. and Suzuki, S. (1984). Diurnal changes in vertical distributions of anchovy eggs and larvae in the western Wakasa Bay. *Bull. Jap. Soc. Sci. Fish.* **50** (8): 1285–1292 (in Japanese with English abstract).

Lane, P. V. Z., Smith, S. L., Urban, J. L., and Biscaye, P. E. (1994). Carbon flux and recycling associated with zooplanktonic fecal pellets on the shelf of the Middle Atlantic Bight. *Deep-Sea Res.* **41**: 437–457.

Last, J. M. (1989). The food of herring, *Clupea harengus*, in the North Sea, 1983–1986. *J. Fish Biol.* **34**: 489–501.

Lebour, M. V. (1920). The food of young fish. No. III (1919). *J. Mar. Biol. Ass. UK* **12**, 261–324.

Lebour, M. V. (1921). The food of young clupeoids. *J. Mar. Biol. Ass. UK* **12**: 458–467.

Leong, R. J. H. and O'Connell, C. P. (1969). A laboratory study of particulate and filter-feeding of northern anchovy (*Engraulis mordax*). *J. Fish. Res. Bd. Can.* **26**: 557–582.

Lewis, R. C. (1929). The food habits of the California sardine in relation to the seasonal distribution of microplankton. *Bull. Scripps Inst. Oceanog., Univ. Calif. (Tech. Ser.)* **2** (3): 155–180.

Lewis, V. P. and Peters, D. S. (1994). Diet of juvenile and adult Atlantic menhaden in estuarine and coastal habitats. *Trans. Am. Fish. Soc.* **123**: 803–810.

Li, X. Y., Kawasaki, T. and Honda, H. (1992). The niches of the far eastern sardine and Japanese anchovy. *Asian Fish. Sci.* **5**: 315–326.

Llanos, A., Herrera, G., and Bernal, P. (1996). Análisis del tamaño de las presas en la dieta de las larvas de cuatro clupeiformes en un área costera de Chile central. *Sci. Mar.* **60**: 435–442 (in Spanish with English abstract).

Lluch-Belda, D., Elourduy-Garay, J., Lluch-Cota, S.E., and Ponce-Díaz, G., eds. (2000). *BAC. Centros de actividad biológica del Pacífico mexicano*. La Paz, B.C.S. México, CIBNOR-CICIMAR-CONACyT., 367 pp. (in Spanish).

Longhurst, A.R. (1971). The clupeoid resources of tropical seas. *Oceanogr. Mar. Biol. Ann. Rev.* **9**: 349–385.

López-Martínez, J. (1991). Alimentación de juveniles y adultos de sardina Monterrey *Sardinops sagax caeruleus* (Girard), en el Norte de Isla Tiburón durante Invierno de 1990. M.Sc. thesis, CICESE, Ensenada, B.C., México, 127 pp. (in Spanish).

Loukashkin, A.S. (1970). On the diet and feeding behavior of the northern anchovy, *Engraulis mordax* (Girard). *Proc. Calif. Acad. Sci. Ser. 4* **37**: 419–458.

Louw, G.G., van der Lingen, C.D., and Gibbons, M.J. (1998). Differential feeding by sardine *Sardinops sagax* and anchovy *Engraulis capensis* recruits in mixed shoals. In *Benguela Dynamics: Impacts of Variability on Shelf-Sea Environments and their Living Resources*, Pillar, S.C., Moloney, C.L., Payne, A.I.L. and Shillington, F.A., eds. *S. Afr. J. Mar. Sci.* **19**: 227–232.

MacKenzie, B.R. and Köster, F.W. (2004). Fish production and climate: sprat in the Baltic Sea. *Ecology* **85**: 784–794.

MacNeill, D.B. and Brandt, S.B. (1990). Ontogenetic shifts in gill raker morphology and predicted prey capture efficiency of the Alewife, *Alosa pseudoharengus*. *Copeia* **1**: 164–171.

Malone, T.C. and Chervin, M.B. (1979). The production and fate of phytoplankton size fractions in the plume of the Hudson River, New York Bight. *Limnol. Oceanogr.* **24**: 683–696.

Mann, F.G. (1954). Vida de los peces en aguas Chilenas. *Inst. Invest. Veter. Santiago de Chile*, 342 pp. (in Spanish).

Massuti, M. and Oliver, M (1948). Estudio de la biometría y biología de la sardina de Mahón (Baleares), especialmente de su alimentación. *Bol. Inst. Esp. Oceanogr.* **3**: 1–15 (in Spanish).

Matarese, A.C., Kendall Jr., A.W., Blood, D.M., and Vinter, B.M. (1989). Laboratory guide to early life history stages of northeast Pacific fishes. *Technical Report No. 80*, US Dept. of Commerce, NMFS, NOAA, Seattle, WA, 652 pp.

Matsumoto, K. and Kohda, M. (2001). Differences in gill raker morphology between two local populations of a benthophagous filter-feeding fish, *Goniistius zonatus* (Cheilodactylidae). *Ichthyol. Res.* **48**: 269–273.

McFarlane, G.A., Smith, P.E., Baumgartner, T.R., and Hunter, J.R. (2002). Climate variability and Pacific sardine population and fisheries. *Am. Fish. Soc. Symp.* **32**: 195–214.

McFarlane, G.A., Schweigert, J., MacDougall, L., and Hrabok, C. (2005). Distribution and biology of Pacific sardines (*Sardinops sagax*) off British Columbia, Canada. *CalCOFI Rep.* **46**: 144–160.

Miller, T.W. (2006). Trophic dynamics of marine nekton and zooplankton in the northern California Current pelagic ecosystem. Ph.D. thesis, Oregon State University, Corvallis, Oregon, USA, 198 pp.

Miller, T.W. and Brodeur, R.D. (2007). Diets of and trophic relationships among dominant marine nekton within the northern California Current ecosystem. *Fish. Bull.* **105** (4): 548–559.

Mitani, I. (1988). Food habits of Japanese anchovy in the shirasu fishing ground within Sagami Bay. *Nipp. Suisan Gakkaishi* **54** (11): 1859–1865 (in Japanese with English abstract).

Mitchell-Innes, B.A. and Pitcher, G.C. (1992). Hydrographic parameters as indicators of the suitability of phytoplankton populations as food for herbivorous copepods. In *Benguela Trophic Functioning*, Payne, A.I.L., Mann, K.H., and Hilborn, R., eds. *S. Afr. J. Mar. Sci.* **12**: 355–365.

Molina-Ocampo, R.E. (1993). Habitos alimenticios de peces pelagicos menores de importancia comercial del Golfo de California, Mexico. M.Sc. thesis, ITESM-Guaymas, Guaymas, Son., México, 107 pp. (in Spanish).

Möllmann, C, and Köster, F.W. (1999). Food consumption by clupeids in the Central Baltic: evidence for top-down control? *ICES J. Mar. Sci.* **56**(suppl.): 100–113.

Möllmann, C., Kornilovs, G., and Sidrevics, L. (2000). Long-term dynamics of main mesozooplankton species in the Central Baltic Sea. *J. Plank. Res.* **22**: 2015–2038.

Möllmann, C., Kornilovs, G., Fetter, M., *et al.* (2003a). The marine copepod, *Pseudocalanus elongatus*, as a mediator between climate variability and fisheries in the Central Baltic Sea. *Fish. Oceanogr.* **12**: 360–368.

Möllmann, C., Köster, F.W., Kornilovs, G., and Sidrevics, L. (2003b). Interannual variability in population dynamics of calanoid copepods in the Central Baltic Sea. *ICES Mar. Sci. Symp.* **219**: 220–230.

Möllmann, C., Kornilovs, G., Fetter, M., and Köster, F.W. (2004). Feeding ecology of central Baltic Sea herring and sprat. *J. Fish Biol.* **65**: 1563–1581.

Möllmann, C., Kornilovs, G., Fetter, M., and Köster, F.W. (2005). Climate, zooplankton and pelagic fish growth in the Central Baltic Sea. *ICES J. Mar. Sci.* **62**: 1270–1280.

Monod T. (1961). *Brevoortia* Gill 1861 and *Ethmalosa* Regan 1971. *Bull. Français d'Afr. Noire* **23**: 506–547.

Muck, P., Rojas de Mendiola, B., and Antonietti, E. (1989). Comparative studies on feeding in larval anchoveta (*Engraulis ringens*) and sardine (*Sardinops sagax*). In *The Peruvian Upwelling Ecosystem: Dynamics and Interactions*, Pauly, D., Muck, P., Mendo, J., and Tsukayama, I., eds. *ICLARM Conf. Proc.* 18: 86–96.

Munroe, T.A. (2002). Herrings. In *Fishes of the Gulf of Maine*, Collette, B.B., and Klein-MacPhee, G., eds. Washington, DC: Smithsonian Institution Press, pp. 111–158.

Munuera, I. (2006). Ecología de la alimentación de las larvas de *Sardina pilchardus* (Walbaum, 1792), *Scomber scombrus* (Linné, 1798) y *Engraulis encrasicolus* (Linné, 1798) en el mar Cantábrico. Ph.D. thesis, Universidad de Oviedo, Oviedo, Spain, 137 pp. (in Spanish).

Muylaert, K. and Sabbe, K. (1999). Spring phytoplankton assemblages in and around the maximum turbidity zone of the estuaries of the Elbe (Germany), the Schelde (Belgium/The Netherlands) and the Gironde (France). *J. Mar. Syst.* **22**: 133–149.

Muzinic, R. (1958). Sur la coincidence et l'alternance dans la pêche de quelques poissons pélagiques. *Rapp. Comm. Int. Mer Médit.* **14**: 313–315 (in French).

Mylius, S.D., Klumpers, K., de Roos, A.M., and Persson, L. (2001). Impact of intraguild predation and stage structure on simple communities along a productivity gradient. *Am. Nat.* **158**: 259–276.

Nakai, Z. (1938). Maiwashi no Saiha no kouzou to shokuji tono kankei ni tsuite [Relationship between gill raker structure and feeding of sardine, *Sardina melanosticta* (T. & S.)]. *Suisan Kenkyusi* **33** (12): 547–561 (in Japanese).

Nakai, Z. (1962). Studies relevant to mechanisms underlying the fluctuation in the catch of the Japanese sardine, *Sardinops melanosticta* (Temminck & Schlegel). *Jap. J. Ichthyol.* **9** (1–6): 1–115.

Nakai, Z., Usami, S., Hattori, S., *et al.* (1955). Progress Report of the Cooperative IWASHI Resources Investigations, April 1949–December 1951. *Fisheries Agency, Tokai Regional Fisheries Research Laboratory, Tokyo*, 116 pp.

Nakai, Z., Honjo, Y., Kidachi, T., *et al.* (1962). Iwashirui kokisigyo no syokuzi to kanyuryo tono kanei [Relationships between food organisms and size of recruitment of iwashi]. *Suisan Sigen ni kansuru kyodo kenkyu suisin kaigi hokokusho, Showa 36 nendo*, 102–121, *Norin Suisan Gijutsu Kaigi, Tokyo* (in Japanese).

Nakata, K. (1988). Alimentary tract contents and feeding conditions of ocean-caught post larval Japanese sardine, *Sardinops melanostictus*. *Bull. Tokai Reg. Fish. Res. Lab.* **126**: 11–24.

Nakata, K. (1995). Feeding conditions of Japanese sardine larvae in and near the Kuroshio examined from their gut contents. *Bull. Natl. Res. Inst. Fish. Sci.* **7**: 265–275 (in Japanese with English abstract).

Nakata, K. and Hidaka, K. (2003). Decadal-scale variability in the Kuroshio marine ecosystem in winter. *Fish. Oceanogr.* **12** (4/5): 234–244.

Navarro, F. D. P. and Navaz, J. M. (1946). Apuntes para la biología y biometría de la sardina, anchoa, boga y chicharro de las costas vascas. *Inst. Esp. Oceanogr. Notas y Resúmenes* **134** (in Spanish).

Nishikawa, T. (1901). Hishiko Chousa Houkoku [A study on the anchovy]. *Suisan Chousa Houkoku (J. Imp. Fish. Bur. Tokyo)* **10** (1): 1–16 (in Japanese).

Nissling, A. (2004). Effects of temperature on egg and larval survival of cod (*Gadus morhua*) and sprat (*Sprattus sprattus*) in the Baltic Sea – implications for stock development. *Hydrobiol.* **514**: 115–123.

Noto, M. and Yasuda, I. (1999). Population decline of the Japanese sardine, *Sardinops melanostictus*, in relation to sea surface temperature in the Kuroshio Extension. *Can. J. Fish. Aquat. Sci.* **56**: 973–983.

O'Connell, C. P. (1972). The interrelation of biting and filtering in the feeding activity of Northern anchovy (*Engraulis mordax*). *J. Fish. Res. Bd Can.* **29** (3): 285–293.

Odate, K. (1994). Zooplankton biomass and its long-term variation in the western North Pacific Ocean, Tohoku Sea Area, Japan. *Bull. Tohoku Natl. Fish. Res. Inst.* **56**: 115–173 (in Japanese with English abstract).

Oliver, M, (1951). La sardina de la costa noroeste española en 1948 y 1949 (Estudio biométrico y biológico). *Bol. Inst. Esp. Oceanogr.* **42**: 1–22 (in Spanish).

Oliver, M. and Navarro, F. D. P. (1952). Nuevos datos sobre la sardina de Vigo. *Bol. Inst. Esp. Oceanogr.* **56**: 25–39 (in Spanish).

Ostrowski, J. and Mackiewicz, A. (1992). Feeding of herring and cod in the southern Baltic in 1991. *ICES CM:1992/***J18**: 1–15.

Oviatt, C. A., Gall, A. L. and Nixon, S. W. (1972). Environmental effects of Atlantic menhaden on surrounding waters. *Ches. Sci.* **13**: 321–323.

Parmanne, R., Rechlin, O., and Sjöstrand, B. (1994). Status and future of herring and sprat stocks in the Baltic Sea. *Dana* **10**: 29–59.

Parr, A. E. (1930). Is the presence of phytoplankton in the stomach contents of the California sardine caused by special pursuit or merely due to incidental ingestion? *Ecology* **11** (2): 465–468.

Patokina, F. A. and Feldman, V. N. (1998). Peculiarities of trophic relations between Baltic herring (*Clupea harengus membras* L.) and sprat (*Sprattus sprattus* L.) in the south eastern Baltic Sea in 1995–1997. *ICES CM:1998/***CC7**, 17 pp.

Pauly, D., Jarre, A., Luna, S., *et al.* (1989). On the quantity and types of food ingested by Peruvian anchoveta, 1953–1982. In *The Peruvian Upwelling Ecosystem: Dynamics and Interactions*, Pauly, D., Muck, P., Mendo, J., and Tsukayama, I., eds. *ICLARM Conf. Proc.* **18**: 109–124.

Peck J. I. (1893). On the food of the menhaden. *Bull. US Fish. Comm.* **13**: 113–126.

Peterson, W. T. (1989). Zooplankton feeding and egg production in comparison to primary production along the west coast of South Africa. In *Proceedings of the Plankton Dynamics Mini-Symposium, May 1989* Anon., ed. *Benguela Ecology Programme Report* **17**, 5 pp.

Plounevez, S. and Champalbert, G. (1999). Feeding behaviour and trophic environment of *Engraulis encrasicolus* (L.) in the Bay of Biscay. *Est. Coast. Shelf Sci.* **49**: 177–191.

Plounevez, S. and Champalbert, G. (2000). Diet, feeding behaviour and trophic activity of the anchovy (*Engraulis encrasicolus* L.) in the Gulf of Lions (Mediterranean Sea). *Oceanol. Acta* **23** (2): 175–192.

Polis, G. A., Myers C. A., and Holt, R. D. (1989). The ecology and evolution of intraguild predation: potential competitors that eat each other. *Ann. Rev. Ecol. Syst.* **20**: 297–330.

Popiel, J. (1951). Feeding and food of herring (*Clupea harengus* L.) in the Gulf of Gdansk and in the adjoining waters. *Rep. Sea Fish. Inst. Gdynia* **6**: 29–56 (in Polish with English abstract).

Radovich, J. (1952). Food of the Pacific sardine, *Sardinops caerulea*, from Central Baja California and Southern California. *Calif. Fish Game* **38** (4): 575–585.

Raid, T. and Lankov, A. (1995). Recent changes in the growth and feeding of Baltic herring and sprat in the north-eastern Baltic Sea. *Proc. Est. Acad. Sci.* **5**: 38–55.

Ramírez-Granados, R. (1957). Aspectos biológicos y económicos de la pesquería de sardina, *Sardinops caerulea*, (Girard, 1854) en aguas del Pacífico Mexicano. Thesis, Escuela Nacional de Ciencias Biológicas-IPN, México, 119 pp. (in Spanish).

Rasoanarivo, R., Folack, J., Champalbert, G., and Becker, B. (1991). Rélation entre les communautés phytoplanctoniques et l'alimentation des larves de *Sardina pilchardus* Walb. dans le Golfe de Fos (Méditerranée Occidentale) influence de la lumière sur l'activité alimentaire des larves. *J. Exp. Mar. Biol. Ecol.* **151**: 83–92 (in French with English abstract).

Rebolledo, H. and Cubillos, L. (2003). Ítems alimentarios de la anchoveta y sardina común en la zona Centro-sur durante el verano del 2003. In *Evaluación hidroacústica del reclutamiento de anchoveta y sardina común entre la V y X Regiones, año 2002. Informe Final Proyecto FIP 2002–13, Inst. Fom. Pesq., Informes Técnicos* **FIP-IT/2002–13**, 203 pp. (http://www.fip.cl) (in Spanish).

Renz, J. and Hirche, H.-J. (2006). Life cycle of *Pseudocalanus acuspes* Giesbrecht (Copepoda, Calanoida) in the Central Baltic Sea: I. Seasonal and spatial distribution. *Mar. Biol.* **148**: 567–580.

Robinson, G. A. (1966). A preliminary report on certain aspects of the biology of the South African anchovy, *Engraulis capensis* (Gilchrist). M.Sc. thesis, University of Stellenbosch, Stellenbosch, South Africa, iii, 61 pp. + 66 tables.

Rojas, E. B. (1953). Estudios preliminares del contenido estomacal de las anchovetas. *Bol. Cient. de la Cía. Adm. del Guano* **1**: 33–42 (in Spanish).

Rojas de Mendiola, B. (1974). Food of the larval anchovy *Engraulis ringens*. In *The Early Life History of Fish*, Blaxter, J. H. S., ed. New York: Springer-Verlag, pp. 277–285.

Rojas de Mendiola, B. (1989). Stomach contents of anchoveta (*Engraulis ringens*), 1953–1974. In *The Peruvian Upwelling Ecosystem: Dynamics and Interactions*, Pauly, D., Muck, P., Mendo, J. and Tsukayama, I., eds. *ICLARM Conf. Proc.* **18**: 97–104.

Rojas de Mendiola, B., and Gómez, O. (1981). Primera alimentación, sobrevivencia y tiempo de actividad de las larvas de anchoveta (*Engraulis ringens* J.). In *Investigación Cooperativa de la Anchoveta y su Ecosistema (ICANE) entre Perú y Canadá*, Dickie, L. M., and Valdivia, J. E., eds. *Bol. Inst. Mar Perú, Vol. Extraordinario*, pp. 72–79 (in Spanish).

Rönkkönen, S., Ojaveer, E., Raid, T., and Viitasalo, M. (2004). Long-term changes in Baltic herring (*Clupea harengus membras*) growth in the Gulf of Finland. *Can. J. Fish. Aquatic Sci.* **61**: 219–229.

Rossi, S., Sabates, A., Latasa, M., and Reyes, E. (2006). Lipid biomarkers and trophic linkages between phytoplankton, zooplankton and anchovy (*Engraulis encrasicolus*) larvae in the NW Mediterranean. *J. Plank. Res.* **28**: 551–562.

Roux, J.-P. (2003). Risks. In *Namibia's Marine Environment*, Molloy, F. J., and Reinikainen, T., eds. Directorate of Environmental Affairs, MET, Windhoek, Namibia, pp. 137–152.

Rudstam, L. G. (1989). Exploring the seasonal dynamics of herring predation in the Baltic Sea: applications of a bioenergetic model of fish growth. *Kieler Meeresforsch. Sonderh.* **6**: 312–332.

Rudstam, L. G., Hansson, S., Johansson, S., and Larsson, U. (1992). Dynamics of planktivory in a coastal area of the northern Baltic Sea. *Mar. Ecol. Prog. Ser.* **80**: 159–173.

Ryther, J. H. (1969). Relationship of photosynthesis to fish production in the sea. *Science* **166**: 72–76.

Sánchez, G., Alamo, A. and Fuentes, H. (1985). Alteraciones en la dieta alimentaria de algunos peces comerciales por efecto del fenómeno El Niño. In *El Niño y su Impacto en la Fauna Marina*, Arntz, W., Landa, A. and Tarazona, J., eds. *Bol. Inst. Mar Perú, Vol. Extraordinario*, pp. 135–142 (in Spanish).

Santander, H., Alheit, J., MacCall, A. D., and Alamo, A. (1983). Egg mortality of the Peruvian anchovy (*Engraulis ringens*) caused by cannibalism and predation by sardines (*Sardinops sagax*). In *Proceedings of the Expert Consultation to Examine Changes in Abundance and Species of Neritic Fish Resources*, San José de Costa Rica, Sharp, G. D. and Csirke, J., eds. *FAO. Fish. Rep.* **291**: 1011–1026.

Schwartzlose, R. A., Alheit, J., Bakun, A., *et al.* (1999). Worldwide large-scale fluctuations of sardine and anchovy populations. *S. Afr. J. Mar. Sci.* **21**: 289–347.

Scofield, E. C. (1934). Early life history of California sardine (*Sardina caerulea*) with special reference to distribution of eggs and larvae. *Fish. Bull. Calif.* **41**: 1–48.

Sears, M. (1941). ¿Qué es el plankton y por qué debemos estudiarlo? *Bol. Cient. Cía. Admin. del Guano* **17** (12): 451–465 (in Spanish).

Shannon, L. J., Nelson, G., Crawford, R. J. M., and Boyd, A. J. (1996). Possible impacts of environmental change on pelagic fish recruitment: modelling anchovy transport by advective processes in the southern Benguela. *Global Change Biol.* **2**: 407–420.

Shannon, L. J., Field, J. G., and Moloney, C. L. (2004). Simulating anchovy–sardine regime shifts in the southern Benguela ecosystem. *Ecol. Model.* **172**: 269–281.

Shen, S. C. (1969). Comparative study of the gill structure and feeding habits of the anchovy, *Engraulis japonica* (Hout.). *Bull. Inst. Zool. Acad. Sinica* **8**: 21–35.

Shimps, E. L., Rice, J. A. and Osborne, J. A. (2005). Hypoxia tolerance in two juvenile estuary-dependent fishes. *J. Exp. Mar. Biol. Ecol.* **325**: 146–162.

Sholto-Douglas, A. D., Field, J. G., James, A. G., and van der Merwe, N. J. (1991). $^{13}C/^{12}C$ and $^{15}N/^{14}N$ isotope ratios in the Southern Benguela ecosystem: indicators of food web relationships among different size-classes of plankton and pelagic fish; differences between fish muscle and bone collagen tissues. *Mar. Ecol. Prog. Ser.* **78**: 23–31.

Siegfried, W. R., Crawford, R. J. M., Shannon, L. V., *et al.* (1990). Scenarios for global warming induce change in the open-ocean environment and selected fisheries of the west coast of southern Africa. *S. Afr. J. Sci.* **86**: 281–285.

Silva, E. (1954). Some notes on the food of the pilchard *Sardina pilchardus* (Walb.), of the Portuguese coasts. *Rev. Fac. Cienc. Lisboa, 2ª Sér.* **4** (2): 281–294.

Sin, Y., Wetzel, R. L., and Anderson, I. C. (2000). Seasonal variations of size fractionated phytoplankton along the salinity gradient in the York River estuary, Virginia (USA). *J. Plank. Res.* **22**: 1945–1960.

Starodub, M. L., Shvetsov, F., and Hoziosky, S. (1992). The feeding of sprat in the eastern Baltic. *ICES CM:1992/***J26**: 1–6.

Sugimoto, T., Kimura, S., and Tadokoro, K. (2001). Impact of El Niño events and climate regime shift on living resources in the western North Pacific. *Prog. Oceanogr.* **9**: 113–127.

Szypula, J., Grygiel, W., and Wyszynski, M. (1997). Feeding of Baltic herring and sprat in the period 1986–1996 in relation to their state and biomass. *Bull. Sea Fish. Inst.* **3**: 73–83.

Takasuka, A. (2003). Growth rate and survival mechanism during early life history stages of Japanese anchovy *Engraulis japonicus*. Ph.D. thesis, University of Tokyo, Tokyo, Japan, 112 pp. (in Japanese).

Takasuka, A., Oozeki, Y., Kimura, R., *et al.* (2004). Growth-selective predation hypothesis revisited for larval anchovy in offshore waters: cannibalism by juveniles versus predation by skipjack tunas. *Mar. Ecol. Prog. Ser.* **278**: 297–302.

Takasuka, A., Oozeki, Y., and Aoki, I. (2007). Optimal growth temperature hypothesis: why do anchovy flourish and sardine collapse or vice versa under the same ocean regime? *Can. J. Fish. Aquat. Sci.* **64**: 768–776.

Tanaka, H. (2006). Comparative study of the feeding ecology of small pelagic fishes with a focus on Japanese anchovy *Engraulis japonicus*. Ph.D. thesis, University of Tokyo, Tokyo, Japan, 189 pp. (in Japanese).

Tanaka, H., Aoki, I. and Ohshimo, S. (2006). Feeding habits and gill raker morphology of three planktivorous pelagic fish species off the coast of northern and western Kyushu in summer. *J. Fish Biol.* **68**: 1041–1061.

Temming, A. (1996). Die quantitative Bestimmung der Konsumption von Fischen. Experimentelle, methodische und theoretische Aspekte. Habilitation thesis, University of Hamburg, Hamburg, Germany, 235 pp. (in German).

Thomas, R. M. and Schülein, F. H. (1988). The shoaling behaviour of pelagic fish and the distribution of seals and gannets off Namibia as deduced from routine fishing reports, 1982–1985. *S. Afr. J. Mar. Sci.* **7**: 179–191.

Tudela, S. and Palomera, I. (1995). Diel feeding intensity and daily ration in the anchovy *Engraulis encrasicolus* in the northwest Mediterranean Sea during the spawning period. *Mar. Ecol. Prog. Ser.* **129**: 55–61.

Tudela, S. and Palomera, I. (1997). Trophic ecology of the European anchovy *Engraulis encrasicolus* in the Catalan Sea (northwest Mediterranean). *Mar. Ecol. Prog. Ser.* **160**: 121–134.

Tudela, S., Palomera, I., and Quilez, G. (2002). Feeding of anchovy *Engraulis encrasicolus* larvae in the north-west Mediterranean. *J. Mar. Biol. Ass. UK* **82**: 349–350.

Ulloa, O., Escribano, R., Hormazabal, S., *et al.* (2001). Evolution and biological effects of the 1997–98 El Niño in the upwelling ecosystem off northern Chile. *Geophys. Res. Lett.* **28**: 1591–1594.

Uotani, I. (1985). The relation between feeding mode and feeding habit of the anchovy larvae. *Bull. Jap. Soc. Sci. Fish.* **51** (7): 1057–1065 (in Japanese with English abstract).

Uotani, I., Izuha, A., and Asai, K. (1978). Food habits and selective feeding of anchovy larvae (*Engraulis japonica*). *Bull. Jap. Soc. Sci. Fish.* **44** (5): 427–434 (in Japanese with English abstract).

Valdés, E. S., Shelton, P. A., Armstrong, M. J., and Field, J. G. (1987). Cannibalism in South African anchovy: egg mortality and egg consumption rates. In *The Benguela and Comparable Ecosystems*, Payne, A. I. L., Gulland, J. A. and Brink, K. H., eds. *S. Afr. J. Mar. Sci.* **5**: 613–622.

Valdés Szeinfeld, E. (1991). Cannibalism and intraguild predation in clupeoids. *Mar. Ecol. Prog. Ser.* **79**: 17–26.

Valdés Szeinfeld, E. S. and Cochrane, K. L. (1992). The potential effects of cannibalism and intraguild predation on anchovy recruitment and clupeoid fluctuations. In *Benguela Trophic Functioning*, Payne, A. I. L., Brink, K. H., Mann K. H., and Hilborn, R., eds. *S. Afr. J. Mar. Sci.* **12**: 695–702.

Valenzuela, V., Balbontín, F., and Llanos, A. (1995). Composición de la dieta y tamaño de las presas de los estadios larvales de ocho especies de peces en la costa central de Chile. *Rev. Biol. Mar., Valparaíso* **30**: 275–291 (in Spanish).

van der Lingen, C. D. (1994). Effect of particle size and concentration on the feeding behaviour of adult pilchard, *Sardinops sagax*. *Mar. Ecol. Prog. Ser.* **109**: 1–13.

van der Lingen, C. D. (1995). Respiration rate of adult pilchard *Sardinops sagax* in relation to temperature, voluntary swimming speed and feeding behaviour. *Mar. Ecol. Prog. Ser.* **129**: 41–54.

van der Lingen, C. D. (1998a). Gastric evacuation, feeding periodicity and daily ration of sardine *Sardinops sagax* in the southern Benguela upwelling ecosystem. In *Benguela Dynamics. Impacts of Variability on Shelf-Sea Environments and their Living Resources*, Pillar, S. C., Moloney, C. L., Payne, A. I. L., and Shillington, F. A., eds. *S. Afr. J. Mar. Sci.* **19**: 305–316.

van der Lingen, C. D. (1998b). Nitrogen excretion and absorption efficiencies of sardine *Sardinops sagax* fed phytoplankton and zooplankton diets. *Mar. Ecol. Prog. Ser.* **175**: 67–76.

van der Lingen, C. D. (1999). The feeding ecology of, and carbon and nitrogen budgets for, sardine *Sardinops sagax* in the southern Benguela upwelling ecosystem. Ph.D. thesis, University of Cape Town, Cape Town, South Africa, vii, 202 pp.

van der Lingen, C. D. (2002). Diet of sardine *Sardinops sagax* in the Southern Benguela upwelling ecosystem. *S. Afr. J. Mar. Sci.* **24**: 301–316.

van der Lingen, C. D., Hutchings, L., and Field, J. G. (2006a). Comparative trophodynamics of anchovy *Engraulis encrasicolus* and sardine *Sardinops sagax* in the southern Benguela: are species alternations between small pelagic fish trophodynamically mediated? *Afr. J. Mar. Sci.* **28** (3/4): 465–477.

van der Lingen, C. D., Shannon, L. J., Cury, P., *et al.* (2006b). Resource and ecosystem variability, including regime shifts in the Benguela Current system. In *Benguela: Predicting a Large Marine Ecosystem*, Shannon, V., Hempel, G., Malanotte-Rizzoli, P., Moloney, C., and Woods, J., eds. Elsevier, Amsterdam, the Netherlands, *Large Marine Ecosystems* **14**: 147–184.

van Khan, N., Drzycimski, I., and Chojnacki, J. (1972). Feeding and food composition of sprat from the Bornholm depth. *Acta Ichthy. Pisc.* **II** (2): 55–66.

Varela, M., Larrañaga, A., Costas, E., and Rodriguez, B. (1988). Contenido estomacal de la sardina (*Sardina pilchardus* Walbaum) durante la campaña Saracus 871 en las plataformas Cantábrica y de Galicia en febrero de 1971. *Bol. Inst. Esp. Oceanogr.* **5**: 17–28 (in Spanish).

Varela, M., Alvarez-Ossorio, M. T., and L. Valdés. (1990). Método para el estudio cuantitativo del contenido estomacal de la sardina. Resultados preliminares. *Bol. Inst. Esp. Oceanogr.* **6**: 117–126 (in Spanish).

Verheye, H. M. (2000) Decadal-scale trends across several marine trophic levels in the southern Benguela upwelling system off South Africa. *Ambio* **29**: 30–34.

Verheye, H. M., Richardson, A. J., Hutchings, L., *et al.* (1998). Long-term trends in the abundance and community structure of coastal zooplankton in the southern Benguela system, 1951–1996. In *Benguela Dynamics. Impacts of Variability on Shelf-Sea Environments and their Living Resources*, Pillar, S. C., Moloney, C. L., Payne, A. I. L., and Shillington, F. A., eds. *S. Afr. J. Mar. Sci.* **19**: 317–332.

Villalobos, H. and Rodríguez-Sánchez, R. (2002). Pattern of increase in gill raker number of the California sardine. *J. Fish Biol.* **60** (1): 256–259.

Villalobos-Ortiz, H. (1998). Morfologia funcional del aparato filtrador de *Sardinops caeruleus* (Girard, 1856) (Pisces: Clupeidae), del noroeste de Mexico. M.Sc. Thesis, CICIMAR-IPN, La Paz, B.C.S., México, 75 pp. (in Spanish).

Vogt, W. (1940). Una depression ecológica de la costa Peruana. *Bol. Cient. Cia. Admin. del Guano* **16** (10): 307–329 (in Spanish).

Voss, R., Köster, F. W., and Dickmann, M. (2003). Comparing the feeding habits of co-occurring sprat (*Sprattus sprattus*) and cod (*Gadus morhua*) larvae in the Bornholm Basin, Baltic Sea. *Fish. Res.* **63**: 97–111.

Voss, R., Clemmesen, C., Baumann, H., and Hinrichsen, H.-H. (2006). Baltic sprat larvae: coupling food availability, larval condition and survival. *Mar. Ecol. Prog. Ser.* **308**: 243–254.

Walker, D. R., and Peterson, W. T. (1991). Relationships between hydrography, phytoplankton production, biomass, cell size and species composition, and copepod production in the southern Benguela upwelling system in April 1988. *S. Afr. J. Mar. Sci.* **11**: 289–305.

Wallace-Fincham, B. P. (1987). The food and feeding of *Etrumeus whiteheadi* Wongratana 1983, off the Cape Province of South Africa. M.Sc. thesis, University of Cape Town, Cape Town, South Africa, 117 pp.

Walsh, J. J. (1981). A carbon budget for overfishing off Peru. *Nature* **290**: 300–304.

Ware, D. M., Rojas de Mendiola, B., and Newhouse, D. S. (1981). Behaviour of first feeding Peruvian anchoveta larvae, (*Engraulis ringens*). In *Investigación Cooperativa de la Anchoveta y su Ecosistema (ICANE) entre Perú y Canadá*, Dickie, L. M. and Valdivia, J. E., eds. *Bol. Inst. Mar Perú*, *Vol. Extraordinario*, pp. 80–87.

Watanabe, Y., and Saito, H. (1998). Feeding and growth of early juvenile Japanese sardines in the Pacific waters off central Japan. *J. Fish Biol.* **52**: 519–533.

Yamashita, H. (1955). The feeding habit of sardine, *Sardinia melanosticta*, in the waters adjacent to Kyushu, with reference to its growth. *Bull. Jap. Soc. Sci. Fish.* **21** (7): 471–475 (in Japanese with English abstract).

Yamashita, H. (1957). Relations of the foods of sardine, jack mackerel, mackerel, and so on, in the waters adjacent to west Kyushu. *Bull. Seikai Nat. Fish. Res. Inst.* **11**: 45–53 (in Japanese with English abstract).

Yasuda, I., Sugisaki, H., Watanabe, Y., *et al.* (1999). Interdecadal variations in Japanese sardine and ocean/climate. *Fish. Oceanogr.* **8**: 18–24.

Yatsu, A. and Kaeriyama, M. (2005). Linkages between coastal and open-ocean habitats and dynamics of Japanese stocks of chum salmon and Japanese sardine. *Deep-Sea Res. II* **52**: 727–737.

Yokota, T., Toriyama, M., Kanai, F., and Nomura, S. (1961). Studies on the feeding habit of fishes. *Rep. Nankai Reg. Fish. Res. Lab.* **14**: 1–234 (in Japanese with English abstract).

Yoneda, Y. and Yoshida, Y. (1955). The relation between the sardine and the food plankton – I. On the food intake by *Sardinops melanosticta*. *Bull. Jap. Soc. Sci. Fish.* **21** (2): 62–66 (in Japanese with English abstract).

Zalachowski, W., Szypula, J., Krzykawski, S., and Krzykawski, I. (1975). Feeding of some commercial fishes in the southern region of the Baltic Sea in 1971 and 1972. *Arch. Polish Hydrobiol.* **22**: 429–448.

Zalachowski, W., Szypula, J., Krzykawski, S. and Krzykawski, I. (1976). Composition and amount of food consumed by sprat, herring and cod in the Southern Baltic in the years 1971–1974. *ICES CM:1976/***P23**: 1–13.

8 Impacts of fishing and climate change explored using trophic models

Lynne Shannon, Marta Coll, Sergio Neira, Philippe Cury, and Jean-Paul Roux

CONTENTS

Summary

Small pelagic fish are termed "wasp-waist" species as they dominate mid trophic levels and comprise relatively few species but attain large abundances that can vary drastically in size. They have been found to exert top-down control on their prey species and bottom-up control on their predators and, in this way, appear to induce unsuspected ecosystem dynamics. Largely based on model results, this chapter explores these effects and associated dynamics, not only illustrating the importance of small pelagic fish in structuring marine ecosystems, but also revealing the consistency of the role of small pelagic fish across various upwelling systems in which they play key roles. The Northern and Southern Benguela, Southern Humboldt, South Catalan Sea and North and Central Adriatic Sea ecosystems are compared in terms of the importance and role of small pelagic fish using information gained from landings and ecological models. Trophic level of the catch, the Fishing-in-Balance (FiB) index and the ratio of pelagic:demersal fish are calculated from reported landings. Sums of all flows to detritus are compared across modelled ecosystems. Models of the Southern Benguela, Southern Humboldt and South Catalan Sea are used to perform two simulations: (1) closure of fisheries on small pelagic fish and (2) collapse of small pelagic fish stocks, to further explore the roles of small pelagic fish in the dynamics of these ecosystems.

Tracking pelagic:demersal fish catch and biomass ratios over time is a means of detecting collapses in the small pelagic fish stocks, and comparing these ratios across ecosystems highlights the greater importance of small pelagic fish in the Humboldt compared to other ecosystems. The trophic level of the catch has strongly increased in the Northern Benguela and Southern Humboldt ecosystems, reflecting collapses in small pelagic and/or other fisheries in these regions in recent years. In the Mediterranean Sea, fluctuations in trophic level of the catch, in conjunction with the FiB index, reflects the collapse of small pelagic fish and disruption of the ecosystems. Models analyzed in this study show how a decrease in small pelagic fish abundance will have detrimental effects on both higher and lower trophic levels of the food web, causing a decrease in predators and the proliferation of other species that are prey or competitors of small pelagic fish, but which do not constitute an alternative pathway for the flow of primary productivity to higher trophic levels. Trophic model simulations consistently suggested that gelatinous zooplankton may increase when small pelagic fish stocks decline and, conversely, a decrease in jellyfish abundance may be expected if pelagic fisheries were to be closed, in agreement with circumstantial evidence for some systems. Simulations of small pelagic fishery collapses in the southern Benguela, Southern Humboldt, and South Catalan Sea all emphasize the dependency of avian, and to a lesser extent more opportunistic mammalian and fish predators on healthy stocks of small pelagic fish. In general, collapses of small pelagic fish seem to disrupt energy flows, which could result in an increase in flows to detritus, demersal processes within the system becoming more important, stronger overall mixed trophic interactions of the pelagic component on the demersal component in more recent periods (usually when fishing

Climate Change and Small Pelagic Fish, eds. Dave Checkley, Jürgen Alheit, Yoshioki Oozeki, and Claude Roy. Published by Cambridge University Press. © Cambridge University Press 2009.

has increased), and often reduced summed impacts of the demersal compartment on the degraded pelagic food web. Examination of summed mixed trophic impacts of detritus on both the pelagic and demersal food webs further supported the hypothesis that detritus becomes increasingly more important as ecosystems are fished and/or degraded. The chapter draws on model results to reveal the ecological role of small pelagic fish within the ecosystem, and to examine how extreme changes in the abundance of small pelagic fish stocks might impact ecosystems. These changes are modeled by examining altered fishing effects using available trophic models, but could similarly be climatically driven. Given the pivotal ecological role of small pelagics in ecosystems, climate change is likely to express itself in the ecosystem via its impacts on small pelagic fish. However, precisely how climate change will be likely to affect abundance, life history, and the spatial distribution of small pelagics (and the food webs they dominate) still requires further study in terms of the mechanisms and processes involved.

Introduction

Small pelagic fish as the wasp-waist of ecosystems

Small pelagic fish, dominating mid-trophic levels and being composed of relatively few species but attaining large abundances that can vary drastically in size, have been named "wasp-waist" species (Rice, 1995; Bakun, 1996). Small pelagic fish have been shown to exert top-down control on their prey species and bottom-up control on their predators, which has been described as a "wasp-waist" control (Cury et al., 2000). The "boom-bust" dynamics of pelagic fish and populations that interact trophically with them suggests a "predator pit" type of dynamics according to Bakun (2006). This special configuration in certain marine food webs and the role and dynamics of pelagic fish appears to induce unsuspected ecosystem dynamics. These effects and associated dynamics need to be explored in detail, not only to reveal the importance of small pelagic fish in structuring marine ecosystems, but also to examine the consistency of their role in various upwelling systems.

Regime shifts or switches in abundance of pelagic fish species may be initiated or sustained by changes in zooplankton availability (Schwartzlose et al., 1999). On the other hand, changes in abundance of small pelagic fish may have top-down effects on the abundance and community structure of zooplankton prey (Schwartzlose et al., 1999). Between the 1950s and the 1990s, the proportion of small cyclopoid copepods off the west coast of South Africa increased, compared to a decrease in the relative abundance of large calanoid copepods, and the biomass of copepods increased tenfold whereas total copepod abundance increased 100-fold (Verheye and Richardson, 1998;

Verheye et al., 1998). Both bottom-up and top-down effects of small pelagic fish on zooplankton in this region have been proposed (Verheye et al., 1998), although Cury et al. (2000) showed that the general rule in upwelling systems, including the Benguela, is that small pelagic fish exert top-down control on their zooplankton prey.

Small pelagic fish have been shown to exert important controls over their predators too. Seabirds, in particular, have been found to respond strongly to changes in pelagic fish abundances. Seabird guano harvests off South Africa and Peru reflected changes in pelagic fish abundance (Crawford and Jahncke, 1999) and several population parameters measured for African penguins, Cape cormorants and swift terns off South Africa have been related to stock fluctuations of anchovy and sardine (Best et al., 1997). These have been verified through changes in seabird diet compositions in response to changes in pelagic fish abundance (e.g. that of the Cape gannet *Morus capensis*; Crawford and Dyer, 1995). Similarly, off Namibia, as sardine abundance decreased, pelagic goby replaced sardine as the important prey item in the diets of Cape gannets, Cape cormorants *Phalacrocorax capensis* and African penguins *Spheniscus demersus* (Crawford et al., 1985). Off Namibia, sardine were an important prey of the Cape fur seal, *Arctocephalus pusillus pusillus*, in earlier years whereas goby, supplemented by myctophids in the southern part of the Northern Benguela, and horse mackerel in the central and northern parts of the Northern Benguela, became the main prey of seals in later years (Mecenero and Roux, 2002; J.-P. Roux, unpublished data). The dominant large pelagic fish predator in the southern Benguela, namely snoek (*Thyrsites atun*), preys non-selectively on pelagic fish (M. Griffiths, personal communication) and diet studies have shown that snoek feed on anchovy and sardine in accordance with their relative abundances in the ocean (M. Griffiths, unpublished data). In the Mediterranean Sea, small pelagic fish have been shown to be important prey for demersal fish, marine mammals and seabirds (Bozzano et al., 1997; Ríos, 2000; Arcos, 2001; Blanco et al., 2001; Stergiou and Karpouzi, 2002).

Empirical studies reveal patterns of interaction with small pelagic fish (e.g. with top predators such as birds and seals or with zooplankton prey) that have been obtained at different spatial and temporal resolutions. Thus the time series used to study the effect of changes in biomass of small pelagic fish on seals or birds are estimated at the level of the ecosystem and on an annual basis, while the time series of plankton have usually been collected locally at a fine temporal scale. These differences in the resolution can substantially bias the way we analyze control versus what might be regarded as a local depletion with no effect on the overall population, and such interactions need to be explored in a broader perspective.

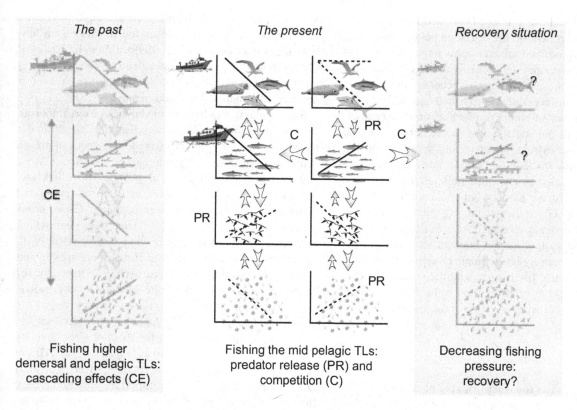

Fig. 8.1. Schematic plot showing expected changes across different trophic levels when the typical wasp-waist ecosystem is disrupted by fishing.

As we have been fishing down marine food webs (Pauly and Palomares, 2005), fishing pressure is increasing on organisms at lower trophic levels, such as small pelagic fish (FAO, 2005). Therefore, their increased exploitation is expected to have a wide impact on the pelagic food web and might provoke the dysfunction of the ecosystem and the further depletion of populations of dependent predators at higher trophic levels. This can cause, as has already been observed when top predators are overfished (Springer *et al.*, 2003; Frank *et al.*, 2005), large trophic cascades and proliferation of species located at lower trophic levels or having shorter life spans (Daskalov, 2002; Cury *et al.*, 2005b). All these changes on exploited marine ecosystems point to the existence of a sequence of flow controls as ecosystems are fished down or as community turnover rates increase, which is of great relevance in terms of recovery of intensely exploited ecosystems (Fig. 8.1). There is thus a need to characterize what changes are occurring where and why, as well as which ecosystem elements favor these changes in relation to fishing pressure, control mechanisms, and environmental variability. An essential element is to analyze to what extent observed ecosystem changes due to the collapse of the pelagic system and changes in flow control are reversible. Our work is based on models that help to uncover the ecological role of small pelagic fish within the ecosystem, to examine how changes in their abundance might have an impact at an ecosystem level. We use altered fishing effects in available trophic models and local information on fishing activities, but changes in small pelagic fish could similarly be climatically driven due to changes in sea water temperature or river runoffs, for example. Given the pivotal ecological role of small pelagics in ecosystems, climate change is likely to express itself in the ecosystem via its impacts on small pelagic fish.

Piscivorous fish (such as large pelagic fish and hake) support fisheries of high commercial value and thus the linkages between small pelagic fisheries and these fisheries can be very important. It is often debated whether there would be value in fishing more lightly on small pelagic fish in order to leave more forage fish for predatory fish, thereby potentially benefiting fisheries on these predatory fish. This is still an open question, to which answers may be provided through model simulations or the implementation of carefully designed experimental fishing scenarios. It is the former option that is very briefly explored in this study, although more rigorous and detailed modeling is recommended.

Chapter outline

In this chapter, we focus on the following ecosystems in which small pelagic fish play important roles, both commercially and within the food web: the Benguela (South

Africa and Namibia), the Southern Humboldt (Chile), the Mediterranean Sea (South Catalan, and the North and Central Adriatic), and to a lesser extent also the Northern Humboldt (Peruvian) system, representing a gradient of exploited ecosystems. We select these ecosystems as key examples because these are the ones for which we have constructed comparable trophic models fitted to time series data, and which we use as a standardized basis from which to examine the ecosystem role of small pelagics.

In a previous study, comparative "static" trophic models of upwelling ecosystems have been considered, and taking into account the exploitation history of these ecosystems, selected indicators were analyzed to test an initial hypothesis. That is, an ecological model representing the South Catalan Sea (Northwestern Mediterranean) would be ranked within the high impacted areas of Namibia and Peru, which were differentiated in terms of ecosystem impact from the moderately impacted regions of Chile and South Africa (Moloney *et al.*, 2005; Coll *et al.*, 2006a).

Inferring from results of comparable dynamic trophic models, fitted to available time series data, this chapter aims to define, in a quantifiable respect, the disruption of food webs (Fig. 8.1) as a result of small pelagic fish collapse[1] (by comparing with situations where small pelagic fish are considered to be sustainably exploited), and to illustrate that small pelagic fish can be used as indicators or warning signs of a future ecosystem collapse or change. By understanding the ecological role of small pelagic fish within the larger ecosystem, and by exploring possible ecosystem implications of severe declines or large increases in small pelagic fish stocks using ecological fishery models, the potential effects of climate change at the level of the ecosystem may be inferred for future situations in which climate change may alter the abundance of small pelagic fish.

Ecosystems examined

Benguela upwelling region

The Benguela upwelling region is subdivided by the permanent Lüderitz upwelling cell around 26° S into two systems, namely the Southern Benguela ecosystem off South Africa and the Northern Benguela ecosystem off Namibia. The Northern Benguela ecosystem extends from the Angola–Benguela front located between 14° S and 16° S (Meeuwis and Lutjeharms, 1990) southwards to Lüderitz. In addition to an upwelling region along the southwestern coast, the Southern Benguela ecosystem is uniquely characterized by a shallow bank region, the Agulhas Bank in the south. This is the reason for the more abundant and ecologically important demersal and benthic biota in the Southern Benguela ecosystem compared to "traditional" (and non-collapsed[2]) upwelling systems (Shannon and Jarre-Teichmann, 1999a, b).

Although fisheries in the Northern and Southern Benguela ecosystems target similar demersal (Cape hake, *Merluccius capensis* and *M. paradoxus*) and pelagic fish species (sardine, *Sardinops sagax*, anchovy, *Engraulis encrasicolus*, horse mackerel, *Trachurus trachurus capensis*), they have experienced different exploitation histories and environmental perturbations, and contrast in their species compositions, community structures and abundance trajectories.

Total catches in the Southern Benguela ecosystem have fluctuated around 0.5 million tonnes since the early 1960s. Since the 1950s and even before then, hake off South Africa have been targeted by bottom trawlers. Catches of hake increased slowly from about 50 000 tonnes to about 140 000 tonnes between 1950 and 1977, when a 200-mile Fishing Zone was proclaimed by South Africa (Payne, 1989), and have remained relatively stable since. Most of the catch was composed of sardine from 1950 until 1965, thereafter the fishing industry used nets of smaller mesh size to target anchovy, which dominated the catches until the mid 1990s. In the early 2000s, significant quantities of both small pelagic species were caught. The sustainability of catches off South Africa is largely attributable to relatively conservative management measures. For example, pelagic fish catches have been maintained well below 25% of estimated annual biomass since the early 1990s (Cury *et al.*, 2005a), although the initialization of the unusually high abundance of both anchovy and sardine off South Africa in the early 2000s has been attributed to a favorable sequence of short-lived environmental events (Roy *et al.*, 2001).

Sardines, anchovy and hake off Namibia have been sequentially exploited and depleted, largely because effective management measures could not be implemented until Namibia's independence and the proclamation of a 200-mile EEZ[3] in 1990. Subsequently, there has been a modest recovery of hake catches, but sardines have remained at very low levels. Catches have been dominated by horse mackerel since the 1970s, although there has been a slight decline since the late 1980s. There has been a steady decline in total catches, which peaked in 1968 at more than 2 million tonnes and decreased to around 0.5 million tonnes in the late 1990s.

Sardine were abundant off Namibia in the early 1960s, but year classes were poor from 1965, whereas catches were increasing from 1967–1969, and sardine biomass thus declined (R. Crawford., unpublished data). Unlike off South Africa, anchovy alone did not replace sardine, rather a suite of planktivorous fish, including horse mackerel, pelagic goby *Sufflogobius bibarbatus* and to a lesser degree anchovy, filled the previously occupied niche from the late 1970s and early 1980s (Boyer and Hampton, 2001). Between 1986 and 1991, sardine and hake biomass increased (R. Crawford, unpublished data). However, two environmental anomalies occurred in the Northern Benguela between 1993 and 1995 (a low oxygen event in 1993/4 and an intrusion of warm water – a Benguela Niño – in 1995), and were linked to distributional shifts in several fish species, poor sardine recruitment, declines to low biomass levels of most commercial species and reduced catch rates of hake (Boyer *et al.*, 2001), which migrated offshore between 1993 and 1995 (Hamukuaya *et al.*, 1998), monkfish (Maartens, 1999) and horse mackerel (Boyer and Hampton, 2001). Concomitant declines were also observed in top predators such as seals, which, due to starvation, declined by a third in the mid-1990s (Roux, 1998), and Bank cormorants, *Phalacrocorax neglectus* (Crawford *et al.*, 1999).

Humboldt upwelling system

The upwelling system off central Chile is located in the southern section of the Humboldt Current System, which is one of the four major eastern boundary ecosystems of the world. This system supports one of the highest levels of primary productivity recorded for the open ocean (1.7 gC m^{-2} per year; Daneri *et al.*, 2000) and globally significant landings (>4.5 million tonnes in 1995). Despite this high biological productivity, the state of exploited stocks is far from healthy (Arancibia and Neira, 2003). In fact, the main target species have been fully exploited or even overexploited for many years, leading to a series of recent stock collapses, for example, of horse mackerel (1998), red squat lobster and yellow squat lobster (1999), and Chilean hake (2004).

The Southern Humboldt system off central Chile, as considered in this chapter, extends from 33° S to 39° S and from the coast line up to 30 nm offshore, covering a total area of approximately 50 000 km^2. This geographic unit corresponds to the "Mediterranean District" and is ecologically independent from the "Peruvian Province" and the "Austral District" located northward and southward, respectively (Camus, 2001). The main oceanographic and biogeographic patterns that characterize the Southern Humboldt are a rather narrow continental shelf (<30 nm), strongly seasonal upwelling (September to March) and high levels of primary productivity (Strub *et al.*, 1998; Daneri *et al.*, 2000; Escribano *et al.*, 2003).

Oceanographically, four main water masses are present, which leads to a very diverse marine environment along the central Chilean coast: Subtropical Surface Water (STSW), Subantarctic Water (SAW), Antarctic Intermediate Water (AAIW), and Equatorial Subsurface Water (ESSW) (for a review of the physical and chemical characteristics of each water mass see Strub *et al.*, 1998). Wind-driven coastal upwelling brings the ESSW to the surface in the coastal zone, causing a persistent and characteristic shallow oxygen minimum zone (< 0.5 ml O$_2$ l^{-1}). Based on the correlation between the low frequency coastal modes and the Pacific and Atlantic sea surface temperature (SST), Montecinos *et al.* (2003) suggest that the coastal SST signal seems to

comprise two main large-scale climate processes affecting the Southern Humboldt. At the inter-annual scale the main source of variability is the El Niño Southern Oscillation cycle. At the long-term scale, an inter-decadal oscillation occurs at a basin-wide, and maybe even global, scale.

In terms of the main biological components, the Southern Humboldt sustains a diverse and productive food web. The phytoplankton group is dominated by large diatoms for most of the year (Avaria and Muñoz, 1982), while the zooplankton is dominated by herbivorous copepods and euphausiids. Jellyfish (Hydrozoa) also constitute an important group in the plankton domain (Palma and Rosales, 1995). Macrocrustaceans are significant benthic components and some species such as red squat lobster (*Pleuroncodes monodon*), yellow squat lobster (*Cervimunida johni*) and pink shrimp (*Heterocarpus reedi*) support important fisheries. The fish community is dominated by pelagic species. Small pelagic fishes such as anchovy (*E. ringens*) and the endemic common sardine (*Strangomera bentincki*) are present at high biomass levels in the Southern Humboldt and dominate landings. These species feed primarily on phytoplankton and secondarily on zooplankton (Arrizaga *et al.*, 1993). The horse mackerel (*Trachurus symmetricus*) is a major fishery resource in the region. This highly migratory species performs large-scale migrations in the Pacific Ocean. Off Chile, horse mackerel feed mainly on euphausiids (Miranda *et al.*, 1998). The demersal fish community is dominated, both in terms of biomass and landings by the Chilean hake (*Merluccius gayi*). Hake inhabits mid-depth waters (200–400 m), feeds on euphausiids, galatheid crustaceans and small pelagic fish (Meléndez, 1984; Arancibia 1989; Cubillos *et al.*, 2003) and is highly cannibalistic (Arancibia *et al.*, 1998).

The Southern Humboldt ecosystem also represents an independent management unit, comprising the main fishing ground for the Chilean purse-seine and trawling fleets, both industrial and small-scale operations, and accounting for approximately 75% of the total landings in Chile (Neira and Arancibia, 2004; Neira *et al.*, 2004). The industrial fishery, based on fish and crustacean species, started in the 1940s, when demersal trawlers targeted Chilean hake (*Merluccius gayi*). However, landings of this fleet were significant only from mid 1950s onwards. By the early 1960s, an industrial pelagic fishery came into operation, targeting small pelagic fish, mainly common sardine (*S. bentincki*) and anchovy (*Engraulis ringens*). At the same time, an industrial fleet operated on a medium-sized pelagic fish, namely horse mackerel (*Trachurus symmetricus*), landings only becoming globally significant from 1975 onwards.

When expanding the trophic analysis to the whole food web, Arancibia and Neira (2005) found a significant decrease in the mean trophic level of the catch, which indicates a likely change in community structure induced by fishing. The current status and basic ecology of top predators such marine birds, sea lions and cetaceans are poorly known, but it is likely that their abundances are very low compared to the period at the start of the industrial fisheries in Central Chile (1940s).

The Northern Humboldt system (Peruvian system) is world famous because it supports one of the most productive fisheries of the world (Ryther, 1969). Specifically, during the 1960s and 1970s, it supported the world's largest single-species fishery, targeting anchovy (*Engraulis ringens*). The Peruvian system is dominated by the dynamics of anchovy and subject to frequent direct environmental perturbations of the El Niño Southern Oscillation (ENSO). The long-term dynamics of the Humboldt Current ecosystem off Peru are controlled by shifts between alternating anchovy (cool) and sardine (warm) regimes that restructure the entire ecosystem, from phytoplankton to top predators (Alheit and Niquen, 2004).

The Peruvian food web, considered to be the northern-central subsystem of the Humboldt, extending from 4°–14° S and west up to 60 nm, and covering approximately 82 000 km², has previously been modeled (Jarre-Teichmann *et al.*, 1998). That model represents the period 1973–1981, corresponding to the period following the collapse of the huge anchoveta fishery of the late 1960s, while sardine biomass was increasing, and did not span any major El Niño events. The model considers the main distribution area of small pelagic fish such as anchovy and sardine, i.e. over the Peruvian continental shelf. This geographic delimitation means that only part of the biomass of horse mackerel, mackerel, and mesopelagic fish were included, since these species are distributed farther offshore (Jarre-Teichmann, 1998).

One of the most remarkable features of the Peruvian system is the rise and collapse of the Peruvian anchovy fishery from 700 thousand tonnes in 1958 to a maximum of 12 million tonnes in 1973 (Walsh, 1981). The evolution in the Peruvian anchovy fishery was accompanied by marked changes in fluxes in the carbon budget of the food web (Walsh, 1981). Similar results can be found in Jarre-Teichmann (1998). The significant drop in anchovy biomass led to a decrease in grazing pressure, allowing plankton biomass to increase. These changes were accompanied by increased stocks of sardine and hake, increased carbon loading and sulfate reduction at the downstream sea bed, and a decline in oxygen and nitrate content of the water column (Walsh, 1981). Trophic flows in the system decreased by one order of magnitude between the periods 1964 and 1971 (anchovy dominated) and between 1973 and 1981 (before the anchovy collapse), and the flow to detritus from the phytoplankton compartment increased by two orders of magnitude between the same periods (Jarre-Teichmann, 1998).

Mediterranean ecosystems of the Catalan and Adriatic seas

The South Catalan Sea (NW Mediterranean Sea) is mainly an oligotrophic area, where enrichment occurs due to regional environmental events, related to wind conditions, the existence of a temporal thermocline and a shelf-slope current and river discharges (Estrada, 1996; Salat, 1996; Agostini and Bakun, 2002). These episodes greatly influence the productivity and fishing activity of the area (Agostini and Bakun, 2002; Lloret *et al.*, 2004), which yields almost half of the total landings of the Catalan coast (Lleonart, 1990) and which is especially important for the reproduction of small pelagic fishes, mainly the European anchovy (*Engraulis encrasicolus*). Moreover, this is also a strategic area for marine vertebrate conservation, sheltering important colonies of terns and gulls (Zotier *et al.*, 1998; Abelló *et al.*, 2003).

Although artisanal gears are still important, most fleets have developed towards a nearly industrial type of activity, fully exploiting continental shelves and upper slopes of the area, associated with the highest landings and mainly composed of bottom trawlers, purse seines and longliners. Small pelagic fish, namely overcoat sardine (*Sardina pilchardus*) and anchovy (*Engraulis encrasicolus*), constitute the principal component of the catch in terms of biomass and are mainly caught by purse seiners and bottom trawlers. There is increasing concern about recruitment overfishing of the north-western Mediterranean anchovy stocks (Papaconstantinou and Farrugio, 2000). Moreover, sardine has decreased steadily since the 1980s, apparently due to fishing and environmental factors (Palomera *et al.*, 2007). Other small and medium-sized pelagic fish (e.g. horse mackerel *Trachurus* spp., mackerel *Scomber* spp., and sardinella *Sardinella aurita*) are exploited to a far lesser degree. The demersal fishery comprises mainly juveniles of several target species, e.g. hake (*Merluccius merluccius*), red mullet (*Mullus barbatus*), and blue whiting (*Micromesistius poutassou*), caught principally by the trawling fleet. Growth overfishing affects some demersal resources because for many species the sizes at first catch are very similar to those at which the fish recruit (Lloret and Lleonart, 2002; Sardà *et al.*, 2005). Large demersal fish (e.g. adult hake) and large pelagic fish (e.g. Atlantic bonito *Sarda sarda*, bluefin tuna *Thunnus thynnus*, and swordfish *Xiphias gladius*) are caught by longline and troll bait fleets and are at low population levels.

Official landings from the area increased dramatically from the early 1960s to the early 1980s, mainly due to governmental aids to the fishing sector. Marked fluctuations in landings occurred thereafter until catches progressively declined from 1994 to the present. Total official landings from 2003 were similar to those attained in the late 1970s. From 1994 to 2003, a decrease of 55% in total official landings has been observed. This reduction is mainly based on the pelagic fraction. Demersal landings have been maintained at similar levels since 1983, with a reduction of 18% in landings, and underwent smaller fluctuations over the period of decline in the pelagic fraction (Coll *et al.*, 2006b).

The Northern and Central Adriatic Sea constitutes the widest continental shelf in the Mediterranean Sea (Pinardi *et al.*, 2006) and is of great value for fishing within the Italian and the European context (Bombace, 1992; Mannini and Massa, 2000). Owing to river runoff and oceanographic conditions, nutrient concentration and production in the Adriatic diminishes from north to south and from west to east (Fonda Umani, 1996; Zavatarelli *et al.*, 1998). The northern parts are characterized by shallow waters, are influenced by river runoff, and are considered eutrophic, whereas the central area is considered to be mesotrophic and the southern region oligotrophic. The area is characterized by highly diverse environmental conditions that translate into high biodiversity (Ott, 1992). Numerous studies describe the distribution and abundance of marine fauna and flora of the Adriatic Sea (e.g. Riedl, 1986; Zupanovic and Jardas, 1989; Jukic-Peladic *et al.*, 2001). The Northern and Central Adriatic is also a strategic area for marine vertebrate conservation, sheltering important seabird populations (Zotier *et al.*, 1998; Baccetti *et al.*, 2002) as well as endangered marine mammals and turtles (Manoukian *et al.*, 2001; Bearzi *et al.*, 2004).

Small pelagic fish, mainly sardine (*Sardina pilchardus*) and anchovy (*Engraulis encrasicolus*), constituted the principal component of the catch in terms of biomass in the 1990s and are mainly caught by purse seiners and mid water trawlers (Arneri, 1996; Mannini and Massa, 2000). The demersal fishery is highly multispecific and mainly comprises juveniles of several target species, e.g. hake (*Merluccius merluccius*) and red mullet (*Mullus barbatus*), caught principally by the trawling fleet. Invertebrates (cephalopods and crabs) also constitute a significant proportion of the catch.

Important changes in landings have been recorded in the Adriatic, with a dramatic increase from the mid 1970s to the mid 1980s, mainly due to the increase of small pelagic fish in the catch. This was followed by marked fluctuations in landings until catches progressively declined from the late 1980s to the present, primarily due to the decrease in small pelagic fish, especially of European anchovy (*Engraulis encrasicolus*) and sardine (*Sardina pilchardus*) (Cingolani *et al.*, 1996; Azzali *et al.*, 2002; Santojanni *et al.*, 2003, 2005). Total official landings from 2000 are lower than those reached in the late 1970s and existing recent data show a significant decrease in fish landings, coupled with an increase of invertebrate landings (Coll *et al.*, 2007). Various target demersal species have been reported to be overexploited (e.g. Papaconstantinou and Farrugio, 2000; Jukic-Peladic *et al.*, 2001), and significant quantities are discarded (Pranovi *et al.*, 2000, 2001; Tudela, 2004). Simultaneously, proliferation of some jellyfish species in the Adriatic Sea since the

1980s has been reported (e.g. Rottini-Sandrini and Stravini, 1981; Zavodnik, 1991; Arai, 2001; Mills, 2001), in parallel with the decrease in small pelagic fish and the increase in eutrophication in the region.

Methods

Data series and selected models available for the Benguela, Humboldt and Mediterranean regions were compared in terms of indicators quantifying the roles of small pelagic fish in the respective ecosystems examined. Models were constructed using the Ecopath with Ecosim software (Box 8.1).

Catch and abundance data series

Recorded landings have been compiled for the selected ecosystems, were used as inputs to the trophic models, and are summarized in the model-generated indicators (i) trophic level[4] of the landed catch and (ii) the Fishing-in-Balance (FiB) index[5]. In addition, the simple data-derived ratio of pelagic:demersal fish in reported landings was also considered. For the Southern Benguela and South-Central Chile, research survey biomass indices of pelagic and demersal species are used to illustrate how the pelagic:demersal fish species ratios in the communities have changed over time.

"Static" ("Snapshot") models

Several available trophic models representing each of the five ecosystems discussed above are used. To facilitate meaningful comparisons between systems, these models were standardized according to the methods described by Moloney *et al.* (2005). Here, with the aim of exploring the effects of small pelagic fish on different ecosystems, we select some of these models as a representative and comparable set capturing the main characteristic periods within each system.

For the Southern Benguela, trophic models of the period 1980–1989 and 1990–1997 are available (Shannon and Jarre-Teichmann, 1999b; Shannon *et al.*, 2003) and an additional, comparable model of the period 2000–2004 has been developed. For the Northern Benguela, several models have been constructed (Roux and Shannon, 2004; Shannon and Jarre-Teichmann, 1999a, b; Heymans *et al.*, 2004). For purposes of this chapter, standardized, updated models have been constructed for the 1960s, 1980s and 1995–1999, and are used to illustrate the main ecosystem changes that have occurred off Namibia. For the Southern Humboldt ecosystem, models have been constructed for 1992 and 1998 (Neira and Arancibia, 2004; Neira *et al.*, 2004) and more recently, also for 2004–2005 (Neira *et al.*, 2006; Neira, 2008). For the South Catalan Sea, trophic models calibrated with time series of data and representing 1978 and 1994 have been used, while in the case of the North and Central Adriatic Sea, models representing the 1980s and 1990s are available (Coll, 2006; Coll *et al.*, 2005, 2006b, 2007, 2008).

Box 8.1. A quick look at Ecopath with Ecosim (EwE)

Ecopath

Mass balance Ecopath models provide a quantitative representation, or snapshot, of marine ecosystems in terms of trophic flows and biomasses for a defined time period, usually a year. Model parameters are expressed in terms of nutrient or energy related currency per unit of surface (frequently expressed as t km^{-2} per year for fisheries applications) (Pauly *et al.*, 2000; Christensen and Walters, 2004).

The ecosystem is represented by functional groups, which can be composed of species, groups of species with ecological similarities or ontogenetic fractions of a species. The model is based on two linear equations that represent the energy balance among groups (production equation) and the energy balance within a group (consumption equation). For each group, the model describes a mass balance situation, where the energy removed from the ecosystem due to, for example, predation or fishing, is balanced by production.

The production (P) of each functional group (i) is divided into predation mortality ($M2_{ij}$) caused by the biomass of the other predators (B_j), exports from the system via fishing activity (Y_i) and other exports (E_i), biomass accumulation in the ecosystem (BA_i) and a baseline mortality or other mortality ($1-EE_i$), where EE_i is the ecotrophic efficiency, or the proportion of the production of group (i) that is exported out of the ecosystem (e.g. by fishing) or consumed by predators:

$$P_i = \sum_j B_j . M2_{ij} + Y_i + E_i + BA_i + P_i .(1 - EE_i)$$

The energy balance within each group is then ensured when consumption by group (i) equals the sum of production by (i), respiration by (i) and food that is unassimilated by (i).

Ecological analyses integrated in Ecopath can be used to examine many ecosystem features based on trophic flows and network analyses, thermodynamic concepts and information theory (Christensen and Walters, 2004; Cury *et al.*, 2005a).

Ecosim

The trophodynamic simulation module Ecosim (Walters *et al.*, 1997) re-expresses the linear equations of the Ecopath model as difference and differential equations that dynamically respond to changes in fishing mortality and biomass. This enables the performance of dynamic simulations at the ecosystem level from the initial parameters of a baseline Ecopath model as follows:

$$\frac{dB_i}{dt} = \left(\frac{P}{Q}\right)_i . \sum Q_{ji} - \sum Q_{ij} + I_i - \left(M_i + F_i + e_i\right) . B_i$$

where dB_i/dt is the growth rate of *i* during the time interval d*t* in terms of B_i, $(P/Q)_i$ is the gross efficiency, M_i is the non-predation natural mortality rate, F_i is the fishing mortality rate, e_i is the emigration rate, I_i is the immigration rate, and $e_i . B_i - I_i$ is the net migration rate. Consumption rates Q are calculated based upon the "foraging arena" theory where the biomass of *i* is divided into a vulnerable and a non-vulnerable fraction and the transfer rate υ between the two fractions is what determines the flow control (Walters *et al.*, 1997; Christensen and Walters, 2004). Default values of υ represent mixed flow control, whilst these values can be modified to represent bottom-up flow control and top-down flow control.

Sum of all flows to detritus derived from the static models was monitored as an indicator of ecosystem change that may be caused by changes in small pelagic fish abundance. In addition, Mixed Trophic Impact (MTI) analysis[6] (Christensen and Walters, 2004) is also used to quantify the variation in benthic–pelagic coupling of trophic interactions in the ecosystems. Two (oldest and most recent) snapshot models from the Southern and Northern Benguela, Southern Humboldt, and the Mediterranean Sea are used and the squared sum of all trophic impacts from the pelagic compartment on the demersal one, and *vice versa*, are computed and compared by means of percentage variation in the squared sum of the MTIs between the early and most

recent modeled periods. The sum of all MTIs of detritus on the demersal and pelagic compartment is also considered.

Models fitted to time series data

Models of the Southern Benguela, the South Catalan Sea and South-Central Chile have been fitted to time series data and are used here to quantify, in ecosystem terms, the changes in these three ecosystems since the 1970s.

A trophic model of the Southern Benguela ecosystem has been fitted to catch and abundance time series data for 1978–2002 with forcing by fishing effort over time, and environmental forcing of primary productivity, using flow control parameters for model tuning (Shannon *et al.*, 2004b).

This model has been updated for revised and additional data series (Shannon *et al.*, 2008). The ecological model developed for the South Catalan Sea was fitted with time series of data from 1978 to 2003 and was able to satisfactorily reproduce the dynamics of several target and top predator species considering fishing, trophic interactions and the environment (Coll *et al.*, 2005, 2008; Coll, 2006). Similarly, an ecosystem model of the Southern Humboldt has been fitted to time series data from 1970–2004 (Neira, 2008).

Fitted models are advantageous in that they make use of available time series data rather than being restricted to the point estimates for one year/few years used in static models, and by means of fitting these series, they constrain estimates of the abundance trajectories of species for which time series data are not available. Since EwE model results have been found to be sensitive to model vulnerability settings (see following section) that quantify flow control interactions (e.g. Shannon *et al.*, 2000), using tuned models, for which best-fit vulnerabilities have been proposed, assists in eliminating potential dynamic model simulations that, when considering the available time series data, would be less likely to be possible ecosystem trajectories in the real world. Fitted models can be used also to quantify likely changes in predation mortality as a result of biomass and fishery fluctuations. Biomass estimates (model outputs) of small pelagic fish and other important groups are plotted over time using the three fitted time-series models.

Simulations

One "snapshot" (static) model was selected for each of the following ecosystems (see section entitled Static ("Snapshot") models): Southern Benguela (1980s), South Catalan (1994), North and Central Adriatic Sea (1990s) and South-Central Chile (1992), and two simple, hypothetical and extreme fishing scenarios were simulated to explore whether the effects of small pelagic fish were likely to be similar across the systems and to try to identify any system specific attributes relating to the role of small pelagic fish in these ecosystems.

All three models fitted to time-series data, as already explained (see prior section, Southern Benguela, South Catalan Sea and Southern Humboldt), suggested that wasp-waist flow control emerges. Thus, in the model scenarios presented here, we adopt a standardized "wasp-waist flow control" assumption, setting vulnerabilities (v, the model parameter describing the availability of a prey species to a predator, see Walters *et al.*, 1997; Christensen and Walters, 2004) for wasp-waist control by small pelagic fish: $v = 1000$ for top down control of small pelagic fish on their plankton prey; $v = 1$ for bottom up control of small pelagic fish on their predators. These vulnerabilities are applied to trophic interactions of the main small pelagic fish in the ecosystems considered: i.e. for Southern Benguela simulations, wasp-waist settings are adopted for anchovy, sardine, "other small pelagic fish" (e.g. flying fish, saury) and redeye; for Southern Humboldt and Mediterranean simulations, wasp-waist settings are assumed for anchovy and sardine only. For all other interacations, v is set to the default of $v = 2$ for mixed flow controls.

Simulations are run for an initial 5-year period, during which no alterations to fishing are simulated (for ease of visualization of model results and extraction of parameter values in year one). Two fishing scenarios are subsequently simulated for an additional 15-year period and model results extracted for the final year of simulation. The following two main simulations are performed.

Simulation 1: Closure of fisheries targeting small pelagic fish

For each of the four models, the fishery on small pelagic fish is closed and model biomass trajectories plotted to examine which species may be likely to benefit or be deleteriously affected. Changes in the ratios of modeled pelagic:demersal fish biomass are tracked over the simulation periods.

Simulation 2: Collapse of small pelagic fish stocks

For each of the four snapshot models selected for simulation purposes, model fishing mortality rates for small pelagics are increased to a level at which anchovy and sardine stocks collapse (are obliterated down to zero or near-zero biomass) by year 16 in model simulations. Changes in model biomass of the main groups responding to the collapse of anchovy and sardine are plotted over the 15-year simulation period. As above, modeled pelagic:demersal fish biomass ratios are compared.

Model results

Catch ratios and abundances

The ratio of small pelagic fish to demersal fish in reported landings ranged between 1 and 9 in the Northern and Southern Benguela, the South Catalan Sea and the North and Central Adriatic, whereas in the Southern Humboldt, the ratio ranges from 1 all the way up to 60 (if horse mackerel, of which a significant portion are caught in the open ocean, are included as pelagic fish) and 18 when horse mackerel are excluded (Fig. 8.2), reflecting the far greater importance of small pelagic fish in the Humboldt compared to other fisheries examined.

The TL of the landed catch increased in the 1990s in the Northern Benguela (Fig. 8.3a), and from 1999 onwards in the Southern Humboldt, as a result of the collapse of small pelagic fish, and thus the increases in the FiB index that

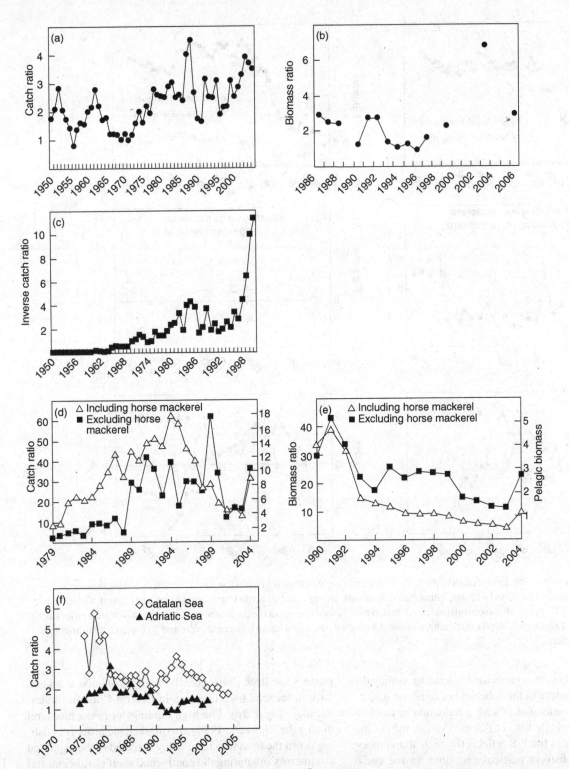

Fig. 8.2. Ratios of pelagic:demersal fish landings in the (a) Southern Benguela, (d) Southern Humboldt, and (f) Mediterranean Sea (South Catalan Sea and North and Central Adriatic Sea). The inverse landings ratio (demersal:pelagic landings) is plotted for the Northern Benguela in (c) to avoid loss of graphic detail on account of the recent collapse of small pelagics off Namibia. (b) Biomass ratio of small pelagic fish to demersal fish and chondrichthyans from research surveys undertaken in the Southern Benguela using hydro-acoustic (small pelagic fish) and swept area trawl (demersal fish and chondrichthyans) surveys. (e) Ratio of surveyed biomass of pelagics (including large horse mackerel) or small pelagics (anchovy, sardine) to hake, in the southern Humboldt. Note that the biomasses for both the Southern Benguela and Humboldt should only be considered in relative terms, not as absolute ratios, as the pelagic and demersal survey estimates are not directly comparable.

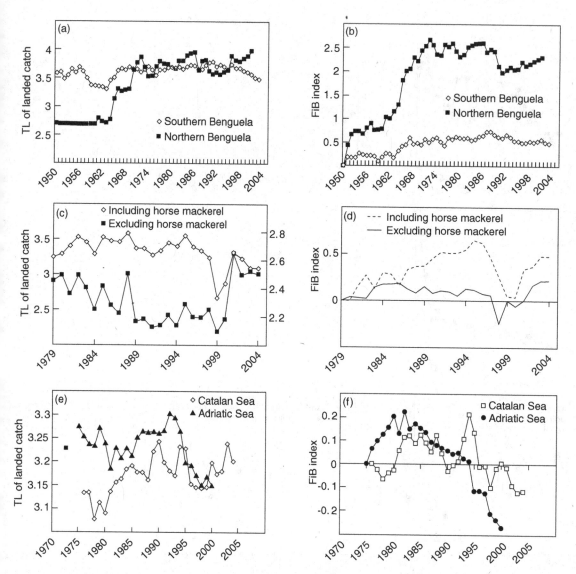

Fig. 8.3. Trophic level (TL) of the landed catch in the (a) Southern and Northern Benguela (after Cury *et al.*, 2005a; van der Lingen *et al.*, 2006), (c) Southern Humboldt (plotted with and without horse mackerel; see text), and (e) South Catalan and North and Central Adriatic Seas, and the corresponding Fishing-in-Balance (FiB) index calculated from landed catches in the (b) Southern and Northern Benguela (after Cury *et al.*, 2005a; van der Lingen *et al.*, 2006), (d) Southern Humboldt (plotted with and without horse mackerel), and (f) South Catalan and North and Central Adriatic Seas.

was observed in these two cases are misleading as they do not reflect an expansion of the fisheries but rather an underlying collapse (Namibia) or at least a reduction in catches (Chile) of small pelagic fish. Off South Africa, the recent short-term decline in the FiB reflects the high abundances of small pelagic fish, in particular sardine, in the early 2000s, rather than the traditional "fishing down the food web" situation.

By the late 1970s, horse mackerel became the most important fishery resource off Central Chile due to the development and later westward expansion of an important offshore fishery, mainly composed of an industrial

purse-seine fleet. Consequently, landings of horse mackerel have influenced total landings in Central Chile in the last decades (Fig. 8.2d). The high landings of horse mackerel during the 1980s and 1990s also masked the process of fishing down the food web from being detected in the exploited community inhabiting the continental shelf (Arancibia and Neira, 2005). Therefore, the indicators are presented with and without including landings of horse mackerel for the Southern Humboldt ecosystem (Fig. 8.3c-d). The TL of the landed catch, excluding horse mackerel, shows an increase in 2001, probably due to a combined decrease in landings of common sardine and anchovy and an increase in landings of hake. The FiB shows an increasing trend until the mid 1990s that reflects the westward expansion of the fleet targeting horse mackerel (Fig. 8.3d). Conversely, FiB excluding horse mackerel shows a decreasing trend since the mid 1980s to minimum (negative) values in the late 1990s (Fig. 8.3c). During the early 2000s, both series of FiB (with and without horse mackerel) increase (Fig 8.3d), probably owing to high landings of hake during this period. However, this trend should not be interpreted as ecosystem recovery since Chilean hake has more recently collapsed and it is expected that the trend in FiB will strongly decline in the next few years.

In the Catalan Sea (Fig. 8.3e-f), TL of the landed catch and the FiB index show the dynamics of small pelagic fisheries, reflecting the expansion of the fishery and subsequent local depletion of pelagic fish stocks (Coll *et al.*, 2006a). In the case of the North and Central Adriatic Sea, data suggest an increase in the TL of the landed catch in the 1980s (due to decrease of small pelagic fisheries and a fishery based on demersal species) and a subsequent decline in TL of landings thereafter due to increased invertebrate landings (Coll *et al.*, 2007). From the 1990s the decrease in the FiB indices reflect deteriorating ecosystems and fisheries.

It is of interest to highlight that simulations of closures of small pelagic fisheries show that, after 10 years, the modeled pelagic:demersal biomass ratios increase from around 1 to over 2.5 in the Southern Benguela, following a steady recovery of model anchovy and sardine over the ten-year simulation period (Fig. 8.4a). By contrast, the modeled pelagic:demersal biomass ratios for the South Catalan Sea, the Adriatic Sea and the Southern Humboldt, which are all higher to begin with (around 5 for Chile and 14 for the Mediterranean), increase following the simulated pelagic fishery closure, as expected, but subsequently decline to original levels or below as a result of increased model hake biomass levels in response to increased availability of model sardine and anchovy prey (Fig. 8.4b-d). Previous model simulations of the Northern Benguela (Shannon *et al.*, 2001; Roux and Shannon, 2004) showed that a moratorium on sardine catches might have some benefit to sardine but likely only small effects on other groups unless abundance of pelagic goby was also reduced. These model results provide an interesting contrast to those for other upwelling systems, suggesting that the Benguela may operate differently than the others examined, possibly because both Southern and Northern Benguela subsystems have an additional, largely unfished and abundant small pelagic fish resource (round herring in the south and pelagic goby in the north, both forage fish species). Alternatively, it is possible that, after the intense exploitation of small pelagic fish in the Southern Humboldt and Mediterranean systems, the subsequent sudden simulated release of fishing pressure allows small pelagic fish to increase very fast initially, whereas at least in the case of the Southern Benguela, small pelagic

fish were not too heavily exploited to begin with. However, in the northern Benguela, the observed pelagic : demersal catch ratio (inversely plotted in Fig. 8.2c) decreased over time as a result of increase in catches of horse mackerel relative to piscivorous hake. Overall, the ratios indicate a shift towards greater importance of demersal fish in the northern Benguela since the 1950s.

When simulating the collapse of small pelagic fish stocks, the modeled pelagic : demersal biomass ratios initially decrease before leveling off or recovering slightly again (Fig. 8.5) due to the decrease of pelagic fish biomass in the models. Decreases of 33% and 46% are estimated in the Catalan and Adriatic simulations respectively, while the Southern Benguela simulation suggests a smaller reduction in the modelled pelagic : demersal biomass ratio of 25%, again perhaps due the presence of a third abundant small pelagic species (redeye round herring). In the Southern Humboldt case, the simulated collapse of small pelagic fish results in a decrease in the model pelagic : demersal biomass ratio with and without horse mackerel of 25% and 100%, respectively (Fig. 8.5b).

Gelatinous zooplankton and proliferation of other species

Model simulations yield relatively consistent results where gelatinous zooplankton tends to increase when small pelagic fish stocks collapse and conversely, there is a decrease in jellyfish abundance modeled when closure of the pelagic fisheries is simulated (Fig. 8.4, Fig. 8.5 and Table 8.1). In Southern Benguela simulations, macro-zooplankton follow similar trajectories to gelatinous zooplankton, increasing by 8% when a collapse of the purse-seine fishery is simulated, and declining to 68% of original levels when the purse seine fishery is closed. Model phytoplankton biomass remains unchanged. In the case of the Catalan Sea, jellyfish increase when a collapse of the small pelagic fishery is simulated, while there is also an increase in model micro-mesozooplankton and macrozooplankton of 21% and 35%, respectively. On the contrary, the model simulation predicts an 8% decrease in phytoplankton biomass. Similarly, in the North and Central Adriatic Sea, simulations suggest an increase in jellyfish biomass when small pelagic fisheries collapse, while micro–meso and macrozooplankton show an increase of 18% and 2%, respectively. Phytoplankton biomass decreases by 5% in the Adriatic model. Similar results were found by Roux and Shannon (2004) when a hypothetical fishery on jellyfish was modeled, in that anchovy and several predatory fish species were favored when 50% of gelatinous zooplankton production was "removed" through "fishing."

Simulations also show the proliferation of other species when small pelagic fish stocks collapse (Fig. 8.5). Both ecological models from the Catalan and Adriatic Sea highlight

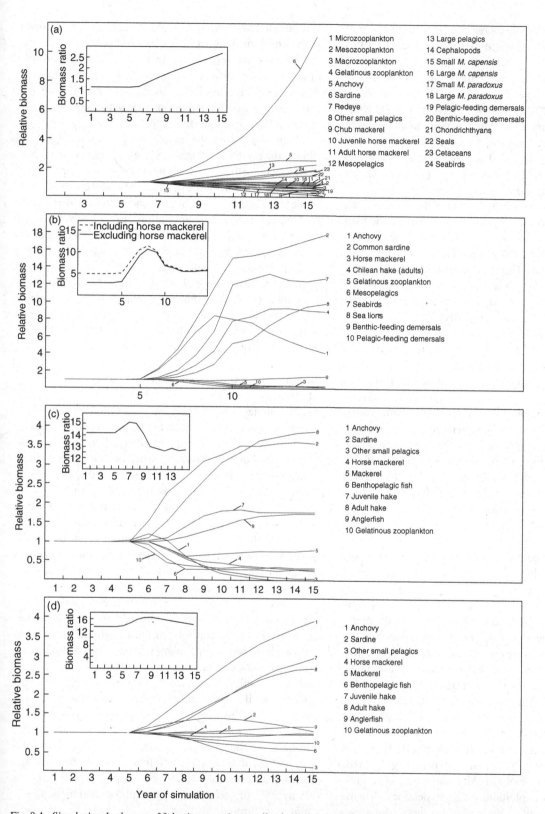

Fig. 8.4. *Simulation 1: closure of fisheries targeting small pelagic fish*. Projected, modeled biomass trajectories (biomass plotted relative to biomass at start of simulation), and the ratio of the biomasses of small pelagic fish to demersal fish (including hake) plotted as inserts, for (a) Southern Benguela (small pelagic fish include anchovy, sardine, round herring and other small pelagic fish), (b) Southern Humboldt (small pelagic fish include only anchovy and sardine), (c) South Catalan Sea (small pelagic fish include anchovy, sardine and other small pelagic fish), and (d) North and Central Adriatic Seas (small pelagic fish include anchovy, sardine and other small pelagic fish).

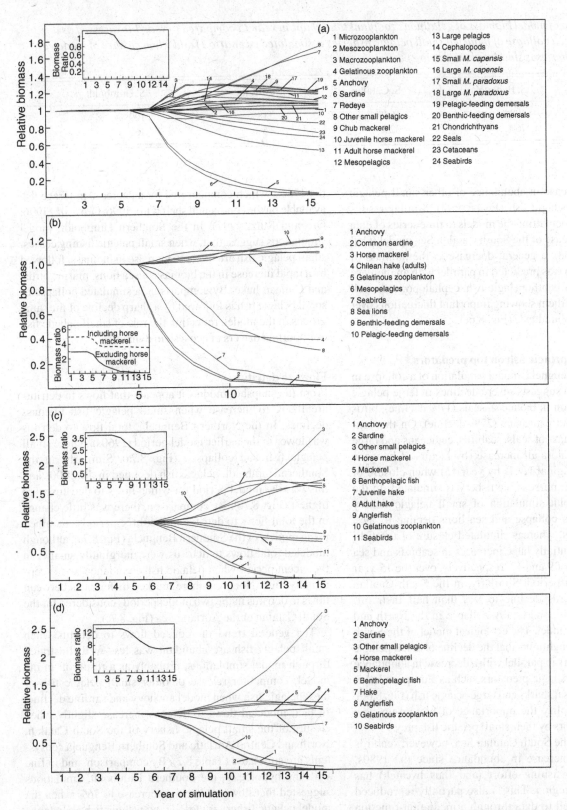

Fig. 8.5. *Simulation 2: collapse of small pelagic fish stocks.* Projected, modeled biomass trajectories (biomass plotted relative to biomass at start of simulation), and the ratio of the biomasses of small pelagic fish to demersal fish (including hake) plotted as inserts, for (a) Southern Benguela (small pelagic fish include anchovy, sardine, round herring and other small pelagic fish), (b) Southern Humboldt (small pelagic fish include only anchovy and sardine), (c) South Catalan Sea (small pelagic fish include anchovy, sardine and other small pelagic fish), and (d) North and Central Adriatic (small pelagic fish include anchovy, sardine and other small pelagic fish).

Table 8.1.   *Ratio of model biomass of gelatinous zooplankton groups in year 15 compared to year 1 (expressed as a percentage) when a collapse of modeled small pelagic fisheries is simulated (scenario 1) and when closure of modeled small pelagic fisheries is simulated (scenario 2)*

Scenario	S. Benguela	S. Catalan	Adriatic	S. Humboldt
1. Collapse	108%	164%	118%	109%
2. Closure	76%	32%	75%	16%

the possible increase in abundance of other small pelagic fish and benthopelagic fish (Fig. 8.5c-d). Similar results were also seen when fitting the models to time series of data (Fig. 8.6). In the case of the South Catalan Sea model (Coll *et al.*, 2005, 2008), a general decrease in the biomass of small pelagic fish was predicted in parallel with an increase of the biomass of benthopelagic fish, cephalopods, as well as jellyfish, all of them showing important fluctuations over time in terms of abundance (Fig. 8.6c).

Effects of small pelagic fish on top predators

In the southern Benguela, model simulation of a collapse in small pelagic fish suggests severe declines in large pelagic fish (48% reduction in biomass), seals (17% decline), birds (33% decline), and cetaceans (27% decline). On the contrary, model biomass of seals, seabirds, cetaceans and large pelagic fish predators all increase (by factors of between 1.5 and 2.2 of original levels by year 15) when closure of the South African purse-seine fishery is simulated. In the Southern Humboldt simulation of small pelagic fishery collapse, seabirds collapse and sea lions decline to 29% of original values, whereas simulated closure of the fishery suggests potentially huge increases in seabirds and sea lions, by a factor of 9 and 12, respectively, over the 15-year model simulation period. Similarly, in the South Catalan Sea model, seabirds decline to less than half their initial model biomass, and totally collapse in the North and Central Adriatic model. The ecological model of the South Catalan Sea also highlights that the decline of small pelagic fish biomass occurs in parallel with a decrease in abundance of demersal and pelagic predators such as European hake, anglerfish, demersal sharks and large pelagic fish (Fig. 8.6c). This could exemplify the importance of bottom-up flow control on predators by their small pelagic fish prey.

In the case of the South Catalan Sea, however, seabirds have shown an increase in abundance since the 1980s, even though the fishing effort (and thus bycatch) has increased in the region. This is also partially reproduced by the model fitted to data through a mediation function (Coll *et al.*, 2005, 2008) (Fig. 8.6c) and it is mainly due to the increase in discards from trawling and the protection of breeding colonies in the area since the beginning of

the 1980s, which compensated for a decrease in forage fish available on the continental shelf (Oro, 1996; Oro *et al.*, 1996; Oro and Ruíz, 1997). In the Southern Humboldt, model simulations suggest that, when small pelagic fishing ceases, small pelagic fish are likely to increase in biomass, followed by a rapid increase in the biomass of sea lions, marine birds, and Chilean hake. By comparison, a simulated collapse of small pelagic fish is followed by a sharp decline of the same groups in the model, reflecting the role of small pelagic fish as important drivers of ecosystem dynamics.

Flows to detritus

From the snapshot models it appears that flows to detritus are likely to increase when small pelagic fish biomass declines. In the Northern Benguela, total flow to detritus was lower in the earlier model period (1960s), before small pelagic fish had collapsed (Fig. 8.7a). Similarly, in the Southern Humboldt, pelagic fish declined in the 2000s and this was the model period for which flows to detritus were highest (Fig. 8.7b). By comparison, there was little change in the total flows to detritus in the three periods examined using models of the Southern Benguela (Fig. 8.7a), although modeled total flows to detritus were marginally smaller in the recent period when pelagic fish abundance was exceptionally high. Nor were there marked differences between flows to detritus in the two model periods considered for the South Catalan or the Adriatic Sea (Fig. 8.7c).

The general trend of reduced flows to detritus when small pelagic fish are abundant was less well illustrated through model simulations, probably as a result of, in the model, competitor release of other small pelagic fish or mesopelagic fish when model anchovy and sardine decline. In fact, modeled flows to detritus increase slightly when closure of the small pelagic fishery of the South Catalan, North and Central Adriatic and Southern Benguela ecosystems are simulated (Table 8.2). By comparison, and in line with the hypothesis, the Southern Humboldt simulations suggested that flows to detritus decrease to 76% when the model pelagic fishery is closed (and pelagic fish abundance increased), and increase slightly when the modeled pelagic fishery collapses anchovy and sardine stocks (Table 8.2). Overall, standardized snapshot models suggested that flows

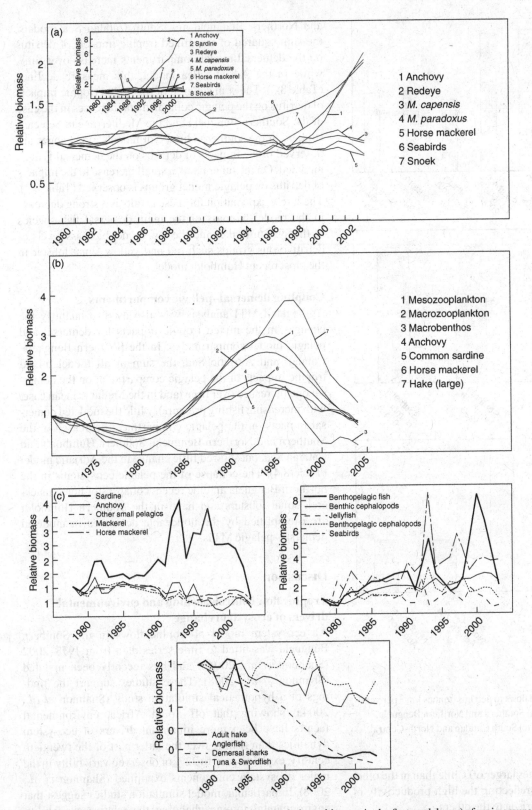

Fig. 8.6. Estimated biomass trajectories (biomass plotted relative to biomass in the first model year) of main groups from (a) Southern Benguela trophic model fitted to time-series data (Shannon *et al.*, 2004b), (b) Southern Humboldt trophic model fitted to time-series data (Neira, 2008), and (c) South Catalan trophic model fitted to time series data (Coll *et al.*, 2008). Because the relative increase in sardine in the Southern Benguela was so large, the plot including sardine is presented as an insert in (a). For the South Catalan Sea, species are grouped according to their scale of relative biomass change over time in (c).

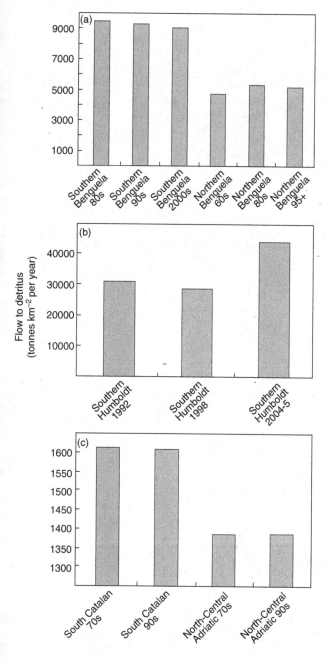

Fig. 8.7. Sum of all modeled flows to detritus (tonnes km⁻² per year) in different periods for the (a) Southern and Northern Benguela, (b) Southern Humboldt, and (c) South Catalan and North-Central Adriatic Seas.

to detritus were three times larger off Chile than in the other ecosystems considered, reflecting the high productivity of this ecosystem in comparison to the others.

Differences between ecosystems are found in the sum squared of all model trophic impacts of detritus on the demersal and pelagic model compartments. In the Southern

and Northern Benguela and South Catalan Sea models, the sum squared of the mixed trophic impacts of detritus on the demersal model compartments increase over time, while in the Adriatic Sea model, this measure declines (Table 8.3). The sum squared of the mixed trophic impacts of detritus on the pelagic compartment increases in the case of the Southern Benguela and the Mediterranean Sea case studies. In the southern Humboldt models, no changes are observed in the impacts of detritus on the demersal groups modeled. On the other hand, a small increase in the impacts of detritus on pelagic model groups is observed (Table 8.3). The likely explanation for these results is a strong decrease in the model biomass of the main demersal fish species in the system (hake) and the low model biomass levels of detritivorous groups such red and yellow squat lobster in the most recent Humboldt model.

Coupling demersal–pelagic compartments

Aggregated MTI analysis was also used to quantify the changes in the mixed trophic impacts for demersal and pelagic model compartments. In the Southern Benguela, Catalan and Adriatic Sea, the sum of all model mixed trophic impacts of the pelagic compartment on the demersal one increased over time (and in the Mediterranean case, with increasing fishing pressure), while the modeled demersal impacts on the pelagic compartment decreased in the Southern and Northern Benguela, Southern Humboldt and Catalan Sea, but showed little change in the Adriatic model (Table 8.4). The collapse of the pelagic ecosystems in the northern Benguela and the recent collapses of horse mackerel, squat lobsters and hake in the Southern Humboldt are exemplified by the noticeable decreases in modeled demersal–pelagic MTIs.

Discussion

Trophic flow controls, fishing and environmental drivers of ecosystem change

An ecosystem model of trophic flows in the Southern Benguela was fitted to time series data from 1978–2002 (Shannon *et al.*, 2004b) and has recently been updated (Shannon *et al.*, 2008). These studies support the findings of a hypothetical simulation study (Shannon *et al.*, 2004a) showing that off South Africa, environmental factors have been more important drivers of ecosystem dynamics than fishing over the latter part of the twentieth century, explaining up to 22% of observed variability in the major ecosystem components examined (Shannon *et al.*, 2008). In particular, model simulation studies suggest that environmental forcing changing the relative availability of mesozooplankton to anchovy and sardine may be a key factor in driving changes in the abundance of small pelagic fish in the southern Benguela (Shannon *et al.*,

Table 8.2. *Ratio of total model flows (all combined) to detritus in year 15, compared to year 1 (expressed as a percentage), when a collapse of modeled small pelagic fisheries is simulated (scenario 1) and when closure of modeled small pelagic fisheries is simulated (scenario 2)*

Scenario	S. Benguela	S. Catalan	Adriatic	S. Humboldt
1. Collapse	99%	95%	97%	102%
2. Closure	100%	113%	108%	76%

Table 8.3. *Variation (% change over time) of the sum (squared) of all modeled trophic impacts of detritus on the demersal and pelagic model components, derived from the Mixed Trophic Impact analysis of snapshot models of Southern and Northern Benguela, Southern Humboldt and Mediterranean Sea case studies*

	S. Benguela (early 2000s compared to 1980s)	N. Benguela (1960s compared to late 1990s)	S. Catalan (1994 compared to 1978)	Adriatic (1990s compared to 1975)	S. Humboldt (2004–05 compared to 1992)
Detritus–demersal	252.5%	52.8%	6.0%	–30.7%	0.0%
Detritus–pelagic	246.6%	–32.7%	7.5%	13.8%	6.7%

Table 8.4. *Variation (% change over time) of the sum (squared) of all model trophic impacts of the pelagic component on the demersal component, and* vice versa, *derived from the Mixed Trophic Impact analysis of snapshot models of Southern and Northern Benguela, Southern Humboldt and Mediterranean Sea case studies*

	S. Benguela (early 2000s compared to 1980s)	N. Benguela (1960s compared to late 1990s)	S. Catalan (1994 compared to 1978)	Adriatic (1990s compared to 1975)	S. Humboldt (2004–05 compared to 1992
Pelagic on demersal	32.2%	–15.2%	4.8%	70.0%	–32.0%
Demersal on pelagic	–18.7%	–50.7%	–19.6	0.5%	–36.0%

2004a). These mechanisms and their physical forcings still need to be explored in more detail. Shannon *et al.* (2004b, 2008) found that model parameterisation describing prey vulnerability to predators accounted for over 40% of the observed variability in the time series examined, and suggest wasp-waist flow control by small pelagic fish as a characteristic of this upwelling system.

A model representing the Southern Humboldt upwelling system in the year 1970 was fitted to available time series of relative biomass, catch and fishing mortality for the period 1970 to 2004 (Neira, 2008; Shannon *et al.*, 2008). Fishing mortality explained 23% of the variability in the times series, whereas vulnerability parameters estimated using EwE explained an additional 16%, and a function affecting primary production a further 21%. The best-fit model indicated that a long-term change in primary production might have affected the dynamics of groups in the Southern Humboldt

system, a result that was reinforced when the primary productivity anomaly proposed by the model was compared with independent time series of sea surface temperature and an upwelling index in the system for the period 1970 to 2000 (Neira, 2008).

When fitting the ecological model of the South Catalan Sea to available time series of data (Coll, 2006; Coll *et al.*, 2005, 2008), similar results were found. Trophic interactions accounted for half of the observed data variability, while fishing and the environment had similar contributions (approximately 7%) to explaining the variability in time-series data. Moreover, small pelagic fish were found to be exerting wasp-waist (sardine) and bottom-up (anchovy) flow control, also in line with previous hypotheses. The function affecting primary production calculated by Ecosim to minimize the sum-of-squares of the time series was significant.

Box 8.2. The importance of understanding ecosystem functioning when considering impacts of climate change on small pelagic fish

Although there are plenty of studies dealing with how environmental variability and small pelagic fish abundance are related, it is difficult to separate direct environmental effects on early life stages of small pelagic fish from the indirect effect of the environment on planktonic prey availability for small pelagics (Cury *et al.*, 2005b) and from anthropogenic effects such as fishing. Nevertheless, by acting at several possible trophic levels (Fig. 8.8), environmental variability or change manifests itself in marine ecosystems as decadal-scale regime shifts or synchronous fish stock fluctuations, or at the shorter time scale, as resource variability such as that related to El Niño events (see final section). The frequency and intensity of El Niño events (Timmerman *et al.*, 1999) and upwelling favorable winds (Bakun, 1990) are expected to increase with global warming. Increased upwelling favorable winds may lead to more nutrients being injected into the photic zone, although adverse offshore dispersal of early life stages of small pelagics and other fish may increase. In addition, fish species are adapted to specific ranges of temperature and other environmental conditions for spawning, recruitment, larval development, etc. Therefore, projected increased upwelling may not necessarily have a positive effect on marine communities. How climate change may affect marine ecosystems will be very much dependent on the plasticity of the species being affected, and on the structure of an ecosystem and the way in which it is functioning.

The concept that ecosystems may be "bottom-up" controlled rests on the assumption that primary production or supply of nutrients regulates groups at higher trophic levels in the food web (Pace *et al.*, 1999). Although bottom-up trophic flow control tends to be the dominant flow control type reported for aquatic ecosystems (Cury *et al.*, 2003), primary producers and herbivores appear to be weakly linked in marine food webs (Cury *et al.*, 2005b), possibly explaining the rarity of effects of changing primary productivity cascading up the food web to consumers of small pelagic fish (Micheli, 1999). In a wasp-waist controlled ecosystem, environmental changes (*sensu* Sinclair, 1988 and Bakun, 1996) directly affect recruitment of small pelagic fish, the effects of which may propagate up and down the food web, at least to a limited extent. Upwelling ecosystems, in particular, have been found to be largely wasp-waist controlled (Cury *et al.*, 2003). However, changing the trophic structure of the marine food web, specifically through overexploitation of fish stocks at higher trophic levels, may change recruitment patterns of commercially valuable species (Beaugrand *et al.*, 2003), thereby potentially enhancing environmental impacts on the ecosystem as a whole; overexploited ecosystems tend to be more likely to be bottom-up controlled (e.g. van der Lingen *et al.*, 2006). It is thus important that the synergistic effects of fishing and climate change be carefully considered (van der Lingen *et al.*, 2006).

Thus, configuration of the trophic web seems to play a major role in explaining the dynamics of marine ecosystems and small pelagic fish, and its modification by fishing (Fig. 8.1) and the environment (Box 8.2) can have important impacts on the whole ecosystem. The synchrony in the dynamics of small pelagic stocks around the world (Kawasaki, 1983; Lluch-Belda *et al.*, 1992) may mean that the link between environmental and ecosystem change could be less complex than initially expected (the environment is acting either directly on small pelagic fish or their main food, and through small pelagic fish these effects are propagated up and down the food web). On the other hand, the effects of overexploiting small pelagic fish could lead to dramatic and possibly even irreversible ecosystem change, since collapse due to overfishing small pelagic fish can exacerbate "natural" fluctuations induced by the environment. For example, there has been a drastic change in the pelagic ecosystem of the Northern Benguela since the 1980s (Bakun and Weeks, 2004). As a result of a series of unfavorable environmental events which have been exacerbated by heavier fishing than was in fact sustainable at the

low stock biomasses at the time, several stocks off Namibia have undergone large changes, including a collapse of the anchovy and sardine stocks and possibly proliferation of jellyfish (although there is some debate regarding reliability of evidence for the latter) (Boyer *et al.*, 2001). Similarly, in the Black Sea (Daskalov, 2002; Gucu, 2002) a regime shift has been described in relation to both overexploitation (causing trophic cascades down the food web), and eutrophication, and a collapse of small pelagic fish has been followed by the proliferation of jellyfish. However, other factors may account for jellyfish outbreaks, such as the beneficial impacts of bottom trawling to benthic-spawning jellyfish. Mills (2001) has reported outbreaks of new non-indigenous jellyfish species in certain areas linked to changes in climatic conditions.

Generic effects of fishing on small pelagic fish

Frank *et al.* (2006) showed a north–south geographic gradation in flow controls in the North Atlantic, with more productive and diverse ecosystems better able to deal with top-down perturbations (such as fishing) and showing

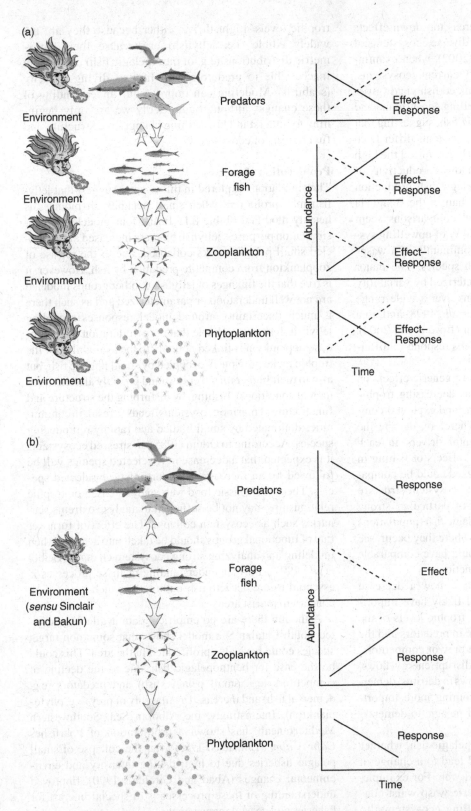

Fig. 8.8. Diagrammatic representation of a simple four-tier marine food chain showing (a) uncertainty in the response (right panel) of the marine ecosystem to climate change, which may act at any or several trophic levels and which will depend on the dominant kind of flow control (left panel), and (b) the possible response of this simplified marine food web (right panel) to climate change directly affecting small pelagic fish, in a wasp-waist flow control dominated marine ecosystem (left panel). Small pelagic fish abundance is hypothetically altered by climate change, affecting the abundance of predators, zooplankton prey and primary producers (modified from Cury *et al.*, 2003).

stronger bottom-up influences, whereas top-down effects are large in less productive and less diverse ecosystems of the northern North Atlantic. Carr (2002), when examining productivity of eastern boundary current ecosystems, found a mismatch between the estimated fish yield (given productivity estimates in these upwelling ecosystems) and the reported landings of small pelagic fish, suggesting that the trophic structure of upwelling ecosystems differ from one another. In particular, Carr (2002) contrasted the high productivity of the Benguela and the lower productivity of the Humboldt, noting that small pelagic fish production was much higher in the Humboldt than in the Benguela ecosystem. Nevertheless, our model comparisons seem to suggest the overarching commonality of upwelling systems tending to be controlled, predominantly in a wasp-waist fashion, by small pelagic fish species. The major coastal upwelling systems are characterized by variability in their chemical–physical conditions over a wide range of spatial and temporal scales (Hill *et al.*, 1998) and, to a lesser extent, also the Mediterranean (Bosc *et al.*, 2004). Individuals, populations and food webs respond in different ways to this variability.

On the other hand, fishing has more generic effects on populations and communities such as decreasing trophic level and size structure of populations and exploited communities (e.g. Pauly *et al.*, 1998). Therefore, despite the importance of different environmental drivers in each system, we expect that the "generic" effects of fishing in driving changes in the systems analyzed could be comparable, given the common wasp-waist ecosystem structure pivoting around small pelagic fish. In particular, strong changes in the abundance of small pelagic fish (populations located in the waist of the food webs where they occur, see Introduction), induced by fishing, could have comparable effects on ecosystem structure and function.

Models analyzed in this study show how a decrease in small pelagic fish abundance will likely have important effects on both higher and lower trophic levels of the food web, possibly causing a decrease in predators and the proliferation of other species that are prey or competitors. Changes in small pelagic fish may disrupt energy flows, which could result in an increase in flows to detritus, demersal processes within the system becoming more important and possible stronger coupling of pelagic to demersal trophic compartments.

In addition, the collapse of small pelagic fish, whether fishery or climatically driven, could lead to a change in the type of trophic control in a food web. For example, a food web could change from generic wasp-waist control, characteristic of several upwelling ecosystems, to being more bottom-up, influenced by the environment. Opportunistic species that benefit from excess plankton production but do not transfer this production to higher

trophic levels might thrive, either because they are not widely edible (i.e. jellyfish) or because their bathymetric distributions (e.g. of mesopelagic fish) make them inaccessible to predators with limited diving capacity (seabirds). Modeling can improve our understanding of these changes although admittedly we are only beginning to understand how fishing changes the structure and functioning of ecosystems.

Proliferation of species

The simulations explored in this chapter suggest that jellyfish may proliferate when small pelagic fish biomass is heavily modified (Table 8.1). In the four models used for simulation purposes, jellyfish biomass increased after modeled small pelagic stocks collapsed, due to the release of zooplankton from compete predation by fish. However, it is true that the linkages of jellyfish to other groups modeled are not well understood or parameterized and as such there is much uncertainty around model responses regarding jellyfish. It seems probable that the proliferation of gelatinous zooplankton is linked not only to the availability of the trophic niche previously occupied by small pelagic fish, but also to their higher turnover rate (and possibly also environmental conditions). Fishing, by disturbing the structure and functioning of marine ecosystems, tends to result in communities dominated by small bodied and rapidly reproducing species. According to Odum (1985), in stressed ecosystems it is expected that a decrease in K-selected species will be followed by an increase in opportunistic r-selected species. Therefore, classic food web analysis based on trophic relationships may not be sufficient to analyze extreme scenarios such as ecosystem collapse. The effect of turnover rate of functional groups should be taken into account when modeling and analyzing strong ecosystem changes (Leibol *et al.*, 1997). Further understanding of these processes is essential to predict jellyfish outbreaks, which are of high concern in tourist areas.

Although there are no empirical data available, the fitted South Catalan Sea model shows that some non-target species could have also proliferated in the area. This could be the case for benthopelagic fish, due to the decline of competitors (e.g. small pelagic fish) and predators (e.g. demersal fish) and increased availability of prey (e.g. phytoplankton). Interestingly, the Alboran Sea, Southwestern Mediterranean, has shown a proliferation of boarfishes *Capros aper* (Linnaeus, 1758) after the collapse of small pelagic fisheries due to high fishing intensity and environmental changes (Abad and Giraldez, 1990). Improved understanding of these processes is of special interest for fisheries and coastal management.

Simulation results also suggest that changes in small pelagic fish abundance have important model impacts on other zooplanktonic species and support the idea that pelagic fish exert top-down control on their zooplanktonic preys (Cury *et al.*, 2000).

Effects of small pelagic fish on top predators

Fluctuations of small pelagic fish have an important impact on predator populations, as can be seen in studies of seabird population dynamics (see later example) and as deduced from model simulations that illustrate the importance of these organisms as prey for demersal and pelagic feeders such as tuna, seabirds, marine mammals, and groundfish predators.

In the southern Benguela, a system in which marine mammals and birds are relatively abundant, a simulated collapse of the small pelagic fish suggests severe declines in large pelagic fish, seals, birds, and cetaceans. In addition to "simple" direct predator–prey effects that have been modeled are other indirect influences; analysis of field data has shown that African penguin breeding success has been found to be significantly related to fish prey abundance (Crawford *et al.*, 2006), and the proportion of penguins choosing to breed in any particular year has also been found to be prey related (Crawford *et al.*, 1999).

In the case of the South Catalan Sea model, top predators feeding on small pelagic fish (e.g. hake, anglerfish, and dolphins) also show decreasing biomass when these prey organisms collapse in the model. However, seabirds (observations) have shown an increase in abundance due to the increase of discards, as already discussed, highlighting the indirect effects of fishing. A moratorium on trawling activities in the Catalan Sea affected populations of seabirds due to reduced food availability during the breeding season (Oro *et al.*, 1996).

However, and not yet considered in ecological modeling studies such as the simulations presented in this chapter, it is not only the quantity of food available to predators that is an important factor affecting trophic mechanisms up the food web, but also the quality of that food. Not only do predators and prey need to overlap in time (Cushing's (1990) match/mismatch hypothesis) and in space, but the quality of the prey needs to be suitable (Beaugrand *et al.*, 2003). For example, in 2004, the common guillemot (*Uria aalge*), which usually feeds on sandeel (*Ammodytes marinus*) during its breeding season, fed sprat (*Sprattus sprattus*) to its chicks; breeding success was poor and the chicks produced were in poor condition (Wanless *et al.*, 2005). Although daily food intake by the chicks in 2004 did not differ from previous years, the nutritional content of sprat was found to be much lower than that of sandeel, pointing to food quality as an important factor in seabird breeding success (Wanless *et al.*, 2005). Changes in the availability of essential fatty acids have been shown to have drastic effects on fish population dynamics, as they contribute to regime shift effects on fish populations (Litzow, 2006).

Environmentally induced decreases in availability of lipid-rich fish to seabirds can also explain dramatic declines in number of breeding kittiwakes (Kitaysky *et al.*, 2006).

Changes in stock sizes of small pelagic fish are often associated with shifts in the geographic distribution of these fish (Lluch-Belda *et al.*, 1989). This has been found to be true of anchovy (Barange *et al.*, 1999) and sardine (J. Coetzee, personal communication) off South Africa; their distributions are expanded at large stock sizes. Thus, small pelagic fish cause, or at least themselves are associated with, shifts in the distributions of several other ecosystem components, in particular their predators. This is particularly well illustrated off southern Africa, where changes in the distribution of small pelagic fish, and therefore also the availability of prey to predators such as seabirds, have been observed. There has been an eastward shift in the distribution of pelagic fish resources off South Africa (van der Lingen *et al.*, 2006), which has led to a large reduction in size of the African penguin colonies along the South African west coast as a result of severely reduced availability of prey to breeders (Crawford, 2006). The eastward distributional shift of small pelagic fish in the southern Benguela in recent years has been matched by a similar southward geographic shift in west coast rock lobster (Mayfield and Branch, 2000), attributed to large scale environmental forcing (Cochrane *et al.*, 2004). Off Namibia, there has been a northward distributional shift of seals and seabirds such as African penguins between the early 1980s and the 2000s, which is concurrent with northward contraction of anchovy and sardine spawning stocks in that region (van der Lingen *et al.*, 2006).

Unfortunately, no biomass/abundance time series for top predators such as marine birds and marine mammals are available off central Chile. Therefore, the effects of changes in the abundance of small pelagics in their populations are practically unknown. Off Peru, changes in population size of guano-producing seabirds (cormorant, *Phalacrocorax bougainvillii*; booby, *Sula variegata*; pelican, *Pelecanus thagus*) could be a response to changes in primary and secondary production of the Peruvian upwelling system, likely mediated by anchovy (Jahncke *et al.*, 2004). Using a modeling approach, these authors indicate that seabirds would have consumed 14.4% of the available anchovy resource during the model period prior to the development of the fishery targeting anchovy. Once the fishery was in full operation, the model estimated that seabirds only accessed 2.2% of the available anchovy resource (Jahncke *et al.*, 2004).

Increased pelagic–benthic coupling and flow to detritus

Benthic–pelagic coupling is viewed as an important process in highly exploited ecosystems. For example, this was found to be important in the Namibian upwelling ecosystem after depletion of stocks due to industrial fishing and environmental events, an important feature that appears to set this system apart from other upwelling ecosystems (Moloney *et al.*, 2005). Dietary studies of jellyfish in Walvis Bay Lagoon, Namibia, highlighted the importance of benthic invertebrates in the diet of jellyfish, suggesting that, by virtue of jellyfish being able to take advantage of increased benthic production, jellyfish biomass in the northern Benguela may have increased after the collapse of small pelagic fish (Flynn and Gibbons, 2007). The Namibian region shows some similarities with the South Catalan Sea (Coll *et al.*, 2006a), and this is also characteristic of highly fished temperate regions (Sánchez and Olaso, 2004). Model results from the North and Central Adriatic Sea model also highlight the importance of coupling between the pelagic and demersal compartments in that ecosystem, mainly due to the link between model detritus and benthic and pelagic invertebrates (Coll *et al.*, 2007). This is likely explained by the shallow waters and the oceanographic features of the area (Ott, 1992).

The Humboldt system, composed of the northern (Peru) and southern (Chile) subsystems, is probably the largest in terms of size when compared to other upwelling systems (Jarre-Teichmann and Christensen, 1998; Neira *et al.*, 2004). In addition, this system has exhibited most pronounced changes in system size after the collapse of the anchovy stock off Peru, and horse mackerel off Chile. After the collapse of anchovy, the Peruvian system became similar to the northern Benguela system, the latter decreasing in size from the mid 1970s to the early 1980s due to the strong decrease of small pelagics (mainly sardine), which was not compensated for by the increased abundance of horse mackerel (Jarre-Teichmann and Christensen, 1998). The simple simulations examined here suggest that reducing small pelagic fish abundance will most likely result in an increase in flows to detritus, probably due to the disruption of energy transfer from low to high trophic levels (i.e. excess primary production that is not efficiently transferred up the food web). These model results are in line with conclusions drawn for the Peruvian ecosystem (Northern Humboldt) (Walsh, 1981) where trophic flows to detritus from 1966 to 1979 were seen to increase due to overfishing of anchovy. Therefore, flows to detritus may be of importance and could be seen as a measure of increasing pelagic–benthic coupling.

The increase in modeled flows to detritus in the Mediterranean was smaller than in other ecosystems analyzed, most probably due to the fact that fluctuations of small pelagic fish abundances are smaller in the Mediterranean and that this area has a history of being heavily fished (Margalef, 1985).

Enhancing catches of commercially valuable predatory fish by reducing small pelagic fish catches?

Model simulation of closed small pelagic fisheries in the Southern Benguela suggests that higher demersal fish

biomasses would be unlikely to be supported despite the likely increased abundances of forage fish (Fig. 8.4a). By comparison, closed small pelagic fisheries in the South Catalan, North and Central Adriatic and Southern Humboldt models clearly suggest that larger stocks of demersal fish, especially hake, would be supported (Fig. 8.4b-d), thus benefiting the demersal fisheries. This was especially noticeable in the Chilean case, where the potential benefits of closing the small pelagic fishery appear to be larger, and projected model hake biomass increases almost nine-fold (compared to 3–4-fold increases in hake biomass in the Catalan and Adriatic models respectively). As the demersal fishery is of greater economic importance in the Mediterranean areas than the small pelagic fisheries, temporal closure of the small pelagic fish fishery could be economically beneficial. However, formal simulations based on sound economic data would need to be carefully designed and examined to analyze this. Specifically, in these simulation results, demersal groups might benefit from an increase in small pelagic fish provided their own fishery is kept at a constant level during the simulation period, suggesting that benefits of increased small pelagic fish abundance may only be reaped if the predator stocks are also sustainably managed (or maintained at viable levels, in the case of charismatic species).

As previously mentioned (section entitled "Catch ratios and abundances"), the Benguela ecosystems (north and south) differ from the others by also supporting large stocks of a third abundant small pelagic, namely round herring in the Southern Benguela and pelagic goby in the Northern Benguela. It is unlikely that round herring and pelagic goby are strong competitors with anchovy and sardine for zooplankton food as they exhibit different feeding behaviors. Tanaka *et al.* (2006) found that feeding behavior of filter-feeding Japanese anchovy, and thus the zooplankton composition in its diet, differs from that of particulate-feeding Pacific round herring and Japanese jack mackerel in the East China Sea, bringing about a certain degree of trophic partitioning in this pelagic ecosystem. Round herring off South Africa feed on different species and sizes of zooplankton to anchovy and sardine (Wallace-Fincham, 1987; C. van der Lingen, personal communication). Diet analyses of pelagic goby off Namibia have recently shown that this species can feed on macrobenthos as well as on zooplankton in the water column (M. Gibbons, personal communication; J.-P. Roux, unpublished data). However, despite the largely limited competition for food amongst these small pelagic fish, model simulations suggest an indirect effect of anchovy and sardine on other small pelagic fish, possibly through predator niche overlaps.

In the closed pelagic fishery simulation using the Southern Benguela model, the round herring stock declines to a mere 13% of its original 1980s level, and the "other small pelagic fish" group is reduced to 28% of starting biomass, in response to competition with the expanding populations of anchovy and sardine. In addition, the Southern Benguela ecosystem also differs from the Catalan, Adriatic, and Humboldt ecosystems by supporting relatively large stocks of large, predatory pelagic fish, in particular snoek. Trophodynamic responses of the latter to a closed purse-seine fishery may explain the apparent lack of favorable effects on demersal predatory fish species when the modeled purse-seine fishery is closed, given that model-projected large pelagic fish biomass increases by a factor of 2.3 over the 10-year simulation period (Fig. 8.4a).

It is also probable that in the Southern Benguela, the simulated biomass of small pelagic fish and thus also some of their predators do not increase substantially when the model purse-seine fishery is closed because small pelagic fishing mortalities were relatively low to begin with (F of around 0.3 y^{-1}), compared to F for anchovy and sardine in the Mediterranean and Chilean models. Another relevant factor to consider is how heavily fished hake are in the various ecosystems. In the Southern Benguela, the exploitation rate of hake was between one-quarter and one-tenth of that in the Chilean and Mediterranean systems, so this would not lend support for the argument that hake are more heavily fished in the Benguela and thus less likely to be able to respond to increased forage fish abundance. A final point of consideration must be the relative importance of small pelagic fish in the diets of hake and other predatory fish in the different modeled ecosystems. In the Southern and Northern Benguela, small pelagic fish comprise around a third of the diet of hake, and even less than this in the 1980s model of the Southern Benguela (before both anchovy and sardine were abundant in the early 2000s). Large pelagic fish such as snoek are estimated to be relying on anchovy and sardine for 30%–40% of their diet off South Africa. By comparison, hake in the Mediterranean are estimated to fill 50%–60% of their diet with small pelagic fish (see Stergiou and Karpouzi, 2002 for a review) and, in the Humboldt, between 30% and 50% of their diet composition with small pelagic fish.

Small pelagic fish as indicators of ecosystem collapse

Ecosystem indicators

We compared ecosystem indicators between systems in different time periods, i.e. when they were considered to have functioned "normally," or at least to have been sustainably fished, and when they were overexploited, or collapsed because of the synergistic effects of the environment and exploitation, to examine the values and trends exhibited by overexploited ecosystems. Ideally, we would have liked to be in a position to propose levels/ranges/limits of indicators

which should be avoided when managing systems that are not yet overexploited/collapsed and when managing in the context of changing climatic conditions. Exploitation rates applicable to certain kinds of fish, such as small pelagic fish, may be one of the better examples available, although these have not been dealt with in details as they are beyond the focus of this chapter. However, most indicators, especially those quantifying trophodynamics, are largely still descriptive and reference points are yet to be agreed upon. Cury *et al.* (2005a) concluded that a suite of trophodynamic indicators is needed to track complex fisheries and ecosystem changes and that, although helpful in understanding ecosystem and fisheries dynamics that arise as a result of interactions between species, these kinds of indicators tend to respond relatively slowly to structural change. Instead of trying to propose values and limits for indicators, García and Staples (2000) propose focusing on how indicators can be implemented in fisheries management to minimize the adverse effects of fishing. There is currently much effort being put into development of "indicator dashboards" that attempt to bring together information and management guidance based on indicators derived from a variety of sources (Shin, 2006).

Ecological modeling is a key tool to monitor and quantify proposed indicators, as has been illustrated by virtue of trophic models considered in this chapter. However, models rely on sufficient high quality data, and the need for real-time sampling integrated at the level of ecosystem-based management of marine ecosystems has been strongly recognized (Shannon *et al.*, 2006).

Combined effect of climate and fishing

The two major drivers of fluctuations and collapses in small pelagic fish stocks, and thus, by virtue of their wasp-waist role in the food web, including the propagation of these effects up and down the food web, are (i) climatic effects on small pelagic fish, and (ii) fishing effects on small pelagic fish. The three models fitted to time-series data and discussed above (section entitled "Trophic flow controls, fishing and environmental drivers of ecosystem change") all illustrated the importance of considering the often synergistic effects of the environment and fishing. Interactions between the environment and marine resources can be synergistic or antagonistic, and van der Lingen *et al.* (2006) point out that a change in the structure of an ecosystem is likely to result in a change in the response of the ecosystem to environmental forcing. In the face of severe environmental perturbations, it may simply not be sufficient to ensure exploitation rates are well below critical levels such as those suggested by Mertz and Myers (1998) or Rochet and Trenkel (2003). According to those guidelines, and considering model results not presented here (L. Shannon, unpublished data), the Northern Benguela would not have been classified as overexploited in the 1980s, despite the collapse of the small pelagic fishery in the 1990s, pointing to the importance of taking into account synergistic effects of environmental conditions and fishing pressure. Sardine in the Catalan Sea would rightfully be considered overexploited in the 1990s (Lleonart and Maynou, 2003), whereas in the Adriatic, exploitation rates of small pelagic fish appear to have been within sustainable levels but large stock declines were observed at the end of 1980s (Azzali *et al.*, 2002; Santojanni *et al.*, 2003, 2005).

Disruption of trophic pathways

Monitoring of pelagic:demersal fish biomass ratios over time shows the general trend since the 1990s to have been a shift towards more demersally dominated ecosystems (Fig. 8.2) after the expansion of pelagic fisheries since the 1950s and the subsequent large declines of small pelagic fish stocks. Proliferation of non-commercial species such as jellyfish, and changes in diet and populations of predators in ecosystems where small pelagic fish are heavily exploited, could be regarded as signs of disruption of the trophic structure. Snapshot model results and dynamic model simulations of severely altered fishing on small pelagic fish in the different ecosystems examined support these hypotheses. Changes in energy transfer can be quantified in terms of increased energy flowing to detritus when pelagic fish are heavily exploited and a higher coupling between pelagic and demersal compartments (see sections entitled "Coupling demersal-pelagic compartments" and "Increased pelagic–benthic coupling and flow to detritus").

No important changes were found in the Southern Humboldt models in terms of the impact of detritus on the pelagic and demersal components. This could likely be attributed to a decrease in the biomass of the demersal fish (i.e. hake) and benthic detritivorous crustaceans in recent years, and thus full advantage is not being taken of the potential increase in flow to detritus. On the other hand, large increases in the sum in the model of the mixed trophic impacts of detritus on both the pelagic and demersal food webs were observed over time in the Southern Benguela, and small increases were observed in the South Catalan Sea model, further supporting the hypothesis that detritus can become increasingly more important as ecosystems are fished and/or degraded, or adversely affected by environmental changes. However, in the Northern Benguela, the near-total collapse of the pelagic ecosystem appears to have weakened detritus–pelagic food web linkages, whereas detritus–demersal linkages seem to have strengthened. The dominant small pelagic fish is now the non-commercial pelagic goby *Sufflogobius bibarbatus*, which, by virtue of being able to feed benthically as well as pelagically, appears to have been able to survive the frequent low oxygen events that have been detrimental to the pelagic food web off the coast of Namibia. In the Adriatic, the decline in the modeled impact of detritus on the demersal compartment could be due to the high rates of exploitation of commercial scallops and other invertebrates after the decrease of fish catches (Coll *et al.*, 2007) and the lower biomasses of these groups in the ecosystem during the 1990s compared to the 1980s.

Comparison of snapshot models also suggests an increase in the pelagic–benthic coupling through stronger overall mixed trophic interactions of the model pelagic component on the model demersal component in more recent periods (usually when fishing has increased), and often reduced summed impacts of the model demersal compartment on the pelagic food web modelled (see sections entitled "Coupling demersal–pelagic compartments" and "Increased pelagic-benthic coupling and flow to detritus"). Although this needs further investigation and more careful quantification, and recognizing that results would depend on whether certain omnivorous groups are categorized as pelagic or demersal, changes in flows between benthic and pelagic compartments may be considered a measure of the impacts of fishing and/or ecosystem change.

Using models to explore the effects of climate change on the ecosystem via small pelagic fish

The chapter has drawn heavily from model results compared across a selection of ecosystems, mainly upwelling systems, for which standardized trophic models are available across several time periods. These, and some simple exploratory model simulations, have been used to reveal the ecological role of small pelagic fish within the ecosystem, and to examine how dramatic changes in the abundance of small pelagic fish stocks might impact ecosystems. The biological characteristics of small pelagic fish render them highly susceptible and responsive to environmental variability, which is reflected in the high amplitude fluctuations observed for small pelagic fish stocks worldwide. Changes in abundances of small pelagic fish, by virtue of their wasp-waist/pivotal position as forage fish within marine food webs, are likely to have some large implications for the ecosystem as a whole, and for certain important fish stocks in particular. An important means by which climatic change is likely to express itself in the ecosystem is thus via small pelagic fish.

The model simulations presented here are purely illustrative ones of extreme situations of collapse or closure of the small pelagic fishery and should not be considered in the quantitative, predictive sense. Robustness tests across ranges of multiple input parameter values would need to be examined to provide such inferences, but that was not the purpose of this contribution, which was to highlight the potential for large ecosystem impacts to prevail if small pelagic fish abundances were to change.

The importance of environmental perturbations as drivers of observed ecosystem changes has already been discussed, where trophic models fitted to time series data are considered. This is a new field still requiring much further effort and expertise, but it is encouraging that models of this kind are enabling these types of effects to be explored. An integral part of climate change effects at the ecosystem level will be on the spatial scale, an aspect which has not been considered in depth in this chapter (although see section entitled "Effects of small pelagic fish on top predators" in relation to spatial dynamics of predators) but which deserves particular mention. Spatial distributions of small pelagic fish spawners, and the environmentally driven processes that determine the transport, survival, growth rate, and development of small pelagic fish eggs, larvae, prerecruits, recruits, and adults, are all likely to be affected by climate change in the years to come. However, until such time that realistic climate change scenarios are available on a regional scale, it will be difficult to go beyond simple "what-if" scenario testing to examine the likely impacts of climate change on small pelagics in particular ecosystems.

"What-if" climate scenario testing was followed using a simple advection "Monte Carlo simulation" model applied off South Africa, to examine the effects on anchovy recruitment of altered advection and spatial distribution of spawners (Shannon *et al.,* 1996). Based on the reduced frequency of south-easterly winds observed in the El Niño period from 1990 to 1993 (Boyd *et al.,* 1998), climate change scenarios were developed of possible changes in advective processes that may develop off South Africa. Reduced westward and northward advection led to a doubling in the mean model year-class strength when compared to the "normal" or enhanced advection situations, and model results suggested that reduced advective losses dampened interannual recruitment variability. On the other hand, enhanced westward and northward model advection, although seemingly beneficial in transporting anchovy eggs and larvae into more productive waters along the South African west coast, carried the disadvantage of severe advective losses into unproductive offshore waters.

More elaborate models such as Individual Based Models (IBMs) have been used to explore transport processes and spatial effects on small pelagic fish. Although beyond the focus of our chapter, these warrant mentioning in terms of future directions for climate change research and highlight the urgent need for climate change scenarios to be identified and interpreted in terms of their implications for ocean processes affecting our marine resources. Again, drawing a comparable example from the Benguela region, IBMs of anchovy and sardine have been coupled to a regional ocean model of the southern Benguela that produces fields for flow, temperature and density (Penven *et al.,* 2001). This has facilitated model studies of the spatial and temporal effects of environmental

variability on anchovy (Mullon *et al.,* 2002; Huggett *et al.,* 2003; Mullon *et al.,* 2003; Parada *et al.,* 2003) and sardine (Miller *et al.,* 2006) recruitment off South Africa.

Under conditions of global warming, it has been hypothesized that upwelling may intensify in the major eastern boundary current systems of the world (Bakun, 1990; Bakun and Weeks, 2004; McGregor *et al.,* 2007), with major implications for those ecosystems. For example, it has been proposed that sardine in the Northern Benguela, by virtue of their filter-feeding strategy and migratory potential, may be able to take advantage of any increase in phytoplankton production that could result if upwelling increases (Bakun and Weeks, 2004). Several general hypotheses have been proposed about possible large-scale impacts of climate change in the Benguela region (van der Lingen *et al.,* 2006): altered wind stress is envisaged, which would enhance coastal upwelling (Bakun, 1990; Bakun, 1992; Shannon *et al.,* 1996); it is proposed that Benguela Niños and the advection of warm tropical water may occur more frequently and be of greater intensity (Siegfried *et al.,* 1990); and it is possible that there would be an increase in sea surface temperature (Roux, 2003). Coupled hydrodynamic and biological models as mentioned above have been identified as promising ways of exploring how climate changes such as those proposed for the Benguela region might affect small pelagic fish and the rest of the ecosystem (van der Lingen *et al.,* 2006).

It is possible that global warming of the eastern equatorial zone of the Pacific Ocean relative to the western zone could move that system into an "elevated El Niño" state similar to that experienced between the mid 1970s and the mid 1980s (Timmermann *et al.,* 1999). These changes may well be felt globally too, via teleconnections between the ocean basins. For example, the extreme El Niño event of 1997–1998 had important ecological, economic and social consequences on both the east and western Pacific and globally (e.g. McPhaden, 1999; McPhaden *et al.,* 2006). Timmermann *et al.*'s (1999) model suggested more frequent El Niño-like conditions and stronger cool events in the tropical Pacific. This would likely affect anchovy and sardine differently, as sardine is better suited to warmer conditions than anchovy (Chávez *et al.,* 2003).

In conclusion, the chapter has examined the potential ecosystem effects of changes in small pelagic fish abundance through models that largely consider altered fishing strategies, but similar changes could be climatically driven. Given the pivotal ecological role of small pelagics in ecosystems, climate change is likely to express itself in the ecosystem via its impacts on small pelagic fish. However, precisely how climate change will be likely to affect abundance, life history and the spatial distribution of small pelagics (and therefore the food webs they dominate) still requires extensive further study and is currently a hot scientific topic of notable concern.

Acknowledgments

The authors would like to thank all researchers involved in ecological models used in this work: Isabel Palmera, Enrico Arneri, Alberto Santojanni, Sergi Tudela, and Frances Sardà (South Catalan Sea and North and Central Adriatic Sea), Astrid Jarre and Coleen Moloney (Southern Benguela upwelling), Hugo Arancibia and Luis Cubillos (Southern Humboldt upwelling). Lynne Shannon's contribution to this chapter was part of her work while employed as fisheries scientist at Marine and Coastal Management, Department of Environmental Affairs and Tourism, South Africa, where she worked until December 2008.

NOTES

1. Collapse is here used to describe severe declines of small pelagic fish stocks (to zero/near-zero biomass) to the extent that fisheries relying on these species are no longer economically or ecologically viable. The collapse of a species alters trophic pathways so that other species within the ecosystem, including predators, prey, and competing species, are dramatically affected in terms of their abundance and/or trophic linkages within the food web.
2. A collapsed ecosystem is considered to be one which no longer comprises functional pelagic and/or demersal food webs supporting viable populations of predators, and for which important changes in energy transfer between food web elements are observed. A collapsed ecosystem has been subjected to adverse effects on its biological productivity that extend beyond natural, reversible variability, leading to alteration of the ecosystem function of key functional groups.
3. Exclusive Economic Zone.
4. The trophic level (TL) of an organism is its position in a food web. Initially, TLs were allocated as integers (Lindeman, 1942) but fractional values are now used (Odum and Heald, 1975): primary producers and detritus are allocated TLs of 1 by default, whereas TLs of consumers are calculated from the weighted average TL of their prey, which can be estimated by means of mass-balance models, analysis of stomach contents or isotope analysis (Stergiou and Karpouzi, 2002).
5. The Fishing in Balance (FiB) index, calculated from catch data over time and the TL of the catch (Pauly *et al.*, 2000), measures to what extent a change in the trophic level of the catch matches theoretical changes in productivity. For example, FiB would be close to zero if a decline in the TL of the catch was matched with an increase in productivity of the ecosystem. FiB <0 could indicate overfishing (a decline in the trophic level of the fishery not matched by enhanced productivity) or that unreported discarding has impaired ecosystem functioning, whereas FiB >0 may indicate an expansion of the fishery or bottom-up effects coming into play.
6. Based on an input–output method for assessing direct and indirect economic interactions (Leontief, 1951), Mixed trophic impact (MTI) analysis quantifies the net effects of direct and indirect trophic interactions among groups, measuring the relative impact of a change in the biomass of one group on each of the others modeled (Ulanowicz and Puccia, 1990). MTI analysis is expressed in the form of matrices of the relative, net impacts, scaled between –1 and 1.

REFERENCES

Abad, R. and Giraldez, A. (1990). Concentrations de Capros aper dans la Mer d'Alboran (Mediterranée Espagnole). *Rapp. Proc. Verb. Comm. Int. Explor. Sci. Mer Medit.* **32** (1): 256.
Abelló, P., Arcos, J.M., and Gil de Sola, L. (2003). Geographical patterns of attendance to trawling by breeding seabirds along the Iberian Peninsula Mediterranean Coasts. *Sci. Mar.* **67** (Suppl. 2): 69–75.
Agostini, V. and Bakun, A. (2002). "Ocean triads" in the Mediterranean Sea: physical mechanisms potentially structuring reproductive habitat suitability (with example application to European anchovy, *Engraulis encrasiclous*). *Fish. Oceanogr.* **11** (3): 129–142.
Alheit, J. and Niquen, M. (2004). Regime shifts in the Humboldt Current System. *Prog. Oceanogr.* **60**: 201–122.
Arai, M.N. (2001). Pelagic coelenterates and eutrophication: a review. *Hydrobiologia* **451**: 69–87.
Arancibia, H. (1989). Distribution patterns, trophic relationships and stock interactions in the demersal fish assemblage off central Chile. Dr. rer. Nat. thesis. University of Bremen, German, 221 pp.
Arancibia H. and Neira, S. (2003). Simulación de cambios en la biomasa de los principales recursos pesqueros de Chile central (V – IX región) bajo el marco de la ley de pesca N° 19.713 y sus posteriores modificaciones. Informe Final. Universidad de Concepción, 53 pp.
Arancibia, H. and Neira, S. (2005). Long-term analysis of the trophic level of fisheries landings in Central Chile. *Sci. Mar.* **69** (2): 295–300.
Arancibia, H., Catrilao, M., and Farías, B. (1998). Evaluación de la demanda de alimento en merluza común y análisis de su impacto en pre – reclutas. *Informe Final Proyecto FIP* **95–17**: 98 pp.
Arcos, J.M. (2001). Foraging ecology of seabirds at sea: significance of commercial fisheries in the Northwestern Mediterranean. Ph.D. thesis. University of Barcelona. 109 pp.
Arneri, E. (1996). Fisheries resources assessment and management in Adriatic and Ionian Seas. FAO Fisheries Rep. **533**: 7–20.
Arrizaga, A., Fuentealba, M., Espinoza, C., *et al.* (1993). Trophic habits of two pelagic fish species: *Strangomera bentincki* (Norman, 1936) and *Engraulis ringens* (Jenyns, 1842), in the littoral of the Biobio Region, Chile. *Bol. Soc. Biol. Concepción (Chile)* **64**: 27–35.
Avaria, S. and Muñoz, P. (1982). Producción actual, biomasas y composición específica del fitoplancton de la bahía de Valparaíso en 1979. *Rev. Biol. Mar. Valparaíso* **18**: 129–157.
Azzali, M., De Felice, A., Luna, M., Cosimi, G., and Parmiggiani, F. (2002). The state of the Adriatic Sea centered on the small pelagic fish populations. *P. S. Z. N. Mar. Ecol.* **23** (Suppl.1): 78–91.
Baccetti, N., Dall'Antonia, P., Magagnoli, P., *et al.* (2002). Risultati dei censimenti degli uccelli acquatici svernanti in Italia: distribuzione, stima e trend delle popolazioni nel

1991–2000. *Instituto Nazionale per la Fauna Selvatica Alessandro Ghigi* **111**: 234.

Bakun, A. (1990). Global climate change and intensification of coastal ocean upwelling. *Science* **247**: 198–201.

Bakun, A. (1992). Global greenhouse effects, multidecadal wind trends, and potential impacts on coastal pelagic fish populations. *ICES Mar. Sci. Symp.* **195**: 316–325.

Bakun, A. (1996). *Patterns in the ocean: ocean processes and marine population dynamics.* University of California Sea Grant, San Diego, California, USA, in cooperation with Centro de Investigaciones Biologicas de Noroeste, La Paz, Baja California Sur, Mexico, 323 pp.

Bakun A. (2006). Wasp-waist populations and marine ecosystem dynamics: navigating the "predator pit" topographies. *Prog. Oceanogr.* **68**: 271–288.

Bakun, A. and Weeks, S. (2004). Greenhouse gas buildup, sardines, submarine eruptions and the possibility of abrupt degradation of intense marine upwelling systems. *Ecol. Lett.* **7**: 1015–1023.

Barange, M., Hampton, I., and Roel, B.A. (1999). Trends in the abundance and distribution of anchovy and sardine on the South African continental shelf in the 1990s, deduced from acoustic surveys. *S. Afr. J. Mar. Sci.* **19**: 367–391.

Bearzi, G., Holcer, D., and Notarbartolo Di Sciara, G. (2004). The role of historical dolphin takes and habitat degradation in shaping the present status of northern Adriatic cetaceans. *Aquat. Conserv. Mar. Freshwater Ecosyst.* **14**: 363–379.

Beaugrand, G., Brander, K.M., Alistair Lindley, J., *et al.* (2003). Plankton effect on cod recruitment in the North Sea. *Nature* **426**: 661–664.

Best, P.B., Crawford, R.J.M., and Van der Elst, R.P. (1997). Top predators in southern Africa's marine ecosystems. *T. Roy. Soc. S. Afr.* **52** (1): 177–225.

Blanco, C., Salomón, O., and Raga, J.A. (2001). Diet of the bottlenose dolphin (*Tursiops truncatus*) in the western Mediterranean Sea. *J. Mar. Biol. Assoc. UK* **81**: 1053–1058.

Bombace, G. (1992). Fisheries of the Adriatic Sea. In *Marine Eutrophication and Population Dynamics. 25th European Marine Biology Symposium.* Olsen and Olsen, Fredensborg, pp. 379–389.

Bosc, E., Bricau, A., and Antoine., D. (2004). Seasonal and interannual variability in algal biomass and primary production in the Mediterranean Sea, as derived from 4 years of SeaWiFS observations. *Glob. Biogeochem. Cy.* **18** (GB1005), 17 pp.

Boyd, A.J., Shannon, L.J., Schulein, F.H., and Taunton-Clark, J. (1998). Food, transport and anchovy recruitment in the southern Benguela upwelling system of South Africa. In *Global Versus Local Changes in Upwelling Systems.* Durand, M.-H., Cury, P., Mendelssohn, R., *et al.*, eds. Orstom Editions, pp. 195–209.

Boyer, D.C. and Hampton, I. (2001). An overview of the living marine resources of Namibia. *S. Afr. J. Mar. Sci.* **23**: 5–35.

Boyer, D.C., Boyer, H.J., Fossen, I., and Kreiner, A. (2001). Changes in abundance of the northern Benguela sardine stock during the decade 1990 to 2000 with comments on the relative importance of fishing and the environment. *S. Afr. J. Mar. Sci.* **23**: 67–84.

Bozzano, A., Recasens, L., and Sartor, P. (1997). Diet of the European hake *Merluccius merluccius* (Pisces: Merluciidae) in the Western Mediterranean (Gulf of Lions). *Sci. Mar.* **61**: 1–8.

Camus, P. (2001). Biogeografía marina de Chile continental. *Rev. Chil. Hist. Nat.* **74**: 587–617.

Carr. E.-M. (2002). Estimation of potential productivity in eastern boundary currents using remote sensing. *Deep-Sea Res. II* **49**: 59–80.

Chávez, F.P., Ryan, J., Lluch-Cota, S.E., and Niquen, M. (2003). From anchovies to sardines and back: multidecadal change in the Pacific Ocean. *Science* **299**: 217–221.

Christensen, V. and Walters, C.J. (2004). Ecopath with Ecosim: methods, capabilities and limitations. *Ecol. Model.* **172**: 109–139.

Cingolani, N., Giannetti, G. G and Arneri, E. (1996). Anchova fisheries in the Adriatic Sea. *Sci. Mar.* **60** (Suppl. 2): 269–277.

Cochrane, K.L., Augustyn, C.J., Cockcroft, A.C., *et al.* (2004). An ecosystem approach to fisheries in the southern Benguela context. *Afr. J. Mar. Sci.* **26**: 9–35.

Coll, M., Palomera, I., Tudela S., and Sardà, F. (2005). Assessing the impact of fishing activities and environmental forcing on a Northwestern Mediterranean ecosystem along the last decades of intense exploitation. *Advances in Marine Ecosystem Modelling Research. International Symposium at Plymouth Marine Laboratory* 27–29 June 2005, Plymouth, UK, 148 pp.

Coll, M., Shannon, L.J., Moloney, C.L., *et al.* (2006a). Comparing trophic flows and fishing impacts of a NW Mediterranean ecosystem with coastal upwellings by means of standardized ecological models and indicators. *Ecol. Model.* **198**: 53–70.

Coll, M., Palomera, I., Tudela S., and Dowd, M. (2008). Food-web dynamics in the South Catalan Sea ecosystem (NW Mediterranean) for 1978–2003. *Ecol. Model.* **217**(1–2): 95–116.

Coll, M., Palomera, I., Tudela, S., and Sardà, F. (2006b). Trophic flows, ecosystem structure and fishing impacts in the South Catalan Sea, Northwestern Mediterranean. *J. Mar. Syst.* **59**: 63–96.

Coll, M., Santojanni, A., Arneri, E., and Palomera, I. (2007). An ecosystem model of the Northern and Central Adriatic Sea: analysis of ecosystem structure and fishing impacts. *J. Mar. Syst.* **67**: 119–154.

Coll, M., Santojanni, A., Palomera, I., and Arneri, E. (in press). Food-web dynamics in the North-Central Adriatic marine ecosystem (Mediterranean Sea) over the last three decades. *Mar. Ecol. Prog. Ser.*

Crawford, R.J.M. (2006). Closure of areas to purse-seine fishing around the St Croix and Dyer island African penguin colonies. Unpublished report **SWG/OCT2006/PEL/06**, Marine and Coastal Management, Department of Environmental Affairs and Tourism, South Africa, 10 pp.

Crawford, R.J.M. and Dyer, B.M. (1995). Responses by four seabirds to a fluctuating availability of Cape anchovy *Engraulis capensis* off South Africa. *Ibis* **137**: 329–339.

Crawford, R. J. M. and Jahncke, J. (1999). Comparison of trends in abundance of guano-producing seabirds in Peru and southern Africa. *S. Afr. J. Mar. Sci.* **21**: 145–156.

Crawford, R. J. M., Cruichshank, R. A., Shelton, P. A., and Kruger, I. (1985). Partitioning of a goby resource amongst four avian predators and evidence for altered trophic flow in the pelagic community of an intense perennial upwelling system. *S. Afr. J. Mar. Sci.* **3**: 215–228.

Crawford, R. J. M., Dyer, B. M., Cordes, I., and Williams, A. J. (1999). Seasonal pattern of breeding, population trends and conservation status of bank cormorants *Phalacrocorax neglectus* off south western Africa. *Biol. Cons.* **87**: 49–58.

Crawford, R. J. M., Barham, P. J., Underhill, L. G., *et al.* (2006). The influence of food availability on breeding success of African penguins *Spheniscus demersus* at Robben Island, South Africa. *Biol. Cons.* **132**: 119–125.

Cubillos, L., Rebolledo, H., and Hernández, A. (2003). Prey composition and estimation of Q/B for the Chilean hake, *Merluccius gayi* (Gadiformes-Merluccidae), in the central-south area off Chile (34°–40° S). *Arch. Fish. Mar. Res.* **50**: 271–286.

Cury, P., Bakun, A. Crawford, R. J. M., *et al.* (2000). Small pelagic fish in upwelling systems: patterns of interaction and structural changes in "wasp-waist" ecosystems. *ICES J. Mar. Sci. Symp.* **57** (3): 603–618.

Cury, P. M., Shannon, L. J., and Shin, Y.-J. (2003). The functioning of marine ecosystems: a fisheries perspective. In *Responsible Fisheries in the Marine Ecosystem*, Sinclair, M., and Valdimarsson G., eds. UK: FAO and CABI Publishing, pp. 103–123.

Cury, P. M., Shannon, L. J., Roux, J.-P., *et al.* (2005a). Trophodynamic indicators for an ecosystem approach to fisheries. *ICES J. Mar. Sci.* **62**: 430–442.

Cury, P., Fréon, P., Moloney, C., *et al.* (2005b). Processes and patterns of interactions in marine fish populations: an ecosystem perspective. In *The Sea: Ideas and Observations on Progress in the Study of the Seas. The Global Coastal Ocean: Multiscale Interdisciplinary Processes*, Robinson, A., and Brink, K., eds. Cambridge, USA: Harvard University Press, pp. 475–554.

Cushing, D. H. (1990). Plankton production and year-class strength in fish populations: an update of the match/mismatch hypothesis. *Adv. Mar. Biol.* **26**: 249–293.

Daneri, G., Dellarossa, V., Quiñones, R., *et al.* (2000). Primary production and community respiration in the Humboldt Current System off Chile and associated oceanic areas. *Mar. Ecol. Prog. Ser.* **197**: 41–49.

Daskalov, G. M. (2002). Overfishing drives a trophic cascade in the Black Sea. *Mar. Ecol. Prog. Ser.* **225**: 53–63.

Escribano, E., Fernández, M., and Aranís, A. (2003). Physical–chemical processes and patterns of biodiversity of the Chilean eastern boundary pelagic and benthic marine ecosystems: an overview. *Gayana* **67**: 190–205.

Estrada, M. (1996). Primary production in the Northwestern Mediterranean. *Sci. Mar.* **60** (Suppl. 2): 55–64.

FAO. (2005). Review of the state of the world marine fishery resources. *FAO Marine Resources Service, Fishery Resources Division. Fisheries Technical Paper* **457**: 246 pp.

Flynn, B. A. and Gibbons, M. J. (2007). A note on the diet and feeding of *Chrysaora hysoscella* in Walvis Bay Lagoon, Namibia during September 2003. *Afr. J. Mar. Sci.* **29** (2): 303–307.

Fonda Umani, S. (1996). Pelagic production and biomass in the Adriatic Sea. *Sci. Mar.* **60** (Suppl. 2): 65–77.

Frank, K. T., Choi, J. S., Petrie, B., and Leggett, W. C. (2005). Trophic cascades in a formerly cod dominated ecosystem. *Science* **308**: 1621–1623.

Frank, K. T., Petrie, B., Shackell, N. L., and Choi, J. S. (2006). Reconciling differences in trophic control in mid-latitude marine ecosystems. *Ecol. Lett.* **9**: 1–10.

García, S. M. and Staples, D. J. (2000). Sustainability reference systems and indicators for responsible marine capture fisheries: a review of concepts and elements for a set of guidelines. *Mar. Freshwater Res.* **51**: 385–426.

Gucu, A. C. (2002). Can overfishing be responsible for the successful establishment of *Mnemiopsis leidyi* in the Black Sea? *Estuar., Coast. Shelf Sci.* **54**: 439–451.

Hamukuaya, H., O'Toole, M. J., and Woodhead, P. M. J. (1998). Observations of severe hypoxia and offshore displacement of Cape hake over the Namibian shelf in 1994. *S. Afr. J. Mar. Sci.* **19**: 57–59.

Heymans, J. J., Shannon, L. J., and Jarre, A. (2004). Changes in the northern Benguela ecosystem over three decades: 1970s, 1980s and 1990s. *Ecol. Model.* **172**: 175–195.

Hill, A., Hickey, B., Shillington, F., *et al.* (1998). Eastern ocean boundaries coastal segment. Chapter 2. In *The Sea: The Global Coastal Ocean, Regional Studies and Synthesis*, Robinson, A., and Brink, K. eds. New York: John Wiley and Sons, Inc., pp. 29–67.

Huggett, J., Fréon, P., Mullon, C., and Penven, P. (2003). Modelling the transport success of anchovy *Engraulis encrasicolus* eggs and larvae in the southern Benguela: the effect of spatio-temporal spawning patterns. *Mar. Ecol. Prog. Ser.* **250**: 247–262.

Jahncke, J., Checkley Jr., D. M., and Hunt Jr., G. L. (2004). Trends in carbon flux to seabirds in the Peruvian upwelling system: effects of wind and fisheries on population regulation. *Fish. Oceanogr.* **13** (3): 208–223.

Jarre-Teichmann, A. (1998). The potential role of mass-balanced models for the management of upwelling ecosystems. *Ecol. Appl.* **8** (1): 93–103.

Jarre-Teichmann, A. and Christensen, V. (1998). Comparative modelling of trophic flows in four Large Marine Ecosystems; Global versus local effects. In *Global Versus Local Changes in Upwelling Systems*, Durand, M.-H., Cury, P., Mendelssohn, R., *et al.* eds. Editions de l'Orstom, 423–443.

Jarre-Teichmann, A., Shannon, L. J., Moloney, C. L., and Wickens, P. A. (1998). Comparing trophic flows in the southern Benguela to those in other upwelling ecosystems. In *Benguela Dynamics: Impacts of Variability on Shelf-sea Environments and their Living Resources*, Pillar, S. C. Moloney, C. L. Payne A. I. L., and Shillington, F. A., eds. *S African Journal of Marine Science*, **19**: 391–414.

Jukic-Peladic, S., Vrgoc, N., Krstulovic-Sifner, S., *et al.* (2001). Long-term changes in demersal resources of the Adriatic

Sea: comparison between trawl surveys carried out in 1948 and 1998. *Fish. Res.* **53** (1): 95–104.

Kawasaki, T. (1983). Why do some pelagic fish have wide fluctuations in their numbers? Biological basis from the viewpoint of evolutionary ecology. *FAO Fisheries Rep.* **291** (3): 1065–1080.

Kitaysky A. S., Kitauskaia E. V., Piatt J. F., and Wingfield, J. C. (2006). A mechanistic link between chick diet and decline in seabirds? *Proc. Roy. Soc. B.* **273**: 445–450.

Leibol, M. A., Chase, J. M., Shurin, J. B., and Downing, A. L. (1997). Species turnover and the regulation of trophic structure. *Ann. Rev. Ecol. Syst.* **28**: 467–94.

Leontief, W. W. 1951. *The Structure of the US Economy.* 2nd edn. New York: Oxford University Press.

Lindeman, R. L. 1942. The trophic–dynamic aspect of ecology. *Ecol.*, **23**: 399–418.

Litzow M. A., Bailey K. M., Prahl F. G., and Heintz, R. (2006). Climate regime shifts and reorganization of fish communities: the essential fatty acid limitation hypothesis. *Mar. Ecol. Prog. Ser.* **315**: 1–11.

Lleonart, J. (1990). La pesquería de Cataluña y Valencia: descripción global y planteamiento de bases para su seguimiento. Informe final del proyecto 1989/3. CE-Dirección General de Pesca (XIV). Vol I-II. Barcelona.

Lleonart, J. and Maynou, F. (2003). Fish stock assessments in the Mediterranean: state of the art. *Sci. Mar.* **67** (Suppl. 2): 37–49.

Lloret, J. and Lleonart, J. (2002). Recruitment dynamics of eight fishery species in the Northwestern Mediterranean Sea. *Sci. Mar.* **66** (1): 77–82.

Lloret, J., Palomera, I., Salat, J., and Sole, I. (2004). Impact of freshwater input and wind on landings of anchovy (*Engraulis encrasiclous*) and sardine (*Sardina pilchardus*) in shelf waters surrounding the Ebro River delta (northwestern Mediterranean). *Fish. Oceanogr.* **13** (2): 102–110.

Lluch-Belda, D., Crawford, R. J. M., Kawasaki, T., *et al.* (1989). World-wide fluctuations of sardine and anchovy stocks: the regime problem. *S. Afr. J. Mar. Sci.* **8**: 195–205.

Lluch-Belda, D., Schwartzlose, R. A., Serra, R., *et al.* (1992). Sardine and anchovy regime fluctuations of abundance in four regions of the world oceans: a workshop report. *Fish. Oceanogr.* **1** (4): 339–347.

Maartens, L. (1999). An assessment of the monkfish resource of Namibia. Ph.D. thesis. Rhodes University: Grahamstown, 190 pp.

Mannini, P. and Massa, F. (2000). Brief overview of Adriatic fisheries landings trends (1972–97). In *Report of the First Meeting of Adriamed Coordination Comité*, Massa, F. and Mannini, P., eds. FAO-MiPAF Scientific Cooperation to Support Responsable Fisheries in the Adriatic Sea. GCP/RER/010/ITA/TD-01, pp. 31–49.

Manoukian, M., Azzali, M., Farchi, C., *et al.* (2001). Sightings distribution and variability in species composition of cetaceans in the Adriatic Sea ecosystem in one decade of study. *Rapp. Proc. Verb. Comm. Int. Explor. Sci. Mer Medit* **36**: 297.

Margalef, R. (1985). Introduction to the Mediterranean. In *Key Environments: Western Mediterranean*, Margalef, R., ed. New York: Pergamon Press, 362 pp.

Mayfield, S. and Branch, G. M. (2000). Interrelations among rock lobsters, sea urchins, and juvenile abalone: implications for community management. *Can. J. Fish. Aquat. Sci.* **57**: 2175–2185.

McGregor H. V., Dima, M., Fischer, H. W., and Mulitza, S. (2007). Rapid 20th-century increase in coastal upwelling off Northwest Africa. *Science* **315**: 637–639.

McPhaden, M. J. (1999). Genesis and evolution of the 1997–98 El Niño. *Science* **283**: 950–954.

McPhaden M. J., Zebiak, S. E., and Glantz, M. H. (2006). ENSO as an integrating concept in earth science. *Science* **314**: 1740–1745.

Mecenero, S. and Roux, J.-P. (2002). Spatial and temporal variability in the diet of a top predator in the northern Benguela. *GLOBEC Rep.* **16**: 62–64.

Meeuwis, J. M. and Lutjeharms, J. R. E. (1990). Surface thermal characteristics of the Angola-Benguela Front. *S. Afr. J. Mar. Sci.* **9**: 261–279.

Meléndez, R. (1984). Alimentación de *Merluccius gayi* (Guichenot) frente a Chile central (32°05 S–36°50 S). *Bol. Mus. Nac. Hist. Nat. Chile*, **54** pp.

Mertz G. and Myers, R. A. (1998). A simplified formulation for fish production. *Can. J. Fish. Aquat. Sci.* **55**: 478–484.

Micheli, F. (1999). Eutrophication, fisheries, and consumer-resource dynamics in marine pelagic ecosystems. *Science* **285**: 1396–1398.

Miller, D. C. M., Moloney, C. L., van der Lingen, C. D., *et al.* (2006). Modelling the effects of physical–biological interactions and spatial variability in spawning and nursery areas on recruitment of sardine in the southern Benguela system. *J. Mar. Syst.* **61** (3–4): 212–229.

Mills, C. E. (2001). Jellyfish blooms: are populations increasing globally in response to changing ocean conditions? *Hydrobiologia*, **41**: 55–68.

Miranda, L., Hernández, A., Sepúlveda, A., and Landaeta, M. (1998). Alimentación de jurel y análisis de la selectividad en la zona centro-sur de Chile. In *Biología y Ecología del Jurel en Aguas Chilenas*, D. Arcos., ed. Instituto de Talcahuano-Chile, Investigación Pesquera, pp. 173–187.

Moloney, C. L., Jarre, A., Arancibia, H., *et al.* (2005). Comparing the Benguela and Humboldt marine upwelling ecosystems with indicators derived from inter-calibrated models. *ICES J. Mar. Sci.* **62** (3): 493–502.

Montecinos, A., Purca, S., and Pizarrro, O. (2003). Interannual-to-interdecadal sea surface temperature variability along the western coast of South America. *Geophys. Res. Lett.* **30** (11): 1570–1573.

Mullon C., Cury, P., and Penven, P. (2002). Evolutionary individual-based model for the recruitment of the anchovy in the southern Benguela. *Can. J. Fish. Aquat. Sci.* **59**: 910–922.

Mullon, C., Fréon, P., Parada, C., *et al.* (2003). From particles to individuals: modeling the early stages of anchovy in the southern Benguela. *Fish. Oceanogr.* **12** (4–5): 396–406.

Neira, S. (2008). Assessing the effects of internal (trophic structure) and external (fishing and environment) forcing factors on fisheries off Central Chile: basis for an ecosystem approach to management. Ph.D. thesis, University of Cape Town. 254 pp. plus 2 appendices.

Neira, S. and Arancibia, H. (2004). Trophic interactions and community structure in the Central Chile marine ecosystem (33° S–39° S). *J. Exp. Mar. Biol. Ecol.* **312**: 349–366.

Neira, S., Arancibia, H., and Cubillos, L. (2004). Comparative analysis of trophic structure of commercial fishery species off Central Chile in 1992 and 1998. *Ecol. Model.* **172** (2–4): 233–248.

Neira, S., Watermeyer, K., Moloney, C., *et al.* (2006). Comparing ecosystem status in the southern Humboldt and the southern Benguela systems before and after heavy fishing. In *Book of extended abstracts, International Conference, The Humboldt Current System: climate, ocean dynamics, ecosystem processes, and fisheries, Lima, Peru, November 27–December 1, 2006*, pp. 164–165.

Odum, E. P. (1985). Trends expected in stressed ecosystems. *BioScience* **35** (7): 419–422.

Odum, W. E. and Heald, E. J. (1975). The detritus-based food web for an estuarine mangrove community. In *Estuarine Research*, Vol. 1., Cronin, L. I., ed. New York: Academic Press.

Oro, D. (1996). Effects of trawler discard availability on egg laying and breeding success in the lesser black-backed gull *Larus fuscus* in the western Mediterranean. *Mar. Ecol. Prog. Ser.* **132**: 43–46.

Oro, D. and Ruíz, X. (1997). Exploitation of trawler discards by breeding seabirds in the north-western Mediterranean: differences between the Ebro Delta and the Balearic Islands areas. *ICES J. Mar. Sci.* **54**: 695–707.

Oro, D., Jover, L., and Ruíz, X. (1996). Influence of trawling activity on the breeding ecology of a threatened seabird, Audouin's gull *Larus audouinii. Mar. Ecol. Prog. Ser.* **139**: 19–29.

Ott, J. (1992). The Adriatic benthos: problems and perspectives. In *Marine Eutrophication and Population Dynamics. 25th European Marine Biology Symposium*. Olsen and Olsen: Fredensborg, pp. 367–378.

Pace, M. L., Cole, J. J., Carpenter, S. R., and Kitchell, J. F. (1999). Trophic cascades revealed in diverse ecosystems. *Trends Ecol. Evol.* **14**: 483–488.

Palma, S. and Rosales, S. (1995). Composición, distribución y abundancia estacional del macroplancton de la bahía de Valparaíso. *Invest. Mar.* **23**: 49–66.

Palomera, I., Olivar, M. P., and Salat, J. (2007). Small pelagic fish in the NW Mediterranean Sea: an ecological review. *Prog. Oceanogr.* **74** (2–3): 377–396.

Papaconstantinou, C. and Farrugio, H. (2000). Fisheries in the Mediterranean. *Mediterr. Mar. Sci.* **1** (1): 5–18.

Parada, C., van der Lingen, C. D., Mullon, C., and Penven, P. (2003). Modeling the effect of buoyancy on the transport of anchovy (*Engraulis capensis*) eggs from spawning to nursery grounds in the southern Benguela: an IBM approach. *Fish. Oceanogr.* **12**: 170–184.

Pauly, D. and Palomares, M. L. (2005). Fishing down marine food web: it is far more pervasive than we thought. *Bull. Mar. Sci.* **76** (2): 197–211.

Pauly, D., Christensen, V., Dalsgaard, J., *et al.* (1998). Fishing down marine food webs. *Science* **279**: 860–863.

Pauly, D., Christensen, V., and Walters, C. (2000). Ecopath, Ecosim, and Ecospace as tools for evaluating ecosystem impact of fisheries. *ICES J. Mar. Sci.* **57**: 697–706.

Payne A. I. L. (1989). Cape hakes. In *Oceans of Life off Southern Africa*, Payne, A. I. L., and Crawford, R. J. M., eds. Cape Town: Vlaeberg, pp. 136–147.

Penven, P., Roy, C., Brundritt, G. B., *et al.* (2001). A regional hydrodynamic model of upwelling in the southern Benguela. *S. Afr. J. Sci.* **97**: 472–475.

Pinardi, N., Arneri, E., Crise, A., *et al.* (2006). The physical, sedimentary and ecological structure and variability of shelf areas in the Mediterranean sea. In *The Sea*, Volume 14, Robinson, A. R., and Brink, K. H., eds. Harvard University Press.

Pranovi, F., Raicevich, S., Franceschini, G., *et al.* (2000). Rapido trawling in the northern Adriatic Sea: effects on benthic communities in an experimental area. *ICES J. Mar. Sci.* **57**: 517–524.

Pranovi, F., Raicevich, S., Franceschini, G., *et al.* (2001). Discard analysis and damage to non-target species in the rapido trawl fishery. *Mar. Biol.* **139**: 863–875.

Rice, J. (1995). Food web theory, marine food webs, and what climate change may do to northern marine fish populations. In *Climate Change and Northern Fish Populations*, Beamish, R. J., ed. *Canadian Special Publication of Fisheries and Aquatic Sciences*, **121**: 516–568

Riedl, R. (1986). *Fauna y Flora del Mar Mediterráneo*. Barcelona: Ediciones Omega.

Ríos, J. (2000). Feeding habits and trophic relationships in a demersal fish community in the Catalan sea (Western Mediterranean). M.Sc. thesis. School of Ocean Sciences, University of Wales, Bangor, 100 pp.

Rochet, M. -J. and Trenkel, V. M. (2003). Which community indicators can measure the impact of fishing? A review and proposals. *Can. J. Fish. Aquat. Sci.* **60**: 86–99.

Rottini-Sandrini, R. L. and Stravisi, F. (1981). Preliminary report on the occurrence of *Pelagia noctiluca* (Semaestomae, pelagidae) in Northern Adriatic. *Rapp. Proc. Verb. Comm. Int. Explor. Sci. Mer Medit.* **27** (7): 147.

Roux, J. -P. (1998). The impact of environmental variability on the seal population. *Namibia Brief*, **Special Issue 18**, Focus on Fisheries and Research: 138–140.

Roux, J-P. (2003). Risks. In *Namibia's Marine Environment*, Molloy, F. J. and Reinikainen, T., eds. Windhoek, Namibia: Directorate of Environmental Affairs, MET, pp. 137–152.

Roux, J. -P. and Shannon, L. J. (2004). Ecosystem approach to fisheries management in the northern Benguela: the Namibian experience. In *Ecosystem Approaches to Fisheries in the Southern Benguela*, Shannon, L. J., Cochrane, K. L., and Pillar, S. C., eds. *Afr. J. Mar. Sci.* **26**: 79–93.

Roy, C. P., Weeks, S. C., Rouault, M., *et al.* (2001). Extreme oceanographic events recorded in the Southern Benguela during the 1999–2000 summer season. *S. Afr. J. Sci.* **97**: 465–471.

Ryther, J. H. (1969). Photosynthesis and fish production in the sea. *Science* **166**: 72–76.

Salat, J. (1996). Review of hydrographic environmental factors that may influence anchovy habitats in the Northwestern Mediterranean. *Sci. Mar.* **60**(2): 21–32.

Sánchez, F. and Olaso, I. (2004). Effects of fisheries on the Cantabrian Sea shelf ecosystem. *Ecol. Model.* **172** (2–4): 151–174.

Santojanni, A., Arneri, E., Barry, C., *et al.* (2003). Trends of anchovy (*Engraulis encrasicolus*, L.) biomass in the northern and central Adriatic Sea. *Sci. Mar.* **67** (3): 327–340.

Santojanni, A., Cingolani, N., Arneri, E., *et al.* (2005). Stock assessment of sardine (*Sardina pilchardus*, Walb.) in the Adriatic Sea, with an estimate of discards. *Sci. Mar.* **69** (4): 603–617.

Sardà, F., Bahamón N., Sardà-Palomera F., and Molí B. (2005). Commercial testing of a sorting grid to reduce catches of juvenile hake (*Merluccius merluccius*, L.) in the western Mediterranean demersal trawl fishery. *Aquat. Living Res.* **18**: 87–91.

Schwartzlose, R. A., Alheit, J., Bakun, A., *et al.* (1999). Worldwide large-scale fluctuations of sardine and anchovy populations. *S. Afr. J. Mar. Sci.* **21**: 289–347.

Shannon, L. J., Christensen, V., and Walters, C. (2004b). Modelling stock dynamics in the southern Benguela ecosystem for the period 1978–2002. In *Ecosystem Approaches to Fisheries in the Southern Benguela*, Shannon, L. J., Cochrane, K. L., and Pillar, S. C. eds. *Afr. J. Mar. Sci.* **26**: 179–196.

Shannon, L. J., Cury, P., and Jarre, A. (2000). Modelling effects of fishing in the southern Benguela ecosystem. *ICES J. Mar. Sci., Symposium Edition* **57** (3): 720–722.

Shannon, L. J., Cury, P. M., Nel, D., *et al.* (2006). How can science contribute to an ecosystem approach to pelagic, demersal and rock lobster fisheries in South Africa? *Afr. J. Mar. Sci.* **28** (1): 115–157.

Shannon, L. J., Field, J. G., and Moloney, C. L. (2004a). Simulating anchovy-sardine regime shifts in the southern Benguela ecosystem. *Ecol. Model.* **172**: 269–281.

Shannon, L. J. and Jarre-Teichmann, A. (1999a). A model of the trophic flows in the northern Benguela upwelling system during the 1980s. *S. Afr. J. Mar. Sci.* **21**: 349–366.

Shannon, L. J. and Jarre-Teichmann, A. (1999b). Comparing models of trophic flows in the northern and southern Benguela upwelling systems during the 1980s. In *Ecosystem approaches for fisheries management*. Fairbanks: University of Alaska Sea Grant, **AK-SG-99–01**, pp. 527–541.

Shannon, L. J., Moloney, C. L., Jarre, A., and Field, J. G. (2003). Trophic flows in the southern Benguela during the 1980s and 1990s. *J. Mar. Syst.* **39** (1–2): 83–116.

Shannon, L. J., Neira, S., and Taylor, M. (2008). Comparing internal and external drivers in the southern Benguela and the southern and northern Humboldt upwelling ecosystems. *Afr. J. Mar. Sci.* **30**(1): 63–84.

Shannon, L. J., Nelson, G., Crawford, R. J. M., and Boyd, A. J. (1996). Possible impacts of environmental change on pelagic fish recruitment: modelling anchovy transport by advective processes in the southern Benguela. *Glob. Change Biol.* **2**: 407–420.

Shannon, L. J., Roux, J. -P., and Jarre, A. (2001). Trophic interactions in the Benguela ecosystem and their implications for multispecies management of fisheries. Report on progress to January 2001: exploring multispecies management options using the completed northern Benguela ecosystem model. *BENEFIT/FAO/Government of Japan Cooperative Programme GCP/INT/JPN, Report* **2.4**, 56 pp.

Shin, Y. (2006). EAF Indicators: a comparative approach across ecosystems. *Report of the Third Workshop: 1–3 February 2006, CRH, Sète, France. EUR-OCEANS Working Group for Tjeme WP6-Ecosystem Approach for Marine Resources*, 9 pp.

Siegfried, W. R., Crawford, R. J. M., Shannon, L. V., *et al.* (1990). Scenarios for global warming induce change in the open-ocean environment and selected fisheries of the west coast of southern Africa. *S. Afr. J. Sci.* **86**: 281–285.

Sinclair, M. (1988). *Marine Populations: An Essay on Population Regulation and Speciation*. Seattle, Washington: University of Washington Press.

Springer, A. M., Estes, J. A., van Vliet, G. B., *et al.* (2003). Sequential megafaunal collapse in the North Pacific Ocean: an ongoing legacy of industrial whaling? *Proc. Natl. Acad. Sci. USA* **100**: 12223–12228.

Stergiou, K. I. and Karpouzi, V. (2002). Feeding habits and trophic levels of Mediterranean fish. *Rev. Fish. Biol. Fisher.* **11**: 217–254.

Strub P., Mesías, J. M., Montrecino, V., *et al.* (1998). Coastal ocean circulation off western South America. In *The Global Coastal Ocean: The Sea 11*, Brink, K. H., and Robinson, A. R., eds. New York: Wiley and Sons, Inc., pp. 273–313.

Tanaka, H., Aoki, I., and Ohshimo, S. (2006). Feeding habits and gill raker morphology of three planktivorous pelagic fish species off the coast of northern and western Kyushu in summer. *J. Fish. Biol.* **68**: 1041–1061.

Timmermann, A., Oberhuber, J., Bacher, A., *et al.* (1999). Increased El Nino frequency in a climate model forced by future green house warming. *Nature* **398**: 694–697.

Tudela, S. (2004). Ecosystem effects of fishing in the Mediterranean: An analysis of the major threats of fishing gear and practices to biodiversity and marine habitats. *General Fisheries Commission for the Mediterranean (FAO). Studies and Reviews*, **74**, 58 pp.

Ulanowicz, R. E. and Puccia, C. J. (1990). Mixed trophic impacts in ecosystems. *Coenoses* **5**: 7–16.

Van der Lingen, C. D., Shannon, L. J., Cury, P., *et al.* (2006). Chapter 8 Resource and ecosystem variability, including regime shifts, in the Benguela Current system. In *Benguela: Predicting a Large Marine Ecosystem*, Shannon, L. V., Hempel, G., Malanotte-Rizzoli, P., *et al.*, eds. Elsevier, USA: Large Marine Ecosystems Series 14, pp. 147–184

Verheye, H. M. and Richardson, A. J. (1998). Long-term increase in crustacean zooplankton abundance in the southern Benguela upwelling region (1951–1996): bottom-up or top-down control? *ICES J. Mar. Sci.* **55**: 803–807.

Verheye, H. M., Richardson, A. J., Hutchings, L., *et al.* (1998). Long-term trends in the abundance and community structure of the coastal zooplankton in the southern Benguela system, 1951–1996. In *Benguela Dynamics: Impacts of Variability on Shelf-Sea Environments and their Living Resources*, Pillar, S. C. Moloney, C. L. Payne A. I. L., and Shillington, F. A. S., eds. *Afr. J. Mar. Sci.* 317–332.

Wallace-Fincham, B.P. (1987). The food and feeding of *Etrumeus whiteheadi* Wongratana 1983, off the Cape Province of South Africa. M.Sc. thesis, University of Cape Town, 117 pp.

Walsh, J.J. (1981). A carbon budget for overfishing off Peru. *Nature* **290**: 300–304.

Walters, C., Christensen, V., and Pauly, D. (1997). Structuring dynamic models of exploited ecosystems from trophic mass-balance assessments. *Rev. Fish. Biol. Fisher.* **7**: 139–172.

Wanless, S., Harris, M.P., Redman, P., and Speakman, J.R. (2005). Low energy values of fish as a probable cause of major seabird breeding failure in the North Sea. *Mar. Ecol. Prog. Ser.* **204**: 1–8.

Zavatarelli, M., Raicich, F., Bregant, D., *et al.* (1998). Climatological biogeochemical characteristics of the Adriatic Sea. *J. Mar. Syst.* **18**: 227–263.

Zavodnik, D. (1991). On the food and feeding in the Northern Adriatic of *Pelagica noctiluca* (Scyphozoa). Jellyfish blooms in the Mediterranean. *Proceedings of the 2nd Workshop on Jellyfish in the Mediterranean Sea. UNEP, MAP Technical Reports Series,* **47**: 212–216.

Zotier, R., Bretagnolle, V., and Thibault, J.-C. (1998). Biogeography of the marine birds of a confined sea, the Mediterranean. *J. Biogeogr.* **26**: 297–313.

Zupanovic, S. and Jardas, I. (1989). *Fauna i flora Jadrana: Jabucka Kotlina.* Logos, Split. Institut za Oceanografiju i Ribarstvo. Vol IV.

9 Current trends in the assessment and management of stocks

Manuel Barange, Miguel Bernal, Maria Cristina Cergole, Luis A. Cubillos, Georgi M. Daskalov, Carryn L. de Moor (formerly Cunningham), José A. A. De Oliveira, Mark Dickey-Collas, Daniel J. Gaughan, Kevin Hill, Larry D. Jacobson, Fritz W. Köster, Jacques Massé, Miguel Ñiquen, Hiroshi Nishida, Yoshioki Oozeki, Isabel Palomera, Suzana A. Saccardo, Alberto Santojanni, Rodolfo Serra, Stylianos Somarakis, Yorgos Stratoudakis, Andres Uriarte, Carl D. van der Lingen, and Akihiko Yatsu

CONTENTS

Summary

The assessment and management of small pelagic fish (SPF) stocks is particularly difficult and uncertain because their short life expectancy, characteristic aggregative behavior, rapid response to climate and environmental signals and large and variable natural mortality make them less tractable through traditional population dynamic models and assumptions. In this review we summarize the assessment and management approaches applied in 29 SPF stocks or management units (12 anchovy, 10 sardine, 4 herring, and 3 sprat). The review demonstrates that the assessment and management of SPF varies substantially in its approach and performance between stocks and regions. Most stocks have a scientific assessment program in place and a management approach that generally takes into account assessment results, but in some stocks management practices deviate substantially from scientific advice and in some, assessment and management processes are largely disconnected. It is concluded that only properly tailored scientific assessment and management programs can provide the speed of response and the flexibility of management that highly variable SPF demand. The most effective monitoring programs are based on fishery-independent surveys (daily egg production or/and hydroacoustics), while analyses based on catch per unit effort offer limited value. Most assessments, defined as what management uses to base its decisions on, rely on catch-at-age or yield per recruit models. Harvest strategies range from those driven by harvest control rules to those derived from outputs of best assessment runs. Some stocks use operating models based on age–structure model outputs or forward VPA[1]. On the issue of scientific uncertainty some practitioners propose reducing it through additional science and measures, while others promote the development of management procedures robust to uncertainty. This difference is particularly evident in relation to the value of recruitment forecasts. Other identified uncertainties include fishing versus natural mortality estimates and fleet catchability estimates. Regarding governance it is suggested that adaptive management practices applied by independent governance structures capable of interacting at ecological, social and economic levels need development for effective stewardship and governance. The review also addresses recent concerns over managing stocks that may be subject to productivity regimes or regime shifts, and whether two-level management strategies are required to address short- and

Climate Change and Small Pelagic Fish, eds. Dave Checkley, Jürgen Alheit, Yoshioki Oozeki, and Claude Roy. Published by Cambridge University Press. © Cambridge University Press 2009.

long-term resource variability. The straddling nature of some stocks adds complexity to management procedures, and it is feared that this could be accentuated by climate-driven changes in stock distribution. Finally, SPF should be managed under ecosystem considerations, to protect their value as forage species to other fish, mammals and birds, and thus to respect the integrity of ecosystems. While the Ecosystem Approach is yet to be successfully applied to the management of any SPF, this may be the single most important driving force in influencing future assessment and management policies.

Introduction

The assessment and management of small pelagic fish stocks is considered particularly difficult and uncertain because these species do not conform to traditional population dynamics models and assumptions. Some of these invalid assumptions include that the size of the unexploited stock (B_∞ or carrying capacity), as well as the catchability coefficient (the probability of a fish being caught) remain constant, and the assumption that the effect of the environment on population parameters is either constant or generates a random noise (Csirke, 1988). Their particular size-selective shoaling behavior poses problems, not only in terms of catch and effort analysis but also in interpreting age and length frequency data from catches.

Small pelagic species are short lived, fast growing, and are characterized by high and often variable levels of natural mortality. As a result, their stock size is very dependent on incoming recruitments, and thus highly variable, unreliable and less responsive to management measures than other longer lived species. Beverton (1983) classified small pelagic fish populations as the most unreliable and vulnerable to unrestrained fishing, making their exploitation a high-risk activity. Small pelagic species have led to the development of very profitable fisheries, but these have suffered well-known collapses and have experienced partial and slow recoveries. Overall, they are fragile enterprises (Pitcher, 1995).

In recent decades we have experienced the development of new and advanced fisheries assessment methodologies that have contributed to more consistent and responsive fisheries management worldwide (although with limited success because of, among others, poor governance). These developments have also involved small pelagic fisheries. For example, the current support for the so-called Ecosystem Approach to marine resources (FAO, 2003) has been influenced by theories such as the possibility of species replacements in the pelagic marine ecosystem (Lluch-Belda *et al.*, 1992), and the idea that the variability of groups of small pelagic fish stocks may be smaller than individual stocks. This chapter intends to describe the state of the art in the assessment and management of small pelagic fisheries by

reviewing current practices in 29 stocks of anchovy, sardine, herring, and sprat, in search of commonalities, successful approaches, and future requirements.

Stock structure, population trends and fishery information

This chapter synthesizes information from 12 anchovy, ten sardine, three sprats, and four herring stocks worldwide (Fig. 9.1). The stock structure, population trends and fishery information are introduced individually, while assessment and management procedures are described and discussed together.

Table 9.1 summarizes information regarding biomass and catch statistics, distribution patterns, and population parameters for the different stocks. Anchovy stocks yield catches between 1200 t (Bay of Biscay) and 7.5 Mt² (Central-South Peru), corresponding to biomass levels between 9000 t and 10 Mt (Table 9.1a). Maximum anchovy length is generally 15–19 cm, corresponding to a maximum age of 2–5 years. Spawning generally takes place in spring–summer (but winter in the case of the Peru–Chile anchovy), and 1 year olds (y.o.) are active spawners (Table 9.1a). Sardine stocks yield catches presently between 4500 t (southern Western Australia) and 350 000 t (Chile), corresponding to biomass levels between 102 000 t (Western Australia) and 1.7 Mt (Chile) (Table 9.1b). Maximum sardine lengths range between 23 and 40 cm, corresponding to 4–10-year-old fish. Spawning is generally year round, many stocks displaying a winter spawning peak (except Brazil, California, and Benguela). Some 1 y.o. fish are reproductively active, but in many stocks the age of first spawning is delayed to ages 2 and 3 (Table 9.1b). Herrings and sprats range in catches between the 30 000 t yield by the Gulf of Riga herring to the 1 Mt of the Arcto-Norwegian spring spawning herring, corresponding to biomasses of 165 000 t and 10 Mt, respectively (Table 9.1c). Maximum fish lengths are 30–40 cm in the case of herrings and 15–18 cm for sprats. Of the examples provided, many stocks spawn in spring and autumn, and the age of first spawning is generally 2–4 y.o. (except for the Black Sea sprat, Table 9.1c).

Anchovy stocks

*Japanese Anchovy (*Engraulis japonicus*) – NW Pacific stock*

Three stocks of Japanese anchovy are distributed around Japan: the NW Pacific stock, the Seto Inland Sea stock and the Tsushima Current stock (Kono and Zenitani, 2005; Ohshimo, 2005; Oozeki *et al.*, 2005). These three stocks are distinguished by distribution and migration patterns. The NW Pacific stock accounts for more than 75% of the Japanese anchovy landings of Japan in recent years, and is the focus of this review. Spawning grounds

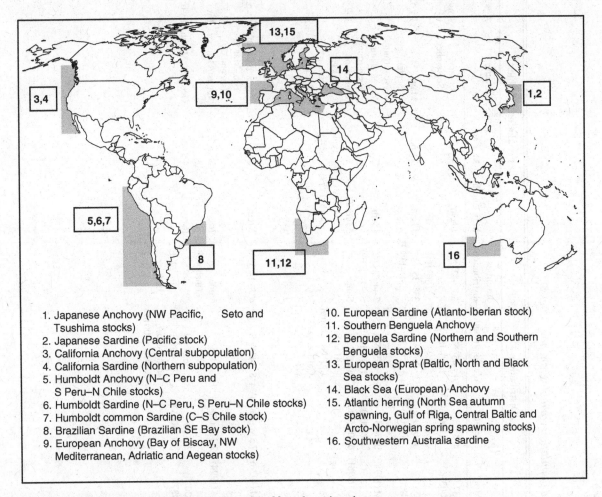

1. Japanese Anchovy (NW Pacific, Seto and Tsushima stocks)
2. Japanese Sardine (Pacific stock)
3. California Anchovy (Central subpopulation)
4. California Sardine (Northern subpopulation)
5. Humboldt Anchovy (N–C Peru and S Peru–N Chile stocks)
6. Humboldt Sardine (N–C Peru, S Peru–N Chile stocks)
7. Humboldt common Sardine (C–S Chile stock)
8. Brazilian Sardine (Brazilian SE Bay stock)
9. European Anchovy (Bay of Biscay, NW Mediterranean, Adriatic and Aegean stocks)
10. European Sardine (Atlanto-Iberian stock)
11. Southern Benguela Anchovy
12. Benguela Sardine (Northern and Southern Benguela stocks)
13. European Sprat (Baltic, North and Black Sea stocks)
14. Black Sea (European) Anchovy
15. Atlantic herring (North Sea autumn spawning, Gulf of Riga, Central Baltic and Arcto-Norwegian spring spawning stocks)
16. Southwestern Australia sardine

Fig. 9.1. Schematic representation of the location of the 29 stocks reviewed.

are confined between the Kuroshio Current and the coast in the southern Japan and extended offshore as east as 170° longitude in the northern Japan (Oozeki *et al.*, 2005). Spawning takes place mainly from February to September. Juveniles are transported in spring by the Kuroshio and Kuroshio Extension Currents to beyond 170° E longitude. The feeding grounds are mainly located in the Oyashio and Kuroshio/Oyashio Transition Zone in summer and autumn. Major fishing grounds of the purse-seine fishery are located along the Pacific coast of northern Japan. Biological minimum size is 8 cm in standard length, which corresponds to age 1, although half of the age 0 group may mature in favorable conditions. Estimated stock biomass was less than 500 000 t until 1988 but has increased since to reach 1.16 Mt in 2004 (Table 9.1a, Fig. 9.2a). Recruitment of the Japanese anchovy was high from 1997 (Fig. 9.2a). Japanese anchovy are landed mainly by purse seiners and also their larvae and juveniles are landed as "shirasu" (white children) by purse nets around the Pacific coast of Japan. Fishing grounds concentrated only in the coastal area, although the distribution of the Japanese anchovy expanded

from the coast of Japan to 160° E as their population size increased. Annual landings have increased from 1990 off the northern Pacific coast of Japan and they now exceed 300 000 t (Table 9.1a). Exploitation rates in recent years were estimated to be around 30%. Although mainly young-of-the-year were historically landed, substantial proportions of age 1 and age 2 fish have been landed since 1989 and 1990, respectively.

California anchovy[3] (Engraulis mordax) – central subpopulation

The central subpopulation of the California anchovy is distributed along the coast from central Baja California, Mexico (30° N) north to central California (37° N) in the United States (PFMC, 1998). It has a preference for SST in the range 12–21.5 °C. The stock is sedentary relative to other small pelagic stocks. Large individuals venture furthest north and offshore (Parrish *et al.*, 1989) and the stock moves north during El Niño events (Pearcy *et al.*, 1985). Anchovy rarely live to more than 4 years of age, although specimens as old as 7 years old have been

Table 9.1a. *Stock distribution and structure, life history, biomass and population variables: anchovy stocks*

Species	Stock	Biomass and Catch				Distribution		Population variables			
		Max biomass (t)	Present biomass (t)	Max catch (t)	Present catch (t)	Adult distribution	Juveniles distribution	Max length	Max age	Spawning season	Age 1st spawning
Japanese anchovy (*Engraulis japonicus*)	NW Pacific	1.48 M (2003)	1.24 M (2004)	415437 (2003)	401158 (2004)	30°–45° N / 130°–180° E	30°–42° N / 130°–160° E	17 cm	4 y.o.	February–September	1 y.o.
	Seto Inland	0.53 M (1985)	0.23 M (2004)	149953 (1985)	66000 (2004)	33°–35° N / 131°–136° E	33°–35° N / 131°–136° E	14 cm	2 y.o.	May–September (year round)	1 y.o.
	Tsushima Current	3.80 M (1998)	0.56 M (2004)	1.75 M (1998)[1]	0.26 M (2004)[2]	20°–45° N / 120°–140° E	?–40° N /? –140° E	15 cm	2 y.o.	March–January (year round)	0.5 y.o.
California anchovy (*Engraulis mordax*)	Central sub-population	1.6 M (1974)	0.39 M (1995)	315000 (1981)	6900 (2004)	Central Baja California (30° N)–central California (37° N)	Same	23 cm *TL* but typically <18 cm *TL*	7 y but typically <4 y	Year round Peaks February–April	1–2 y.o.
Humboldt anchovy (*Engraulis ringens*)	North-Central Peru	22.8 M (1970)	10.5 M (2005)	10.9 M (1970)	7.5 M (2005)	4° S–14°59'S	4° S–14°59'S	20 cm	4 y.o.	Aug/Sep (main) Feb–Mar (secondary)	1 y.o.
	South Peru–North Chile	10 M (2001)	5.7 M (2003)	2.75 M (1994)	2.3 M (2003)	15°–24° S	Same (inshore)	20 cm	5 y.o.	August/October	1 y.o.
Benguela anchovy (*Engraulis encrasicolus*)	Southern (South African)	7.93 M (2001)	3.06 M (2005)	596000 (1987)	283000 (2005)	32°–36.5° S / 17°–27° E Inshore to depths >200 m	29°–36° S / 17°–22° E Inshore	16 cm	4 y.o.	October–March	1 y.o.

Species	Region										
European anchovy (*Engraulis encrasicolus*)	Bay of Biscay	93 000 (1998)	9200 (2005)	83 600 (1965) 40 500 (2001)	1200 (2005)	South of 47°N and East of 5°W (continental shelf and shelf edge)	South of 47°N and East of 5°W	18.5 cm	4 y.o.	April to August	1 y.o.
	NW Mediterranean	GL: 112 000 (2001) CS: 27 000 (2003)	GL: 18 000 (2004) CS: 26 700 (2004) [3]	25 030 (1994)	9379 (2004)	38.5°–43.5°N/0°–6.5°E	38.5°–43.5°N/0°–6.5°E	19 cm	4 y.o.	April–October	1 y.o.
	Adriatic Sea	350 000 (1978) VPA	114 000 (2004) VPA	60 000 (1980)	30 000 (2004)	10–200 m	Inshore	19 cm	6 y.o.	March–October	< 1 y.o.
	Aegean Sea	48 000 (2003) [7]	46 000 (2004) [4]	27 227 (1998)	22 007 (2003) [5]	37.5°–41° N/ 22.5°–27° E 10–200 m	37.5°–41° N/ 22.5°–27° E inshore	18 cm	4 y.o.	May–September	1 y.o.
	Black Sea	708 000 (1979)	387 556 (2001)	468 807 (1988)	322 711 (2004)	All Black sea – maximum over shelf	All Black sea – max. in shelf	16 cm	5 y.o.	Summer (Jun–Aug)	1 year

Notes:

[1] Including Korea and China.
[2] Including Korea.
[3] Acoustics estimates.
[4] Acoustics in Greek waters.
[5] Greece and Turkey.

Table 9.1b. *Stock distribution and structure, life history, biomass and population variables: sardine stocks*

Species	Stock	Biomass and catch				Distribution		Population variables			
		Max biomass (t)	Present biomass (t)	Max catch (t)	Present catch (t)	Adult distribution	Juvenile distribution	Max length	Max age	Spawning season	Age 1st spawning
Japanese sardine (*Sardinops melanostictus*)	Pacific	19.5 M (1987)	0.1 M (2004)	2.9 M (1987)	48 000 (2004)	29°–54°N / 130°E–160°W (high-stock period), shrink in low-stock period	29°–38°N / 130°E–160°W (high-stock period)	23 cm	7 y.o.	October–May	1 y.o. (high-stock period), 3 y.o. (low-stock period)
California sardine (*Sardinops sagax caerulea*)	Northern sub-population	3.63 M (1934)	1.06 M (2005)	718 000 (1936)	135 000 (2004)	29°–59°N 115°–136°W Over depths 0–50 m	29°–59°N 115°–136°W Inshore 0–30 m	41 cm	14 y.o.	April–August	1 y.o.
Humboldt sardine (*Sardinops sagax*)	North-Central Peru	11.8 Mt (1984)	0.2 Mt (2000)	3.4 Mt (1988)	0.001 M (2003)	4°S–14°59'S	4°S–14°59'S Inshore 0–30 m	39 cm	8 y.o.	Aug/Sep (Main) Jan–Feb (Secondary)	3–4 y.o.
	South Peru–North Chile	9.1 M (1980)	<0.1 M (1996)	3.0 M (1985)	2201 (2003)	15°–24°S	Same (inshore)	40 cm	11 y.o.	Main:August/September Secondary: February/March	5 y.o.
Humboldt common sardine (*Strangomera bentincki*)	Central-South Chile	3.09 M (1995/96)	1.65 M (2003/04)	693 833 (1998/99)	353 952 (2003/04)	34°S–40°S	34°S–40°S	20 cm	4 y.o.	August/September	1 y.o.
Brazilian sardine (*Sardinella brasiliensis*)	Brazilian Southeast Bight	1.2 M (1977)	0.05 M (2007)	228 000 (1973)	17 000 (2000)	22°–29°S Over depths <100 m	23°–26°S inshore	27 cm	3.8 y.o.	October–March	1 y.o.
Benguela sardine (*Sardinops sagax*)	Southern (South African)	4.14 M (2002)	0.96 M (2005)	410 000 (1962)	247 000 (2005)	32°–36.5°S / 17°–32°E Over depths 10–500 m	29°–36°S / 17°–32°E Inshore	25 cm	10 y.o.	Year round but mainly August–March	1–3 y.o.
	Northern (Namibian)	11.14 M (1964)	0.27 M (2005)	1.40 M (1968)	25 300 (2005)	15°–25°S / 13°–15°E Over depths 10–200 m	15°–25°S / 13°–15°E Inshore	25 cm	10 y.o.	Year round but mainly August–March	1–3 y.o.
European sardine (*Sardina pilchardus*)	Atlanto-Iberian	0.65 M (2001)	0.4 M (2004)	250 000 (1964)	95 000 (2004)	36°–44°N / 1°–9°W below 100 m	39°–41°N (north Portugal) 6°–7°W (Gulf of Cadiz) inshore	25 cm	10 y.o.	October–April	1–2 y.o.
Australian sardine (*Sardinops sagax neopilchardus*)	Southern Western Australia	0.14 M (1990)	0.10 M	8000 (1988 from one region)	4500 (=TAC, only 1800 tonnes landed)	Part of a continuous distribution around southern half of Australia	Same (inshore)	23 cm	9 y.o.	Summer–winter, variable peak	2 y.o.

Table 9.1c. *Stock distribution and structure, life history, biomass and population variables: herrings and sprats*

Species	Stock	Biomass and catch				Distribution		Max length	Population variables		
		Max biomass (t)	Present biomass (t)	Max catch (t)	Present catch (t)	Adult distribution	Juveniles distribution		Max age	Spawning season	Age 1st spawning
European sprat (*Sprattus sprattus*)	Black Sea	0.58M (1975)	0.58M (1999)	0.10M (1989)	49446 (2004)	All Black sea with maxima in shelf waters and the northwestern part	All Black sea Upper water layer	14.5cm	5 y.o.	All year with max in Nov–Mar	1 year old
	Baltic Sea	3.1M (1995)	2.0M (2005)	0.53M (1997)	405000 (2005)	54°–63°N / 10°–29°E	54°–63°N / 10°–29°E depends on drift	16cm	>10 y.o.	March–June	2 y.o.
	North Sea	n/a	n/a	0.36M (1995)	0.21M (2005)	North Sea and English Channel 51–60°N 0–12°E	E.N. Sea and Kattegat, more coastal 51–58°N 0–12°E	18cm	10 y.o.	February to September	2 y.o.
Atlantic herring (*Clupea harengus*)	North Sea autumn spawning stock	2.18M (1963)	1.70M (2005)	1.17M (1965)	0.66M (2005)	North Sea and English Channel 51–62°N 0–10°E	E.N. Sea and Kattegat, more coastal 51–60°N 0–12°E	38–39cm	17–20 y.o.	July to January	2–3 y.o.
	Arcto-Norwegian spring spawning stock	16.2M (1950)	10.1M (2005)	1.96M (1969)	1.0M (2005)	Norwegian Sea, NE Atlantic 62–78°N 10°W–20°E	Norwegian Sea, Barents Sea 62–79°N 10°E–35°E	42cm	20 y.o.	March to April	4–5 y.o.
	Central Baltic	3.05M (1974)	0.93M (2005)	0.37M (1974)	92000 (2005)	54°–63°N / 15°–29°E coastal (spring spawner)	54°–63°N / 15°–29°E coastal (spring spawner)	35cm	>10 y.o.	March–April (spring spawner) Sept.–Oct. (autumn spawner)	2–3 y.o.
	Gulf of Riga	0.17M (2001)	0.16M (2005)	41000 (2003)	32000 (2005)	57°–58.5°N / 22°–24°E	57°–58.5°N / 22°–24°E	30cm	>10 y.o.	March–April	2 y.o.

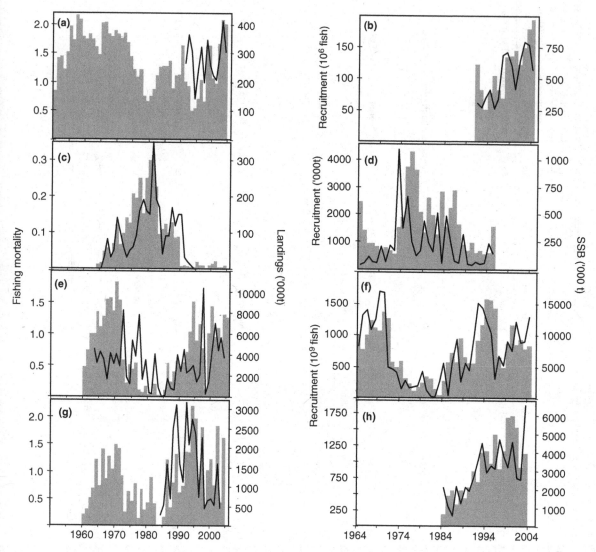

Fig. 9.2a. Time series of fishing mortality (y^{-1}), landings ('000 t), recruitment and spawning stock biomass (SSB, '000 t) for Japanese anchovy (a, b), California anchovy (c, d), and Humboldt anchovy (Central Peru stock, e, f; South Peru–North Chile stock, g, h). Solid lines: fishing mortality and recruitment; grey bars: landings and spawning stock biomass.

recorded (Table 9.1a). Spawning is opportunistic and occurs year round in the main spawning areas off northern Mexico and the southern United States, with peaks during February–April. Maturity is reached by age 2 and the proportion mature at age 1 depends on water temperatures (Methot, 1989; Table 9.1a).

Annual landings in recent years indicated the highest level during the last 25 years (Fig. 9.2a). Anchovy biomass (ages 1+) averaged 326 000 t during 1964–1970, increased rapidly to 1.6 Mt in 1974, and then declined to range 153–392 000 t during 1991–1995 (Jacobson *et al.*, 1995; Table 9.1a). No information is available after 1995. Total catches in the US and Mexico peaked at about 315 000 t during 1981 and then declined to about 10 000 t tonnes during 1991–2005 (PFMC, 2005). Relatively low catch

levels since the 1990s have been due to poor market conditions. Anchovy are harvested mainly by purse seines in US waters off California and in Mexican waters off Baja California. They are landed and sold for reduction to fish meal, human consumption, live and dead bait used in recreational fisheries, and as an ingredient in pet food. Live bait, dead bait, and pet food are the most important and economically important uses.

Humboldt anchovy[4] (Engraulis ringens)
In the Humboldt Current System anchovy is distributed from Zorritos (4°30 S) in northern Peru to about Chiloé (42°30 S) in southern Chile. Three main discrete stocks can be identified (Serra, 1983; Alheit and Ñiquen, 2004). The most productive is located in north-central Peru (4°–15° S),

followed by the Peruvian–Chilean stock (16°–24° S), and the most austral and less abundant located off–central–southern coast of Chile (33°–42° S; not considered in this review). The separation between the three stocks is supported by meristic, morphometric, spawning grounds and/or tagging studies (Serra, 1983; Pauly and Tsukayama, 1987; Mendo, 1991). Anchovy is restricted to 100–120 nm offshore, but in spring and summer is limited to 30 nm offshore. Fishing takes place within 60 nm from the coast by an industrial fleet of purse seiners.

Anchovy is a short-lived species with maximum longevity and length of 4 years and 20 cm in total length, respectively. Natural mortality is 0.8–1.0 y^{-1} (Csirke *et al.*, 1994).

For the North-Central Peru (N/C Peru) stock the main spawning areas are between 8–9.5° S and 12–14° S, while for the Peruvian–Chilean (SP/NCh) stock the main spawning area extends from about 17° S to 23° S (Oliva *et al.*, 2001; Braun *et al.*, 2005). Anchovy spawn almost all year round but the main spawning seasons are between August and September (winter) with a second spawning in late summer (February–March) (Table 9.1a). Spawning is in batches and the population fecundity is indeterminate. Length at first maturity occurs at 11.5 cm in winter and 12.5 cm total length in summer (Simpson and Gil, 1967). It is assumed that all fish at age 1 are mature. The recruitment to the fishery tends to occur from mid November to April (late spring and summer), with a second, less important recruitment period in winter.

Peruvian anchovy fluctuates dramatically in relation to the El Niño phenomenon. In the N/C Peru stock estimated biomass dropped from an average of 14 Mt in 1967–1971 to 4 Mt in 1972 (Tsukayama, 1983), and took years to recover. However, the stock recovered well after the El Niños of 1982–83 and 1998 (Fig. 9.2a). Acoustic estimates agree with monthly VPA analysis (Pauly and Palomares, 1989). Catches have fluctuated with biomass, peaking at 12 Mt in 1970. Current catches are around 7–8 Mt. For the SP/NCh stock the biomass is estimated by stock assessment models. Biomass increased since 1984 and after 1992 it has fluctuated between 6 and 10 Mt. Similar trends were followed by catches, but important variations after 1993 without correlation with the stock biomass are observed (Fig. 9.2a). The catch history of this stock shows coherent and synchronous long-term changes alternating with sardine (*S. sagax*) catches.

Biological information for the assessment of the N/C Peru stock is obtained at all landing sites and, since 1964, through the EUREKA project (Villanueva, 1970). This consists of using the commercial fleet to obtain biological and fishery information in real time. In addition all vessels are equipped with VMS[5], and geo-referenced data are used in the management. From 1965 the assessment consists mainly of spatially separated CPUE from the industrial fleet (from 1965) and fishery independent surveys. For the assessment of the SP/NCh stock the time series from 1984 to 2003 is considered. Catch-at-age matrices from the Chilean and the southern Peru fisheries are computed monthly and grouped by year, and then added to obtain an overall annual matrix for the combined stock. Weight-at-age was obtained by converting the mean length-at-age to weight by an overall length–weight relationship (Serra *et al.*, 2004; GTE, 2003).

Benguela anchovy *(Engraulis encrasicolus) – southern Benguela*

Benguela anchovy occur from northern Namibia (around 16° S) to South Africa's east coast (around 27° E), but within this range are thought to comprise two distinct stocks separated by a permanent zone of intense coastal upwelling located at Lüderitz (26° S). Anchovy has been the dominant component of catches made off South Africa but is less important to the Namibian fishery, particularly over the past decade (van der Lingen *et al.*, 2006a), hence only southern Benguela anchovy are discussed here.

Spawning of southern Benguela anchovy occurs over the Agulhas Bank (south of the African continent), peaking between October and December (van der Lingen and Huggett, 2003). Historically, spawning was concentrated on the western Agulhas Bank, but in recent years spawners have been concentrated to the east of Cape Agulhas (van der Lingen *et al.*, 2002; Roy *et al.*, 2007). Eggs and larvae are transported from the south coast spawning grounds to the west coast nursery areas by a shelf-edge jet current, and juvenile fish undergo a return migration to reach the spawning grounds at an age of around 1 year (Hutchings *et al.*, 1998), by which time they are sexually mature. Anchovy has a maximal recorded age of 4 years (Melo, 1984; Table 9.1a).

The South African pelagic fishery began to target anchovy in the early 1960s following the drastic decline in sardine catches. Anchovy catches increased steadily, peaking around 600000 t in 1987 and 1988 (De Oliveira, 2003; Fig. 9.2b). Landings decreased to a minimum of 40000 t in 1996, and then increased sharply to about 287000 t in 2001. All catches are taken by purse seiners and reduced for fish meal. The fishery commences in January, and operates sometimes well into October/November, with over 80% (by mass and number) of the catch each year comprising juvenile fish of around 6 months old (De Oliveira, 2003). Anchovy spawner biomass was initially estimated using VPA, which is considered to have substantially underestimated stock size; more recent monitoring via acoustic and DEPM[6] surveys has shown large fluctuations in recruitment and spawner biomass, with record levels over the period 2000–2003 (Fig. 9.2b).

As annual age-length keys are not available for the commercial anchovy landings, assumptions about the age structure of the catch are made based on the life-history

200

M. Barange, M. Bernal, M. C. Cergole et al.

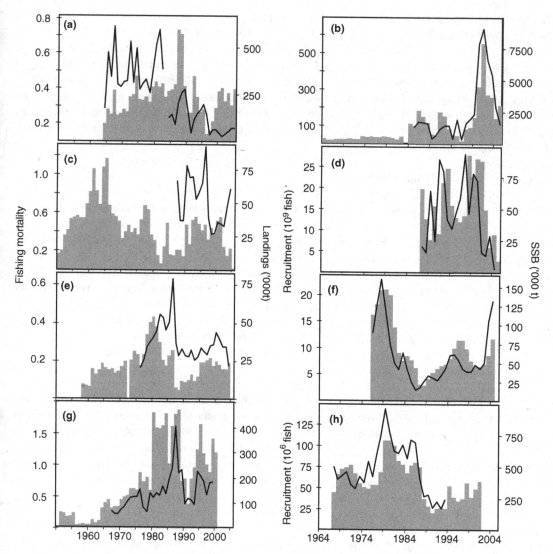

Fig. 9.2b. Time series of fishing mortality (y⁻¹), landings ('000t), recruitment and spawning stock biomass (SSB, '000t) for southern Benguela anchovy (a, b), Bay of Biscay anchovy (c, d), Adriatic Sea anchovy (e, f) and Black Sea anchovy (g, h). Notation as in Fig. 9.2a.

characteristics of the stock and general fishing patterns (De Oliveira, 2003). It is assumed that all anchovy landed between November (when spawning is modeled to occur) and March are 1 year olds, and those landed during the remainder of the year are recruits. Corresponding mean masses-at-age are available. Two-year-old and older fish hardly appear in the catch and are not taken into account in the assessment.

European anchovy (Engraulis encrasicolus) –
Bay of Biscay
The European anchovy occurs from the Bay of Cadiz (Southern Spain) in the south to the North Sea and the western Baltic Sea in the north (Reid, 1966; Beare *et al.*, 2004). Anchovy in the Bay of Biscay is considered to be

a stock isolated from the small populations either to the north or to the south. Although there is evidence for some heterogeneity inside the Bay, the evident interconnection of the fisheries and homogenous recruitment led ICES to consider the anchovy in the Bay of Biscay as a single unit for assessment and management (ICES, 2006a). Spawning takes place in spring, mainly in the south-east corner, particularly in front of the Adour and Gironde estuaries. Fish mature at age 1, and age-groups 1 and 2 constitute the bulk of the catches and the population. Occasionally, fish of 4 or even 5 years old are recorded, but these are very rare. The anchovy population undergoes seasonal migrations, moving northwards in summer and autumn, after spawning, being matched by the different fisheries in the Bay (Uriarte *et al.*, 1996).

Landings show a high degree of inter-decadal variability, with catches peaking from the mid 1950s to the early 1970s (peak of 83 600 t in 1965, Fig. 9.2b, Table 9.1a). Catches progressively diminished since, followed by a similar decline in the Spanish purse-seine fishery. In the late 1980s French pelagic trawlers entered the fishery and now catch as much as the Spanish purse-seine fleet. During the 1990s catches reached 40 000 t, but recently these have declined to historical minima until the recent fishery collapse and subsequent regime of closure periods (ICES, 2006a; Fig. 9.2b). Since 1989, the stock is annually assessed by ICES, surveyed by Spain and France.

The resource is shared between Spain and France: the Spanish fleet is composed of about 211 purse seines that operate in the SE corner of the bay mainly in spring during the anchovy's spawning period. The main anchovy catches from the French fleet are taken in the central-east part of the bay in the first half of the year and the NE during the second half. The fleet consists of 20–30 pair trawlers and about 30 purse seines. No standardization of effort or any series of CPUE is available for the fishery as they are considered unreliable indicators of abundance for small pelagic fishes (Ulltang, 1980; Csirke, 1988; Pitcher, 1995; Mackinson *et al.*, 1997). Total landings are being reported by France and Spain to ICES. Catches are monitored by national institutes for length and age composition. Catches at age are inferred by otolith sampling and serve as input for the assessment. Since 1989, the stock is annually assessed by ICES, supported by the direct monitoring of the resource and the fishery made by Spain and France.

Mediterranean Sea

The European anchovy is the most important pelagic fish resource in the Mediterranean (Lleonart and Maynou, 2002). Three major genetically distinct (Magoulas *et al.*, 2006) stocks exist, with reduced gene exchanges: the NW Mediterranean (Catalan Sea and Gulf of Lions) stock, the Adriatic Sea stock and the northern Aegean Sea stock (Somarakis *et al.*, 2004). Other areas in the Mediterranean are inhabited by smaller and highly fluctuating stocks.

In the NW Mediterranean, anchovy forms a genetically panmictic stock which is distributed and reproduces over the continental shelf areas associated with the runoff from the Ebre river in the Catalan Sea (CS) and the Rhône river in the Gulf of Lions (GL). In the CS, spawning takes place from late April to September, but the reproductive period is shorter in the GL (May–August) (Palomera, 1992). Fish mature at a length of about 11 cm (Palomera *et al.*, 2003), i.e. at the first year of life (Table 9.1a). Maximum length recorded is 19 cm and maximum age, 4 years (Pertierra and Lleonart, 1996; Torres *et al.*, 2004).

In the NW Mediterranean the stock is mainly caught by the Spanish fleet of purse seiners and bottom trawlers and to a lesser extent by a French fleet of mid-water trawlers operating in the GL. Catches show a decreasing trend in the CS from the early 1990s to present (from about 20 000 to 5000 t), while in the GL, catches increase from the mid 1980s to early 1990s and fluctuate thereafter from 1500 to 10 000 t. Mean biomass values are presently at around 15 000 t in the Catalan Sea and 50 000 t in the GL (Table 9.1a).

In the Adriatic Sea, anchovy is distributed all over the northern and central parts, influenced by river runoff (Russo and Artegiani, 1996). Spawning takes place mostly in the western part of the Adriatic, from March to October with a peak in June–July (Regner, 1996). Fish mature at 8 cm, i.e. at less than one year old (Rampa *et al.*, 2005) and young-of-the-year recruit to the fishery during spring. Most individuals fished belong to age classes 0, 1 and 2 (Cingolani *et al.*, 1996) but specimens up to 6 years old have been recorded (Table 9.1a). Most of the catches are taken by the Italian fleet of pelagic trawlers and purse seiners. Biomass was at its highest levels in the late seventies (more than 300 000 t; Santojanni *et al.*, 2003) and is presently estimated to be over 100 000 t (Cingolani *et al.*, 2005; Table 9.1a). Catches reached a peak of about 60 000 t in 1980, decreased steadily thereafter, and now fluctuate between 20 000 and 30 000 t (Fig. 9.2b).

In the Aegean Sea, anchovy is mainly distributed and spawns over continental shelf areas of the northern part of the basin (Somarakis *et al.*, 2005a), influenced by the Black Sea water advection and discharge from large rivers (Somarakis *et al.*, 2002a; Isari *et al.*, 2006). Spawning takes place during late spring/summer (usually from May to September) when the stock is also mainly fished by the purse-seine fleet (Stergiou *et al.*, 1997). Fish mature at a length of about 10.5 cm (Somarakis *et al.*, 2005b), i.e. at the first year of life. Maximum length recorded is 18 cm and maximum age, 4 years (Nikoloudakis *et al.*, 2000; Table 9.1a). The combined Greek–Turkish catch showed an increasing trend from the early 1970s to mid 1980s and fluctuates thereafter between about 15 000 t and 25 000 t. Biomass estimates indicate a stock of about 40 000–50 000 t.

The NW Mediterranean and the Aegean Sea anchovy stocks are not currently being assessed. However, catch records have been collected by the National and Regional Statistical Services of Spain since the early 1940s, France since the early 1970s and Greece since the early 1960s (Stergiou *et al.*, 1997). Additional data on catch-per-day of the involved purse-seine fleet are available in the central database of the Hellenic Center for Marine Research since 1995 (Kapadagakis *et al.*, 2001). Since 2002 the framework of the National Programs for Fisheries Data collection (co-funded by the EC), collects length and age information on a regular basis in both the NW Mediterranean and the Aegean Sea, in order to derive catch-at-age estimates. The first attempt to assess the stock of anchovy in the Catalan Sea has been presented to the General Fisheries Council for the Mediterranean (GFCM; Torres *et al.*, 2004).

In contrast to the NW Mediterranean and the Aegean, data on catch, fishing effort and age composition have been routinely collected in the northern and central Adriatic since 1975 (Fig. 9.2b). The activity involves the estimation of length frequency distribution of catches, the ageing of a subsample to obtain catch-at-age numbers and the standardisation of the fishing effort for certain ports using Generalized Linear Models (Santojanni *et al.*, 2005).

Black Sea anchovy

Black Sea anchovy is distributed throughout the Black sea but is subject to seasonal migrations. In October–November it moves to the wintering grounds along the Anatolian and Caucasian coasts in the southern Black Sea. In these areas it forms dense wintering concentrations in November–March, which are subject to intensive commercial fishery. The rest of the year it occupies its usual spawning and feeding habitats across the sea with some preference for the shelf areas and the northwestern part of the sea. This area is characterized by the largest shelf area and high productivity due to abundant river runoff (Faschuk *et al.*, 1995; Daskalov, 1999).

Two subpopulations of anchovy exist in the Black Sea: the Black Sea and the Azov Sea stocks (Ivanov and Beverton, 1985). The latter reproduces and feeds in the Azov Sea and hibernates along the northern Caucasian and Crimean coast of the Black Sea. The Black Sea stock is of bigger ecological and commercial importance and the information below concerns only this stock.

Anchovy first spawns at about age 1 but precocious maturation and spawning (at age 2–3 months) has been reported in years of stock collapse (Mikhailov and Prodanov, 2002; Table 9.1a). It spawns in the surface layer of warm and stratified areas in summer, the main feeding season (Arkhipov, 1993; Fashchuk *et al.*, 1995). A large convergence zone is formed on the northwestern and the western shelf (the main anchovy spawning area) due to the river Danube inflow, which favors offspring retention.

Anchovy is the object of both an artisanal (with coastal trapnets and beach seines) and commercial purse-seine fishery on the wintering grounds. Decadal fluctuations of abundance are observed and likely to be related to changes in climate (Daskalov, 1999, 2003; Fig. 9.2b). The increasing trend in biomass started in the 1970s and 1980s promoted the expansion of powerful purse-seine fishing fleet and a steady increase in fishing effort (Gucu, 1997). Maxima of catch and fishing mortality were recorded in the late 1980s parallel to the decrease in exploited biomass following

recruitment failures in the previous years. Sharp reductions in biomass and catch in the early 1990s were described as stock collapses. In the recent decade the stock partially recovered and catches reached levels of 300 000–400 000 t (Fig. 9.2b, Table 9.1a).

Sardine stocks

Japanese sardine (Sardinops melanostictus) – *Pacific stock*

The Japanese sardine fishery targets two stocks, the Pacific stock and the Tsushima Current stock, which are distinguished by distribution and migration patterns (Nishida *et al.*, 2006). The Pacific stock, the focus of this review, is distributed along the Pacific coast of Japan; its eastern boundary extends to 160° W during periods of high abundance but in periods of low abundance adults are confined to west of ca. 155° E longitude (Yatsu *et al.*, 2003; Yatsu and Kaeriyama, 2005). The western boundary coincides with the spawning ground which at high stock abundance extends from the southern coast of Kyushu Island to the offshore area beyond the Kuroshio axis and to northern Honshu Island (Zenitani and Yamada, 2000).

Spawning takes place from October to May and juveniles are transported in spring by the Kuroshio and Kuroshio Extension Currents to beyond 170° E longitude (Watanabe and Nishida, 2002). The feeding grounds are mainly located in the Oyashio and Kuroshio/Oyashio Transition Zone (KOTZ) in summer and autumn. Major fishing grounds of the purse-seine fishery are located along the Pacific coast of northern Japan. Longevity is about 7 years (Nishida *et al.*, 2006). Sexual maturation is in general attained at age 2 (Table 9.1b).

Japanese catch statistics indicate two historic peaks in the 1930s and 1980s (Noto and Yasuda, 2003; Fig. 9.3a). Recruitment increased in the early 1970s and peaked in the mid 1980s, while SSB[7] lagged according to the age of maturity. Consecutive recruitment failures were observed during 1988–1991 when SSB achieved its historical peak (Watanabe *et al.*, 1995). The current catch level (48 000 t in 2004) approaches the historic minimum (7000 t in 1965). Purse seine is the major gear used for catching Japanese pelagic fish, including sardine, chub mackerel, anchovy, and jack mackerel, which are targeted alternately according to availability and market demand (Nishida *et al.*, 2006; Yatsu *et al.*, 2005). During the 1970s and early 1980s, the purse-seine fleets expanded their fishing effort to compensate for the declined global supply of fish meal (Yatsu, 2005). Because of the lower commercial value of anchovy and the cost of adopting anchovy-mesh nets, the purse-seine fishery has been adapted to targeting age-0 and age-1 chub mackerel and sardine since the early 1990s, when chub mackerel and anchovy began to increase (Yatsu *et al.*, 2003). In recent years continued intensive Japanese sardine fishing mortality has prevented recovery of sardine despite good recruitment per spawner (RPS) biomass (Yatsu and Kaeriyama, 2005).

California sardine (Sardinops sagax caerulea)
California sardine comprise three groups: a "northern" (northern Baja California to Alaska), a "southern" (outer Baja California peninsula to southern California), and a "Gulf of California" subpopulation (for review see Smith, 2005). Vrooman (1964) proposed the three subpopulations based on serological evidence, supported by temperature at capture and otolith morphometrics (Felix-Uraga *et al.*, 2004, 2005). Geographic ranges of the three subpopulations partially overlap seasonally, but the degrees of mixing and relative contributions to productivity remain undetermined. The northern stock likely grows larger and lives longer than its counterparts. When the population is large, California sardine are abundant from the Gulf of California to southeastern Alaska. When abundance is low, sardine do not occur north of 34.5° N (southern California). Spawning typically occurs from 50 to 200 nm offshore (Lo *et al.*, 2005), but sardine have been captured 300 nm offshore (Macewicz and Abramenkoff, 1993). Sardine migrate north in the late spring and south in the fall. Movements are more extensive for larger sardine and during El Niño conditions. Spawning occurs year-round in the southern stock (17–21°C), peaks April through August in the northern stock (13–15°C), and January through April in the Gulf of California (22–25°C, Table 9.1b). At low biomass they mature at age 1, at high biomass only some 2-year-olds are mature. They are oviparous multiple-batch spawners with indeterminate and age/size-dependent fecundity (Macewicz *et al.*, 1996). Two-year-olds spawn six times per year and old sardine may spawn up to 40 times in a year (Butler *et al.*, 1993). Maximum age is 14 years, but most commercially caught sardine are less than 5 years old (Table 9.1b). Size-at-age and composition vary regionally, with both increasing in northern and offshore areas (Phillips, 1948; Hill *et al.*, 2006). Recruits first appear in the California fishery between October and April at age 6–12 mo. Adult natural mortality was estimated at $M = 0.4$ yr^{-1} (Murphy, 1966), but M varies by subpopulation.

Paleo-oceanography indicates dramatic changes in California sardine over the past two millennia (Baumgartner *et al.*, 1992, revised in MacFarlane *et al.*, 2002), with biomass peaking from 5 to 6 Mt. At the peak of the historical fishery, biomass was 3.6 Mt (1934) (Murphy, 1966) and the fishery captured 718 000 t (1936). Fisheries progressively collapsed from British Columbia to northern Baja California, and the biomass was only 10 000 t in 1965. Minor fisheries remained in southern Baja California (Radovich, 1982; Lluch-Belda *et al.*, 1989) and a substantial fishery eventually developed in the Gulf of California

(Cisneros-Mata *et al.*, 1995). The northern subpopulation began recovering in the early 1980s, and the population grew from 5000 t in 1983 to 1.49 Mt by 1996. The latest assessment indicates current biomass to be 1.06 Mt (Hill *et al.*, 2006). The recovery has fueled fishery redevelopment from Baja California to British Columbia, with a current combined harvest of up to 142 000 t year^{-1} (PFMC, 2006).

Biological data for assessment purposes have been collected since 1919. Fishery sampling resumed for most areas with the onset of the population recovery, and biological data are available from the California and Baja California ports beginning in the early 1980s and from Pacific Northwest ports since the year 2000. The current stock assessment for US management (Hill *et al.*, 2006) includes age composition and weight-at-age data from three fisheries (northern Baja California, California, and Pacific Northwest) for the period 1982–83 to present, aggregated by season (July–June).

Humboldt sardine (Sardinops sagax) – *South Peru/ North Chile stock*
Three stock units of sardine exist along the Humboldt Current: from north-central Peru to about 15° S; from southern Peru to northern Chile (15–24° S, the focus of this review) and a third stock off central-south Chile (30–42°S). The latter is only obvious in the expanded phase of sardine's distribution. The Galapagos Islands may have a separate stock (Serra, 1983; Parrish *et al.*, 1989). When sardine is abundant, its distribution ranges from Ecuador (0°S) and the Galapagos Islands down to south-central Chile (42°S) (Serra, 1983; Serra and Tsukayama, 1988; Parrish *et al.*, 1989). When scarce its distribution is from 5–27° S. Its offshore distribution exceeds 200 nm but juveniles are found close to the coast (Serra and Tsukayama, 1988). They may reach over 40 cm in total length and sizes over 35 cm were frequent in the Chilean fishery. The maximum longevity of the sardine is 11 years (Table 9.1b). The natural mortality rate has been estimated to be $M = 0.3$ yr^{-1} (Serra and Tsukayama, 1988).

Off Chile the sardines have a long spawning season with two peaks, in winter and summer (Serra, 1983; Serra and Tsukayama, 1988). Sardines are multi-batch spawners with an indeterminate annual fecundity. The size at first spawning is generally 24 cm, equivalent to age 3. All fish > 6 y.o. are mature. Catches are sustained by fish from 5 to 8 years old. In northern Chile, sardines start to be recruited to the fishery at age 1, but are not fully recruited before age 6.

The sardine biomass started to increase in the early 1970s and has declined since 1981 due to intensive exploitation and poor recruitment (Fig. 9.3a) (GTE, 2003; Serra *et al.*, 2004). Meanwhile, catches continued to grow reaching 3 Mt in 1985, decreasing to present levels of about 100 000 t. The catch of this stock shows

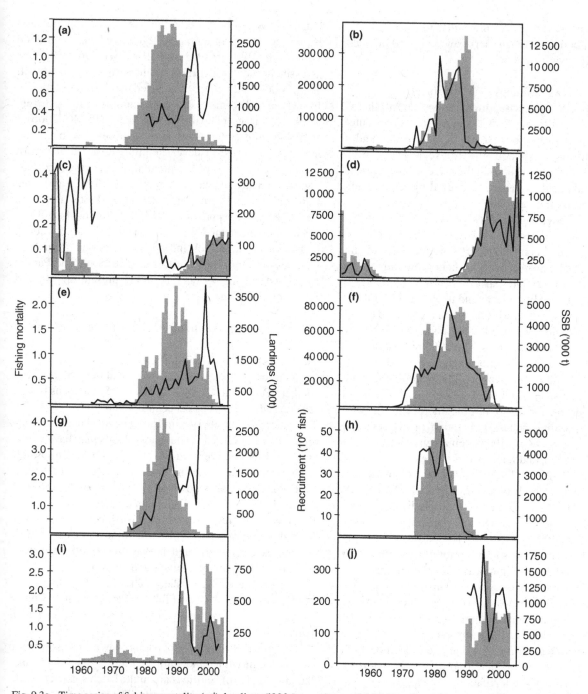

Fig. 9.3a. Time series of fishing mortality (y⁻¹), landings ('000 t), recruitment (106 fish) and spawning stock biomass (SSB, '000 t) for Japanese sardine (a, b), California sardine (c, d), Humboldt sardine (North/Central Peru stock, e, f; South Peru/North Chile stock, g, h) and Chilean common sardine (i, j). Notation as in Fig. 9.2a.

coherent and synchronous long-term changes alternating with anchovy catches (Serra, 1991; Lluch-Belda *et al.*, 1992; Schwartzlose *et al.*, 1999; Figure 9.2a).

The Chilean pelagic fish fishery is monitored by a program of observers and samplers on fixed landing sites. Biological sampling on the landings for species composition, size, weight, sex, maturity and otoliths for age studies

has also been taken since 1964. For the assessment of the shared sardine stock between Peru and Chile only the time series from 1974 to 1996 has been considered, as subsequent abundance indices are not reliable. Catch-at-age matrices from the Chilean and the Peru–Chile fisheries are computed monthly and grouped by year, and then added to obtain an overall annual matrix for the combined stock.

Weight-at-age was obtained converting the mean length-at-age to weight by an overall length–weight relationship (Serra *et al.*, 2004; GTE, 2003).

Chilean common sardine (Strangomera bentincki)

The common sardine is a Chilean endemic species, distributed from northern Coquimbo (29° S) to Puerto Montt (42° S; Arrizaga, 1981). Despite morphometric and meristic differences between individuals from different landing ports (Cortes *et al.*, 1996) it constitutes a single genetic population (Galleguillos *et al.*, 1997).

In the central-south area off Chile, two areas of high abundance of common sardine can be identified between 34°30–37°10 S and between 38–40° S (Castillo *et al.*, 2005; Cubillos *et al.*, 2005). Eggs, recruits and adults concentrate between the coast and 10–20 nm offshore. The main bays (Golfo de Arauco, Bahía de Concepción, and Bahía Coliumo) are important spawning zones probably through a combination of retention and concentration processes.

The spawning season extends from July to September, with a peak between August and September (Cubillos *et al.*, 1999, 2001). Relative fecundity has been estimated by Cubillos *et al.* (2005), ranging between 433 and 535 oocytes per female body weight. Three to four months after spawning, juveniles (5–6 cm total length) recruit to the population (Cubillos *et al.*, 2001, 2002). The fishery is heavy dependent on the annual pulse of recruitment, which concentrates in bays and gulfs from January to March (Cubillos *et al.*, 1998). The common sardine is a short-lived species and attains a maximum age of 4 years. Size at first maturity is 10 cm total length (1 y.o.; Cubillos *et al.*, 1999; Table 9.1b).

Biomass has been estimated by stock assessment models for the period 1990–2003 (Canales *et al.*, 2004). Total biomass was highest (3 Mt) at the beginning of the 1995/96 fishing season (Fig. 9.3a).

Total catch and catch per unit effort data from the industrial purse-seine fleet are used in stock assessment from 1991. In addition, catch-at-age and weight-at-age matrices are used in stock assessment models. For the common sardine stock, all age groups are defined to be born on July 1, and the fishery information is pooled by fishing seasons starting the July 1 and ending the June 30 of the following year.

Brazilian sardine (Sardinella brasiliensis)

The Brazilian sardine inhabits the Brazilian Southeast Bight, between 22 and 29° S. Morphometry, seasonality, and biochemical studies (Saccardo and Rossi-Wongtschowski, 1991) suggest that the species does not constitute a single stock unit. However, it is considered as a single stock for management purposes, as there is not enough information to characterize the different stocks. This species spawns in batches, from October to March, with maximum intensity in December–January (Saccardo and Rossi-Wongtschowski, 1991). Spawning occurs at night in the upper layers of the water mass on the continental shelf (Matsuura, 1983, 1996, 1998; Saccardo and Rossi-Wongtschowski, 1991). Individuals up to 4 y.o. have been recorded and maximum size is 27 cm (Cergole and Valentini, 1994; Table 9.1b). Females mature at approximately age 1, and the whole population is matured by age 2 (Vazzoler, 1962; Isaac-Nahum *et al.*, 1983; Isaac-Nahum *et al.*, 1988; Wenzel *et al.*, 1988; Cergole and Valentini, 1994). Maximum recruitment to the adult stock occurs around July (Cergole, 1993, 1995).

Abundance estimates through indirect methods suggested periods of high and low abundance (Fig. 9.3b): the first period (1977 to 1986) showed a total mean biomass of 668 000 t (SSB 255 000). From 1986 the stock showed two significant declines in 1990 and 2000 (Cergole, 1993, 1995; Cergole *et al.*, 2002). First records of sardine catches were collected in 1964, reaching 228 000 t in 1973 before initiating a downward trend until the 1990s. As far back as 1988, it was recognized that the stock was collapsing. Severe management recommendations (Rossi-Wongtschowski *et al.*, 1995; SUDEPE/PDP, 1989) led to some signs of recovery, with catches of 118 000 t in 1997. Since then, a new decline took the total catch to 17 000 t in 2000 (IBAMA, 2005).

The Brazilian sardine purse-seine fishery suffered a 50% reduction by the year 2000, but fishing effort has not decreased because the larger boats with larger fishing power remained in the fishery. From the mid 1990s the fleet has diversified into a multispecies fishery targeting sardines, mackerels and other pelagics.

Benguela sardine (Sardinops sagax)

Benguela sardine are distributed from southern Angola (approximately 15° S) to South Africa's east coast (around 30° E), and comprise separate northern and southern stocks (Beckley and van der Lingen, 1999). Benguela sardine have historically supported large catches off Namibia and South Africa.

Sardine have a protracted spawning period in both the northern and southern subsystems (although with very low egg concentrations in the austral winter) peaking in September/November and February/March (Beckley and van der Lingen, 1999; van der Lingen and Huggett, 2003). The location of the primary spawning grounds has shown substantial spatial shifts through time in both subsystems (van der Lingen *et al.*, 2006a). In the northern Benguela, spawning has moved north with the decline in population size. Off South Africa, intense sardine spawning has been confined to the south coast in recent years (van der Lingen *et al.*, 2005). Sardine recruitment in the northern Benguela occurs primarily inshore of major spawning sites, whereas the South African west coast is the principal nursery ground

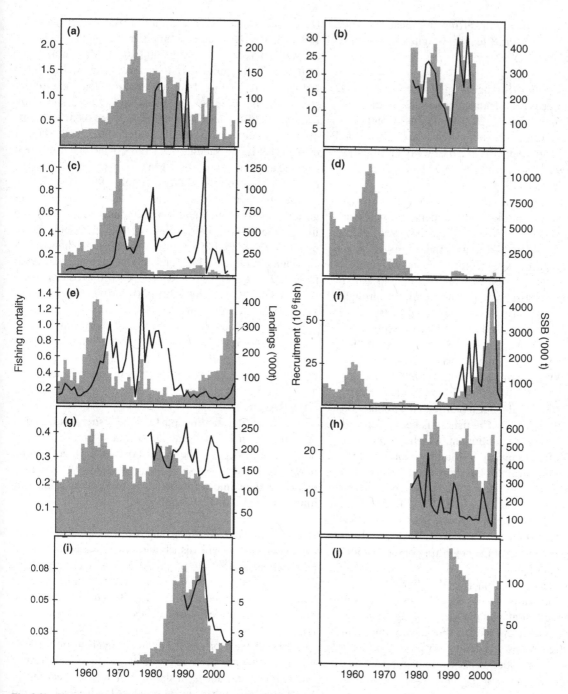

Fig. 9.3b. Time series of fishing mortality (y⁻¹), landings ('000t), recruitment (10⁶ fish) and spawning stock biomass (SSB, '000t) for Brazillian sardine (a, b), northern Benguela sardine (c, d), southern Benguela sardine (e, f), European sardine (g, h) and SW Australian sardine (i, j). Notation as in Fig. 9.2a.

of southern Benguela sardine (Barange *et al.*, 1999). The size at sexual maturity of sardine has varied in both the northern (Thomas, 1986) and southern (Armstrong *et al.*, 1989; Akkers *et al.*, 1996; Fairweather *et al.*, 2006) subsystems, possibly in a density dependent response to stock size (van der Lingen *et al.*, 2006b). Sardine is a relatively short-

lived species, with few adults older than 8 years observed and a maximum age of 10 (Le Clus *et al.*, 1988).

Commercial fishing operations targeting South African sardine commenced in 1943 and landings rose dramatically during the late 1950s and early 1960s, peaking at 410000t in 1962 and declining rapidly thereafter (De Oliveira,

2003; Fig. 9.3b). The collapse of the southern Benguela sardine stock was ascribed to over-fishing, expansion of fishing grounds, and variable recruitment (Beckley and van der Lingen, 1999), and resulted in increased fishing effort in the northern Benguela, where annual sardine catches rose rapidly to a maximum of 1.4 Mt in 1968 (Boyer and Hampton, 2001; Fig. 9.3b). Thereafter, sardine landings declined and have remained below 100 000 t since the early 1980s. As in the southern Benguela, the collapse of the northern Benguela sardine population was primarily attributed to overfishing, although poor sardine recruitment resulting from adverse environmental conditions exacerbated the decline, both during the 1960s (Boyer and Hampton, 2001) and more recently (Boyer *et al.*, 2001).

Sardine spawner biomass in the southern Benguela has been increasing since the start of the current acoustic monitoring program in 1984, reaching a peak in 2002. A unique phenomenon for this stock is the KwaZulu-Natal "sardine run" (Armstrong *et al.*, 1991), in which large shoals of sardine move up the South African east coast during winter (June) each year, reaching Durban and occasionally further north. Catches (typically < 1000 t) are small in comparison to those on the west and south coasts and are not included in the catch statistics for assessment. It has been proposed that the "sardine run" occurs in response to an expansion of the environment suitable for sardine on the east coast during the cooler water conditions of winter (Armstrong *et al.*, 1991). Off Namibia sardine spawner biomass has remained at a very low level since the start of the current acoustic monitoring program in 1990.

Monthly age–length keys have been compiled for the southern Benguela sardine from 1980 to 1999, giving monthly catch-at-age data (De Oliveira, 2003), and corresponding mean masses-at-age are available. Age readings since 1999 remain scarce at the moment and thus average age–length keys have been used to estimate catch-at-age from 2000 for the most recent assessment. Although a less than ideal scenario, this has enabled a temporary update of the assessment in the absence of such ageing information (Cunningham and Butterworth, 2004b). The catch of juvenile sardine taken between November (when spawning is modeled to occur) and the beginning of the recruitment survey is also utilized in the assessment. The bycatch of juvenile sardine caught in the anchovy directed fishery has also been recorded since 1987, and is used in constructing the management procedure. Length frequencies are generated for sardine caught in the northern Benguela and, as fish are not aged, division into age classes is done according to length, with fish ≤ 16.5 cm TL[8] assigned to the 0-group, 16.6–22.4 cm TL the 1-group, and ≥22.5 cm TL the 2+ fish (Boyer *et al.*, 2001).

European sardine (Sardina pilchardus) – Atlanto-Iberian stock

The European sardine is widely distributed in shelf waters along the northeast Atlantic (NEA) coast (from 15–20° N to 52–58° N) and the Mediterranean Sea, with residual populations off the Azores, Madeira and the Canary Islands. A recent study integrating morphometrics with genetics (allozymes and microsatellite DNA) revealed five genetic stocks: Azores, Madeira, Mediterranean Sea, and two Atlantic stocks in the area from the North Sea to Mauritania (with an internal boundary possibly at the Bay of Agadir) (Y. Stratoudakis, pers. comm.). For assessment purposes, sardine in European Atlantic waters has always been considered to belong to a single stock, but its geographic limits have changed over time. Sardine assessment under ICES began in 1978 (ICES, 1978), when the stock covered the area from the north of France to southern Iberia. The northern border was revised in 1980, based on a mixture of biological evidence and administrative needs (ICES, 1980). This new delimitation gave rise to what is currently known as the Atlanto-Iberian stock, from the France/Spain border in the inner Bay of Biscay to the Strait of Gibraltar.

Sardine is an indeterminate, batch spawner with high relative fecundity. Spawning activity extends for many months, with regionally varying local peaks (Coombs *et al.*, 2006). Sexual maturation is attained during the first two years of life (Silva *et al.*, 2006). Although sardines up to 14 years of age have occasionally been reported, most fish do not exceed 6–8 years (Table 9.1b). Spawning probably takes place near the bottom towards dusk. Adult fish are widely distributed within the continental shelf, forming characteristically dense schools with marked diel variations in vertical position, shape and integrity. Recruitment demonstrates high spatial fidelity, creating a complex mosaic of adjacent subpopulations of different ages. In general, European sardine shows less pronounced migration patterns than other pelagic species. Sardine maximum size, longevity, and growth are highest at the northern extreme (northern France and English Channel) and lowest in the southern Iberian Peninsula. Upper temperature tolerance to spawning seems also to increase with decreasing latitude. Within the NEA, the largest and most productive stock is situated off Morocco (stock biomass 1–5 Mt; catches around 600 000 t in recent years; FAO, 2003). In the Atlantic waters of the Iberian Peninsula (stock biomass around 500 000 t; ICES, 2005a), catches peaked at 250 000 t in the mid 1960s, but have declined during the past 15 years, and are currently at 100 000 t (ICES, 2005a; Fig. 9.3b). Sardine catches north of the Iberian Peninsula are comparatively small. A dedicated fishery with annual catches of some 10–15,000 t exists in the French waters of the Bay of Biscay, where fisheries interest seems to be increasing. A small seasonal fishery is found in the western

English Channel (3000 – 5000 t), while in recent years sardine distribution is reported to have extended further north, well into the North Sea.

Iberian catches are almost exclusively destined for human consumption, through a complex marketing circuit involving intermediate sales that progressively increase the cost. The average first sale price is 0.61 euros kg^{-1} (2004 in Portugal), with a marked seasonality (price peaking in summer months when sardine are fatter and fresh consumption is highest).

Systematic biological monitoring of sardine catches started in the late 1940s by Spain and Portugal. The first ICES sardine assessment meeting was held in 1978, reviewing existing biological information and sampling plans to provide input data to future assessments (ICES, 1978). Since then, regular sampling of fish catches (length distribution, macroscopic biological information and age-length keys) has been performed, progressively providing a more homogeneous coverage of the Iberian Peninsula and increasing the sampling frequency and intensity. Length distribution samples are obtained fortnightly, biological samples monthly and age readings trimestrally or semestrally. Reviews on sampling and data availability can be found in Pestana (1989), Carrera and Porteiro (2003), Jardim *et al.* (2004), Silva *et al.* (2006).

Southwestern Australia sardine (Sardinops sagax neopilchardus)

Sardine are continuously distributed around the southern Australian coast from approximately 26° S on both the eastern and western coasts. Highest abundance, and the more significant fisheries, are along the southern coast, with a center of distribution in the state of South Australia, where the fishery has reached annual total catches of 30 000–40 000 t. The fishery along the southwestern region, in the state of Western Australia (WA), has reached an annual total catch of around 12 000 tonnes.

Sardines in WA consist of two breeding stocks: one along the west coast and another along the south coast. The remainder of this summary focuses on the south coast (WA) breeding stock, which has had the longest period of research and monitoring for sardine in Australia. The south coast stock is partitioned into three management units (zones) because of limited alongshore movement of the mature biomass (Edmonds and Fletcher, 1997; Gaughan *et al.*, 2002); each management unit has its own TAC[9]. Uncertainty regarding spatial dynamics of recruits has been managed via a "recruit pool" hypothesis, which suggests that each unit of the stock is important for recruitment to the broader breeding stock.

Sardines in Western Australia live to 8–9 years (Fletcher and Blight, 1996) both maturing and entering the fishery at approximately 2 years but full recruitment into the fishery

may not occur until 4 or 5 years of age. South coast sardines spawn for most of the year with major peak in winter and a smaller peak in summer; spawning occurs across much of the continental shelf seaward of ~30 m depth (Fletcher and Sumner, 1999; Gaughan *et al.*, 2004). These general patterns vary regionally and interannually, depending on size of the SSB and oceanographic conditions. Larvae from the main (winter) spawning are transported east, potentially up to 1000 km (Gaughan *et al.*, 2001), requiring a return migration by juveniles.

Purse-seine fisheries on the south coast expanded in the mid to late 1980s at which time the maximum annual catch (8000 t) was recorded. The fleet of relatively small vessels typically fish within 10–15 km of port. Biological assessment and monitoring began in 1988. SSB declined in the 1990s due to poor recruitment, fishing pressure in some regions and a mono-specific mass mortality, but has grown since 1999. However, uncertainty in the estimates of SSB preclude setting harvest rates higher than 10% of SSB.

Herring and sprats

Atlantic herring (Clupea harengus) *– North Sea autumn spawning stock*

North Sea herring is made up of a number of spawning components, including Shetland, Buchan, Banks, and Downs (Heinke, 1898; Redeke and van Breemen, 1907; Cushing, 1955, 1992; Zijlstra, 1958). The current stock definition only covers the autumn and winter spawning herring and spring spawners are not included in the stock assessment or the management agreement (Cushing, 1967; ICES, 2006d). Herring spawn benthic eggs and for this stock, spawning occurs in the western North Sea along the coast of the UK (Boeke, 1906; Cushing and Burd, 1957; Burd and Howlett, 1974). Atlantic herring are spatial repeat spawners (McQuinn, 1997) and this behavior is either caused by natal returns to the "home" spawning bed or adopted behavior (Harden Jones, 1968; Wheeler and Winters, 1984; McQuinn, 1997, and references cited therein).

The majority of the larvae drift in an easterly direction towards the German Bight and Skagerrak (Munk and Christensen, 1990; Bartsch, 1993). It is during this life stage that year class strength is determined (Nash and Dickey-Collas, 2005). The juveniles stay in the east until maturity, upon which they then join the adults. Recruits from one spawning will not necessarily mature in synchrony (McQuinn, 1997; Brophy and Danilowicz, 2003). The adults feed in the central and northern North Sea. After feeding, the herring migrate to the spawning grounds. As the majority of North Sea herring are autumn and winter spawners, they exhibit a different energy strategy than Norwegian spring spawning herring (Isles, 1984; Winters and Wheeler, 1996; Slotte, 1999). Genetically, the stock shows no major differentiation between the spawning components, but

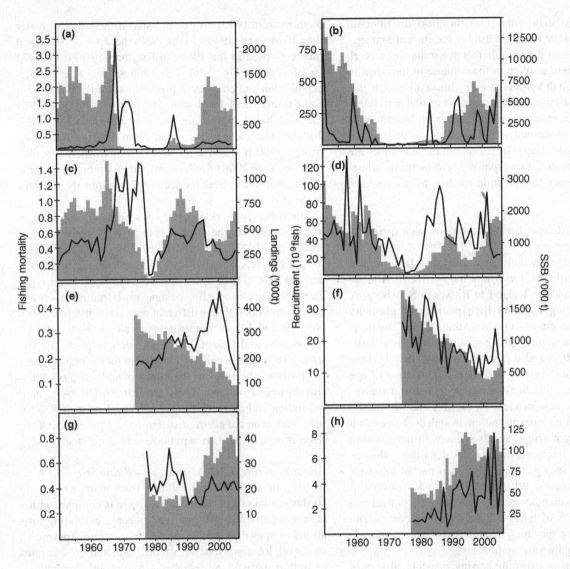

Fig. 9.4a. Time series of fishing mortality (y⁻¹), landings ('000t), recruitment (10^9 fish) and spawning stock biomass (SSB, '000t) for Arcto-Norwegian spring spawning herring (a, b), North Sea autumn spawning herring (c, d), Central Baltic herring (e, f) and Gulf of Riga herring (g, h). Notation as in Fig. 9.2a.

there is drift with distance between spawning grounds in the genetic make-up of the herring (Mariani *et al.*, 2005). There are also strong density dependent effects in the population characteristics (Cushing and Bridger, 1966; Hubold, 1978; Winters and Wheeler, 1996; ICES, 2006d).

In the 1960s, the spawning stock biomass of North Sea herring was over 2 Mt (ICES, 2006d, Fig. 9.4a). However, the stock collapsed in the 1970s through overfishing (Burd, 1985; Cushing, 1992; Nichols, 2001; Simmonds, 2007). The stock slowly recovered to above 1 Mt by the 1990s and a management agreement was brought in to reduce fishing effort. The stock then grew to approximately 1.8 Mt in 2004 (Table 9.1c). This high spawning biomass, however, has not prevented 4 years of poor recruitment (2003 to 2006, at

age 0). In 2006, the stock began to decrease in size again in response to this.

Regulated fisheries exist for both the adult and juvenile herring. Juvenile herring are caught as a bycatch in the North Sea industrial fisheries, which usually target sandeel, sprat, and Norway pout. The fisheries on the adults takes place both whilst the adults are feeding in the summer, and during the autumn and winter spawning aggregations. The only comprehensive fishery information collected at present for the management of North Sea herring are the numbers and weights of herring, by age in the catch. These are estimated by each catching nation and combined under the auspices of ICES. Each nation uses different methods, some correct for misreporting and others include estimates of discarded fish. No information on the effort, distribution, or efficiencies of the fleets are used in the current management of the stock, as it is thought that over-reliance on catch per unit effort data resulted in the collapse of the stock in the 1970s. Although VMS is now on almost all vessels that fish North Sea herring, these data are not available to scientists. The species make-up of the catches of the industrial fisheries is also monitored and used to determine the total catch of juvenile herring. The herring in these samples are also monitored to determine spawner type (autumn, winter or spring) to ensure that juvenile catches are allocated to the correct stock.

Atlantic herring (Clupea harengus) – Norwegian spring spawning stock

Norwegian spring spawning herring exhibits large migrations and is distributed throughout the north east Atlantic (from the Shetlands to Iceland to Russia to Spitsbergen). Like other herring it shows high phenotypic plasticity (Jennings and Beverton, 1991; McQuinn, 1997). The maximum size of Norwegian spring spawning herring is large (>40 cm and 700 g) and is considered longer lived (Holst et al., 2004). These herring mature at 4 to 6 years of age (Table 9.1c). The migrations of Norwegian spring spawning herring change, dependent on the size and age profile of the stock and changes in the environment. Norwegian spring spawning herring can also mix with neighboring stocks. This has impacted on the stock definition; whereas Icelandic spring spawners were classed as part of the stock in the past (Johansen, 1919), since the stock recovery in the 1980s the Icelandic spring spawners are considered a separate stock (which has still yet to fully recover). The Norwegian spring spawning herring is also called Atlanto-Scandian herring by some authorities.

Norwegian spring spawning herring spawn benthic eggs on gravel. The spawning grounds are along the Norwegian coast and, unlike North Sea herring, as the adults age they spawn on grounds further and further south along the Norwegian coast (Slotte, 2000; Slotte et al., 2000), i.e. natal homing is thought to be weak. Spawning appears heavily influenced by environmental conditions, with large-scale atresia of the ovaries in some years (Oskarsson et al., 2002). Spring spawning herring exhibit a different strategy in terms of feeding and energy utilization than autumn spawning herring, in that their condition is based on the feeding season of the previous year (Isles, 1984; Winters and Wheeler, 1996; Slotte, 1999). The summer feeding and overwintering areas appear to vary with time; the stock overwintered offshore during the 1950s and 1960s and then moved inshore in the 1970s and 1980s. In the last decade some fish have begun to overwinter offshore again (see Holst et al., 2004).

This stock has had a well-documented collapse and recovery (Toresen and Østvedt, 2000) with an increase in biomass from 1900 onwards to a spawning biomass greater than 15 Mt in the 1940s. In the 1950s the stock declined and then collapsed in the 1960s to less than 20000t by 1972 (ICES, 2005b; Fig. 9.4a). The collapse was due to over-exploitation and changes in productivity of the Norwegian Sea (Toresen and Jakobsson, 2002). From 1980 onwards the stock has slowly rebuilt, and is now above 5 Mt (ICES, 2005b). The catches increased from 1900 to the 1960s (from 200000t to above 1.5 Mt) and then by the 1970s they were less than 7000t. Under a program for rebuilding of the stock catches have been reduced to approximately 800000t in the mid 2000s.

The fishery now occurs in two main locations: in international waters to the northwest of Norway in the summer and along the Norwegian coast in the winter. The fishery includes vessels from Norway, Russia, Iceland, the Faroes, and the EU[10]. These fleets operate a range of vessel types and fishing gears including pelagic trawls, paired trawls and purse seine, operating in different seasons in different areas. Norway and Russia dominate the catches with approximately 60% and 15% of the annual catch, respectively. The fishery information collected at present for the management of North Sea herring are the numbers and weights of herring, by age in the catch. These are estimated by each catching nation and combined under the auspices of ICES. No information on the effort, distribution or efficiencies of the fleets is used in the current management of the stock.

Atlantic herring (Clupea harengus) – Baltic Sea

Herring inhabits the entire Baltic from marine to nearly freshwater habitats. The stock structure is complex with a number of different spawning components, exhibiting variations in spawning period (spring vs. autumn spawners), spawning locations (coastal vs. offshore) and growth rates as well as meristic, morphometric and otolith characteristics (e.g. Ojaveer, 1981; Parmanne et al., 1994). Altogether ICES (2001b) has identified 11 herring stocks in the Baltic. Spring spawning herring dominate in abundance since the early 1970s, with the reasons for the decline in autumn spawner abundance still being unclear (Parmanne et al., 1994). Various spring spawning components mix during their feeding period in summer and autumn in open sea areas of the Baltic (Aro, 1989), which makes stock separation at this time of the year difficult.

Baltic herring is at present assessed in five different stock units (ICES, 2006c), the Western Baltic (ICES Subdivision 22–24 with prolonged feeding migrations into the Kattegat, Skagerrak and the North Sea), the Central Baltic (Subdivision 25–29 inclusive of the Gulf of Finland, i.e. Subdivision 32), the Gulf of Riga (eastern part of Subdivision 28), the Bothnian Sea (Subdivision 30) and the Bothnian Bay (Subdivision 31). In focus of the present review are the Central Baltic and the Gulf of Riga stocks.

Baltic herring is fished by a variety of fishing fleets, for human consumption in trawl, trap-net and gill-net fisheries and in a mixed trawl fishery with sprat for industrial purposes (see Parmanne *et al.*, 1989; ICES, 2006c). Central Baltic herring landings and spawning stock biomass decreased from 1980s to 2000 by 60–70% (Fig. 9.4a). Stock abundance decreased as well until the mid 1990s, but less drastically. The decline in landings and biomass was partly driven by a decline in weight-at-age from early 1980s to mid 1990s (ICES, 2006c). Herring in the Gulf of Riga showed an opposite development, with tripled landings and more than doubled spawning stock biomass in the last 20 years (Fig. 9.4a). SSB and stock abundance increased especially until the mid 1990s, while the continued increase in landings also in later years was accomplished by increasing fishing mortalities (ICES, 2006c).

The drastic decline in weight-at-age of the Central Baltic herring has been explained by (1) a reduction in size selective feeding by cod, preying predominantly on smallest individuals within a herring age group (Beyer and Lassen, 1994), (2) different developmental success in sub-stocks exhibiting different growth rates (Sparholt, 1994a), and (3) limitation in food supply (e.g. Cardinale and Arrhenius, 2000). Both the Gulf of Riga herring and the Baltic sprat showed similar reductions in weight-at-age in the absence of predation by cod or size-selective predation, respectively, indicating that the first two hypotheses can not explain the observed changes in weight-at-age alone. In turn, zooplankton data suggest that individual prey availability for both clupeid species declined concurrently with their weight-at-age. Especially the decline of *Pseudocalanus acuspes* affected the nutrional status of herring negatively, while sprat utilizes also other copepods, e.g. *Temora longicornis* and *Acartia* spp., and thus sustained a good nutritional status until density dependent processes started to act as a consequence of the drastic increase in stock size in early 1990s (Möllmann *et al.*, 2005). A potential impact of low condition on reproductive success of herring has been hypothesized, but not addressed in specific studies. Herring sexually mature between age 2 and 3, with substantial interannual variability. Yearly maturity ogives have been compiled (ICES, 2002a) but are presently not used in the assessment.

Predation by cod is a major source of clupeid mortality in the Baltic, with especially juvenile herring being preyed upon intensively (Sparholt, 1994b). To account for this, the stock assessment uses age- and year-specific predation mortalities estimated by a Multispecies VPA (ICES, 2005c) as input. A drastic decline in the cod stock throughout the 1980s as the major predator in the system caused a substantial reduction in predation (Köster *et al.*, 2003a). While open sea herring of the Central Baltic should have benefited most from the release of predation pressure, this is not obvious from above-described stock dynamics.

Recruitment of herring stocks in the Gulf of Riga, Gulf of Bothnia and Gulf of Finland is highly correlated, but largely decoupled from recruitment variability of the Central Baltic stock (Kornilovs, 1995). Also the longer-term trends are different: while the latter showed a decline in recruitment from mid 1980s to 2000, the reproductive success of Gulf herring, e.g. in the Gulf of Riga, increased (Fig. 9.4a), with recruitment at age 1 being significantly related to SSB, temperature in April and zooplankton abundance in May (Kornilovs, 1995; Barange, 2003). Similarly, winter/spring temperature is correlated with the recruitment of herring in the Bothnian Sea and Bay. The mechanisms affecting year class formation of herring in the Central Baltic are not fully understood. A separate assessment for three popultions performed by ICES (2003a) revealed a similar pattern in recruitment dynamics; however, variability was higher for the southern coastal herring population.

European Sprat (Sprattus sprattus) – *Black Sea*

Sprat is distributed in the whole Black sea with maximum abundance in the northwestern part and shelf waters. In spring, schools migrate to coastal waters for feeding and in the summer they stay under the seasonal thermocline forming dense near-bottom aggregations during the day and rising to the surface at night (Ivanov and Beverton, 1985).

Black Sea sprat forms a self-sustaining stock (Ivanov and Beverton, 1985). It is one of the most abundant species in the area with importance for the commercial fishery as well as a food for fish and mammals (Daskalov, 2002).

Sprat matures at age 1 and reproduces during the whole year with a maximum between November and March (Table 9.1c). Spawning can be associated with the winter divergence and spring plankton blooms (Daskalov, 1999). The reproductive niche is limited to offshore subsurface (10–50 m) layers, which are stabilized by the permanent pycnocline. Horizontally, sprat eggs and larvae are concentrated above the shelf edge, and in the central cyclonic areas (Arkhipov, 1993; Fashchuk *et al.*, 1995).

Sprat is an object of both artisanal and commercial midwater trawl fisheries. Decadal fluctuations of abundance are observed and likely to be related to changes in climate (Daskalov, 1999, 2003). Maxima of recruitment and biomass were observed in the mid 1970s and mid 1980s (Fig. 9.4b). Maximum catch was in 1989, followed by a stock collapse. In the recent decade the stock partially recovered, with catches of up to 40 000–50 000 t.

European Sprat (Sprattus sprattus) – *North Sea*

Sprat is found mostly in the east and south of the North Sea, and also in the coastal areas and lochs of the east of the British Isles. No major studies on stock definition have

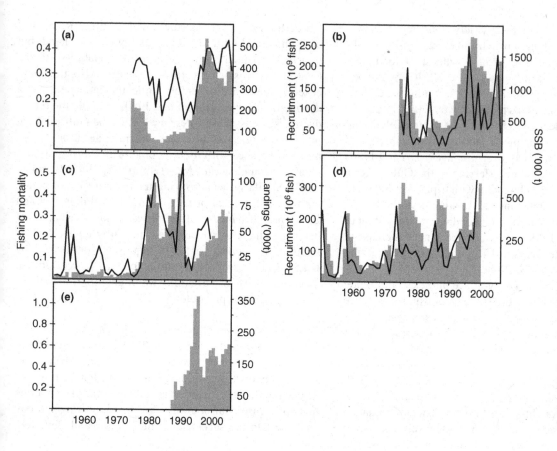

Fig. 9.4b. Time series of fishing mortality (y⁻¹), landings ('000t), recruitment and spawning stock biomass (SSB, '000t) for Baltic Sea sprat (a, b), Black Sea sprat (c, d), and North Sea sprat (e). Notation as in Fig. 9.2a

taken place, but at present it is assessed as a separate stock from that in the Skagerrak and the English Channel. In the North Sea, sprat spawns from spring through to late summer (Table 9.1c). This causes problems in aging the fish, as a variable part of each year's new young may not metamorphose that year, and hence a blurring of year classes or cohorts may occur. Sprat spawn pelagic eggs and, in general, larva drift is in a northeasterly direction towards the German Bight and Skagerrak, or towards the coast. Year-class strength appears to be determined during the larval phase, although as this is a short-lived species in the North Sea, there are not many cohorts in the population to support analysis of the consistency of year-class strength. The sprat are not thought to make large migrations, with both feeding and spawning occurring in similar areas. Estimates of absolute biomass are not available. However time series of catch (from 1985 onwards) and research surveys suggest that in 2000–2005, the stock biomass was high compared to earlier in the series (by at least three-fold). As this is a short-lived species in the North Sea, the biomass is strongly dependent on the incoming year-class strength. There are considerable fluctuations in total landings, from a peak in 1975 of 641 000 t to a low in 1986 of around 20 000 t (Fig. 9.4b). Since 1994, landings have varied from ca. 100 000 t (1997) to ca. 300 000 t (1994).

Denmark, Norway, and UK trawlers and purse seiners exploit North Sea sprat. Most of the catch (98%) is used to create fish meal (industrial fishery) and the remainder, mostly from UK and Norwegian catches, is for human consumption. Juvenile herring are also caught as a bycatch. The sprat fishery occurs in the second half of each year. The only fisheries data collected are the total landings, per quarter, from each country and samples of the age and weight of fish in the catch in the industrial fishery. The catch is usually dominated by 1-year-old fish. The bycatch of species other than sprat is monitored in the industrial fishery.

European Sprat (Sprattus sprattus) – *Baltic Sea*
Sprat is distributed throughout the entire Baltic Sea, with the exception of the Bay of Bothnia, and it is assessed and managed as a single stock. However, morphometric studies, otolith microstructure and genetic analyses suggest that at least two stocks are present in the western and central to northern Baltic (Parmanne *et al.*, 1994). The stock identity of the latter component has been discussed controversially (e.g. Ojaveer, 1989; Sjöstrand, 1989), and the evidence of further separation into different stocks is not generally accepted.

The Baltic Sea is located near the northernmost limit of sprat's geographic distribution (Parmanne *et al.*, 1994).

Thus, sprat shows a preference for warmer water layers in winter, specifically the bottom water layers of the deep Baltic basins, for which the pronounced stratification prevents a vertical convection during winter. Spawning takes place in March to June (Table 9.1c) as well in the deep Basins (Parmanne *et al.*, 1994), while the distribution in summer and autumn is more widespread, covering coastal as well as deep water areas (Aro, 1989).

Depending on the fishing mortality and predation pressure by cod, sprat of 8–10 years of age can be abundant in the stock. Sexual maturity is reached at the age of 2 (Elwertowski, 1960), but age-group 1 fish sometimes contribute to spawning in significant proportions (ICES, 2002a). The mechanism affecting maturation is not yet clear, but low temperature delays sexual and gonadal maturation.

The stock size of sprat declined during the 1970s, remained at a stable low level throughout the 1980s and increased in the first half of the 1990s to historic high values, reaching more than 3 Mt biomass in 1995 (Fig. 9.4b). This positive stock development was enabled by a reduction in predation pressure by the main predator cod and increasing reproductive success at relatively low fishing mortalities (Köster *et al.*, 2003a).

The reproductive success of sprat during the last 15 years is related to high winter temperatures having a positive impact on egg production and survival (Köster *et al.*, 2003b; Nissling, 2004), as well as the strong preference of the larvae for the copepod *Acartia* spp. (Voss *et al.*, 2003), which has increased since the 1990s in parallel to the increase in temperature (Möllmann *et al.*, 2000). This may have led in general to higher survival of larvae, being the critical early life stage in Baltic sprat (Köster *et al.*, 2003b). The drastic decline of weight-at-age observed during the 1990s (see section on Atlantic herring in the Baltic Sea) had obviously no impact on the reproductive success of the stock.

As a consequence of the drastic stock increase during the 1990s, an industrial fishery on sprat developed with 300 000–500 000 t catch annually resulting in a reduction in stock size in most recent years. Recruitment has been highly variable since the mid 1990s, but outstanding year classes, on average every second year, sustain the stock on a relatively high level (ICES, 2006c).

Most of the catches are taken by pelagic single and pair trawling, but some demersal trawling exists. The main fishing season is in the first half of the year, but in the northern part of the Baltic ice cover is a limiting factor. Sprat is fished mostly for industrial purposes, but also for human consumption and animal feed (ICES, 2006c). Sprat is also caught as bycatch in some herring fisheries and in turn herring bycatch occurs in several of the sprat fisheries. The latter is regulated by bycatch ceilings and, in fact, use of the TAC for herring has stopped the sprat fishery from fishing out its TAC in recent years. The species composition of the mixed catches is defined from logbooks and, partly, by observers on board of larger vessels. Misreporting of herring as sprat occurs, but the magnitude is unknown. Commercial catch rates are available for some fisheries but not used in the assessment (ICES, 2006c).

Fishery independent monitoring surveys

Until the 1960s most pelagic fish stocks were not assessed or their assessment relied on analysis of data from commercial fisheries. Since then the use of catch data has been compromised by its limitations; CPUE from commercial fleets have proven to be a poor proxy for stock abundance because pelagic fish schooling habits may result in increased catchability as stock abundance decreases rather than the expected opposite (Csirke, 1988). Consequently, the demand for fishery independent estimates of population size increased, either as input to management plans or to provide auxiliary information in catch-at-age analyses.

The main survey programs currently applied in the assessment of pelagic fish stocks are egg and larvae surveys and acoustic surveys. Bottom-trawl and aerial surveys have also been used sporadically (for review, see Gunderson, 1993).

Egg production surveys

Ichthyoplankton sampling started in the North Sea in 1895 (Hensen and Apstein, 1897), initially to identify and describe the biology of eggs and larvae. Currently, most egg and larvae programs are aimed at understanding the processes that determine recruitment fluctuations and to estimate recruitment strength, while some are used to estimate the size of the parental stock. The latter is based on one of three methods: the Annual Egg Production (AEP, Gunderson, 1993) and the Daily Fecundity Reduction (DFR, Lo *et al.*, 1992a) methods, both applicable to determinate spawners, and the DEPM (Parker, 1980; Lasker, 1985), which can be used on determinate as well as indeterminate spawners. The DEPM has been most successfully applied to pelagic fish, and is based on daily measurements of egg production and fecundity during a single survey (Alheit, 1993; Hunter and Lo, 1997; Stratoudakis *et al.*, 2006). A brief summary of the egg survey programs currently applied in the assessment of anchovy and sardine stocks follows. For sprat, the potential for the application of the egg production method has been demonstrated as well (Kraus and Köster, 2004), but at present the method is not used in any of the assessments.

Anchovy stocks

Monthly egg production surveys have been conducted along the Pacific coast of Japan from 1949 in support of the assessment and management of the Japanese anchovy (e.g.

Kubota *et al.*, 1999), using NORPAC-net vertical samples from 150 m depth (or near bottom) to the surface. Monthly egg production is calculated on each 30×30 latitudinal square.

California anchovy are monitored but not actively managed in the US because catches have been relatively low in recent years (PFMC, 1998). The most recent stock assessments used biomass estimates from forward-projecting models (Methot, 1989; Jacobson *et al.*, 1994) with spawning biomass estimates from DEPM surveys, two egg production indices from CalCOFI ichthyoplankton surveys, aerial fish spotter data (see below), and sonar survey data to measure spawning biomass during 1963–1995 (Jacobson *et al.*, 1995).

In the Humboldt anchovy egg surveys have been undertaken in northern Chile since 1992 to assess the spawning biomass applying the DEPM in the area from the border with Peru to 26°S. The Peruvian portion of the anchovy stocks is monitored acoustically twice a year (see below), but the winter survey also includes DEPM estimates of biomass.

Southern Benguela anchovy spawner biomass was estimated using combined DEPM and acoustic summer surveys between 1984 and 1993 (Armstrong *et al.*, 1988; Shelton *et al.*, 1993). The DEPM estimates were considered unbiased and were used to scale acoustic estimates at a time when there was no information on the accuracy or applicability of the acoustic target-strength expression used (Barange *et al.*, 1996; Hampton, 1996). DEPM estimates are no longer provided, and fishery independent monitoring is limited to a program of acoustic surveys (see below).

Similarly, the Bay of Biscay anchovy stock is monitored by annual DEPM spring surveys since 1987 (Santiago and Sanz, 1992; Motos *et al.*, 2005) and acoustics (regularly since 1989, although surveys were also conducted in the early 1970s and 1980s; Massé, 1988, 1994, 1996). Both surveys provide spawning biomass and population-at-age estimates. The DEPM is taken as an absolute indicator of biomass, while the acoustic as a relative one. This survey based monitoring system provides population estimates by the middle of the year, when about half of the annual catches have already been taken; and provides no prediction of the state of the stock in the next year, since the bulk of it will consist of 1-year-old fish being born at the time the surveys take place.

The DEPM has been applied on the NW Mediterranean anchovy stock in 1993 and 1994 (Garcia and Palomera, 1996), and experimentally in the monitoring of the southern Adriatic fishery (e.g. Casavola, 1998). However, neither program is applied routinely. Direct biomass estimates for anchovy in the Aegean Sea through acoustic and/or DEPM surveys have been obtained within the framework of various European and National projects (Machias *et al.*, 1997,

2000, 2001; Somarakis, 2005; Somarakis *et al.*, 2002b, 2004), but lack a regular (in time and space) time series. Since 2003, the Greek National Program for Fisheries data collection monitors the Aegean Sea anchovy stock, using the DEPM and acoustics, which are applied concurrently during the June spawning peak (Somarakis *et al.*, 2005a). Experimental DEPM surveys have also been performed on the Black Sea anchovy for the period 1987–1991 by the USSR, Bulgaria and Romania (Arkhipov *et al.*, 1991). Regular Black Sea anchovy pre-recruit surveys have also been carried out by the former USSR (now Ukrainian) institute YugNIRO, Kerch from early 1960s to 1993 (Tkacheva and Benko, 1979; Arkhipov, 1993).

Sardine stocks

Egg census and pre-recruit surveys are used as predictors of Japanese sardine SSB and recruitment, respectively (Nishida *et al.*, 2006). Monthly egg production census surveys have been conducted since 1949. From these surveys the extension of the sardine spawning area is calculated as a measure of SSB by integrating the areas in which early developmental stage eggs are present (Zenitani and Yamada, 2000). Since 1996, pre-recruit surveys have been carried out in May–June in KOTZ,[11] which is assumed to be the key area of recruitment (Watanabe and Nishida, 2002). A temperature-weighted pre-recruit index corresponds well to the VPA-derived recruitment estimate.

The California Cooperative Oceanic Fisheries Investigations (CalCOFI) survey has collected ichthyoplankton, including California sardine eggs and larvae, since 1951. The present CalCOFI sampling grid is smaller and less frequently sampled than the original design – quarterly surveys now span from San Diego to Avila Beach, California. However, the range is expanded northward each April to sample a significant portion of the sardine's spawning habitat. A NOAA synoptic survey from British Columbia to the US–Mexico border was conducted in April–May, 2006, while Mexico sampled areas off the Baja California coast, in the first tri-national stock assessment cruise. Sampling included oblique bongo and vertical egg net tows, Continuous Underway Fish Egg Sampler (CUFES; Checkley *et al.*, 1997), and trawling for data on reproductive parameters and age composition. Data are used in the DEPM (Lo *et al.*, 1996, 2005). Adult reproductive parameters have been averaged across years, a source of bias caused by limited adult sardines in the sampling. A time series of DEPM-biomass estimates is available from 1986–1988 and 1994–2006, and is updated annually (Hill *et al.*, 2006). The IMECOCAL program (Investigaciones Mexicanas de la Corriente de California) was initiated off Baja California in 1997. Similar to CalCOFI, IMECOCAL conducts quarterly cruises at historic (pre-1985) CalCOFI stations. Sardine are a major focus of this program; however, the ichthyoplankton data are not yet incorporated into the US stock assessment. Lack of adult samples (i.e. reproductive parameters) from the Baja California region has thus far hindered reliable application of the DEPM approach.

The Chilean IFOP[12] monitors the distribution of eggs and larvae of the main pelagic fish stocks. A time series of a Humboldt sardine larval index has been constructed for the winter season, which has been used in stock assessment (Braun *et al.*, 2002). In addition, a conventional DEPM has been applied on the Humboldt common sardine from 2002 (Cubillos *et al.*, 2005). However, because of the short length of the series, DEPM estimates of spawning biomass are not yet integrated into common sardine stock assessment models.

Egg and larvae surveys (Matsuura, 1983, 1996, 1998) and DEPM (Rossi-Wongtschowski *et al.*, 1994; Rossi-Wongtschowski *et al.*, 1995) have been used to monitor the Brazilian sardine stock, but they are not applied in management (Saccardo and Rossi-Wongtschowski, 1991). Instead, catch data have provided the basis for the application of Surplus Production Models and Yield-per-Recruit Models, defining maximum catches since 1974 (Cergole, 1993). Virtual Population Analysis was applied for the period 1977–1997 (Cergole, 1993, 1995; Cergole *et al.*, 2002), including oceanographic and meteorological variables (Sunyé and Servain, 1998; Sunyé, 1999; Jablonski, 2003; Jablonski and Legey, 2004, 2005). Sporadic hydroacoustic surveys have also been conducted (Johanneson, 1975; Rijavec and Amaral, 1977; Castello *et al.*, 1991; Madureira and Rossi-Wongtschowski, 2005).

The first application of the DEPM on the European sardine stock was conducted in 1988, although the method did not become a regular tool until 1997. Since 1999 it has been performed triennially (ICES, 2004, 2006e). As for the case of acoustic surveys, the different national sampling efforts are coordinated within the ICES framework (ICES, 2006e). Since 2002, CUFES is used as an auxiliary egg sampler, in addition to the standard station net sampling. Adult sampling is carried out opportunistically. DEPM-based spatially explicit estimates of SSB are experimentally obtained by spatial modeling as the population is known to show large spatial variability which can produce bias if unaccounted for (ICES, 2006e). Spatial estimates are generally consistent with traditional DEPM-based estimates, although their use in assessment is pending a general revision of the methods and data series.

DEPM has periodically been applied to each of the three sardine management units off southern Western Australia since the early 1990s (e.g. Fletcher *et al.*, 1996; Gaughan *et al.*, 2004). Scarcity of adult samples for some surveys has required the application of rules (e.g. sex ratio not to exceed 70%) and the use of reproductive data obtained outside of the survey period. Despite that Fletcher and Sumner (1999)

determined the appropriate spatial scales for sampling sardine eggs; patchiness of eggs also remains problematic. Imprecision of SSB estimates has therefore been a problem, particularly since even acceptably low CV[13] can occur by chance. Nonetheless, the time series of DEPM-based SSB estimates have been crucial for managing sardine in Western Australia.

Hydroacoustic surveys

Hydroacoustic winter surveys targeting anchovy recruits have been conducted off Japan in the offshore area $35°-37°$ N / $141°-145°$ E to detect and estimate recruitment. From 2004 a second series of hydroacoustic surveys has been conducted to assess the stock biomass outside the fishing ground $35°-45°$ N / $145°-170°$ E in the spring season.

The Peruvian anchovy stock is monitored acoustically twice a year, in summer and winter, while in Chile hydroacoustic surveys have been conducted to assess the Humboldt sardine from 1981 to 1995. The surveys estimate the spawning biomass in winter during the spawning process (Castillo *et al.*, 1995; Castillo and Robotham, 2004). Since 1999 annual recruitment surveys are conducted in January for common sardine assessment (Castillo *et al.*, 2005).

Acoustic surveys form the cornerstone of the South African anchovy and sardine assessment program and have been in place since 1984 to monitor the biomass of the spawner stocks (combined with DEPM for anchovy until 1993) in November and recruitment in May/June (Hampton, 1992; Barange *et al.*, 1999). Coetzee *et al.* (2008) and de Moor *et al.* (2008) have updated the SSB time series, accounting for receiver saturation in old-generation echo sounders, more accurate target strength expressions, and acoustic signal attenuation by dense sardine schools. The recruitment survey is conducted at the earliest time possible to provide a reliable survey index of the abundance of the incoming anchovy and sardine recruits for that year, but this estimate is available only 2–3 months after fishing has commenced. The November surveys have also facilitated the collection of data on mean mass-at-age since 1990 for anchovy and since 1988 for sardine, and age–length keys are currently available for surveys conducted between 1992 and 1995 for anchovy and since 1988 for sardine. For other years, a combined 1992–1995 age–length key for anchovy has been used to obtain the proportion (by number) of 1-year-olds in the November surveys (De Oliveira, 2003).

In the northern Benguela hydroacoustic sardine surveys are conducted in autumn (February–April) to estimate adult abundance, and in spring (October–December) to both assess adult stock and provide an index of recruitment. Occasionally surveys have extended into Angolan waters since as much as 50% of the sardine adult stock has been found there in winter (Boyer *et al.*, 2001). In contrast to the southern Benguela where a survey grid of randomly stratified cross-shelf acoustic transects are surveyed by the research vessel only, surveys in the northern Benguela use a two-stage adaptive strategy, whereby systematic zig–zag transects are followed by pelagic fishing vessels that locate sardine school groups which are then assessed using closely spaced parallel transects surveyed by the research vessel.

In addition to the program of acoustic estimation of the size of the Bay of Biscay anchovy stock in spring mentioned above (Massé, 1988, 1994, 1996), an experimental program to survey anchovy juveniles was conducted in 1998 and 1999 in the Bay of Biscay (Uriarte *et al.*, 2001, 2002; Carrera *et al.*, 2006). This concluded that acoustic surveys at the end of summer and early autumn could provide a good index of pre-recruit abundance. Thus, since 2003 acoustic surveying of anchovy juveniles takes place annually in September–October in order to assess the biomass of anchovy juveniles that will enter the fishery the next year (Boyra *et al.*, 2005; Boyra and Uriarte, 2005). However, the time series is too short to allow a proper evaluation of its performance as a recruitment predictor and it is therefore not yet used for the management of the population.

Routine acoustic surveys of the Mediterranean anchovy have been carried out along the NW Mediterranean coast since 1993 (Abad *et al.*, 1998; Liorzou *et al.*, 2004; Alemany *et al.*, 2002) directed towards recruits in the Catalan Sea (autumn) and to adults in the Gulf of Lions (summer). From 2002, these surveys have been incorporated in the Spanish and French National Programs for Fisheries Data collection. In the Adriatic Sea, anchovy has been acoustically monitored since 1975, covering the Italian territorial waters, i.e. the western half of the Adriatic (Azzali *et al.*, 2002). The eastern side has not been covered regularly, but this has improved recently with the participation of Croatian scientists (Tičina *et al.*, 2005). Despite this, monitoring of the Adriatic Sea stock has been based mainly on analytical assessments using fishery data. As mentioned above, an acoustic monitoring program for the Aegean Sea anchovy was implemented in 2003.

In the Black Sea regular hydroacoustic or mid-water trawl surveys were performed by the former USSR (in collaboration with Bulgaria and Romania) from 1980 to 1992. Regular pre-recruit surveys have been carried out by the former USSR (now Ukrainian) institute YugNIRO (Kerch, Crimea) from early 1960s to 1993 (Tkacheva and Benko, 1979; Arkhipov, 1993). Sprat biomass has been assessed using mid-water trawl surveys in 1967–1993 (Ivanov and Beverton, 1985; Prodanov *et al.*, 1997). The anchovy stock has been monitored using hydroacoustics in Turkish waters (southern Black Sea) for the period 1990–1994.

Acoustic surveys dedicated to the estimation of the European sardine distribution and abundance started in Spain and Portugal during the early 1980s. Off northern Spain, surveys have been performed annually in spring and continue to the present day. Off Portugal, acoustic surveys were carried out every 6 months (spring and autumn) during the period 1984–1988, were interrupted until 1992, and from 1995 onwards are always performed in spring and in most years during autumn. Survey plans and methods from Spain and Portugal have been coordinated in different ICES Planning and Working groups (ICES, 1998b; ICES, 2006e), and since 2000 are also coordinated with the French survey in the Bay of Biscay (ICES, 2006e). In summary, acoustic sampling is carried out following a systematic transect-based survey sampled only during daytime (since the mid 1990s). Opportunistic adult sampling is conducted to identify eco-traces, species biological and demographic characteristics, age–length keys, and so on. Adult sampling is performed using mainly pelagic (Spain) or both pelagic and bottom trawls (Portugal), although opportunistic samples from purse seiners are sometimes used to gather additional information. Following general trends within ICES, acoustic surveys in the Iberian Peninsula have gradually evolved into pelagic community surveys, and a series of pelagic species are now monitored concurrently.

The ICES-coordinated North Sea acoustic surveys began in the late 1980s to estimate the abundance of herring and sprat during summer. The surveys were originally targeted at herring, but from 1996 onwards records of sprat were also processed. However, the coverage of the surveys did not include the complete distribution of sprat, and so they were modified in 2003. These surveys appear to give consistent results in terms of the numbers of sprat aged 1+ and herring 2+, but estimates of the younger groups show very high year effects. The time series of sprat is at present too short to be used in any assessment, but herring estimates are used.

Baltic Sea sprat and herring abundances are derived from annual autumn hydroacoustic surveys targeting both species (ICES, 2006b). The hydroacoustic surveys cover the entire stock areas, and all age groups. The open sea survey is internationally standardized; calibration of equipment and ship intercalibrations are performed at each survey. To differentiate between sprat and herring, approximately 100–150 identification trawl sets are carried out. Other abundance indices not used in the present assessment of sprat are hydroacoustic surveys conducted in May/June since 2001 (ICES, 2006b) and egg surveys (Köster *et al.*, 2003b) covering both the central stock distribution area. Herring larvae and juvenile surveys have also been conducted more or less regularly in specific areas of the Baltic, but none is used in the assessment.

Finally, the Norwegian spring spawning herring stock is one of the most acoustically monitored pelagic resources worldwide. Surveys are conducted in the overwintering areas (November–December 1992–2004, but also in January 1991–1999), the spawning grounds

(February–March 1988–2005) and the feeding areas (April–June 1996–2005). These surveys target the adult stock (age classes >3–15+). There are also acoustic surveys for the 1–2 y.o. conducted in the Barents Sea in May–June (1991–2005) and August–September (1974–2004) (ICES, 2005b). In addition, a mid-water trawl survey is conducted in August–September (1974–2004) in the Barents Sea to estimate the size of the 0 age class, and an extensive tagging program started in 1975 (ICES, 2005b). A time series of acoustic estimates of 0 group herring from the Norwegian fjords and coastal areas, running from 1975 to 2004, is available, but this is not used in the stock assessment.

Other monitoring surveys

Some pelagic fish, such as herring and sprat, are also surveyed through the ICES coordinated Bottom Trawl Surveys in the North Sea (IBTS) and the Baltic (BITS). The IBTS was originally set up after the collapse of North Sea herring to monitor the recovery of the stock (Heessen *et al.*, 1997) but it covers the distribution of sprat as well. The survey uses the GOV (Grande Overture Verticale) trawl in the first and third quarter of the year. The results are used for both following the trajectory of the herring stock, and for catch forecasts of sprat for within-year management. However, catchability fluctuates as a function of hydrographic condition, which is also the reason why catch rates of Baltic herring and sprat available from first and fourth quarter BITS (ICES, 2006b) are presently not used in stock assessment.

Surveys of newly hatched larvae are also used in the assessments of North Sea herring and Norwegian spring spawning herring. The time series run from 1973 and 1981, respectively. Both series are used as indices of spawning stock potential in the assessments. Conversely, surveys of post-larval herring (1992 to present) are used as indicators of recruiting year classes in the North Sea (see Nash and Dickey-Collas, 2005).

In addition to egg production and acoustic surveys, additional surveys are in place to monitor abundance and biology of adult Japanese sardine: driftnet surveys during spring–summer in the Hokkaido area since 1994, longitudinal transects at 155°E, 175°E and 180°E since 1982 (Yatsu *et al.*, 2003), and surface trawl surveys during winter and spring in the Kuroshio, Oyashio and KOTZ since 2001 (Nishida *et al.*, 2006).

Aerial fish spotting data have been used in California and Namibia in recent decades. California's fishing fleet uses spotter pilots to locate pelagic fish schools. Data for each flight have been provided to the US National Marine Fisheries Service (NMFS) since 1962 (Squire, 1972). Spotter logbooks contain data on species, school size (tonnes), effort, and location (10×10 nm statistical areas). Lo *et al.* (1992b) developed a standardized index of relative abundance, estimating year effects using delta lognormal linear models. The current index for California sardine includes data from 1986 through 2005. After the year 2000, there was a rapid decline in both the number of active pilots and total logbooks returned, as well as a southward shift in effort to offshore areas of Baja California in response to an increase in the tuna net-pen fishery. To remedy this problem and continue the time series, NMFS has contracted professional spotter pilots to survey the Southern California Bight region in 2004 and 2005 and the newly available data have been included in the index. Since the spotter pilots operate in the area of the fishery, i.e. within 30 nm from shore, the index is assumed to represent relative abundance of younger sardine (ages 0–2) and is treated as such in the stock assessment model.

A preliminary program of aerial pelagic fish spotting using aircraft-borne remote sensors was also applied in Namibia in the 1970s, driven by concerns that the extreme patchiness of the shoals and their tendency to avoid vessels may invalidate acoustic surveys of the stock. The program localized and measured shoals at night-time through bioluminescence, while a vessel made synchronous measurements of shoal thickness and packing density (Cram and Hampton, 1976). The expected result was the provision of unbiased absolute estimates of pelagic fish, but it did not survive the experimental phase.

Stock assessment, modeling, and harvest strategies

The assessment and management of small pelagic fish stocks does not fit well into the traditional population dynamics models and assumptions. Beverton (1990) concluded that small shoaling pelagic fish species are the most unreliable and vulnerable to unrestrained fishing. This is mostly because their high natural mortality (around 1 y^{-1} or larger), short lifespan and dependency on annual recruitment pulses results in high population variability and a tendency to go from boom to bust in a short time span. In addition, variability in catchability coefficients means that catch information cannot be reliably used to estimate biomass (Csirke, 1988), thus limiting assessment options. On the other hand, pelagic fisheries are highly targeted, and thus generally do not have the management problems that mixed fisheries pose. In this section we reviewed present approaches to outline the state of the art in small pelagic assessment and management.

Japanese stocks

Japanese anchovy and sardine stock assessments are conducted annually in support of the fishery management process. VPA and survey results are used in this assessment (Table 9.2a,b).

The TAC for the Pacific stocks of Japanese anchovy and sardine for year t are determined in November of year $t–1$, on the basis of allowable biological catches (ABC) and socio-economic factors. ABC in year t is recommended on the basis of projected stock abundance, SSB, recruitment (age-0 fish) for year t and either limiting or targeting fishing mortalities, i.e. F_{limit} or F_{target} for both species. Abundances-at-age in year t are predicted using a forward VPA from year $t–2$ with prognoses of SSB and recruitment in years $t–1$ and t, assuming fishing mortality in year $t–1$. In the case of anchovy the relationship between SSB and recruitment is significant and its linear relationship is used to forecast future recruitments. In the case of sardine Recruitment-Per-Spawner (RPS) is negatively correlated with sea surface temperature (SST) in the Kuroshio Extension South Area (KESA) in winter (Noto and Yasuda, 2003). An extended Ricker model has been proposed by including winter SST in the KESA (Yatsu *et al.*, 2005). Suda and Kishida (2003) developed a recruitment model of Japanese sardine by incorporating transportation, prey condition, inter-species competitions and predation. These models, however, have not been applied in management. The sardine management target in recent years has been set to 222 000 t SSB, a magnitude at which relatively stable recruitment has been observed (Nishida *et al.*, 2006). The abundance estimate of pre-recruits from spring surveys in the KOTZ is used to predict recruitment level. As 8–10-month-old recruits are targeted together with older fish by the fishery, these estimates are included in catch prognosis and quota determination (Watanabe and Nishida, 2002). Owing to uncertainties in SSB and recruitment predictions, the TACs are reviewed within each season when additional information from surveys and commercial catch becomes available.

Japanese catch statistics indicate a sardine-anchovy replacement, with historical peaks of anchovy around 1960s and the present, and 1930s and 1980s in the case of sardine. The current catch level of anchovy is the maximum during the last 25 years, but given its low landing price, it is unlikely that the Japanese anchovy will become the major target species for the purse-seine industry. The current sardine catch level is approaching the historic minima.

California sardine and anchovy

The California anchovy fishery is monitored at this time but not actively managed because harvest levels and demand are low (< 10 000 t per year during 1991–1995; PFMC, 2005). Under these conditions, a "default" ABC level of 25 000 t is allowed annually in US waters (Table 9.3a). Active management will be required if catch levels rise above the ABC.

In the 1980s, just after stock biomass and fishery landings declined from record highs, the anchovy fishery was managed actively using a harvest control rule that was the first of its kind (PFMC, 1983). The rule allowed very low catches at spawning biomass levels below a pre-specified cut-off level. As spawning biomass increased above the cut-off, catches were allowed to increase up to pre-specified maximum. DEPM spawning biomass estimates were used initially to set harvest levels but proved too expensive to carry out on an annual basis. In lieu of annual DEPM estimates, forward-casting stock assessment models, including the original "Stock Synthesis" model (Methot, 1989), were developed and used to estimate spawning biomass based on a diverse range of data (Jacobson *et al.*, 1994).

Murphy (1966) developed the age-based cohort analysis now known as VPA, first applying it to California sardine. Deriso *et al.* (1996) applied the forward-projecting approach (CANSAR[14]) to contemporary sardine data; Hill *et al.* (1999) modified CANSAR into a two-area model. The population model currently used for assessment and management of the US fishery is called "ASAP" (Age-Structured Assessment Program; Legault and Restrepo, 1999), a forward simulation approach (Table 9.3b). The population dynamics and statistical underpinnings of ASAP are well established (Fournier and Archibald, 1982; Deriso *et al.*, 1985). The current sardine ASAP model includes catch and age compositions for three fisheries (Ensenada, California, Pacific Northwest), as well as the aerial spotter and DEPM indices of relative abundance described above (Hill *et al.*, 2006). Modeled time series begin in 1982–83. The current model does not include SST as a term in the S–R model (as was done with CANSAR): however, the next modeling platform for sardine, Stock Synthesis 2 (Methot, 2005), will likely resume including environmental data.

ASAP's estimation approach is that of a flexible forward simulation allowing for the efficient and reliable estimation of a large number of parameters using the maximum likelihood method. The current ASAP model for sardine includes nine likelihood components and a few penalties: Selectivity in 1st Year, F_{mult} in 1st Year, Catchability in 1st Year, Stock-Recruitment Relationship, Steepness, F_{mult} Deviations, Recruitment Deviations, and Selectivity Deviations.

Periods of warm SST in the California Current Ecosystem are associated with good recruitment and higher productivity for California sardine (Ahlstrom, 1960; Jacobson and MacCall, 1995; Jacobson *et al.*, 2001). Sea-surface temperature (SST-SIO) measured at the Scripps Institution of Oceanography pier (La Jolla, California) since 1916 is a good proxy for environmental conditions influencing positive or negative surplus production in California sardine. The US management approach is unique in that it uses a harvest control rule that depends on a 3-year running average of SST–SIO (PFMC, 1998). In addition, the same SST series has been directly incorporated in the S–R function of the stock assessment model (Deriso *et al.*, 1996; Hill *et al.*, 1999). The ASAP assessment for sardine uses projected catch for the coming season in order to calculate population abundance at the start of the final time step. This provides estimation of B_t for calculating H_{t+1} with the harvest control rule (see OMP[15]).

The harvest control rule used by the Pacific Fishery Management Council (PFMC, 1998) has the following form: $H_{t+1}=(B_t-E)\,U*f$, with $H_{t+1} < H_{MAX}$; where H_{t+1} is the harvest guideline for year $t+1$, B_t is the stock biomass estimate at the beginning of the previous year t, E is a minimum escapement level (150000 t), U is the exploitation rate at F_{MSY}, f is the fraction of the stock assumed in US waters, and H_{MAX} is the maximum allowable harvest level (200000 t). In an attempt to make the control rule responsive to environmental forcing, control rules were constructed with U_t for each year based on a regression function relating U_{MSY} to a range of average SST values (Jacobson and MacCall, 1995). U_{MSY} is constrained to range $0.05–0.15$ y^{-1}.

Humboldt fisheries

In Peru anchovy assessments consist of direct methods (acoustics and DEPM) as well as VPAs that incorporate fishery data. VPAs have been applied in the N/C Peru stock for the period 1953–1985 (Pauly and Palomares, 1989), 1960–1994 (Csirke *et al.*, 1996), and for the SP/NCh stock for the period 1984–2001 (GTE, 2002; Table 9.2a). For the assessment of the Peruvian–Chilean anchovy stock a statistical catch-at-age model was developed (Serra *et al.*, 2004, Table 9.3a). The objective function of this model is the sum of the log-likelihood terms for the catch in tonnes, catch-at-age, spawning biomass, CPUE and penalizations of the parameters of the selectivity model and recruitment. Parameter estimates are found by minimizing the residual sum of squares for each terms of the objective function. The management of the anchovy fishery off Peru aims at maintaining a spawning biomass of 5 Mt at the beginning of the spawning periods in August and February. The management is flexible and short term and comprises disaggregated TAC in time and space, closures during spawning periods, and short-term closures to protect areas with a high proportion of juvenile fish. Anchovy in northern Chile is managed under quota since 2002 through a "maximum catch limit per ship owner" and also by closed periods during the spawning and recruitment seasons. The TAC is estimated under a constant F policy and medium term projections with uncertainty and risk analysis of different scenarios for recruitment (Serra *et al.*, 2004; Table 9.3a).

The South Peru–North Chile sardine stock has been assessed by age-structured Sequential Population Analysis, in which an extended version of ADAPT was used (Serra *et al.*, 2004; Gavaris, 1988; Conser and Powers, 1990; Table 9.3b). The abundance indices used for the assessment were standardized CPUE and a larval index. The CPUE time series was obtained through a GLM[16] following the approach proposed by Stefánsson (1996). The larval index is from winter surveys from 1985 to 2001 and was obtained from Braun *et al.* (2002). This fishery historically operated under an open access policy. The Chilean fishery in Northern Chile was managed through closed season to protect the spawning stock and size restriction. A catch quota was applied in 1982–1983 but was resisted by the industry. The criterion to estimate the quota was a constant F policy. Since 2002 a catch quota has been applied on the Chilean fishery.

The stock assessment model for common sardine consists of a general population dynamics model to predict catch and age composition, as well as catch-per-unit effort and hydroacoustic biomass (Table 9.2b); likelihood functions for the observed data; priors and penalties to constrain parameterization, and parameter estimation by minimization of an objective function (Table 9.3b). Catch data are expressed by weight and must be available for each time period. In terms of the population dynamics component, standard survival and catch equations are used to describe changes in the age structure of the population. The parameterization used in the models allows for a separation assumption in fishing mortality. Selectivity at age is allowed to vary, but treated by block of years in which they remain constant. The harvest strategy consists of computing a TAC by using a fixed fishing mortality rate defined at the level of 60% of the spawning biomass per recruit. Short-term projections of abundance are conducted under uncertainty, and the performance function is a ratio between the spawning-stock biomass at the end of the projection period and the current spawning-stock biomass. Risk is defined as the probability that fishing mortality in the short term will exceed a target fishing mortality for a range of alternative and equally probable quota options in the short term. A level of 10% of risk is usually taken into account. A second review of the TAC is carried out after the hydroacoustic survey, because of uncertainty in the short-term recruitment projection (Table 9.3b).

The Benguela operational management procedure (OMP) approach

Age-structured production models are used to assess southern Benguela anchovy and sardine stocks. These assume lognormal distributions of recruitment about a hockey-stick (single-sloped) stock–recruitment relationship (base case operating model). Beverton–Holt and Ricker stock–recruitment curves were used in alternative operating models (or assessments), on which the management procedure was tested. The most recent assessment, conducted using data from 1980/81 up to 2003, was Bayesian (Cunningham and Butterworth, 2004a; Table 9.3a,b).

As mentioned above, acoustic estimates of anchovy and sardine SSB are considered relative, while DEPM estimates of anchovy SSB are considered absolute. A probability density function (pdf) for the overall bias in the November survey estimate of sardine SSB has been calculated and the median was used for the base case run, suggesting a 28% underestimation of the stock.

A key feature of the South African pelagic fishery is the problem of juvenile sardine taken as bycatch in the anchovy fishery. This problem arises because the two species school together as juveniles (Crawford *et al.*, 1980) and means that catches of the two species cannot be simultaneously maximized. For this reason, a joint OMP is used for the sardine and anchovy fisheries (Table 9.3a,b). The anchovy fishery is managed through a TAC, with an initial TAC set at the start of the fishing season based on the biomass estimated from the preceding November (SSB) survey, and assuming forthcoming recruitment will be the median of observed values (De Oliveira and Butterworth, 2004). Because the bulk of the coming year's catch consists of incoming recruits, the initial TAC is revised as soon as an estimate of recruitment becomes available from the mid year recruit survey. The computation of the revised or final TAC replaces the median recruitment assumed earlier with the actual estimate from the recruit survey. An "additional sub-season" extending from August until the end of the year was introduced into the anchovy fishery in 1999, with the goal of targeting anchovy that no longer shoal with juvenile sardine. Targeting such "clean" anchovy would minimize the juvenile sardine bycatch, thereby optimizing long-term sardine catch. This sub-season TAC is set using the same data and at the same time as the revised TAC for the normal season, after the mid-year recruit survey.

The southern Benguela sardine fishery is managed by TAC and Total Allowable Bycatch (TAB). The TAC is set at the start of the fishing season based on the biomass estimated from the preceding November (SSB) survey (De Oliveira and Butterworth, 2004). Since anchovy and sardine shoal together during their first few months of life, directed fishing for anchovy is accompanied by a bycatch of juvenile sardine. The initial sardine TAB is based on a fixed tonnage for bycatch (mainly adults) linked to a small fishery for round herring (*Etrumeus whiteheadii*), and a component proportional to the initial anchovy TAC, using a conservative estimate of the ratio of sardine to anchovy juvenile fish. As for the anchovy TAC, the sardine TAB is revised after the recruit survey using an updated estimate of the ratio of sardine to anchovy juveniles. A TAB for the "additional sub-season" (see preceding section on southern Benguela anchovy) from August onwards is proportional to the additional sub-season anchovy TAC, subject to a maximum of 2000t.

The South African anchovy stock is a worthy candidate for investigating the use in management procedures of environmental indices that predict recruitment, because it is a resource that sustains a recruit fishery for which a measurement of recruitment is not available until after substantial fishing on that recruitment has already taken place (Butterworth *et al.*, 1993). Recent analyses have concluded that, under existing management constraints, an environmental index needs to explain roughly 50% or more of the total variation in anchovy recruitment before a management procedure which takes account of such information starts to show benefits in terms of risk and/or average catch (De Oliveira and Butterworth, 2005).

Assessments for northern Benguela sardine are conducted using an age-structured model in which age classes are either assigned from length frequency data or following cohort analysis, and a Ricker stock–recruitment curve is fitted to the data (Fossen *et al.*, 2001). More recently, simple age-structured models have been investigated, covering a range of uncertainties regarding key assumptions, and these have been proposed as operating models to form the basis for future development of OMPs for Namibian sardine (De Oliveira *et al.*, 2007). Currently though, management of the northern Benguela sardine fishery off Namibia is via a TAC based on a projected fishing mortality ($F = 0.2$ y^{-1}) of the fishable stock (fish >16cm; Boyer and Hampton, 2001). Fishing is also temporally limited, with the fishing season generally opening in March and closing at the end of August (Boyer *et al.*, 2001). Sardine bycatch taken in fishing directed at anchovy or horse mackerel (*Trachurus trachurus*) was not previously included as part of the TAC but has been included since 1990, a year when a bycatch of 29000t was taken in addition to a TAC of 60000t. Socioeconomic concerns (i.e. keeping processing factories operating) resulted in TACs being set in the mid 1990s that were higher than had been scientifically recommended, and this, together with permission being granted from 1994 onwards for Namibian vessels to catch unlimited quantities of sardine from Angolan waters (and outside of the closed season), meant that catches of northern Benguela sardine were at unsustainably high levels on several occasions during the past decade (Boyer *et al.*, 2001). This resulted in further decreases, both in TACs set and catches attained (Fig. 9.3b), and culminated in the setting of a zero TAC in 2002, although small TACs of 20000–25000t have since been set.

ICES stocks

ICES advice is based on the work done by research organizations in member countries, which contribute to ICES through data collection and analysis in the institutes and through participation in ICES expert groups. The outcomes of these analyses are translated into operational advice by ICES advisory committees which include scientists from all ICES member countries (ICES, 2005d). The stocks considered in this review are assessed using models in the VPA family. Three (Bay of Biscay anchovy, European pilchard, and North Sea herring) are managed through Integrated-Catch-at-Age models (ICA, Patterson and Melvin, 1996). In contrast to conventional VPA, which assumes that catches at age are measured without error, ICA models consider catches at age to be measured with an error that has an independent lognormal distribution. One advantage of this type of model is that approximate estimates of the variance of many of the estimated parameters can be calculated. ICA also allows catch information to be run against biomass (non-aged) survey data, such as egg, larvae, and total acoustic biomass indices. Baltic Sea sprat and herring are assessed by the Extended Survivor Analysis model (XSA; Shepherd, 1999), a method that partially solves the sensitivity to observation error in the final year when tuning virtual population analysis using abundance indices, while it also makes efficient use of previous estimates of year-class strength. XSA allows for the simultaneous analysis of several sets of abundance indices and for non-linear relationships between abundance indices and population size for the younger age groups. North Sea sprat is assessed using Catch-Survey Analysis (CSA; Mesnil, 2003), an assessment method that aims at estimating absolute stock abundance given a time series of catches and of relative abundance indices, typically from research surveys, by filtering measurement error in the latter through a simple two-stage population dynamics model. Finally, Arcto-Norwegian spring spawning herring stock uses SeaStar, a model specifically developed to use both survey and tagging data to tune long-time catch series.

Bay of Biscay anchovy

The current ICA provides a Maximum Likelihood Estimator (MLE) of recruits and fishing mortality by tuning catch-at-age and direct biomass and population-at-age estimates from surveys (DEPM and acoustics surveys; Table 9.2a). The acoustic estimates are treated as relative and DEPM as absolute. The assessment assumes a constant natural mortality of $1.2\ y^{-1}$ and a separable model of fishing mortality by age, applied over a period of the last 15 years. The most abundant age classes (catches at ages 1 and 2) receive higher weighting factors than other ages. The operating model of this assessment is therefore an age-structured model with the classical survivors and catch equations defining its dynamics within the frame of a separable model of fishing mortality (Table 9.3a). The assessment with ICA is heavily influenced by the surveys because the short lifespan of anchovy does not allow for a proper VPA convergence. In recent years, in addition to ICA, a simpler biomass delay-difference model (Schnute, 1987), based on the model applied to squid by Roel & Butterworth (2000), is being applied with successful results via Bayesian approach (Ibaibarriaga *et al.*, 2005). The model seeks to

estimate recruitment at age 1 at the beginning of each year (in mass) accounting for the signals of inter-annual biomass variations obtained from the direct surveys (DEPM and acoustics) and the level of total catches (in tonnes) produced each year. The ICES working group considered that the biomass model is as appropriate as ICA (with less risk of over-parameterization) and it is intended to be adopted as the standard assessment for this species in a next future.

Although a stock/recruitment relationship is not explicitly used, for management purposes it is assumed that a link exists below 21 000 t. This value coincides with the minimum spawning biomass level B_{lim} (set at 21 000 t, the lowest observed biomass in the 2003 assessment) below which the dynamics of this stock is unknown. Management is aimed at keeping biomass above B_{lim} (ICES, 2003b). This led ICES to propose a precautionary level of spawning Biomass (B_{pa}) around 33 000 t once assessment uncertainty and natural variability are accounted for.

No proper management plan or harvest control rules are adopted for this fishery. At the end of the year, when the TAC is set and the fishery starts, direct estimates of recruitment are not available. Therefore advice on the short-term development of the fishery has been based on assumptions rather than assessments of the strength of recruitment. Managers have not followed this advice which was usually based on assumed recruitment levels (usually precautionary – low levels of recruitment).

The development of the fishery until the 1970s was conducted without any management. Since 1979 a precautionary TAC for anchovy of about 30 000–33 000 t has always been agreed for the international catches of anchovy (except at the start of 2000 and 2006). Recent scientific advice warning about continuous recruitment failures since 2001 has not been considered. In 2005 the complete failure of the fishery and low levels of spawning biomass led to the closure of the fishery.

In the late 1990s the possibility of forecasting anchovy recruitment on the basis of environmental indices was explored, thus potentially enhancing the TAC setting at a time when year-class strength is still unknown. Using historical data Borja *et al.* (1996, 1998) concluded that recruitment can be favored by the occurrence of spring wind-driven upwelling along the French and Spanish coasts, close to anchovy spawning grounds. The statistical link was supported by Allain *et al.* (2001) using a 3D hydrodynamic model of the Bay in which upwelling intensity was deduced from modeled vertical water velocities rather than from geostrophic wind calculations. Using results from the 3D hydrodynamic model, Allain *et al.* (2001) observed that strong wind events during summer (June–July) provoke a breakdown of the water-column stratification, potentially reducing recruitment. Allain *et al.* (2001, 2004) proposed an environmental model that combines upwelling intensity

Table 9.2a. *Monitoring information and data input: anchovy stocks*

Species	Stock	Fishery data			Fishery-independent data		
		Parameter	Time series	Frequency	Parameter	Time series	Frequency
Japanese anchovy (*Engraulis japonicus*)	NW Pacific	Length frequencies	1978–2004	Monthly	Hydroacoustic surveys	2002–2004	2×year
		Landings	1978–2004	Monthly	Egg surveys	1978–2004	Monthly
		CPUE	1978–2004	Monthly	SST	1978–2004	Monthly
	Seto Inland Sea	Length frequencies	1981–2004	Monthly	Egg surveys	1981–2004	Monthly
		Landings	1955–2004	Monthly	SST	1981–2004	Monthly
		CPUE	1993–2004	Monthly	Hydroacoustic surveys	1981–2004	1×year
	Tsushima Current	Length frequencies	1991–2004	Monthly	Egg surveys	1991–2004	Monthly
		Landings	1960–2004	Monthly	SST	1991–2004	Monthly
		CPUE	1991–2004	Monthly			
California anchovy (*Engraulis mordax*)	Central sub-population	Length frequencies	1963–2005	Annually	Hydroacoustic surveys	1968–1984	Annually
		Landings			Egg and larval surveys	1963–2005	Various
					SST (Scripps Pier)	1963–present	Daily
Humboldt anchovy (*Engraulis ringens*)	North-Central Peru	Length frequencies	1953–2005	Monthly	Hydroacoustic surveys	1980–2005	2×year
		Age-length keys	2000–2005	Quarterly	Egg surveys	1999–2005	1×year
		Landings	1950–2005	Monthly	SST	1925–2005	Monthly
		CPUE	1960–2005	Annually			
	South Peru–North Chile	Length frequencies	1984–2005	Monthly	Egg surveys	1992–2005	1×year
		Age-length keys	1984–2005	Quarterly			
Benguela anchovy (*Engraulis encrasicolus*)	Southern Benguela	Length frequencies	1988–2005	Monthly	Hydroacoustic surveys	1984–2005	2×year
		Landings	1981–2005	Annually	Egg surveys	1984–1993	1×year

Species / Region	Data type	Time period	Frequency
European anchovy (*Engraulis encrasicolus*)			
Bay of Biscay	Catches at age	1987–2005	Quarterly
	Landings	1987–2005	Monthly
	SSB Hydroacoustic surveys	1989–2005	1×year
	Egg surveys (DEPM)	1987–2005	1×year
	Acoustic survey on juveniles (not used in assessment)	2003–2005	1×year
NW Mediterranean	Length frequencies	2001–2005	Monthly
	Hydroacoustic surveys	GL(summer): 1993–2005 CS (autumn): 1996–2005	Monthly
	Landings	1985–2005	Monthly
	CPUE	2000–2004	Monthly
	Egg surveys	1993–1994	1×year
Adriatic Sea	Length frequencies	1975–present	Monthly
	Landings	1975–present	Monthly
	CPUE	1975–present	Daily for one port
	Hydroacoustic surveys	1976–present	1×year
Aegean Sea	Length frequencies	2003–2006	Monthly
	Landings	1964–2006	Monthly
	Discards	2003–2006	Quarterly
	CPUE	1996–2006	Monthly
	Hydroacoustic surveys	2003–2006	1×year
	Egg surveys	2003–2006	1×year
Black Sea	Size and age	1949–2005	Monthly
	Landings	1925–2005	Annually
	CPUE	1984–2005	
	Hydroacoustic surveys	1980–1994	1×year
	Egg surveys	1987–1991	1×year
	Juvenile fish surveys	1980–1993	1×year
	SST	1925–2005	Monthly

and stratification breakdown events, which explained about 75% of the inter-annual variability between 1986 and 1997. In 1999 the ICES assessment working group used Borja's upwelling index to predict a failure of recruitment at age 1 for 2000. However the predictions failed and the reduction of the initial TAC adopted by managers turned out to be unnecessary. The TAC was raised to 33000 t on July 1, after direct estimates of SSB from surveys became available. This experience caused intensive debates among the management bodies and the scientific advisers, not only about the use of environmental indices to predict recruitment but also about the nature of the advice required for this short-living species. The ICES working group decided to abandon the use of the environmental indices given the limited predictive power of the simple model used. In a recent review (ICES, 2005f) the performance of the environmental models up to 2003 were again revisited and concluded that the available models (both Borja's and Allain's) offer poor predictive power.

Of late ICES has proposed a two-stage TAC regime based on a preliminary TAC at the beginning of the year (from an analytic assessment in the fourth quarter), and a revised TAC according to measurements of the stock by acoustic and DEPM surveys in May–June (Table 9.3a). The preliminary TAC at the beginning of the year was aimed at keeping the stock safely above B_{lim} even if the incoming year class is poor (ICES, 2001a, 2003a). However, the two-step TAC regime was not followed by managers before the collapse of the fishery in 2005.

European sardine

Routine assessment of the Atlanto-Iberian stock of sardine has been conducted annually since 1982, currently based on a data series since 1978. In the late 1990s, assessment was based on the ICA model (Table 9.3b), using age-disaggregated catch data from the fishery (0 to 6+ age group), three CPUE series from Galicia (abandoned in 1999 due to difficulties in estimation of changes of effort), three incomplete series of acoustic estimates (relative index of stock abundance from spring series from Spain and Portugal, autumn series from Portugal) and a few point estimates of spawning biomass from the DEPM (absolute index of abundance from combined Iberian spawning biomass estimates; Table 9.2b). A natural mortality of 0.33 (Pestana, 1989) has been used as a fixed value for all ages and years. During the late 1990s, survey indices (acoustics and DEPM) and catch data from Spain and Portugal provided contradictory indications, showing a marked stock decline in Spanish but no indication of change in Portuguese waters.

Although the assessment model used at the time (ICA) allowed for a change in exploitation pattern within the study period, it did not allow for the existence of independent biological units within the stock area or for changes in

the relative distribution and abundance within the study period. Finally, there was a bulk of evidence about wider environmental changes in the western coast of the Iberian Peninsula during the 1990s, which in some cases were linked to larger scale inter-decadal changes in hydrological and climatic conditions in the North Atlantic. These issues were reviewed in a dedicated ICES Workshop in 1998 (ICES, 1998c) and as part of the ICES Working Group on sardine assessment in 1999 (ICES, 2000) but with inconclusive results.

In the 2002 assessment, a new and more flexible catch-at-age model and software (Assessment Model Combining Information – AMCI) was used to explore the input data and assumptions underlying the ICA assessment. Results from the exploratory analysis detected violations of assumptions regarding survey catchability and fishery selectivity, which pointed to structural uncertainties in the stock assessment. AMCI provided a description of the stock dynamics which was more consistent perceived spatio-temporal variability, and so AMCI was adopted as the standard assessment model in 2003.

Structural uncertainties in the models do not change perceptions of the present level of the stock, but question the evolution of the biomass from the late 1980s to the present. This uncertainty has prevented the setting of reference points for management purposes, as well as the setting of official harvest control rules. However, national management measures have been taken since the decreasing SSB trend observed in the late 1990s (Fig. 9.3b). These national management measures mainly considered the control of fleet size, a reduction in fishing effort and the implementation of stricter catch limits (in many areas translated to a maximum daily catch per vessel).

Catch predictions for management purposes are only carried out for the short term, using a deterministic projection assuming the parameters estimated by the chosen assessment model, with corrections for the estimation of last year. recruitment and with subsequent recruitments estimated as the geometric mean of the time series. Nevertheless, results from the predictions are not considered very reliable, both due to uncertainties in the recruitment prediction and to the inadequacy of any equilibrium assumption. Unlike most other European stocks, assessment advice is not translated into TAC and national quotas, but is left to the governments of Portugal and Spain to define annual catch limits for the national fleets and distribute them among producer organizations and non-associated vessels.

North Sea autumn spawning herring stock

The stock assessment is carried out by the ICES herring assessment working group. In 2006, a comprehensive review of the assessment took place (ICES, 2006d), and concluded that the assessment was robust to scrutiny, had little

Table 9.2b. *Monitoring information and data input: sardine stocks*

Species	Stock	Fishery data — Parameter	Time series	Frequency	Fishery-independent data — Parameter	Time series	Frequency
Japanese sardine (*Sardinops melanostictus*)	Pacific	Length frequencies	1976–2004	Monthly	Egg surveys	1978–2004	Monthly
		Landings	1976–2004	Monthly	SST	1978–2004	Monthly
		Immature fish wintering abundance index	1976–2004	1×year	Surface trawl survey	1996–2004	2×year
California sardine (*Sardinops sagax caerulea*)	Northern sub-population	Landings	1982–2005	Monthly	Driftnet survey	1994–2004	5×year
		Age composition	1982–2005	Monthly	Egg survey	1985–1987	Each April
		Weight-at-age	1982–2005	Monthly	Aerial survey	1995–2006	Year-round
					SST	1986–2005	Daily
Humboldt sardine (*Sardinops sagax*)	North–Central Peru	Length frequencies	1968–2005	Monthly	Hydroacoustic surveys	1975–2005	2×year
		Age length keys	2000–2005	Quarterly	Egg surveys	1985–2000	1×year
		Landings	1960–2005	Monthly	SST	1925–2005	Monthly
		CPUE	1970–2005	Annually			
	South Peru–North Chile	Length frequencies	1974–2005	Monthly	Hydroacoustic surveys	1981–1995	1×year
		Age–length keys	1974–2005	Quarterly	Egg surveys	1985–2005	4×year
		Landings		Annually	Larval surveys		
					SST		
Humboldt common sardine (*Strangomera bentincki*)	Central-South Chile	CPUE	1986–2005	Quarterly	Egg surveys	2002–2005	1×year
		Age–length keys	1991–2005	Annually	Larval surveys		
		Landings		Annually	SST		
Brazilian sardine (*Sardinella brasiliensis*)	Brazilian Southeast Bight	Age–length keys	1981–1995	Annually	Egg surveys	1970–2008	2×year
		CPUE	1964–2005	Annually	Larval surveys	1970–2008	2×year
		Landings	1964–2005	Monthly	Hydroacoustic surveys	1970–1990s	2×year
		Length frequencies	1968–2005	Monthly	SST	1977–1983	Mean values (Dec–Jan)
Benguela sardine (*Sardinops sagax*)	Southern (South African) stock	Length frequencies	1980–2005	Monthly	Hydroacoustic surveys	1984–2005	2×year
		Landings	1980–2005	Annually			
	Northern (Namibian) stock	Length frequencies	1980–2005	Monthly	Hydroacoustic surveys	1990–2005	2–4×year
		Landings	1980–2005	Annually			
European sardine (*Sardina pilchardus*)	Atlanto-Iberian stock	Length frequencies	1978–2005	Monthly	Hydroacoustic surveys	1984–1988, 1995 onwards	2×year
		Landings	1978–2005	Annually	Egg surveys	1988, 1997, 1999 – onwards	1 every 3 years
Australian sardine (*Sardinops sagax neopilchardus*)	Southwestern Australia (3 management units)	Length frequencies	1989–2005	Monthly	Egg survey	1993–2005	Every few years for each of 3 management units
		Weight, sex, maturity	1989–2005	Monthly			
		Age composition (age–otolith weight relationship)	1989–2005	Monthly			
		Landings	1975–2005	Monthly			

Table 9.2c. *Monitoring information and data input: herrings and sprats*

Species	Stock	Fishery data			Fishery-independent data		
		Parameter	Time series	Frequency	Parameter	Time series	Frequency
European sprat (*Sprattus sprattus*)	Black Sea	Size and age composition	1945–2005	Monthly	Mid-water trawl survey	1967–1993	1×year
		Landings	1925–2005	Monthly	Juvenile fish surveys	1980–1993	1×year
		CPUE	1978–2005	Monthly	SST	1915–2005	Monthly
	Baltic Sea	Length and age frequencies	1974–2005	Depends on catches	Hydroacoustic surveys	Autumn: 1983–2005	2×year
						Spring 2001–2005	
		Landings	1974–2005	Annually	Egg surveys	1974–2005	3×year
					Cod stomach contents	1977–1993	Quarterly
					Winter NAO	1973–2006	Annually
	North Sea	CPUE	1995–2005	Log-book per trip	Hydroacoustic survey	2003–2005	1×year
		Length frequencies	1984–2005	Quarterly			
Atlantic herring (*Clupea harengus*)	North Sea autumn spawning stock	Landings	1984–2005	Quarterly	Trawl survey	1984–2006	1×year
		Length frequencies	1960–2005	Quarterly	Hydroacoustic survey	1989–2005	1×year
		Landings	1960–2005	Annually	Larvae	1973–2005	4×year
		Discards	1996–2005	Quarterly- not all nations	Post larval surveys	1992–2006	1×year
	Arcto-Norwegian spring spawning stock	Length frequencies	1950–2005	Quarterly	Trawl survey	1984–2006	1×year
		Landings	1950–2005	Annually	Hydroacoustic surveys	1988–2005	5×year
	Central Baltic	Length and age frequencies	1974–2005	Depends on catches	Larvae	1981–2005	1×year
		Landings	1974–2005	Annually	0 group surveys trawl	1974–2004	1×year
					Tagging	1975–2004	1×year
					Hydroacoustic surveys	Autumn: 1982–2005	1×year
	Gulf of Riga	Length and age frequencies	1977–2005	Depends on catches	Cod stomach contents	1977–1993	Quarterly
		Landings	1977–2005	Annually	Hydroacoustic surveys	July–August 1999–2005	1×year
		CPUE (trap-net)	1982–2005	Depends on catches	SST	1977–2005	1×year (April)
					Zooplankton abundance	1977–2005	1×year (May)

Table 9.3a. Assessment and management: anchovy stocks

Species	Stock	Assessment			Harvest strategy				Operating model	
		Type	Inputs	Outputs	Type	Inputs	Decision criteria	Frequency	Type	Inputs
Japanese anchovy (Engraulis japonicus)	NW Pacific	Annual VPA	Annual landings, age structure, commercial CPUEs	F, SSB, R, N-at-age	Harvest Control Rules (HCRs) under stable high stock biomass condition	HCRs based on results of annual VPA	$F_{limit}=F_{sim}$ $F_{target}=1.0\times F_{limit}$ F_{sim} can continue the lowest SSB level during the recent five years	Annual	n.a	n.a
	Seto Inland	Monthly VPA	Monthly landings, age structure, commercial CPUEs	F, SSB, R, N-at-month	HCRs under stable middle stock biomass condition	HCRs based on results of monthley VPA	$F_{limit}=F_{sim}$ $F_{target}=0.8\times F_{limit}$	Annual	n.a.	n.a.
	Tsushima Current	Monthly VPA	Monthly landings, age structure, commercial CPUEs	F, SSB, R, N-at-month	HCRs under increasing middle stock biomass condition	HCRs based on results of monthley VPA	$F_{limit}=F_{current}$ $F_{target}=0.8\times F_{limit}$	Annual	n.a.	
California anchovy (Engraulis mordax)	Central Sub-population	Monitored only	Landings	Trends in landings	Low constant quota	Landings	None	Annual	None	None
Humboldt anchovy (Engraulis ringens)	North-Central Peru	Acoustic	Length–frequency sampling, L-W relationship	Length–frequency weighted to biomass, Adult biomass, juvenile biomass	Short-term projections under uncertainty	Adult biomass, juvenile biomass, Fishing and natural mortality	TAC on the basis of minimum spawning biomass	6-monthly	None	None
	South Peru-North Chile	Statistical catch-at-age analysis	Landings, commercial CPUE, and a larval indice	F, SSB, R, N-at-age	Constant F policy and short-term projections under uncertainty	Based on results of annual assessment	TAC based on SSB/R criteria	Annual	None	None

Table 7.3.1. (cont.)

Species	Stock	Assessment Type	Inputs	Outputs	Harvest strategy Type	Inputs	Decision criteria	Frequency	Operating model Type	Inputs
Benguela anchovy (*Engraulis encrasicolus*)	Southern Benguela anchovy	Model-free			OMP – Harvest Control Rules (HCRs), simulation tested on alternative operating models (alternative "states of nature" for the stock)	HCRs based directly on survey indices of abundance	Evaluate summary performance statistics (e.g. risk, average catch, interannual catch variability, stock depletion)	TAC. Initial in January, revised in June. OMP. Multi-annual (3–5 y)	Age-structured production model; stock-recruit model differs between operating models, (Hockey-Stick stock-recruit model assumed for base case)	Catch, survey indices, age composition of catches and survey indices
European anchovy (*Engraulis encrasicolus*)	Bay of Biscay	ICA, (Patterson and Melvin, 1996)	Catches at age, Spawning biomass and population at age indices from Acoustics and DEPM	F, SSB, R, N-at-age	No HCR is adopted. But in practice Fixed TAC of 33000t, but within season closure if SSB falls below B_{lim}	In season closure depending on survey indices of biomass at spawning time	No HCR. No projection of the population and the fishery available. In season closure of the fishery if SSB falls below B_{lim}	Annual	No formal operating model is adopted. Draft testing of HCR include age structure dynamic models and biomass Dealy models	n.a
	NW Mediterranean	n.a.	n.a.	n.a.	n.a.	n.a.	n.a.	n.a.	n.a.	n.a.
	Adriatic Sea	Laurec-Sheperd Tuned VPA	Landings, commercial CPUEs	F, SSB, R, N-at-age	n.a.	n.a.	n.a.	Annual	n.a.	n.a.
	Aegean Sea	n.a.	n.a.	n.a.	n.a.	n.a.	n.a.	n.a.	n.a.	n.a.
	Black Sea	VPA Modified Baranov method	Catch at age	N-at-age, F-at-age, SSB, R,	Deterministic short-term projections	Output from "best" assessment	$F \leq F_{pa}$ (F_{msy}) Catch $\leq$ TAC	Annual	n.a.	n.a

Table 9.3b. Assessment and management: sardine stocks

Species	Stock	Assessment			Harvest strategy				Operating model	
		Type	Inputs	Outputs	Type	Inputs	Decision criteria	Frequency	Type	Inputs
Japanese sardine (*Sardinops melanostictus*)	Pacific	Annual VPA	Annual landings, age structure, trawl survey index, egg abundance	F, SSB, R, N-at-age	Harvest Control Rules (HCRs) under low stock biomass condition	HCRs based on results of annual VPA	$F_{limit}=F_{recovery}$ $F_{target}=0.8 \times F_{limit}$ $F_{recovery}$ can recover SSB to the target SSB (1996 level) in 2015	Annual	n.a.	n.a.
California sardine (*Sardinops sagax caerulea*)	Northern sub-population	'ASAP' age-structured forward-projected model, with Beverton-Holt stock-recruit model	Catches, age composition of catches, survey indices	F, SSB, R, N-at-age,	OMP – Harvest Control Rule; simulation tested on alternative states of nature (SST-based F_{msy}) using different thresholds of minimum biomass and maximum catch	Age 1+ biomass output from "best" assessment	F<=Fpa, B>=Bpa	Annual	'ASAP' age-structured forward-projected model, with Beverton-Holt stock-recruit model	Catches, age composition of catches, survey indices
Humboldt sardine (*Sardinops sagax*)	North-Central Peru	Acoustic	Length-frequency sampling, L–W relationship	Length-frequency weighted to biomass, Adult biomass, juvenile biomass	Short-term projections under uncertainty	Adult biomass, juvenile biomass, Fishing and natural mortality	TAC on the basis of minimum spawning biomass. Since 2000, low abundance by environmental effect	Semestral	n/a	n/a
	South Peru–North Chile	ADAPT catch-at-age analysis	Landings, commercial CPUE, and a larval indice	F, SSB, R, N-at-age	n.a	n.a	n.a	n.a	n.a	n.a
Humboldt common sardine (*Strangomera bentincki*)	Central-South Chile	Statistical catch-at-age analysis	Landings, commercial CPUEs, acoustic biomass indices	F, SSB, R, N-at-age	Short-term projections under uncertainty	Output from the age-structured assessment	TAC on the basis of 10% of risk, at F60% fishing mortality reference point.	Annual, reviewed after the acoustic survey to take into account recruitment	n.a	n.a

Species	Stock	Assessment			Harvest strategy				Operating model	
		Type	Inputs	Outputs	Type	Inputs	Decision criteria	Frequency	Type	Inputs
Brazilian sardine (*Sardinella brasiliensis*)	Brazilian Southeast Bight	Surplus Yield and Yield/Recruit Models, VPA	Landings, commercial CPUEs, age structure, survey indices	Estimated CPUE trends, F, SSB, R, N-at-age	Alternative "states of nature" for the stock	Survey indices of abundance (hydroacoustic, eggs and larvae)	Evaluate summary performance statistics (average catch, interannual catch variability, stock depletion)	Multi-annual (~3 years)	Age-structured model, assuming a Beverton-Holt stock-recruit model	Catch, age-composition of catches and survey indices
Benguela sardine (*Sardinops sagax*)	Northern Benguela	Simple age-structured models, accounting for uncertainties related to key assumption	Catch survey indices, length frequencies of surveys and catches	Current stock status	TACs based on projected outcomes from a range of model hypotheses		Reference points are available, but decisions have been heavily influenced by socio-economic factors in recent years of low biomass	TAC: annual	n.a	n.a
	Southern Benguela	Model-free			OMP – Harvest Control Rules (HCRs), simulation tested on alternative operating models (alternative "states of nature" for the stock)	HCRs based directly on survey indices of abundance	Evaluate summary performance statistics (e.g. risk, average catch, interannual catch variability, stock depletion)	TAC: annual TAB: Initial in January, revised in June OMP: Multi-annual (3–5 years)	Age-structured production model; stock-recruit model differs between operating models. (Hockey-Stick stock-recruit model assumed for base case)	Catch, survey indices, age-composition of catches and survey indices
European sardine (*Sardina pilchardus*)	Atlanto-Iberian	AMCI	Landings, survey indices	F, SSB, R, N-at-age	Deterministic short-term projections	Output from "best" assessment	$F \leq F_{pa}$, $B \geq B_{pa}$	Annual	n.a	n.a
Australian sardine (*Sardinops sagax neopilchardus*)	Southern Western Australia (3 management units)	Age-structured model. Simple analysis of age composition	Landings, age-structure, SSB from egg surveys	SSB Mean age	Harvest control rules: SSB relative to SSB and mean age provide a two-way table to objectively set a harvest rate of 0–10% of SSB. Apply a carrying-capacity conceptual model	SSB output from the age-structured assessment. Note: May discard use of model and use annual landings and mean age as HCR indicators	Minimum threshold for SSB; max and min thresholds for mean age. To set a harvest rate of 10% (i.e. maximum allowed) SSB must be >0.4 of SSB_∞ and mean age must be 4–5 years	Annual	Age-structured model, assuming a Beverton-Holt stock-recruit model	Catches, age-composition of catches, SSB estimates from egg surveys

Table 9.3c. *Assessment and management: herrings and sprats*

Species	Stock	Assessment			Harvest strategy				Operating model	
		Type	Inputs	Outputs	Type	Inputs	Decision criteria	Frequency	Type	Inputs
European sprat (*Sprattus sprattus*)	Black Sea	XSA, ICA Absolute biomass from mid-water trawl survey	Catch ate age commercial CPUEs, survey indices	N-at-age, F-at-age, SSB, R	Deterministic short-term projections	Output from "best" assessment	$F \leq Fpa$ (Fmsy) Catch $\leq$ TAC	Annual	n.a	n.a
	Baltic Sea	Extended Survivor Analysis (XSA)	Landings, survey indices, M2 from MSVPA	F, SSB, R, N-at-age	Deterministic short-term projections	Output from "best" assessment, 0-group from hydroacoustic survey, winter NAO	$F \leq Fpa$, $B \geq Bpa$	Annual	n.a	n.a
	North Sea	Catch Survey analysis (CSA)	Landings, surveys – IBTS	Relative biomass, relative recruits	None. No clear management objectives or evaluated strategy	Output from "best" assessment	None, based on Council of Ministers decision and advice from ICES and STECF	Annual, within year	n.a	n.a
Atlantic herring (*Clupea harengus*)	North Sea autumn spawning stock	Integrated catch age age (ICA)	Landings, misreported catch, discards, surveys	F, SSB, R, N-at-age	Harvest control rule (EU Norway agreement)	Output from "best" assessment	SSB $>1.3 \times 10^6$ $F_{2-6}=0.25$, $F_{0-1}=0.12$ SSB $<0.8 \times 10^6$ $F_{2-6}=0.10$, $F_{0-1}=0.04$ If SSB betw. 0.8–1.3×10^6 then linear decrease in F	Annual	n.a	n.a
	Arcto-Norwegian spring spawning stock	SeaStar	Landings, surveys, tagging recoveries	F, SSB, R, N-at-age	Harvest control rule (Iceland Russia Faroes agreement)	Output from "best" assessment	SSB $>5.0 \times 10^6$ $F_{5-14}=0.125$ SSB-2.5×10^6 $F_{5-14}=0.05$ If SSB betw. 2.5–5.0×10^6 then linear decrease in F	Annual	n.a	n.a
	Central Baltic stock	Extended Survivor Analysis (XSA)	Landings, survey indices, M2 from MSVPA	F, SSB, R, N-at-age	Deterministic short-term projections	Output from "best" assessment, 0-group from acoustic survey		Annual	n.a	n.a
	Gulf of Riga stock	Extended Survivor Analysis (XSA)	Landings, survey indices, CPUE	F, SSB, R, N-at-age	Deterministic short-term projections	Output from "best" assessment, SST in April, Zooplankton abundance in May	$F \leq Fpa$, $B \geq Bpa$	Annual	n.a	n.a

retrospective change between years and performed well. A survey-based SURBA[17] model (Needle, 2004), a separable period-based ICA model (Patterson, 1998), a VPA-based model XSA (Darby and Flatman, 1994; Shepherd, 1999) and a simple two-stage population model CSA (Mesnil, 2003) gave similar perceptions of the trends in mortality and spawning stock biomass. Due to the higher precision in the estimation of fishing mortality in the terminal year, and the lower retrospective bias between years, the assessment model ICA is used to provide the basis for advice.

The assessment feeds directly into the management agreement between the EU and Norway for Norway Sea herring. The agreement was adopted in 1997 and amended in 2004. According to the agreement, efforts should be made to maintain the SSB of North Sea Autumn Spawning herring above 800 000 t. The agreement states that an SSB of 1.3 Mt acts as a trigger. In years with an SSB above this trigger, the following years TAC will be based on an $F = 0.25\,y^{-1}$ for adult herring (mean ages 2–6) and $F = 0.12\,y^{-1}$ for juveniles (mean ages 0–1). If the SSB falls below 1.3 Mt, the fishing mortality will have to be linearly reduced. If the SSB is below 0.8 Mt a TAC equivalent to $F = 0.1\,y^{-1}$ for adult herring and $0.04\,y^{-1}$ for juveniles should be set. Simulations suggest that this management agreement results in a sustainable fishery within the precautionary approach.

The agreement also states that a TAC deviation of more than 15% between two subsequent years should be avoided, unless the parties consider this appropriate. The regulations on the bycatch of herring in the industrial fishery tend to change between years, but all bycatch should be counted against the TAC.

Arcto-Norwegian spring spawning herring stock

The stock assessment is carried out through the ICES northern pelagic and blue whiting fisheries working group (ICES, 2005b). The SeaStar model is used (www.assessment.imr.no), a statistically based model designed to minimize the use of weighting of time series. It also allows for the inclusion of tagging data into the assessment and only "tunes" the assessment on the bigger year classes. This allows the year-class signals, which are very strong in this stock, to have maximum influence on the assessment.

The assessment feeds directly into the long-term management agreement between the EU, Norway, Faroes, Iceland and Russia (called the Coastal States). The agreement states that every effort should be made to maintain SSB above 2.5 Mt. The TAC should be restricted to ensure that fishing mortality (F) remains less than 0.125 y^{-1}. Should SSB fall below 5 Mt, the F should be reduced from 0.125 y^{-1} to allow a safe and rapid recovery of SSB to above 5 Mt. The basis of the reduced F should be at least a linear reduction in F from 0.125 y^{-1} at 5 Mt to 0.05 y^{-1} at 2.5 Mt. So basically the fishery is managed with a target F

of 0.125 y^{-1}, and a trigger biomass of 5 Mt. This agreement is dependent on the advice from ICES, and simulations by ICES suggest that this management agreement results in a sustainable fishery within the precautionary approach. Despite having this management agreement, the coastal states have not been able to agree on an annual quota for Norwegian spring spawning herring since 2003 but have, however, still kept their quotas close to the scientific advice coming from ICES.

North Sea sprat

The ICES herring assessment working group conducts assessments of North Sea sprat based on indicators derived from a research survey and on a two-stage Catch-Survey Analysis (CSA). The CSA model (Mesnil, 2003) assumes that the population consists of two stages: the recruits and the fully recruited ages. The model results are highly dependent on the assumptions about natural mortality and the IBTS is the survey used with the commercial catch data. The assessment is only used to indicate stock trends. There are no reference points for this stock (either precautionary or target).

ICES has, for many years, recommended "within year" management for North Sea sprat, with the advice based on the results of surveys. The current advice is based on a regression between the annual survey index (IBTS) and annual catches, suggesting a probable catch for that year based on the quarter 1 IBTS, and extrapolations of the CSA assessment with a range of TACs into the forthcoming year. Neither of these methods has been evaluated and can be considered reliable.

Currently, North Sea sprat is managed by a TAC and a bycatch quota, but the scientific rational for the TAC is unclear. Other management measures include that Norwegian vessels are not allowed to fish in the second and third quarters and are not allowed to fish in the Norwegian zone until the quota has been taken in the EU zone. Recently, the sprat fishery has not reached its TAC and the bycatch of herring has been declining (now at <10% of the catch).

Baltic Sea sprat

The standard stock assessment model deployed is an Extended Survivor Analysis (XSA) (Shepherd, 1999) using Multispecies Virtual Population Dynamics (MSVPA) derived predation mortalities as input (ICES, 2005c). Alternatively, ICAs (ICES, 1997a) and multispecies stock production models (Horbowy, 1996) have been tested, revealing similar stock development trends.

The multispecies assessment separates the western and eastern sprat stocks (ICES, 2003c), but as the western stock is comparatively small, the single species assessment uses predation mortalities from the eastern multispecies assessment. Additionally, MSVPA runs for single statistical regions containing major spawning areas. They are presently not used in assessment, but are used in stock recruitment studies (Köster *et al.*, 2003b).

Short-term catch and stock predictions are conducted to recommend TACs, with management constraints being a precautionary fishing mortality (F_{pa}) and a precautionary spawning stock biomass (B_{pa}), below which action is taken to ensure recovery of the stock above B_{pa} and reduction of the fishing mortality below F_{pa}. Furthermore, there exists a biomass limit reference point B_{lim} which if violated result in severe fisheries restrictions, i.e. in case of a reduction below B_{lim} the fishery is stopped. These reference points are either based on stock-recruitment relationships (e.g. F_{pa} equals F_{med}) or medium-term projections (ICES, 1998a).

The short-term prediction conducted in year *t* for the year *t*+1 is based on assessment data from year *t*–1 yielding the population size at the beginning of year *t*, and requires estimation of recruitment for age-group 1 in years *t* and *t*+1. Traditionally, a geometric mean recruitment over a range of recent years has been used as recruitment estimate for all years. However, considering a temperature (or NAO)–recruitment relationship outperformed the standard procedure (MacKenzie and Köster, 2004) and consequently the winter NAO index was incorporated into the short-term prediction by ICES (2006c), together with the 0-group abundance index from the hydroacoustic survey, to predict recruitment in 2006 (year *t*), while recruitment in 2007 (year *t*+1) was still assumed to be the geometric mean. In an exploratory analysis MacKenzie *et al.* (2008) investigated the effect of additionally replacing the geometric mean recruitment for 2007 by an estimate from a NAO–recruitment relationship. The sensitivity of the spawner biomass at the end of the prediction period to the most likely range in NAO values (10th–90th percentiles) was found to be ca. 27%. For the specific case of the NAO in 2008, MacKenzie *et al.* (2008) found a 15% difference in spawner biomass.

The stochastic medium-term projection allows for uncertainty in stock size at the beginning of the projection period, the stock recruitment relationships and weight-at-age. However, the applied stock-recruitment relationship is environmentally insensitive, despite a clear dependence of recruitment on environmental conditions, i.e. temperature (Köster *et al.*, 2003b), NAO (Mackenzie *et al.*, 2008), and transport (Baumann *et al.*, 2006) and a limited variability in recruitment explained by the model. Mean weight-at-age and maturity ogives are kept constant as average over a number of most recent years in both short- and medium-term predictions, although clear density dependent relationships exist for weight-at-age and would, in fact, allow prediction independent of stock size. In single species predictions, natural mortality is kept constant, which appears to be reasonable for short-term predictions, while

medium-term projections should include cod as predator, with the methodology of multispecies projections being implemented (ICES, 2005c).

Baltic Sea herring

Herring in the Central Baltic has been assessed by ICES in varying stock units (Sjöstrand, 1989), but these were combined in 1990 due to practical reasons regarding input data. In 2002 preliminary assessments were conducted separating the main Central Baltic stock and the Gulf of Riga herring (ICES, 2002b) and these separate assessments form the basis for the scientific advice given by ICES since 2003.

Advice on both the Central Baltic and the Gulf of Riga stock is based on standard ICES short-term catch and stock predictions against the background of limiting biological reference points. The assessment model deployed is an XSA, tuned by hydroacoustic survey indices and for the Gulf of Riga herring additionally by CPUE from the trap-net fishery (ICES, 2006c). While for the Central Baltic herring MSVPA derived predation mortalities are used as input, the cod abundance in the Gulf of Riga has been low for over two decades and thus natural mortality (M) has been taken as constant at $0.2y^{-1}$ since 1984.

Standard ICES short-term catch and stock predictions are conducted to recommend TACs (see section on Baltic sprat). For the Central Baltic herring recruitment at age 1 in year *t* is estimated from hydroacoustic derived 0-group abundance, while Gulf of Riga recruitment is predicted from a multiple regression of historical recruitment on water temperature in April and zooplankton abundance in May of the year of birth (ICES, 2006c). As the annual assessment is conducted before environmental data for the assessment year are available, recruitment in year *t*+1 is estimated as geometric mean over a range of year classes, similar to the Central Baltic herring stock.

In stochastic medium-term projections stock-recruitment relationships are used for both stocks, for the Central Baltic herring assuming a linear increase in recruitment from origin to a breakpoint, estimated by ICES (2003d), and constant recruitment above this breakpoint. For the Gulf of Riga herring, a Beverton and Holt stock-recruitment relationhip (ICES, 2006c) is applied. Weight-at-age, maturity and natural mortality are kept constant, but uncertainty is considered for stock-recruitment relationships, initial stock sizes, and weight-at-age.

Management constraints for the Gulf of Riga herring include a defined precautionary fishing mortality (F_{pa}) and a precautionary spawning stock biomass level (B_{pa}) as well as a limit spawning stock biomass reference point (B_{lim}). For the Central Baltic herring only F_{pa} is defined. The reference points are either based on stock-recruitment relationships (F_{pa} equals F_{med} for the central Baltic herring, B_{pa} equals

M_{BAL} and B_{lim} derived from B_{pa} for the Gulf of Riga herring; see ICES, 1996) or derived from medium-term predictions (F_{pa} for Gulf of Riga herring; see ICES, 2003e). Other management measures include international (EU and Russia) and national technical regulations. National regulations by Estonia and Latvia for the Gulf of Riga include a limitation on the number and power of trawlers operating in the Gulf, a summer ban in the Estonian part of the Gulf from mid June to September and a 30-day ban for all trawl fishery during the main spawning migrations of herring in April–May.

Mediterranean and Black Sea pelagic fisheries

In the Adriatic Sea, stock assessment has been carried out by means of VPA (Darby and Flatman, 1994). Commercial catch data collection includes Italy, Slovenia and Croatia. Age–length keys are applied to the annual, catch-weighted length frequency distributions to obtain the corresponding age distribution. Age classes used in the VPA are 0 to 3 with the 4 used as a plus group. In most recent years, estimates of fishing mortality rates at age are obtained by Laurec–Shepherd tuning (Laurec and Shepherd, 1983) based on the CPUE at age obtained from the Porto Garibaldi fleet (which lands 25% of total landings). Natural mortality (M) is assumed to be 0.6 y^{-1} according to the age distribution of the catch and the observed maximum longevity of the species in the Adriatic Sea. However, additional VPA runs with M = 0.8 y^{-1} are usually performed. Recent work points out the importance of environmental factors in determining recruitment strength and the potential of using stock-recruitment models incorporating environmental variables (Santojanni et al., 2006). However, such models have not yet been included in the assessment procedure. Preliminary trials to tune VPA using fishery independent data (echosurveys) are also currently being conducted, showing positive results.

There is no established harvest strategy nor any quota system enforced. The management of the Mediterranean small pelagic fisheries is rather independent from scientific advice.

International stock assessments of Black Sea sprat and anchovy are based on catch-at-age models (Prodanov et al., 1997). Pre-recruit and adult stock abundance indices were used by Daskalov et al. (1996) and Daskalov (1998) for tuning catch-at-age assessment models of sprat. In recent years research surveys are performed more rarely because of financial problems of the research institutes. A quasi-decadal pattern dominates the abundance series of both species. Abundance of both stocks increased during the 1970s and 1980s when the highest catches were recorded. A combination of low recruitment and excessive fishing led to stock collapses in the early 1990s (Daskalov et al., 2008), which are illustrated by the peaks in fishing mortality in the late 1980s to early 1990s. At the same time the accidentally introduced ctenophore *Mnemiopsis leidyi* contributed to the decline in recruitment (Grishin et al., 1994). During the 1990s the anchovy and sprat stocks recovered despite the growing *Mnemiopsis* population and catches increased again.

The management of Black Sea stocks has been done separately by the surrounding countries despite the existence of a Joint Commission for the Fisheries in the Black Sea until 1991, which included Bulgaria, Romania and the former USSR, but not Turkey. Studies on stock assessment and management of anchovy and sprat have been published, e.g. by Tkatcheva and Benko (1979), Ivanov (1983), Ivanov and Beverton (1985), Domashenko et al. (1985). After the political changes in Eastern Europe, the Commission ceased its function.

At present, there is no routine international assessment and management of Black Sea stocks, although the stocks are shared. In the early 1990s a group of international experts performed stock assessments of 18 commercially important stocks including sprat and anchovy (Prodanov et al., 1997) and the GFCM recommended this study as a background paper for future stock assessment work. In the early 1990s the countries decided to prepare and sign a new convention for the fisheries in the Black Sea under the auspices of which international working groups will prepare assessment and management advice. However, the Convention has not been ratified. At the moment some internationally coordinating functions are carried out by the Commission on the protection of the Black Sea Against Pollution (http://www.blacksea-commission.org) in conjunction with GFCM. Meanwhile, each national government applies its own methodology and legislation in stock assessment and management. Before the collapse of the fisheries in the early 1990s (Daskalov et al., 2008), small pelagics in the Black Sea were considered as under-exploited and no special measures for their protection were adopted. Since 1989, Turkey (which catches 80%–90% of the total sprat and anchovy production) applies restrictions on minimum body size (9 cm total body length for anchovy), minimum mesh size, minimum discard rate, closed season and areas (near shore area of less than 16 m of depth). Regular (annual) stock assessments are used as base for short-term projections (e.g. in Bulgaria, Russia and Ukraine) and F_{msy} and TAC are used as limit reference points (e.g. by Russia and Ukraine). Target harvest is usually recommended as some percentage of the TAC related to the state of the stock. Other regulations on minimum body size, closed areas and seasons are applied on a regular or *ad hoc* basis related to the status of the stock.

Daskalov (1999) used generalized additive modeling (GAM) in exploring non-linear relationships in four Black Sea fish species (sprat, anchovy, horse mackerel and whiting) and different environmental variables (SST, SLP[18], wind, run-off). Despite relatively strong fish–environment linkages found, environmental indices have not been used in operational stock assessment. Pre-recruit surveys have been used for short-term forecasts of recruitment and SST data have been used to predict timing of formation of fishable concentrations of anchovy and sprat by the YugNIRO, Kerch (A. Mihaylyuk, pers. comm.).

Southwestern Australia sardine

The southwestern sardine fishery is managed using TACs and ITQs.[19] TACs for each of three management units are set (recommended) by a formal management advisory committee. An age-structured model with spatial components was developed initially for one region (Fletcher, 1992), with DEPM-based SSB estimates and relative recruitment strength forming the assessment advice for the other two regions. Fletcher's model was later replaced by a more sophisticated age-structured model (Hall, 2000) designed to encompass the entire southern WA sardine breeding stock but incorporating differing parameters for, e.g. recruitment and fishing mortality, between the three regional management units. This model fitted to monthly catches, fishery independent SSB estimates, and age composition. The log-likelihoods for the observed SSB estimates from DEPM surveys, age composition data, and recruitment deviations were combined, together with the penalty functions to form the overall log-likelihood. These objective functions were maximized simultaneously for each region to obtain maximum likelihood estimates of the parameters for that region. Parameters were estimated for each of the three regions including the initial recruitment, estimates of natural mortality (the two parameters of the selectivity curve), and annual recruitment deviations. A Bayesian procedure was used to determine estimates of uncertainty of the mature biomass using Markov-Chain Monte Carlo procedure.

Although the assessment model has successfully tracked trends in SSB, the levels of uncertainty (e.g. in the spatial dynamics of recruits) have negatively impacted the use of the models for providing definitive or "stand-alone" stock assessment advice. Along with the comparatively low natural productivity of the pelagic ecosystem off southern WA, the lack of precise point estimates of SSB has necessitated a conservative harvest strategy, and one that may eventually be based on fishery performance (e.g. annual landings) and age structure.

The current aim of management is to limit the harvest rate to less than 10% of SSB; harvest rate is set objectively using a two-way matrix that considers spawning biomass (0%–100% SSB_∞[20]) and the average age (from 1–8+ years). The age component provides an indication of the health of the stock relative to past and recent recruitment levels. The harvest rate control rules allow the maximum exploitation rate (i.e. 10%) only if SSB > 0.4 SSB_∞ and mean age is 4–5 years.

A mono-specific mass mortality killed 70% of Australia sardines in 1998/99 (Gaughan *et al.*, 2000); SSB has recovered but in WA recovery of the commercial fisheries has been hampered by failure of market-size (i.e. mature, >3+ years) sardines to return to the traditional fishing grounds. Current research activities are focused on bycatch mitigation, particularly quantifying and minimizing interactions with seabirds and dolphin.

Discussion

The present review demonstrates that the assessment and management of small pelagic fish varies substantially in its approach and performance. Most of the stocks reviewed have some sort of scientific assessment program in place and a management approach that takes into account this assessment, albeit to different degrees. California sardine, Humboldt anchovy, southern Benguela, and Baltic stocks are examples where the advice based on scientific assessments is respected reasonably in the management process, while Namibian sardine and Bay of Biscay anchovy are examples of management practices that have deviated substantially from scientific advice on the basis of political or poorly explained socio-economic reasons. Considering aspects not included in biologically based scientific assessments is both understandable and to be expected. For example, issues such as employment, industry and social stability or revenue income should be an intrinsic part of the management process. However, these factors should be subject to the same type of analysis applied in determining biological risk if we are to demonstrate a consistent approach to management (Cochrane, 1999). For example, there was considerable deviation from the scientific advice provided for the southwestern Australia sardine fishery in the mid 1990s, with TACs set higher than recommended for a number of years (which were not filled), which contributed to a stock crash in one region. The development of an objective decision table helped overcome this. The southern Benguela OMP (De Oliveira and Butterworth, 2004) approach appears to have broadened its management objectives without compromising its quantitative, analytical approach. However, it is more common to have rigorous, data-intensive scientific advice modified to account for vague, subjective and untested socio-economic reasons. Such a two-tier approach cannot demonstrate true resource stewardship and use, nor does it allow the implementation of management performance measures.

Surprisingly, this review has demonstrated that many stocks have assessment and management processes that are disconnected, or have scientifically based management advice that is often overlooked. The Bay of Biscay anchovy, for example, has been managed by fairly fixed (30 000– 33 000 t for the period 1979–2004) annual TACs set independently of the scientific advice given. Annual catches have been well below this fixed TAC in recent years, indicating that the TAC was unable to constrain the fishery. The main reason for this is that the anchovy TAC responded more to European Union-wide politics, in which quotas can be exchanged between countries for products (e.g. quotas for other fisheries, agriculture products) completely unrelated with anchovy management issues. In the case of the Mediterranean anchovy, as in other fisheries in this basin, management is limited to technical measures on effort and gear regulations that disregard the state of the stock. Despite dwindling transboundary stocks, governments appear to discourage the collection of fisheries data and the development of effective, stock-wide assessment procedures (Lleonart and Maynou, 2002). In another interesting example, the North Sea sprat is managed through an assessment methodology that relies heavily on assumptions about natural mortality and using a historical time series of catches where herring was misreported as sprat. All parties involved accept that there is no reliable scientific basis for the specific numeric TAC advice given. Whilst this is not problematic when exploitation is light and stock productivity high, it is likely to haunt managers if demand increases and/ or productivity declines require a more effective management approach.

In conclusion, in scientific circles it is widely accepted that when science is credible, it has proven to be a strong basis for advice and management of pelagic fish stocks. The examples reviewed here show that the most successfully managed stocks rely on a scientific program of advice, suggesting that science-based management cannot be replaced, and that only a properly tailored scientific assessment and management program can provide the speed of response and the flexibility of management that highly variable small pelagic fisheries demand. However, while reliable science provides the best set of principles, this prerequisite is not sufficient to ensure sustainable long-term use and conservation of fishery resources.

Regarding assessment standards, the most effective monitoring programs are based on fishery-independent surveys that provide precise information of the state of the stocks. Certainly, for short-lived species, catches-at-age analysis based on VPA-like assessments offer little value, given the poor VPA convergence for these species and the poor prognosis of next evolution of the population they can provide. Therefore, direct surveying of the population offers the best alternative to such methods, providing current and ready to apply knowledge about the status of the population. Preferred surveys are either based on ichthyoplankton sampling techniques (e.g. Daily Egg Production Method), hydroacoustics, or both. There are also suggestions that pelagic fish assessment may benefit from incorporating the spatial structure of the population into the estimation

procedure. This, for example, has been attempted for southwestern Australia sardine, but the spatial complexities or non-linearities (e.g. interactions between management units change over time) hampered fitting of the model, hence the move back to a simpler management system. The use of catch per unit effort (CPUE) information, particularly in the absence of fishery-independent methods, is questioned because the schooling habits of pelagic fish suggest CPUE may not reflect stock abundance (Ulltang, 1980; Csirke, 1988; Pitcher, 1995; Mackinson et al., 1997). However, changes in spawning stock biomass estimated by VPA have been compared to those estimated using annual egg production methods in Japan, indicating similar trends and almost the same stock level although variability in the VPA estimate was lower (Y. Oozeki, unpublished data). As a result, the latter is adopted to estimate the Allowable Biological Catch (ABC). There is also agreement that biological sampling programs directed at the acquisition of catch volumes, length frequency and age data (one of the major sources of uncertainty) must be protected, and that this information should be spatially segregated to allow for spatial management rules.

Overall, scientific uncertainty and lack of governance are the most cited reasons to explain fisheries management failure. Regarding scientific uncertainty two approaches have dominated fisheries science in recent times. One argued that uncertainty needs to be reduced through additional science and measures (Ulltang, 1996, 2003; King et al., 2001), while the other assumes that uncertainty cannot be significantly reduced and thus that the development of management procedures robust to it should be promoted (Walters and Collie, 1988). Evidence of both approaches can be found in this review, for example, in relation to recruitment forecasting. Short-lived small pelagic fish species rely on their annual recruitment to secure a healthy adult biomass the following year. In addition, many small pelagic fisheries target these recruits. As a result recruitment estimation has been one of the most debated, conflictive and productive areas of pelagic fish research in recent years (e.g. Shepherd et al., 1984; Myers, 1998). Survey-based recruitment estimation in South Africa, for example, is only available after sometimes substantial anchovy catches have been landed (Butterworth et al., 1993) on the basis of an initial TAC that assumes median recruitment. Indirect early recruitment estimations based on environmental proxies have been explored because the average annual catch could be increased by up to 48% if a very precise prediction could be made before fishing commences (Cochrane and Starfield, 1992). On the basis of simulation work, De Oliveira and Butterworth (2005) concluded that an environmental index needed to explain 50% or more of the total variation in recruitment before an Operational Management Procedure that uses the index as a recruitment proxy starts to show benefits in terms of long-term catches and risk. However, removing constraints on the extent of TAC changes in the South African OMP would relax the amount of variation that the index needs to explain for comparable benefits, but at the cost of larger interannual variability in TAC. This is because constraints on TAC changes limit the response of an Operational Management Procedure to the environmental index.

In the case of the Bay of Biscay, science cannot produce the advice required by managers because of the lack of a procedure for forecasting the anchovy population and the fishery for the management year. Management needs a reliable indicator of the latest year-class strength prior to its recruitment into the fishery, but the failure of recruitment predictors based on statistical relationships (Borja *et al.*, 1998) led to their abandonment in TAC determination. De Oliveira *et al.* (2005) investigated under what circumstances incorporating environmental-based recruitment indices would lead to management improvements. Results show that precautionary approaches may better ensure successful management than consideration of uncertain or moderate-to-weak (in terms of the relationship to recruitment) environmental effects. Scientists in charge of this stock now consider that any future attempts to use environmental estimates of recruitment would require prior definition of criteria against which success or failure would be judged, a process that should apply to other stocks and species. In the meantime plans are in place to revise the current management regime so as to take into account recruitment fluctuations. This may be achieved by developing a decision rule using directly the information from the existing time series of spring surveys or by taking into consideration the results of a recently established autumn juvenile survey program.

Environment-based recruitment estimators, however, are explicitly used in the management of the California sardine (Jacobson and MacCall, 1995; PFMC, 1998). Justification is based on the observation that Maximum Sustainable Yield (MSY) and stock biomass for MSY (B_{MSY}) are dependent on habitat area (Jacobson *et al.*, 2005), and thus monitoring habitat area (proxied by SST) can help anticipate periods of high and low productivity and adjust management accordingly (Jacobson *et al.*, 2005). The value of this index has been consistently proven (Myers, 1998). In recent years more examples have been provided to justify the incorporation of environment–recruitment relationships (e.g. Yatsu *et al.*, 2005), suggesting that this area of work will continue, particularly in relation to developing medium to long term exploitation strategies in the context of global climate change.

Mackenzie and Köster (2004) consider that incorporating a recruitment estimate based on indices for atmospheric NAO and ice coverage outperforms the current assumption of medium recruitment in the assessment of the Baltic Sea sprat and first trials of implementation in assessment are being conducted (ICES, 2006c). The estimate of recruitment in North Sea herring is provided by the survey of post larvae, which explains over 70% of the variance in 0-group fish. This direct method is thus relatively robust and prior to the onset of fishing pressure in this relatively longer-lived species.

Discussing the value of pre-season forecasting, Walters (1989) concluded that improvements in management performance depended strongly on the average productivity of the stock concerned, and on the flexibility of the in-season regulatory system used to manage the stock. The fact is that there is not a single approach to be recommended on the subject of scientific uncertainty, because factors such as the dynamics of the fishery (targeting recruits vs. adults), the type of assessment (monthly, annual, or multi-annual) and the subsequent management measures to be taken, would determine whether resolving uncertainty is cost effective. However, a major stumbling block to using recruitment estimators is the lack of understanding of the dynamics of recruitment, the links between nursery and recruitment grounds, and growth and feeding ecology of young fish, among others. Several recruitment proxies are used, not without problems. For example, the proportion of 2-year-olds in the catch is used as an indicator of southwestern Australia sardine recruitment in different subregions, a figure that is critical in the scientific advice. Indirect methods, in particular, suffer from poor process understanding linking trends in proxy variables with recruitment variability (see De Oliveira and Butterworth, 2005; ICES, 2006f). These aspects are major scientific challenges that are expected to remain research priorities in the future (ICES, 2006e,f).

Another important source of uncertainty relates to the actual levels of fishing vs. natural mortality, as small pelagics are major forage species for other fish, birds and mammals (see Box 9.1). This aspect is influenced by catchability assumptions in the DEPM, bottom trawl surveys and VPAs, posing questions over management strategies based on fishing mortality levels. The catchability of fishing fleets also deserves further attention, particularly in relation to patterns and seasonality, which can be used to assess the effect of different management options and harvest control rules. This is particularly relevant in the case of species that are harvested using both trawlers and purse-seiners, and is linked to behavioral changes in the stocks themselves, as perceived in the Bay of Biscay anchovy of late.

Box 9.1. How is science contributing to the ecosystem approach to South African fisheries?

The ecosystem approach to fisheries (EAF) requires a sound scientific basis to provide the means of assessing the eco-system effects of fishing and the effectiveness of management options in response to identified risks. South Africa, by virtue of several decades of multidisciplinary studies and monitoring, and a relatively flexible management approach with some stakeholder participation, is well placed for a test case for EAF. Ecological risk assessment (Fletcher, 2005) has been applied to identify and prioritize ecosystem issues in three main fisheries: hake, small pelagics and west coast rock lobster. An example of a high priority issue is given in the table below for small pelagics (SPF) to illustrate how South Africa is moving towards an EAF by identifying required research and/or monitoring, indicators and management actions to address each issue (Shannon *et al.,* 2006). Indicators derived from biological or catch data facilitate the monitoring of ecosystem responses to management actions, implemented to optimize economic and social objectives while ensuring the ecological sustainability. Careful interpretation of these indicators is required such that a decrease in abundance is not misinterpreted as poor management (e.g. catch set too high) if it is rather a consequence of normal fluctuations in the stock (e.g. poor recruitment). Implementation of EAF is regarded as an ongoing process, composed of the following main needs:

(i)   Identification of the current status of the resource.
(ii)   Examination of concerns regarding single-species, community or ecosystem-based approaches (such as spatial issues or species interactions not currently accounted for in management) and expression of these as ecosystem objectives.
(iii)   Identification of indicators related to ecosystem objectives.
(iv)   Translation of ecosystem indicators into decision criteria (for example, by defining limit reference points to be avoided).
(v)   Identification of research and monitoring needs.
(vi)   Development of appropriate management actions to be taken with stakeholder participation.
(vii)   Development of evaluation criteria for adopted management measures.

A number of the above needs are already implemented as part of the Operational Management Procedure in South Africa (see Section "The Benguela operational management procedure (OMP) approach", above), while others (points (ii) to (iv)), are specific to EAF.

Issue	*Indicators*	*Research approaches*	*Management actions under consideration*
Impacts of removal of SPF on seabirds bound to breeding sites on land.	Bird population sizes; breeding success (fledgling weight, fledglings raised per breeding pair, breeding proportion); seabird diet composition; spatial indicators (e.g. overlap of seabird foraging and SPF fisheries).	Underway: Routine monitoring of seabird colonies; satellite tracking to assess foraging ranges; minimum realistic models; spatialized models of SPF around seabird colonies; quantification and formalization of the link between the SPF fishery and seabirds; quantification of functional responses of seabirds to SPF and identification of thresholds below which there are serious negative implications for seabirds.	Avoid seabird populations falling below limit reference points set according to IUCN[21] conservation criteria by reducing TACs or closing areas within foraging ranges; allow sufficient escapement of SPF for predators; avoid threshold levels of SPF below which the risk to seabirds is unacceptable

Whether investment is on more complex science, or on more flexible, robust and adaptive management systems (Walters, 1997), or both, a problem most of the fisheries reviewed here have suffered from, at some point in their exploitation history, is inadequate governance. Governance is the sum of the legal, social, economic, and political arrangements used to manage fisheries. It has local, national, and sometimes international dimensions and includes legally binding rules as well as customary social arrangements. Global reviews point out that prevailing systems of fisheries governance have often been ineffective. Governance failures are caused by managers trying to address multiple objectives in the absence of a formal analytical framework that places resource sustainability in its appropriate context. For example, the management of the Bay of Biscay anchovy has been driven by quota trade agreements between countries rather than resource sustainability. Another example of governance failure is the fact that scientific advice is more readily accepted when it implies an increase in fishing effort than when it requires a reduction. Failure to take sufficient action in the face of clear stock declines has been implicated in the collapses of the California and Namibian sardine, for example, even if overfishing was not necessarily the catalyst for the initial stock productivity declines. The impact of governance is clearly shown when comparing North Sea herring in the 1970s and 1990s. From the same biomass (0.5 Mt) failure to act on the advice resulted in a collapse of the stock in the 1970s and action in line with the advice resulted in a recovery in the 1990s (Simmonds, 2007). An example of an adequate governance framework is the South African OMP. This is anchored on fishery-independent surveys, leading to annual TACs which are revised twice through the year and on ground-breaking ongoing consultation with stakeholders to determine the most effective and valuable management solutions. These solutions are taken on the basis of trade-offs and exploitation patterns that are simulated and negotiated (De Oliveira and Butterworth, 2004). Although initially TACs were often overruled or adjusted by political authorities (Cochrane, 1999), over the last decade OMP recommendations have been respected (Plagányi *et al.*, 2007). The South African OMP example provided the basis for the development of objective management criteria in the southwestern Australia sardine fishery, a process for which industry members actively participated. Developing adaptive management practices applied by independent governance structures capable of interacting at ecological, social and economic levels is a necessity for effective stewardship and governance.

The reliance of small pelagic fish populations on annual recruitment explains the boom and bust nature of their fisheries, but there is also increasing appreciation of the existence of low frequency, multidecadal productivity regimes (or regime shifts) underlying their long-term dynamics. Several authors have suggested that these regimes may be synchronized across basins (Kawasaki 1983; Lluch-Belda *et al.*, 1989, 1992; but see Freon *et al.*, 2003). Multi-decadal productivity cycles are poorly understood, although they are observable in centennial records (Baumgartner *et al.*, 1992) as well as in modern catch statistics (Klyashtorin 1998). Among the examples reviewed here, productivity cycles have been detected in the Brazilian and Californian sardine stocks, Humboldt sardine and anchovy, Black Sea anchovy and sprat, Baltic Sea sprat and Arcto-Norwegian spring spawning herring, among others. The existence of regimes poses substantial problems to fisheries managers because the effects of the fishery and its management cannot be separated from natural cycles. They also suggest that, for management decisions to be effective, they must occur in synchrony with these cycles (Fréon *et al.*, 2005). For example, there are current concerns over whether sardine in the northern Benguela can recover from its present low level when the structure and functioning of the system nowadays seems to differ substantially from the way it operated in the 1970s, when the population was at its peak (Heymans *et al.*, 2004), particularly as natural mortality appears to have increased in current times (Fossen *et al.*, 2001). Fréon *et al.* (2005) advocates a two-level management strategy that would combine conventional fisheries management to deal with short-term fluctuations, and a long-term management strategy (for example, driving fleet capacity and investment cycles) driven by decadal fluctuations. Because productivity regimes synchronistically affect different species in opposite ways, single-species management systems are more vulnerable than those that consider the dynamics of competing resources. For example, Baltic sprat and cod have dominated successive productivity cycles in recent decades, with consequences also for associated fish species such as herring (Köster *et al.*, 2003a). Simulations have demonstrated that consideration of species interactions stabilizes the assessment of sprat (ICES, 1997b).

King (2005) and Polovina (2005) conclude that the most appropriate approach to managing fisheries under scenarios of productivity cycles is through the application of regime-specific harvest rates (RSHR). These harvest rates should be part of a decision rule framework, associated with timeframes for management response triggered when there are indications that a regime shift has occurred. Improved results could be achieved even if the switch in harvest rates was delayed some years after the regime shift (MacCall, 2002) either because of detection delays or because stakeholder pressure and institutional stiffness delays implementation. Jacobson *et al.* (2001) noted that the instantaneous rate of surplus production was effective in detecting regime shifts, and therefore useful in adapting management practices in real time. Katsukawa and Matsuda (2003) suggested a management procedure based on non-parametric target switching, i.e. prohibiting catch of the most depleted species in favour of alternating species, which would bring more long-term yield and stable biomass than those obtained from a single-species point of view. However, the different value of the targets and the relationship between value and volume requires further investigation before target switching could be successfully applied.

De Oliveira (2006) investigated different harvesting strategies in the South African fishery for sardine and anchovy and concluded that management procedures designed under the assumption of out-of-phase sinusoidal trends in species abundance could be more effective than traditional modeling approaches. The author observed that, comparing regime shift estimators, a 6-year running mean of the estimator outperforms a precise knowledge of the actual position in the cycle. The performance of these management procedures shows rapid deterioration as the amplitude of regime cycles increases. In the case of marine ecosystems in which exploited small pelagic species play a key role, Fréon *et al.* (2005) recommend specific statistical and socio-economic studies on the feasibility of a two-level management strategy that would combine adaptive management (short-term) and fleet capacity control (long-term) and would take into account ecosystem considerations. In western Australia the current approach to dealing with biomass fluctuations has been to suggest a small pelagic carrying-capacity, which equates to an upper biomass level against which industry should consider investment decisons but also noting that SSB can be much less than the carrying capacity (Gaughan *et al.*, 2004). Certainly, the development of exploitation strategies and levels in low and high abundance regimes must be a future research priority.

Regime-specific harvest rates, however, have a potential problem with the long-term investment in fisheries, while regime longevity is uncertain. For example, the Japanese purse-seine fishery – which is the major fishery for sardine, anchovy and mackerels – started its investments in new fishing fleets in the 1980s during the high stock period of Japanese sardine, resulting in over-capitalization when the fishery regime turned unproductive after 1988 (Yatsu, 2005; Yatsu *et al.*, 2005). This mismatch in the longevity of fishing fleets vs. productivity regimes is a fundamental problem for sustainable fishing. Fleets expand following increasing trends in fish biomass, usually some time after the start of a new high productivity regime, and the length of this regime is generally shorter than the lifespan of the new fishing fleet, thus leading to overcapacity at the start of a subsequent low productivity regime. Based on a simulation study of a hypothetical sardine-like stock, Harada and Nishiyama (2005) argued that constant harvest rate strategies (CHRS), such as F_{MSY} with a precautionary approach, are preferable to

RSHR because (1) CHRS is robust to estimation errors of stock abundance, (2) the resulting long-term catch is only slightly smaller than RSHR, and (3) CHRS is essentially constant for investment and thus can avoid overcapitalization derived from the above mismatch. CHRS also brings less year-to-year variation of catch and biomass compared to RSHR, hence it is preferable for the fishes and probably the surrounding fishery community. The performance of CHRS can be improved by incorporating catch prohibition of immature fish (Harada and Nishiyama, 2005). A CHRS that incorporates annual catch as an indicator of fishery performance is currently being investigated for the southwestern Australian sardine fishery in recognition of both the low value of this fishery and the uncertainty regarding point estimates of SSB.

The assessment and management of small pelagic fish is also affected by their straddling nature across political boundaries. This review has noted how the management of Mediterranean and Black Sea fish stocks has been hampered by the international nature of the fisheries and the lack of arrangements for the development of common assessment and management tools. Some of the anchovy stocks in the Mediterranean are shared between Spain and France, others between Greece and Turkey, and others by Italy, Slovenia and Croatia. Each country has its own set of technical management measures, with different degrees of enforcement. Regional assessments and management (as proposed for the Black Sea; Prodanov *et al.*, 1997; Zengin, 2003) are essential.

The California sardine is exploited by Mexico, USA, and Canada. Despite the recent creation of a Tri-National Sardine Forum, transboundary management has not been considered. In fact, USA and Mexico have no record of agreeing on a single management system or mechanism for setting or allocating harvest levels. Coordination and data sharing seem to be the most important initial steps, involving annual synoptic surveys, coordinated stock assessments, formalized data exchange, and agreement on the relative contributions of each country to fishing mortality. At a later stage these may allow consistent management guidelines which currently range from the USA's annual assessments with environmentally based harvest control rules, to Canada's limit based on a constant fraction of the USA's harvest guideline, to Mexico's fish size (but no total harvest) limits.

Discrepancies in trans-boundary management approaches often reflect national perceptions. During one of the latest crises in the management of the NE Atlantic sardine, additional management measures were applied to the Spanish and Portuguese fleets. The former did not contest the measures because the reduced abundance was already reflected in lower catches and revenues. In Portugal, however, the reduction in catches was buffered by price increases and as a result they did not see the need for concerted action.

Box 9.2. Management of fisheries under climate change

Climate change will affect, and is already likely to be affecting, fish populations worldwide. There are three generic impacts to be considered:

(a) changes in species distributions due to warming of the oceans and changes in circulation patterns,

(b) changes in species composition as a result of the previous impact, resulting in more tropical species expanding towards the subpolar regions, and

(c) changes in population parameters, such as growth rates, timing and success of reproduction. The latter will be affected by changes in the composition of the food web, temporal mismatch between fish larvae and their prey, and disruptions in the connectivity between spawning and recruitment grounds due to circulation changes.

There is no doubt that improved management of fisheries and marine ecosystems will play an important role in adapting to the impacts of climate change. Many of the improvements require new science and understanding, but particularly require the development of acceptable, effective responsive social institutions and instruments for achieving adaptive management (see this volume, Chapter 11). In the context of climate change, additional and continuous investment in monitoring exploited populations is needed to ensure adequate model parameterization. Most models on which management relies reproduce the dynamics of the stocks using specific population parameters such as average recruitment, growth and natural mortality rates or weight-at-age keys. As these will change with a changing climate, science must be responsive to this reality, for example, by using trends in specific variables rather than averages of a recent past. Well-designed and reliable monitoring programs of exploited stocks and their associated environment remain essential in order to detect changes and give advance warning of alterations in the productivity and structure of the marine ecosystem. In a changing environment, management advice needs to include complete and transparent information on risks and uncertainties which arise from data quality and assessment models' structural deficiencies.

Climate change is another stress on exploited fish resources, which compounds the impacts of overfishing, making the populations less resilient to unfavorable environmental conditions and more vulnerable to excessive exploitation. Under climate change, management will have to be even more precautionary, paying particular attention to uncertainties, and developing structures and institutions capable of applying adaptive management measures on behalf of all stakeholders.

The ICES system is a model for the generation of cross-boundary assessments. Its scientific advice to management bodies (e.g. European Union, Northeast Atlantic Fisheries Commission) is based on the work of ICES expert groups that include government scientists from ICES countries. Government laboratories contribute data and analysis, and the outcomes of their common work are translated into operational advice by ICES advisory committees. The management result is, however, variable. The North Sea herring, which is fished by many nations, assessed by ICES and managed by an EU/Norway management agreement, is an example of a well-managed stock that has recently achieved Marine Stewardship Council (MSC) accreditation, as recognized by FAO. However, ICES advice is not always followed, as noted in the case of the Bay of Biscay anchovy discussed above. In the case of the Norwegian spring spawning stock the lack of international agreement on quotas has maintained fishing mortality above the target value of $0.125\,\mathrm{y}^{-1}$, resulting in catches of approximately 100 000 t more than required in the management plan. This species has a well-documented tendency to change its distributions and behavior, which complicates shared resource usage, particularly under threats of climate change. Hannesson (2006) studied the complex sharing of this stock between exploiting nations, and the role that straddling a hole of international waters can play in determining management strategies. Whether to adopt a competitive versus a cooperative multi-national management solution has substantial consequences. Climate-driven regime shifts can also play a significant role in destabilizing cooperation in the management of transboundary stocks. Shifts of productivity or geography require adaptive and resilient management institutions, based on innovative models of co-operative games in which the players anticipate the possibility of a regime shift (or climate change-driven geographic displacement, Perry *et al.*, 2005) that radically alters the strength of their relative bargaining positions (Miller, 2007).

In recent years nations increasingly have accepted the obligation to consider the impacts of their policies on marine ecosystems. Ecosystem-based fisheries management has been defined as an approach that takes all major ecosystem components and services – both structural and functional – into account in managing fisheries. The EAF (Ecosystem Approach to Fisheries) is now widely accepted (although not yet widely implemented). As a result of the World Summit for Sustainable Development (WSSD), fisheries need to develop and implement an ecosystem approach to their management by 2012. The need to apply ecosystem considerations to fisheries management is unarguable. 30% of the world's marine primary productivity is used directly to sustain current catch levels (Pauly and Christensen, 1995) and about 25% of the catch is discarded (Alverson *et al.*, 1994). The widespread application of single-species MSY policies may have contributed to the evident deterioration in the structure of many ecosystems, in particular regarding the loss of top predator species (Walters *et al.*, 2005). Managing small pelagic fish species under ecosystem considerations would be aimed at protecting their value as forage species to other fish, mammals, and birds, and thus to respect the integrity of ecosystems, a need increasingly recognized by fisheries managers (PFMC, 1998). To accomplish this requires quantitative understanding of marine food webs and their dynamics not only with respect to fisheries impact removing large quantities of predators and prey, but also environmental drivers altering structure and functioning of food webs. It has been argued that, while there may be insufficient understanding of the processes involved, a qualitative understanding may suffice to make management arrangements that recognize the pivotal role of small pelagics in marine foodwebs. In the case of the southwestern Australia sardine fishery the carrying capacity concept (including consideration of CHRS) has been applied specifically to account for ecosystem needs, with qualitative descriptions of the importance of small pelagics used as ancillary scientific advice when setting TACs. Besides trophic interactions, direct interactions also need to be managed. Due to political concerns, research activites for sardines in southwestern Australia are currently focused on mitigating interactions with seabirds and dolphins, rather than on stock assessment.

Of the species considered in this review only a small minority are managed under some ecosystem considerations. Baltic Sea clupeids are partially assessed using multispecies models, one of the few multispecies assessments currently implemented. However, assessments neither consider predatory interactions affecting fish stock recruitment nor intra- and interspecific competition, both being evident as important processes affecting Baltic fish stock dynamics. The South African combined anchovy and sardine OMP covers some requirements of an ecosystem approach by including certain risk criteria for both species. These risk criteria relate to low probabilities of depleting each resource below specified levels, and incidentally secure relatively high escapement on average for these species at levels (relative to average pre-exploitation abundances), which are similar to those adopted by CCAMLR[22] to take appropriate account of the needs of natural predators (of krill in CCAMLR's case). Ecological issues that are pertinent to South Africa's small pelagic fishery in terms of an EAF have been discussed by Shannon *et al.* (2006), and the OMP for this fishery is currently being revised to allow for explicit incorporation of the effects of small pelagic fish–predator (African penguin, *Spheniscus demersus*) interactions (Cunningham and Butterworth, 2006). In general, however, the Ecosystem Approach is still to be successfully applied to the management of small pelagic fisheries, and this may be the single most important driving force in influencing future assessment and management policies .

Acknowledgments

The authors would like to thank Kevern Cochrane (FAO) and an anonymous reviewer for their useful and thorough reviews of the original version, and the editors of the book for their encouragement and support.

NOTES
  1 Virtual Population Analysis.
  2 Million tonnes.
  3 Northern anchovy.
  4 Peruvian anchovy.
  5 Vessel Monitoring System.
  6 Daily Egg Production Method.
  7 Spawning Stock Biomass.
  8 Total Length.
  9 Total Allowable Catch.
 10 European Union.
 11 Kuroshio-Oyashio Transition Zone.
 12 Instituto de Fomento Pesquero.
 13 Coefficient of Variation.
 14 Catch-at-age Analysis for Sardine.
 15 Operational Management Procedure.
 16 General Linear Model.
 17 Survey-Based.
 18 Sea Level Pressure.
 19 Individual Transferable Quota.
 20 Maximal Spawning Stock Biomass.
 21 International Union for the Conservation of Nature and Natural Resources.
 22 Commission for the Conservation of Antarctic Marine Living Resources.

REFERENCES

Abad, R., Miquel, J., Iglesias, M., and Alvarez, F. (1998). Acoustic estimation of abundance and distribution of anchovy in the NW Mediterranean. *Sci. Mar.* **62**: 37–43.

Ahlstrom, E. H. (1960). Synopsis on the biology of the Pacific sardine (*Sardinops caerulea*). *Proc. World Sci. Meet. Biol. Sardines and Related Species*, Rome, FAO, **2**: 415–451.

Akkers, T. R., Melo, Y. C. and Veith, W. (1996). Gonad development and spawning frequency of the South African pilchard *Sardinops sagax* during the 1993–1994 spawning season. *S. Afr. J. Mar. Sci.* **17**: 183–193.

Alemany, F., Bigot, J.L., Giráldez, A. *et al.* (2002). *Preliminary results on anchovy shared stock in the Gulf of Lions.* Working document presented in the Working Group on small pelagic species. SAC-GFCM, ftp://cucafera.icm.csic.es/pub/scsa/.

Alheit, J. (1993). Use of the daily egg production method for estimating biomass of clupeoid fishes: a review and evaluation. *Bull. Mar. Sci.* **53**: 750–767.

Alheit, J. and Ñiquen, M. (2004) Regime shift in the Humboldt Current ecosystem. *Progr. Oceanogr.* **60**: 201–222.

Allain, G., Petitgas, P., and Lazure, P. (2001). The influence of mesoscale ocean processes on anchovy (*Engraulis encrasicolus*) recruitment in the Bay of Biscay estimated with a three-dimensional hydrodynamic model. *Fisher. Oceanogr.* **10**: 151–163.

Allain, G., Petitgas, P., and Lazure, P. (2004). Use of a biophysical larval drift growth and survival model to explore the interaction between a stock and its environment: anchovy recruitment in Biscay. *ICES cm:2004/J:***14**.

Alverson, D. L., Freeberg, M. H., Murawski, S. A., and Pope, J. G. (1994). *A Global Assessment of Fisheries by Catch and Discards.* Rome, Italy: FAO.

Arkhipov, A. G. (1993). Estimation of abundance and peculiarities of distribution of the commercial fishes in the early ontogeny. *Vopr. Ihktiologii* **33** (4): 511–522. (In Russian)

Arkhipov, A. G., Andrianov, D. P., and Lisovenko, L. A. (1991). Application of Parker's method for estimating the spawning biomass of batch-spawning fish as exemplified by the Black Sea anchovy *Engraulis encrasicolus ponticus. Vopr. Ikhtiol.* **31** (6): 939–950.

Armstrong, M., Shelton, P., Hampton, I., *et al.* (1988). Egg production estimates of anchovy biomass in the southern Benguela system. *CalCOFI Rep.* **29**: 137–157.

Armstrong, M. J., Roel. B. A., and Prosch, R. M. (1989). Long-term trends in patterns of maturity in the southern Benguela pilchard population: evidence for density-dependence? *S. Afr. J. Mar. Sci.* **8**: 91–101.

Armstrong, M. J., Chapman, P., Dudley, S. F. J., *et al.* (1991). Occurrence and population structure of pilchard *Sardinops ocellatus*, round herring *Etrumeus whiteheadi* and anchovy *Engraulis capensis* off the east coast of Southern Africa. *S. Afr. J. Mar. Sci.* **11**: 227–249.

Aro, E. (1989). A review of fish migration patterns in the Baltic. *Rapp. P. -v. Réun. Cons. Int. Explor. Mer* **190**: 72–96.

Arrizaga, A. (1981). Nuevos antecedentes biológicos para la sardina común, *Clupea* (Strangomera) *bentincki* Norman 1936. *Bol. Soc. Biol. Concepción* **52**: 5–66.

Azzali, M., De Felice, A., Cosimi, G., *et al.* (2002). The state of the Adriatic sea centred on the small pelagic fish populations. *P.S.Z.N. Mar. Ecol.* **23** (Suppl. 1): 78–91.

Barange, M., ed. (2003). Report of the 2nd meeting of the SPACC/IOC Study Group on "Use of environmental indices in the management of pelagic fish populations," 9–11 December 2002, Paris, France. *GLOBEC Spec. Cont.* **6**: 156 pp.

Barange, M., Hampton, I., and Soule, M. A. (1996). Empirical analysis of the in situ target strength of three loosely-aggregated pelagic fish species. *ICES J. Mar. Sci.* **53**: 225–232.

Barange, M., Hampton, I., and Roel, B. A. (1999). Trends in the abundance and distribution of anchovy and sardine on the South African continental shelf in the 1990s, deduced from acoustic surveys. *S. Afr. J. Mar. Sci.* **21**: 367–391.

Bartsch, J. (1993). Application of a circulation and transport model system to the dispersal of herring larvae in the North Sea. *Cont. Shelf Res.* **13**: 1335–1361.

Baumann, H., Hinrichsen, H. -H., Malzahn, A., *et al.* (2006). Sprat recruitment in the Baltic Sea: the importance of temperature and transport variability during the late larval and early juvenile stages. *Can. J. Fish. Aquat. Sci.* **63**: 2191–2201.

Baumgartner, T., Soutar, A., and Ferriera-Bartrina, V. (1992). Reconstruction of the history of pacific sardine and northern anchovy populations over the past two millenia from sediments of the Santa Barbara Basin, California. *Calif. Coop. Oceanic Fish. Invest. Rep.* **33**: 24–40.

Beare, D. J., Burns, F., Greig, A., *et al.* (2004). An increase in the abundance of anchovies and sardines in the north-western North Sea since 1995. *Global Change Biol.* **10** (7): 1209–1213.

Beckley, L. E. and van der Lingen, C. D. (1999). Biology, fishery and management of sardines (*Sardinops sagax*) in southern African waters. *Mar. Freshwat. Res.* **50**: 955–978.

Beverton, R. J. H. (1983). Science and decision-making in fisheries regulations. In *Proceedings of the Expert Consultation to Examine Changes in Abundance and Species Composition of Neritic Fish Resources*, Sharp, G. D. and Csirke, J., eds. *FAO Fish Rep.* **291** (3): 919–936.

Beverton, R. J. H. (1990). Small marine pelagic fish and the threat of fishing; are they endangered? *J. Fish Biol.* **37** (Suppl. A): 5–16.

Beyer, J. E. and Lassen, H. (1994). The effect of size-selective mortality on the size-at-age of Baltic herring. *Dana* **10**: 203–234.

Boeke, J. (1906). Eier und Jugendformen von Fischen der südlichen Nordsee. *Verhandelingen u.h. Rijksinstituut v.h. Onderzoaek der Zee* **1** (4): 3–35.

Borja, A., Uriarte, A., Motos, L. and Valencia, V. (1996). Relationship between anchovy (*Engraulis encrasicolus* L.) recruitment and the environment in the Bay of Biscay. *Sci. Mar.* **60** (Suppl. 2): 179–192.

Borja, A., Uriarte, A., Egaña, J., Motos, L., and Valencia, V. (1998). Relationship between anchovy (*Engraulis encrasicolus* L.) recruitment and environment in the Bay of Biscay. *Fish. Oceanogr.* **7**: 375–380.

Boyer, D. C. and Hampton, I. (2001). An overview of the living marine resources of Namibia. *S. Afr. J. Mar. Sci.* **23**: 5–35.

Boyer, D. C., Boyer, H. J., Fossen, I., and Kreiner, A. (2001). Changes in the abundance of the northern Benguela sardine stock during the decade 1990–2000, with comments on the relative importance of fishing and the environment. *S. Afr. J. Mar. Sci.* **23**: 67–84.

Boyra, G. and Uriarte, A. (2005). Informe de campaña / Survey Report: Campaña acústica de juveniles de anchoa en 2005. Acoustic survey on juvenile anchovy in 2005: "JUVENA 2005" *Working Document to STECF meeting 06–10 November 2005 at Brussels.*

Boyra, G., Arregi, I., Cotano, U., Alvarez, P., and Uriarte, A. (2005). Acoustic surveying of anchovy juveniles in the Bay of Biscay: JUVENA 2003 and 2004: preliminary biomass estimates. In *Working Document to the ICES Working Group on the assessment of mackerel, horse mackerel, sardine and anchovy, held in Vigo from 06 to 15 September 2005.*

Braun, M., Reyes, H., Osses, J., *et al.* (2002). Monitoreo de las condiciones bio-oceanográficas en la I y II regiones. *Informe Final,* **FIP** 2001–01, 432 pp.

Braun, M., Claramunt, G., Valenzuela, V., *et al.* (2005). Evaluación del Stock Desovante de Anchoveta en la I y II Regiones, año 2004. *Informe Final*, **FIP**, 244 pp.

Brophy, D. and Danilowicz, B. S. (2003). The influence of pre-recruitment growth on subsequent growth and age at first spawning in Atlantic herring (*Clupea harengus* L.). *ICES J. Mar. Sci.* **60**: 1103–1113.

Burd, A. C. (1985). Recent changes in the central and southern North Sea herring stocks. *Can. J. Fish. Aquat. Sci.* **42** (Suppl 1): 192–206.

Burd A. C. and Howlett, G. J. (1974). Fecundity studies on North Sea herring. *J. Cons. Perm. Int. Explor. Mer.* **35** (2): 107–120.

Butler, J. L., Smith, P. E., and Lo, N. C. H. (1993). The effect of natural variability of life-history parameters on anchovy and sardine population growth. *Calif. Coop. Oceanic Fish. Invest. Rep.* **34**: 104–111.

Butterworth, D. S., De Oliveira, J. A. A., and Cochrane, K. L. (1993). Current initiatives in refining the management procedure for the South African anchovy resource. In *Proceedings of the International Symposium on Management Strategies for Exploited Fish Populations*, D. M. Eggers, R. J. Marasco, C. Pautzke and T. J. Quinn II, eds. G. Kruse, *Alaska Sea Grant College Program Report* **93–02**, Fairbanks: University of Alaska, pp. 439–473.

Canales, M., Saavedra, J. C., Böhm, G., and Aranis, A. (2004). Investigación CTP de sardina común, zona centro-sur. *Informe Final*, **IFOP/SUBPESCA**, 41 pp.

Cardinale, M. and Arrhenius, F. (2000). Decreasing weight-at-age of Atlantic herring (*Clupea harengus*) from the Baltic Sea between 1986 and 1996: a statistical analysis. *ICES J. Mar. Sci.* **57**: 882–893.

Carrera, P. and Porteiro, C. (2003). Stock dynamics of the Iberian sardine (*Sardina pilchardus*, W.) and its implication on the fishery off Galicia. *Sci. Mar.* **67** (Suppl. 1): 245–258.

Carrera P., Churnside, J. H., Boyra, G., *et al.* (2006). Comparison of airborne lidar with echosounders: a case study in the coastal Atlantic waters of southern Europe. *ICES J. Mar. Sci.* **63**: 1736–1750.

Casavola, N. (1998). Daily egg production method for spawning biomass estimates of anchovy in the south-western Adriatic during 1994. *Rapp. P. –v. Réun. Cons. Int. Explor. Mer.* **35** (2): 394–395.

Castello, J. P., Habiaga, J. C., and Lima, I. D. Á., Jr. (1991). Prospecção hidroacústica e avaliação da biomassa de sardinha e anchoita, na região sudeste do Brasil (outubro/novembro de 1988). *Publções esp. Inst. Oceanogr., S Paulo* **8**: 15–30.

Castillo, J. and Robotham, H. (2004). Spatial structure and geometry of schools of sardine (*Sardinops sagax*) in relation to abundance, fishing effort and catch in northern Chile. *ICES J. Mar. Sci.* **61**: 1113–1119.

Castillo, J., Barbieri, M. A., Parker, U., *et al.* (1995). Evaluación hidroacústica de los stocks de sardina española, anchoveta y jurel en la zona norte, I a IV regiones. *Fondo de Investigación Pesquera, FIP. Informe Final*, 356 pp.

Castillo, J., Saavedra, A., and Gálvez, P. (2005). Evaluación acústica de la biomasa, abundancia, distribución espacial y caracterización de cardúmenes de anchoveta y sardina común durante el periodo de reclutamiento. Zona Centro-Sur. Verano 2005. In *Evaluación hidroacústica del reclutamiento de anchoveta y sardina común entre la V y X Regiones, año 2004. Informes Técnicos FIP – IT/2004–05*, 205 pp. + anexos (www.fip.cl).

Cergole, M. C. (1993). Avaliação do estoque da sardinha, *Sardinella brasiliensis*, da costa sudeste do Brasil, período 1977 a 1990. Ph.D. thesis. Universidade de São Paulo, Instituto Oceanográfico, 245 pp.

Cergole, M. C. (1995). Stock assessment of the Brazilian sardine, *Sardinella brasiliensis*, of the southeastern Coast of Brazil. *Sci. Mar.* **59** (3–4): 597–610.

Cergole, M. C. and Valentini, H. (1994). Growth and mortality estimates of *Sardinella brasiliensis* in the southeastern Brazilian bight. *Bol. Inst. Oceanogr., S Paulo* **42** (1/2): 113–127.

Cergole, M. C., Saccardo, S. A. and Rossi-Wongtschowski, C. L. D. B. (2002). Fluctuations in the spawning stock biomass and recruitment of the Brazilian sardine (*Sardinella brasiliensis*): 1977–1997. *Rev. Bras. Oceanogr.* **50**: 13–26.

Checkley, D. M. Jr., Ortner, P. B., Settle, L. R., and Cummings, S. R. (1997). A continuous, underway fish egg sampler. *Fish. Oceanogr.* **6** (2): 58–73.

Cingolani, N., Giannetti, G., and Arneri, E. (1996). Anchovy fisheries in the Adriatic Sea. *Sci. Mar.* **60** (Suppl. 2): 269–277.

Cingolani, N., Santojanni, A., Arneri, E., *et al.* (2005). Anchovy (*Engraulis encrasicolus*, L.) stock assessment in the Adriatic Sea: 1975–2004. *Working document presented in the Working Group on small pelagic species. SAC-GFCM*, ftp://cucafera.icm.csic.es/pub/scsa/

Cisneros-Mata, M. A., Hammann, M. G. and Nevarez-Martinez, M. O. (1995). The rise and fall of the Pacific sardine, *Sardinops sagax caeruleus* Girard, in the Gulf of California, Mexico. *Calif. Coop. Oceanic Fish. Invest. Rep.* **36**: 136–143.

Cochrane, K. and Starfield, A. (1992). The potential use of predictions of recruitment success in the management of the South African anchovy resource In *Benguela Trophic Functioning*, Payne, A. I. L., Brink, K. H., Mann K. H., and Hilborn, R., eds. *S. Afr. J. Mar. Sci.* **12**: 891–902.

Cochrane, K. L. (1999). Complexity in fisheries and limitations in the increasing complexity of fisheries management. *ICES J. Mar. Sci.* **56**: 917–926.

Coetzee, J. C., Merkle, D., de Moor (formerly Cunningham), C. L., *et al.* (2008). Refined estimates of South African pelagic fish biomass from hydro-acoustic surveys: quantifying the effects of target strength, signal attenuation and receiver saturation. *Afr. J. Mar. Sci.* **30**(2): 205–217.

Conser, R. J. and Powers, J. E. (1990). Extensions of the ADAPT VPA tuning method designed to facilitate assessmenmt work on tuna and swordfish stocks. *ICCAT Coll. Vol. Sci. Pap.* **32**: 461–467.

Coombs, S. H., Smyth, T. J., Conway, D. V. P., *et al.* (2006). Spawning season and temperature relationships for sardine (*Sardina pilchardus*) in the eastern North Atlantic. *J. Mar. Biol. Assoc. UK*, **86**: 1245–1252.

Cortes, N.A., Oyarzun, C. and Galleguillos, R. (1996). Diferenciación poblacional en sardina común, *Strangomera bentincki* (Norman, 1936) II: Análisis multivariado de la morfometría y merística. *Rev. Biol. Mar. Valparaiso* **31** (2): 91–105.

Cram, D.L. and Hampton, I. (1976). A proposed aerial/acoustic strategy for pelagic fish stock assessment. *J. Cons. Perm. Int. Explor. Mer.* **37** (1): 91–97.

Crawford, R.J.M., Shelton, P.A., and Hutchings, L. (1980). Implications of availability, distribution and movements of pilchard (*Sardinops ocellata*) and anchovy (*Engraulis capensis*) for assessment and management of the South African purse-seine fishery. *Rapp. P. –v. Réun. Cons. Int. Explor. Mer.* **177**: 355–373.

Csirke, J. (1988). Small shoaling pelagic fish stocks. In *Fish Population Dynamics*, Gulland, J.A., ed. 2nd edn. John Wiley & Sons Ltd.

Csirke, J.R., Guevara, G., Cardenas, M., *et al.* (1996). Situación de los recursos anchoveta (*Engraulis ringens*) y sardina (*Sardinops sagax*) a principios de 1994 y perspectivas para la pesca en el Perú, con especial referencia a la región norte-centro de la costa peruana. *Boletín Inst. Mar Perú* **15**: 11–23.

Cubillos, L., Canales, M., Hernández, A., *et al.* (1998). Poder de pesca, esfuerzo de pesca y cambios estacionales e interanuales en la abundancia relativa de *Strangomera bentincki* y *Engraulis ringens* en el área frente a Talcahuano, Chile (1990–97). *Invest. Mar. Valparaíso* **26**: 3–14.

Cubillos, L., Canales, M., Bucarey, D., Rojas, A., and Alarcón, R. (1999). Época reproductiva y talla media de primera madurez sexual de *Strangomera bentincki* y *Engraulis ringens* en la zona centro-sur de Chile en el período 1993–1997. *Invest. Mar. Valparaíso* **27**: 73–86.

Cubillos, L., Arcos, D.F., Bucarey D.A., and Canales, M.T. (2001). Seasonal growth of small pelagic fish off Talcahuano, Chile (37° S, 73° W): a consequence of their reproductive strategy to seasonal upwelling? *Aquat. Living Resour.* **14**: 115–124.

Cubillos, L.A., Bucarey, D.A., and Canales, M. (2002). Monthly abundance estimation for common sardine *Strangomera bentincki* and anchovy *Engraulis ringens* in the central-south Chile (34–40° S). *Fish. Res.* **57**: 117–130.

Cubillos, L., Castro, L., and Oyarzún, C. (2005). Evaluación del stock desovante de anchoveta y sardina común entre la V y X Región, año 2004. *Informes Técnicos* **FIP-IT/2004–03**: 130 pp.

Cunningham, C.L. and Butterworth, D.S. (2004a). *Base Case Bayesian Assessment of the South African Anchovy Resource.* Unpublished MCM document, **WG/APR04/PEL/01**.

Cunningham, C.L. and Butterworth, D.S. (2004b). *Base Case Bayesian Assessment of the South African Sardine Resource.* Unpublished MCM document, **WG/APR04/PEL/02**.

Cunningham, C.L. and Butterworth, D.S. (2006). *Update: Proposals for Issues to be Addressed in the Revision of the Pelagic OMP.* Unpublished MCM document, **SWG/OCT2006/PEL/04**.

Cushing, D.H. (1955). On the autumn-spawned herring races in the North Sea. *J. Cons. Int. Explor Mer* **21** (1): 44–59.

Cushing D.H. (1967). The grouping of herring populations. *J. Mar. Biol. Ass. UK* **47**: 193–208.

Cushing, D.H. (1992). A short history of the Downs stock of herring. *ICES J. Mar. Sci.* **49**: 437–443.

Cushing, D.H. and Bridger, J.P. (1966). The stock of herring in the North Sea, and changes due to fishing. *Fish. Invest. London, Ser II* **25** (1): 1–123.

Cushing, D.H. and Burd, A.C. (1957). On the herring of the southern North Sea. *Fish. Invest., London, Ser II* **20** (11): 1–31.

Darby, C.D. and Flatman, S. (1994). Virtual population analysis: version 3.1 (Windows/DOS) user guide. Lowestoft: Directorate of Fisheries Research. *MAFF Information Technology Series* **1**.

Daskalov, G. (1998). Using abundance indices and fishing effort data to tune catch-at-age analyses of sprat *Sprattus sprattus*, whiting *Merlangius merlangus* and spiny dogfish *Squalus acanthias* in the Black Sea. *Options Medit.* **35**: 215–228.

Daskalov, G. (1999). Relating fish recruitment to stock biomass and physical environment in the Black Sea using generalized additive models. *Fish. Res.* **41** (1): 1–23.

Daskalov, G.M. (2002). Overfishing drives a trophic cascade in the Black Sea. *Mar. Ecol. Prog. Ser.* **225**: 53–63.

Daskalov, G.M. (2003). Long-term changes in fish abundance and environmental indices the Black Sea. *Mar. Ecol. Prog. Ser.* **255**: 259–270.

Daskalov, G., Prodanov, K., Shlykhov, A., and Maxim, K. (1996). Stock assessment of sprat *Sprattus* L. in the Black Sea during 1945–1993 using international fishery and research data. *Izv. IRR, Varna* **24**: 67–93.

Daskalov, G.M., Prodanov, K. and Zengin, M. (2008). The Black Seas fisheries and ecosystem change: discriminating between natural variability and human-related effects: In: *Proceedings of the Fourth World Fisheries Congress: Reconciling Fisheries with Conservation.* Nielsen, J., Dodson, J., Friedland, K., *et al.*, eds. American Fisheries Society Symposium 49, AFS, Bethesda, MD, pp. 1645–1664.

Deriso, R., Quinn, T.J., and Neal, P.R. (1985). Catch-age analysis with auxiliary information. *Can. J. Fish. Aquat. Sci.* **42** (4): 815–24.

Deriso, R.B., Barnes, J.T., Jacobson, L.D., and Arenas, P.J. (1996). Catch-at-age analysis for Pacific sardine (*Sardinops sagax*), 1983–1995. *Calif. Coop. Oceanic Fish. Invest. Rep.* **37**: 175–187.

de Moor (formerly Cunningham), C.L., Butterworth, D.S., and Coetzee, J.C. (2008). Revised estimates of abundance of South African sardine and anchovy from acoustic surveys adjusting for echosounder saturation in earlier surveys and attenuation effects for sardine. *Afr. J. Mar. Sci.* **30**(2): 219–232.

De Oliveira, J.A.A. (2003). The development and implementation of a joint management procedure for the South African pilchard and anchovy resources. Ph.D. thesis. University of Cape Town, South Africa. iv, 319 pp.

De Oliveira, J.A.A. (2006). Long-term harvest strategies for small pelagic fisheries under regime shifts: the South African pilchard-anchovy fishery. In *Climate Change* and the Economics of the World's Fisheries: Examples of Small Pelagic Stocks, Hannesson, R., Barange, M., and Herrick, S. F., Jr., eds. *New Horizons in Environmental Economics*. Cheltenham, UK: Edward Elgar Publishing, pp. 151–204.

De Oliveira, J. A. A. and Butterworth, D. S. (2004). Developing and refining a joint management procedure for the multispecies South African pelagic fishery. *ICES J. Mar. Sci.* **61**: 1432–1442.

De Oliveira, J. A. A. and Butterworth, D. S. (2005). Limits to the use of environmental indices to reduce risk and/or increase yield in the South African anchovy fishery. *Afr. J. Mar. Sci.* **27** (1): 191–203.

De Oliveira, J. A. A., Uriarte, A., and Roel, B. A. (2005). Potential improvements in the management of Bay of Biscay anchovy by incorporating environmental indices as recruitment predictors. *Fish. Res.* **75**: 2–14.

De Oliveira, J. A. A., Boyer, H. J., and Kirchner, C. H. (2007). Developing age-structured stock assessment models as a basis for management procedure evaluations for Namibian sardine. *Fish. Res.* **85**: 148–158.

Domashenko, G. P., Mikhajlyuk, A. N., Chashchin, A. K., *et al.* (1985). Contemporary state of commercial stocks of anchovy, scad, sprat and whiting in the Black Sea. In *Oceanological and Fisheries Investigations in the Black Sea*. Tr. VNIRO, pp. 87–100.

Edmonds, J. S. and Fletcher, W. J. (1997). Stock discrimination of pilchards *Sardinops sagax* by stable isotope ratio analysis of otolith carbonate. *Mar. Ecol. Prog. Ser.* **152**: 241–247.

Elwertowski, J. (1960). Biologische Grundlagen der Sprottenfischerei in der östlichen und mittleren Ostsee. *Fisch. Forsch.* **3** (4): 1–19.

Fairweather, T. P., van der Lingen, C. D., Booth, A. J., *et al.* (2006). Indicators of sustainable fishing for the South African sardine (*Sardinops sagax*) and anchovy (*Engraulis encrasicolus*). *Afr. J. Mar. Sci.* **28** (3–4): 661–680.

FAO (2003). The ecosystem approach to marine capture fisheries. *FAO Technical guidelines for Responsible Management*, **4** (Suppl. 2): 112 pp.

FAO (2004). The State of World Fisheries and Aquaculture. Rome, Italy: FAO, 153 pp.

Faschuk, D. Ya., Arkhipov, A. G., and Shlyakhov, V. A. (1995). Concentration of the Black Sea mass commercial fishes in different ontogenetic stages and factors of its determination. *Vopr. Ihktiologii* **35** (1): 34–42. (In Russian)

Félix-Uraga, R., Gómez-Muñoz, V. M., Quiñónez-Velázquez, C., *et al.* (2004). On the existence of Pacific sardine groups off the west coast of Baja California and Southern California. *Calif. Coop. Oceanic Fish. Invest. Rep.* **45**: 146–151.

Felix-Uraga, R., Gómez-Muñoz, V. M., Quiñónez-Velázquez, C., *et al.* (2005). Pacific sardine stock discrimination off the west coast of Baja California and southern California using otolith morphometry. *Calif. Coop. Oceanic Fish. Invest. Rep.* **46**: 113–121.

Fletcher, W. J. (1992). Use of a spatial model to provide initial estimates of stock size for a purse seine fishery on pilchards (*Sardinops sagax neopilchardus*) in Western Australia. *Fish. Res.* **14**: 41–57.

Fletcher, W. J. (2005). The application of qualitative risk assessment methodology to prioritize issues for fisheries management. *ICES J. Mar. Sci.* **62**: 1576–1587.

Fletcher, W. J. and Blight, S. J. (1996). Validity of using translucent zones of otoliths to age the pilchard *Sardinops sagax neopilchardus* from Albany, Western Australia. *Mar. Freshwat. Res.* **47**: 617–624.

Fletcher W. J. and Sumner, N. R. (1999). Spatial distribution of sardine (*Sardinops sagax*) eggs and larvae: an application of geostatistics and resampling to survey data. *Can. J. Fish. Aquat. Sci.* **56**: 907–914.

Fletcher, W. J., Lo, N. C. H., Hayes, E. A., *et al.* (1996). Use of the daily egg production method to estimate the stock size of Western Australian sardines (*Sardinops sagax*). *Mar. Freshwat. Res.*, **47**: 819–25.

Fossen, I., Boyer, D. C., and Plarre, H. (2001). Changes in some key biological parameters of the northern Benguela sardine stock. *S. Afr. J. Mar. Sci.* **23**: 111–121.

Fournier, D. A. and Archibald, C. P. (1982). A general theory for analyzing catch at age data. *Can. J. Fish. Aquat. Sci.* **39**: 1195–1207.

Fréon, P., Mullon, C., and Voisin, B. (2003). Investigating remote synchronous patterns in fisheries. *Fish. Oceanogr.* **12** (4/5): 443–457.

Fréon P., Cury, P., Shannon, L., and Roy, C. (2005). Sustainable exploitation of small pelagic fish stocks challenged by environmental and ecosystem changes: a review. *Bull. Mar. Sci.* **76** (2): 385–462.

Galleguillos, R., Troncoso, L., Monsalve, J., and Oyarzún, C. (1997). Diferenciación poblacional en la sardina chilena *Strangomera bentincki* (Pices:Clupeidae): análisis de variabilidad proteínica. *Rev. Chil.Hist. Nat.* **70**: 351–361.

García, A. and Palomera, I. (1996). Anchovy early life history and its relation to its surrounding environment in the Western Mediterranean basin. *Sci. Mar.* **60** (Suppl. 2): 155–166.

Gaughan, D. J., Mitchell, R. W., and Blight, S. J. (2000). Impact of mortality, possibly due to *Herpesvirus* sp., on pilchard, *Sardinops sagax*, stocks along the south coast of Western Australia 1998–1999. *Mar. Freshwat. Res.* **51**: 601–612.

Gaughan, D. J., White, K. V. and Fletcher, W. J. (2001). Links between functionally distinct adult assemblages of *Sardinops sagax*: larval advection across management boundaries. *ICES J. Mar. Sci.* **58** (3): 597–606.

Gaughan, D. J., Fletcher, W. J., and McKinlay, J. P. (2002). Functionally distinct adult assemblages within a single breeding stock of the sardine, *Sardinops sagax*: management units within a management unit. *Fish. Res.* **59**: 217–231.

Gaughan, D. J., Mitchell, R. W. D., Leary, T. I., and Wright, I. W. (2004). A sudden collapse in distribution of Pacific sardine (*Sardinops sagax*) in southwestern Australia enables an objective re-assessment of biomass estimates. *Fish. Bull.* **102**: 617–633.

Gavaris, S. (1988). An adaptative framework for the estimation of population size. *CAFSAC Res. Doc.* **88/29**.

Grishin, A.N., Kovalenko, L.A. and Sorokolit, L.K. (1994). Trophic relations in plankton communities in the Black Sea before and after *Mnemiopsis leidyi* invasion. The main results of YugNIRO complex researches in the Azov-Black Sea region and the World ocean in 1993, *Tr. YugNIRO, Kerch* **40**: 38–44.

GTE (2002). *Informe Técnico de Evaluación Conjunta de los Stocks de Sardina y Anchoveta del Sur de Perú y Norte de Chile*. Grupo de trabajo IFOP-IMARPE sobre Pesquerías de pequeños Pelágicos. Octavo Taller, Lima, 23–30 de setiembre de 2002.

GTE (2003). Evaluación Conjunta de los stocks de anchoveta y sardine del sur del peru y norte de Chile. Grupo de Trabajo IMARPE-IFOP sobre pesquerías de Pelágicos Pequeños. *Informe Técnico del Noveno Taller*. Callao 17 al 21 de noviembre de 2003.

Gucu, A.C. (1997). Role of fishing in the Black Sea ecosystem. In *Sensitivity to Change: Black Sea, Baltic Sea and North Sea*, Özsoy, E., and Mikaelyan, A., eds. NATO ASI Series 2. Environment vol. 27. Dordrecht the Netherlands: Kluwer Academic Publishers.

Gunderson D.R. (1993). *Surveys of Fisheries Resources*. New York, USA: John Wiley and Sons Inc., 248 pp.

Hall, N.G. (2000). Modelling for fishery management, utilising data for selected species in Western Australia. Ph.D. thesis, Murdoch University.

Hampton, I. (1992). The role of acoustic surveys in the assessment of pelagic fish resources on the South African continental shelf. *S. Afr. J. Mar. Sci.* **12**: 1031–1050.

Hampton, I. (1996). Acoustic and egg production estimates of South African anchovy biomass over a decade: comparisons, accuracy and utility. *ICES J. Mar. Sci.* **53**: 493–500.

Hannesson, R. (2006). Sharing the herring: fish migrations, strategic advantage and climate change, pp. 66–99. In *Climate Change and the Economics of the World's Fisheries*, Hannesson, R., Barange, M., and Herrick, S.F., eds. Cheltenham, UK: Edward Elgar Publishing Ltd.

Harada, Y. and Nishiyama, M. (2005). Management strategy robust to environmental variation. In *Regime Shift and Fisheries Stock Management*, Aoki, I., Nihira, A., Yatsu, A., and Yamakawa, T., eds. Tokyo: Koseisya-Koseikau Publ. Co. (in Japanese)

Harden Jones, F.R. (1968). *Fish Migration*. London: Edward Arnold Ltd., 325 pp.

Heessen, H.J.L., Dalskov, J., and Cook, R.M. (1997). The international bottom trawl survey in the North Sea, the Skagerrak and Kattegat. *ICES cm:1997/Y:*31: 23 pp.

Heincke, F. (1898). Naturgeschichte des Herings. *Abhandl. Deutschen Seefisch Ver II.*

Hense, V. and Apstein, C. (1897). Dir Nordsee-Expedition 1895 des Desutschen Seefischereivereins. Über die Eimenge de rim Winter laichenden Fische. *Wiss. Meresunters. Helgoland* **2** (2): 1–101.

Heymans, J.J., Shannon, L.J., and Jarre, A. (2004). Changes in the northern Benguela ecosystem over three decades: 1970s, 1980s and 1990s. *Ecol. Model.* **172**: 175–195.

Hill, K.T., Jacobson, L.D., Lo, N.C.H., *et al.* (1999). Stock assessment of Pacific sardine for 1998 with management recommendations for 1999. *Calif. Dept. Fish. Game. Marine Region Admin. Rep.* **99**–4: 92 pp.

Hill, K.T., Lo, N.C.H., Macewicz, B.J., and Felix-Uraga, R. (2006). Assessment of the Pacific sardine (*Sardinops sagax caerulea*) population for US management in 2006. *NOAA Tech. Mem.*, **NOAA-TM-NMFS-SWFSC-386**, 85 pp.

Holst, J.C., Røttingen, I., and Melle, W. (2004). The herring. In *The Norwegian Sea Ecosystem*, Skjoldal, H.R., ed. Trondheim: Tapir Academic Press, pp. 203–226.

Horbowy, J. (1996). The dynamics of Baltic fish stocks on the basis of a multispecies stock-production model. *Can. J. Fish. Aquat. Sci.* **53**: 2115–2225.

Hubold, G. (1978). Variations in growth rate and maturity of herring in the Northern North Sea in the years 1955–1973. *Rapp. P.-v Réun. Cons. Int. Explor. Mer* **172**: 154–163.

Hunter, J.R. and Lo, N.C.H. (1997). The daily egg production method of biomass estimation: some problems and potential improvements. *Ozeanografika* **2**: 41–69.

Hutchings, L., Barange, M., Bloomer, S.F., *et al.* (1998). Multiple factors affecting South African anchovy recruitment in the spawning, transport and nursery area. In *Benguela Dynamics: Impacts of Variability on Shelf-Sea Environments and their Living Resources*, Pillar, S.C., Moloney, C.L., Payne, A.I.L., and Shillington, F.A., eds. *S. Afr. J. Mar. Sci.* **19**: 211–225.

Ibaibarriaga, L., Uriarte, U., and Roel, B. (2005). More on harvest control rules for the Bay of Biscay anchovy. In *Working Document to the ICES Working Group on the assessment of mackerel, horse mackerel, sardine and anchovy,* held in Vigo from 06 to 15 September 2005.

Ibama (2005). *Estatística Pesqueira Nacional: 2003.* Brasília, junho de 2005.

ICES (1978). Rapport du groupe de travail pour l´évaluation des stocks de reproducteurs de sardines et autres clupeides au sud des Isles Britanniques. *ICES cm:1978/***H:5**.

ICES (1980). Rapport du Groupe de Travail pour l' évaluation des stocks de sardines dans les Divisions VIIIc et IXa. *ICES cm:1980/***H:53**.

ICES (1996). Report of the Baltic Fisheries Assessment Working Group. *ICES cm:1996/***Assess:13**.

ICES (1997a). Report of the Baltic Fisheries Assessment Working Group. *ICES cm:1997/***Assess:12**.

ICES (1997b) Report of the Study Group on Multispecies Model Implementation in the Baltic. *ICES cm:1997/***J:2**.

ICES (1998a). Report of the Study Group on Management Strategies for Baltic Fish Stocks. *ICES cm:1998/***Assess:11**.

ICES (1998b). Report of the planning group for acoustic surveys in ICES sub-areas VIII and IX. *ICES cm:1998/***G:2**.

ICES (2000). Report of the Working Group on the assessment of Mackerel, Horse Mackerel, Sardine, and Anchovy. *ICES cm:2000/***ACFM:5**.

ICES (2001a). Report of the ICES Study Group on the further development of the precautionary approach to Fisheries Management. *ICES cm:2001/***ACFM:11**.

ICES (2001b). Report of the Study Group on Herring Assessment Units in the Baltic Sea. *ICES cm:2001/***ACFM:10**.

ICES (2002a). Report of the Study Group on Baltic Herring and Sprat Maturity. *ICES cm :2002/***ACFM:21**.

ICES (2002b). Report of the Baltic Fisheries Assessment Working Group. *ICES cm:2002/***ACFM:17**.

ICES (2003a). Report of the Study Group on Herring Assessment Units in the Baltic Sea. *ICES cm:2003/***ACFM:11**.

ICES (2003b). Report of the ICES Study Group on the further development of the precautionary approach to fishery management. *ICES cm:2003/***ACFM:09**.

ICES (2003c). Report of the Study Group on multispecies assessments of Baltic Fish. *ICES cm:2003/***H:03**.

ICES (2003d). Study Group on Precautionary Reference Points For Advice on Fishery Management. *ICES cm:2003/***ACFM:15**.

ICES (2003e). Report of the Baltic Fisheries Working Group. *ICES cm:2003/***ACFM:21**.

ICES (2004). The DEPM estimation of spawning stock biomass for sardine and anchovy. *ICES Coop. Res. Rep.* **268**, 91 pp.

ICES (2005a). Report of the Working Group on the Assessment of Mackerel, Horse Mackerel, Sardine and Anchovy. *ICES cm:2005/***ACFM:08**.

ICES (2005b). Report of the Northern Pelagic and Blue Whiting Fisheries Working Group (WGNPBW). *ICES cm:2006/***ACFM:05**.

ICES (2005c). Report of the Study Group on Multispecies Assessments of Baltic Fish. *ICES cm:2005/***H:06**.

ICES (2005d). Report of the ICES Advisory Committee on Fishery Management, Advisory Committee on the Marine Environment and Advisory Committee on Ecosystems, 2005. *ICES Advice.* Volumes **1–11**.

ICES (2005e). Report of the Study Group on Regional Scale Ecology of Small Pelagic Fish. *ICES cm:2005/***G:06**.

ICES (2005f). Report of the Working Group for Regional Ecosystem Description (WGRED). *ICES cm:2005/***ACE:1**.

ICES (2006a). Report of the Working group on the assessment of mackerel, horse mackerel, sardine and anchovy. (Vigo Spain from 6–15 September 2005). *ICES cm:2006/***ACFM:08.**

ICES (2006b). Report of the Baltic International Fish Survey Working Group. *ICES cm:2006/***LRC:07**.

ICES (2006c). Report of the Baltic Fisheries Working Group. *ICES cm:2006/***ACFM:24**.

ICES (2006d). Report of the Herring Assessment Working Group South of 62° N (HAWG). ICES Advisory Committee on Fishery Management, *ICES cm:2006/***ACFM:20**.

ICES (2006e). Report of the working group on acoustic and egg surveys for sardine and anchovy in ICES areas VIII and IX (WGACEGG). *ICES cm:2006/***LRC:01**.

ICES (2006f). Incorporation of process information into stock-recruitment models. *ICES Coop. Res. Rep.* **282**: 152 pp.

Isaac-Nahum, V. J., Vazzoler, A. E. A. de M., and Zaneti-Prado, E. M. (1983). Estudos sobre a estrutura, ciclo de vida e comportamento da *Sardinella brasiliensis* (Steindachner, 1879), na área entre 22° Se 28° S, Brasil. 3 – Morfologia e histologia de ovários e escala de maturidade. *Bolm. Inst. Oceanogr. São Paulo* **32** (1): 1–16.

Isaac-Nahum, V. J., Cardoso, R. de D., Servo, G. J. de M. and Rossi-Wongtschowski, C. L. D. B. (1988). Aspects of the spawning biology of the Brazilian sardine, *Sardinella brasiliensis* (Steindachner, 1879), (Clupeidae). *J. Fish. Biol.* **32** (3): 383–396.

Isari, S., Ramfos, A., Somarakis, S., *et al.* (2006). Mesozooplankton distribution in relation to hydrology of the Northeastern Aegean Sea, eastern Mediterranean. *J. Plankton Res.* **28**: 241–255.

Isles, T. D. (1984). Allocation of resources to gonad and soma in Atlantic herring *Clupea harengus* L. In *Fish Reproduction*, Potts, G. W. and Wootton, R. J., eds. London: Academic Press, pp. 331–348.

Ivanov, L, (1983). Population parameters and limiting methods of sprat *Sprattus sprattus* L. catches in the Western Black sea. *Izv. IRR, Varna*, **20**: 7–46. (In Bulgarian)

Ivanov, L. and Beverton, R. J. H. (1985). The fisheries resources of the Mediterranean, part two: Black Sea. *FAO Stud. Rev.*, **60**, 135 pp.

Jacobson, L. D. and MacCall, A. D. (1995). Stock-recruitment models for Pacific sardine (*Sardinops sagax*). *Can. J. Fish. Aquat. Sci.* **52**: 566–577.

Jacobson, L. D., Lo, N. C. H., and Barnes, J. T. (1994). A biomass based assessment model for northern anchovy *Engraulis mordax*. *Fish. Bull. US* **92**: 711–724.

Jacobson, L. D., Lo, N. C. H., , S. F., Jr., and Bishop, T. (1995). Spawning biomass of the northern anchovy in 1995 and status of the coastal pelagic species fishery during 1994. *National Marine Fisheries Service, Southwest Fisheries Science Center Admin. Rep.* **LJ-95**-11.

Jacobson, L. D., De Oliveira, J. A. A., Barange, M., *et al.* (2001). Surplus production, variability, and climate change in the great sardine and anchovy fisheries. *Can. J. Fish. Aquat. Sci.* **58**: 1891–1903.

Jacobson, L. D., Bograd, S. J., Parrish, R. H., *et al.* (2005). An ecosystem-based hypothesis for climatic effects on surplus production in California sardine (*Sardinops sagax*) and environmentally dependent surplus production models. *Can. J. Fish. Aquat. Sci.* **62**: 1782–1796.

Jablonski, S. (2003). Modelos Não Paramétricos e Sistemas Especialistas na Avaliação da Influência de Fatores Ambientais sobre Recursos Pesqueiros: O Caso da Sardinha Verdadeira (*Sardinella brasiliensis*). Tese de Doutorado. Programa de Planejamento Energético, Universidade Federal do Rio de Janeiro, COPPE, 152 pp.

Jablonski, S. and Legey, L. F. L. (2004). Environmental effects on the recruitment of the Brazilian sardine (*Sardinella brasiliensis*) (1977–1993). *Sci. Mar.* **68** (3): 385–398.

Jablonski, S. and Legey, L. F. L. (2005). Towards the development of an environmental rule based model for predicting recruitment in Brazilian sardine *Sardinella brasiliensis*, (1977–1993). *Afr. J. Mar. Sci.* **27** (3): 539–547.

Jardim, E., Trujillo, V., and Sampedro, P. (2004). Uncertainties in sampling procedures for age composition of hake and sardine in Iberian Atlantic waters. *Sci. Mar.* **68** (4): 561–569.

Jennings, S. and Beverton, R. J. H. (1991). Intraspecific variation in the life history tactics of Atlantic herring (*Clupea harengus* L.) *ICES J. Mar. Sci.* **48**: 117–125.

Johanneson, K. A. (1975). Relatório preliminar das observações acústicas quantitativas sobre tamanho e distribuição dos recursos de peixes pelágicos ao largo do Brasil. *SUDEPE/ PDP, Sér. Doc. téc.*, **10**: 1–10.

Johansen, A.C. (1919). On the large spring spawning herring in the north-west European waters. *Medd. Komm. Hauundersökeler, serie Fisk.* Bd **5** (8).

Kapadagakis, A., Machias, A., Somarakis, S., and Stergiou, K. I. (2001). Patterns and propensities in Greek fishing effort and catches. *Final report, DG XIV Project* 00/018.

Katsukawa, T. and Matsuda, H. (2003). Simulated effects of target switching on yield and sustainability of fish stocks. *Fish. Res.* **60**: 515–525.

Kawasaki, T. (1983). Why do some pelagic fishes have wide fluctuations in their numbers? Biological basis of fluctuation from the viewpoint of evolutionary ecology. *FAO Fish. Rep.* **291** 1065–1080.

King, J.R., ed. (2005). Report of the Study Group on Fisheries and Ecosystem Responses to Recent Regime Shifts. *PICES Sci. Rep.* **28**: 162 pp.

King, J.R., McFarlane, G.A., and Beamish, R.J. (2001). Incorporating the dynamics of marine systems into the stock assessment and management of sablefish. *Prog. Oceanogr.* **49** (1–4): 619–639.

Klyashtorin, L.B. (1998). Long-term climate change and main commercial fish production in the Atlantic and Pacific. *Fisher. Res.* **37**: 115–125.

Kono, N. and Zenitani, H. (2005). Stock assessment and evaluation for the Seto Inland Sea stock of Japanese anchovy (fiscal year 2005). In *Marine Fisheries Stock Assessment and Evaluation for Japanese waters (fiscal year 2005/2006)*. Fisheries Agency and Fisheries Research Agency of Japan. pp. 629–656. (in Japanese)

Kornilovs, G. (1995). Analysis of Baltic herring year-class strength in the Gulf of Riga. *ICES cm:1995/*J:10.

Köster, F.W., Möllmann, C., Neuenfeldt, S., *et al.* (2003a). Fish stock development in the Central Baltic Sea (1976–2000) in relation to variability in the physical environment. *ICES Mar. Sci. Symp.* **219**: 294–306.

Köster, F.W., Hinrichsen, H.-H., Schnack, D., *et al.* (2003b). Recruitment of Baltic cod and sprat stocks: identification of critical life stages and incorporation of environmental variability into stock-recruitment relationships. *Sci. Mar.* **67** (Suppl. 1): 129–154.

Kraus, G. and Köster, F.W. (2004). Estimating Baltic sprat (*Sprattus sprattus balticus* S.) population sizes from egg production. *Fish. Res.* **69**: 313–329.

Kubota, H., Oozeki, Y., Ishida, M., *et al.* (1999). *Distributions of Eggs and Larvae of Japanese Sardine, Japanese Anchovy, Mackerels, Round Herring, Jack Mackerel and Japanese Common Squid in the Waters around Japan, 1994 through 1996.* National Research Institute of Fisheries Science, 352 pp.

Lasker, R. (1985). Introduction: an egg production method for anchovy biomass assessment. In *An Egg Production Method for Estimating Spawning Biomass of Pelagic Fish: Application to the Northern Anchovy, Engraulis mordax,* ed. R. Lasker. *NOAA Technical Report NMFS*, Springfield, VA, USA: US Department of Commerce, pp. 1–4.

Laurec, A. and Shepherd, J.G. (1983). On the analysis of catch and effort data. *J. Cons. Int. Explor. Mer* **41**: 81–84.

Le Clus, F., Melo, Y.C., and Cooper, R.M. (1988). Impact of environmental perturbations during 1986 on the availability and abundance of pilchard and anchovy in the northern Benguela system. *Colln. Scient. Pap. Int. Commn SE. Atl. Fish.* **15**: 49–70.

Legault, C.M. and Restrepo, V.R. (1999). A flexible forward age-structured assessment program. *ICCAT Coll. Vol. Sci. Pap.* **49** (2): 246–253.

Liorzou, B., Bigot, J.L. and Guennegan, Y. (2004). Evolution des stocks de sardine et d'anchois dans le golfe du Lion. Working document presented in the Working Group on small pelagic species. SAC-GFCM, ftp://cucafera.icm. csic.es/pub/scsa/.

Lleonart, J. and Maynou, F. (2002). Fish stock assessments in the Mediterranean: state of the art. *Sci. Mar.* **67** (Suppl. 1): 37–49.

Lluch-Belda, D., Crawford, R.J.M., Kawasaki, T., *et al.* (1989). World-wide fluctuations of sardine and anchovy stocks: the regime problem. *S. Afr. J. Mar. Sci.* **8**: 195–205.

Lluch-Belda, D., Schwartzlose, R.A., Serra, R., *et al.* (1992). Sardine and anchovy regime fluctuations of abundance in four regions of the world oceans: a workshop report. *Fish. Oceanogr.* **1** (4): 339–347.

Lo, N.C.H., Hunter, J.R., Moser, H.G., *et al.* (1992a). A daily fecundity reduction method: a new procedure for estimating adult biomass. *ICES J. Mar. Res.* **49**: 209–215.

Lo, N.C.H., Jacobson, L.D. and Squire, J.L. (1992b). Indices of relative abundance from fish spotter data based on delta-lognormal models. *Can. J. Fish. Aquat. Sci.* **49**: 2515–2526.

Lo, N.C.H., Green Ruiz, Y.A., Cervantes, M.J., *et al.* (1996). Egg production and spawning biomass of Pacific sardine (*Sardinops sagax*) in 1994, determined by the daily egg production method. *Calif. Coop. Oceanic Fish. Invest. Rep.* **37**: 160–174.

Lo, N.C.H., Macewicz, B.J., and Griffith, D.A. (2005). Spawning biomass of Pacific sardine (*Sardinops sagax*), from 1994–2004 off California. *Calif. Coop. Oceanic Fish. Invest. Rep.* **46**: 93–112.

MacCall, A.B. (2002). Fishery management and stock rebuilding prospects under conditions of low frequency variability and species interactions. *Bull. Mar. Sci.* 70: 613–628.

Macewicz, B.J. and Abramenkoff, D.N. (1993). Collection of jack mackerel, *Trachurus symmetricus*, off southern California during 1991 cooperative US–USSR cruise. *Southwest Fisheries Science Center, National Marine Fisheries Service, Admin. Rep.* **LJ-93–07**, 13 pp.

Macewicz, B.J., Castro-Gonzalez, J.J., Cotero Altamirano, C.E., and Hunter, J.R. (1996). Adult reproductive parameters of Pacific sardine (*Sardinops sagax*) during 1994. *CalCOFI Rep.* **37**: 140–151.

MacKenzie, B.R.M. and Köster, F.W. (2004). Fish production and climate: sprat in the Baltic Sea. *Ecology* **85**: 784–794.

MacKenzie, B.R. M, Horbowy, J., and Köster, F.W. (2008). Incorporating environmental variability in stock assessment: predicting recruitment, spawner biomass and landings of sprat (*Sprattus sprattus*), in the Baltic Sea. *Can. J. Fish. Aquat. Sci.* **65**: 1334–1341.

McQuinn, I. H. (1997). Metapopulations and the Atlantic herring. *Rev. Fish Biol. Fisher.* **7**: 297–329.

Machias, A., Somarakis, S., Giannoulaki, M., *et al.* (1997). Estimation of the northern Aegean anchovy stock in June 1995 by means of hydroacoustics. *Proceedings of the 5th Panhellenic Symposium on Oceanography and Fisheries*, Vol. **2**, pp. 47–50.

Machias, A., Somarakis, S., Drakopoulos, P., *et al.* (2000). Evaluation of the Southern Greek anchovy stocks. *Final Report, DG XIV Project* **97–0048**, 105 pp.

Machias, A., Giannoulaki, M., Somarakis, S., *et al.* (2001). Estimation of the anchovy (*Engraulis encrasicolus*) stocks in the central Aegean and Ionian Seas by means of hydroacoustics. *Proceedings of the 10th Panellenic Conference of Ichthyologists*, pp. 57–60.

Mackinson, S., Sumaila, U.R., and Pitcher, T.J. (1997). Bioeconomics and catchability: fish and fishers behavior during stock collapse. *Fish. Res.* **31**: 11–17.

Madureira, L.S.P. and Rossi-Wongtschowski, C.L.D.B. (2005). Prospecção de recursos pesqueiros pelágicos na Zona Econômica Exclusiva da Região Sudeste-Sul do Brasil: hidroacústica e biomassa. São Paulo, Instituto Oceanográfico-US, 144 pp.

Magoulas, A., Castilho, R., Gaetano, S., *et al.* (2006). Mitochondrial DNA reveals a mosaic pattern of phylogeographical structure in Atlantic and Mediterranean populations of anchovy (*Engraulis encrasicolus*). *Molec. Phylogenet. and Evol.* **39**: 734–746.

Mariani, S., Hutchinson, W.F., Hatfield, E.M.C., *et al.* (2005). North Sea herring population structure revealed by microsatellite analysis. *Mar. Ecol. Prog. Ser.* **303**: 245–257.

Massé, J. (1988). Utilisation de l'echo-integration en recherche halieutique (analyse des campagnes effectuées dans le Golfe de Gascogne de 1983 à 1987). Direction des Ressources Vivantes, IFREMER.

Massé, J. (1994). Acoustic Surveys DAAG 90, DAAG 91, DAAG 92. In *Annex to Improvement of Stock Assessment by Direct Methods, its Application to the Anchovy (Engraulis encrasicholus) in the Bay of Biscay*, Orestes ed. Report of the EC FAR Project (1991–1993) (Contract No. MA 2495 EF), 90 pp. + annexes.

Massé, J. (1996). Acoustics observation in the Bay of Biscay: schooling, vertical distribution, species assemblages and behavior. *Sci. Mar.* **60** (Suppl. 2): 227–234.

Matsuura, Y. (1983). Estudo comparativo das fases iniciais do ciclo de vida da sardinha-verdadeira, *Sardinella brasiliensis* e da sardinha-cascuda, *Harengula jaguana* (Pisces: Clupeidae), e nota sobre a dinâmica da população da sardinha-verdadeira na região sudeste do Brasil. Universidade de São Paulo, Instituto Oceanográfico. 150 pp. (Tese de Livre-docência).

Matsuura, Y. (1996). A probable cause of recruitment failure of the Brazilian sardine *Sardinella aurita* population during the 1974/75 spawning season. *S. Afr. J. Mar. Sci.* **17**: 29–35.

Matsuura, Y. (1998). Brazilian sardine (*Sardinella brasiliensis*) spawning in the southeast Brazilian Bight over the period 1976–1993. *Rev. Bras. Oceanogr.* **46** (1): 33–43.

McFarlane, G.A., Smith, P.E., Baumgartner, T.R., and Hunter, J.R. (2002). Climate variability and Pacific sardine populations and fisheries. *Am. Fish. Soc. Symp.* **32**: 195–214.

Melo, Y.C. (1984). Age studies on anchovy *Engraulis capensis* Gilchrist off South West Africa. *S. Afr. J. Mar. Sci.* **2**: 19–31.

Mendo, J. (1991). Stock identification of Peruvian anchoveta (*Engraulis ringens*): morphometric, tagging/recapture, electrophoretic and ecological studies. Ph.D. dissertation. Alfred-Wegener Institute für Polar- und Meeresforschung, Bremerhaven, Germany.

Mesnil, B. (2003). The Catch-Survey Analysis (CSA) method of fish stock assessment: an evaluation using simulated data. *Fish. Res.* **63**: 193–212.

Methot, R.D. (1989). Synthetic estimates of historical abundance and mortality for northern anchovy. *Am. Fish. Soc. Symp.* **6**: 66–82.

Methot, R.D. (2005). *Technical Description of the Stock Synthesis II Assessment Program. Version 1.17 – March 2005*. Seattle, WA: NOAA Fisheries, 54 pp.

Mikhailov, K. and Prodanov, K. (2002). Early sex maturation of the anchovy *Engraulis encrasicolus* (L.) in the northwestern part of the Black Sea. In *3rd EuroGOOS Conference: Building the European Capacity in Operational Oceanography*, Athens, Greece: National Centre for Marine Research.

Miller, K.A. (2007). Fish stew: uncertainty, conflicting interests and climate regime shifts. In Bjørndal, T., Gordon, D.V., Arnason, R., and Sumaila, U.R., eds. *Advances in Fisheries Economics: Festschrift in honour of Professor Gordon R. Munro*. Blackwell, Oxford, UK: Chapter 13, pp. 207–221.

Möllmann, C., Kornilovs, G., and Sidrevics L. (2000). Long-term dynamics of main mesozooplankton species in the Central Baltic Sea. *J. Plank. Res.* **22**: 2015–2038.

Möllmann, C., Kornilovs, G., Fetter, M. and Köster, F.W. (2005). Climate, zooplankton and pelagic fish growth in the Central Baltic Sea. *ICES J. Mar. Sci.* **62**: 1270–1280.

Motos L., Uriarte, A., Prouzet, P., *et al.* (2005). Assessing the Bay of Biscay anchovy population by DEPM: a review 1989–2001. In *Report of the SPACC Meeting on Small Pelagic Fish Spawning Habitat Dynamics and the Daily Egg Production Method (DEPM)*, Castro, L. R., Fréon, P., van der Lingen, C. D., and Uriarte, A., eds. *GLOBEC Rep.* **22**: xiv, 107 pp.

Munk, P. and Christensen, V. (1990). Larval growth and drift pattern and the separation of herring spawning groups in the North Sea. *J. Fish Biol.* **37**: 135–148.

Murphy, G.I. (1966). Population biology of the Pacific sardine (*Sardinops caerulea*). *Proc. Calif. Acad. Sci.* **34** (1): 1–84.

Myers, R. A. (1998). When do environment-recruitment correlations work? *Rev. Fish Biol. Fisher.* **8**: 285–305.

Nash, R. D. M. and Dickey-Collas, M. (2005). The influence of life history dynamics and environment on the determination of year class strength in North Sea herring (*Clupea harengus* L.). *Fish. Oceanogr.* **14**: 279–291.

Needle, C. L. (2004). Absolute abundance estimates and other developments in SURBA. In *Working Document to the ICES Working Group on Methods of Fish Stock Assessment, Lisbon, 11–18 February.*

Nichols, J. H. (2001). Management of North Sea herring and prospects for the new millennium. In *Herring: Expectations for a New Millennium.* pp. 645–665. Fairbanks, AK, USA: Alaska Sea Grant Coll. Program. Lowell Wakefield Fisheries Symposium Series, **18**.

Nikoloudakis, G., Machias, A., Somarakis, S., *et al.* (2000). Comparison of growth in two anchovy stocks. In *Proceedings of the 6th Panhellenic Symposium on Oceanography and Fisheries*, Vol. **2**: 104–108.

Nishida, H., Yatsu, A., Ishida, M., Noto, M., and Katsukawa, Y. (2006). Stock assessment and evaluation for the Pacific stock of Japanese sardine (FY 2005). In *Marine Fisheries Stock Assessment and Evaluation for Japanese Waters FY2005/2006*, Fisheries Agency and Fisheries Research Agency of Japan, pp. 11–45. (in Japanese)

Nissling, A. (2004). Effects of temperature on egg and larval survival of cod (*Gadus morhua*) and sprat (*Sprattus sprattus*) in the Baltic Sea – implications for stock development. *Hydrobiologia* **514**: 115–123.

Noto, M. and Yasuda, I. (2003). Empirical biomass model for the Japanese sardine, *Sardinops melanostictus*, with surface temperature in the Kuroshio Extension. *Fish. Oceanogr.* **12**: 1–9.

Ohshimo, S. (2005). Stock assessment and evaluation for the Tsushima Current stock of Japanese anchovy (fiscal year 2005). In *Marine Fisheries Stock Assessment and Evaluation for Japanese Waters (fiscal year 2005/2006)*, Fisheries Agency and Fisheries Research Agency of Japan, pp. 657–674. (in Japanese)

Ojaveer, E. (1981). *Fish Fauna of the Baltic.* In The *Baltic Sea*, Voipio, A., ed. Elsevier Oceanography Series. pp. 275–292.

Ojaveer E. (1989). Population structure of pelagic fishes in the Baltic. *Rapp. P.-v. Réun. Cons. int. Explor. Mer*, **190**: 17–21.

Oliva, J., Montenegro, C., Braun, M., *et al.* (2001). Evaluación del stock desovante de anchoveta en la I y II regiones, año 2000. Informe Final. FIP-IT 2000–06. 189 pp.

Oozeki, Y., Kubota, H., Takasuka, A., *et al.* (2005). Stock assessment and evaluation for the northwestern pacific stock of Japanese anchovy (fiscal year 2005). In *Marine Fisheries Stock Assessment and Evaluation for Japanese Waters (fiscal year 2005/2006)*, Fisheries Agency and Fisheries Research Agency of Japan, pp. 604–628. (in Japanese)

Oskarsson, G. J., Kjesbu, O. S., and Slotte, A. (2002). Predictions of realised fecundity and spawning time in Norwegian spring-spawning herring (*Clupea harengus*). *J. Sea Res*, **48** (1): 59–79.

Palomera, I. (1992). Spawning of anchovy *Engraulis encrasicolus* in the northwestern Mediterranean relative to hydrographic features in the region. *Mar. Ecol. Prog. Ser.* **79**: 215–223.

Palomera, I., Tejeiro, B., and Alemany, X. (2003). Size at first maturity of the NW Mediterranean anchovy. *Working Document Presented in the Working Group on Small Pelagic Species*. SAC-GFCM, ftp://cucafera.icm.csic.es/pub/scsa/.

Parker, K. (1980). A direct method for estimating northern anchovy, *Engraulis mordax*, spawning biomass. *Fish. Bull. US* 78541–78544.

Parmanne, R., Rechlin, O., and Sjöstrand, B. (1994). Status and future of herring and sprat stocks in the Baltic Sea. *Dana* **10**: 29–59.

Parrish, R. H., Serra, R., and Grant, W. S. (1989). The monotypic sardines, *Sardina* and *Sardinops*: their taxonomy, distribution, stock structure and zoogeography. *Can. J. Fish. Aquat. Sci.* **46** (11): 2019–2036.

Patterson, K. R. (1998). Integrated catch at age analysis version 1.4. *Scottish Fisheries Research Report*, **38**: Aberdeen: FRS.

Patterson, K. R. and Melvin, G. D. (1996). Integrated catch at age analysis version 1.2. *Scottish Fisheries Research Report*, **56**: Aberdeen, FRS.

Pauly, D. and Christensen, V. (1995). Primary production required to sustain global fisheries. *Nature* **374**: 255–257.

Pauly, D. and Palomares, M. L. (1989). New estimates of monthly biomass, recruitment and related statistics of anchoveta (*Engraulis ringens*), 1953 to 1981. In *The Peruvian Upwelling Ecosystem: Dynamics and Interactions. ICLARM Conference Proceedings*, **18**. Pauly, D., Muck, P., Mendo, J., and Tsukuyama, I., eds. pp. 189–206.

Pauly D. and Tsukayama, I. (1987). On the implementation of management-oriented fishery research: the case of the Peruvian anchoveta. In *The Peruvian Anchoveta and its Upwelling Ecosystem: Three Decades of Changes. ICLARM Studies and Reviews*, **15**: Pauly, D. and Tsukayama, I., eds. Instituto del Mar del Perú (IMARPE), Callao, Perú; Deutsche Gesellschaft für Technische Zusammenarbeit (GTZ), Eschborn, Federal Republic of Germany; and International Center for Living Aquatic Resources Management (ICLARM), Manila, Philippines, pp. 1–13.

Pearcy, W., Fisher, J., Brodeur, R., and Johnson, S. (1985). Effects of the 1983 El Niño on coastal nekton off Oregon and Washington. In El Niño North: Niño Effects in the Eastern Subarctic Pacific Ocean, Wooster, W. S. and Fluharty, D. L., eds. Washington Sea Grant Program, Seattle, USA: University of Washington, pp. 188–204.

Perry, A. L., Low, P. J., Ellis, J. R., and Reynolds, J. D. (2005). Climate change and distribution shifts in marine fishes. *Science* **308**: 1912–1915.

Pertierra, J. P. and Lleonart, J. (1996). NW Mediterranean anchovy fisheries. *Sci. Mar.* **60** (Suppl. 2): 257–267.

Pestana, G. (1989). *Manancial Ibero-Atlântico de Sardinha (Sardina pilchardus Walb.), sua Avaliação e Medidas de Gestão.* INIP, 192 pp.

PFMC (1983). *Northern Anchovy Fishery Management Plan (Amendment No. 5)*. Portland, OR, USA: Pacific Fishery Management Council.

PFMC (1998). *Amendment 8 (to the Northern Anchovy Fishery Management Plan) Incorporating a Name Change to: the Coastal Pelagic Species Fishery Management Plan*. Portland, OR, USA: Pacific Fishery Management Council.

PFMC (2005). *Status of the Pacific Coast Coastal Pelagic Species Fishery and Recommended Acceptable Biological Catches. Stock assessment and fishery evaluation 2005*. Portland, OR, USA: Pacific Fishery Management Council.[23]

PFMC (2006). *Status of the Pacific Coast Coastal Pelagic Species Fishery and Recommended Acceptable Biological Catches. Stock Assessment and Fishery Evaluation 2005*. Portland, OR, USA: Pacific Fishery Management Council.

Phillips, J. B. (1948). Growth of the sardine, *Sardinops caerulea*, 1941–42 through 1946–47. *Calif. Div. Fish Game Fish Bull.* **71**, 33 pp.

Pitcher, T. J. (1995). The impact of pelagic fish behavior on fisheries. *Sci. Mar.* **59** (3–4): 295–306.

Plagányi, É. E., Rademeyer, R. A., Butterworth, D. S., et al. (2007). Making management procedures operational – innovations implemented in South Africa. *ICES J. Mar. Sci.* **64** (4): 626–632.

Polovina, J. J. (2005). Climate variation, regime shifts, and implications for sustainable fisheries. *Bull. Mar. Sci.* **76**: 233–244.

Prodanov, K., Mikhaylov, K., Daskalov, G., et al. (1997). Environmental management of fish resources in the Black Sea and their rational exploitation. *Studies and Reviews, GFCM*, **68**. Rome: FAO. 178 pp. http://www.fao.org/DOCREP/006/W5020E/W5020E00.HTM.

Radovich, J. (1982). The collapse of the California sardine fishery: what have we learned? *Calif. Coop. Oceanic Fish. Invest. Rep.* **23**: 56–78.

Rampa, R., Arneri, E., Belardinelli, A., et al. (2005). Length at first maturity of the Adriatic anchovy (*Engraulis encrasicolus*, L.) *Working document presented in the Working Group on small pelagic species. SAC-GFCM*, ftp://cucafera.icm.csic.es/pub/scsa/.

Redeke, H. C. and van Breemen, P. J. (1907). Die Verbreitung der planktonischen Eier und Larven einiger Nützfische in der südlichen Nordsee. *Verhand. Rijksinstituut Onderzoek Zee*, **II** (2): 3–37.

Regner, S. (1996). Effects of environmental changes on early stages and reproduction of the anchovy in the Adriatic Sea. *Sci. Mar.* **60** (Suppl. 2): 167–177.

Reid, J. L. (1966). Oceanic environment of the genus *Engraulis* around the world. *CalCOFI Rep.* **11**: 29–33.

Rijavec, L and Amaral, J. C. (1977). Distribuição e abundância de peixes pelágicos na costa sul e sudeste do Brasil (resultados da pesquisa com ecointegrador). *SUDEPE/PDP* **24**: 1–55.

Roel, B. A. and Butterworth, D. S. (2000). Assessment of the South African chokka squid *Loligo vulgaris reynaudii*. Is disturbance of aggregations by the recent jig fishery having a negative impact on recruitment? *Fish. Res.*, **48**: 213–228.

Rossi-Wongtschowski, C. L. D. B., Saccardo, S. A., Miranda, L. B., et al. (1994). Aplicação do Método de Produção de Ovos ao estoque desovante de sardinha, *Sardinella brasiliensis*: resultados do cruzeiro oceanográfico de janeiro de 1988. (manuscript)

Rossi-Wongtschowski, C. L. D. B., Saccardo, S. A., and Cergole, M. C. (1995). Situação do estoque da sardinha (*Sardinella brasiliensis*) no litoral sudeste e sul do Brasil. *IBAMA/CEPSUL, Coleção Meio Ambiente. Série Estudos Pesca* **17**: 1–45.

Roy, C., van der Lingen, C. D., Coetzee, J. C., and Lutjeharms, J. R. E. (2007). Abrupt environmental shift links with changes in the distribution of Cape anchovy (*Engraulis encrasicolus*) spawners in the southern Benguela. *Afr. J. Mar. Sci.* **29**: 309–319.

Russo, A. and Artegiani, A. (1996). Adriatic Sea hydrography. *Sci. Mar.* **60** (Suppl. 2): 33–43.

Saccardo, S. A. and Rossi-Wongtschowski, C. L. D. B. (1991). Biologia e avaliação do estoque da sardinha *Sardinella brasiliensis*: uma compilação. Atlântica, Rio Grande, **13** (1): 29–43.

Santiago, J. and Sanz, A. (1992). Egg production estimates of the Bay of Biscay anchovy, *Engraulis encrasicolus* (L.), spawning stock in 1987 and 1988 (Estimaciones de la producción de hevos de 1 stock reproductor de anchoa, *Engraulis encrasicolus* (L.), del golfo de Vizcaya en 1987 y 1988). *Bol. Inst. Esp. Oceanogr.* **8** (1): 225–230.

Santojanni, A., Arneri, E., Barry, C., et al. (2003). Trends of anchovy (*Engraulis encrasicolus*, L.) biomass in the northern and central Adriatic Sea. *Sci. Mar.* **67**: 327–340.

Santojanni, N., Cingolani, E., Arneri, G., et al. (2005). Stock assessment of sardine (*Sardina pilchardus*, Walb.) in the Adriatic Sea, with an estimate of discards. *Sci. Mar.* **69**: 603–617.

Santojanni, A., Arneri, E., Bernardini, V., Cingolani, N., Di Marco, M., and Russo, A. (2006). Effects of environmental variables on recruitment of anchovy in the Adriatic Sea. *Clim. Res.* **31**: 181–193.

Schnute, J. 1987. A general fishery model for a size-structured fish population. *Can. J. Fish. Aquat. Sci.* **44**: 924–940.

Schwartzlose, R. A., Alheit, J., Bakun, A., et al. (1999). Worldwide large-scale fluctuations of sardine and anchovy populations. *S. Afr. J. Mar. Sci.* **21**: 289–347.

Serra, J. R. (1983). Changes in the abundance of pelagic resources along the Chilean coast. In *Proceedings of the Expert Consultation to Examine Changes in Abundance and Species Composition of Neritic Fish Resources. San José, Costa Rica, 18–29 April* 1983, Sharp, G. D. and Csirke, J., eds. *FAO Fish. Rep.* **291** (2): 255–284.

Serra R. (1991). Long-term variability of the Chilean sardine. In: *Proceedings of the International Symposium on the Long-Term Variability of Pelagic Fish Populations and their Environment*, Kawasaki, T., Tanaka, S., Toba, Y., and Taniguchi, A., eds. New York: Pergamon Press, pp. 165–172.

Serra, R. and Tsukayama, I. (1988). Sinopsis de datos biológicos y pesqueros de la sardina *Sardinops sagax* (Jenyns, 1842) en el Pacifico Suroriental. *FAO Sinopsis sobre la pesca*, **13** (Rev.1): p. 59.

Serra, R., Canales, C. and Böhm, G. (2004). Investigación CTP anchoveta norte, 2004. *Informe Final. IFOP/SUBPESCA.*

Shannon, L. J., Cury, P. M., Nel, D., *et al.* (2006). How can science contribute to an ecosystem approach to pelagic, demersal and rock lobster fisheries in South Africa? *Afr. J. Mar. Sci.* **28**: 115–157.

Shelton, P. A., Armstrong, M. J., and Roel, B. A. (1993). An overview of the application of the daily egg production method in the assessment and management of anchovy in the southeast Atlantic. *Bull. Mar. Sci.* **53**: 778–794.

Shepherd, J. G. (1999). Extended survivors analysis: an improved method for the analysis of catch-at-age data and abundance indices. *ICES J. Mar. Sci.* **56**: 584–591.

Shepherd, J. G., Pope, J. G. and Cousens, R. D. (1984). Variations in fish stocks and hypotheses concerning their links with climate. *Rapp. P.-v. Réun. Cons. Int. Explor. Mer* **185**: 255–267.

Silva, A., Santos, M. B., Caneco, B., *et al.* (2006). Temporal and geographic variability of sardine maturity at length in the north-eastern Atlantic and the western Mediterranean. *ICES J. Mar. Sci.* **63**: 663–676.

Simmonds E. J., (2007). Comparison of two periods of North Sea herring stock management: success, failure, and monetary value. *ICES J. Mar. Sci.* **64** (4): 686–692.

Simpson, J. G. and Gil, E. (1967). Maduración y desove de la anchoveta (*Engraulis ringens*) en Chile. *Bol. Cient. Inst. Fom. Pesq., Santiago, Chile*, **4**, p. 55.

Sjöstrand, B. (1989). Assessment reviews: exploited pelagic fish stocks in the Baltic. *Rapp. P.-v. Réun. Cons. int. Explor. Mer* **190**: 235–252.

Slotte, A. (1999). Effects of fish length and condition on spawning migration in Norwegian spring spawning herring (*Clupea harengus* L). *Sarsia* **84**: 111–127.

Slotte, A (2000). Factors influencing location and time of spawning in Norwegian spring-spawning herring: an evaluation of different hypotheses. In: *Herring. Expectations for a New Millennium. Lowell Wakefield Fisheries Symposium Series*, **18**, pp. 255–278.

Slotte, A., Johannessen, A., and Kjesbu, O. S. (2000). Effects of fish size on spawning time in Norwegian spring-spawning herring. *J. Fish Biol.* **56** (2): 295–310.

Smith, P. E. (2005). A history of proposals for subpopulation structure in the Pacific sardine (*Sardinops sagax*) population off western North America. *Calif. Coop. Oceanic Fish. Invest. Rep.* **46**: 75–82.

Somarakis, S. (2005). Marked inter-annual differences in reproductive parameters and daily egg production of anchovy in the northern Aegean Sea. *Belg. J. Zool.* **135**: 247–252.

Somarakis, S., Drakopoulos, P. and Filippou, V. (2002a). Distribution and abundance of larval fishes in the northern Aegean Sea – eastern Mediterranean – in relation to early summer oceanographic conditions. *J. Plankt. Res.* **24**: 339–357.

Somarakis, S., Koutsikopoulos, C., Machias, A. and Tsimenides, N. (2002b) Applying the daily egg production method to small stocks in highly heterogeneous seas. *Fish. Res.* **55**: 193–204.

Somarakis, S., Palomera, I., Garcia, A., *et al.* (2004). Daily egg production of anchovy in European waters. *ICES J. Mar. Sci.* **61**: 944–958.

Somarakis, S., Machias, A., Giannoulaki, M., *et al.* (2005a). Ichthyoplanktonic and acoustic biomass estimates of anchovy in the Aegean Sea (June 2003 and June 2004. A. *Working Document Presented in the Working Group on Small Pelagic Species. SAC-GFCM*, ftp://cucafera.icm. csic.es/pub/scsa/.

Somarakis, S., Machias, A., Giannoulaki, M., *et al.* (2005b). Lengths-at-maturity of anchovy and sardine in the central Aegean and Ionian Sea. *Working document Presented in the Working Group on Small Pelagic Species. SAC-GFCM*, ftp://cucafera.icm.csic.es/pub/scsa/.

Sparholt, H. (1994a). Growth changes of herring in the Baltic. *Tema Nord 1994*, **532**.

Sparholt, H. (1994b). Fish species interactions in the Baltic Sea. *Dana*, **10**: 131–162.

Squire, J. L. Jr. (1972). Apparent abundance of some pelagic marine fishes off the southern and central California coast as surveyed by an airborne monitoring program. *Fish. Bull.*, **70**: 1005–1019.

Stefánsson, G. (1996). Analysis of groundfish survey abundant data: combining the GLM and delta approaches. *ICES J. Mar. Sci.* **53**: 577–588.

Stergiou, K. I., Christou, E. D., Georgopoulos, D., Zenetos, A., and Souvermezoglou, C. (1997). The Hellenic Seas: physics, chemistry, biology and fisheries. *Oceanogr. Mar. Biol. Ann. Rev.* **35**: 415–538.

Stratoudakis Y., Bernal, M., Ganias, K., and Uriarte, A. (2006). The daily egg production method: recent advances, current applications and future challenges. *Fish Fisher.* **7**: 35–37.

Suda, M. and Kishida, T. (2003). A spatial model of population dynamics of the early life stages of Japanese sardine, *Sardinops melanostictus*, off the Pacific coast of Japan. *Fish. Oceanogr.* **12**: 85–99.

SUDEPE/PDP (1989) Relatório da reunião técnica do grupo permanente de estudo sobre sardinha (07 a 10 de novembro de 1989). Instituto de Pesquisa e Desenvolvimento Pesqueiro, 39 pp.

Sunyé, P. S. (1999). Effet de la variabilité climatique régionale sur la pêche de la sadinelle le long de la côte sud-est du Brésil (1964–1993). These de Doctorat. Bregtane, Université de Bretagne Occidentale, Institut Universitaire Européen de la Mer, 130 pp.

Sunyé, P. S. and Servain, J. (1998). Effects of seasonal variations in meteorology and oceanography on the Brazilian sardine fishery. *Fish. Oceanogr.* **7** (2): 89–100.

Thomas, R. M. (1986). The Namibian pilchard: the 1985 season, assessment for 12952–1985, and recommendations for 1986. *Colln scient. Pap. Int. Commn SE. Atl. Fish.* **13**: 243–269.

Tičina, V., Katavić, I., Dadić, V., Grubišić, L., Franičević, M. and Tičina, V. E. (2005). Acoustic estimates of small pelagic fish stocks in the eastern part of the Adriatic Sea: September 2004. *Working Document Presented in the Working Group on Small Pelagic Species. SAC-GFCM*, ftp://cucafera.icm. csic.es/pub/scsa/.

Tkatcheva, K. S. and Benko, Yu. K., ed. (1979). Resources and raw materials in the Black Sea. Moskow: AztcherNIRO, Pishtchevaya promishlenist (In Russian).

Toresen, R. and Jakobsson, J. (2002). Exploitation and management of Norwegian spring-spawning herring in the 20th century. *ICES Mar. Sci. Symp.* **215**: 558–571.

Toresen, R. and Østvedt, O. J. (2000). Variation in abundance of Norwegian spring spawning herring (*Clupea harengus*, Clupeidae) throughout the 20th century and the influence of climatic fluctuations. *Fish Fisher.* **1**: 231–256.

Torres, P., Giráldez, A., González, M., *et al.* (2004). Anchovy (*Engraulis encrasicolus*) stock assessment in the GFCM Geographical sub-area 06 North (2001, 2002 and 2003). *Working document presented in the Working Group on small pelagic species. SAC-GFCM*, ftp://cucafera.icm. csic.es/pub/scsa/.

Tsukayama, I. (1983). Recursos Pelágicos y sus pesquerías en Perú. *Rev. Com. Perm. Pacífico Sur* (**13**): 25–63.

Ulltang, O. (1980). Factors affecting the reaction of pelagic fish stocks to exploitation and requiring a new approach to assessment and management. *Rapp. Procès-Verb. Réun. Cons. Int. Explor. Mer*, **177**: 489–504.

Ulltang, O. (1996). Stock assessment and biological knowledge: can prediction uncertainty be reduced? *ICES J. Mar. Sci.* **53** (4): 659–675.

Ulltang, O. (2003). Fish stock assessments and predictions: integrating relevant knowledge. An overview. *Sci. Mar.* **67**: 5–12.

Uriarte A., Prouzet, P., and Villamor, B. (1996). Bay of Biscay and Ibero Atlantic anchovy populations and their fisheries. *Sci. Mar.* **60** (Suppl. 2): 237–255.

Uriarte, A., Sagarminaga, Y., Scalabrin, C., *et al.* (2001). Ecology of anchovy juveniles in the Bay of Biscay 4 months after peak spawning: do they form part of the plankton? *ICES cm:2001/W:20*.

Uriarte, A., Roel, B. A., Borja, A., *et al.* (2002). Role of environmental indices in determining the recruitment of the Bay of Biscay anchovy. *ICES cm:2002/O:25*.

van der Lingen, C. D., Coetzee, J. C., and Hutchings, L. (2002). Temporal shifts in the spatial distribution of anchovy spawners and their eggs in the Southern Benguela: implications for recruitment. In van der Lingen, C. D., Roy, C., Fréon, P., *et al.* eds. Report of a GLOBEC-SPACC/IDYLE/ENVIFISH workshop on upwelling areas. *GLOBEC* **Rep. 16**: 46–48.

van der Lingen, C. D. and Huggett, J. A. (2003). The role of ichthyoplankton surveys in recruitment research and management of South African anchovy and sardine. In *The Big Fish Bang: Proceedings of the 26th Annual Larval Fish Conference*, Browman, H. I. and Skiftesvik, A. B., eds. Bergen, Norway: Institute of Marine Research, ISBN 82-7461-059-8: 303–343.

van der Lingen, C. D., Coetzee, J. C., Demarcq, H., *et al.* (2005). An eastward shift in the distribution of southern Benguela sardine. *GLOBEC Int. Newsl.* **11**: 17–22.

van der Lingen, C. D., Shannon, L. J., Cury, P., *et al.* (2006a). Resource and ecosystem variability, including regime shifts, in the Benguela Current system. In *Benguela: Predicting a Large Marine Ecosystem, Large Marine Ecosystems*, Shannon, L. V. Hempel, G., Malanotte-Rizzoli, P., eds. Elsevier Large Marine Ecosystems Series **14**: 147–185.

van der Lingen C. D., Fréon, P., Fairweather, T. P., and van der Westhuizen, J. J. (2006b). Density-dependent changes in reproductive parameters and condition of southern Benguela sardine *Sardinops sagax*. *Afr. J. Mar. Sci.* **28** (3–4): 625–636.

Vazzoler, A. E. A. de M. (1962). Sobre a primeira maturação sexual e destruição de peixes imaturos. *Bolm Inst. Oceanogr. São Paulo* **12** (2): 5–58.

Villanueva, R. (1970). The peruvian Eureka programme rapid acoustic surveys using fishing vessels. *Technical Conference on Fish Finding, Purse Seining and Aimed Trawling. FAO, FII: F70/5*, 12pp.

Voss, R., Köster, F. W., and Dickmann, M. (2003). Comparing the feeding habits of co-occurring sprat (*Sprattus sprattus*) and cod (*Gadus morhua*) larvae in the Bornholm Basin, Baltic Sea. *Fish. Res.*, **63**: 97–111.

Vrooman, A. M. (1964). Serologically differentiated subpopulations of the Pacific sardine, *Sardinops caerulea*. *J. Fish. Res. Bd. Can.* **21**: 691–701.

Walters, C. J. (1989). Value of short-term forecasts of recruitment variation for harvest management. *Can. J. Fish. Aquat. Sci.* **46** (11): 1969–1976.

Walters, C. J. (1997). Challenges in adaptive management of riparian and coastal ecosystems. *Conserv. Ecol.* **1** (2): 1 [http://www.ecologyandsociety.org/vol1/iss2/].

Walters, C. J. and Collie, J. S. (1988). Is research on environmental factors useful to fisheries management? *Can. J. Fish. Aquat. Sci.* **45** (10): 1848–1854.

Walters, C. J., Christensen, V., Martell, S. J., and Kitchell, J. F. (2005). Possible ecosystem impacts of applying MSY policies from single-species assessment. *ICES J. Mar. Sci.* **62**: 558–568.

Wantanabe, Y., Zenitani, H. and Kimura, R. (1995). Population decline of the Japanese sardine *Sardinops melanostictus* owing to recruitment failures. *Can. J. Fish. Aquat. Sci.* **52**: 1609–1616.

Watanabe, C. and Nishida, H. (2002). Development of assessment techniques for pelagic fish stocks: applications of daily egg production method and pelagic trawl in the Northwestern Pacific Ocean. *Fish. Sci.* **68** (Suppl. 1): 97–100.

Wenzel, M. S. M. T., Cardoso, R. de D., Servo, G. J. de M., and Braga, B. S. (1988). PIEBS – Programa Integrado de Estudos Biológicos sobre a Sardinha. III – Comprimento médio de primeira maturação sexual, época e local de desova. In *Simpósio da FURG sobre Pesquisa Pesqueira, Rio Grande, 05–08/12:69*.

Wheeler, J. P. and Winters, G. H. (1984). Homing of Atlantic herring (*Clupea harengus*) in Newfoundland waters as indicated by tagging data. *Can. J. Fish. Aquat. Sci.* **41**(1): 108–117.

Winters, G. H. and Wheeler, J. P. (1996). Environmental and phenotypic factors affecting the reproductive cycle of Atlantic herring. *ICES J. Mar. Sci.* **53** (1): 73–88.

Yatsu, A. (2003). ABC determination rules and TAC System. *Mizu-Joho (Aquatic Information)* **23**: 8–12. (in Japanese)

Yatsu, A. (2005). Decadal-scale variability in ocean productivity and fishery management of small pelagic fishes. *Nippon Suisan Gakk.*, **71**: 854–858. (in Japanese)

Yatsu, A. and Kaeriyama, M. (2005). Linkages between coastal and open-ocean habitats and dynamics of Japanese stocks of chum salmon and Japanese sardine. *Deep-Sea Res. II* **52**: 727–737.

Yatsu, A., Nagasawa, K., and Wada, T. (2003). Decadal changes in abundance of dominant pelagic fishes and squids in the Northwestern Pacific Ocean since the 1970s and implications on fisheries management. *Amer. Fish. Soc. Symp.* **38**: 675–684.

Yatsu, A., Watanabe, T., Ishida, M., *et al.* (2005). Environmental effects on recruitment and productivity of Japanese sardine *Sardinops melanostictus* and chub mackerel *Scomber japonicus* with recommendations for management. *Fish. Oceanogr.* **14**: 263–278.

Zengin, M. (2003). The current status of Turkey's Black Sea fisheries and suggestions on the use of those fisheries. In *Workshop on Responsible Fisheries in the Black Sea and Azov Sea and evaluation of demersal fish resources.* BSERP. 15–17/4/2003 Istanbul, Turkey http://www.black-sea-environment.org/text/default.htm.

Zenitani, H. and Yamada, S. (2000). The relation between spawning area and biomass of Japanese pilchard, *Sardinops melanostictus*, along the Pacific coast of Japan. *Fish. Bull.* **98** (4): 842–848.

Zijlstra, J. J. (1958). On the herring races spawning in the southern North Sea and English Channel. *Rapp. P.-v. Réun. Cons. Int. Explor. Mer.* **143** (2): 134–145.

10 Global production and economics

Samuel F. Herrick, Jr., Jerrold G. Norton, Rognvaldur Hannesson, U. Rashid Sumaila, Mahfuzuddin Ahmed, and Julio Pena-Torres

CONTENTS

Summary

Global production and trade in small pelagic fish (SPF) are affected by complex interactions between physical, ecological and economic systems, which give rise to relatively long-term, asynchronous cycles in SPF abundance and distribution. These cycles can have serious impacts on local SPF fisheries' production, but because they tend to be counterbalancing, global production of SPF tends to remain relatively stable. Nevertheless, recent patterns of landings indicate that most SPF are being harvested at or near their maximum yield levels, which in the face of increasing demand is expected to result in rising prices in supply-limited markets. Adding to these concerns are the uncertainties of climate change, which leads us to consider important economic issues related to SPF fisheries production, starting with how the redistribution of SPF resources affects respective rates of resource utilization, particularly when SPF move between independently managed fishing zones. This entails an associated issue, the time preferences for experiencing the range of benefits from SPF resources among nations sharing access to these resources. Because the ecological and economic impacts of climate change will extend well beyond directed SPF fisheries, we consider the economic impact of a climate–SPF regime shift from an ecosystem perspective. Of interest here is the full range of economic benefits SPF resources provide; not only their commercial value, but as prey for commercially valuable predators, and for recreational and non-commercial predators. In this context we examine the socially optimum use of these resources, balancing the benefits from commercially harvesting SPF with those from leaving them in the ocean ecosystem. Lastly, we address the impact of climate change on SPF-dependent developing coastal communities, which rely heavily on SPF resources for livelihoods and food security making them particularly vulnerable to unfavorable climate-related changes in the abundance and distribution of these resources. Overall, our investigation suggests that developed nations would likely favor a long-term, cooperative approach towards international stewardship of SPF resources that focuses on the full range of potential economic benefits, while developing nations would probably be more concerned with capturing the commercial benefits from SPF resources now in order to address more immediate social objectives.

Introduction

Changes in climate affect the productivity, abundance, and distribution of predominantly planktivorous small pelagic fish (SPF) resources worldwide. These resources characteristically exhibit relatively long-term boom and bust cycles related to their geographic abundance. They may be extremely plentiful in one area for a period of time while virtually absent from another, only for the pattern to reverse itself in response to some significant climatic or environmental change. Moreover, SPF may vary in opposite phases over this cycle, as is typically the case for sardine and anchovy. These climate–SPF regime shifts can have important ecological and economic implications for fisheries that harvest SPF directly, as well as fisheries that harvest species which prey on SPF. At the regional level, unfavorable climate change can mean the virtual disappearance of small pelagic and related fisheries, and the net national income and economic activity therefrom. Conversely, favorable climate conditions can result in population explosions and extraordinary economic benefits at the regional level. Internationally, the economic impacts are primarily in terms of international conservation and management of transboundary SPF fishery resources, and maintaining SPF commodity flows in international markets. Although

Climate Change and Small Pelagic Fish, eds. Dave Checkley, Jürgen Alheit, Yoshioki Oozeki, and Claude Roy. Published by Cambridge University Press. © Cambridge University Press 2009.

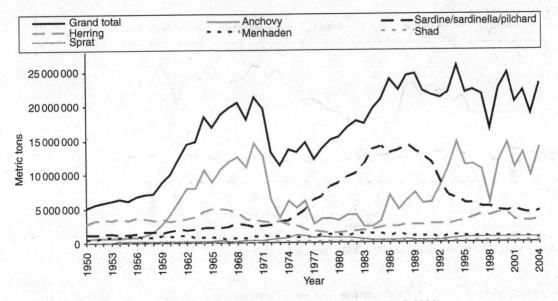

Fig. 10.1. Annual global harvests of SPF by species category, 1950–2004. (Source: FAO, 2006.)

climate change can persist over varying time scales, in this chapter we primarily focus on the economic impacts of climate–SPF regime shifts, which occur rather infrequently, on decadal scales or longer.

The chapter is organized as follows: first we review trends in global harvests of SPF fisheries, with an eye towards prospective SPF fisheries production in the light of increasing demand for SPF commodities, particularly as regards projected growth in aquaculture production. In the next section we present several economic issues relating to climate change and SPF fisheries production, starting with how the redistribution of SPF resources, in response to short-term, long-term, and seemingly permanent climate change, can have significant conservation and management implications when they move between sovereign fishing zones. Such circumstances will typically evoke strategic behavior by the nations sharing an SPF resource, concerning respective rates of resource utilization. This leads to a discussion of a related issue: the time preferences for experiencing the range of benefits from SPF resources among nations sharing access to these resources.

The economic impact of climate–SPF regime change will extend well beyond directed SPF fisheries. We next consider the total economic impact of a regime shift from an ecosystem perspective, which will include not only the economic value of SPF in terms of commodity production, but also value as forage for higher trophic level commercial fisheries. In this context we examine the economic tradeoff between commercially harvesting SPF and leaving them in the ocean ecosystem. The last issue deals with the impact of climate change on developing coastal communities, whose reliance on SPF resources for livelihoods and food security may make them particularly vulnerable to unfavorable

climate-related changes in the abundance and distribution of SPF resources. In the final section a brief summary and conclusion are provided.

Global harvests, production and trade

Based on statistics compiled by the Food and Agriculture Organization of the United Nations (FAO), SPF averaged 33% of marine fish harvests, in tonnes, over the 1950–2004 period. There was an upward trend in the global harvest of SPF from 1950 through 2004, with anchovies, sardines, and herring contributing the most to annual global harvests of SPF during that time, averaging 36%, 31% and 24% respectively of the total (Fig. 10.1). However, while the global harvest of SPF was increasing over this period, the relative contributions by the two major species categories, anchovy and sardine, tended to offset each other. Anchovy dominated global SPF harvests during the years 1960–1976 and 1992–2004 (averaging 56% of annual global SPF harvests over both periods); sardine during the intervening years (averaging 64% of global SPF harvests). Herring and related SPF (shad, sprat and menhaden) harvests remained fairly stable, at relatively low levels, throughout the period. From an economic standpoint, this trend in the overall availability of SPF is apt to offset volatility in the availability of a particular species since individual SPF species are frequently substituted for each other in the global market.

Most of the annual global SPF (from here on, SPF will collectively refer to anchovy, sardine (includes pilchards) and herring unless otherwise noted) harvest came from the Pacific Ocean over the 1950–2004 period (Fig. 10.2). The pattern of global SPF harvests during the period was predominantly influenced by harvests of anchovy and sardine from the Pacific Ocean. The increase in Pacific

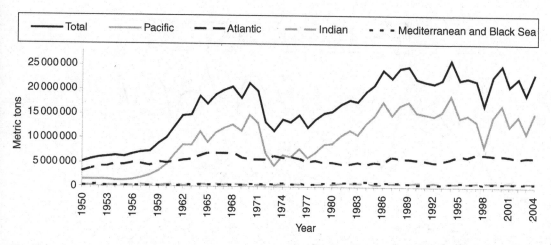

Fig. 10.2. Annual global harvest of SPF by oceanic region, 1950–2004. (Source: FAO, 2006.)

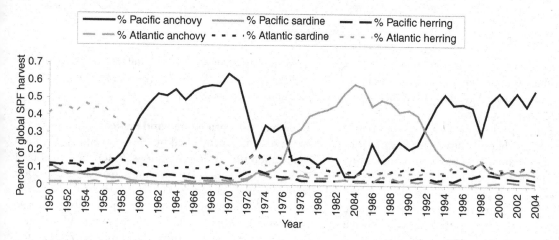

Fig. 10.3. Fraction of anchovy, herring and sardine harvests from the Atlantic and Pacific Oceans, of global SPF harvests, 1950–2004. (Source: FAO, 2006.)

harvests followed from the rapid development of SPF fisheries for anchovy and then sardine in Peru and Chile during the 1950s. Since the early 1970s interannual fluctuations in Pacific SPF harvests have been common, but the overall trend has leveled off, suggesting that the limits of sustainability for these species have been attained. Pacific anchovy harvests averaged 32% of global SPF harvests for the entire 1950–2004 period, but were over 50% of global SPF harvests from the 1960s through early 1970s, and from the mid 1990s on, falling to less than 30% in between these periods. Sardine harvests from the Pacific averaged 17% of global SPF harvests for the entire 1950–2004 period but dominated global SPF harvests from the mid 1970s through the early 1990s, with more than 40% of the global total. Respective harvests from the Atlantic were lower, but tended to be much more stable over the entire period, suggesting that Atlantic harvests had reached their maximum sustainable yield (MSY) levels; Atlantic anchovy harvests

averaged 2% of global SPF harvests for the period, Atlantic sardine harvests 11%. The proportion of the Atlantic herring harvest of the global SPF harvest was higher than the proportion for the Pacific, averaging 17% for the period compared to 5% for the Pacific (Fig. 10.3).

The major SPF harvesting countries during the 1950–2004 period were (in descending order of average annual global harvest share) Peru, Japan, Chile, the USSR/Russia, Norway, South Africa, and Canada. On average, these countries accounted for over 65% of the annual global SPF harvest for the period (Fig. 10.4). There were 117 other countries that also harvested SPF during the period. Peru was the leading harvester throughout most of the period, averaging 29% of the annual global SPF harvest. Peru thoroughly dominated anchovy harvests from the Pacific over the period, switching to sardines when the anchovy fishery was depressed from the mid 1970s through the early 1990s. Japan averaged 11% of the annual global SPF harvest

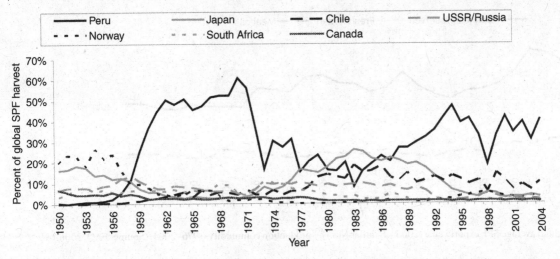

Fig. 10.4. Major harvesting countries based on average share of global annual SPF harvest, 1950–2004. (Source: FAO, 2006.)

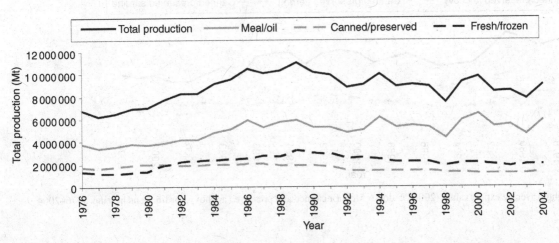

Fig. 10.5. Annual processed commodity production from global SPF harvests, 1976–2004. (Source: FAO, 2006.)

for the period, mainly due to its prodigious harvests of sardines from the Pacific in the mid 1970s through the early 1990s. Chile's annual harvests of anchovy, sardine and, to a lesser extent, herring from the Pacific averaged 7% of annual global SPF harvests for the period. The annual harvests of the USSR/Russia also averaged 7% of annual global SPF harvests for the period, consisting mainly of herring and sardines from the Atlantic through the 1970s, and sardines from the Pacific and the Atlantic from the mid 1970s through the early 1990s. Norway's annual share of the global SPF harvest, almost exclusively herring from the Atlantic, averaged 6% for the period; Norway was the leading global SPF harvester during most of the 1950s, but its share slipped to less than 1% from 1974–84 with the collapse of its herring fishery. South Africa averaged 5% of the global SPF harvest over the period; mainly sardines through the mid 1970s and then again in the late 1990s, and anchovy and herring from

the mid 1960s on, all from the Atlantic. Canada's combined harvests of herring from the Atlantic and the Pacific averaged 2% of the global SPF harvest over the period.

Harvests of SPF are converted into various commodities through value-added production processes. Based on FAO fisheries commodities, production, and trade data from 1976–2004, most of the reported SPF commodities production was in the fishmeal and fish oil category,[1] which averaged 57% of total annual SPF commodities production over the period. Fresh and frozen SPF was the next most important commodity category averaging 24% of total annual production for the period, followed by canned and preserved, averaging 19% of total annual production for the period. Production in all three commodity categories exhibited a gradual increase up until about 1990 and then started to decline with a slight uptick in 2004 (Fig. 10.5).

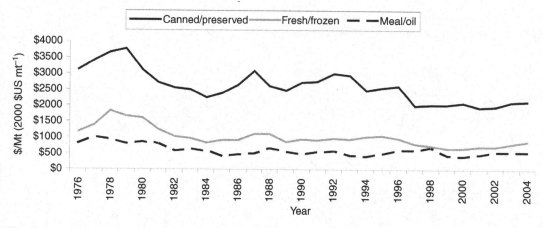

Fig. 10.6. Real weighted average SPF export value (2000 US dollar Mt⁻¹) by production commodity, 1976–2004. (Source: FAO, 2006.)

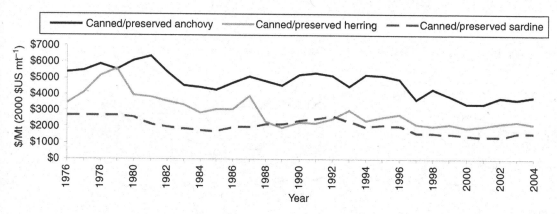

Fig. 10.7. Real weighted average export values (2000 US dollar Mt⁻¹) for canned and preserved anchovy, sardines and herring, 1976–2004. (Source: FAO, 2006.)

Canned and preserved SPF generated the highest per unit export values based upon (real $2000) global-weighted average export prices ($ Mt⁻¹ US), which averaged $2611 over the 1976–2004 period. Fresh and frozen SPF was the next most valuable commodity category followed by meal and oil (Fig. 10.6). Anchovies in the canned and preserved, and fresh-chilled and frozen categories were the most valuable SPF exports, with average unit export values of $4746 and $1795, respectively for the period. This compared to an average unit export value of $2911 for canned and preserved herring and $2024 for canned and preserved sardine, and an average unit export value of $807 for fresh, frozen, and chilled herring and $582 for fresh, frozen, and chilled sardine (Fig. 10.7 and 10.8).

Of the global fishmeal and fish oil exports for the 1976–2004 period, 98% were reported as being produced from species in the FAO oily fish category. Except for minuscule amounts produced from sardine, the balance of SPF meal and oil exports for the period were produced from herring; there were no exports of meal and oil specifically identified as being produced from anchovy. Because they were conspicuously lacking in the FAO fishmeal and fish oil production and trade statistics for the 1976–2004 period, it is reasonable to assume that large amounts of anchovy and sardine were included in the "oily fish not elsewhere included" category (FAO, 2006) during the times when they dominated global harvests (Fig. 10.1). The average unit price of oily fish, meal, and oil exports for the period was $604 (Fig. 10.9).

For both the canned or preserved commodity category, and the fresh, chilled, or frozen commodity category, herring and sardine export unit values were relatively close in recent years, while anchovy unit values were comparatively higher. These patterns of export unit values suggest that, while herring and sardines might readily substitute for each other in the production of canned or preserved, and fresh, chilled, or frozen SPF commodities, anchovy appears to be more exclusive in this regard. Because anchovies are

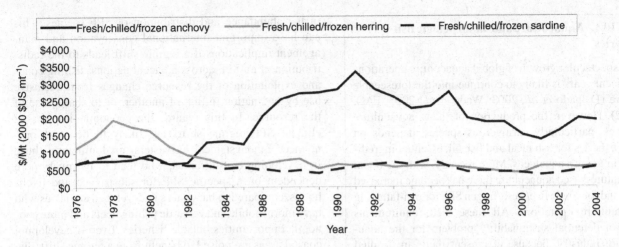

Fig. 10.8. Real weighted average export values (2000 US dollar Mt⁻¹) for fresh, chilled or frozen anchovy, sardines and herring, 1976–2004. (Source: FAO, 2006.)

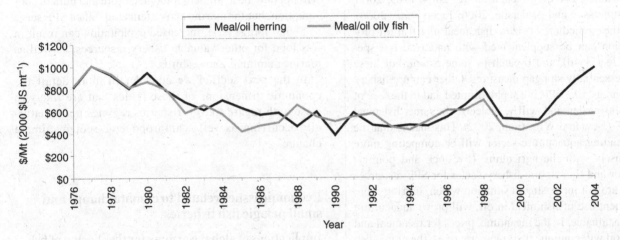

Fig. 10.9. Real weighted average export values (2000 US dollar Mt⁻¹) for meal and oil from herring and "oily fish not elsewhere included", 1976–2004. (Source: FAO, 2006.)

predominantly used for fish meal, anchovy products for human consumption would be quite unique and therefore have few substitutes.

The similar patterns of export unit values for meal and oil produced from herring and for meal and oil produced from species in the oily fish category suggest that SPF species can be easily substituted for each other in the production of fish meal and oil.

The outlook for SPF fisheries production is one of increased demand for resources that are approaching or have already reached their maximum sustainable harvest levels. Increased demand for SPF commodities, mainly driven by rapid expansion of aquaculture and corresponding growth in the use of fishmeal and fish oil in aquafeeds

(Box 10.1), is expected to result in rising relative prices, in supply-limited markets, before too long. Unless price pressures induce innovations in substituting other ingredients for fish meal and fish oil in aquafeeds, as well as in more traditional agrifeeds, we can expect to see increasing amounts of the global SPF harvest being diverted from human consumption into reduction. Demand for SPF is also expected to increase as a more health and nutritional conscious society increases its consumption of omega-3 based nutraceuticals. Yet, while these expanding uses of SPF may be increasing their value, transforming relatively low value food into much higher valued food and other commodities does not bode well for those that depend on SPF for food security.

Box 10.1. Aquaculture and small pelagic fish fisheries

The spectacular growth of global aquaculture operations in recent years is likely to continue into the foreseeable future (Delgado *et al.*, 2003; Wada *et al.*, 2003; FAO, 2002). However, the production of many aquaculture species, particularly carnivorous species, depends on SPF fisheries for fish meal and fish oil, essential ingredients in today's aquafeeds. More recently, with the boom in capture-based aquaculture there has become increased demand for whole live/fresh/frozen SPF for pen-fattening aquaculture operations. All these feed requirements pose a potential sustainability problem for the aquaculture industry, because at present there are limited opportunities to replace SPF, either in fresh or in fish meal form, with cost-effective protein substitutes, which greatly increases the risk of overfishing these resources (Naylor *et al.*, 2000; New and Wijkstrom, 2002; Brugere and Ridler, 2004; Ottolenghi *et al.*, 2004; FAO, 2006; Kristofersson and Anderson, 2006; Tacon *et al.*, 2006). On the immediate horizon, fish meal and fish oil production can be supplemented with underutilized species (e.g. krill), and bycatch to some extent, but these are essentially stopgap measures. Other capture fishery resources, like SPF, are supply limited and in the face of increasing demand, will be subject to the same "fish meal trap" (New and Wijkstrom, 2002). This means that the expanding aquaculture sector will be competing more intensely with the agriculture (livestock and poultry) sector and the capture fishery sector for SPF resources that are not increasing; a situation which, barring more efficient use of fish meal in oil, will prove aquaculture unsustainable. In the meantime, prices for fish meal and fish oil will continue their meteoric rise – the global fish meal price reached an all-time high of US $1400 per Mt in May of 2006, more than double that from a year earlier (Globefish, 2006) – and wild fish stocks will be increasingly threatened with overfishing while we wait for a technological breakthrough in terms of finding substitutes for fish meal and fish oil. This also means making tradeoffs among the many ecosystem services SPF provide. For instance, SPF are forage for numerous finfish species targeted by higher trophic level fisheries, as well as economically valuable seabirds and marine mammals (see below). More intense use of SPF for fish meal and fish oil can therefore incur significant economic costs in terms of foregoing many of the alternative ecosystem services provided by SPF resources.

Adding to these concerns are the uncertainties of climate change. Based on historical evidence, it is likely that a climate–SPF regime shift will lead to significant changes in the abundance and distribution of SPF species. This can have important international conservation and management implications if a regime shift leads to the redistribution of an SPF across adjacent nations' fishing zones and exploitation of the resource changes from exclusive use by one nation to that of another, or to shared use of the resource. In this regard, the economic impacts of climate–SPF regime shifts are likely to be more pronounced amongst local harvesters, predominantly those in developing coastal areas, who tend to be much more dependent on a specific SPF for subsistence and livelihoods, compared to harvesters in developed countries who have more mobile and versatile fishing fleets and more economic opportunities outside fisheries. Even if developing coastal areas are poised to benefit from a regime shift, due to an increase in SPF availability, chances are that inadequate conservation and management policies are likely to encourage intense exploitation, especially by distant water fleets. Overfishing of SPF is a probable result which will not only be a problem for aquaculture and human consumption, but for marine ecosystems too. Since SPF serve as a vital forage base, increased exploitation can result in less food for other valuable fishery resources, as well as marine mammals and seabirds.

In the next section we consider in more detail the economic dimensions of these issues that are likely to arise with regard to SPF fisheries regimes under naturally occurring, as well as anthropogenic-rooted, climate change.

Economic issues related to climate change and small pelagic fish fisheries

Implications of global warming for the sharing of fish stocks

Many, if not most, fish stocks migrate between the exclusive economic zones of different countries. Efficient management of such stocks requires co-operation between the countries involved. That co-operation revolves around two critical decisions: (i) how much to harvest from the stock, and (ii) how to divide the harvest between the countries involved. Both questions can be addressed indirectly by deciding the total effort to apply in the fishery and how it should be distributed among the countries involved, rather than deciding on a global harvest quota and its division among the countries. Biological and other circumstances dictate which method is most appropriate.

The migrations of the fish stocks have a bearing on how they are shared among the countries involved. This is obvious if we consider the fallback position of a country in case co-operation does not materialize. The country

Box 10.2. Climate change and global fisheries production

Fluctuations in SPF abundance and distribution are increasingly regarded as a biological response to climate–ocean variations, and not just as a result of overfishing and other anthropogenic factors. Shifts in SPF regimes resulting from climate change can have various effects on fisheries: some fisheries are likely to be positively affected, and others negatively, although the net global effect may be negligible. However, now that the coastal ocean has been divided up into different national exclusive zones, the fisheries of countries in whose economic zones these stocks are located could be affected in vastly different ways if an SPF resource undergoes a major bio-geographic redistribution. While an SPF stock is in transition from one regime to another, any existing agreements on how it is shared will probably come under strain and be renegotiated. Countries from whose zones the stock is receding may have little incentive to protect the stock, but may be in a position to negotiate favorable concessions based on enhanced conservation incentives on the part of neighboring countries into whose waters the stock is transitioning. Whether or not the countries involved are able to negotiate a mutually beneficial conservation and management agreement will among other things depend on: (1) their expectations concerning the duration of the regime shift; (2) their time preferences for the stream of benefits SPF resources are capable of providing; and, (3) the relative importance they place on the full range of benefits provided by SPF resources. In regard to the latter, the fact that SPF are important prey for species targeted by higher trophic level fisheries introduces ecosystem considerations into the negotiations. International, and domestic, conservation and management decisions concerning SPF will also be influenced by how these – typically more commercially and recreationally valuable – higher trophic level species are affected by the regime shift. Another consideration relates to the differences in attitudes towards these factors that we have noted between developed and developing countries. In this context, developing countries would tend to be more concerned about providing for the current generation's needs in terms of food security and income, which means greater emphasis on immediate, consumptive use of SPF resources. In contrast, developed countries, having already fulfilled these necessities, will be more inclined to place greater emphasis on future use and the non-consumptive benefits associated with SPF resources. How far neighboring countries have evolved in this regard will be a major factor in terms of negotiating a mutually beneficial transboundary conservation and management agreement in response to a climate–SPF regime shift.

involved would spare fish that occurs within its zone only to the extent that this would improve its future fishing possibilities. The fact that any fish spared by a given country may end up in the economic zone of another obviously weakens the incentive for any single country to leave any fish uncaught within its own zone. This is the fundamental reason why non-co-operation leads to solutions that typically are vastly inferior to cooperative solutions.

In order that a country be willing to participate in a co-operative solution, it must be offered an outcome that is at least as good as it would get in the absence of cooperation. The critical share of the co-operative solution clearly depends on the migrations of the fish stock; a country whose zone only has a small share of the stock obviously cannot expect a large share of the co-operative solution, since it cannot lay its hands on a large share of the stock anyway. But it is worth noting that a country with a minor interest must typically be offered a larger share of the co-operative solution than corresponds to its share of the stock. The reason is that a country like that has a weak incentive to spare any of the fish inside its zone, because it would mainly benefit countries with a larger share of the stock. Hence, in the absence of co-operation, the country with a minor interest would take a disproportionally large fraction of the stock appearing in its zone, and maybe all of it, and would thus come out relatively strong in the absence of co-operation.[2]

Fish migrations are governed by two things. Many fish have a "homing instinct" to where they were originally spawned. The salmon are a famous case; somehow they find their way over thousands of miles to their home rivers, but even species like cod seem to be driven by a similar instinct; the Arctic cod migrate every year from the Barents Sea or the Norwegian Sea to Lofoten and other places along the Norwegian coast to spawn. Second, fish migrate in search of food. Since blooming of plankton and availability of feed fish is influenced by ocean currents, any change in the critical currents will affect fish migrations and the distribution of fish between the economic zones of different countries. It is well known that such changes occur, both from year to year and on longer time scales. Uncertainty about such changes is bound, therefore, to affect international agreements on shared fish stocks.

It is useful to distinguish between three types of fluctuations, (i) repetitive but uncertain fluctuations on a short time scale (year to year or so); (ii) regime change, which may be repetitive but occurs on a long time scale (decadal or more), and (iii) secular change such as global warming, which nevertheless may be characterized by variations around a trend.

Repetitive but uncertain fluctuations probably pose the least problem. Sometimes a country will be disadvantaged so that the fish stay out of its economic zone, and it might not even be able to harvest the fish it is entitled to, unless there is some reciprocity agreement among the parties that would allow the disadvantaged country to do so. The EU-Norwegian fisheries agreement allows for reciprocity in this regard; Norwegian boats can take some of their harvests in the EU zone, and *vice versa*. If the changes in fish migrations are repetitive on a sufficiently short time scale, these fluctuations will even out; in the past it has happened that the Northeast Arctic cod has mainly been in the Norwegian economic zone, and a part of the Russian quota has then been transferred to Norway by a mutual agreement. But a careful analysis of this situation shows that the outcome can be unexpected and counterintuitive; lack of information about fish migrations can give some protection to stocks in the absence of co-operation, and information can sometimes be disadvantageous for the party that holds it (Golubtsev and McKelvey, 2007).

Changes that occur infrequently, on a decadal scale or longer, will be more problematic, even if they are in some sense repetitive. Such changes that have occurred in the recent past have turned out to be problematic. One such is the regime shift in the North Pacific that changed the salmon migrations and upset the existing treaty between the United States and Canada on the sharing of the salmon (Miller and Munro, 2004). In a recent paper, Miller (2007) discusses how management agreements for highly migratory species like tuna can be made more resilient against fluctuations in migration and argues for transferability of harvest and effort quotas among the nations involved. How well this would work for changes that take place on a decadal scale or longer is debatable, as the question of whether this is a permanent change is likely to arise.

This brings us to the last type of change, a permanent change in stock habitat or migration, in the sense of a trend but with short-term fluctuations. This can give rise to both counterintuitive and unpredictable effects. Unexpectedly perhaps, the bargaining position of a country that sees its share of a stock dwindle may in fact be strengthened – up to a point. The reason is that the incentive to conserve the stock will be weakened while the country can ride for free on the conservation incentives of other parties, which will have been strengthened.[3]

A permanent change in migrations or habitat would be likely to sooner or later upset existing sharing agreements based on outdated stock distribution or migrations. Such a change could also increase the risk of extinction of a fish stock. With a climate change such that a stock is slipping permanently out of a country's economic zone into that of another, the former country has little or no incentive to protect the stock and might just as well elect to over-harvest it while the stock is still in its zone. Whether or not the countries involved would be able to come to a mutually beneficial agreement about saving the stock is an open question. Some uncertainty about what is happening and a slow enough revision of beliefs about the behavior of the stock

could save the stock while perfect and timely information could be disastrous (Hannesson, 2007b).

Discount rates and the optimal time path of SPF resource exploitation

Another issue related to the economics of climate change and conservation and management of transboundary SPF resources concerns the optimal time path of a SPF stock's exploitation. Because society is typically the owner of the resource stock, its overarching objective is to manage the resource sustainably for the overall benefit of both current and future generations (Sumaila, 2005). In the case of a shared resource fishery, this social objective is difficult to achieve since it is often at odds with the outcome from what would be considered rational behavior on the part of private fishing firms exploiting the publicly held resource for individual financial gain. Private fishing firms would be inclined to use the resource in a manner that maximizes the net value (ex-vessel revenues minus fishing costs) of their directed commercial harvests over time, whereas for other members of society, including those with non-consumptive use and non-use[4] interests in the resource (e.g. recreational fishers, environmentalists), there are non-market values provided by the resource (e.g. value as forage). Therefore, from contemporary society's view, the total economic value – market and non-market from both the use and non-use of the resource – would be taken into account when considering the optimal use of its resource.

With respect to a country's fisheries conservation and management policy, society is concerned with the full array of benefits provided by the resource over time, and the time preference rate at which these benefits are received. To account for people's time preferences for receiving benefits when determining the optimal use of SPF resources, discounting is used to convert anticipated future values into present values, where the choice of discount rate reflects the weight placed on receiving benefits now in contrast to the uncertainty of deferring benefits into the future.[5] All else being equal, the optimal time path is that which maximizes the present value of the projected total net benefit stream provided by the resource; that is, its net present value.

Each resource interest group will apply a discount rate that is highly influenced by its time preference for receiving the benefits it expects to derive from the resource; those preferring to sacrifice future for immediate benefits will have a higher discount rate than those placing a higher preference on benefits realized in the future. In this context, one would expect that the discount rate for privately owned commercial fishing firms would be higher than that for society as a whole. Profit maximizing fishing firms exploiting a common property resource under great uncertainty, are driven to pursue a harvest strategy that favors current harvests over future harvests (Brennan, 1997). If unchecked, this will likely result in a rate of commercial exploitation which jeopardizes society's ability to manage the resource in a sustainable manner. On the other hand, society's emphasis on providing for future generations and on non-market values would typically favor a more conservative harvest strategy that defers current use to maintain a higher stock level. In general society has a longer time horizon, so that the social discount rate would be comparatively lower than the private discount rate (Krutilla, 1967; Goulder and Kennedy, 1997).

In the case of publicly held fishery resources the choice of social discount rate will reflect the degree of sustainability that policy makers ascribe to the use of the resource, and therefore it serves as a policy variable when evaluating the economic benefits of long-term fishery conservation and management strategies. However, the time preference rate for the use of a fishery resource will also be greatly influenced by overall economic conditions within a country. For instance, high levels of unemployment, poverty and indebtedness will press a country to apply higher discount rates to the flow of benefits from fishery resources (Sumaila, 2005).

These circumstances can then lead to a situation where the resource could become biologically overfished, but not economically overfished.[6] Sumaila and Stephanus (2006) addressed this prospect by exploring the question of whether the Namibian pilchard had been both biologically and economically overfished. They came to the conclusion that pilchard had indeed been biologically overfished. However, they also found that, from the perspective of a private economically rational fishing operation, it was not obvious that the pilchard had been overfished economically.

The steep decline in the harvest of the Namibian pilchard over the 1960–2002 time period was used to support the argument that the pilchard had been overfished biologically. To explore whether pilchard had been economically overfished, Sumaila and Stephanus (2006) addressed the question: would a rationally behaving private economic agent or sole owner of the resource had chosen the observed harvest pattern of Namibian pilchard over a more sustainable harvest profile? The results from their analysis indicated that the observed harvest time path would be preferred by the sole owner over the calculated average sustainable yield time path for a discount rate equal to or greater than 3%. This was because the net present value of the observed commercial harvest over the 1960–2002 period was greater than the net present value of average sustainable harvest profile calculated for the same period when the discount rate was at least 3% (Fig. 10.10). For the period under consideration, a discount rate of less than 3% for Namibia is simply unrealistic, implying that for all practical purposes it cannot be said that the Namibian pilchard had been economically overfished; that is, that the net present value is not maximized.

Even though the outcome for the Namibian pilchard may have been perfectly rational from the private economic

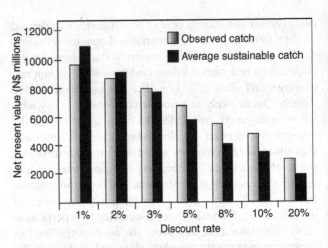

Fig. 10.10. Total discounted present values from the observed and average sustainable harvests of Namibian pilchard for the period 1960 to 2002 (Sumaila and Stephanus, 2006).

agent's perspective, it may have been at variance with Namibia's social objective of exploiting its fishery resources sustainably through time (Sumaila and Walters, 2005; Berman and Sumaila, 2006). If so, this means that private fishing firms tended to undervalue the resource relative to the rest of Namibian society by ignoring the non-market benefits of leaving more fish in a dynamic ecosystem. Hence from society's standpoint, the present value of the projected total net benefit stream provided by the resource would not have been maximized under the observed harvest profile for the 1960–2002 time period.

When dealing with the problem of climate change and the shared use of marine ecosystem resources among nations, differences in attitudes towards the time preference for benefits, the relative importance of market and non-market values and the treatment of risk and uncertainty, at the level of society, will bear on each country's willingness to participate in a co-operative solution. Differences in these attitudes are reflected in each affected country's social discount rate which, as pointed out in the previous section, will be affected by how permanent it believes a redistribution of SPF resource stock shares, brought about by global climate change, is likely to be. If there is relative uncertainty concerning the redistribution of the stock, then each country may be willing undertake a collective more precautionary sustainable utilization strategy, reflected in a comparatively low discount rate. On the other hand, if a country expects to experience a permanent loss in its resource share, then it may reasonably accelerate its exploitation of the SPF resource reflecting a relatively high social discount rate (Herrick *et al.*, 2007).

As we shall discuss in the next section, when climate change impacts are considered at the ecosystem level, the total economic value of the SPF resource, which includes its non-market value as a prey species, becomes the focus. In this sense, the success of co-operative ecosystem-based conservation and management of SPF fisheries will depend on the relative importance of market and non-market values to the countries involved. Countries that place greater emphasis on non-market values tend to be low discount rate countries and *vice versa*. Low discount rate countries are therefore more likely to advocate co-operative ecosystem-based management, and undertake costly investment in the resource now in order to realize future benefits. Conversely, high discount rate countries would deem these investments too costly in terms of the benefits foregone from not harvesting the resource now (Sumaila, 2005). In this context, Berman and Sumaila (2006) provide a nice discussion of the twin issue of valuation and discounting as it pertains to ecosystem restoration, and stress the need, at the level of society, to incorporate amenity values and apply new discounting approaches (Weitzman, 2001; Settle and Shogren, 2004; Sumaila and Walters, 2005) to ensure that the full range of ecosystem benefits to be enjoyed by future generations are adequately captured in economic valuations of SPF resources.

Ecosystem, economic and management implications of climate change

Historically, much of the debate on the failure of SPF fisheries worldwide has focused on overfishing as the primary cause. This has resulted in the development of SPF fisheries conservation and management measures to prevent such future occurrences. However, continuing observation has shown the importance of environmental variability in determining SPF biomass and research has been directed towards incorporating environmental variables into ecosystem-oriented conservation and management analyses (Jacobson and MacCall, 1995; Jacobson *et al.*, 2005; Herrick *et al.*, 2007; PFMC, 2006).

As a case in point, we use historical California commercial fish landings records (CACom) and environmental information to examine the influence of climate-scale changes in the fishes and fisheries ecological and economic systems, respectively, and their interconnections. Previous studies have described the data sets and examined linkages between species and large-scale environmental variables (Norton and Mason, 2004, 2005) as well as between phylogenetic and ecological groups and ex-vessel value (Mason, 2004). We extend these investigations to link changes in SPF ecological systems to landings compositions, ex-vessel values and investment in the California fishing industry. Then we discuss the total economic value of SPF in terms of their directed harvest value, along with their value as food for other organisms in the ecosystem, and use these empirical premises to develop a conceptual model to indicate the socially optimum harvest level of a SPF based on a broad spectrum of economic benefits.

Analysis of historic CACom data show that the availability of SPF species in the waters off California has varied with an apparently complete environmental cycle during the 1930–2000 period (MacCall, 1996; Norton and Mason, 2003, 2004, 2005; Herrick *et al.*, 2007). This cycle, diagrammed in Fig. 10.11, shows that the periods of species abundance maxima depend on the progression of environmental conditions through the 1930–2000 period. Because of the relationship of the ecological space represented in Fig. 10.11 to the physical environment (Norton and Mason, 2005) each quadrant represents the ecological and fisheries results of varying physical environmental conditions.

The fluctuating availability of various species and corresponding economic effects have caused the fishing industry and state and federal conservation and management agencies to alter their operations following environmental changes. The three species of predominant interest in this respect, sardine, anchovy and herring, have unique, well-separated, positions in Fig. 10.11. Sardine is at the right middle of Fig. 10.11, and anchovy and herring are on the left side of Fig. 10.11 – upper left and lower left, respectively. These positions in the diagram show that each of the three SPF species has unique environmental conditions that facilitate maximum abundance and consequent landings maxima. All three SPF species have been present in the waters off California throughout the 1930–2000 period; however, their availability to commercial fishers varies over several orders of magnitude and each species has been sufficiently low in abundance or in market value that it has been landed only as bycatch at some time during this 70-year period. Associated wetfish[7] species, such as Pacific mackerel, jack mackerel, and market squid are also distributed about the environmental space (Fig. 10.11) showing the association of their maximum catch and abundance with varying physical environmental conditions.

By inference, the spatial separation in SPF species in the ecological space (Fig. 10.11) also suggests differences in the quality and quantity of forage available to upper trophic levels such as predatory fish, birds and mammals, which may be valued based on their non-consumptive or existence value, rather than their value as food. White sharks (*Carcharodon carcharias*), brown pelicans (*Pelecanus occidentalis Californicus*), and gray whales (*Eschrichtius robustus*) are examples of California fish, bird and mammal species with non-consumptive and existence value. From the perspective of ecosystem-based management, which includes SPF and the predators that depend upon them directly or indirectly, the ecosystem–economic impacts of an environmental–SPF regime shift will be much more profound than the collapse of a single fishery.

The environmentally controlled cycle of SPF abundance has been modified by four exogenous factors associated with the development of the California fishing industry.

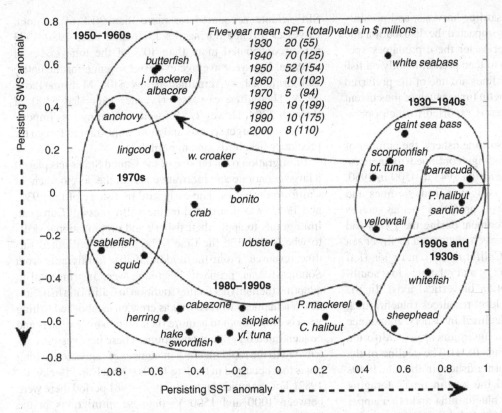

Fig. 10.11. The progression of ecosystem changes, as measured by the variation in California commercial landings weight of 29 species captured through 1930–2000 is indicated by the closed loops, which show species having abundance maxima during successive decades. Each species is plotted according to loadings in the empirical orthogonal function explaining the most (EOF1, horizontal) and second most (EOF2, vertical) variance. These EOF-axes provide an ecological space that displays the fluctuation of fisheries abundance. EOF1 and EOF2 time varying coefficients (principal components) have been found to correlate with temporal variation indices of sea surface temperature (SST) in the California Current region (increasing to the right on the horizontal scale, with $r > 0.9$) and southward wind stress (increasing downward, more negative, on the vertical scale, with $r > 0.8$). Decadal values of 5-year mean ex-vessel revenue are shown in the upper middle of the figure; with the decade, corresponding mean ex-vessel revenues for all the SPF landings, and mean ex-vessel revenue from the landings of all 29 species, are shown in the first and second columns and in parentheses, respectively. Ex-vessel revenue changes significantly as the ecosystem and species available for commercial capture changes with ocean climate.

First, fisheries for alternative species, and invertebrate species have emerged as the fishery has evolved through the 1930–2000 period (Mason, 2004). Second, immigration has been important, as the immigrants have brought fishing knowledge and willingness to work for low initial return to California's commercial fisheries. Third, legislation by the United States (US) Congress in the 1960s and 1970s provided tax incentives and guaranteed loans for fishing boat improvement and construction. Fourth, regulation of fishery harvests by state and federal agencies has increased, particularly after 1976 (PFMC, 1998). Empirical orthogonal function analysis of California commercial fish landings records suggest that management and other social and economic factors may account for as much as 50% of the interannual landing variability in California landings

through the 1930–2000 period (Norton and Mason, 2003, 2004).

Overall, the California fisheries grew rapidly from a 5-year average total ex-vessel revenue[8] slightly more than \$50M in 1928 to more than \$170M over 1943–1946. Sardines were more than 80% of the total (all species) landings weight and 55% of the total landings value during the 1936–1946 sardine fishery. The total California ex-vessel revenue declined in the late 1940s as sardine availability declined (MacCall, 1979; Jacobson and MacCall, 1995) and total ex-vessel fishery value of \$170M was not equaled for another 30 years. The early California fishery is shown in the middle right hand section of Fig. 10.11. Bluefin tuna, barracuda, yellowtail, and salmon were commercial species that fed on sardines during the 1930s and 1940s and it is probable that the large sardine harvests, and possibly bycatch of other SPF species, impacted the distribution and possibly the growth of fisheries for these predatory species. Bluefin tuna are known to feed predominantly on fish (Pinkas et al., 1971), and sardines are one of the preferred provisions for pen-reared bluefin tuna. Much of the current California harvest of sardines is exported to such operations in Australia.

During the decline of the sardine fishery, the number of vessels with commercial landings continued to increase from about 1500 in 1934 to 2700 in 1943 to 4500 in 1950, and California fishers were switching from sardines and other SPF to higher trophic level, predatory species such as albacore. The physical environment during the 1950s and 1960s (Parrish et al., 2000) was characterized by lower sea surface temperatures in the California current region (left on horizontal axis in Fig. 10.11), and relatively low southward windstress (upper portion of vertical axis). In the 1950s several upper trophic level resources (bluefin tuna, yellowtail, and barracuda) declined in fishery abundance with the change in the ocean environment off California (middle right to upper left, Fig. 10.11). The decline of the sardine availability to California fishers in the late 1940s and 1950s probably affected the bluefin tuna's foraging range and the availability of bluefin tuna and other upper trophic level predators to California fishers.

California fisheries for high-valued species (e.g. swordfish, albacore, salmon, mackerel) became more fully developed in the 1950s, 1960s, and 1970s. Other species (e.g. yellowfin and skipjack tuna) were available during high frequency climate perturbations, particularly in the late 1950s and the early 1980s. Fisheries for these species developed in response to the environmental shifts (right to left in Fig. 10.11). Good year classes of anchovy resulted in peak anchovy landings during the 1972–1976 period. Anchovy landings in 1973 earned about $21 M in ex-vessel revenue and accounted for nearly 65% of the total California fishery landings weight. Anchovy harvest for reduction continued through 1982. Anchovy, herring, Pacific mackerel, jack mackerel, and other small fish landings contributed 28% of the total to the overall ex-vessel revenue maximum that occurred in the early 1980s.

The albacore fishery produced as much as $40 M annually for California fishers. Albacore fisheries were the most important fishery for upper trophic level species off California from 1947 through 1980, contributing an average of $15 M to $20 M annually to total ex-vessel revenues. Where they co-exist, anchovy are among the preferred prey of albacore (Pinkas et al., 1971; Bernard et al., 1985). As the environmental cycle progressed through the 1970s and into the 1980s (lower left to center in Fig. 10.11), albacore decreased in the California landings. Although SPF landings continued at 50% or more of total landings by weight,

SPF seldom accounted for more than 20% of the total ex-vessel revenue during the 1970s and after 1980–1983 rarely contributed more than 10% of the total ex-vessel revenue. Five-year average ex-vessel revenue for the entire California fishery remained below $100 M from 1961–1970. Thereafter, ex-vessel revenues exceeded $100 M through 2004. However, SPF were becoming less important as major target species and more important as forage for predatory fish, birds, and mammals.

Immigration from outside of the United States has played a large economic and innovative part in the development of California fisheries. The early sardine fishery of the 1930s and 1940s was dominated by the willingness of European immigrants to apply their fishery and organizational skills to what seemed at the time to be a nearly unlimited sardine resource. From the middle 1970s, immigrants from Southeast Asia, primarily Vietnam, were instrumental in adapting their gill net fishing tradition to California fisheries. The Vietnamese worked first as crew on established fishing vessels, then began to acquire their own fishing vessels and concentrated on gill net operations. These new vessels contributed to the rapid increase in number of registered fishing boats that occurred in the late 1970s (Orbach and Beckwith, 1982). For example, during the 1975–1981 period there were between 1000 and 1500 Vietnamese immigrants in the Monterey Bay area, in central California. As many as 10% of these immigrants were boat owners and fishers, operating about 40 fishing boats out of Monterey Bay Harbors (Orbach and Beckwith, 1982). Vietnamese also entered the fishing business in other California ports. Later, Latin American immigrants supplied a low-wage labor force in California fisheries (Jacobson and Thomson, 1993).

Passage of the US Fishery Conservation and Management Act (P.L. 94–265) in 1976 and creation of the federal EEZ established a favorable institutional and economic environment for the expansion of existing California fisheries, as well as the development of new California fisheries (e.g. herring roe and urchins). The anticipation of greater profits from a larger harvest area and the exclusion of foreign harvesters from the EEZ led to an increased interest by federal and private lenders to facilitate investment in fishery operations (Dunnigan, 1999). This contributed to the rapid expansion of the California fishing fleet. The number of registered fishing vessels increased to more than 6600 during 1977–1980 and declined steadily thereafter to 2000 in 2004. California average ex-vessel revenue exceeded $190 M during 1979–1981. This maximum was achieved through fisheries diversification that included SPF; gillnet-caught tuna, swordfish and sharks; jack and Pacific mackerel; demersal fishes; and invertebrate species including lobster, crab, urchins, shrimp and squid (Mason, 2004). Squid have supported a major California fishery since 1990, regularly contributing more than 20% of the total landings and 10%–20% of the total ex-vessel revenue. In addition to harvesting previously under-utilized species, it is probable that this transitional environmental period of increasing ocean temperatures (from left to right across the bottom of Fig. 10.11) made a variety of fishing opportunities (e.g. skipjack, yellowfin tuna, swordfish) more available to California fishers. Fisheries characteristic of the lower left and the lower to middle-right were simultaneously available to California fishers during the transitional environmental period from the mid 1970s into the 1990s.

From 1976 to 1999 the herring-roe fishery was an important contributor to California SPF harvests, contributing from 5% to 10% of the total ex-vessel revenue. After being nearly absent from the fishery from 1966 to 1976, due to environmental conditions (MacCall, 1996; Norton and Mason, 2004), the Pacific mackerel fishery re-emerged in the late 1970s at ex-vessel revenue levels comparable to those seen in the 1930s and 1940s. As the Pacific mackerel fishery declined after 1990, the sardine fishery started anew, suggesting that the environmental and SPF harvest cycles had returned to the approximate ecological state of the 1930s and 1940s (Fig. 10.11). However, there were important differences between the sardine fishery that reached a maximum in the late 1990s and the earlier fishery (Norton and Mason, 2004; Herrick *et al.*, 2006). First, there were no regulations limiting sardine harvests in the 1930s and 1940s, but since the 1980s the fishery has been managed to preserve viable reproductive populations and a forage base for birds and mammals as well as predatory fish. Second, the management method for sardines makes the harvest quotas dependent on an environmental index (Jacobson and MacCall, 1995; PFMC, 1998). Third, sardines contributed 25% to 40% of the total harvest by weight and 5% to 15% of the total ex-vessel revenue during 1990–2000, whereas in the historical fishery sardines contributed 80% to 90% of the total landings by weight and about 50% of the total ex-vessel revenue. Fourth, the largest portion of the sardines landed in Washington, Oregon and California since 1990 have been sold fresh for food or exported in frozen blocks for aquaculture feed, bait and human consumption. In the historical California sardine fishery, fish meal, canned sardines, and fish oil were the main products. None of these products is important in today's west coast sardine fishery.

The rejuvenated California sardine fishery had a 5-year average value of about $5M in 2002 as compared to more than $90M in 1943 on landings of about 58000 t and 453000 t, respectively. In 2004, 44000 t of sardines were landed in California which earned about $4M ex-vessel. Total California fisheries ex-vessel revenue in 2004 was $121M. Sardine fisheries recommenced in Oregon and Washington in 2000, a decade following renewal of the California fishery. In 2004, the Oregon fishery harvested 36000 t of sardines and the Washington fishery landed 8800 t. The possible loss of these northern fisheries will probably mark the next climate shift, from right to left in Fig. 10.11, as in the 1940s, 1950s and 1960s (Herrick *et al.*, 2007). However, it is not impossible for continued heating of the ocean and atmosphere to shift the cycle toward the right and provide environments consistently favorable to sardine reproduction and growth from the 26° N, off northern Mexico, to 57° N off southeastern Alaska. In this case, the harvestable biomass available to US, Mexican and Canadian fisheries might exceed that of the 1930s and 1940s by several fold. Climate changes have impacted California fisheries landings and ex-vessel value over the 1930–2000 period, and climate changes will continue to have an effect on economic opportunities in California fisheries and present conservation and management challenges throughout the twenty-first century.

Climate changes are propagated into the ecosystem, which reorganizes tropic relationships and relative species composition. The impact of these ecosystem reorganizations, or biological regime shifts, extends beyond fisheries. SPF are also the primary prey of many marine mammals and seabirds. A change in the availability of a particular forage species may affect the distributions and population levels of these predators. This can have significant implications in terms of the economic benefits these predator species provide, particularly with regard to the increasing importance of the non-consumptive use and non-use benefits they impart (Tisdell, 1991; Kuronuma and Tisdell, 1993; Sumaila, 2005).

Some marine mammals, which have become important as symbols of environmental health and as ecotourism attractions (marine revenue producers), feed heavily on SPF. Individuals and businesses may obtain economic benefits from management strategies that allow ample SPF forage for marine mammals to thrive and to congregate offshore within reach of reasonably short ecotourism excursions. From an ecosystem-based management perspective, the economic benefits that marine mammals provide must be taken into account when evaluating the total economic impact of climate change on the abundance and distribution of SPF. In this context, the economic benefits to society that SPF provide as forage for marine mammals – and other natural predators (e.g. fish and birds) – need to be balanced against the economic benefits society gains from harvesting SPF in determining socially optimal harvest levels. Accordingly, the socially optimal harvest (forage) level is that which equates the respective marginal net benefits from each use, B*, in Fig. 10.12 (Haraden *et al.*, 2004).

Large fluctuations in the availability of SPF are not only a concern from the standpoint of directed fisheries and as forage for numerous less conspicuous naturally occurring marine species – which may be key in the maintenance

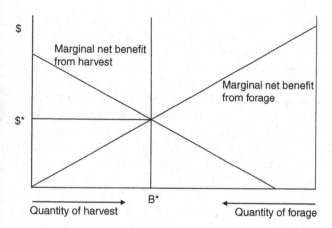

Fig. 10.12. The socially optimal SPF harvest level (B*). The vertical axis is net value in dollars from the quantity of the SPF left in the ocean for forage and the quantity extracted as commercial landings, which are shown along the horizontal axis. Harvesting will occur as long as the net benefit from harvesting is greater than that from forage, and the converse. Hence, the social optimum occurs where the respective marginal net benefit curves intersect: a harvest level of B* and marginal net benefit of $*. The slopes of the net benefit curves will vary depending on net harvest value of the fish and the net non-market value of providing forage for predators, both of which would be affected by climate change.

of a robust ecosystem – they are also a major constituent in feeds used in aquaculture and other fish rearing operations. With the expected increase in global aquaculture production, demand for SPF based fish food will also increase (see aquaculture box above), so that impacts of climate change on SPF will likely affect the economics of aquaculture as well.

The California commercial fish landings data indicate that as the environment becomes unfavorable for sardine, they seem to be replaced in the CCE by mackerel, northern anchovy, or market squid as the predominant forage species. Generally, the new regime is as diverse and ecologically acceptable as the one it replaced, even if it does represent a major change in economic value. Environmental signals portend ecosystem changes which are likely to alter the species compositions of commercial landings and of recreational harvests. This suggests that ecosystem-based fisheries conservation and management can be greatly enhanced by modeling and monitoring decadal scale changes in biological and physical indices. Extrapolation of decadal scale signals in SPF landings will then allow prediction of the rise and fall of fisheries and local economies. A comprehensive environmental–ecological–economic-based SPF conservation and management approach would also incorporate knowledge about the total economic value of SPF. This will depend on the marginal valuations of different SPF species for use as natural forage and other non-commercial purposes, requiring information which may be quite difficult to obtain.

Implications of climate change for developing coastal communities

In developing countries, the risk of climate change poses an additional challenge to the livelihoods and food security of poor households dependent on SPF resources.[9] The problem is particularly acute in areas with rapidly growing populations such as developing Southeast Asia where countries are likely to share their fishery resources with each other, suggesting a number of the international resource use issues raised earlier. FAO estimates that around 22 million people in Asia depend on fishing for a livelihood. The bulk of these are small-scale or artisanal fishers in coastal areas, for whom SPF represent an important contribution to their overall harvests (Briones *et al.*, 2006). On the consumption side, fish contributes more than one-quarter of animal protein intake in Asia (FAO, 2002). Fish consumption is concentrated on low value species (Kent, 1998), which includes most of the SPF. Overfishing and other human activities have already reduced SPF resources in Asia to a precarious state. Any undesirable climate change poses an external threat to the productivity of SPF fisheries. The threat is more alarming in developing Asia, given the dependence of poor households on SPF stocks.

Bioeconomic analyses have largely established that coastal fishery resources throughout the region are experiencing overfishing, both biological and economic. While the countries in the region collectively contribute to overfishing, the complexity and intensity of the surrounding issues means that there will be no single conservation and management strategy for their resolution (Silvestre *et al.*, 2003). The most important strategies are those needed to promote sustainability, which will be useful and practical even in the absence of climate change. Most of the countries in the region have fisheries policies to varying degrees, which are principally intended to achieve sustainability. However, given the overfished state of most fisheries resources in the region, the inadequacy of these policies is evident. In most cases the fundamental reason for inadequate fisheries policy is a lack of the information needed to determine appropriate levels of fisheries exploitation. Without systematic, comprehensive, bioeconomic data collection for these fisheries and subsequent analyses, there exists a shaky foundation for the development of sound fisheries policy, which will ultimately result in a "tragedy of the commons" (Hardin, 1968) throughout the neighborhood.

Meanwhile, the downward spiral continues as diminishing resources inspire more productive means of harvesting what remains. This situation gives rise to increased competition and conflict within and among the region's commercial and artisanal fishery sectors, where, in general, owners of large-scale commercial vessels earn large sums while artisanal (small scale) fishers, those largest in number, and for the most part already deeply impoverished, barely earn a living. These differences in the existing socio-economic structure within each fishery sector have important implications for the sustainable management of shared SPF resources as well as for maintaining livelihoods in the coastal communities of the region.

Whether international, inter- or intrasectoral, conflicts essentially occur over rights to fishery resources and are likely to intensify as resources become less available, resulting in significant social and economic impacts. These impacts will likely be borne disproportionately by artisanal fishers having the fewest alternative livelihoods. Given these circumstances, it is the artisanal fishers who are likely to be the most vulnerable to unfavorable climate-SPF regime shifts. They are vulnerable economically, in terms of access to alternative earnings opportunities, and also vulnerable nutritionally, because as poor people, who are highly reliant on fish for consumption, they will be forced to substitute inferior foods. Regional governments may find it prohibitively costly to respond to these circumstances when they arise, underscoring the need for proactive fishery conservation and management policies.

Naturally occurring short- and long-term climate change cycles will affect the availability of SPF to coastal fishers in south and Southeast Asia, compounding the overfishing problems already beleaguering the region. Enhancing systematic data collection, including biophysical data, in these fisheries will be important towards identifying biophysical indicators that forewarn of a climate–SPF regime change. This information can be incorporated into harvest policies for the region's fisheries, which will improve the ability of the region to prepare for and deal with such events. Of current concern are the specific impacts of human activity on climate change. If the atmosphere and oceans continue to warm under the influence of anthropogenic forces, the plight of developing coastal communities is exacerbated by the threat of being inundated by rising sea levels. Managing such a calamity will extend far beyond the realm of fisheries policy; cutting to the core of each nation's basic social welfare policies. In the meantime, dealing effectively with short-term natural climate change offers a good learning opportunity.

Conclusions

The outlook for SPF fisheries production is one of increased demand for resources that are approaching or have already reached their maximum sustainable harvest levels. Harvests of SPF have been highly variable in terms of shifting species availability over time and area. Global production of SPF commodities is relatively much more stable, suggesting that there is a greater degree of substitutability of raw SPF as inputs.

In the major eastern boundary current ecosystems, SPF fisheries have exhibited boom and bust cycles on a decadal scale, with the cycles for particular SPF (e.g. anchovy and sardine) characteristically out of phase. Historic fishery collapses were initially ascribed to overfishing and more recently changes in the environment have been recognized as important factors. Recovery under a favorable climate–SPF regime has been partly tempered by conservation and management to prevent overfishing, and reduced commercial interest due to the emergence of new fisheries.

The economic and social consequences of climate–SPF regime shifts are likely to be greater among local harvesters, particularly those in developing countries where there tends to be widespread reliance on SPF for food security and income. Conservation and management in developing countries may therefore have a much greater socioeconomic emphasis, particularly in terms of trying to avoid undesirable community impacts (this volume, Chapter 11).

Historical evidence from a number of areas suggests that a climate regime shift is likely to result in substantial changes in the abundance and distribution of SPF species. This can have important international conservation and management implications when an SPF are redistributed across adjacent nations' fishing zones. Exploitation of the resource may change from exclusive use to shared use, and the country from which the SPF are leaving may have no incentive to protect the stock, choosing instead to overharvest it while it is still in its zone. The decision whether to co-operate in international conservation and management will, to a great extent, depend on what is known about climate–SPF regime shifts. The conundrum here is that perfect information may work to the detriment of the resource, while some uncertainty about what is happening may lead to a more precautionary and co-operative approach.

Problems of climate change and the shared use of SPF resources among nations, when they are distributed across national fishing zones, reflect differences the time preferences for the economic benefits that can be derived from SPF resources, as well as the relative importance of the full range of benefits that these resources are capable of providing. These differences are reflected in the discount rate each country uses to evaluate the time stream of net economic benefits under various resource conservation and management policies. High discount rates are often characteristic of developing nations and the converse holds for developed nations. Developing nations, for whom a primary social objective is to accumulate monetary wealth, would therefore be more interested in using SPF resources to provide for the current generation's needs, while developed countries, having already greatly achieved this objective, would be more apt to undertake costly investment in SPF resources now in order to realize more extensive economic benefits in the future. This suggests that, in terms of climate change, developed nations might be more inclined to advocate cooperative international conservation and management that focuses on the full range of economic values for the benefit of future generations, whereas developing nations would be most interested in capturing the commercial benefits from SPF resources now in order to satisfy internal social objectives.

Besides being the target species of directed fisheries, SPF are key components of marine food webs, and constitute the forage base for many commercially and recreationally targeted species as well as marine mammals and seabirds. This means that the potential economic impacts of a climate–SPF regime shift will be much more widespread than those associated with directed fisheries. Therefore, the impacts of climate change on the availability of SPF would take into account the full range of ecosystem services that SPF provide and the corresponding consumptive, non-consumptive and non-use economic value from these services. From an analytical standpoint, this means a shift from traditional single species, bioeconomic analysis to a multispecies ecosystem-based approach towards evaluating the total economic impacts of a climate–SPF regime shift.

Artisanal and/or subsistance fishers are likely to be the most vulnerable to unfavorable climate–SPF regime shifts. For the most part, these are people with limited economic means who rely heavily on SPF for livelihoods and food security. Under these circumstances, diminished SPF resources mean reduced incomes and a need to find alternative food sources. With reduced incomes and limited access to alternative livelihoods, these people will become further impoverished and forced to substitute inferior foods for reduced amounts of SPF, as well as the species for which SPF are a critical forage source. Ideally, regional governments would have proactive fishery conservation and management policies in place to be ready for these circumstances. To respond to them after the fact would probably be much more costly than being prepared.

Acknowledgments

The authors express their gratitude to Dave Checkley, Kathleen Miller, and an anonymous reviewer for their constructive comments on an earlier version of this chapter, and to Janet Mason and Cindy Bessey for their contributions to the preparation of this chapter.

NOTES

1. In the case of fish meal and fish oil production only, SPF includes anchovy, sardine, herring and the "oily fish not elsewhere included" category in the FAO Fisheries commodities production and trade 1976–2004 database (FAO, 2006).
2. On this, see Hannesson (2007a).

3. For speculations along those lines, applied to the Northeast Arctic cod, see Hannesson (2006).
4. Non-consumptive use values occur in the case where one person's enjoyment does not prevent others from enjoying the same resource (e.g. my viewing of wildlife does not prevent others from enjoying the same resource). Non-use value is also described as existence value, the value conferred by humans on an environmental resource even though there is no personal interaction with it. An environmental resource may be valuable simply because one is happy knowing that it exists (Sumaila, 2005). These values are not directly observable through market prices and therefore have to be estimated using non-market valuation methods.
5. In this context, discounting accounts for the forgone opportunities of undertaking a particular resource conservation policy (the opportunity cost of investing in the resource); that is, the fact that \$X in the future is worth less than \$X now, because \$X now could be invested and earn interest, whereas \$X received in the future cannot. According to the discrete period discounting formula: Present Value = Future Value$_t*(1+i)^{-t}$; where t is the number of periods from the present when the future value will occur, and the discount rate (i) is the rate of interest at which the streams of cash outflows (costs) and inflows (benefits) associated with a multi-period economic activity are to be discounted. The choice of discount rate is a means of reflecting the opportunity cost of investing in the resource.
6. In general, biological overfishing occurs when effort in the fishery exceeds the level corresponding to MSY from the fishery. Economic overfishing is generally associated with the long-run equilibrium in open access, common property commercial fisheries, where fishing effort has increased to the point where the average cost of effort equals the average ex-vessel revenue generated by the effort, so that there is no net economic benefit from use of the fishery resource, resulting in a "tragedy of the commons" according to Hardin (1968). It is usually the case that the MSY harvest level will be exceeded before economic overfishing occurs as in the Namibian pilchard fishery, where private economic expectations dictated applying effort in excess of that corresponding to MSY.
7. The term "wetfish," as used here, includes anchovy, mackerel, sardine, squid, and is an expression that has its origins in how these fish were processed in canneries (California Seafood Council, 1997).
8. All dollar values are inflation adjusted (2000 \$), and a change in a 5-year average total ex-vessel revenue of more than \$20 M is statistically significant by the standard t-test at $P < 0.05$.
9. While we recognize the potential gains that some developing coastal communities may experience from a change in the abundance and distribution of SPF resources due to climate change, our focus here is on the potential downside of climate change in these areas.

REFERENCES

Berman, M. and Sumaila, U. R. (2006). Let's start a conversation on the role of discounting and amenities on the economic viability of marine ecosystem restoration. *Mar. Resour. Econ.*

Bernard, H.J., Hedgepeth, J.B. and Reilly, S.B. (1985). Stomach contents of albacore, skipjack and bonito off southern California during summer 1983. *Calif. Coop. Ocean. Fish. Invest. Rep.* **26**: 175–182.

Brennan, T.J. (1997). Discounting the future: economics and ethics. In *The RFF Reader In Environmental and Resource Management*, Oates, W.E., ed. Washington, DC. Resources for the Future, pp. 28–35.

Briones, R., Garces, L., and Ahmed, M. (2006). Climate change and small pelagic fisheries in developing Asia: the economic impact on fish producers and consumers. In *Climate Change and the Economics of the World's Fisheries*, Hannesson, R., Barange, M., and Herrick, S., eds. UK: Edward Elgar, pp. 215–235.

Brugere, C. and Ridler, N. (2004). *Global Aquaculture Outlook in the Next Decades: An Analysis of National Aquaculture Production Forecasts to 2030*, FAO Fisheries Circular. N. 1001. Rome: Food and Agriculture Organization of the United Nations.

California Seafood Council. (1997). California's wetfish industry: Its importance… past, present and future. Available at: http://ca-seafood.ucdavis.edu/news/wetfish/index.htm.

Delgado, C.L., Wada, N., Rosegrant, M.W., *et al.* (2003). *Fish to 2020: Supply and Demand in Changing Global Markets*, Washington, DC: International Food Policy Research Institute, and Penang, Malaysia: Worldfish Center.

Dunnigan, J.H. (1999). Federal fisheries investment task force report to Congress, Atlantic States Marine Fisheries Commission for the National Atmospheric and Oceanic Administration. 229 pp. Available at: http://www.nmfs.noaa.gov/sfa/itfcover.pdf.

FAO (2002). *State of the World Fisheries and Aquaculture*: Rome: Food and Agriculture Organization of the United Nations.

FAO (2006). *State of World Aquaculture: 2006*, Rome: Food and Agriculture Organization of the United Nations.

FAO Fishery Information, Data and Statistics Unit. (2006). Fisheries commodities production and trade 1976–2004. FISHSTAT plus – universal software for fishery statistical time series [online or CD-ROM], Rome: Food and Agriculture Organization of the United Nations. Available at: http://www/fao.org/fi/statist/FISOFT/FISHPLUS. asp.

Globefish (2006). *Fishmeal Market Report*, http://www.globefish.org/index.php?id=3367.

Golubtsov, P.V. and McKelvey, R. (2007). The incomplete-information split-stream fish war: examining the implications of competing risks. *Natural Res. Modeling* **20**: 263–300.

Goulder, H. and Kennedy, D. (1997). Valuing ecosystem services: philosophical bases and empirical methods. In *Nature's Services: Societal Dependence on Natural Ecosystems*, Daily, G.C., ed. Washington, DC: Island Press, pp. 23–48.

Hannesson, R. (2006). Sharing the northeast arctic cod: possible effects of climate change. *Natural Res. Modeling* **19**: 633–54.

Hannesson, R. (2007a). Incentive compatibility of fish-sharing agreements. In *Festschrift to Gordon Munro*, Bjørndal, T., Arnason, R., and Gordon, D., eds. Oxford: Blackwell.

Hannesson, R. (2007b). Global warming and fish migrations. *Nat. Res. Modeling* **20**: 301–320.

Haraden, J., Herrick, S.F., Squires, D., and Tisdell, C.A. (2004). Economic benefits of dolphins in the United States eastern tropical Pacific purse-seine tuna fishery. *Environ. Resource Econ.* **28**: 451–468.

Hardin, G. (1968). The tragedy of the commons. *Science* **162**: 1243–1248.

Herrick, S.F., Hill, K., and Reiss, C. (2006). An optimal harvest policy for the recently renewed United States Pacific sardine fishery. In *Climate Change and the Economics of the World's Fisheries*. Hannesson, R., Barange, M. and Herrick, S. eds. UK: Edward Elgar, pp. 126–150.

Herrick, Jr. S.F., Norton, J.G., Mason, J.E., and Bessey, C. (2007). Management application of an empirical model of sardine-climate regime shifts. *Mar. Policy* **31**: 71–80.

Jacobson, L.D. and MacCall, A.D. (1995). Stock-recruitment models for Pacific sardine (*Sardinops sagax*). *Can. J. Fish. Aquat. Sci.* **52**: 566–567.

Jacobson, L.D. and Thomson, C.J. (1993). Opportunity costs and the decision to fish for northern anchovy. *N. Am. J. Fish. Manage.* **13**: 27–34.

Jacobson, L.D., Bograd, S.J., Parrish, R.H., *et al.* (2005). An ecosystem-based hypothesis for climatic effects on surplus production in California sardine (*Sardinops sagax*) and environmentally dependent surplus production models. *Can. J. Fish. Aquat. Sci.* **62**: 1782–1796.

Kent, G. (1998). Fisheries food security, and the poor. *Food Policy* **22**: 393–404.

Kristofersson, D. and Anderson, J. (2006). Is there a relationship between fisheries and farming? Interdependence of fisheries, animal production and aquaculture. *Mar. Policy* **30**: 721–725.

Krutilla, J. (1967). Conservation reconsidered. *Amer. Econ. Rev.* **57**: 787–796.

Kuronuma, Y. and Tisdell, C. (1993). Institutional management of an international mixed good: the IWC and socially optimal whale harvests. *Mar. Policy* **17**: 235–250.

MacCall, A.D. (1979). Population estimates for the waning years of the Pacific Sardine Fishery. *Calif. Coop. Ocean. Fish. Invest. Rep.* **20**: 72–82.

MacCall, A.D. (1996). Patterns of low-frequency variability in the California Current. *Calif. Coop. Ocean. Fish. Invest. Rep.* **37**: 100–110.

Mason, J.E. (2004). Historical patterns from 74 years of commercial landings from California waters. *California Coop. Ocean. Fish. Invest. Rep.* **45**: 180–190.

Miller, K.A. (2007): Climate variability and tropical tuna: management challenges for highly migratory fish stocks. *Mar. Policy* **31**: 56–70.

Miller, K. A. and Munro, G. R. (2004). Climate and cooperation: a new perspective on the management of shared fish stocks. *Mar. Resour. Econ.* **19**: 367–393.

Naylor, R. L., Goldberg, R. J., Primavera, J. H., *et al.* (2000). Effect of aquaculture on world fish supplies. *Nature* **405**: 1017–1024.

New, M. B. and Wijkstrom, U. N. (2002). *Use of Fishmeal and Fish Oil in Aquafeeds: Further Thoughts on the Fishmeal Trap*, FAO Fisheries Circular. N. 975. Rome: Food and Agriculture Organization of the United Nations.

Norton, J. G. and Mason, J. E. (2003). Environmental influences on species composition of the commercial harvest of finfish and invertebrates off California. *Calif. Coop. Ocean. Fish. Invest. Rep.* **44**: 123–133.

Norton, J. G. and Mason, J. E. (2004). Locally and remotely forced environmental influences on California commercial fish and invertebrate landings. *Calif. Coop. Ocean. Fish. Invest. Rep.* **45**: 136–145.

Norton, J. G. and Mason, J. E. (2005). Relationship of California sardine (*Sardinops sagax*) abundance to climate-scale ecological changes in the California current system. *Calif. Coop Ocean. Fish. Invest. Rep.* **46**: 83–92.

Orbach, M. K. and Beckwith, J. (1982). Indochinese adaptation and local government policy: An example from Monterey. *Anthropol. Quart.* **55**: 135–145.

Ottolenghi, F., Silvestri, C., Giordano, P., *et al.* (2004). *Capture-Based Aquaculture. The Fattening of Eels, Groupers, Tunas and Yellowtails.* Rome: Food and Agriculture Organization of the United Nations.

Parrish, R. H., Schwing, F. B., and Mendelssohn, R. (2000). Mid-latitude wind stress: the energy source for climatic shifts in the North Pacific Ocean. *Fish. Oceanogr.* **9**: 224–238.

PFMC (1998). *Amendment 8 (To The Northern Anchovy Fishery Management Plan) Incorporating a Name Change to: The Coastal Pelagic Species Fishery Management Plan.* Portland, OR: Pacific Fishery Management Council.

PFMC (2006). *Status of the Pacific Coast Coastal Pelagic Species Fishery and Recommended Acceptable Biological Catches. Stock Assessment and Fishery Evaluation – 2006.* Portland, OR: Pacific Fishery Management Council.

Pinkas, L., Oliphant, M. S., and Iverson, I. L. K. (1971). Food habits of albacore, bluefin tuna and bonito in California waters. *State of California, The Resources Agency. Fish. B.*: **152**: 1–84.

Settle, C. and Shogren, J. F. (2004). Hyperbolic discounting and time inconsistency in a native-exotic species conflict. *Resour. Energy Economics* **26** (Special Issue): 255–274.

Silvestre, G., Garces, L., Stobutzki, I., *et al.* (2003). Assessment, management and future directions for coastal fisheries in Asian countries. *Worldfish Center Conference Proceedings* **67**: 1. Malaysia: Worldfish Center.

Sumaila, U. R. (2005). Differences in economic perspectives and the implementation of ecosystem based management of marine resources. *Mar. Ecol. Prog. Ser.* **300**: 279–282.

Sumaila, U. R. and Stephanus, K. (2006). Declines in Namibia's pilchard catch: the reasons and consequences. In *Climate Change and the Economics of the World's Fisheries*, Hannesson, R., Barange, M., and Herrick, S., eds. UK: Edward Elgar.

Sumaila, U. R. and Walters, C. (2005). Intergenerational discounting: A new intuitive approach. *Ecol. Econ.* **52**: 135–142.

Tacon, A. G. J., Hasan, M. R., and Subasinghe, R. P. (2006). Use of fishery resources as feed inputs for aquaculture development: trends and policy implications, *FAO Fisheries Circular.* N. 1018. Rome: Food and Agriculture Organization of the United Nations.

Tisdell, C. A. (1991). *Economics of Environmental Conservation.* Amsterdam: Elsevier.

Wada, N., Delgado, C., and Tacon, A. (2003). Outlook for fishmeal and fish oil demand: the role of aquaculture. Paper presented at the International Institute of Fisheries Economics and Trade (IIFET) 2002 Conference, Wellington, New Zealand. Bruce Shallard and Associates, Wellington, New Zealand.

Weitzman, M. L. (2001). Gamma discounting. *Am. Econ. Rev.* **91**: 260–272.

11 Human dimensions of the fisheries under global change

Rosemary E. Ommer, Astrid C. Jarre, R. Ian Perry, Manuel Barange, Kevern Cochrane, and Coleen Moloney

CONTENTS

Summary

The human dimensions of fisheries for small pelagic fishes represent an interface of natural and social sciences, mainly concerning management and governance issues, but not limited to these. This chapter provides a brief overview of the scope of interactions between these domains, as a guide to future studies. We highlight important social, economic, institutional, and cultural issues arising from other chapters in this volume, such as power, resource access, equity, property rights, scales, globalized markets, ethics, interactive political agendas between developed and developing countries and technology, and refer to results of existing studies focusing on SPACC (Small Pelagic Fish and Climate Change) regions. We outline how scientists from these disciplines can work together, and with stakeholders such as resource users and managers, towards improved stewardship of the local and global marine ecosystems that are faced with significant global changes (including climate change), of which small pelagic fisheries and humans are an important interactive part.

Introduction: coasts under stress!

This chapter is a preliminary exploration of the human dimensions of small pelagic fisheries under global change. It examines the interface between the natural and social sciences and humanities, mainly (but not entirely) concerning management, use and governance issues in small pelagic fisheries. This is particularly important at a time when national and local fisheries are being forced to cope with globalization, major global climatic environmental stresses, and the interactions between these two external driving forces. Under such conditions, answers to governance issues will not be clarified by looking at fisheries (individually or globally) from a purely bio-economic perspective. The need is for a multi-disciplinary, multi-scale, and interactive "worldwide system of pelagic fisheries" (Mullon and Fréon, 2006) analysis, one that identifies the feedback processes between climate and other global changes, but that extends to include people and human community survival (which includes adaptation to change), thus reaching far beyond purely economic concerns.

The range of people involved in the capture of small pelagic fishes is large, from individual artisans to members of large, industrial fisheries, but these nonetheless comprise a very specific group. Of the world's fish catches 75% are taken by seven countries, which have diverse economies and management systems: Peru, Chile, Japan, USA, China, Norway, and Russia (FAO, 2002). Almost 50% of the catch is reduced to fish meal and fish oil, mostly from small pelagic fishes, and marketed as a global commodity. By 2010 it is predicted that 50% of the fish meal and 80% of fish oil will be used in the global aquaculture feed industry. The majority of small pelagic fisheries are therefore, in one sense, part of a single global community in that they are united by the common commodity they are producing. The Small Pelagic Fish and Climate Change (SPACC) program of GLOBEC has focused primarily on the large industrial fisheries, especially those of the California, Humboldt, Benguela, European Atlantic, Canary, and Kuroshio Current systems. Other stocks of interest include those in the Mediterranean, off Australia, and in the Baltic Sea.

The human dimensions of these fisheries, both industrial and artisanal, have not been a focus of SPACC, except as part of its economic context (e.g. Hanesson *et al.*, 2006). For artisanal fisheries in particular, there is a paucity of good data on all aspects of community involvement with the fishery. We recognize, however, the importance of small pelagic fish populations to fishers, to related persons and indeed the entire community (e.g. Ferraris *et al.*, 1998; Brookfield, *et al.*, 2005; Hamilton *et al.*, 2006; Kennedy, 2006) – we know they merit attention.

Current discussions of global management and capture fisheries tend to wrestle with issues of overcapacity; illegal, unregulated and unreported fishing; overexploitation of many stocks and scarcity of new resources, and their impacts on ecosystems and markets, as well as uncertainty over globalization and climate change. There is, however, an urgent need for greater attention to cross-disciplinary issues, including:

(1) the problems of temporal, spatial, and organizational scale and how they relate to the ability to define appropriate inter-disciplinary research questions;
(2) the way in which social and economic power, institutional structures and paradigms, and science and management practices interact, including issues of equity and ethics;
(3) the problems of preserving people's livelihoods in the face of resource depletion (Ommer, 2006); and
(4) the nature and quality of the information available to fisheries science through collaborations with industrial fish harvesters for accurate stock assessments, partly as a result of very rapid changes in fishing efficiency and practices (e.g. Hamilton *et al.*, 2006; Ommer *et al.*, 2007) and human responses to management regulations and policies.

These issues are at the intersection of the natural and social sciences with respect to small pelagic fisheries. We examine them, and provide examples of the types of inter-disciplinary questions and research that need to be conducted for a more complete understanding of these coupled social–ecological systems and their responses to global changes.

The problems of scale

Human communities dependent on small pelagic fisheries are exposed to problems over a wide range of scales, with scale issues of central importance in such matters as data acquisition, fish capture, fisheries management

and the survival of communities of fish and of all fishers, not just those who are employed in industrial, large-scale fishing fleets (Perry and Ommer, 2003). Global drivers of environmental change include climate variability and change, internal ecosystem dynamics, geographic shifts of resources, overfishing and pollution. Global drivers of changes in human systems include environmental changes plus globalization of trade, information and capital, increases in population, technological change, changing public tastes and preferences, corporate concentration and education. Small pelagic fisheries, with their generally highly fluctuating volumes, high turnover rates, and significance to developed and developing countries, are exposed to all these pressures, which are further aggravated by changes in the destination of the catch product following the development of aquaculture and its demand for fish meal (Kristofersson and Anderson, 2006), and by discrepancies between the management systems of developed and developing countries (Garcia and Leiva-Moreno, 2003). Globalization is a particular concern that threatens the livelihoods, societies, cultures, and the very existence of those coastal communities that are dependent upon small pelagic fisheries. How they respond to these stressors is important because fisheries-based coastal communities have a vital relationship with their environment: marine resources are a matter of livelihood and thus of immediate and direct concern.

Small pelagic fish stocks pose specific problems to the fishers exploiting them, and managers and policy makers need to be aware of those problems in the broader human, not just capture and marketing, context. Some of these problems are caused by factors such as the short lifespan of these species, their seasonality and their high mobility, which are inherent in their biology and are strategies for dealing with environmental variability. The short lifespan of small pelagics means that they can reproduce in large numbers, and quickly, and this in turn leads to high variability in abundance on interannual to decadal scales. Whereas the economics of industrial fisheries on such species have been well studied (e.g. Hanneson *et al.*, 2006), relatively little is known about the impacts of this variability on the artisanal fishers of these species. Overall, however, it is likely to be substantial (see, for example, Sumaila and Stephanus, 2006). The biological patterns of small pelagics force human capture and management strategies to be flexible and responsive to shifting abundance (see Cochrane *et al.*, 1998).

What is also needed is consideration of whether or not the ramifications of such strategies are appropriate in terms, not only of the conservation of the stocks and the maintenance of adequate fish biomass, but also of the survival potential of the human communities that depend on these fisheries. How have human communities dealt with large fish catch fluctuations in the past? Studies pertaining to small pelagic fisheries include that of Sumaila and Stephanus (2006); this describes diversification of economic activities as one strategy. We provide another example below (see Box 11.1). These and other – notably social and cultural – human adaptations to change need to be considered and embedded in the context of past and current natural systems. Of course, adaptation does not mean that the status quo – preserving fishing communities – is always the goal. Indeed, in some cases this may not be as appropriate as facilitating their adaptation to changing conditions.

Scale mismatches have, to date, been a challenge for interdisciplinary studies that look at the interrelationships between climate, ecosystem, and human concerns. Scale effects are found throughout such inter-relationships, from physical to biological coupling, to fishing and social processes. Spatial-scale mismatches occur when activities appropriate to one geographic level are applied without due consideration at another, or when a regulatory process is wrong and decisions are made at one level that pertain to another without the people at that other level (usually, but not always, lower down) being consulted. Some government policies, for example, may be directed at "the individual," when "the community" is actually the more appropriate focus for them; again, national-level management decisions concerning quotas may not be spatially subdivided in ways that make sense at the regional level. Temporal-scale mismatches occur when the rate of fishing does not fit with the life cycle of the fish species concerned. They also occur, from a human perspective, when major management changes are introduced too fast (or too slowly) with the result that problems arise in human communities as they attempt to deal with overly rapid, or insufficiently rapid, change. Herrick *et al.* (this volume, Chapter 10) point out potential mismatches of temporal scale in their discussion of discounting, discount rates and optimal time paths for exploitation by small pelagic fisheries, and how attitudes towards risk and uncertainty on the part of resource beneficiaries can affect their preferred path for the exploitation of these resources.

Organizational-scale mismatches occur when, for example, activities appropriate at the level of the firm are applied to government or community organizations or when, biologically speaking, simultaneous fishing occurs on a predator and its prey without taking into account these trophic interactions. Different types of mismatching tend to go together: decisions made at the wrong level run the risk, not only of not applying at a different spatial scale, but also of not matching at the temporal scale (Ommer, 2006; Ommer *et al.*, 2007).

Moreover, the scales at which scientific observations are made can limit perception of a problem. The high mobility of most small pelagic fish populations makes them susceptible to scientific investigations of stock fluctuations at the wrong scale: hence the potential confusion of distinguishing between stock collapses and stock changes in distribution and/or migration patterns. For example, it is unclear whether observed increases in the number of small fish in the Chilean jack mackerel catch were due to collapse of the stock from overfishing or to changes in distribution and increased availability of younger age classes due to environmental change (Arcos *et al.*, 2001). The collapse of the Namibian pilchard (sardine) in the 1960s provides another example. At the local scale, the job loss as a result of this collapse for the largely seasonal community work force was serious while, at the scale of the industry, the impact was minor. Although processing plants (and thus jobs) were lost, the internationalization of the industry meant that vessels and machinery owned by international capital could be redeployed to other areas and firms could continue on a "business as usual" basis. At the national level the economic impact on Namibia was minimal, since the development of other fisheries cushioned the collapse (Sumaila and Stephanus, 2006). However, although most fishers were seasonal workers, and thus no full-time jobs appeared to have been lost, the loss of part-time work, which may have been the only wage employment some workers had, must have had serious social consequences. In South Africa, the increases in pelagic fish biomass in the 2000s have masked the social and industrial changes created by a shift in resource distribution from the traditional west coast to less dependent south and south-east coastal communities (van der Lingen *et al.*, 2006).

It is, therefore, important that we understand at which scales marine ecosystems, fisheries resources, coastal communities and production industries interact, and to what extent coastal communities are affected by scale mismatches in and between these components. This means that we need to be aware of how different groups or users perceive and use "scale." What are the different scales at which oceanographers, biologists, fisheries managers, economists, business, politicians, and communities (which may be further split into such categories as aboriginal communities and commercial fishing communities) operate? We need also to be constantly aware of the possible mismatches of scale between the functioning of environmental change and that of human vulnerability and adaptation, understood as the combination of exposure, sensitivity, and adaptive capacity to change. Finally, there is the related problem of "down-scaling": how to move between global-scale concerns and locally based responses.

In inter-disciplinary work, obvious methodological matches tend to dominate, as in the case of quantitative large-scale social science studies (often using modeling techniques), which fit standard natural science approaches to analysis: such is the fit which results in bio-economic modeling, for example. This kind of work makes valuable contributions to our understanding of human and biological issues in small pelagic fisheries: see, for example, Hannesson *et al.* (2006). However, such methodological techniques can, at best, provide only partial understanding, since they cannot capture the complexities of dynamic interactions between a society and the environment in which it is embedded. Cochrane *et al.* (1998), for example, identified the failure to confront underlying conflicting objectives in the South African small pelagic fishery – which at a superficial level appeared to have a common view on the issues of concern – as a primary cause of management malfunctions in the 1990s.

Human dimension contexts of research on small pelagic fisheries

Several chapters in the present volume consider issues that link to the human dimensions issues introduced in the previous section. MacCall (this volume, Chapter 12) expands on the history of the exploitation of small pelagic fish. Indeed, for centuries small pelagic fisheries have been an important component of many coastal communities. In Portugal for example, sardines are a staple of local diets, as are herring in Scandinavia and the Baltic. Globally, however, the highest catches are taken from small pelagic fisheries that have only developed since the middle of the last century. The anchoveta fishery in Peru developed in the 1960s and early 1970s (FAO, 2005). The first fishery for small pelagic fish in South Africa started in 1935, becoming established as a commercial operation in the 1940s as a result of demand for canned products during the Second World War (e.g. Cochrane *et al.*, 1997). In Namibia, the purse-seine fishery, which targeted sardine, also developed during the 1940s and increased rapidly during the 1960s, in part as a result of the introduction of factory vessels (Roux and Shannon, 2004). The relatively recent development of these fisheries in no way diminishes the importance of their human dimensions, however. Whether the current fishers are second- or third-generation members of small pelagic fishing families or communities, or have found their way into a new occupation, what matters is their degree of dependence on the fishery and other resources and their vulnerability to the vagaries of globalization, climate change, and policy.

In this volume, Chapters 5 and 9, authors examine interdecadal variability as well as assessment and management of small pelagic fish. At the interface of these two topics, it appears to be advantageous to manage fisheries differently when populations are increasing rather than when decreasing, as recently introduced for the management

of the sardine fishery off California (Herrick *et al.*, 2006, and references therein). Such work needs to be tied to social science analyses of human community sustainability under such differing fishing conditions: what happens to employment, including that of seasonal workers? How do those affected adapt to, and cope with, variability and uncertainty? Unless such impacts are appreciated, and responses to them comprehended and taken into account, policy makers who devise safeguards for the sustainability of fish stocks may well find that fishing communities are forced to ignore bureaucratic management decisions and resort to overfishing since they have no choice other than to protect their livelihoods and that of their families. In this respect, it is important to remember that economies (and hence economics) are part of societies, not distinct from them. Chapter 10, by Herrick, for example, while concentrating on macroeconomics, does so without ignoring this fact, although the chapter has a clear economics focus. The present chapter is the obverse of that of Herrick. It recognizes the vital importance of economics to fishing nations and communities, but focuses on the social side of human impacts in small pelagic fisheries.

In this volume, in Chapter 7 and 8, authors elaborate on feeding interactions of small pelagics and their importance for ecosystems. Interesting in terms of sustainability and the interactivity of humans and the ecosystems of which they are part is the fact that there are foodweb complexities that reach beyond the ocean onto the shore and into the sky. One such example is that of "guano scraping" as a related income-generating activity when populations of small pelagics are sufficiently high to sustain high populations of guano-producing birds such as cormorants, gannets, and pelicans in Peru or Namibia (Muck and Pauly, 1987; Best *et al.*, 1997; Jahncke *et al.*, 2004). A modern case in point is seabirds and marine mammals as attractions for ecotourism, and the related benefits for coastal communities. Considering the trade-offs between different societal objectives is an integral part of an ecosystem approach to fisheries (e.g. Shannon *et al.*, 2004), but such an approach can only be effective if the human dimensions are well understood and the different stakeholders are consulted, involved, and able to reap the benefits from the policy and management processes (e.g. Degnbol and Jarre, 2004).

Checkley and other authors, in this volume, Chapters 3 and 9, emphasize that small pelagic fishes comprise fast-growing populations in variable habitats. How do fisheries and fishing communities adapt? Are there lessons to be learned from the past? Can we tie fish population fluctuations to the fate of local human communities in the distant past (e.g. by working with archeologists, paleoceanographers, and climatologists;

this volume, Chapter 4)? What about more recent and present-day fluctuations in fish populations? The issue of human diet will be vital here, allowing rich collaboration between natural, social, and health scientists which could also be used to analyze human preparation and consumption of small pelagic fishes in aboriginal communities, in other coastal communities, and under industrialized conditions.

Perspectives on collaboration of social and natural scientists

From the preceding discussion, it is clear that there is a need for greater insight into the complexities of the human dimensions of small pelagic fisheries. The outcomes will be more effective if such work is carried out by social scientists in partnership with natural scientists. It is possible to adopt a "spotlight and searchlight" technique in which natural and social science case studies at various scales are conducted in tandem. That would allow natural and social scientists to determine which issues are appropriate for interdisciplinary research, and to agree on methodologies that will work. This may mean identifying data gaps that can be filled by local fishers from their own observations and then doing joint interdisciplinary field work to collect those data. This would meet the needs of marine scientists to understand, in scientific terms, what is being communicated, and the needs of social scientists to identify contexts and analyze interactions, stresses, and concerns (Ommer *et al.*, 2007). At other times, data collection of various kinds, which documents different aspects of a problem, will need to be done separately, and then be followed by joint discussion of the meaning of the various data sets and mono-discipline initial findings. One such example, from Ghana, illustrates the point (see Box 11.1).

Consider, at the same scale, the manager's dilemma with the pelagic fisheries of Chile. There, the large, international, globally footloose industrial fishing fleets pursue horse mackerel both under license inside the 200-mile limit and, unlicensed, outside it as well. As a consequence, stocks are scarcer for small-scale indigenous and local fishers who are driven to engage in a "race for fish," which is counterproductive over the long term with respect to conservation. Managers are faced with the dilemma of having to give a quota which is considered sufficient for their needs by the large-scale fleets whose contribution to national income through licensing fees is significant, and still have enough quota left to sustain the stocks and the small fishing communities whose livelihoods are at stake. Managers have found (R. Quiñones, personal communication) that conservation measures are impossible without the cooperation of local

---

Box 11.1. Marine ecosystem variability and human responses in West Africa

Variability in the marine ecosystem of Ghana, West Africa, exists on several temporal and spatial scales. Perry and Sumaila (2007) discuss how human communities using this ecosystem respond to this variability to cope socially and economically. Ghanaian marine waters are part of an upwelling system with strong seasonal and inter-annual variability. Much of the variability at these time scales appears to be forced at large spatial scales, with El Niño–Southern Oscillation events in the Pacific Ocean influencing interannual variability of sea surface temperature and subsequent pelagic fish landings off Ghana. At decadal scales, Ghanaian marine waters experienced cool sea temperatures and low fishery landings during the 1960s, rapid warming and increases in fishery landings during the late 1970s and 1980s, and variable temperatures and fishery landings during the 1990s. In the late 1990s, pelagic and demersal fish populations appeared to be declining, partly due to overfishing, although the per capita supply (domestic production plus net imports) of fish was kept high by increased imports. As expected, artisanal fishers and fishing communities in Ghana have devised strategies to deal with variability on seasonal and interannual scales. These livelihood strategies include (i) exploiting marine and terrestrial natural resources more intensively, initially at local scales but expanding to regional scales; (ii) ensuring multiple and diversified income sources; (iii) investing in social relationships and communities for support; and (iv) undertaking seasonal or permanent migrations, sometimes outside of Ghana depending on the scales of the environmental change. In addition, the national government imports fish to deal with shortages. However, these strategies may be less adapted to variability at decadal scales, and may not be sustainable when viewed at the larger scales of environmental change (Perry and Sumaila, 2007).

---

unions and fishers, and they are seeking to find an equitable way in which to manage important fisheries at two radically different scales and with respect to two currently conflicting sets of interests. A linked natural–social science study here by marine biologists, political economists, anthropologists, and sociologists could produce a path-breaking analysis of the interface between marine science, management, and international and community requirements in small pelagic fisheries.

At a trans-national scale there is the study of small pelagic fisheries in the Americas. Pacific sardine was one of the largest fisheries in western North America from the mid 1920s until its collapse in the late 1940s. Those employed in fishing and processing sardines had to find alternative employment; some people and capital equipment moved into the newly developing anchovy fishery off Peru (Radovich, 1981). By the 1990s, however, this western North American stock of Pacific sardines had rebuilt and McFarlane and Beamish (2001) suggest that this return was driven by environmental and ecosystem changes – this had happened before in several northeast Pacific marine systems (Francis and Hare, 1994). The ability of the northeast Pacific ecosystem to produce sardines, and the importance of sardines to the local fisheries and processing sectors, will be perceived to be either positive or negative, depending on which year is chosen as the start of the trend analysis. If analysis starts (sets analytical year 0) at a time when the stock is at low abundance, forecasts of biological production will not look promising, nor will economic and employment potential. However, the opposite will be the case if the trend analysis sets year 0 at a time of high abundance. This is referred to as "the shifting temporal baseline problem" (Pauly, 1995), and it is important for management, since assumptions about biomass and stock vulnerability depend on having an analysis which takes into account sufficient data on the past to allow proper assessment of future trends (a different kind of scale problem). Indeed, the sardine industry is a fine example of the potential for interdisciplinary analysis at various scales ranging from the local (ecosystem and human system linked activities) to global (capital shifts and the survival of industrial capital assets at a global level in the face of ecosystem change) (Perry and Ommer, 2003). We need to understand the interdependence of ecosystem, social and economic resilience in the face of environmental variability much better, which means dealing with climatological variability, marine ecosystem variability, questions of overfishing, community vulnerability and resilience, institutional responses to change, management challenges, and the behavior of international capital. Similar combinations of historical and marine biological research are now being conducted on the herring fisheries of the North and Baltic Seas (cf. Alheit and Hagen, 1997; Holm et al., 2001; Poulsen, 2005). Moving to the global scale, a general case can be made that global warming will alter the production, harvesting and marketing of fisheries products significantly over the twenty-first century. In schooling species, such as small pelagic fish, a climate-driven change in biomass growth will unequivocally lead to changed harvesting policies (e.g. Arnason, 2006), while the resultant changes in fish distribution would aggravate the balance between co-operative and competitive management

solutions in the case of shared resources (see e.g. Hanesson, 2006) with dramatic consequences for the fishing communities exploiting them. Changes of fish distributions as a result of climate change will also become an international issue, as fish cross borders and expand into perhaps new locations, while possibly shrinking in the old. Rodriguez-Sánchez et al. (2002) show how small pelagic fish populations have shifted in the California Current System across the US and Mexican border associated with regime shifts during the twentieth century, which might be considered as a proxy for (short-lived) climate variations. Although not related to a "small" pelagic fish species, Miller (2000) describes the conflict and governance issues that can arise when dealing with cross-border fisheries issues (in this example relating to Pacific salmon). Now that we know we live under the threat of global environmental change, the development of adaptive, flexible management solutions with multiple objectives (biological, social and economic) increasingly is being recognized as being essential to protect our communities of fish and our fishing communities with, once again, adaptation rather than preservation being the most likely outcome. However, there is cause for concern when we consider that coastal community capacity to adapt has, in some areas, been shrinking (Ommer et al., 2007) at the same time that there has been an increase in the rate of biological change, brought about by major environmental crises. For the fishing communities, given the heavy fishing and sometimes excessive pressure being exerted on many stocks, the solutions may not be found entirely within the fishing sector and alternative livelihoods, part-time or full-time, will frequently need to be a part of the answer.

Future social–ecological research in small pelagic fisheries

Beyond the above topics, there are many other areas that would benefit from interdisciplinary research on small pelagic fisheries. Related to human diet, health, and economic well-being is the major issue of the impact of fishing pressure on both the ecosystem and human communities. A fascinating study is to be made of the impact of different catching technologies (canoes vs. purse seiners) on the well-being of both fish and fishing communities. Another important topic is that of ethics and equity in fisheries (Coward et al., 2000): who benefits from modern fishing methods, who loses, how, and why? What value issues are involved, and what different kinds of valuation (market, cultural, ecosystem, and so on) are embedded in different analyses and management systems? These are studies that need the combined efforts of biologists, ethicists, and humanists. This ties in tightly to the study of management systems and the advice on which managers rely. All too often in the past managers have used bio-economic modeling as their primary source of scientific advice when seeking to predict fish populations. Rigorous social science studies which employ methodologies that move beyond economic models to get at human motivations and socio-cultural concerns are all too rarely employed when making decisions on management measures for national or regional economies. Bio-economic modeling is inadequate, because such models fail to understand what drives the lives of ordinary fishing people – community, family, and household economics can look very different from averages that depict regional patterns (scale again) – while non-rigorous informal impressions of social issues are (at best) incomplete and are often inaccurate. Such inadequate sources result in suboptimal management decision making and have been the basis for many fishery protests (be those political or in indirect forms such as data fouling, illegal fishing, and so on).

Then there are the institutional aspects of management. The uncertainty that is inherent in scientific knowledge of marine species needs to be recognized by policy makers, managers, and fishers. It may be the result of poor data or incomplete research (over which we may have some control), but it may be the inevitable consequence of the complexity of dynamic natural systems. In either case, the challenge of having to base regulations for heavily exploited ecosystems on limited knowledge requires considerable implementation of risk assessment and management, while taking due consideration of the precautionary approach. This is a joint task for political scientists, sociologists, economists, and marine biologists, because regulations of scarce resources have immediate implications for the lives of fishing families in both industrial and artisanal fisheries, and the input into decision making must recognize the human complexities involved if regulations are to be respected. Such work will have to take into account the thorny issue of access to fish,

---

Box 11.2. Climate impacts on human communities

Climate-related changes, both short-term variability and long-term climate change, have serious consequences for everyone, especially coastal communities for many of whom, as we note in this chapter, their capacity to endure further change is already seriously tested. Climate change will not have an impact on communities fishing small pelagics that will be notably different from other communities whose resource bases are already damaged by such things as environmental and social restructuring. "Global climate changes will have direct population and community health impacts as exposure to highly variable climate conditions increases, and there will also be direct environmental impacts as local ecologies adjust, thus increasing the uncertainty of resource quality and quantity, including community access and availability" (H. Dolan and R. Ommer, personal communication). How such climate-related changes will affect coastal communities is dependent, in part, on their social and environmental adaptive capacity, which is best thought of interactively as the idea of social-ecological health (Berkes *et al.,* 2003; Ommer *et al.,* 2007). Coastal communities globally are now considered to be vulnerable to climate change, given their location and isolation, exposure to extreme climate variability, and their dependence on environmental resources for continued community health and well being (E. H. Allison *et al.* 2004, unpublished data). The insightful discussion of resource-dependent communities' vulnerability to climate change in the 2004 Allison *et al.*'s report also points to the absolute need to ensure adaptive capacity in resource-based communities (especially those in the developing world) and cites the risks of increased water levels, lost corals, damaged mangrove swamps, and serious issues of nutrition that accompany climate change-induced ecosystem variability, destruction, and change.

The real problem is that climate change will interact with a vast array of other factors so that, for example, the response of the global marketplace to this new stressor will affect fishing communities, and that will also interact with resource shifts and changes, political responses to new stressors, human migrations, and a vast array of other factors, which taken together form a global interactive network whose behavior cannot be adequately predicted. That is why the human dimensions of climate change tend to be broken down into discrete topics such as, for example, "the implications of potential future impacts on the health of populations via simulations and modeling that draw a direct link between a change in an environmental variable (e.g. extreme heat) and health outcome (e.g. increased hospitalizations, increased respiratory illnesses). Prescribed adaptations, whether governmental or individual, are then based on such projected health consequences" (H. Dolan and R. Ommer, personal communication). However, that approach, while necessary, cannot get at the network of interactivity that is at the heart of social-ecological research. This chapter stresses the complexity of these inter-linkages and recommends thinking about small pelagic-dependent communities in terms of the network of global change stressors that surrounds them. Case by case, small-scale studies can then begin to identify pathways between stressors and community responses, paving the way for a holistic new manner of examining such matters.

and so there will also be a need for legal expertise. Findings will then need to be linked back to the biology, and hence sustainability of the fish to which access has been assigned to some group or groups. Decision makers may well also have to rely in part on local and traditional knowledge of inheritance and property practices, as well as highly localized information about stocks. Historical resource rights, issues of equity, and issues of biological resilience are all crucial in such a scenario (e.g. Isaacs *et al.*, 2005; Isaacs, 2006; Fairweather *et al.*, 2006; Raakjer Nielsen and Hara, 2006). That is, there is a direct line of thought from the issue of access to fish, to the current pattern of global production of small pelagics, to the economics of their exploitation, the appropriate scale of exploitation and of fishing technology, and the societal implications. Within the context of global change, all this needs to happen within an adaptable framework, which allows for rapid response to changes, as and when needed.

Finally, co-management practices need to be developed and protected from external threats. These must be well informed about both the needs of communities of fish and of fishers if they are to be successful (Wilson *et al.*, 2003; Ommer *et al.*, 2007). This is the best way to achieve local equity, stewardship of resources, and ecosystems, biomass resilience and human community well-being. Implicit in such goals is the need to educate managers, local communities, industries, and academics of all types for the future, so that the next generation thinks in ecosystem terms – and consequently recognizes the need to think and work within the context of a range of interdependencies. Without such a perspective, small fishing communities and the species they fish are likely to become endangered. Over the past decade, the north–south divide in some marine science endeavors has been bridged through the activities of SPACC, resulting in an overall strengthening of scientific understanding of small pelagic fisheries on both sides of the divide. This knowledge transfer and augmentation of research capabilities provides both a foundation and a platform for enhancing such education activities.

Conclusions

Important social, institutional, and cultural issues arise from the topics dealt with in the biological chapters of this volume: issues of power, resource access, equity, property rights, scales, ethics, and technology, although none of these has been part of the SPACC analysis. Our goal in this chapter has been to point the way for future interdisciplinary work in small pelagic fisheries within a context of increasing global changes (of which climate change is a significant component), and to suggest methodologies and approaches that would be fruitful. In tying biological and oceanographic science to social science, we note that scale is the

major challenge that will confront all such research. While we have pointed to a few examples of interdisciplinary work that has been done with respect to fish and fisheries, much still remains to be done, and the current generation of young social and natural science marine researchers needs to be aware that new ways of confounding the old idea that there are fundamentally incompatible belief systems do exist. They should be encouraged to explore these efforts and indeed pioneer new ways of practicing their professions which will include mutual social–natural science interactions that will actively seek to heal past schisms. The extensive social science research on industrial fisheries needs to be informed by marine science of various kinds, while also maintaining a balance between industrial and small-scale fisheries, encompassing a continuum from the global, spatially footloose, and economically rooted to the local, spatially fixed, and culturally rooted. Because small-scale fisheries commonly have to compete directly with large-scale industrial fisheries for the same fish stocks, and for government quotas and subsidies, small-scale fisheries set urban and rural labor against one another in the competition for jobs. This is a major problem of equity, and one that needs to be recognized and redressed in a globalized economy where the prevailing international laws and practices have so far failed to achieve even the minimum acceptable standards.

Finally, we insist that, in the search for true interdisciplinarity in fisheries science (writ large), there is a great need for integrated work that brings together the social sciences and humanities with the natural sciences, not as add-on, but as full partners in the research endeavor. This will entail bringing in the social and cultural questions (framed as social scientists and humanists would lay them out) from the beginning, doing the same with the natural science questions, selecting those which address the most urgent problems and on which interdisciplinary work will provide new and necessary light on both sides of the present intellectual divide, and then creating new interdisciplinary methodologies to fit the resultant studies. We hope that this brief essay starts what can be, and needs to be, a richly productive dialog in the future.

Acknowledgments

The authors would like to thank Dave Checkley, Sam Herrick, and our anonymous reviewers for their helpful and constructive suggestions.

REFERENCES

Alheit, J. and Hagen, E. (1997). Long-term climate forcing of European herring and sardine populations. *Fish. Oceanogr.* **6**: 130–9.

Allison, E. H., Adger, W. N., Badjeck, M.-C., *et al.* Project R4778J: Effects of climate change on the sustainability of capture and enhancement fisheries important to the poor: analysis of the vulnerability and adaptability of fisherfolk living in poverty. Summary report, Draft, November 2004.

Arcos, D. F., Cubillos, L. A., and Núñez, S. P. (2001). The jack mackerel fishery and El Niño 1997–98 effects off Chile. *Progr. Oceanogr.* **49**: 597–617.

Arnason, R. (2006). Global warming, small pelagic fisheries and risk. In *Climate change and the Economics of the World's Fisheries: Examples of Small Pelagic Stocks*, Hannesson, R., Barange, M. A., and Herrick. S. F., eds. Cheltenham, UK: Edward Elgar Publishing Ltd., pp. 1–32.

Berkes, F., Colding, J., and Folke, C. (2003). *Navigating Social–Ecological Systems: Building Resilience for Complexity and Change.* Cambridge, UK: Cambridge University Press.

Best, P. B., Crawford, R. J. M., and van der Elst, R. P. (1997). Top predators in southern Africa's marine ecosystems. *Trans. Roy. Soc. S. Afr.* **52**: 177–225.

Brookfield, K., Gray, T., and Hatchard, J. (2005). The concept of fisheries-dependent communities. A comparative analysis of four UK case studies: Shetland, Peterhead, North Shields and Lowestoft. *Fish. Res.* **72**: 55–69.

Cochrane, K. L., Butterworth, D. S., and Payne, A. I. L. (1997). South Africa's offshore living marine resources: the scientific basis for management of the fisheries. *Trans. Roy. Soc. S. Afr.* **52**: 149–176.

Cochrane, K. L., Butterworth, D. S., de Oliveira, J. A. A., and Roel, B. A. (1998). Management procedures in a fishery based on highly variable stocks and with conflicting objectives: experiences in the South African pelagic fishery. *Rev. Fish Biol. Fisher.* **8**: 177–214.

Coward, H., Ommer, R. E., and Pitcher, T. (2000). *Just fish. Ethics and Canadian Marine Fisheries.* St John's, Newfoundland: ISER Books.

Degnbol, P. and. Jarre, A. (2004). Indicators in fisheries management: a development perspective. *Afr. J. Mar. Sci.* **26**, 303–326.

Fairweather T. P., Hara, M., van der Lingen, C. D., *et al.* (2006). The knowledge base for management of the capital-intensive fishery for small pelagic fish off South Africa. *Afr. J. Mar. Sci.* **28**: 645–660.

FAO (2002). *The State of the World Fisheries and Aquaculture.* Rome: FAO.

FAO (2005). Review of the state of world marine fishery resources: B15 Southeast Pacific. FAO *Fisher. Tech. Pap.* **457**: 130–143.

Ferraris, J., Koranteng, K. A., and Samba, A. (1998). Comparative study of the dynamics of small-scale marine fisheries in Senegal and Ghana. In *Global Versus Local Changes in Upwelling Systems*, Durant, M.-H., Cury, P., Mendelsson, R., *et al.*, eds. Paris, France: Editions ORSTOM, pp. 447–464.

Francis, R. C. and Hare, S. R. (1994). Decadal-scale regime shifts in the large marine ecosystems of the North-east Pacific: a case for historical science. *Fish. Oceanogr.* **3**: 279–291.

Garcia, S. and de Leiva Moreno, I. (2003). Global overview of marine fisheries. In *Responsible Fisheries in the Marine Ecosystem*, Sinclair, M. and Valdimarsson, G., eds. Wallingford, UK: CAB International, pp. 103–123.

Hamilton, L., Otterstad, O., and Ögmundardóttir, H. (2006). Rise and fall of the herring towns: impacts of climate and human teleconnections. In *Climate Change and the Economics of the World's Fisheries: Examples of Small Pelagic Stocks*, Hannesson, R., Barange M. A., and Herrick, S. F., eds. Cheltenham, UK: Edward Elgar Publishing Ltd., pp. 100–125.

Hanneson, R. (2006). Sharing the herring: fish migrations, strategic advantage and climate change. In *Climate Change and the Economics of the World's Fisheries: Examples of Small Pelagic Stocks*, Hannesson, R., Barange, M. A., and Herrick, S. F., eds. Cheltenham, UK: Edward Elgar Publishing Ltd., pp. 66–99.

Hannesson, R., Barange, M., and Herrick, S. F. (eds.) (2006). *Climate Change and the Economics of the World's Fisheries: Examples of Small Pelagic Stocks.* Cheltenham, UK: Edward Elgar Publishing Ltd.

Herrick, S. F., Jr., Hill, K., and Reiss, C. (2006). An optimal harvest policy for the recently renewed United States Pacific sardine fishery. In *Climate Change and the Economics of the World's Fisheries: Examples of Small Pelagic Stocks*, Hannesson, R., Barange, M. A., and Herrick, S. F., eds. Cheltenham, UK: Edward Elgar Publishing Ltd., pp. 126–150.

Holm, P., Smith, T. D., and Starkey, D. J. (eds.). (2001). *The Exploited Seas: New Directions for Marine Environmental History.* St. John's, Newfoundland: International Maritime Economic History Association, Census of Marine Life.

Isaacs, M. (2006). Small-scale fisheries reform: expectations, hopes and dreams of "a better life for all". *Mar. Pol.* **30**: 51–59.

Isaacs, M., Hara, M., and Raajkær Nielsen, J. (2005). South African fisheries reform – past, present and future? *PLAAS Policy Brief No. 16, Programme for land and agrarian studies*, University of the Western Cape, Cape Town, South Africa. Available from www.plaas.wcape.ac.za.

Jahncke, J., D. M. Checkley, Jr., and G. L. Hunt, Jr. (2004). Trends in carbon flux to seabirds in the Peruvian upwelling system: effects of wind and fisheries on population regulation. *Fish. Oceanogr.* **13**: 208–223.

Kennedy, J. C. (2006). *Island voices: fisheries and community survival in Northern Norway.* The Netherlands: Eburon Academic Publishers.

Kristofersson, D., and Anderson, J. L. (2006). Is there a relationship between fisheries and farming? Interdependence of fisheries, animal production and aquaculture. *Mar. Policy.* **30** (6): 721–725.

McFarlane, G. A. and Beamish, R. J. (2001). The reoccurrence of sardine off British Columbia characterize the dynamic nature of regimes. *Prog. Oceanogr.* **49**: 151–165.

Miller, K. (2000). Pacific salmon fisheries: climate, information and adaptation in a conflict-ridden context. *Clim. Change* **45**: 37–61.

Muck, P. and Pauly, D. (1987). Monthly anchoveta consumption of guano birds, 1953–1982. In ICLARM Studies and

Reviews 15: *The Peruvian Anchoveta and its Upwelling Ecosystem: Three Decades of Change*. Pauly D. and Tukayama, I., eds. Manila, Philippines: International Center for Living Aquatic Resources Management, pp. 219–233.

Mullon, C. and Fréon, P. (2006). Prototype of an integrated model of the worldwide system of small pelagic fisheries. In *Climate Change and the Economics of the World's Fisheries: Examples of Small Pelagic Stocks*, Hannesson, R., Barange, M. A., and Herrick, S. F., eds. Cheltenham, UK: Edward Elgar Publishing Ltd., pp. 262–295.

Ommer, R. E. (2006). *Coasts Under Stress: Understanding Restructuring and Social–Ecological Health. Policy Implications.* St. John's, Newfoundland: ISER Books

Ommer, R. E., *et al.* (2007). Coasts Under Stress: understanding restructuring and social-ecological health, Chapter 3. McGill-Queen's University Press.

Pauly, D. (1995). Anecdotes and the shifting baseline syndrome of fisheries. *Trends Ecol. Evol.* **10**: 430.

Perry, R. I. and Ommer, R. (2003). Scale issues in marine ecosystems and human interactions. *Fish. Oceanogr.* **12**: 513–522.

Perry, R. I. and Sumaila, U. R. (2007). Marine ecosystem variability and human community responses: the example of Ghana, West Africa. *Mar. Pol.* **31**: 125–134

Poulsen, B. (2005), Climate change and catch rates in the North Sea herring fisheries, c. 1600–1850. Paper presented at the Conference on *OCEANS PAST – Multidisciplinary Perspectives on the History of Marine Animal Populations HMAP Open Science Conference*, 24–27 October 2005, Kolding, Denmark (www.hmapcoml.org).

Raakjær Nielsen J. and Hara, M. (2006). Transformation of South African industrial fisheries. *Mar. Pol.* **30**: 43–50.

Radovich, J. (1981). The collapse of the California sardine industry: what have we learned? In *Resource Management and Environmental Uncertainty: Lessons from Coastal Upwelling Fisheries (Advances in Environmental Science and Technology V)*. Glantz, M., ed. New York: John Wiley, pp. 107–136.

Rodríguez-Sánchez, R., Lluch-Belda, D., Villalobos, H., and Ortega-García, S. (2002). Dynamic geography of small pelagic fish populations in the California Current System on the regime time scale (1931–1997). *Can. J. Fish. Aquat. Sci.* **59**: 1980–1988.

Roux, J-P. and Shannon, L. J. (2004). Ecosystem approach to fisheries management in the northern Benguela: the Namibian experience. *Afr. J. Mar. Sci.* **26**: 79–93.

Shannon, L. J., Cochrane, K. L., Moloney, C. L., and Frèon, P. (2004). Ecosystem approaches to fisheries management in the southern Benguela: a workshop overview. *Afr. J. Mar. Sci.* **26**: 1–8.

Sumaila, U. R. and Stephanus, K. (2006). Declines in Namibia's pilchard catch: the reasons and consequences. In *Climate Change and the Economics of the World's Fisheries: Examples of Small Pelagic Stocks*, Hannesson, R., Barange, M., and Herrick, S. F., eds. Cheltenham, UK: Edward Elgar Publishing Ltd., pp. 205–214.

van der Lingen, C. D., Shannon, L. J., Cury, P., *et al.* (2006). Resource and ecosystem variability, including regime shifts, in the Benguela Current System. In *Benguela: Predicting a Large Marine Ecosystem.* LME Series Vol. 14. Shannon, V., Hempel, G., Malanotte-Rizzoli, P., *et al.*, eds. Amsterdam, the Netherlands: Elsevier, pp. 147–184.

Wilson, D. C., Degnbol, P., and Raakjær Nielsen, J. (2003). *The Fisheries Co-management Experience: Accomplishments, Challenges and Prospects.* Amsterdam, the Netherlands: Kluwer Academic Publishers.

12 Mechanisms of low-frequency fluctuations in sardine and anchovy populations

Alec D. MacCall

CONTENTS

Summary

The hypothesized mechanisms reviewed in this chapter have been selected as having potential for generating low frequency variability in sardine (*Sardinops* spp.) and anchovy (*Engraulis* spp.) populations such as those experienced in the California Current, Humboldt Current, and the Benguela and Kuroshio Current systems. No generally accepted theory yet exists. An initial framework for such a theory is proposed, in which sardine productivity is linked to low frequency variability in boundary current flows, which is also related to the characteristic sea surface temperature anomalies associated with sardine productivity in these systems. During periods of weaker flow, planktonic sardine larvae are able to gain swimming capacity before being flushed from the system, allowing sardines to inhabit the main body of the boundary current. During periods of stronger flows, successful sardine reproduction is restricted to coastal waters, and productivity is relatively low. Anchovies are always restricted to coastal waters, and are more influenced by upwelling and coastal productivity; these characteristics tend to be correlated with boundary current fluctuations, giving rise to a tendency (but not requirement) of sardine and anchovy alternations. The Japanese system lacks coastal upwelling, but the cold, nutrient-rich Oyashio Current provides an analogous function.

A wide variety of mechanisms can be added to this framework as appropriate to individual systems. Physical processes include patterns of boundary current flow, including current meandering and formation of cyclonic eddies. A latitudinal shift in the source water coming from the North Pacific has been identified in the California Current, and contributes to the characteristic temperature and nutrient anomalies. Important biological mechanisms include indeterminate fecundity, strong age-dependent increases in fecundity, temperature preferences, and a regime-dependent plasticity in life history (sardines become longer lived during productive periods). Behavioral mechanisms include possible imprinting on regions or waters where a successful cohort is spawned, and possible learning of favorable migratory paths from older individuals in the population. Patterns of distribution may be abundance dependent, due to density dependent habitat selection, and during periods of low abundance, individuals may mix in schools of other species, whether this is to their advantage or to their disadvantage. Multispecies mechanisms include trophic dynamic influences from both higher and lower trophic levels, as well as competition from other species at similar levels. There is evidence of cyclic dominance of pelagic species in some systems, but it is unclear whether it is driven biologically by competition and predation, or by long-term cyclic properties of the physical environment. Fisheries on sardine and anchovy populations have tended to be intense, and have resulted in rapid declines in abundance when productivity rates decrease. Fisheries also affect demographics, including reduced average lifespan and net lifetime fecundity, as well as introducing geographic variability in survivorship. Fisheries may increase the contrast between favorable and unfavorable regimes.

Introduction

This review focuses on possible mechanisms associated with sardine (*Sardinops* spp.) and anchovy (*Engraulis* spp.) "regimes" – the very large, low frequency fluctuations and alternations exhibited by these small coastal pelagic species around the world. For a complementary review of environmental effects on populations of small pelagic fishes, the reader is referred to Fréon *et al.* (2005).

Climate Change and Small Pelagic Fish, eds. Dave Checkley, Jürgen Alheit, Yoshioki Oozeki, and Claude Roy. Published by Cambridge University Press. © Cambridge University Press 2009.

Box 12.1. Was it due to overfishing, or was it natural?

Collapses of small pelagic fisheries consistently raise the question, what caused them? In their simplest form, the two standard alternative hypotheses are H_A: Collapse was a natural result of environmental fluctuations, and H_B: Collapse was due to overfishing. Despite the rather obvious issue that these hypotheses are not mutually exclusive, the question continues to be posed in these overly simplistic terms. The usual conclusion to the debate is that significant decline was inevitable, due to an adverse change in environmental conditions. However, the speed and severity of decline that merit the term "collapse" were due to intense fishing. A disturbing corollary is that the modern ideal of "sustainable" development may in fact be impossible for these fisheries.

A wide variety of mechanisms have been proposed, but these mechanisms tend not to be exclusive of each other. Thus, both mechanism "A" and mechanism "B" may be acting simultaneously, and we can easily be misled by the results of conventional hypothesis tests of H_A vs. H_B (Box 12.1). The difficulties of studying these mechanisms are further compounded by differences among stocks in different geographic regions: mechanism "A" may predominate in one region, and mechanism "B" may predominate in another region. Even within a single region, mechanism "A" may predominate for several decades, and then the system may suddenly change behavior to favor mechanism "B". Thus, during periods of low sardine productivity, it may not be possible to observe or study mechanisms associated with periods of high sardine productivity: The critical mechanisms associated with high productivity may not be operating during a 20- to 30-year regime of low sardine productivity. This helps to explain why it has been so difficult to make progress toward a general understanding of fluctuations in sardine–anchovy systems (also see Alheit et al., 2008; MacCall et al., 2008).

Most of the associated coastal ecosystems have been monitored for little more than a single cycle of sardine or anchovy abundance, and the high serial correlation of physical and biological characteristics within a regime further reduces effective sample sizes. Strong co-variability of those characteristics can make it difficult to isolate causal factors from non-causal co-varying factors. Consequently, existing historical data tend to provide low statistical power to distinguish among competing hypotheses.

A long list of mechanisms can be compiled with regard to explaining annual fluctuations in recruitment (e.g. Bakun

1996), but the mechanisms listed in this chapter are selected as having potential for generating high contrast at interdecadal time scales. The evidence for these mechanisms is drawn mostly from four oceanic sardine–anchovy regions: The California Current (but excluding the Gulf of California in this discussion), the Humboldt Current, the Benguela Current, and the Kuroshio Current (excluding the waters west of Japan). The Canary Current (where the resident sardine is of the genus *Sardina* rather than *Sardinops*) could have been included, but that region is poorly documented and would contribute little to this review. These four regions share many oceanographic features, and share remarkably similar guilds of pelagic fishes (Parrish et al., 1983). However, each region also has unique physical and biological properties. Thus, some of the proposed mechanisms may apply to several systems, while other mechanisms may be unique to one region. It is the presence of similar mechanisms operating in multiple regions that provides the best opportunity for achieving a general understanding of sardine–anchovy regimes.

An initial framework

Shifts in average sea surface temperature may be one of the most prominent features associated with sardine–anchovy regimes, but the preoccupation with temperature also may have been misleading: As Chavez et al. (2003) observed "It remains unclear why sardines increase off Japan when local waters cool and become more productive, whereas they increase off California and Peru when those regions warm and become less productive." MacCall (2002) identified flow, rather than temperature, as the likely unifying feature associated with worldwide regime shifts of small pelagic fish production. Warm temperatures in coastal currents off California, Peru–Chile, and South Africa (where the source water is cold), and cold temperatures off Japan (where the source water is warm) all occur when the flow weakens. Patterns of flow have long been associated with sardine productivity in the Kuroshio Current System (Kondo, 1980). However, researchers outside Japan seem previously to have regarded flow patterns as an issue uniquely important to the highly energetic Kuroshio system, which is the only western boundary current exhibiting a major sardine–anchovy system. MacCall's flow hypothesis extended some aspects of the Japanese view to the contrasting eastern boundary current sardine–anchovy systems. In all cases, patterns of flow are strongly associated not only with region-specific characteristic shifts in ocean temperature, but with shifts in the physical and biological structure of entire ecosystems. As will be shown below, the flow hypothesis also provides an explanation for the mysterious near-synchrony of sardine–anchovy regime shifts on a worldwide scale (Kawasaki, 1983).

Before listing individual mechanisms, I propose the following general physical and biological framework of sardine–anchovy regimes (the "initial framework"), reflecting elements that seem to apply to most of the systems. It must be emphasized that despite the long history of research on this subject (see Chapter 2) there is still no generally accepted theory of sardine–anchovy fluctuations. The following initial framework is proposed by this author as a candidate working theory that links various mechanisms that have been previously identified by a number of scientists working on this problem.

The initial framework consists of alternating physical patterns of ocean circulation or boundary current flow, adaptive behavioral and physiological responses, and fundamental underlying constraints imposed by differences among species. These elements are the following.

Physical variability

There is an interdecadal alternation of flow conditions so that, during favorable conditions, retention of offspring spawned in the major boundary currents is enhanced because planktonic offspring have a higher probability of growing to free-swimming nektonic life stages before being swept out of the system. This can take several forms, but a common form is slower flow of the primary boundary current. For convenience, this set of conditions will be referred to as "weak flow."

1. In eastern boundary currents nutrient enrichment co-varies with strong and weak flow conditions due to correlated intensity of advective transport and coastal upwelling.
2. In the non-upwelling western boundary current system off Japan, Kuroshio Current and Kuroshio Current Extension flow conditions (including patterns of meandering) co-vary with penetration of cold, nutrient-rich Oyashio Current water into the Kuroshio Current system.

Sardines

Sardines occupy alternative habitats during productive and unproductive phases. During productive periods, sardines often migrate upstream and/or offshore, to take advantage of conditions associated with improved net fecundity, including both improved survival of offspring, and enhanced production of gametes. Because these locations are favorable only during periods of weak current flow, sardine productivity is primarily associated with flow conditions in the major currents.

Anchovy

Adult anchovies are restricted to the nearshore region due to their smaller size, and anchovy productivity is influenced by coastal nutrient fluctuations, primary productivity, and water column stability, all of which are related to coastal upwelling.

Evidence for synchronous interdecadal fluctuations in ocean circulation

Coastal sea level is a well-established indicator of integrated boundary current flow (Chelton *et al.*, 1982). I obtained long-term records of sea level from six coastal locations on the Pacific Rim and four coastal locations on the Atlantic Ocean (University of Hawaii Sea Level Center, http://uhslc. soest.hawaii.edu). Data series covered the period 1901 to 2001, with more complete coverage in the latter half of the time period. The data were converted to a 7-year moving average to reduce influences of El Niño events. Thus the analysis reflects patterns of low frequency variability on a coarse, near-decadal time scale. The dominant pattern of co-variability given by the first empirical orthogonal function (a.k.a. principal component), accounted for 52% of the total variance (Fig. 12.1). The loadings for all six locations from the Pacific (two each from California, Peru–Chile, and Japan) were similar, suggesting a Pacific-wide tendency toward synchrony of fluctuations in sea levels and associated boundary current strengths. In the eastern Atlantic Ocean, fluctuations in sea level at Vigo, Spain (at the northern end of the fishery for *Sardina pilchardus*) also tended to be synchronous with those in the Pacific. Sea levels and current strengths off Namibia appear to vary inversely with those in the Pacific Ocean, and indeed, the South African sardine fishery has tended to fluctuate inversely to the Pacific *Sardinops* fisheries (Lluch-Belda *et al.*, 1989; Schwartzlose *et al.*, 1999). Sea levels on the northwestern Atlantic coast also appear to vary inversely with those in the Pacific. The physical basis for the pattern seen in Fig. 12.1 is not well

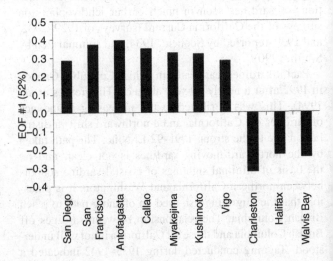

Fig. 12.1. Empirical orthogonal function (principal component) loadings of 7-year averaged sea level associated with the largest eigenvalue.

understood, but Bakun (1995) anticipated the importance of flow to understanding sardine–anchovy and related shifts, and described a possible mechanism that links gyral circulation strength in the North and South Pacific through decadal scale coupled ocean–atmospheric dynamics in the equatorial region.

Sardine movement upstream and utilization of offshore habitat

Kobayashi and Kuroda (1991) provide a clear illustration of the progressive offshore and upstream (southward) shift of Japanese sardine spawning between pre-regime conditions in the mid 1960s and the abundance peak in the late 1980s. There was also a large post-spawning downstream expansion of the Japanese sardine feeding migration: By the mid 1980s, the sardine range had expanded over 4000 km eastward into the Kuroshio Extension (Wada and Kashiwai, 1991). The favorable period ended abruptly in 1988, and due to heavy exploitation, estimates of parental spawning biomass dropped by 98% from 1988 to 1994 (Wada and Jacobson, 1998).

A similar offshore and upstream expansion of the California Current sardine occurred during the early 1990s. The California offshore population was not suspected, and its discovery in 1991 came as a surprise. While conducting exploratory trawling for jack mackerel (*Trachurus symmetricus*) in international waters, the Russian survey vessel *Novodrutsk* encountered abundances of sardines at the farthest edge of the range previously covered by standard CalCOFI ichthyoplankton surveys, and outside the US 200-mile Exclusive Economic Zone (Macewicz and Abramenkoff, 1993). Discovery of this offshore segment of the California sardine population helped explain the puzzling offshore historical distributions of sardines seen in much earlier ichthyoplankton surveys of the California Current (surveys of 1929–1932, and 1939, reported by Scofield, 1934, and summarized by Smith, 1990).

Pacific sardines reappeared in British Columbia, Canada in 1992 after a nearly 40-year absence (Hargreaves *et al.*, 1994). This was attributed to the increasing abundance of sardines off California, and a northward shift in distribution due to the strong 1991–92 El Niño. The path taken by the northward-moving sardines is not clear, and on the basis of minimal sightings of coastal sardine schools between northern California and Washington, it is plausible that the migrating fish used an offshore path to reach British Columbia. The relationship between sardines off British Columbia and those off California is not well understood. Tagging conducted during 1935–1942 indicated a low rate of movement from Central California to British Columbia, and consistent differences in length-at-age suggested restricted mixing (Clark and Marr, 1955).

An analogous situation appears to have occurred when, in 1973, *Sardinops sagax* suddenly appeared in Talcahuano, Chile (37° S.), more than 1000 km south (upstream) of traditional sardine fishing grounds near Iquique, Chile (Serra, 1983). The new spawning area in Talcahuano was thought to be a "colonization" event as described by Sharp (1980). The author proposes a modified hypothesis that the Talcahuano colonization was made possible by an offshore path (similar to that hypothesized for North America) that connected the Iquique habitat to the newly favorable upstream habitat at Talcahuano, and that sardines in these offshore waters were not observed by surveys due to lack of sampling in sufficiently offshore areas: all of the existing samples were taken within ca. 100 nm of the coast. However, there is circumstantial evidence in Zuzunaga's (1985) observation that, before 1972–73, sardines were found outside the coastal area occupied by the anchovy. Also, there are anecdotal reports that sardines were taken by the offshore Russian fishery for jack mackerel (M. Niquen, personal communication). Unlike the sardines off Canada, the Talcahuano sardines do not appear to be larger than those caught downstream in northern Chile or Peru (R. Serra, personal communication), suggesting that the Talcahuano segment may be reproductively independent.

There is also some historical evidence for an offshore segment of sardines off the Western Cape, South Africa. Crawford *et al.* (1983) observed that older fish formerly spawned off St. Helena Bay "in warm water outside the oceanic front" but that few fish later attained these ages due to heavy exploitation. In the recent productive period off South Africa, sardine spawning initially overlapped with anchovy spawning in the Agulhas Bank area east of Cape Point, but in the 1990s briefly shifted to a non-overlapping area west of Cape Point (van der Lingen *et al.*, 2001). However, at the end of the 1990s, the South African sardines inexplicably shifted their spawning distribution over 500 km eastward to the vicinity of Port Elizabeth in the Eastern Cape, completely abandoning the western spawning area (van der Lingen *et al.*, 2005; Coetzee, 2007). Perhaps surprisingly, the eastward shift is easily reconciled with the initial framework: the South African system is unique in that it has two alternative "upstream" regions, one extending off the Western Cape into the source of the Benguela Current in the open South Atlantic, and the other extending toward the Indian Ocean, up the Agulhas Current and the Eastern Cape. It appears that the South African sardines may be exploring both "upstream" possibilities. As van der Lingen *et al.* (2005) note, the eastward shift may have been influenced by the intense exploitation of sardines in the Western Cape. The shift was too sudden to have a genetic basis. Alternative mechanisms such as the "entrainment hypothesis" are being considered (Coetzee, 2007).

The relationship between the South African and Namibian (Walvis Bay) sardine stocks is problematic. These two stocks are now generally considered to be independent, though the potential analogy to the California–British Columbia and Iquique–Talcahuano cases may be worth consideration. The nearshore region north of St. Helena Bay is too cold for sardine reproduction, and is considered to be a barrier to migration between the Western Cape and Walvis Bay. An extensive tagging study conducted from 1957 to 1966 (Newman, 1970a,b) showed a very low amount of movement from Walvis Bay to the Western Cape (and no movement in the opposite direction), which appeared to be evidence that the two stocks were independent. However, much of the tagging study was conducted during the historical collapse of Western Cape sardines; estimated migration rates were based on results for years 1963 to 1966 (Newman, 1970b). A sequence of recruitment failures began in the early 1960s, and estimated spawning biomass declined by 95% from 1961 to 1967 (Butterworth, 1983), suggesting that the favorable regime had already ended by 1963, and if so, the hypothesized favorable offshore habitat required for migration past the cold nearshore region would no longer have existed by the time the tagging program was conducted. This is a case where mechanisms associated with a productive regime cannot be studied during an unproductive regime. In the author's opinion, existing information is insufficient either to support or to disprove the hypothesized possibility of at least an occasional offshore connection between the Western Cape and Walvis Bay during warm Benguela Current regimes of high sardine productivity and abundant older fish.

The mechanisms associated with these changes in life history are poorly understood. The sardine switch to migratory behavior and subsequent migratory patterns were long assumed to have a genetic basis, but the behavioral "entrainment hypothesis" (described on p. 293) is a more flexible alternative, and is quickly gaining support (Coetzee, 2007; Lo, 2007).

Anchovy restriction to nearshore habitat
Anchovies are generally not found far offshore in any system. A possible exception is that of juvenile anchovies occurring far offshore in the Shatsky Rise area of the Kuroshio Extension (Komatsu *et al.*, 2002), but these originate from coastally spawned eggs rather than from local offshore spawning. The fate of those offshore juveniles is unknown.

Size constraints
Bakun (1993) noted that anchovies have less swimming capacity than sardines, and speculated that anchovies may be dominant in habitats where good feeding conditions and good spawning conditions coincide. MacCall (2002) similarly speculated that smaller body size (relative to

sardines) and associated limitation in swimming capacity could restrict anchovies to nearshore areas. Presumably their small size renders them more susceptible to predation and less able to maintain their position in a persistent current, and they may therefore be unable to utilize the offshore habitat that episodically becomes available to sardines. Anchovies in eastern boundary currents are highly dependent on nearshore plankton production associated with upwelling (Parrish *et al.*, 1983), and their relative abundance in the various systems is generally (but not at annual time scales) correlated with the regional intensity of upwelling-driven productivity: The ecological importance of anchovies in the four systems ranges from relatively low significance in Japan (which is not an upwelling-driven system) to ecological dominance in the Humboldt Current, which is the strongest of the three upwelling systems.

The circumstantial basis of sardine–anchovy "alternations"
Periods of weakened boundary current flow are associated with reduced upwelling productivity and reduced plankton abundance in the three eastern boundary currents, e.g. Muck (1989) for the Humboldt Current, and Roemmich and McGowan (1995) for the California Current. Abundance of anchovy stocks tends to decline during these periods. Conversely, upwelling productivity increases during periods of stronger eastern boundary current flow, and anchovy stocks tend to increase in abundance. Under the initial framework hypothesis, any inverse relationship of anchovy and sardine productivity is largely due to the tendency for offshore flow conditions favoring sardines to coincide with nearshore productivity conditions that disfavor anchovies, and *vice versa*. Thus, in contrast to most previous attempts to explain anchovy and sardine variability, this initial framework contains very little direct linkage or interaction between sardines and anchovies: the impression of sardine–anchovy alternation is circumstantial, and should not be expected to be a consistent feature of these systems. This hypothesis predicts that there should be oceanic conditions under which both species may be abundant or scarce simultaneously. Such examples exist: in recent years, sardines and anchovies have been simultaneously abundant in the southern Benguela system (this volume, Chapter 9). In the paleosedimentary record, the formerly puzzling lack of correlation between California's anchovy and sardine scale deposition rates (Soutar and Isaacs, 1974, and confirmed by Baumgartner *et al.*, 1992) is consistent with this initial framework.

Physical processes

Retention and critical development time
Retention of planktonic eggs and larvae is a critical component of Bakun's (1996) "fundamental triad" of conditions

favoring recruitment success. In the relatively open-ended circulation of eastern boundary currents as well as the Kuroshio Current system, the downstream end of the system is a lethal sink for offspring, where there is effectively "a point of no return." The problem of retention in these open-ended systems can be thought of as a developmental time threshold. This critical length of time is the duration of the planktonic phase of a fish's life cycle, i.e. the time from release of a spawned egg to when the offspring achieve sufficient size to become nektonic and capable of actively maintaining their position in a current system. This critical length of time can be related to numerous properties of a current system: its speed, its path (straight or meandering), the upstream distance of spawning from the downstream "point of no return," larval food densities, and temperature (development tends to be faster at higher temperatures). These aspects are treated individually later in this chapter. Because of the threshold nature of the critical development time, relatively small persistent physical changes in an open-ended current system can result in distinct shifts between periods with and without sufficient retention of offspring, thereby creating regime-like contrasting periods of high and low fish production.

Sardine productivity has been associated with warm conditions in the eastern boundary currents, and anchovy productivity has similarly been associated with cool conditions, though no clear causal relationship has been identified, and species fluctuations in the Kuroshio Current system appear to show the opposite relationship to sea surface temperature (McFarlane *et al.*, 2002). The contrary relationship of the Kuroshio Current is resolved if we consider the issue from the standpoint of flow rather than temperature. A faster flow will tend to retain the temperature characteristics of the source water, and a slower flow will tend to show temperature anomalies opposite to that of the source water. Thus slower flow is associated with warm temperature anomalies in the eastern boundary currents with their cold, high latitude source water, and with colder anomalies in the Kuroshio Current which receives warm, low latitude source water.

Although it is convenient to think of the boundary current as being "strong" or "weak," the associated temperature anomalies reflect a wider suite of properties, and that suite is closely related to the above-listed properties that are associated with planktonic retention. The characteristic temperature anomalies would be seen if a larval fish, or equivalently a parcel of water, takes a shorter or longer time to move through the system, which is again related to speed, trajectory, and origin of the source water. One of the simplest mechanisms is that the same volume of water can be transported at reduced velocity if the current becomes wider, or deeper. For example, the thermocline depth increased in both Japan and Peru ca. 1969, at the beginning

of productive regimes for sardines and the characteristic temperature shifts in both areas (this volume, Chapter 5).

Kuroshio Current meandering

The pattern of Kuroshio Current meandering has long been associated with fluctuations in pelagic fish productivity off Japan (Nakai, 1949, 1962). Notably, it is the most extreme or "A-type" meandering that is thought to be associated with sardine productivity and the associated cold temperatures (Hayasi, 1983). A meandering path increases the length of time needed to transport a water parcel from one end of the system to the other, enhancing larval retention by increasing the available time for larval development. However, meandering may have additional effects such as extending the length of frontal regions that may be important feeding areas for both adults and offspring. Increased admixing of cold, nutrient-rich Oyashio Current water is beneficial to sardine production, and seems to be enhanced by some patterns of Kuroshio meandering (Kondo, 1980).

Cyclonic eddies

Logerwell and Smith (2001) have shown that offshore mesoscale cyclonic eddies in the California Current are associated with elevated densities of sardine larvae, and that the periphery of anticyclonic eddies may be zones of enrichment and entrainment of coastal waters. Logerwell *et al.* (2001) modeled the spatial bioenergetics of such an offshore eddy, and concluded that they are likely to be a significant source of sardine recruitment. This mechanism has not yet been confirmed by direct quantitative comparison of eddy activity and resulting recruitment strengths estimated by stock assessments. With respect to mechanisms associated with productivity regimes, an important oceanographic question is whether cyclonic eddy formation is enhanced during periods of weak boundary current flow, as was postulated by MacCall (2002).

Latitudinal shifts in source water

Parrish *et al.* (2000) analyzed North Pacific atmospheric observations and concluded that the North Pacific regime shift in 1976 was associated with distinct changes in wind patterns influencing the North Pacific Current. They found that the total eastward transport actually increased, but that the portion of the water entering the California Current had a more equatorward source than before 1976. This not only accounts for the post-1976 warming of the California Current, but also accounts for the decrease in nutrients and plankton production observed by Roemmich and McGowan (1995). This post-1976 North Pacific wind pattern also was associated with an extreme cooling in the Kuroshio–Oyashio mixing area that Japanese sardines used as nursery habitat, and may have contributed to nutrient enrichment in that area. A comparable analysis of South

Pacific atmospheric observations (if sufficient data exist) would be worthwhile, given the very similar post-1972 Peruvian pattern of declining zooplankton and increased offshore temperatures (Muck, 1989). Importantly, latitudinal shifts in source water can result in characteristic temperature and nutrient anomalies without a weakening of the boundary current. Also, latitudinal shifts are in a class of precession-like rotational mechanisms that can generate periodic variability at low frequencies.

Biological mechanisms

Life history and reproduction
Indeterminate fecundity and access to forage
Unlike herring, anchovies and sardines are indeterminate spawners: their seasonal fecundity is not fixed at the beginning of a spawning season, and they have the capacity to generate additional new oocytes and convert current food intake into successive batches of gametes as long as abundant forage is available (Blaxter and Hunter, 1982). This is an important aspect of their life history, and is undoubtedly a key to their productivity. Hunter and Goldberg (1980) used incidence of post-ovulatory follicles to infer that anchovies spawn at intervals of 6–8 days during a protracted period of active spawning. Hunter and Leong (1981) calculated that a typical female anchovy in southern California may produce 20 batches of eggs in a year, only about two-thirds of which could be attributed to fat stored during the pre-spawning period. Similarly, Le Clus (1989) estimated that large sardines off Namibia may produce 80 batches of eggs in a year.

This direct linkage of plankton production to net fecundity is especially important for sardines. Richard Parrish (personal communication) has hypothesized that regime conditions favorable to sardines in the California Current provide the spawning population access to the high production of plankton at the upstream end of the system at a time when the California Current otherwise exhibits a substantial decline in primary and secondary production (Roemmich and McGowan, 1995). Abundant food for sardines is nearly always present at the upstream end of the California Current, but during periods of strong flow the upstream temperature is too cold for sardine larvae to develop, and sardines abandon the northern area. However, during periods of weaker flow, upstream temperatures are warm enough to allow larval development and successful reproduction. Sardines extend their feeding range northward where they can convert the abundant forage supply into a substantial increase in fecundity. This provides the enrichment and concentration aspects of Bakun's (1996) "fundamental triad." The third component of the triad, which is retention, is provided by the combination of upstream spawning and slower transport of eggs and larvae through the system (MacCall, 2002).

Table 12.1. *Age-specific per capita fecundity of northern anchovy*

Age	Number of spawnings	Eggs/g per year	Avg wt. (g)	Annual fecundity (10^6)
1	5.3	2803	11.6	0.17
2	11.9	6550	16.5	1.29
3	19.2	11434	20.7	4.54
4+	23.5	13861	26.6	8.66

From Parrish *et al.* (1986).

Age-dependent fecundity and reproductive age structure
Although spawning biomass has long been used as a measure of stock reproductive potential, it is an inappropriate measure of reproductive potential for indeterminate spawners such as anchovies and sardines which increase their fecundity with age. Parrish *et al.* (1986) estimated that a 4-year-old female anchovy produces about 50 times the egg output of a 1–year-old (Table 12.1). Le Clus (1989) estimated that Namibian sardines increased their annual number of spawning batches by 50% to 70% for each 1 cm increase in length. Murphy (1967) warned that risk of stock collapse is greatly increased by the tendency of fisheries to reduce the number of reproductive cohorts in a population. The age-related pattern of increasing fecundity reported by Parrish *et al.* (1986) and Le Clus (1989) suggests that, for anchovies and sardines, the risk may be even greater than was envisioned by Murphy.

Life history plasticity
Many fish species exhibit latitudinal trends in life histories: individuals in the more poleward segments of fish populations tend to grow larger, live longer, and mature at a larger size and/or older age. In the California Current, the northern anchovy (*E. mordax*) strongly exhibits this latitudinal structure (Mais, 1974; Parrish *et al.*, 1985). Mais (1981) reported a very sudden change in southern California anchovy age and size composition, beginning in 1977. Where fishery catches had previously been mostly 2- and 3-year-old fish, suddenly catches were mostly 0- and 1-year-old fish. In hindsight, this demographic change coincided with what has since become a well-known California Current regime shift in 1976, at which time ocean temperatures increased markedly. After 1976, anchovies in southern California suddenly exhibited the smaller sizes, earlier maturity and shorter life spans that were previously typical of warmer-water Baja California fish (Table 12.2), suggesting a temperature-based trigger for the shift in anchovy life history. Lasker and MacCall (1983) found paleosedimentary evidence for a similar demographic shift in the mean sizes of anchovy scales taken from

Table 12.2. *Comparison of California Current anchovy and sardine life histories during respective productive and unproductive ocean conditions*

	Regime	
	Productive	Unproductive
Sardine	Warm	Cool
~ Age of maturity	2–3	1–2
~ Maximum age	>13	4
Anchovy	Cool	Warm
~ Age of maturity	2	1
~ Maximum age	6	4

Table 12.3. *Temperature (°C) associated with peak abundance of eggs and larvae*

Region	Sardines	Anchovies
Japan	16.5	22
California	14	14
So. Baja California	24	25

the Santa Barbara Basin, and concluded that anchovies had about half the average body weight during periods of sardine abundance as compared with periods of sardine scarcity. Because of the age-dependent fecundity pattern described previously, changes in lifespan may strongly influence population productivity.

The California Current's sardines also show a regime-dependent change in their life table. Murphy (1966) reports that, in the early years of the historical fishery, about half of the 2-year-olds spawned, but that in the late 1950s, all of the 2-year-olds spawned. Anecdotal accounts suggest that 1-year-old sardines were actively spawning in the 1960s and early 1970s. In the early 1930s, at the beginning of the historical fishery, sardines achieved maximum ages of at least 13 years (the plus group reported by Mosher and Eckles, 1954), seen both in Central California and in British Columbia. In the early 1970s, 4-year-olds were rare, despite a harvest moratorium. Murphy (1966) concluded that the natural mortality rate of sardines approximately doubled in the 1950s, at a time when sardine abundance was declining to very low levels.

Remarkably, during cool regimes unfavorable to sardines, in which California Current anchovies and sardines occupy the same coastal habitats, the two species have very similar life tables. However, during regimes favorable to sardines, in which the two species tend to occupy different habitats, the life tables of the two species diverge sharply. Anchovies become short-lived and mature younger, while sardines become relatively long-lived and mature later (Table 12.2).

Kawasaki (1993) reports similar life history plasticity in the Japanese sardine: periods of high sardine abundance are associated with slower growth rate and older age of maturity than are seen during periods of low abundance. A similar shift may have occurred in South Africa. Fairweather *et al.* (2006) document a 53-year history of sardine length at 50% maturity covering two separate regimes favorable to sardines. There is a strong correlation between length at maturity and spawning abundance. However, the length at maturity increased in the mid to late 1980s in advance of

the population growth. This phase relationship allows the possibility that the change in length at maturity was not only associated with abundance, but may have been a more direct response to the new environmental conditions that allowed subsequent population growth.

Temperature optima

Sardine and anchovy fluctuations are strongly correlated with temperature fluctuations (e.g. Chavez *et al.*, 2003), but it is difficult to distinguish direct influences of temperature from effects of correlated variables. Nonetheless, Butler *et al.* (1993) analyzed stage-based sardine and anchovy life tables, and concluded that temperature-related changes in the duration of egg and early larval stages could have a strong influence on population productivity. Associations of spawning activity and temperature have been described for sardines and anchovies in the California Current system (Lluch-Belda *et al.*, 1991), and in the Japanese system (Takasuka and Aoki, 2006; A. Takasuka, personal communication), and are summarized in Table 12.3. Eggs and larvae in the California Current system have a bimodal temperature distribution, corresponding to separate historical spawning populations off California and off Southern Baja California. Temperature associations in Japanese waters show a separation between spawning habitats of sardines and anchovies, and Takasuka and Aoki (2006) have demonstrated that growth rate of anchovy larvae has a peak value at about 22 °C independently of the availability of food. In contrast to Japan, California Current sardines and anchovies tend to spawn at similar temperatures, but show a difference among stocks. The temperature associations of more recent sardine spawning in the far northern California Current were not covered by the data available to Lluch-Belda *et al.*, and that relatively cold region is not represented in Table 12.3.

Behavioral mechanisms
Behavioral imprinting

Cury (1994) hypothesized that small pelagic species may exhibit a form of behavioral imprinting – a "natal homing" tendency for an individual spawning fish to return to habitats or environmental conditions similar to those experienced in its own early life history. If locations of good reproductive habitat are serially correlated in time, this behavioral tendency would amplify species productivity, and also is consistent with the geographic shift of sardine spawning during favorable regimes. While this remains a viable hypothesis, it is partially superseded by the entrainment hypothesis described in the next section.

The entrainment hypothesis

An International Council for Exploration of the Sea (ICES) Working Group has explored an "entrainment hypothesis" that recruits to spawning migrations learn the navigational path from experienced older fish (Petitgas *et al.*, 2006; ICES, 2007). The importance of learning migratory paths from older experienced individuals is well established in bird migration (Able and Able, 1998) and strengthens the plausibility of this hypothesis for fish migration. Coetzee (2007) examined the evidence for entrainment in the South African sardine population, and Lo (2007) examined the evidence in the sardine population in the California Current. The ICES Working Group concluded that both cases supported the hypothesis (ICES, 2007).

With regard to the hypothesized tendency for sardines to occupy offshore/upstream habitats during favorable regimes, the entrainment hypothesis provides a mechanism, and also helps account for the delay between the shift in ocean conditions and subsequent utilization of the newly available habitat by the fish stocks (Lo, 2007). Important considerations include the contrast between optimal habitat locations in unfavorable and favorable regimes, and the extent to which older "knowledgeable" fish remain in the population at the time the regime shift occurs. Because fishing pressure is often geographically localized, and also may target older fish in particular spawning areas, the impact of intense fishing on the transmission of migratory knowledge may contribute to shifts in spawning migrations such as were seen off South Africa in the early 2000s (Coetzee, 2007).

Mixed-species schooling and the "school trap"

Bakun and Cury (1999) cited widespread observations that, during unproductive periods, species experiencing low abundance often join schools of other more abundant fish species of similar size or swimming speed. Unlike Radovich (1979), who hypothesized that the scarce species benefits from the protection offered by the schools of the abundant species, Bakun and Cury (1999) hypothesized that the scarce species could suffer disadvantages from mixed schooling, and this mechanism could provide the kind of dynamic instability that allows prolonged dominance of one species or the other.

"School-mix feedback"

Bakun's (2001) "school-mix feedback" hypothesis is an extension of Bakun and Cury's "school trap" hypothesis that adds elements of behavioral imprinting (see preceding) and predator–prey dynamics. The result is a predicted tendency for multispecies cycling. This behavioral tendency would contribute to regime-like population behavior, but it is difficult to quantify the expected phasing, periodicity and amplitude or to explain the tendency toward synchronization of the resulting population cycles such as between the South American and Japanese sardine stocks. However, this hypothesis could be easily extended to include the influence of environmental conditions that could strongly influence phasing and periodicity.

"Basin model" of habitat selection

MacCall (1990) proposed a "basin model" of habitat selection mechanism to account for the tendency of fish populations to expand and contract with changes in abundance. A related hypothesis also based on the "ideal free distribution" was proposed by Wada and Kashiwai (1991). Because relative density near the center of the distribution tends to vary less than population abundance as a whole, this population behavior renders fish stocks more vulnerable to centrally foraging predators and fishers during stock declines, and less vulnerable during stock increases, potentially increasing contrast between favorable and unfavorable periods.

The basin model originally described the population size-dependent distribution of anchovies off California, but is occasionally applied to sardines by analogy. The basin model predicts that spawning area should vary approximately as the square or cube root of abundance. In the case of the Japanese sardine, Wada and Kashiwai (1991) and Kobayashi and Kuroda (1991) concluded that spawning was subject to density dependent habitat selection. Zenitani and Yamada (2000) found that a power function was indeed the best mathematical model relating abundance to spawning area, and that the exponent of abundance was in the range of 0.32 to 0.46, in accordance with a basin model. However, Barange *et al.* (1999) concluded that unlike anchovies, the distribution of South African sardines did not conform to the basin model, and that sardines and anchovies "may have different strategies of occupying space." Smith (1990) showed that the California Current sardine spawning area has been approximately proportional to spawning biomass (i.e. a power function exponent of 1.0) over the full range of historically observed sardine abundance, which again does not agree with behavior predicted by a basin model. Whereas MacCall's basin model would suggest that Smith's (1990) offshore expansion of the sardine spawning grounds (considered as "effect") was in response to the increase in abundance (considered as "cause"), the initial framework proposed here reverses the interpretation of cause and effect, so that the improvement of offshore spawning habitat (now the "cause") is itself the source of the increased reproductive rate and consequently increased abundance (now the "effect"). Thus the initial framework freely allows Smith's proportional relationship, and does not assume that a basin model necessarily applies to the productive phase of sardine populations. It is possible that a basin model fits some sardine populations, as in Japan, but it may be best suited to describing more coastally constrained populations such as anchovies, and perhaps sardines during their unfavorable periods (MacCall, 1990; Barange et al., 1999).

Multispecies mechanisms
Trophic dynamics
Biological oceanography has a long history of research on trophic control of marine ecosystems, specifically whether productivity is controlled by the primary production rate or by predation from higher trophic levels. Abundant small pelagic fishes may even exert controlling influences on both higher and lower trophic levels ("wasp-waist" control; Rice, 1995). This is a large area of research, and is reviewed in this volume, Chapters 7 and 8.

Bottom-up control of sardines in the eastern boundary current systems is unlikely. In those systems, primary and secondary production tends to be lower during productive sardine regimes than during regimes of low sardine productivity (e.g. Muck, 1989; Roemmich and McGowan, 1995). Unlike the eastern boundary currents, the source of Japan's Kuroshio Current is tropical nutrient-poor water. Sardines benefit from conditions of increased mixing with nutrient rich Oyashio Current water (Kondo, 1980), allowing a possible bottom-up trophic influence on Japanese sardines. In contrast, eastern boundary current production of anchovies is dependent on nutrient enrichment by coastal upwelling, and anchovies may experience a bottom-up trophic influence. Importantly, the trophic dynamics may be substantially different in regimes favoring sardines or anchovies, respectively.

Species competition
Classic competition was the first multispecies mechanism proposed to explain the shift from sardines to anchovies. Murphy (1966) reviewed specific mechanisms for sardine–anchovy competition in the California Current, which included direct competition among larvae for food, and adult feeding on the competitor's eggs and larvae, but neither was supported by evidence. Stander and LeRoux (1968) similarly suggested that competition from anchovies was a factor in the decline of South African sardines. In both cases, the main evidence was circumstantial: anchovy abundance increased after sardines declined, and sardines showed little ability to recover from their low abundance. In California, Murphy and others strongly advocated an "experimental fishery" on anchovies that was designed to reverse the apparent equilibrium and bring back the sardines (McEvoy, 1986). The experiment was not conducted. However, it is interesting to imagine the result of such an experiment: because we now know that California's sardines began increasing after 1976, the experiment would have appeared successful whether or not it had any influence on actual events.

Predator–prey dynamics
On a predator–prey basis similar to the earlier competition hypothesis, Muck (1989) similarly argued for increasing the Peruvian fishery for mackerel (Scomber japonicus) and horse mackerel (Trachurus murphyi), on the premise that these predators were preventing the anchoveta from recovering. This "ecosystem-oriented" policy was not implemented, but in hindsight it too would have appeared to be successful: the anchoveta increased rapidly in abundance after 1988.

Cyclic dominance of pelagic species
Matsuda et al. (1991, 1992) observed a historical sequence of three dominant pelagic species groups off Japan, and proposed a cyclic dominance model (a.k.a. the "rock–paper–scissors" game, where A defeats B, B defeats C, and C defeats A) to explain the temporal pattern. Their species groups were sardines (Sardinops melanosticta), chub mackerels (mostly Scomber japonicus), and a combined group of anchovies (Engraulis japonica), horse mackerels (Trachurus japonica and Decapturus muroadsi) and saury (Cololabis saira). Matsuda et al. (1992) and Matsuda and Katsukawa (2002) developed a simple biologically driven asymmetric competition model, and identified parameter values that approximately could reproduce the observed species replacement cycle. They acknowledged that environmental variability could influence some aspects of the pattern, but considered biological interactions to be the primary force in the system.

Documented cases of cyclic dominance exist for behavioral interactions among individuals, but to the author's knowledge no cases have been documented for this pattern of interactions at the population level. Evolutionary theory indicates that a single strategy should prevail unless each species of the trio has a fitness advantage when rare (Maynard-Smith, 1982; Sinervo and Lively, 1996). It is difficult to envision any of these species having a fitness advantage during their periods of low abundance, and, as reviewed earlier, Bakun and Cury's (1999) "school trap" hypothesis argues that these pelagic species may suffer a disadvantage when scarce.

MacCall (1996) observed a similar sequence of dominant pelagic species off California, where at least four groups of species were present. As in Japan, the cycle follows a characteristic order: sardine (Sardinops spp.)–jack mackerel (*Trachurus* spp.) –anchovy (*Engraulis* spp.) –chub mackerel (*Scomber japonicus*)–sardine. Note that MacCall splits the multispecies anchovy–horse mackerel group of Matsuda *et al.*, but that their combined position in the sequence remains the same. Thus the classically accepted (though perhaps not as well established in fact) pattern of sardine–anchovy alternations may be a subset of a larger multispecies sequential pattern. The timing of species transitions is closely linked to physical regime shifts: sardines and anchovies become abundant during respectively favorable environmental regimes, while jack mackerel and chub mackerel become abundant during the transitions between those environmental regimes. MacCall also noted that the two regime-transition fishes (*Scomber* and *Trachurus*) are piscivorous and are potential predators of the preceding planktivorous species (*Engraulis* and *Sardinops*) in the cycle, providing some circumstantial biological support for the cyclic dominance mechanism of Matsuda *et al.* However, MacCall favored the hypothesis that the cycle of sequential dominance is primarily a response to a consistent pattern of decadal shifts in the physical system.

MacCall (1996) also noted a loose synchrony in the timing of similar species transitions in Japan and California. The initial framework proposed here suggests that trans-Pacific synchrony may result mainly from the strong response of sardines to mutually experienced decadal fluctuations in current flow. The timing of transitions and peak abundances of the other pelagic species may result from a combination of biological interactions (competition and predation) and from other biological and physical properties of the environment that are associated with interdecadal fluctuations of flow patterns and upwelling.

Predator outbreaks

A speculative addition to this list of mechanisms is the possibility that predator outbreaks may hasten the end of sardine regimes in eastern boundary current systems where persistent warm water may allow an influx of sub-tropical predators. During the present decade, the jumbo squid, *Dosidicus gigas*, has become unusually abundant in the California and Humboldt Currents, whereas it was formerly found mainly at the tropical ends of these systems (Field *et al.*, 2007). There has been plausible speculation that this recent outbreak of piscivorous squid contributed to a massive decline in hake (*Merluccius gayi*) abundance off Chile (Contreras, 2005), and it is quite possible that these squid could contribute substantially to offshore sardine mortality off both North and South America. *Dosidicus* was previously abundant off California in the 1930s, at a time when the sardine population was declining rapidly (Croker, 1937).

In a review of early faunal surveys in California, Hubbs (1948) noted that, in 1880, ichthyologists Jordan and Gilbert encountered substantial abundances of subtropical predators in Monterey Bay, central California. The year 1880 was near the end of an extraordinarily warm period, inferred by Hubbs from faunal range extensions including those of several prominent predators. The paleosedimetary sardine record of Baumgartner *et al.* (1992) shows a very sharp decline in sardine scale deposition rate from the 1870s to the 1880s (see Chapter 4). As is frequently the case, these relationships are circumstantial, and further work is needed to distinguish causal relationships from coincidental associations with environmental conditions.

Environmental "loopholes"

Bakun and Broad (2003) proposed that transient conditions of poor ocean productivity, such as El Niño, could impact predator populations so that production of small pelagics is enhanced in subsequent years. This mechanism could account for interannual variability, but it is more difficult to accept Bakun and Broad's argument that it can account for large amplitude interdecadal variability. While the "loophole hypothesis" is consistent with sudden beginnings of regimes, it is less consistent with their sudden endings.

Impacts of fisheries

Most small pelagic species are now subject to fisheries, and exploitation may play an influential role in multiple aspects of their population dynamics. Classical fishing theory describes how stock status is strongly impacted by the rate of removal by fishing relative to the rate of production (recruitment). However, there may also be important fishery impacts on fish distribution, and especially migration behaviors.

Exploitation rate relative to production rate

Intense exploitation following the end of favorable regimes has, in several cases, depleted the reproductive potential of the resource within a very few years. Subsequent increases in abundance when conditions again become favorable follow a pattern of exponential growth. In some systems such as Japan and South America, the exponential growth rate is sufficiently high that the initial abundance may be unimportant. In other systems such as the California Current, a lower growth rate makes initial conditions much more important.

Inter-regime fluctuations in sardine productivity in Japan and California provide a useful comparison. The recruitment per spawning biomass of Japanese sardines is 20-fold higher during a favorable regime (Wada and Jacobson, 1998). This allows an intense fishery to be sustained even during the early period of population growth. However, when the system suddenly becomes unfavorable for Japanese sardines, intense fishing quickly depletes the remaining stock. In contrast, the California sardine experiences only a doubling of recruitment per spawning biomass during favorable regimes (Jacobson and MacCall, 1995). In the absence of fishing, the population growth rate during favorable conditions off California is about 40% per year. At this relatively low growth rate, even low fishing rates and the initial size of the California stock have strong influences on the growth trajectory – following the beginning of favorable conditions in 1976, the stock required 20 years to approach its asymptotic abundance. In contrast to the rapid post-1988 depletion of the Japanese and South American sardine stocks, recent sardine fishing rates in California have been low, which may have contributed to prolonged abundance.

Demographic and behavioral effects of fishing

Fisheries also cause truncation of the age structure and may differentially favor some migratory patterns over others. Older, larger individuals play an important role in many of the hypothesized mechanisms presented in this review, both directly, through their high fecundity, and indirectly through their behavior, especially migratory behaviors in the case of sardines.

Optimal harvesting

It is doubtful that sufficient understanding of pelagic fish demographics and dynamics will soon be gained to identify a generic optimal harvesting policy for these small pelagic fishes. The first anchoveta fishery off South America lasted perhaps 15 years, and the present anchoveta fishery is now about that old. Both recent sardine fisheries in Japan and South America lasted little more than 20 years under heavy exploitation, as did the Japanese and Californian fisheries during the 1930s and 1940s. The present South African sardine fishery is still less than 20 years old, and appears viable. However, the heavily exploited stock has strangely abandoned its traditional spawning grounds, which may be cause for concern.

The California Current's intentionally small sardine fishery is exceptional, in that it is still viable after nearly 30 years (assuming it began in the late 1970s, though little fishing was allowed for much of the first two decades in order to allow maximal population growth). This raises the question of whether this 30-year record has been the lucky result of continuing favorable environment, or whether it has been the earned result of unusually conservative management that seeks to maintain older individuals in the population and to provide sufficient abundance to endure the inevitable shift to a prolonged unfavorable regime. The most recent stock assessment suggests that the abundance of the California Current's sardine population has been declining since 2001, and that recruitment rates have recently fallen to their lowest levels since the 1980s (Hill *et al.*, 2006).

Conclusions

The sardine–anchovy puzzle has still not been solved, despite the passage of half a century during which it has motivated some of the most extensive marine research programs in history. One sign of progress is that we no longer expect that some single cause or factor is likely to be identified as the key to it all. Sardines (*Sardinops* spp.) and anchovies (*Engraulis* spp.), as well as a characteristic suite of other pelagic fishes and marine fauna, necessarily must find some critical aspects sufficiently similar in the four major systems that they are able to reside in those systems. In a sense, science has been given a "hint" regarding the solution to the puzzle. Yet our inability to use that particular hint in solving the puzzle indicates that we still don't know very much about what aspects of an ecosystem are critical to the creatures that live there. The Kuroshio Current system is profoundly different from the three eastern boundary currents considered in this chapter, yet sardines live there too. South Africa's Agulhas Current seems to have no analog in any of the other systems, and yet it appears to be important to the sardines' life strategy in the Western Cape. The Namibian segment of the Benguela sardines superficially seems to occupy a linear coastal habitat more like the Humboldt or California Current, and yet its sardines have not responded to the recent favorable Benguela Current conditions being experienced by its southern neighbors in the Western Cape.

Any attempt at a general theory of sardine–anchovy fluctuations will encounter many objections based on the way things work in one or another of these ecosystems. Rather than simply responding to an objection by abandoning that particular explanatory framework, it would be appropriate to first ask what aspects of those systems are the causes of the exceptions. While not guaranteeing a solution to the anchovy–sardine puzzle, this incremental comparative approach (which has long been championed by Bakun, Parrish, and others) has an arguably greater chance of making scientific progress than trying to develop independent theories for each individual system. We still have not experienced more than two well-documented episodes of sardine or anchovy abundance in any of these systems, which is an absurdly small sample size by any statistical standard. It will most likely require at least a third episode, presumably some time in the next few decades, for science to make further progress. While such issues as global climate change may introduce new aspects to the sardine–anchovy puzzle, it seems safe to say that studying the response of sardines and anchovies to perhaps new and different oceanic circumstances should provide valuable insights. It is the author's opinion that these species have sufficient flexibility to survive nearly any conceivable future climate scenario, and they will long survive our attempts to understand them as well as to harvest them.

REFERENCES

Able, K. P. and Able, M. A. (1998). The roles of innate information, learning rules and plasticity in migratory bird orientation. *J. Navigation* **51**: 1–9.

Bakun, A. (1993). The California Current, Benguela Current, and southwestern Atlantic shelf ecosystems: a comparative approach to identifying factors regulating biomass yields. In *Large Marine Ecosystems – Stress, Mitigation, and Sustainability*, Sherman, K., Alexander, L. M. and Gold. B., eds. Washington, DC: American Association for the Advancement of Science, pp. 199–224.

Bakun, A. (1995). A dynamical scenario for simultaneous "regime-scale" marine population shifts in widely separated LMEs of the Pacific. In *The Large Marine Ecosystems of the Pacific Rim – A Report of a Symposium held in Qingdao, Peoples Repiblic of China, 8–11 October, 1994*, Tang, Q. and Sherman, K., eds. Gland, Switzerland: IUCN, World Conservation Union, pp. 47–72.

Bakun, A. (1996). *Patterns in the Ocean: Ocean Processes and Marine Population Dynamics*. San Diego, CA, USA: University of California Sea Grant, in cooperation with Centro de Investigaciones Biologicas de Noroeste, La Paz, Baja California Sur, Mexico, 323 pp.

Bakun, A. (2001). "School-mix feedback": a different way to think about low frequency variability in large mobile fish populations. *Prog. Oceanogr.* **49**: 485–511.

Bakun, A. and Broad, K. (2003). Environmental "loopholes" and fish population dynamics: comparative pattern recognition with focus on El Niño effects in the Pacific. *Fish. Oceanogr.* **12**: 458–473.

Bakun, A. and Cury, P. (1999). The "school trap": a mechanism promoting large-amplitude out-of-phase population oscillations of small pelagic fish species. *Ecol. Lett.* **2**: 349–351.

Barange, M., Hampton, I., and Roel, B. A. (1999). Trends in the abundance and distribution of anchovy and sardine on the South African continental shelf in the 1990s, deduced from acoustic surveys. *S. Afr. J. Mar. Sci.* **21**: 367–391.

Baumgartner, T. R., Soutar, A., and Ferreira-Bartrina, V. (1992). Reconstruction of the history of Pacific sardine and northern anchovy populations over the past two millennia from sediments of the Santa Barbara Basin, California. *Calif. Coop. Oceanic Fish. Invest. Rep.* **33**: 24–40.

Blaxter, J. H. S. and Hunter, J. R. (1982). The biology of clupeoid fishes. *Adv. Mar. Biol.* **20**: 1–223.

Butler, J. L., Smith, P. E. and Lo, N. C. (1993). The effect of natural variability of life-history parameters on anchovy and sardine population growth. *Calif. Coop. Oceanic Fish. Invest. Rep.* **34**: 104–111.

Butterworth, D. S. (1983). Assessment and management of pelagic stocks in the southern Benguela region. In *Proceedings of the Expert Consultation to Examine Changes in Abundance and Species Composition of Neritic Fish Resources, San José, Costa Rica, 18–29 April 1983*, Sharp, G. D., and Csirke, J., eds. *FAO Fish. Rep.* **291** (2): 329–405.

Chavez, F. P., Ryan, J., Lluch-Cota, S. E., and Ñiquen, M. (2003). From anchovies to sardines and back: Multidecadal change in the Pacific Ocean. *Science* **299**: 217–221.

Chelton, D. B., Bernal, P. A., and McGowan, J. A. (1982). Large-scale interannual physical and biological interaction in the California Current. *J. Mar. Res.* **40**: 1095–1124.

Clark, F. N. and Marr, J. C. (1955). Population dynamics of the Pacific sardine. *Calif. Coop. Oceanic Fish. Invest. Rep.* **4**: 11–48.

Coetzee, J. (2007). The South African sardine population – evidence of entrainment? In *Report of the Workshop on Testing the Entrainment Hypothesis (WKTEST), 4–7 June 2007, Nantes, France. ICES CM:2007/LRC:10*, pp. 63–72.

Contreras, I. P. (2005). Chilean hake (*Merluccius gayi gayi*) stock assessment in 2005. Insto. de Fomento Pesquero, Chile, manuscript reviewed at Chilean Hake Stock Assessment Workshop, Valparaiso, November 2005.

Crawford, R. J. M., Shelton, P. A., and Hutchings, L. (1983). Aspects of variability of some neritic stocks in the southern Benguela system. In *Proceedings of the Expert Consultation to Examine Changes in Abundance and Species Composition of Neritic Fish Resources. San José, Costa Rica, 18–29 April 1983*, G. D. Sharp and J. Csirke, eds. *FAO Fish. Rep.* **291**: 407–448.

Croker, R. S. (1937). Further notes on the jumbo squid, *Dosidicus gigas*. *Calif. Fish Game* **23**: 246–247.

Cury, P. (1994). Obstinate nature: an ecology of individuals. Thoughts on reproductive behavior and biodiversity. *Can. J. Fish. Aquat. Sci.* **51**: 1664–1673.

Fairweather, T. P., van der Lingen, C. D., Booth, A. J., *et al.* (2006). Indicators of sustainable fishing for South African sardine *Sardinops sagax* and anchovy *Engraulis encrasicolus*. *Afr. J. Mar. Sci.* **28**: 661–680.

Field, J. C., Baltz, K., and Phillips, A. (2007). Range expansion and trophic interactions of the jumbo squid, *Dosidicus gigas*, in the California Current. *Calif. Coop. Oceanic Fish. Invest. Rep.* **48**: 131–146.

Freon, P., Cury, P., Shannon, L., and Roy, C. (2005). Sustainable exploitation of small pelagic fish stocks challenged by environmental and ecosystem changes. *Bull. Mar. Sci.* **76**: 385–462.

Hargreaves, N. B., Ware, D. M., and McFarlane, G. A. (1994). Return of Pacific sardine (*Sardinops sagax*) to the British Columbia coast in 1992. *Can. J. Fish. Aquat. Sci.* **51**: 460–463.

Hayasi, S. (1983). Some explanation for changes in abundances of major neritic-pelagic stocks in the northwestern Pacific Ocean. In *Proceedings of the Expert Consultation to Examine Changes in Abundance and Species Composition of Neritic Fish Resources. San José, Costa Rica, 18–29 April 1983*, Sharp, G. D. and Csirke, J., eds. *FAO Fish. Rep.* **291**: 37–55.

Hill, K. T., Lo, N. C. H., Macewicz, B. J., and Felix-Uraga, R. (2006). Assessment of the Pacific sardine (*Sardinops sagax caerulea*) population for US management in 2007. *NOAA Tech. Memo.* NOAA-TM-NMFS-SWFSC-396, 79pp.

Hubbs, C. L. (1948). Changes in the fish fauna of western North America correllated with changes in ocean temperature. *J. Mar. Res.* **7**: 459–482.

Hunter, J. R. and Goldberg, S. R. (1980). Spawning incidence and batch fecundity in northern anchovy, *Engraulis mordax*. *US Fish. Bull.* **77**: 641–652.

Hunter, J.R. and Leong, R. (1981). The spawning energetics of female northern anchovy, *Engraulis mordax*. *US Fish. Bull.* **79**: 215–230.

ICES (2007). *Report of the Workshop on Testing the Entrainment Hypothesis (WKTEST)*, 4–7 June 2007, Nantes, France. *ICES CM:2007/***LRC:10**, 111 pp.

Jacobson, L.D. and MacCall, A.D. (1995). Stock-recruitment models for Pacific sardine (*Sardinops sagax*). *Can. J. Fish. Aquat. Sci.* **52**: 566–577.

Kawasaki, T. (1983). Why do some pelagic fishes have wide fluctuations in their numbers? – Biological basis of fluctuation from the viewpoint of evolutionary ecology. In *Proceedings of the Expert Consultation to Examine Changes in Abundance and Species Composition of Neritic Fish Resources. San José, Costa Rica, 18–29 April 1983*, Sharp, G.D. and Csirke, J., eds. *FAO Fish. Rep.* **291**: 1065–1080.

Kawasaki, T. (1993). Recovery and collapse of the Far Eastern sardine. *Fish. Oceanogr.* **2**: 244–253.

Kobayashi, M. and Kuroda, K. (1991). Estimation of main spawning grounds of the Japanese sardine from a viewpoint of transport condition of its eggs and larvae. In *Long-term variability of pelagic fish populations and their environment*, Kawasaki, K., Tanaka, S., Toba, Y. and Taniguchi, A., eds. Pergamon Press, pp. 109–116.

Komatsu, T., Sugimoto, T., Ishida, K., *et al.* (2002). Importance of the Shatsky Rise Area in the Kuroshio Extension as an offshore nursery ground for Japanese anchovy (*Engraulis japonicus*) and sardine (*Sardinops melanostictus*). *Fish. Oceanogr.* **11**: 354–360.

Kondo, K. (1980). The recovery of the Japanese sardine – The biological basis of stock size fluctuations. *Rapp. P.-v. Reun. Cons. Int. Explor. Mer* **177**: 332–354.

Lasker, R. and A. D. MacCall. (1983). New ideas on the fluctuations of the clupeoid stocks off California. In *Proceedings of the Joint Oceanographic assembly 1982 – General Symposia*. Ottawa, Canada: Canadian National Committee/Scientific Committee on Oceanic Reasearch, pp. 110–120.

Le Clus, F. (1989). Size-specific seasonal trends in spawning of pilchard *Sardinops ocellatus* in the northern Benguela system, 1973/74. *S. Afr. J. Mar. Sci.* **8**: 21–31.

Lluch-Belda, D., Crawford, R.J.M., Kawasaki, T., *et al.* (1989). World-wide fluctuations of sardine and anchovy stocks: the regime problem. *S. Afr. J. Mar. Sci.* **8**: 195–205.

Lluch-Belda, D., Lluch-Cota, D.B., Hernandez-Vazquez, S., *et al.* (1991). Sardine and anchovy spawning as related to temperature and upwelling in the California Current system. *Calif. Coop. Oceanic Fish. Invest. Rep.* **32**: 105–111.

Lo, N. (2007). Pacific sardine (*Sardinops sagax*) off west coast of American continent. In *Report of the Workshop on Testing the Entrainment Hypothesis (WKTEST), 4–7 June 2007, Nantes, France. ICES CM:2007/***LRC:10**, pp. 48–62.

Logerwell, E. and Smith, P. (2001). Mesoscale eddies and survival of late stage Pacific sardine (*Sardinops sagax*) larvae. *Fish. Oceanogr.* **10**: 13–25.

Logerwell, E., Lavaniegos, B., and Smith, P. (2001). Spatially-explicit bioenergetics of Pacific sardine in the Southern California Bight: are mesoscale eddies areas of exceptional prerecruit production? *Prog. Oceanogr.* **49**: 391–406.

MacCall, A.D. (1990). *Dynamic Geography of Marine Fish Populations*. Seattle, WA: University of Washington Press, 153 pp.

MacCall, A.D. (1996). Patterns of low-frequency variability in fish populations of the California Current. *Calif. Coop. Oceanic Fish. Invest. Rep.* **37**: 100–110.

MacCall, A.D. (2002). An hypothesis explaining biological regimes in sardine-producing Pacific Boundary Current Systems (South America, North America and Japan): implications of alternating modes of slow, meandering flow and fast linear flow in the offshore region. In *Climate and Fisheries. Interacting Paradigms, Scales, and Policy Approaches*, Bakun, A. and Broad, K., eds. Palisades, New York: International Research Institute for Climate Prediction, Columbia University, pp. 39–42.

Macewicz, B.J. and Abramenkoff, D.N. (1993). Collection of jack mackerel, *Trachurus symmetricus*, off southern California during 1991 cooperative U.S.–U.S.S.R. cruise. *NMFS SWFSC Admin. Rep.*, **LJ-93–07**.

Mais, K.F. (1974). Pelagic fish surveys in the California Current. *Calif. Fish Game Fish Bull.* **162**: 1–79.

Mais, K.F. (1981). Age-composition changes in the anchovy, *Engraulis mordax*, central population. *Calif. Coop. Oceanic Fish. Invest. Rep.* **22**: 82–87.

Matsuda, H. and T. Katsukawa. (2002). Fisheries management based on ecosystem dynamics and feedback control. *Fish. Oceanogr.* **11**: 366–370.

Matsuda, H., Wada, T., Takeuchi, Y., and Matsumiya, Y. (1991). Alternative models for species replacement of pelagic fishes. *Res. Popul. Ecol.* **33**: 41–56.

Matsuda, H., Wada, T., Takeuchi, Y., and Matsumiya, Y. (1992). Model analysis of the effect of environmental fluctuation on the species replacement pattern of pelagic fishes under interspecific competition. *Res. Popul. Ecol.* **34**: 309–319.

Maynard-Smith, J. (1982). *Evolution and the Theory of Games*. Cambridge, UK: Cambridge University Press, 226 pp.

McEvoy, A.F. (1986). *The Fisherman's Problem: Ecology and Law in California Fisheries 1850–1980*. New York: Cambridge University Press, 368 pp.

McFarlane, G.A., Smith, p. E., Baumgartner, T.R., and Hunter, J.R. (2002). Climate variability and Pacific sardine populations and fisheries. *Am. Fish. Soc. Symp.* **32**: 195–214.

Mosher, K. and Eckles, H.H. (1954). Age determination of Pacific sardines from otoliths. *US Fish. Wild. Serv., Res. Rep.* **37**: 1–40.

Muck, P. (1989). Major trends in the pelagic ecosystem off Peru and their implications for management. In *The Peruvian Upwelling System: Dynamics and Interactions*, Pauly, D., Muck, P., Mendo, J., and Tsukayama, I., eds. *ICLARM Conference Proceedings* **18**: 386–403.

Murphy, G.I. (1966). Population biology of the Pacific sardine (*Sardinops caerulea*). *Proc. Calif. Acad. Sci.* **34** (1): 1–84.

Murphy, G.I. (1967). Vital statistics of the California sardine and the population consequences. *Ecology* **48**: 731–735.

Nakai, Z. (1949). Iwashi wa naze torenai? (Why cannot the sardine be caught?). *Suisan Kaikan* **2**: 92–101.

Nakai, Z. (1962). Studies relevant to mechanisms underlying the fluctuation in the catch of the Japanese sardine, *Sardinops melanosticta* (Temminck and Schlegel). *Japan J. Ichth.* **9**: (1–6), 1–115.

Newman, G. G. (1970a). Stock assessment of the pilchard *Sardinops ocellata* at Walvis Bay, South West Africa. *Investl. Rep. Div. Sea Fish.*, **85**, 13 pp.

Newman, G. G. (1970b). Migration of the pilchard *Sardinops ocellata* in southern Africa. *Investl. Rep. Div. Sea Fish.* **86**: 6 pp.

Parrish, R. H., Bakun, A., Husby, D. M., and Nelson, C. S. (1983). Comparative climatology of selected environmental processes in relation to eastern boundary current pelagic fish reproduction. In *Proceedings of the Expert Consultation to Examine Changes in Abundance and Species Composition of Neritic Fish Resources. San José, Costa Rica, 18–29 April 1983*, G. D. Sharp and J. Csirke. eds. *FAO Fish. Rep.* **291**: 731–777.

Parrish, R. H., Mallicoate, D. L., and Mais, K. F. (1985). Regional variations in the growth and age composition of northern anchovy, *Engraulis mordax*. *US Fish. Bull.* **83**: 483–496.

Parrish, R. H., Mallicoate, D. L., and Klingbeil, R. A. (1986). Age dependent fecundity, number of spawnings per year, sex ratio, and maturation stages in northern anchovy, *Engraulis mordax*. *US Fish. Bull.* **84**: 503–517.

Parrish, R. H., Schwing, F. B., and Mendelssohn, R. (2000). Mid-latitude wind stress: the energy source for climatic shifts in the North Pacific Ocean. *Fish. Oceanogr.* **9**: 224–238.

Petitgas, P., Reid, D., Planque, B., *et al.* (2006). The entrainment hypothesis: an explanation for the persistence and innovation in spawning migrations and life cycle spatial patterns. *ICES CM:2006/**B:07**, 9 pp.

Radovich, J. (1979). Managing pelagic schooling prey species. In *Predator–Prey Systems in Fisheries Management*, Clepper, H., ed. Washington, DC: Sport Fishing Institute.

Rice, J. (1995). Food web theory, marine food webs, and what climate change may do to northern marine fish populations. In *Climate Change and Northern Fish Populations*, Beamish, R. J., ed. *Can. Spec. Pub. Fish. Aquat. Sci.* **121**: 561–568.

Roemmich, D. and McGowan, J. (1995). Climatic warming and the decline of zooplankton in the California Current. *Science* **267**: 1324–1326.

Scofield, E. C. (1934). Early life history of the California sardine (*Sardina caerulea*), with special reference to distribution of eggs and larvae. *Calif. Dep. Fish and Game Fish Bull.* **41**: 1–48.

Schwartzlose, R., Alheit, J., Bakun, A., *et al.* (1999). Worldwide large-scale fluctuations of sardine and anchovy populations. *S. Afr. J. Mar. Sci.* **21**: 289–347.

Serra, J. R. (1983). Changes in the abundance of pelagic resources along the Chilean coast. pp. 255–284 In *Proceedings of the Expert Consultation to Examine Changes in Abundance and Species Composition of Neritic Fish Resources. San José, Costa Rica, 18–29 April 1983*, ed. G. D. Sharp and J. Csirke. *FAO Fish. Rep.* **291** (2): 1–553.

Sharp, G. D. (1980). Colonization: Modes of opportunism in the ocean. In *Workshop on the Effects of Environmental Variation on the Survival of Larval Pelagic Fishes*. Paris: IOC/UNESCO. *IOC Workshop Rep.* **28**: 143–166.

Sinervo, B. and Lively, C. M. (1996). The rock-paper-scissors game and the evolution of alternative male strategies. *Nature* **380**: 240–243.

Smith, P. E. (1990). Monitoring interannual changes in spawning area of Pacific sardine (*Sardinops sagax*). *Calif. Coop. Oceanic Fish. Invest. Rep.* **31**: 145–151.

Soutar, A. and Isaacs, J. D. (1974). Abundance of pelagic fish during the 19th and 20th centuries as recorded in anaerobic sediment off the Californias. *US Fish. Bull.* **72**: 257–273.

Stander, G. H. and LeRoux, P. J. (1968). Notes on fluctuations of the commercial catch of the South African pilchard (*Sardinops ocellata*) 1950–1965. *Investl. Rep. Div. Sea Fish* **65**: 14 pp.

Takasuka, A. and Aoki, I. (2006). Environmental determinants of growth rates for larval Japanese anchovy *Engraulis japonicus* in different waters. *Fish. Oceanogr.* **15**: 139–149.

van der Lingen, C. D., Hutchings, L., Merkle, D., *et al.* (2001). Comparative spawning habitats of anchovy (*Engraulis capensis*) and sardine (*Sardinops sagax*) in the southern Benguela upwelling system. In *Spatial Processes and Management of Marine Populations*, Kruse, G. H. Bez, N., Booth, T., *et al.* eds. Fairbanks: University of Alaska Sea Grant, AK-SG-01–02, pp. 185–209.

van der Lingen, C. D., Coetzee, J. C., Demarcq, H., *et al.* (2005). An eastward shift in the distribution of southern Benguela sardine. *GLOBEC Int. Newsl.* **11** (2): 17–22.

Wada, T. and Jacobson, L. D. (1998). Regimes and stock-recruitment relationships in Japanese sardine (*Sardinops melanostictus*), 1951–1995. *Can. J. Fish. Aquat. Sci.* **55**: 2455–2463.

Wada, T. and Kashiwai, M. (1991). Changes in growth and feeding ground of Japanese sardine with fluctuation in stock abundance. In *Long-Term Variability of Pelagic Fish Populations and their Environment*, Kawasaki, K., Tanaka, S., Toba, Y., and Taniguchi, A., eds. Pergamon Press, pp. 181–190.

Zenitani, H. and Yamada, S. (2000). The relation between spawning area and biomass of Japanese pilchard, *Sardinops melanistictus*, along the Pacific coast of Japan. *US Fish. Bull.* **98**: 842–848.

Zuzunaga, J. (1985). Cambios del equilibrio poblacional entre la anchoveta (*Engraulis ringens*) y la sardina (*Sardinops sagax*), en el sistema de afloramiento frente al Perú. In "*El Niño," su impacto en la fauna marina*, Arntz, W., Landa, A., and Tarazona, J., eds. *Bol. Insto. del Mar del Perú-Callao*, Vol. Extraordinario: 107–111.

13 Research challenges in the twenty-first century

Andrew Bakun

CONTENTS

Summary

A selection of scientific issues is presented as being likely keys to (1) effectively managing the future of small pelagic fisheries under the combined stresses of climate change and ever-increasing demand for fishery products, and (2) understanding the role that small pelagic fishes may play in maintaining the health, integrity and resilience of the large marine ecosystems in which they function. The concept of a wasp-waist ecosystem is presented, according to which a highly variable small pelagic fish component, largely responding to its own internal dynamics, may significantly drive the operation of its entire ecosystem. On the basis of this conceptual framework, a number of vital unresolved issues are briefly elaborated. Several types of nonlinear feedback mechanisms, capable of leading to abrupt instabilities and durable regime shifts, are presented together with the concept of breakout thresholds that may trigger a shift from a favorable phase of feedback operation that supports resilience to an adverse phase leading to abrupt system changes. Proposed mechanisms underlying distributional dynamics are appraised; a suggested connotation is that changes in distributional aspects of key populations may be timely harbingers of unwelcome regime shifts. Several alternative interpretations of apparent density dependent growth are offered, each presenting rather different implications with respect to sustainability. The mechanisms behind apparent niche replacements and species alternations are pondered. It is suggested that SPACC-type research be extended to similar small pelagic species and issues in other more tropical and/or more oceanic ecosystems. A need for technological advances is stressed, as is the need to allow our minds to move beyond conventional dogmas and terrestrial-based "common-sense" notions, so as to be able to conceive and address mechanisms that may be specifically marine in nature and operating on scales beyond those accessible to direct human marine-based sensory experience.

Introduction

The prospect of climate change proceeding at an unprecedented rate, combined with continuing pressure for ever heavier fishery exploitation, is causing great uncertainty regarding the future of marine ecosystems. Biodiversity developed over eons of evolution is even now almost certainly being irrevocably lost. Productive fish communities have already been displaced on scales as large as entire regional large marine ecosystems by massive incursions of jellyfish and similar taxa (Brodeur *et al.*, 1999; Boyer and Hampton, 2001; Mills, 2001; Lynam *et al.*, 2006). Legendary fish stocks such as northern cod, that had been sources of enormous economic wealth for centuries and had once been thought to be inexhaustible, have nearly disappeared from traditional grounds, although other exploited fish stocks continue to appear sustainable (e.g. Sibert *et al.*, 2006).

Many marine scientists entered the field due in substantial degree to fascination with the beautifully intricate ecosystems and communities of the ocean world. Ensuring the future viability of these ecosystems and of the economic, cultural and esthetic values that we have come to expect from them must be a major goal. But identifying areas of scientific work that would be truly telling, and at the same time feasible, is not simple (it is never easy to describe very specifically what one may not yet know). However, one that well may qualify is developing a real understanding

Climate Change and Small Pelagic Fish, eds. Dave Checkley, Jürgen Alheit, Yoshioki Oozeki, and Claude Roy. Published by Cambridge University Press. © Cambridge University Press 2009.

of the role that small pelagic fishes play in maintaining the basic health and integrity of the "complex adaptive system" (Levin, 1998, 1999) that underlies the dynamics of a large marine ecosystem (LME).

Wasp-waist ecosystems

The most typical trophic configuration of a marine ecosystem features many species at the lower (e.g. planktonic) trophic levels, many at the top (e.g. predatory fishes, large coelenterates, seabirds, marine mammals), but very few (often only one) dominant small pelagic species at a mid-level in the trophic hierarchy (Rice, 1995). Ecosystems of this form have been referred to as wasp-waist ecosystems (Bakun, 1996, 2005a, 2006; Cury et al., 2000, 2004), the name wasp-waist referring to the "choke point" occurring at this "small pelagic planktivore" trophic level that may control the volume and efficiency of the entire trophic flow from the lower trophic levels to the upper trophic levels of the ecosystem. Thus variability in the trophic dynamics of wasp-waist ecosystems tends to be largely dominated by variations in its mid-trophic-level small pelagic populations (Rice, 1995). These wasp-waist species typically have complex life cycles containing notable "weak links" (Bakun, 1996) causing their populations to vary radically in size in a manner that can be more or less independent of other community components. In addition to direct physical effects (e.g. transport anomalies and effects of turbulence on feeding success), any "top down" or "bottom up" trophic effects acting to affect survival rates at the larval or other "weak link" stages of wasp-waist species may thus be explosively amplified. These amplified variations may then cascade upward and downward in the trophic system. In such cases, the major control in these ecosystems may be neither "bottom-up" nor "top-down" but rather "both upward and downward from the middle."

Moreover, the wasp-waist level is generally the lowest trophic level that is mobile in the sense that it has the capability to expand, contract, or relocate its area of operation according to its own internal dynamic and behavioral responses, which may or may not be keyed to environmental changes. Furthermore, these small pelagic fishes tend to have shorter generation cycles than the larger, longer-lived fishes that prey upon them. This gives them a distinctive advantage in that, to the extent that the rapidity of the adaptive processes leading to the changes in geographic distribution may be generation cycle dependent, they may take advantage of their shorter life cycles to shift their population away from their major predators (Bakun, 2001, 2005b), while their own prey, being planktonic, lack the built-in mobility to react in kind.

Feedback instabilities

Importantly, wasp-waist species often consume the eggs and larvae of their locally spawning predators that are their major sources of natural mortality. This is a very different situation from that usually encountered in terrestrial ecosystems. It represents a potential for extreme instability within marine ecosystems for which terrestrial experience offers little preparation. Thus the wasp-waist populations of small pelagic fish represent not only a temporally variable, spatially shifting "choke point" in the trophic web, but may be the source of radical gyrations not only in their own abundance, but also in the abundance of other major component populations trophically linked to them. This primary feedback loop, produced by consumption by prey fishes of early stages of their predators, has been called the "prey-to-predator", or "P2P" feedback loop by Bakun and Weeks (2006) and the "cultivation effect" by Walters and Kitchell (2001).

Another potential non-linear feedback is the predator pit loop (Bakun, 2006). Shorter-lived prey populations generally vary more rapidly than do the populations of their longer-lived predators. Thus, as soon as a prey population exceeds an abundance level sufficient to entirely satiate its predators, any further increases act to steadily lower its rate of predation mortality per unit biomass. Thus, once this point is passed, other factors being equal, the population would grow exponentially. Conversely, the mortality rate per unit biomass increases steadily as abundance falls, causing the prey population to collapse in an accelerating manner ("fall into the pit"), and to continue to do so until it reaches its "refuge level" where numbers of individuals are so low that predators cease to hunt them, or seemingly even to notice them.

A third potential non-linear feedback has been called "school-mix feedback" by Bakun (2001, 2005b). Its mechanism relies on the fact that behavior of a fish school is influenced by the behavioral propensities (whether inherent, imprinted, or learned) of its members in a manner that increases with the relative proportion within the school of members possessing those propensities. Members of a school containing relatively high proportions of individuals possessing propensities favorable for survival and/or reproductive success will tend on average to survive and/or reproduce more successfully than those in other schools in which those favorable propensities are less prevalent. This will act on the inter-generational time scale to amplify tendencies in the population that are currently successful. When these tendencies are geographic in nature, the process could potentially bring about adaptive changes in characteristics such as reproductive grounds and areas of suitable habitat occupied. It could even potentially allow fish populations to, on inter-generational time scales, withdraw away from heavy fishing pressure, even if no individual would ever have been aware of danger posed by the fishery until already caught (for diagrammatic cartoons of how these sequences could hypothetically proceed, see Bakun, 2005b).

Such unstable feedbacks could provide an explanation for the plentiful documented experiences of rather sudden massive breakouts in abundance of wasp-waist populations, and their even more abrupt, often durable, collapses. For example, the ctenophore, *Mnemiopsis leidyi*, is an exotic "jelly predator" species that is believed to have been introduced into the Black Sea ecosystem in discharges of ballast water from ships coming from the American side of the ocean. It was first noted in the Black Sea in 1982 (Pereladov, 1988), and it is unclear how much earlier it may actually have been introduced. But it was only following an apparent triggering event, namely the fishery-associated collapse of the anchovy population in 1988–89, that *Mnemiopsi* suddenly exploded to overwhelm the Black Sea ecosystems with a biomass that has been estimated to have reached as high as a billion tons. Another example is the collapse of the population of *Sardinella aurita* (a more tropical sardine analog) off Ghana following particularly heavy exploitation in 1972, which appears to have unleashed the famous triggerfish explosion that spread along the West African coast in the 1970s (Caverivière, 1991). A third example is the collapse in 1989 of the heavily exploited sardine resource (genus: *Sardinops*) in the Gulf of California, whereupon anchovies (*Engraulis mordax*) suddenly appeared in abundance, a fish that in the knowledge of the current inhabitants had never before existed there. Actually, the very first time an anchovy was ever reliably recorded in the Gulf seems to have been in 1986 (Hammann and Cisneros-Mata, 1989), just as the earlier increasing trend in sardine population abundance had evidently reversed abruptly and been followed by a decreasing trend.

A better understanding of how exceeding a threshold may trigger a shift in the direction of one or more crucial dynamic feedbacks from a favorable to an adverse phase of operation would clearly be a boon to management of these tricky systems. It might enable us to avoid potentially devastating regime shifts, or even to design innovative actions for reversing unwelcome regime shifts that may already have taken place. Consider the precipitous declines of cod stocks off the US East Coast (Mayo *et al.*, 1998), in the Baltic Sea (Köster *et al.*, 2003), and in other areas of the North Atlantic in the late 1980s and the 1990s. With some slight lag, local stocks of clupeid prey favored by cod (herring off the US East Coast, sprat in the Baltic) exploded, such that in both systems there are currently more than 50 times the biomasses of herring, per unit cod biomass, than were evidently sufficient to support the nutritional needs of the respective cod stocks prior to their collapses. Both herring and sprat prey heavily on cod eggs (Köster and Möllmann,

2000). Both cod stocks have not recovered. While there is sentiment in both regions favoring environmental explanations rather than fishery-related reversals in predatory advantage (i.e. a "P2P feedback" loop), it is reasonable to ponder whether actions to trim the clupeids down to a level nearer to the earlier prey–predator ratios might help to raise the productivity of the cod stocks nearer to earlier levels.

Improved understanding of the action and effect of such feedbacks operating within the ecosystem could not but help management to achieve a preferable outcome. But a sufficiently improved level of understanding is vital. It appears that multiple feedbacks can interact in quite deleterious ways (Bakun and Weeks, 2006) and simplistic approaches based on a sketchy understanding of a plausible feedback interaction might lead to highly unintended results that may be nearly impossible to reverse, such as the failed attempt to enhance the sardine resource in the Northern Benguela LME by deliberate overfishing of anchovy, its assumed competitor (Butterworth, 1983; Bakun and Weeks, 2006).

Distributional dynamics

Changes in small pelagic fish abundance have been generally coupled to changes in their distribution (Bakun, 1996, 2005a; Schwartzlose *et al.*, 1999). It is reasonable to think that distributional transformations at the wasp-waist may, in fact, be key elements of the dynamical sequences that may lead to regime shifts or to durable ecosystem disruption. When a change of distribution (e.g. an apparent shift in reproductive habitat selection) is identified, it would be a great asset to have capability to (a) interpret the reasons, (b) make an informed decision as to whether the change may presage an unwelcome result, and if so (c) devise management actions to interrupt and reverse the shift.

We have little understanding of the factors determining basic zoogeography of small, pelagic fish. Obviously, zoogeography must represent a degree of historical accident in that any species must have arisen in some geographic location and then have radiated from there to colonize any other locations where it occurs (e.g. Grant and Bowen, 1998; Lecomte *et al.*, 2004). For example, it has been assumed (e.g. Parrish *et al.*, 1981) that the reason that sardines of the genus *Sardinops* do not occur in the North Atlantic, even though they are prodigious migrators and currently occur in all other major temperate upwelling regions as well as south Australia and Japan, is that none has never managed to take advantage of the acceptable temperature conditions in the thermocline to travel beneath the prohibitively warm surface waters of the Atlantic equatorial zone in order to reach the temperate North Atlantic. But why should we assume that to have necessarily been the case? Consider the sudden appearance of anchovies in the Gulf of California coinciding with the collapse of the sardine resource.

Although there are no historical reports of anchovy in the Gulf, analysis of fish scale deposits in anaerobic sea floor sediments (Holmgren-Urba and Baumgartner, 1993) show that they had in fact dominated the fish biomass in the Gulf for most of the nineteenth century. Therefore, with respect to the absence of *Sardinops* from the North Atlantic, a more important question than "why did they never get there?" might be "what are the factors that may have prevented their establishment when they may have arrived, perhaps repeatedly, over geologic time?"

Small pelagic fish stocks clearly do shift geographically, sometimes over great distances. In the 1920s, for example, northern Morocco was the reported southern limit of the distribution of sardines (*Sardina*) in the North Atlantic. However, by the 1950s, sardines had become abundant as far south as Mauritania (Belvese and Erzini, 1983), and by the mid 1970s they were fairly common even in Senegal (Conand, 1975; Fréon, 1988). Later, the range again contracted northwards for a time, sardines disappearing nearly entirely from Mauritanian waters in 1982–84. Subsequently, the range again expanded southward, with sardines being caught in Senegal during the 1990s (Binet *et al.*, 1998).

Fish distributional limits often appear to follow movements of isotherms (Perry *et al.*, 2005); it is not unexpected that extents of zones of acceptable temperatures would tend to limit distributions for fish, just as they may for zooplankton (Beaugrand *et al.*, 2002). And indeed, certain aspects of the multi-decadal movements of *Sardina* along the coast of northwestern Africa have corresponded to low frequency temperature trends, notably a trend of decreasing SST in the region from the 1950s to the mid 1970s coinciding with an increasing trend in upwelling-favorable winds. However, the latest extension southwards appears to have occurred in the opposite situation of increasing temperature trend and decreasing upwelling-favorable wind trend (Binet *et al.*, 1998; Demarque, 1998; Kifani, 1998).

These large multi-decadal latitudinal movements of southern limits of *Sardina* correspond to similar movements of northern limits of its warmer water analog, *Sardinella*, i.e. the southward movements of *Sardina* appearing to "push" *Sardinella* ahead of it, or alternatively, it might have been southward movements of *Sardinella* that vacated habitat for *Sardina* to occupy. Again, the mechanisms by which one group manages to wholly or partially exclude the other need to be much better understood. Of course, a component of the movements could be attributed merely to inter-decadal sea temperature trends, with each group tending to operate in zones of characterized by temperatures nearer to its optimum. But, as mentioned earlier, the movements have not corresponded in a sufficiently consistent manner to temperature itself to support the belief that simple tracking of optimal temperatures provides a sufficient explanation.

Other examples could be cited where major isotherm movements (e.g. during an El Niño event in the eastern Pacific) have caused incursions of population segments to areas where they earlier had been absent. Such incursions have sometimes remained established for substantial periods after the isotherms retreated back to their earlier positions. Evidently there is a combination of effects operating that include (1) a short-term opportunistic tracking of zones of perceived favorable environmental conditions and (2) a contrary "obstinate" (Cury, 1994) tendency to replicate previously successful geographic behavior. Moreover, this "obstinancy" appears to operate on a much longer time scale (i.e. inter-generational) than does the opportunistic tracking of conditions, suggesting that the mechanism underlying it must be something more elaborate than mere learned behavior by individuals.

Clearly, there are a substantial number of key issues yet to be understood concerning the distributional dynamics of even the best studied small pelagic fish groups, temperate sardines (genera: *Sardinops*, *Sardina*) and anchovies (genus: *Engraulis*). But it is becoming evident (1) that major changes in population dynamics are generally accompanied by distinct changes in geographic distribution, and (2) that changes in spawning habitat selection appear to presage major changes at the key wasp-waist level of an ecosystem that have been perhaps the most consistent characteristics of major ecosystem regime shifts (Bakun, 2005a). Thus, the ability to properly interpret observed distributional changes and assess their cause and as well as their potential effect, be it positive or negative, might yield powerful new management capabilities. For example, if spatially biased fishing pressure were deemed to be the cause of what is projected to be a developing negative progression potentially leading to an unwelcome outcome, management might be able to mandate the application of fishing pressure in a pattern that may serve to reverse whatever original distortion may be deemed to have initiated the changes (for a heuristic example, see Bakun, 2005b).

Density dependence

Numerous fish species exhibit evidence of density-dependent growth, i.e. of decreased average size at age with increased abundance (Ware, 1980). The standard explanation for such density-dependent growth is that at high population sizes, individuals must be experiencing increasing competition for food, i.e. that the population is cropping down its available food resource to a substantial degree. This provides a major argument for the assumption of strong compensatory density dependence that is often a feature of conventional resource population models. Because belief in the existence of strong density dependence leads to confidence in efficacy of adaptive fishery management (i.e. to the idea that fishing pressure may be incrementally increased to a point where stock productivity begins to decline noticeably and then relaxed to a point where productivity is restored to a value near its earlier maximum), it is worthwhile continuing to examine the underlying argument. And there are available at least two plausible alternative hypotheses to the one that envisions a population substantially eating up its available habitat-scale food resources. These are (1) size-dependent predation and (2) increased school sizes at higher abundance.

Prey populations tend to vary in abundance more rapidly than do their longer-lived predators. Thus, when prey populations are declining, the ratio of predators to prey tends to increase. Conversely, when they are increasing, the ratio tends to decrease. Smaller, slower-growing individuals may experience relatively greater predation due to the fact that smaller individuals are slower and easier to catch and also spend more time in passing through the various size-dependent early life stages characterized by very high predation mortality (Hunter, 1981; Healy, 1982; Folkvord and Hunter, 1986; Hargreaves and LeBrasseur, 1986; Miller *et al.*, 1988). Thus, when predation is relatively intense (when the prey population is lower), the average size-at-age will increase, due to the biased cropping of smaller, slower growing individuals (Parker, 1971).

If mean school size increases with population abundance, this could also explain evidence of density-dependent growth. Food and oxygen must be transported to the interior of the school, where individuals evidently prefer to be (see previous section), through its outer surface area, which is roughly proportional to the second power ("square") of its characteristic linear dimension. The number of individuals that must be nourished is proportional to the volume of the school which is roughly proportional to the third power ("cube") of its linear dimension. Thus, as schools get larger, the amount of food and oxygen (oxygen too is needed to support growth; see Pauly, 1981) available to each individual declines.

These issues have received relatively little attention in recent years. They are important as each of the mechanisms leads to a different interpretation of the evidence for density dependent growth in terms of ecology and potential management policy. The "school size"-related mechanism appears to be a true compensatory density dependence, but without the requirement for a major effect of the population on its food resources on the habitat scale, but only on the scale of the school volume itself. On the other hand, if size-dependent predation is the mechanism behind the evidence, it implies the much more dangerous alternative of increasing mortality as abundance decreases (i.e. depensation). Because of their natures, both of these latter two mechanisms need to be addressed on fairly large time and space scales to produce useful inferences. Thus the comparative

method (either in time or space) appears to be the relevant approach (see the discussion in the "Concluding Remarks" section of this chapter).

Time series correlation studies are one way to conduct comparative analysis in the time domain and in fact have been the dominant approach in fisheries science for addressing ecosystem interactions and linkages. It appears that negative (inverse) correlations between zooplankton and pelagic fish abundances may be more common in the literature than are positive correlations (Cury *et al.*, 2000), suggesting a "top-down" effect of pelagic fish abundance on their planktonic food supplies, and this offers some support to the idea of intra-species competition for available food and thus to the possibility of density-dependent population control. But, if any fish population should have been able to graze down its food supply, it should have been the Peruvian anchoveta, which prior to its collapse in the early 1970s, may have been the most densely packed mass of fish ever observed essentially to cover an entire large marine ecosystem. For example, Lasker (1989) estimated that consumption by the anchoveta population, at a pre-collapse biomass level of 20 million tonnes, would account for about one-fourth of the entire primary production of that particularly rich ecosystem. But, while Palomares *et al.* (1987) did indeed find size-at-age of anchoveta to have increased since the population collapse, there is evidence that the zooplankton biomass in the Peru Current system following the anchoveta collapse, rather than increasing as might be expected, actually decreased markedly (Loeb and Rojas, 1988; Carrasco and Lozano, 1989; Alheit and Bernal, 1993). An analysis of more recent IMARPE/Peru survey data (1983–2005) by Ayón *et al.* (unpublished data) has yielded similar positive relationships in the time domain. In sum, the Peruvian anchoveta results would seem to support a "bottom-up" trophic effect of zooplankton abundance on fishery resource productivity and thus perhaps to argue against the idea of strong density dependent control, even in this intermittently most abundant and densely concentrated of all pelagic fish populations. Interestingly, Ayón *et al.* also incorporated the spatial domain in their analysis by comparing samples at neighboring locations. Proceeding this way produced inverse relationships, implying that on very local scales the anchoveta, where highly concentrated, may indeed be paring down its food supply (in agreement with some of the studies in other regions cited by Cury *et al.*, 2000).

Interpretations of such studies may be complicated by the fact that the zooplankton data used are generally gathered with fairly coarse-meshed net samplers, which may lose smaller zooplankton size classes that pass through the net meshes. Moreover, they are mainly measured as settling volumes. Even when large coelenterates are selectively removed from the samples before the volumes are determined, smaller coelenterates and other predatory organisms that may not represent attractive food for anchoveta might swell the measured volumes in which they are particularly abundant, while perhaps having reduced through their predation the food organisms that would be more attractive anchoveta food (i.e. such that large volumes may not reliably denote high fish-food densities). These issues are not easily addressed, but the importance of the notions of density dependence and carrying capacity to management decisions is crucial, understanding is inadequate, and scientific progress is much needed.

So far, the Peruvian system has consistently rebounded from radical population fluctuations that have occurred under heavy fishing. However, a period of catastrophic collapse of heavily fished small pelagics in the northern Benguela LME in the 1970s appears to have durably transformed that system to a chronic situation of greatly reduced fishery productivity (Bakun and Weeks, 2006). We need a much clearer notion of the "tipping points" that may act to trigger such durable, very regrettable regime shifts. Obtaining that clearer notion will require ingenuity and innovative approaches. Otherwise, management of small pelagic fishery resources through the coming period (1) of increasing scarcity and value of food fish, in which (2) explosive aquaculture development may be rapidly increasing the commodity value of fish meal, and in which (3) the background climatic state may meanwhile be changing rapidly due to climate change, will remain largely a matter of guesswork and polemics.

Niche replacement

A remarkable feature of many regional marine systems is an evident tendency for domination of the wasp-waist level of the ecosystem to alternate on multi-annual time scales among two different species groups.

An example is the widely observed pattern of alternations in abundance of anchovies and sardines in a wide variety of diverse ocean settings (Schwartzlose *et al.*, 1999; Bakun and Broad, 2003). Interestingly, sardines are species obviously adapted to highly productive ocean conditions (e.g. upwelling areas), but they seem often to experience better reproductive success, at least in the eastern Pacific, during anomalous seasonal warm events, such as El Niño episodes, which are characterized by lowered primary productivity (Bakun and Broad, 2003). They also have markedly increased their population sizes during the El Niño-dominated period from the early 1970s to mid 1980s. In contrast, in the same areas, the anchovy reproduces most successfully in highly productive, cool interannual events such as La Niña. Thus we have the seemingly paradoxical situation in which these two wasp-waist species alternate in abundance, in a wide variety of very diverse ocean settings

(Bakun and Broad, 2003), while substantial direct interactions, such as preferential predation on the other's eggs and larvae and competition for limiting resources, appear to be lacking.

Sea temperature has sometimes been hypothesized to be the controlling factor, with warmer temperatures favoring sardines and cooler temperatures favoring anchovies. But, if temperature were the causative agent, why would the same relationship not also hold in the western Pacific, where sardines have done best in relatively cool periods? And, if temperature preferences were the answer, why would not anchovies simply occupy the cooler sides, and sardines the warmer sides, of each regional ecosystem, as in other situations where two similar species with differing temperature preferences co-exist? On the contrary, the two species groups tend to directly replace one another as the dominant wasp-waist component throughout the entire regional large marine ecosystem that they share.

Other mechanisms have been suggested. Since sardines are able to filter much smaller particles than anchovies can (van der Lingen, 1994), anchovies might increase when particle size increases. But sardines also take large particles. In fact, they seem to prefer large particles, being observed to cease filtering and deviate from their paths in order to take opportune large particles by biting (van der Lingen, 1994). Thus a simple difference in minimum consumable particle size would, by itself, seem to be an incomplete explanation.

Bakun and Broad (2003) suggested that, on the one hand, the particular attributes of the sardine (e.g. very fine gill raker filters, strong migratory capability, exploratory spawning behaviors) may make it particularly able to survive disrupted or poorly productive ecosystem conditions (such as in El Niño) while taking good reproductive advantage of resulting severe suppressive effects on predator populations that normally wreak havoc on its early life stages. On the other hand, relative efficiencies of the anchovies' life style may make possible extremely rapid population growth during more productive conditions even in the face of resulting heavy predation.

Clearly, the relative tradeoffs of nutrition for both adults and larvae vs. production or attraction of larval predators in poorly productive versus highly productive conditions might resolve some prominent paradoxes and would seem to be worthy of focused research attention.

Whether one particular species, rather than its intermittent alternative, may be dominating the wasp-waist at any particular time may have important ecosystem-wide effects. Sardines, being far more migratory than anchovies, tend to spread the productivity of an upwelling system over much wider geographic areas. For example, when sardines dominate off western South America, Chile may harvest a much larger portion of small pelagic biomass than when anchovy dominate, which are more concentrated within

Peruvian waters. In addition, the feeding mode of the current wasp-waist species may be very important to the basic ecosystem structure. The loss of sardine, with its ability to swim into strongly divergent upwelling zones and directly filter phytoplankton, may have been a key to the transformation of the wasp-waist of the northern Benguela LME to domination by low-valued medusas and pelagic gobies, its consequent drastic decline in fishery resource productivity, and to serious degradation in other customary ecosystem goods and services (Bakun and Weeks, 2004).

Function of small pelagic fish in less-studied ecosystems

Even as we begin to accumulate significant understanding of temperate sardine/anchovy-dominated boundary current systems, as well as of selected boreal shelf–sea systems, we still know very little of small pelagic function in many other more tropical and/or oceanic ecosystems.

Ordinarily, near the low-latitude limit of a temperate sardine distribution, a lower latitude distribution of one or more species of the genus *Sardinella* commences. Sardinellas appear to be rather close analogs to sardines, although they would seem to be less migratory (than *Sardinops*, at least). In some regional cases, e.g. the eastern tropical Atlantic, there may be a pairing of a more migratory species (e.g. *S. aurita*) and a less migratory species (e.g. *S. maderensis*) of the same genus. To what degree this may represent an analogy to a temperate sardine–anchovy pairing is unclear, but the pattern is certainly suggestive. Some tropical systems appear to have a rather full range of analogs to the pelagic fish communities of temperate upwelling systems. For example, in the upwelling region that is spread along the northern coastal arc of the Arabian Sea, major species are the Indian oil sardine (*Sardinella longiceps*), Indian mackerel (*Rastrelliger kanagurta*), and anchovies (*Stolephorus* spp.). Such situations appear to offer rich opportunities for drawing comparative instances, although, in some cases, information may be limited.

Tunas are among the most evolutionarily advanced fishes. Many are voracious, high-energy predators with voracious, high-energy larvae. Although as adults they may feed in relatively rich ocean areas, they may migrate long distances to spawn in relatively poorly productive areas of the ocean. An explanation may be that large tunas may be chronic victims of a P2P feedback loop. They are indeed oceanic "top predators," but only as adults. Earlier in life, they are definitely prey of smaller fish species. As a result, adult tuna may be forced to migrate long distances to spawning grounds where conditions are so poor that even the predictable appearance of tuna early stages may not be sufficient to make it worthwhile for potential predators to concentrate their feeding within those areas at those times.

The situation thus may be rather similar to that for the sardine where particularly poor conditions may impact predators on larval or juvenile stages to the degree that the benefit of decreased predation may outweigh detrimental effects of poor feeding conditions.

The long-term decline in the Southern Oscillation Index during the mid 1970s to mid 1980s period signaled a decadal-scale relaxation of the trade wind circulation in the near-equatorial zone of the Pacific. Correspondingly, one expects that the equatorial upwelling that is a major factor in the relative productivity of the near-equatorial latitude band should slow down, resulting in poor conditions at the low end of the food chain and resulting poor food densities for tuna larvae and early juveniles. Nonetheless, this is a period when the Pacific yellowfin and Pacific skipjack tuna populations, which spawn in the near-equatorial band (Cole, 1980; Foresberg, 1980), apparently experienced extended upward population trends (Lehodey *et al.*, 2003). Frigate mackerels (genus: *Auxis*; also called frigate tunas) are small-sized, fast-swimming, voraciously feeding pelagic fishes, but of a scombroid rather than a clupeoid body form. They may represent a wasp-waist analog in tropical open-ocean "blue water" ecosystems. Although they are not extensively fished, *Auxis* are apparently enormously abundant in the tropical oceans (Richards, 1984; Fonteneau and Marcille, 1993; Colette and Aadland, 1996). While they are favored prey of adult tunas, they are likely to be major predators on young tunas. If low productivity during this period caused the local *Auxis* populations, or whatever other species groups may exert particular predation mortality on tuna early life stages, to collapse or, alternatively to shift their distributions away from the near-equatorial zone, this could perhaps have enabled enhanced reproductive success of yellowfin and skipjack through this period. Notably, the albacore tuna, which spawn to the north and south of the equatorial zone, experienced declines during the same period. We know almost nothing about variability in distribution, abundance, or behavior of *Auxis*. If valuable tuna populations are subject to P2P feedbacks, it would seem worthwhile to try to improve our knowledge of this perhaps most abundant of all pelagic predators (which, in the oceanic habitat of large tunas, might be classed as "small pelagics").

New technologies and approaches

How will we obtain the information to investigate these various hypothetical possibilities, or even to expand the comparative approach to less studied systems as suggested above? So far, it has been necessary to rely largely on information constructed either from a combination of mathematical assumptions and fisheries landings statistics, from shipboard acoustic assessments, or from filtration of a tiny sample of the ecosystem through various types of towed net samplers. How can we get the new information we will need to make the forward leaps in understanding that we seek?

Whenever a new technology offers a way to view a phenomenon from an entirely new perspective, rapid leaps forward in scientific knowledge and understanding invariably follow. The recent development of electronic archive tags is one example. The use of genetic markers and tracers is another. We can be certain that there are patterns of distribution and interaction that are potentially rife with meaning and enlightenment, but are currently invisible to us. We need access to new technologies that will let us see them.

For example, changes in small pelagic fish distribution, notably in reproductive habitat selection, have generally corresponded to ecosystem regime shifts and durable collapses of traditional fishery resource stocks. Distributional changes appear to be both a cause and an effect and thus a potentially revealing signal of underway dynamic changes. However, distributional information is available only in a few selected situations. And in those few available cases, that information is usually quite sketchy and incomplete. But imagine if one were actually able to follow in some detail the changes in reproductive habitat selection by major species in a large marine ecosystem over a number of years. As one watched the patterns developing and unfolding, one could perhaps begin to interpret the causes and the effects. If an evolving pattern seemed to be headed in an unfavorable direction, tactical management measures might be devised to interrupt the processes. Further, imagine if one were able to follow such developments in a number of large marine ecosystems simultaneously, multiplying the learning opportunities, as well as gaining a perspective from which to perhaps distinguish among local, basin-scale, and global effects.

Would it be feasible, for example, to develop an automated "super" egg pump (Checkley *et al.*, 1997), suitable for continuous operation on ships or fishing boats of opportunity, that could sample and filter large amounts of water and count and identify eggs of a selected set of important species? Fish eggs are nearly all nearly the same size, and are of similar spherical or oblate shape (Cury and Pauly, 2000), and thus perhaps relatively easy to separate from other particles and to count automatically even in relatively large volumes of water. In the process, a specially designed set of electronic chip-based genetic probes could perhaps identify the major species in a subsampled portion. If the engineering problems were surmountable, it would be a fascinating tool for marine ecosystem research and a potential boon to establishment of true operational marine ecosystem management.

Others, of course, may have other, better ideas than this. The microelectronics revolution is opening mind-boggling possibilities for innovative approaches toward pushing our science forward. Instruments and sensors that would have been inconceivable a few years ago are already in use today. Examples are sensor arrays that can track the movements of individual fish that have been tagged with transmitting microelectronic tags, as well other arrays that can sense untagged fish moving through their sensory fields. The possibilities for finally seeing clearly how pelagic fish, was well as their predators and prey, move and behave in their natural habitat settings seem limited only by ingenuity, vision, and, of course, available funding.

A similarly amazing revolution has taken place in molecular genetics. Genetic techniques may offer ways to answer such key questions as effective population sizes and inter-population connectivities. Such answers will provide crucial insights into the essential mechanisms of population stability and adaptive movement (e.g. Poulin *et al.*, 2004). Because of the sheer power of the results, and because it appears that different practitioners may sometimes arrive at different conclusions on crucial issues, it will be important to enlist an involved expert community that can provide continuing well-founded peer review and influx of ideas. Genetic techniques also offer a possibility to efficiently identify, for example, fish eggs, partially digested gut contents (e.g. Smith *et al.*, 2005).

The micro-electronics revolution has also made feasible very powerful mathematical modeling approaches. Modeling offers a very efficient way to test whether various mechanistic assumptions actually do add up to observed reality. Model simulations can point out key issues to be tested and thus can be extremely helpful in planning and focusing process-oriented field investigations. A computer model's results can only reflect the knowledge that was incorporated in its construction, and so cannot by itself generate, or substitute for, new observational information. However, components of a model that have been adequately verified as yielding consistently realistic results can often indeed contribute inputs that can, with proper caution, be validly considered to be analogs of real data. In recent years, very good computer-model representations of regional small pelagic fish habitats have been implemented (e.g. Penven *et al.*, 2001), generating important insights regarding processes and mechanisms controlling reproductive success and population dynamics (e.g. Parada *et al.*, 2003; Lett *et al.*, 2006).

In addition, in order to gain additional realizations on the longer time scales on which we know that major changes occur, we must continue to try, in every way possible, to extend our knowledge and data series deeper into the past. Additional well-founded paleosedimentary (e.g. Holmgren-Urba and Baumgartner, 1993) and archeological investigations will undoubtedly be extremely useful; ingenuity in discovering and assembling informative proxy time series (e.g. Pinnegar *et al.*, 2006) will be at a premium. Moreover, commercial and artisanal fishers undoubtedly possess a wealth of meticulously acquired knowledge (e.g. Johannes, 1981). Effective ways to access, record, and utilize this knowledge need to be found.

Box 13.1. Climate change and complex eco-dynamics

Just as we are beginning to fully appreciate the nature of marine ecosystems as complex adaptive systems, in which steady-state concepts such as carrying capacities and sustainable yield may be illusory or at least temporary, we are faced with the prospect of rapid climate change that threatens to remove any reasonable degree of long-term stationarity even in the basic physical context within which these systems operate.

Complex systems are characterized by nonlinearities in dynamical mechanisms and responses that can lead to chaotic solutions in mathematical model formulations. However, certain strong regularities, such as large-scale population synchronies and fairly dependable patterns of alternation among dominant species, indicate that chaos does not overwhelm the dynamics of large marine ecosystems. This implies strong constraining mechanisms that ought to be amenable to scientific investigation and elaboration. Developing a useful level of understanding of the nature and function of these constraining mechanisms may be the key to effective efforts to minimize disruptions and losses of marine ecosystem qualities and fishery resource productivities in the coming turbulent decades.

From the point of view of science itself, climate change will not necessarily be a detriment. The inability to exert experimental controls on marine ecosystems, at least on the scales of large mobile marine populations, has heretofore made the classical experimental scientific method largely unavailable to marine ecosystem science. But the coming steep clines in the climatic context, developing according to comparatively well-understood physical laws and dynamics, will change this, offering essentially an analog to an ever-continuing series of externally controlled experiments performed on the scales of entire populations, each with multiple comparative quasi-replications occurring simultaneously in similarly configured ecosystems throughout the world's oceans. Thus the coming decades should be exciting ones in terms of sorting out resistant scientific issues, even if rather desperate ones in terms of preserving our global marine patrimony.

The wasp-waist level of an ecosystem, at which small pelagic fish stocks operate, exerts major effects on the more speciose trophic levels above and below it. It is also the level at which the promising regularities cited above have been most particularly evident. Moreover, the small pelagic fishes operating at this level are known to be notably subject to climate effects. Accordingly, small pelagic fish studies stand out as a natural scientific entry point for making fortuitous use of the psuedo-experimental attributes of climate change to try unravel many of the outstanding mysteries of complex marine ecosystem dynamics. (Thus, one could perhaps say that, with respect to marine ecosystems and resources, climate change might show us the way to survive climate change.)

Concluding remarks

Some of the ideas presented in the earlier sections of this chapter sketch out a somewhat more complex problem than often assumed. Undoubtedly, the notion that we could satisfactorily account for the ecosystem dynamics by simply correlating deviations from a simple steady-state, density dependent stock–recruitment model to some environmental property such as temperature has been oversimplistic. It appears that uncritical reliance on linear "bottom-up" or "top-down" trophic flow models for the same purpose may be inadequate. If marine ecosystems operate as complex adaptive systems, the dynamics of which may be substantially controlled by unstable interactions, then addressing them as simple stationary systems will not work. Such approaches are undoubtedly useful in exploring various aspects of the problem, as well as in many cases being all that the available data may support. Indeed, they may be very good approximations of responses on short time scales. Yet, they are not a complete answer in themselves, and likely will not predict the most radical outcomes, i.e. the "tails" of the distributions of potential consequences, that may be most crucial to understand and avoid if we are

to preserve the economic, cultural and esthetic values traditionally associated with marine resources and ecosystems. It is possible that linear, stationary approaches may, in fact, describe fairly acceptably the "well behaved" periods with positive (i.e. constructive) feedbacks that promote and maintain the current structure. But abruptly, when some threshold may be exceeded, a tipping point occurs (Scheffer *et al.*, 2001) and, once triggered, may run on to a durably altered ecosystem state. It is this unstable, non-linear phase, and the "tipping point" that initiated it, that are most crucial to understand.

In such a non-linear context, highly specific prediction and skilled forecasting are difficult. To a large degree, we may have to give up the traditional narrow focus on a particular regional system of interest, and try learn as much as can be learned from what may have already happened in other systems. It would seem that the job of a SPACC scientist, whether in GLOBEC, IMBER, or other follow-on scientific programs, is to keep generating and sharing information and to collaborate in recognizing and interpreting the informative patterns in that information. The great evolutionary ecologist, Ernst Mayr, has called the experimental method and the comparative method the "two great methods of science" (Mayr, 1982). The comparative method offers an alternative to the experimental method in situations in which the ability to impose experimental controls may be lacking, which is almost always the case when addressing ecosystem mechanisms at the level of large populations of mobile marine organisms. What could be more fascinating than to study the beautifully intricate ecosystems of the world's oceans in their full global diversity? And if receptive minds remain fascinated, new patterns will emerge, and with them new understanding.

Working in the non-stationary context of progressive global climate change due to continuing build-up of atmospheric greenhouse gases will add obvious difficulties. But, from a science prospective, it may also offer distinct opportunities. For one thing, greenhouse-mediated climate change in itself represents the type of large-scale experiment that we ordinarily have considered to be unavailable to us. We possess a scientific basis to more or less understand the action of climate change in affecting the physical dynamics of various types of marine ecosystems (e.g. Bakun, 1990; Vecchi *et al.*, 2006). Following the progress of such events, population movements, and shifts in phenology in different similarly and non-similarly affected regional ecosystems as the next decades unfold will constitute a simultaneous application of the comparative and experimental approaches on the scales of whole populations and larger. This cannot but be revealing in ways that otherwise might be unavailable.

As we proceed, we may to some degree have to free our minds from the conventional dogmas that admittedly have been extremely effective in underpinning a period of very rapid progress in fisheries science half a century ago. At that earlier time, little was actually known of large-scale mechanisms that are particularly "marine" in nature. And so, the great early innovators of fisheries science understandably based their conceptual framework on terrestrial and small-scale aquatic analogies. In the ensuing decades, information and experiences that are uniquely marine in nature have been steadily accumulating. We are accumulating powerful new tools and approaches. We are already seeing repeatable patterns that appear to be rife with meaning. While these are leading us to the realization that the problem may be more complex than we earlier may have hoped, one can now hope that they are also inevitably leading us toward the truly relevant answers that, as concerned scientists and good citizens of our planet, we must keep seeking.

Acknowledgments

Dave Checkley contributed very helpful suggestions as well as substantive scientific inputs to the preparation of this chapter. The author's effort in preparing the chapter was supported by the PIOS "Adverse Eco-Feedbacks" Project, funded by the Pew Institute for Ocean Science.

REFERENCES

Alheit, J. and Bernal, P. (1993). Effects of physical and biological changes on the biomass yield of the Humboldt Current ecosystem. In *Large Marine Ecosystems – Stress, Mitigation, and Sustainability*, Sherman, K., Alexander, L.M. and Gold, B., eds. Washington, DC: American Association for the Advancement of Science, pp. 53–68.

Bakun, A. (1990). Global climate change and intensification of coastal ocean upwelling. *Science* **247**: 198–201.

Bakun A. (1996). *Patterns in the Ocean: Ocean Processes and Marine Population Dynamics*. San Diego, CA: University of California Sea Grant, in cooperation with Centro de Investigaciones Biológicas de Noroeste, La Paz, Baja California Sur, Mexico.

Bakun, A. (2001). "School-mix feedback": a different way to think about low frequency variability in large mobile fish populations. *Prog. Oceanogr.* **49**: 485–511.

Bakun, A. (2005a). Chapter 24 – Regime Shifts. In *The Sea, Vol. 13*, Robinson, A.R., McCarthy, J., and Rothschild, B.J., eds. Cambridge, MA: Harvard University Press, pp. 971–1018.

Bakun, A. (2005b). Seeking a broader suite of management tools: the potential importance of rapidly-evolving adaptive response mechanisms (such as "school-mix feedback"). *Bull. Mar. Sci.* **76**: 463–483.

Bakun, A. (2006). Wasp-waist populations and marine ecosystem dynamics: Navigating the "predator pit" topographies. *Prog. Oceanogr.* **68**: 271–288.

Bakun, A. and Broad, K. (2003). Environmental loopholes and fish population dynamics: comparative pattern recognition with particular focus on El Niño effects in the Pacific. *Fish. Oceanogr.* **12**: 458–473.

Bakun, A. and Weeks, S.J. (2004). Greenhouse gas buildup, sardines, submarine eruptions, and the possibility of abrupt degradation of intense marine upwelling ecosystems. *Ecol. Lett.* **7**: 1015–1023.

Bakun, A. and Weeks, S.J. (2006). Adverse feedback sequences in exploited marine systems. *Fish Fisher.* **7**: 316–333.

Beaugrand, G., Reid, P.C., Ibanez, F., *et al.* (2002). Reorganization of North Atlantic marine copepod biodiversity and climate. *Science* **296**: 1692–1694.

Belvese, H. and Erzini, K. (1983). Influence of hydroclimatic factors on the availability of sardine (*Sardina pilcardus* Walb.) in the Moroccan fisheries in the Atlantic. In *Proceedings of the Expert Consultation to Examine Changes in Abundance and Species Composition of Neritic Fish Resources*, Sharp, G.D. and Csirke, J., eds. FAO Fish. Rep., **291**: pp. 285–328.

Binet, D., Samb, B., Sidi, M.T., *et al.* (1998). Sardine and other pelagic fisheries changes associated with multi-year trade wind increases in the southern Canary Current. In *Global Versus Local Changes in Upwelling Systems*, Durand, M.-H., Cury, P., Mendelssohn, R., *et al.*, eds. Paris: ORSTOM Editions, pp. 211–233.

Boyer, D.C. and Hampton, I. (2001). An overview of the living marine resources of Namibia. *S. Afr. J. Mar. Sci.* **23**: 5–35.

Brodeur, R.D., Mills, C.E., Overland, J.E., *et al.* (1999). Evidence for a substantial increase in gelatinous zooplankton in the

Bering Sea, with possible links to climate change. *Fish. Oceanogr.* **8**: 296–306.

Butterworth, D. S. (1983). Assessment and management of pelagic stocks in the southern Benguela region. In *Proceedings of the Expert Consultation to Examine Changes in Abundance and Species Composition of Neritic Fish Resources*, Sharp, G. D. and Csirke, J., eds. *FAO Fish. Rep.* **291**: 329–405.

Carrasco, S. and Lozano, O. (1989). Seasonal and long-term variations of zooplankton volumes in the Peruvian sea, 1964–1987. In *The Peruvian Upwelling Ecosystem: Dynamics and Interactions*, Pauly, D., Muck, P., Mendo, J. and Tsukayama, I., eds. Manila: International Center for Living Aquatic Resources Management (ICLARM), pp. 82–85.

Caverivière, A. (1991). L'explosion démographique du baliste (*Balistes carolinensis*) en Afrique de l'Ouest et son évolution en relation avec les tendaces climatique. In *Pêcheries Ouest-Africaines Variabilité, Instabilité et Changement*, P. Cury and C. Roy, ed. Paris: ORSTOM Editions, pp. 354–367.

Checkley, D. M., Jr., Ortner, P. B., Settle, L. R., and Cummings, S. R. (1997). A continuous, underway fish egg sampler. *Fish. Oceanogr.* **6**: 58–73.

Cole, J. S. (1980). Synopsis of biological data on the yellowfin tuna, *Thunnus albacares* (Bonneaterre, 1788) in the Pacific Ocean. *Inter-American Tropical Tuna Commission Special Report* **2**: 71–150.

Collette, B. B. and Aadland, C. R. (1996). Revision of the frigate tunas (Scombridae, *Auxis*), with descriptions of two new subspecies from the eastern Pacific. *Fish. Bull. US* **94**: 423–441.

Conand, F. (1975). Distribution et abondance des larves de clupeids au large des côtes du Sénégal et de la Mauritanie en septembre, octobre et novembre 1972. *ICES CM:1975/J:4*.

Cury, P. (1994). Obstinate nature: an ecology of individuals – thoughts on reproductive behavior and biodiversity. *Can. J. Fish. Aquat. Sci.* **51**: 1664–1673.

Cury, P. and Pauly, D. (2000). Patterns and propensities in reproduction and growth of marine fishes. *Ecol. Res.* **15**: 101–106.

Cury, P., Bakun, A., Crawford, R. J. M., *et al.* (2000). Small pelagics in upwelling systems: patterns of interaction and structural changes in "wasp-waist" ecosystems. *ICES J. Mar. Sci.* **210**: 603–618.

Cury, P., Fréon, P., Moloney, C. L., *et al.* (2004). Chapter 14 – Processes and Patterns of Interactions in Marine Fish Populations: An Ecosystem Perspective. In *The Sea, Vol. 13*, Robinson, A. R., McCarthy, J. and Rothschild, B. J., eds. Cambridge, MA: Harvard University Press, pp. 475–553.

Demarque, H. (1998). Spatial and temporal dynamics of the upwelling off Senegal and Mautrtania: local change and trend. In *Global Versus Local Changes in Upwelling Systems*, Durand, M.-H., Cury, P., Mendelssohn, R., *et al.*, eds. Paris: ORSTOM Editions, pp. 149–165.

Folkvord, A. and Hunter, J. R. (1986). Size-specific vulnerability of northern anchovy (*Engraulis mordax*) larvae to predation by fishes. *Fish. Bull., US*, **84**: 859–869.

Fonteneau, A. and Marcille, J. (1993). Resources, fishing and biology of the tropical tunas of the Eastern Central Atlantic. *FAO Fish. Tech. Pap.*, 292.

Forsberg, E. D. (1980). Synopsis of biological data on the skipjack tuna, *Katsuwanus pelamis* (Linaeus, 1758) in the Pacific Ocean. *Inter-American Tropical Tuna Commission Special Report* **2**: 295–360.

Fréon P. (1988). Réponses et adaptation des stocks de clupeids d'Afrique de l'ouest à la varibilité du milieu et de l'exploitation. Analyse et reeflexion à partir de l'example du Sénégal. *Coll. Etudes et Thèses*, Paris: ORSTOM.

Grant, W. S. and Bowen, B. W. (1998). Shallow population histories in deep evolutionary lineages of marine fishes: Insights from sardines and anchovies and lessons for conservation. *J. Heredity* **89**: 415–426.

Hammann, M. G. and Cisneros-Mata, M. F. (1989). Range extention and commercial capture of the northern anchovy, *Engraulis mordax* Girard, in the Gulf of California, Mexico. *Cal. Dep. Fish Game Fish. Bull.* **75**: 49–53.

Hargreaves, N. B. and LeBrasseur, R. J. (1986). Size-selectivity of coho (*Onchorhynus kisutch*) preying on juvenile chum salmon (*O. keta*). *Can. J. Fish. Aquat. Sci.* **43**: 581–586.

Healy, M. (1982). Timing and relative intensity of size-selective mortality of juvenile chum salmon (*Onchorhynus keta*) during early sea life. *Can. J. Fish. Aquat. Sci.* **39**: 952–957.

Holmgren-Urba, D. and Baumgartner, T. R. (1993). A 250-year history of pelagic fish abundances from the anaerobic sediments of the central Gulf of California. *CalCOFI Reports* **34**: 60–68.

Hunter, J. R. (1981). Feeding ecology and predation of marine fish larvae. In *Marine Fish Larvae: Morphology, Ecology and Relation to Fisheries*, Lasker, R., ed. Seattle: University of Washington Press, pp. 33–77.

Johannes, R. E. (1981). *Words of the Lagoon*. Berkeley, CA: University of California Press.

Kifani, S. (1998). Climate dependent fluctuations of the Moroccan sardine and their impact on fisheries. In *Global Versus Local Changes in Upwelling Systems*, Durand, M.-H., Cury, P., Mendelssohn, R., *et al.*, eds. Paris: ORSTOM Editions, pp. 235–245.

Köster, F. W. and Möllmann, C. (2000). Trophodynamic control by clupeid predators on recruitment success in Baltic cod? *ICES J. Mar. Sci.*, **57**: 311–323.

Köster, F. W., Möllmann, C., Neuenfeldt, S., *et al.* (2003). Fish stock development in the central Baltic Sea (1974–1999) in relation to variability in the environment. *ICES Mar. Sci. Symp.* **219**: 294–306.

Lasker, R. (1989). Food chains and fisheries: an assessment after 20 years, In *Toward a Theory of Biological–Physical Interaction in the World Ocean*, Rothschild, B. J., ed. Dordrecht: Kluwer Academic Press, pp. 173–182.

Lecomte, F., Grant, W. S., Dodson, J. J., *et al.* (2004). Living with uncertainty: Genetic imprints of climate shifts in east Pacific anchovy (*Engraulis mordax*) and sardine (*Sardinops sagax*). *Mol. Ecol.* **13**: 2169–2182.

Lehodey, P., Chai, F., and Hampton, J. (2003). Modeling the climate-related fluctuations of tuna populations with a coupled ocean-biogeochemical–population dynamics model. *Fish. Oceanogr.* **12**: 483–494.

Lett, C., Roy, C., Levasseur, A., van der Lingen, C. D., and Mullon, C. (2006). Simulation and quantification of enrichment and retention processes in the southern Benguela upwelling ecosystem. *Fish. Oceanogr.* **15**: 363–372.

Levin, S. A. (1998). Ecosystems and the biosphere as complex adaptive systems. *Ecosystems* **1**: 431–436.

Levin, S. A. (1999). *Fragile Dominion: Complexity and the Commons*. Reading, MA: Perseus Books.

Loeb, V. J. and Rojas, O. (1988). Interannual variation of ichthyoplankton composition and abundance relations off northern Chile, 1964–83. *Fish. Bull. US* **84**: 1–24.

Lynam, C. P., Gibbons, M. J., Axelsen, B. E., *et al.* (2006). Jellyfish overtake fish in a heavily fished ecosystem. *Curr. Biol.* **16**: R492–R493.

Mayr, E. (1982). *The Growth of Biological Thought*. Cambridge, MA: Harvard University Press.

Mayo, R. K., O'Brien, L. and Wiggley, S. E. (1998). Assessment of the Gulf of Maine cod stock for 1998. *NOAA Northeast Fisheries Science Center Reference Document 98–13*.

Miller, T. J., Crowder, L. B., Rice, J. A., and Marschall, E. A. (1988). Larval size and recruitment mechanisms in fishes: toward a conceptual framework. *Can. J. Fish. Aquat. Sci.* **45**: 1657–1670.

Mills, C. E. (2001). Jellyfish blooms: are populations increasing globally in response to changing ocean conditions? *Hydrobiologia* **451**: 55–68.

Palomares, M. L., Muck, P., Mendo, J., *et al.* (1987). Growth of Peruvian anchoveta (*Engraulis ringens*), 1953 to 1982. In *The Peruvian Upwelling Ecosystem: Dynamics and Interactions*, Pauly, D., Muck, P., Mendo, J. and Tsukayama, I., eds. Manila: International Center for Living Aquatic Resources Management (ICLARM), pp. 117–141.

Parada, C., van der Lingen, C. D., Mullon, C., and Penven, P. (2003). Modelling the effect of buoyancy on the transport of anchovy (*Engraulis capensis*) eggs from spawning to nursery grounds in the southern Benguela: an IBM approach. *Fish. Oceanogr.* **12**: 170–184.

Parker, R. R. (1971). Size-selective predation among juvenile salmonid fishes in a British Columbia inlet. *J. Fish. Res. Board Can.* **28**: 1503–1510.

Pauly, D. (1981). The relationships between gill surface area and growth performance in fish: a generalization of von Bertalanffy's theory of growth. *Meeresforschung* **28**: 251–282.

Penven, P., Roy, C., Brundrit, G. B., *et al.* (2001). A regional hydrodynamic model of the Southern Benguela upwelling. *S. Afr. J. Sci.* **9**: 472–475.

Pereladov, M. V. (1988). Some observations for biota of Sudak Bay of the Black Sea. In *The Third All-Russian Conference on Marine Biology*. Kiev: Naukova Dumka, pp. 237–238. (in Russian)

Perry, A. L., Low, P. J., Ellis, J. R. and, Reynolds, J. D. (2005). Climate change and distribution shifts in marine fishes. *Science* **308**: 1912–1915.

Pinnegar, J. C., Hutton, T. P., and Placenti, V. (2006). What relative seafood prices can tell us about the status of stocks. *Fish Fisher.* **7**: 151–229.

Poulin, E., Cárdenas, L., Hernández, C. E., Kornfield, I., and Ojeda, F. P. (2004). Resolution of the taxonomic status of Chilean and Californian jack mackerels using mitochondrial DNA sequence. *J. Fish Biol.* **65**: 1160–1164.

Rice, J. (1995). Food web theory, marine food webs, and what climate change may do to northern fish populations. In *Climate Change and Northern Fish Populations*, Beamish, R. J., ed. *Can. Spec. Pub. Fish. Aquat. Sci.* **121**: 561–568.

Richards, W. J. (1984). Kinds and abundances of fish larvae in the Caribbean Sea and adjacent areas. *US Dep. Comm. NOAA Tech. Rep. NMFS-SSRF-776*.

Scheffer, M., Straile, D., van Nes, E. H., and Hosper, H. (2001). Climatic warming causes regime shifts in lake food webs. *Limnol. Oceanogr.* **46**: 1780–1783.

Schwartzlose, R. A., Alheit, J., Bakun, A., *et al.* (1999). Worldwide large-scale fluctuations of sardine and anchovy populations. *S. Afr. J. Mar. Sci.* **21**: 289–347.

Sibert, J., Hampton, J., Kleiber, P., and Maunder, M. (2006). Biomass, size, and trophic status of top predators in the Pacific ocean. *Science* **314**: 1773–1776.

Smith, P. J., Mcveagh, S. M., Allain, V., and Sanchez, C. (2005). DNA identification of gut contents of large pelagic fishes. *J. Fish Biol.* **67**: 1178–1183.

Van der Lingen, C. D. (1994). Effect of particle size and concentration on the feeding behavior of adult pilchard, *Sardinops sagax. Mar. Ecol. Prog. Ser.* **119**: 1–13.

Vecchi, G. A., Soden, B. J., Wittenberg, A. T., Heid, I. M., Leetma, A., and Harrison, M. J. (2006). Weakening of tropical Pacific atmospheric circulation due to anthropogenic forcing. *Nature* **441**: 73–76.

Walters, C. and Kitchell, J. F. (2001). Cultivation/depensation effects on juvenile survival and recruitment: implications for the theory of fishing. *Can. J. Fish. Aquat. Sci.* **58**: 39–50.

Ware, D. M. (1980). Bioenergetics of stock and recruitment. *Can. J. Fish. Aquat. Sci.* **37**: 1012–1024.

14 Conjectures on future climate effects on marine ecosystems dominated by small pelagic fish

Pierre Fréon, Francisco Werner, and Francisco P. Chavez

CONTENTS

Summary

Direct effects of humans on the environment (agricultural practices, fishing) have been evident for hundreds of years. Some of these direct effects (increased carbon dioxide in the atmosphere) are now spilling over into the climate system creating uncertainty regarding the future of marine ecosystems. Here, we review possible scenarios of climate change (CC) and physical oceanography in the SPACC context. Three predicted avenues of ecological change are discussed: (1) changes in productivity and composition of lower trophic levels; (2) distributional changes of marine organisms; and (3) changes in circulation and their effects on recruitment processes. Research gaps are identified with special attention to current limitations of available data, models, and projected scenarios. We identify significant gaps in the knowledge of processes and interactions between changes in climate and other ecosystem stressors. These other stressors, such as ocean acidification, eutrophication, and overfishing, constitute additional anthropogenically induced components of global change but are not the focus of this review. The main conclusion is that, although more information is needed before the scientific community is able to make reliable predictions regarding the future state of marine ecosystems, there is already evidence of sensitivity of pelagic species and pelagic ecosystems to CC and of decreased resilience of natural ecosystems caused by overexploitation.

Introduction

Fishers have known for centuries that climate fluctuations are both important and normal. Climate varies on daily, weather-system, and seasonal time scales, and over the past several decades scientists have learned about longer-term fluctuations that occur within the bounds of "natural" variability (e.g. El Niño, Pacific Decadal Oscillation, and ice ages). Long-term records from ice, tree rings, and sediment cores all show that climate has varied considerably and continuously on interannual to centennial time scales. Additionally, it is now known that humans can also affect climate through the emissions of greenhouse gases and the associated global warming (IPCC, 2001; Steffen *et al.*, 2004; Houghton, 2005; IPCC, 2007). And, while there is still some debate over the source and magnitude of twentieth-century warming (e.g. Lomborg, 2001) the evidence for a relation of global increases of temperatures with increasing levels of greenhouse gases continues to mount (Levitus *et al.*, 2003; Barnett *et al.*, 2005; Field *et al.*, 2006). The latest IPCC[1] report states that "it is extremely unlikely (<5%) that the global pattern of warming observed during the past half century can be explained without external forcing" and that there is "very high confidence that the global average net effect of human activities since 1750 has been one of warming, with a radiative forcing of +1.6 [+0.6 to +2.4] $W\,m^{-2}$" (IPCC, 2007). For the next two decades and for a range of emission scenarios, model projections include a warming of about 0.2 °C per decade, sea level rise, a reduction in ice cover, and other global and regional signals. The ecological impacts of a warming climate are already evident in both terrestrial and marine ecosystems, with clear responses of both the flora and fauna, from the species to the community levels (Walther *et al.*, 2002; Parmesan and Yohe, 2003; ACIA, 2004; Harley *et al.*, 2006). The continued influence of humans on climate will result in further changes in abundance and distribution of marine species with strong implications for fisheries, tourism and "goods and services" provided by the marine ecosystems (Costanza *et al.*, 1997). As such, there is a need to estimate the nature and severity of the possible ecological consequences to be able

Climate Change and Small Pelagic Fish, eds. Dave Checkley, Jürgen Alheit, Yoshioki Oozeki, and Claude Roy. Published by Cambridge University Press. © Cambridge University Press 2009.

to develop strategies that will allow us to adapt and better manage our resources (Field *et al.*, 2002; Clark, 2006).

Estimating the effects of climate change (CC) on marine ecosystems is complicated because of the likely differences in which it will be manifested in different parts of the world's oceans, a point that is often overlooked or misunderstood. Using sea surface temperature as a proxy, analyses of global *in situ* or satellite data sets show patterns of variability in the Pacific and Indian Oceans that exceed those in the Atlantic Ocean (Enfield and Mestas-Nunez, 2000). While similar variability in spatial and temporal structures exists in the Atlantic (the Atlantic El Niño, North Atlantic Oscillation, Atlantic Multidecadal Oscillation), their effects are not as pronounced in global data sets (Enfield and Mestas-Nunez, 2000). The western Pacific and eastern Indian Oceans harbor the largest area of warm surface water in the world and the effects that this warm water pool exerts on inter-annual (El Niño) and multi-decadal (El Viejo) time scales can result in significant variations in primary production, fish abundance, and ecosystem structure at a Pacific basin scale (Chavez *et al.*, 2003). Based on the above, different responses to climate change have been projected in the Atlantic and Pacific Oceans (Fréon *et al.*, 2006). The question of whether global warming will increase or decrease the natural variability is a matter of continued research. Some studies expect stronger and more frequent El Niños as a result of global warming (e.g. Timmerman *et al.*, 1999; Hansen *et al.*, 2006), while others suggest that the evidence is still inconclusive (Cane, 2005; IPCC, 2007).

Another important and complicating consideration is the interaction of human-induced climate change with the natural variability reviewed in Chapters 3 and 4 of this volume. Scientific discussions today still debate the relative importance of natural climate variability vs. human-induced disturbance/change in determining the observed declines or increases in populations (Barange, 2003; Barange *et al.*, 2003; Pauly, 2003). Because these two interact, it is difficult at times to ascertain which is the dominant effect, and as above, their relative importance may be regionally dependent. For example, fishing can lead to changes in age structure, spatial contraction, and alteration of life history traits in populations, making them more sensitive to climate variability (Perry *et al.*, in press).

In this chapter we discuss how effects related to global warming, e.g. changes in physical forcing and related bottom-up processes, will affect regions where small pelagic fish such as anchovies and sardines are numerically dominant (SPACC[2] regions). These include, among others, the four major eastern boundary upwelling systems (i.e. the Benguela, California, Canary and Humboldt currents), the Kuroshio Current system, the Bay of Biscay, the Baltic Sea, the Barents Sea, and the Norwegian Sea. When appropriate, we use examples from non-SPACC regions and species

to illustrate possible CC effects on marine ecosystems. Political and scientific concerns over CC are driven by potential negative impacts. SPACC regions produce abundant concentrations of small pelagics, with significant economic returns (Hannesson *et al.*, 2006a). These regions are areas with high levels of primary productivity (Ryther, 1969; Carr, 2002) and are susceptible to climatic perturbations (e.g. Yasuda *et al.*, 1999; Chavez *et al.*, 2003; Hannesson *et al.*, 2006b; Lehodey *et al.*, 2006; Herrick *et al.*, 2007). Our focus is on the direct effect of anthropogenic CC on SPACC systems. We will focus less so on other anthropogenically induced components of global change such as ocean acidification (e.g. Raven, 2005), eutrophication (such as low oxygen events, e.g. Grantham *et al.*, 2004; Monteiro *et al.*, 2006) or overfishing (e.g. Boyer *et al.*, 2001).

The chapter is divided into three main sections. First, scenarios of change in physical oceanography are summarized. Second, predicted avenues of ecological change are described, with some examples. Finally, research gaps are identified.

Future scenarios of physical oceanographic climate change and variability

The physical forcing of a warmer climate is hypothesized to result in several important changes that can impact biological productivity directly: (1) wind patterns and intensities may be significantly modified (e.g. Bakun, 1990; Snyder *et al.*, 2003; Vecchi *et al.*, 2006) impacting the strength of the ocean currents, turbulence levels, and mixing, and the frequency and magnitude of occurrences of phenomena such as El Niño (Cane, 2005; Hansen *et al.*, 2006); (2) stratification in the surface layers can increase with warmer surface ocean (e.g. McGowan *et al.*, 2003) or increased freshwater fluxes (the latter primarily at high latitudes, e.g. Rodhe and Winsor, 2002; Carmack and Wassman, 2006; Doney, 2006; Royer and Grosch, 2006), thereby suppressing the mixing of nutrients into the euphotic zone and decreasing the fertilizing effects of mixing and upwelling (Behrenfeld *et al.*, 2006); (3) changes in precipitation coupled with changes in wind may alter dust transport, and therefore the delivery of iron and other micro-nutrients to the oceans (e.g. Jickells *et al.*, 2005; McTainsh and Strong, 2006). Figure 14.1 summarizes how physical forcings and bottom-up effects might affect populations of small pelagics.

Effects of changing wind fields

Different views have been offered on possible changes of the wind fields and their impact on ocean dynamics under global warming. Bakun (1990) suggests that, because land warms more quickly than water, upwelling-favorable sea breezes should strengthen, resulting in stronger upwelling (also see Auad *et al.*, 2006), resulting in higher

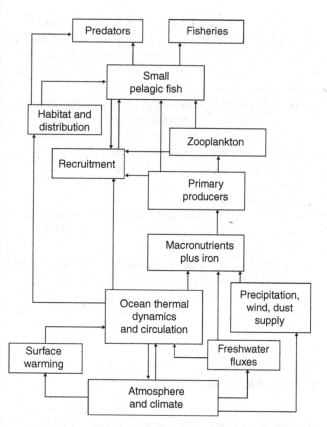

Fig. 14.1. Atmospheric forcing can alter warming, freshwater fluxes
and winds which can directly modify ocean thermal dynamics
and circulation. By thermal dynamics we mean changes in the
horizontal (i.e. El Niño, PDO) or vertical (depth of mixed layer)
distribution of heat. Changes in ocean circulation include the limbs
of the subtropical and polar gyres, including strength and location.
These changes affect fish populations in at least four ways: (1) by
changing the supply of nutrients and consequently overall ecosystem
production – the classical bottom-up forcing; (2) by physically (i.e.
mixing and advection) affecting recruitment processes (as opposed
to bottom-up biological effects); (3) by physical modification of
adult habitat, resulting in changes in distribution; and (4) by physical
modification of habitat either restricting or allowing greater access
by predators. The multiplicity of physical effects on fish populations
complicates future prediction as does the feedback from the ocean
back to climate. Further, the links between physical and biological
systems are decidedly non-linear.

phytoplankton productivity, cooler surface ocean temperatures, and possibly increasing the occurrence of fog,
thus affecting the light field in the water column. Evidence
for such changes in atmospheric conditions over the past
30 years is provided in the observations of wind-driven
upwelling along the California coast by Schwing and
Mendelssohn (1997) and Mendelssohn and Schwing (2002).
Similarly, a recent study of core sediments from the Canary
Current (Cape Ghir) reveals a negative trend in temperature during the whole twentieth century, consistent with
increased upwelling (McGregor *et al.*, 2007).

Using a regional climate model study, Snyder *et al.*
(2003) estimated changes in wind-driven upwelling under
increased CO_2 concentrations, and projected an intensified upwelling season, with some changes in seasonality of
upwelling. It is suggested that the intensification may lead
to enhanced productivity along the coast of California and
possibly counteract increases in sea surface temperature
due to atmospheric warming. In a study complementary
to Snyder *et al.* (2003) it was found by Diffenbaugh *et al.*
(2004) that biophysical land-cover–atmosphere feedbacks
induced by CO_2 radiative forcing enhance the radiative
effects of CO_2 on the land–sea thermal contrast, resulting in
changes in total seasonal upwelling and upwelling seasonality along eastern boundary currents. Such changes were
also found to be spatially heterogeneous, e.g. land-cover–
atmosphere feedbacks were found to lead to a stronger
increase in peak- and late-season near-shore upwelling in
the northern limb of the California Current and a stronger
decrease in peak- and late-season near-shore upwelling in
the southern limb (Diffenbaugh *et al.*, 2004). Generally
speaking, all recent models that incorporate the coupling of
the carbon cycle to CC show a positive feedback effect due
to a limitation of land and ocean uptake of CO_2 by warming, but the strength of this feedback effect varies markedly
among models (IPCC, 2007).

Auad *et al.*'s (2006) study of a region in the California
Current found that a 36% increase in atmospheric CO_2 concentration leads to increased sea surface and near-surface
temperatures, and to increased stratification along the coast.
However, the predicted increase in stratification was found
not to be strong enough to overcome the effect of increased
upwelling favorable winds (see discussion on stratification in the section that follows below), and mild oceanic
subsurface cooling is forecast below the 70 m depth. Auad
et al. (2006) predict a 30% increase in near-surface vertical
velocities early in the upwelling season, as well as anomalous offshore transports in most of the coastal areas, and an
annual mean decrease in the eddy kinetic energy along the
California Current main path. The upward displacement of
isopycnal surfaces, in the northern half of their study area,
is found to lead to an increase in the concentration of nutrients in the subsurface. The forecast increased upwelling
velocities are predicted to result in nutrient richer waters
in the surface and near-surface ocean off California, with
local currents transporting these nutrients (and associated
plankton and larvae) offshore.

In contrast to the above studies that suggest increases
in local upwelling winds, large-scale changes in the wind
fields may be different. Using an ensemble of coupled climate model simulations, Diffenbaugh (2005) finds potentially important changes in the large scale forcing of eastern
boundary current forcing over the next century. Among
these are a relaxation of the strength and variability of

peak-season equatorward wind forcing in eastern boundary current regions. In another study, Vecchi *et al.* (2006) suggest that, because the poles will warm more dramatically than the tropics, the trade wind system which also drives upwelling-favorable winds should weaken, thus resulting in weaker upwelling, less phytoplankton productivity, warmer surface ocean temperatures, and less fog. Vecchi *et al.* (2006) note that atmosphere–ocean conditions in the equatorial Pacific have changed since the mid nineteenth century. In particular, they point to a slowdown of atmospheric circulation that, in turn, has driven a measurable response in ocean circulation predicted by modeling studies. They note that the observed trend in the Pacific surface zonal sea level pressure gradient is unlikely to be due to natural variability alone, and that much of the long-term trend reproduced in model simulations is related to anthropogenic effects on the global radiative budget. Vecchi *et al.* (2006) conclude that the agreement between the theoretical, observed, and model-produced changes in strength of atmospheric circulation increases confidence in the model-projected reduction in the strength of tropical circulation during the twenty-first century, and that, on the basis of climate model simulations, this weakening may be of the order of 10% by the end of this century.

We note that the differences in the predicted changes of the wind fields at different scales are not mutually exclusive, and highlight the simultaneous changes in oceanic conditions that may arise from increased local winds (e.g. due to increased land–ocean temperature gradients) and a reduction in larger scale winds resulting from a decrease in temperature gradients between high and low latitudes. Their combined effects may contribute to a downwelling favorable wind stress curl (stronger wind stress close to shore, weaker offshore). The implications could be an enrichment of nutrients and higher productivity towards shore with reduced offshore losses due to advection. Gaps and uncertainties in our ability quantitatively to anticipate the likelihood of these effects are such that only time and continued observation will tell which, if either, scenario is correct (e.g. Fréon *et al.*, 2006). Irrespective of the sign, e.g. increased upwelling vs. downwelling, it seems clear that SPACC regions whose productive ecosystems are set-up by wind forcing will be modified.

Increased stratification of the ocean surface layers

SPACC systems fall into two primary categories of physical bottom-up forcing: (1) at high latitudes (e.g. off Japan and the Nordic Seas), typically in winter, deep mixing erodes the thermocline downwards while nutrients are mixed upwards (e.g. Sverdrup, 1953; Kasai *et al.*, 1997; Skjoldal, 2004; Doney, 2006; Hashioka and Yamanaka, 2007), with phytoplankton blooms occurring after the winter mixing season during the onset of stratification; (2) at lower latitudes and

along eastern boundaries (Humboldt, Benguela, Canary, California), the dynamics of the wind-driven surface flow cause the thermocline and the nutricline to upwell towards or into the euphotic zone (Pennington *et al.*, 2006). While the effects of climate change on these two regimes will be different, both will be affected by changes in atmospheric circulation (wind speed/direction) and stratification due to surface warming.

Based on more than 50 years of measurements, McGowan *et al.* (2003) show that significant ecosystem changes have taken place in the California Current System including a large, decadal decline in zooplankton biomass, along with a rise in upper-ocean temperature. Specifically, they note the abrupt temperature change that occurred around 1976–77, concurrent with other Pacific basin-wide changes associated with an intensification of the Aleutian Low pressure system. Associated with this shift were changes in the patterns of net surface heat fluxes, turbulent mixing, and horizontal transport. Three hypotheses for the mechanism(s) behind the decline of plankton populations and biomass have been considered: changes in coastal upwelling (Bakun, 1990), interannual variations in horizontal input of cooler, fresher, nutrient-rich water from the north (Chelton *et al.*, 1982) and, the long-term post-1977 warming and deepening of the mixed layer, which led to an increase in stratification of the water column, and a change in the source and the nutrient content of the waters introduced into the surface layers (Roemmich and McGowan, 1995). Based on Sverdrup's (1953) ideas, such a deepening of the mixed layer relative to the "critical depth" for net phytoplankton production can also lead to a decline in production. McGowan *et al.*'s (2003) results are consistent with their hypothesis that the recent warming in the California Current system, which they associated with the 1976–77 regime shift, led to an increase in water column stability. However, they point out that more than one mechanism can be invoked to explain the surface-intensified warming and the deepening thermocline. Possible explanations include: altered air–sea heat fluxes that warmed the surface mixed layer after the shift (Miller *et al.*, 1994) with changes in wind stress curl have forced the thermocline downward (Miller *et al.*, 1998) in the offshore California Current region, and/or remotely forced signals propagating along the eastern Pacific boundary from the tropics due to a preponderance of tropical warm (El Niño) waters following the regime shift (Clarke and Lebedev, 1999). Neither mechanism could be demonstrated or ruled out, and they may also be acting concurrently. McGowan *et al.*'s (2003) results are consistent with the "optimal stability window" hypothesis (Gargett, 1997), wherein increased water column stability along the eastern boundary of the North Pacific would reduce (enhance) biological production at southern (northern) latitudes, where productivity is nutrient (light) limited. Regardless of the mechanism, McGowan *et al.* (2003) propose variability in thermal structure of the upper water column as the dominant mechanism responsible for the stability increase in the California Current system. In contrast, at higher latitudes, salinity changes were found to be potentially more important (e.g. Freeland *et al.*, 1997; Overland *et al.*, 1999).

Hashioka and Yamanaka (2007) simulated the effect of a future climate/global warming scenario in the oceanic region off the NW Pacific (near Japan) using a 3-D ecosystem–biogeochemical model forced by simulated fields from the IS92a global warming scenario (CO-AGCM developed by CCSR/NIES[3]). They found that, for the last decade of this century (2090–2100), the average sea surface temperature (SST) would increase by 2 °C to 3 °C in all regions, relative to present values, except in the Sea of Okhotsk. Similarly, predicted annually averaged maximum sea surface salinity (SSS) increased by 0.7 in the subtropical region due to a decrease in fresh water flux and decreased by 0.3 in the subarctic region due to an increase in the freshwater flux, except off northern Japan. A salinity front is formed in the Kuroshio extension region, which is the transition region between subarctic and subtropical gyres. The change in the maximum mixed layer depth (MLD) is largest in the subarctic–subtropical transition region where maximum MLD is deepest. In the global warming scenario, the MLD becomes shallower by 50 to 100 m along the southern part of the Kuroshio Current due to the increase in surface water temperature. The predicted northward shift of the separation latitude of the Kuroshio Current near 42°N can, in these regions, result in future MLDs to be either 100 m shallower or 200 m deeper than present values over short horizontal distances. The effect of these and other changes in water column structure in Hashioka and Yamanaka's (2007) global warming scenarios on lower trophic level productivity and its timing are discussed in sections that follow.

Effects of freshwater input on water column stratification through Gargett's (1997) "optimal stability window" hypothesis were mentioned above and additional global considerations are given in the section that follows. However, specific to high latitudes, observations and predictions of increased freshwater input and the associated stratification due to melting Arctic regions (as well as related changes in the hydrologic cycle, e.g. precipitation) are becoming robust (ACIA, 2004; IPCC, 2007; Schrank, 2007). Most ocean climate models predict increased stability of the surface mixed layer, reduction in salt flux, less ocean convection, and less deepwater formation (e.g. Stenevik and Sundby, 2007) which could lead to a prolonged reduction in thermohaline circulation and ocean ventilation, particularly the meridional overturning circulation in the Atlantic (Bryden, 2005; IPCC, 2007). If the thermohaline circulation is reduced, this could impact the conditions in Norwegian waters where herring, capelin, and other small pelagics are found (Sissener and Bjørndal, 2005). The inflow of Atlantic water into the Nordic Seas, which depends in part on the thermohaline circulation, could be reduced following a weakened thermohaline circulation. The result would be cooling which could act counter to the projected anthropogenic heating. However, the effect of an expected increase in the winds and the increased temperature flux from atmosphere to ocean during summer could, in turn, exceed the effects of the reduction in thermohaline circulation. One possible outcome is that the increased wind transport of warm Atlantic water may compensate for the reduction in the thermohaline circulation, and hence the net effect will be a warmer Nordic region (Stenevik and Sundby, 2007). However, uncertainties remain in terms of which process will dominate, and the results discussed should be considered as first steps in our depiction of future scenarios.

Predicted avenues of ecological change

We focus on three major processes below: (1) changes in productivity and zooplankton, and subsequent effects on the trophic web; (2) distributional changes and subsequent effects on exploitation; and (3) changes in circulation and their effects on recruitment processes. For each process we will first describe the mechanism(s) and second illustrate it with some examples of already observed or suspected interdecadal changes.

Changes in productivity and composition of lower trophic levels

Bottom-up climatic effects resulting from changes in nutrient supply are among the most studied (e.g. McGowan *et al.*, 2003). Predicted global changes in ocean physics suggest generally higher temperatures, increased stratification, and a slowing down of the deep water formation (Houghton *et al.*, 2001; Meehl *et al.*, 2007). The ocean's primary productivity will also change from possible modifications in cloud cover, light availability, and micro-nutrient supply, although the IPCC (2007) concludes that there is insufficient evidence to determine whether trends exist in tornados, hail, lightning, and dust storms at small spatial scales. While warming and light supply affect photosynthesis directly, increased vertical stratification and water column stability will reduce the nutrient availability to the euphotic zone and thus possibly reduce primary (Behrenfeld *et al.*, 2006) and secondary (Roemmich and McGowan, 1995) production. However, Doney (2006) proposes an opposite scenario in which an increased stratification will increase the productivity because the residence time of particles in the euphotic zone will increase, assuming the nutrient supply remains the same. This might not be the case in SPACC areas, where the increased stratification will decrease the nutrient input, and despite the phytoplankton cells remaining longer in the euphotic zone, they may still not have enough nutrients to take advantage of the increased exposure to light. Predicted global changes in ocean physics are also expected to induce shifts in phytoplankton species composition (Boyd and Doney, 2002; Hashioka and Yamanaka, 2007). Nonetheless, past climate changes suggest that, despite these shifts, the goods and services provided to the ecosystem by phytoplankton should not be strongly affected due to the genetic diversity and high functional redundancy of this trophic level (Behrenfeld *et al.*, 2006).

Bottom-up processes

Quantitative projections of future ecological scenarios are in their early stages, with several key limitations including uncertainties in the biological responses (e.g. the range of primary production algorithms such as those of Behrenfeld and Falkowski, 1997 and Carr, 2002), and the range in predicted physical changes (Cubasch *et al.*, 2001; Doney *et al.*, 2004). Nevertheless, and with these caveats in mind, the study of Sarmiento *et al.* (2004) summarized possible mid twenty-first century responses of ocean ecosystems, based on estimates from six climate models. In general, ice retreats occur in both hemispheres, and sea surface temperature warms globally, with southern high latitudes warming less than northern high latitudes. Changes in salinity in response to the enhancement of the hydrologic cycle are such that subtropical regions (high evaporation and salinity) become saltier, whereas waters at higher latitudes and the tropics become fresher as a consequence of greater rainfall. Overall, the ocean's surface density is reduced. At most locations the resulting increase in vertical stratification reduces the vertical supply of nutrients and decreases the maximum mixed layer depth. Note, however, as previously mentioned, that, while warmer temperatures can increase water column stratification and thus result in a shallowing of the thermocline, changes in wind-stress curl can deepen the thermocline. In the models studied by Sarmiento *et al.* (2004) there was no discernible change in the upwelling patterns except for a slight reduction near the equator. The coastal and equatorial upwelling within 15° on each side of the equator drops by 2%–15%. The predicted magnitude of these changes is subject to revision with models of increased spatial and temporal resolution. In the sub-Arctic areas of the Eastern-North Atlantic, where a 2 °C–4 °C increase is expected, associated with sea-ice retreat, an increase in the primary and secondary production is expected to favor pelagic fish species such as, for instance, the herring in the Norwegian Sea (Skjoldal, 2004).

Sarmiento *et al.* (2004) also considered the possible areal changes of existing biomes (or biogeographic provinces). Relevant to the SPACC low-latitude upwelling biomes, the changes found were modest with increases in the Northern Hemisphere on the order of 1%, and slightly greater decreases in the Southern Hemisphere of ~2%, although estimates of primary production are not robust due to the sensitivity of the results to existing algorithms mentioned above. Primary production in lower latitude biomes was found to decrease or increase depending on the algorithm used, with potentially large differences. Part of the difficulty lies in the different capabilities of the algorithms to include direct effects due to warming as well as the simultaneous indirect (e.g. stratification) effects.

Hashioka and Yamanaka's (2007) model results for the NW Pacific under a global warming scenario also show increases in vertical stratification due to rising temperatures. They predict decreases in nutrient and chlorophyll *a* concentrations in the surface water at the end of the twenty-first century. In the global warming scenario, the onset of the diatom spring bloom is predicted to take place one half-month earlier than in the present-day simulation due to the stronger stratification. The maximum biomass in the spring bloom is predicted to decrease significantly compared to present conditions due to decreases in nutrient concentrations. In contrast, the biomass maximum of the other small phytoplankton at the end of the diatom spring bloom is the same as the present, because their ability to adapt to low nutrient conditions due to their small half-saturation constant. Therefore, a change in the dominant phytoplankton group appears noticeably at the end of spring bloom. Hashioka and Yamanaka (2007) find that changes due to warming are not predicted to occur uniformly in all seasons, but that they may occur most noticeably at the end of the spring and in the fall bloom.

Phenology and the match–mismatch hypothesis

The above shifts in the timing of the phytoplankton blooms can be related to the match–mismatch hypothesis initially proposed by Cushing (1969, 1990) and can be extended to effects of CC. One such case study is that of Beaugrand *et al.* (2003) and Reid *et al.* (2003) who showed that fluctuations in plankton abundance in the North Sea due to CC affected larval cod survival due to a mismatch between the size of prey (calanoid copepods) and cod larger than 30 mm after the mid 1980s. In addition to the conventional match–mismatch hypothesis, there is a need to consider the amplitude of the peak (Fig. 14.2) and possibly a threshold effect (Stenseth and Mysterud, 2002; Durant *et al.*, 2005). Durant *et al.* (2005) demonstrated that, in the case of the herring/puffin match–mismatch, the abundance of herring was structuring the match between predator and prey.

If, under the influence of CC, the phenology of a species shifts at a different rate from that of its prey, this will lead to a mismatch (Stenseth and Mysterud, 2002; Abraham and Sydeman, 2004; Edwards and Richardson, 2004; Visser and Both, 2005). Durant (Joël Durant,

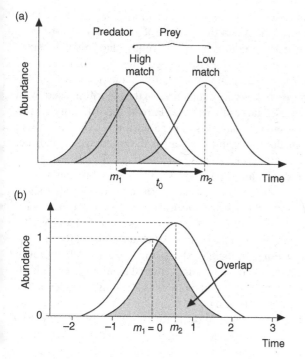

Fig. 14.2. Interaction between two trophic levels: (a) conventional match–mismatch hypothesis. A high match is represented by a temporal overlap of the predator and its prey. An increase of the time-lag (t_0) between the two populations leads to a low match: a small or nonexistent overlap; (b) determination of the overlap of two normally distributed populations. The higher trophic level distribution is the left curve. The overlap (gray shade) depends on the degree of mismatch ($m_1 - m_2$) and the difference in abundances (Durant *et al.*, 2005; permission granted by Blackwell Publishing).

University of Oslo, Norway, pers. commun.) identified three theoretical effects of CC on phenological mismatch: a change in the mean relative timing of predators and prey; a change in the level of prey abundance; or a change in the amplitude of year-to-year variations in prey timing for regions where interannual variability in temperature is expected to increase. This new vision of match–mismatch seems relevant for high latitude SPACC systems where a single spring plankton bloom occurs. In lower latitude and in pulsed upwelling systems, a similar process may still occur but on a shorter time scale, that is at the scale of upwelling events. Under the hypothesis of increased winds favorable to upwelling, one can speculate that the matching events will increase.

Food web considerations
The direct effect of temperature on cod recruitment in different areas around the North Atlantic has been reinterpreted by Sundby (2000) who suggests that, in addition to its direct effect, temperature was likely to be a proxy for zooplankton abundance which, in turn, has a major effect

on cod larvae survival. But, what caused this change in zooplankton? Sundby (2000) argues that, at least in the Barents Sea, the cause may be the advection of warm and zooplankton-rich Atlantic water from the Norwegian Sea. In the Norwegian Sea itself, temperature could directly control the growth of copepods, especially *Calanus finmarchicus*. Additionally, Sundby (2000) suggests that the abundance of the zooplankton population also depends on the abundance of its prey, phytoplankton. In the end, the optimal temperature window observed for cod abundance by Planque and Frédou (1999) could result from the combination and interaction of a direct effect of temperature on cod, but also through indirect effects of temperature on the foodweb modulated by advective processes that, depending on the flux direction, will associate prey abundance for cod with cool or warm temperature (Fig. 14.3). Because copepods are also a main prey of small pelagic fish such as herring and capelin, Sundby's finding is likely to apply to these species, in the North Sea and elsewhere. Indeed, Robinson and Ware (1994) and Robinson (1994) show that high temperatures enhance phytoplankton production and trophic transfer from phytoplankton to zooplankton to fish (Pacific hake and Pacific herring) in the coastal upwelling system off southern Vancouver Island, British Columbia.

Under scenarios of increased frequencies of El Niño occurrences and intensities in relation to CC, sardine regimes might be favored in the Humboldt ecosystem, and possibly in the Californian ecosystem. In these ecosystems, the coastal food web is dominated by large phytoplankton, typically colonial centric diatoms, which supports a food web with large zooplankton (i.e. euphausiids) and small pelagic fish that graze directly on phytoplankton. During cool conditions this ecosystem occupies a large area (~200 km offshore extension) and there is a spatial and temporal separation of sources (nutrients, phytoplankton) and sinks (phytoplankton, zooplankton). An "oceanic" ecosystem is found offshore and is dominated by picophytoplankton whose grazers are protists with similar growth rates creating an efficient recycling system. A complex food web evolves with a smaller proportion of the primary production reaching upper trophic levels (Fig. 14.4, upper panel). During warm years the productive area is reduced and the oceanic ecosystem impinges on the shore. In the reduced productive habitat spatial separations are less evident. In this scenario, warm conditions result in reduced productivity, however global warming may enhance upwelling and production (Fig. 14.4, lower panel). Although the processes determining such regime shifts are not fully understood (see this volume, Chapter 12), the most commonly advocated explanation is a change in food chains, possibly due to weakened current flows, favoring smaller plankton species that are more easily filtered and more efficiently assimilated by sardines than by anchovy (van der Lingen *et al.*, 2008).

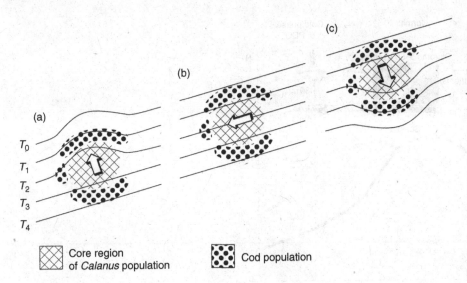

Fig. 14.3. Conceptual figure of spatial distributions of Atlantic cod stocks around the core regions of *Calanus finmarchicus* in the North Atlantic Subpolar Gyre. The temperature distribution is indicated by isotherms of increasing temperature from T_0 to T_4. The increased volume flux (arrows) from the core region of *C. finmarchicus* to cod stocks can be (a) along a negative temperature gradient at lower ambient temperature, (b) at approximately equal ambient temperature, or (c) along a positive temperature gradient at higher ambient temperature (from Sundby, 2000, permission granted by Elsevier).

Furthermore, increased occurrences of El Niños, should they occur, may also enhance the connections between the different stocks of sardine and favor gene exchanges and faster recovery from collapses.

In Eastern Boundary Ecosystems, global warming in offshore regions may result in a warming of surface waters, increased stratification, and lower overall primary production. In contrast, in inshore regions we may expect stronger upwelling, cooler waters, and increased primary production. Such patterns would lead to an increase in cross-shelf gradients (including hydrographic and biological fronts), as well as affect the distribution and abundance of pelagic species with sardine being favored offshore and near frontal zones, and anchovy favored closer to shore.

Changes in community structures of plankton and fish

Boyd and Doney (2002) consider possible floristic shifts induced by climate variability. They find differences in chlorophyll fields that show regional patterns of both higher and lower stocks in temperate and high latitudes and in coastal upwelling areas, as well as a global reduction in new production between 5% and 10% associated mainly with the decreased availability of subsurface nutrients. Boyd and Doney (2002) conclude that regional floristic shifts can be as important as changes in bulk productivity. Hashioka and Yamanaka's (2007) model results for the NW Pacific predict shifts in the dominant phytoplankton group from diatoms to other small phytoplankton. Changes in seasonal variations of biomass and the dominant phytoplankton group in the subarctic–subtropical transition region associated with the global warming are large in all regions. These shifts in phytoplankton could be another, or an additional, explanation for the above-mentioned large-scale changes in the biogeography of zooplankton (Beaugrand *et al.*, 2002) in response to shifts in climatic zones. In order to explore

these issues, detailed coupled models of zooplankton dynamics including the developmental stages of different species are required, but their development is still limited (e.g. Carlotti and Radach, 1996; Carlotti *et al.*, 2000).

Brodeur *et al.* (1999) found that the biomass of large medusae in the eastern Bering Sea shelf increased gradually from 1979 to 1989, followed by a dramatic increase in the 1990s. The authors found that several large-scale winter/spring atmospheric and oceanographic variables in the Bering Sea displayed concomitant changes beginning around 1990 and speculate on whether the increase in biomass of medusae resulted from anthropogenic perturbations or is a manifestation of natural ecosystem variability. However, beginning in 2000 the peaks of gelatinous zooplankton in the Bering have declined markedly and have remained low (see North Pacific Ecosystem Status Report, PICES, 2005) and hence, whether the bloom of the 1990s is associated with a long-term trend, or isolated events remains unknown. Harley *et al.* (2006) state that direct climatic impacts on one or a few leverage species could drive the response of an entire system. Hence, if the future presence of increased concentrations of gelatinous zooplankton is more persistent, we may expect profound shifts in ecosystem structure.

In the Bay of Biscay, many fish species are at the southern or northern limit of their distributional range and recent climate changes have resulted in a change of the fish community structure. Abundance data from experimental ground fish surveys performed during the period 1973–2002 identified two groups of species. Within the group displaying a negative trend, one-third (seven) of the species are characterized by a northerly distribution and a narrow latitudinal distribution range. Their biomass declined during the study period. The biomass of the other group, comprised mainly of transition species, displayed no long-term trend. Within

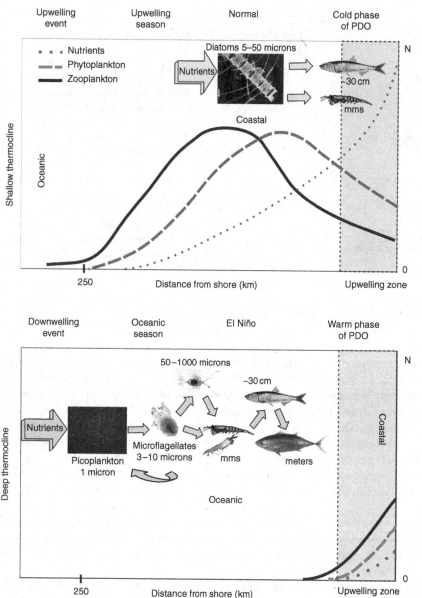

Fig. 14.4. Conceptual model of the spatial distribution of nutrients, phytoplankton, zooplankton and food web structure during cool (upwelling, normal, cool phase of the PDO) and warm (upwelling relaxation, El Niño, warm phase of the PDO) conditions in a coastal upwelling ecosystem. (Modified from Chavez *et al.*, 2002, with permission from Elsevier.)

the group exhibiting an increasing trend in abundance, one-third (12) of the species are characterized by a southern distribution and wide range of latitude (Poulard and Blanchard, 2005).

Distributional changes

Climate change is expected to drive most terrestrial and marine species ranges toward the poles (Southward *et al.*, 1995; Parmesan and Yohe, 2003), as was the case during the Pleistocene–Holocene transition (for review see Fields *et al.*, 1993), although the amplitude might be different. Because temperature is a key factor in the distribution of marine animals, both due to direct (e.g. metabolism, tolerance especially for planktonic larval stages) or indirect effects (e.g.

proxy for other environmental conditions, predator–prey distribution) it is expected that CC will affect the location and geographic range of these organisms. There are already several examples of such changes that manifest as either latitudinal shifts or habitat extensions/contractions, especially for mobile species like fish (e.g. Beaugrand and Ibáñez, 2004).

Perry *et al.* (2005) showed that nearly two-thirds of North Sea demersal species shifted their mean latitude, depth, or both, over 25 years in response to recent increases in sea temperature. Half of the species with northerly or southerly range margins in the North Sea have displayed northward boundary shifts with warming (Fig. 14.5), except one moving southward. The authors tested the hypothesis that observed

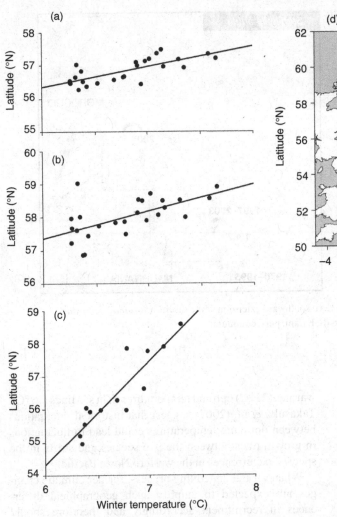

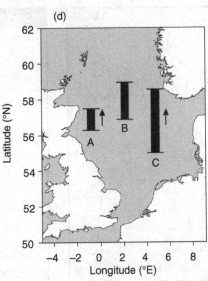

Fig. 14.5. Relationships between mean latitude and 5-year running mean winter bottom temperature for (a) cod, (b) anglerfish, and (c) snake blenny are shown. In (d), ranges of shifts in mean latitude are shown for (a), (b), and (c) within the North Sea. Bars on the map illustrate only shift ranges of mean latitudes, not longitudes. Arrows indicate where shifts have been significant over time, with the direction of movement. (From Perry *et al.*, 2005; reprinted with permission from *Science*.)

differences in rates of movement among the species result from differential rates of population turnover. As expected, the shifting species have faster life cycles and smaller body sizes than non-shifting species. Perry *et al.* (2005) also discuss the effect of further temperature increases on commercial fisheries targeting cod, sole, blue whiting, or redfish through continued shifts in distribution but also alterations in food-web structures, leading to eventual ecosystem-level changes. For example, differential rates of shift could result in altered spatial overlap among species, thereby disrupting interactions, especially trophic interactions between small pelagic fish and their predators. In addition, these changes can also potentially interfere with the decoupling effects of climate-driven changes in phenology.

In a study of small pelagic fish, Beare *et al.* (2004) reported a sudden increase of warm water pelagic species, anchovy (*Engraulis encrasicolus*) and sardine (*Sardina pilchardus*) in the northwestern North Sea after 1995, in the first quarter of each year. The authors attributed these long-term changes to CC, although the causal mechanism is not clear. These observations are confirmed by recent GLOBEC studies from 2003–2005 showing that anchovies and sardines are now spawning in the southern North Sea (J. Alheit, pers. commun.). Sardines have appeared off British Columbia, Canada in the late 1990s after an almost 50-year absence (McFarlane and Beamish, 2001), but this expansion could be related more to the recovery of the sardine population off California rather than to warmer waters enabling sardine to migrate further north.

An example of the direct effect of temperature on recruitment is given by Planque and Frédou (1999), who considered the recruitment of cod in different areas around the North Atlantic. In areas that are cooler than optimal for cod, recruitment is better in warmer years, but in areas that are warmer than optimal, recruitment is better in cooler years. This nicely illustrates the principle of an optimal

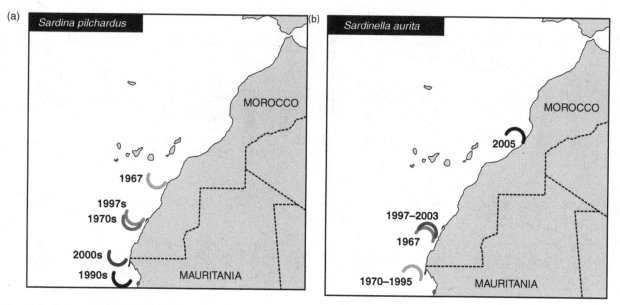

Fig. 14.6. Changes in the distribution of exploited pelagic fish stocks. Left: southward extension of the sardine (*Sardina pilchardus*) stock. Right, northward extension of sardinella (*Sardinella* spp.) stock. (S. Kifani, pers. commun.)

environmental window, although an alternative explanation related to temperature as a proxy for zooplankton abundance has been proposed subsequently (Sundby, 2000; see previous section).

Drinkwater (2005) estimated the response of cod stocks in the North Atlantic using predicted changes in temperature fields and the known observed responses of cod to temperature variability. Stocks in the Celtic and Irish Seas are suggested to possibly disappear under predicted temperature changes by the year 2100, while those in the southern North Sea and Georges Bank will decline. Northward spreads are likely along the coasts of Greenland and Labrador and will occupy larger areas of the Barents Sea, with possible extensions on to some of the continental shelves of the Arctic Ocean. Spawning sites are estimated by Drinkwater (2005) to be established further north than presently found, with spring migrations occurring earlier, and fall returns occurring later. Consistent with other attempts to project changes, it is pointed out that these responses to future climate changes are highly uncertain, and will depend on the changes to other climate and oceanographic variables besides temperature, such as plankton production, the prey and predator fields, and industrial fishing.

Recently, Takasuka *et al.* (2007) proposed an "optimal growth temperature" hypothesis to explain the sardine and anchovy regime shifts. They suggest that the shifts are caused by differential optimal temperatures for growth rates during early life stages, with anchovy having

warmer (22 °C) optimal temperatures than sardines (16 °C). Takasuka *et al.* (2007) suggest that historical fluctuations between optimum temperatures could lead to fluctuations in growth rates between the two species and shifts in the species' occurrences in the western North Pacific.

Whatever the underlying process, further climatic changes are expected to amplify such geographical differences in recruitment accordingly and therefore should result in shifts in the distribution of the species. Although we do not have fully documented examples to cite, individual-based models of the early life history of anchovy and sardine in the southern Benguela indicate substantial mortality of larvae due to contact with cold waters (Mullon *et al.*, 2003; Miller *et al.*, 2006) assuming larval mortality below a threshold of 13 °C, as observed *in vitro* (King, 1977). This result suggests that an increased upwelling intensity could result in lower recruitment success.

Latitudinal shifts in the Canary Current

Quero *et al.* (1998) and Brander *et al.* (2003) noticed a northward shift of commercial and non-commercial fish species distributions from southern Portugal to northern Norway that has been occurring since the late 1980s. According to Quero *et al.* (1998), 16 species of tropical fishes, unnoticed in commercial and scientific records prior to 1950, are now present in European Atlantic waters, mostly (68%) on the upper slope, between approximately 200 m and 600 m. Furthermore, since 1963, at least six species from the upper

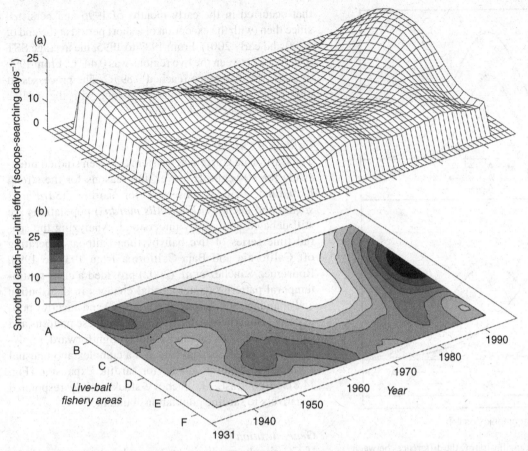

Fig. 14.7. Geographical variability of Pacific sardine (*Sardinops sagax*) abundance along California and Baja California for a 67-year period. The annual catch-per-unit-effort (CPUE) values are smoothed. Area A is off California and area F is in the mouth of the Gulf of California. (a) A three-dimensional view and (b) its projection in two dimensions. (Rodríguez-Sánchez *et al.*, 2002; reprinted with permission from National Research Press.)

slope displayed northward range extensions from southern Portugal to north-western Ireland during time periods varying from 30 years for *Cyttopsis roseus* to 6 years for *Phoeroides pachygaster*. The northward range extension of upper slope species and the higher frequency of records of continental shelf species from the southern part of the Bay of Biscay are concurrent with the 1 °C per decade warming of the north–south current in the upper slope of northern Spain (Le Cann and Pingree, 1995) and of the southern French Atlantic continental shelf (Koutsikopoulos *et al.*, 1998), respectively. This finding is confirmed by more recent studies also based on bottom species and Myctophidae in northwest Spain (e.g. Bañón *et al.*, 2002) and even further north in the Bay of Biscay (Blanchard and Vandermeirsch, 2005). Nonetheless, recent changes in the distribution of two major pelagic species in the region is troublesome. While *Sardinella aurita* expansion is predominantly poleward, as expected under a warming scenario, *Sardina pilchardus* expansion is mostly equatorward (Fig. 14.6), which can be

related to the finding of an upwelling intensification in this region (McGregor *et al.*, 2007).

Europe

Latitudinal shifts are not limited to fish species. Using Continuous Plankton Recorder data from 1960 to 1999, Beaugrand *et al.* (2002) documented a major large-scale reorganization of the plankton communities, especially the calanoid copepod crustaceans, in the eastern North Atlantic Ocean and European shelf seas. A northward extension of more than 10° in latitude occurred for warm-water species over the last four decades associated with a decrease in the number of colder water species and were related to both the increasing trend in Northern Hemisphere temperature and the North Atlantic Oscillation. Beaugrand *et al.* (2003) showed that, in addition to the effects of overfishing, these fluctuations in plankton abundance have resulted in long-term changes in cod recruitment in the North Sea through three bottom-up control processes (changes in mean size of

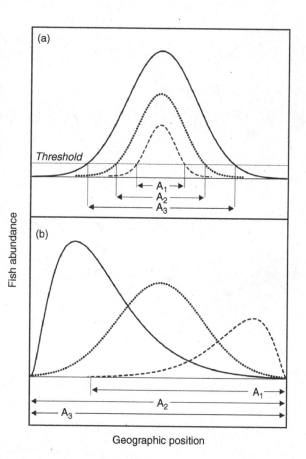

Fig. 14.8. Simplified diagram illustrating the differences between a homogeneous spread resulting from the simple expansion–contraction model of range changes with population abundance increase–decrease and the spatial process described by Rodríguez-Sánchez *et al.* (2002) for the *Sardinops sagax* population in the California Current System. (a) Expansion–contraction model, with the implicit assumption of a geostationary population, usually underlying conventional analyses of long data series.
(b) A non-geostationary dynamic model proposed for Pacific sardine on a regime scale. In both models, A_1 represents the refuge area occupied when the population size is at a low level, A_2 is that when the population size is growing or diminishing, and A_3 is the largest extension occupied when the population size is at a high level. (Reproduced from Rodríguez-Sánchez *et al.*, 2002; reprinted with permission from National Research Press.)

prey, seasonal timing and abundance) that will be described in the next sections.

Anchovy spawning grounds in the Benguela Current
Roy (2007) and others report an eastward shift of the spawning grounds of anchovy in the Southern Benguela from acoustic and egg data collected during the period 1984–2005. This shift occurred in 1996 and it was linked with the pronounced shift in the difference of SST observed between the western and central part of the Agulhas Bank

that occurred in the early months of 1996 and persisted since then (with the exception of a short period at the end of 2000 and early 2001). From 1987 to 1995, the average SST difference between the two regions was 0.44 °C. From 1996 to 2005, the difference reached 0.89 °C. The processes (if any) linking this environmental change with the sardine eastward shift are not elucidated.

California Current
The California Current ecosystem has been studied intensively for 80 years. However, the reasons for the strong latitudinal changes observed for sardine (*Sardinops sagax*) and anchovy (*Engraulis mordax*) populations are still debated and poorly understood. Analyzing the spatial time series of live bait by tuna baitboats operating off California and Baja California from 1931 to 1998, Rodríguez-Sánchez *et al.* (2002) provided a clear spatio-temporal pattern of interdecadal changes in distribution and abundance of the sardine (Fig. 14.7) and anchovy populations. Nonetheless, the link between these patterns and CC proposed by the authors is not straightforward, implying unexplained decadal lags and a complex and unusual asymmetric dynamic model for sardine expansion (Fig. 14.8). In contrast, the anchovy stock mainly responded negatively to sardine population abundance.

Generalization
If CC simply results in uniform global warming of the oceans, one could readily predict that many species distributions will shift toward the poles (Parmesan and Yohe, 2003) and a retreat or extinction of Arctic (Beaugrand *et al.*, 2002) and Antarctic species. In upwelling areas, this should translate into the intrusion of subtropical species into the upwelling ecosystems, especially during summer, and the poleward shift of the species usually found in these ecosystems. For instance one could expect the following from a simplistic point of view.

1. In the Benguela ecosystem, an increased intrusion of *Sardinella* spp. from Angolan to Namibian waters, associated with southward retreat of *Sardinops sagax*, *Trachurus* spp., and *Engraulis encrasicolus*.
2. In the California ecosystem, a northward displacement of the sardine and anchovy populations.
3. In the Canary ecosystem, an increased northward intrusion and residence time of *Sardinella maderensis*, *Pomadasys* spp., *Chloroscombrus chrysurus* and *Trachurus trecae* from Guinea to Senegalese and Mauritanian waters, associated with a northward retreat of *Sardina pilchardus*, *Caranx rhonchus*, *Trachurus* spp., and *Engraulis encrasicolus*, particularly in Morocco.
4. In the Humboldt ecosystem, a southward shift of the anchovy and sardine stocks. Anchovy could completely

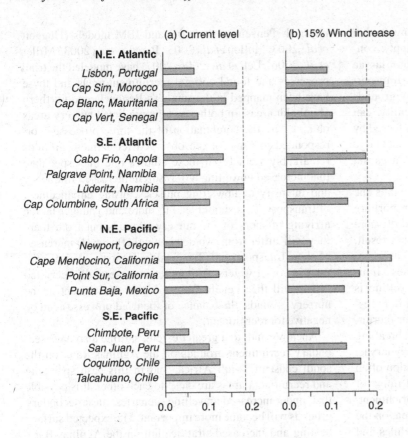

Fig. 14.9. Comparisons of indices of rates of water renewal (units: fraction of total upper layer volume replaced per day), for prominent eastern ocean coastal upwelling zones. (a) Estimates characterizing the late twentieth century. (b) The same data adjusted for a hypothetical 15% wind increase at each location. (Reproduced from Bakun and Weeks, 2004; reprinted with permission from Blackwell Publishing.)

disappear from Ecuador but extend further south into Chile.

But, the manifestations of CC are not that simple, particularly in upwelling areas where CC is expected to induce opposite effects on water temperature, at least according to some authors: an increase due to global warming of the oceans and consequently warmer surface waters advected by the dominant currents (Benguela, California, Canary and Humboldt currents), but a decrease due to the intensification of trade winds (Snyder *et al.*, 2003; Diffenbaugh *et al.*, 2004). Auad *et al.* (2006) forecast an increased inshore transport in the southern part of the Southern California Bight during the summer season, and suggest that the sardine fishing industry/fleet based in southern California could benefit from these forecast oceanic changes. In contrast, in the northern limit of their model domain, the model forecasts an offshore transport that would be detrimental to the concentration of nutrients and larvae in the coastal ocean. As expressed earlier, other authors (e.g. Vecchi *et al.*, 2006) suggest that the upwelling intensity will decrease offshore despite enhanced wind close to shore, and that the combination of increased coastal turbulence and weaker upwelling would negatively affect populations of small pelagics.

Similar to the previous California current example, the northward extension of sardinella and southward extension of sardine in the Canary current ecosystem (Fig. 14.6)

do not display a clear trend and are not clearly correlated with changes in trade wind intensity or SST time series (but are more related to changes in abundance; see this volume, Chapter 5). Furthermore, not all species will respond in the same way to CC, due to their differences in trophic level, tolerance to environmental changes, and latitudinal distance from their optimal habitat. For instance, Crawford *et al.* (1991), based on an empirical analysis of trends in sardine catches in three regions of the Pacific, suggest that *Sardinops* spp. are more susceptible to CC than are other small pelagic fish, an opinion consistent with the study of Rodríguez-Sánchez *et al.* (2002). Another reason preventing straightforward generalization of distributional shifts induced by CC is the interactions among species at the community level that also influence range boundaries (Davis *et al.*, 1998). Finally, biogeographic boundaries in coastal oceans are also set by current-mediated dispersal limitation, regardless of the existence of suitable habitats beyond the dispersal barrier (Gaylord and Gaines, 2000).

Effects of changes in circulation on recruitment processes
Climate changes and Bakun's triad
Most marine organisms have a pelagic phase in their early life history associated with high mortality which, in turn, results in high fecundity to compensate for it. Bakun (1998)

hypothesized that three essential processes are involved in recruitment success: enrichment of nutrient supply, concentration of larvae and their prey, and retention of larvae and recruits in the nursery area. In some cases, transport of ichthyoplankton from the spawning to nursery grounds can be a key process. CC, including its variability, can result in modifications of the circulation patterns forced by changes in the coastal wind fields or other changes in large scale ocean forcing through El Niño-like Kelvin waves and coastal trapped waves.

If we consider scenarios of intensified coastal winds due to CC, then increased Ekman pumping and transport are expected, resulting in increased productivity and offshore losses of ichthyoplankton through advection. As a result, one expects favorable effects on recruitment in areas where upwelling strength is weak and therefore constitutes a limiting factor. However, in other areas where upwelling is already optimal, increased upwelling may result in excessive offshore losses of larvae and therefore poorer recruitment. Increased winds can also affect turbulence, breaking down the thermocline and dispersing food for early larvae. It is important to note that enrichment and retention often work in opposite directions so that an optimal range of upwelling strength is achieved by balancing enrichment, on the one hand, with retention in the nursery area, on the other (Cury and Roy, 1989). It is likely that sardines and anchovies have evolved serial spawning in response to fluctuations in upwelling strength during the spawning season that can be seen as a bet-hedging strategy to cope with the triad of enrichment, concentration, and retention.

Considerations from empirical models of the effect of CC on the triad

In eastern boundary upwelling systems (EBUS), maximum rates of upwelling are observed roughly near the center of each region (Bakun and Weeks, 2004). In some upwelling regions such as the Lüderitz upwelling cell of the Benguela Current (BC), this area of peak upwelling is unfavorable for primary production and retention because of deep turbulent mixing and strong advection. In the scenario of increased upwelling with CC, it is likely that this area will expand, moving the more favorable areas on either side northwards and southwards. Similar phenomena are likely to occur in other EBUS (Fig. 14.9).

The southern Benguela Current is an archetype for the separation of the major spawning and nursery grounds of sardine and anchovy, the two being on average 400 km apart and connected through a coastal jet that transports the ichthyoplankton from one to the other (Hutchings *et al.*, 1998; van der Lingen *et al.*, 2008). There is, of course, a reverse migration of the recruiting adults. Therefore, transport is here a key element of the triad. These processes have been investigated thoroughly using coupled 3-D physical (Penven *et al.*, 2001) and IBM models (Huggett *et al.*, 2003; Mullon *et al.*, 2003; Parada *et al.*, 2003; Miller *et al.*, 2006). Lett *et al.* (2006, 2007) have modeled the triad processes and found optimal recruitment areas and these have been mapped in the southern Benguela and northern Humboldt areas. In both cases the predicted nursery areas obtained by the combination of the three processes corresponded to some of the observed recruitment grounds remarkably well. From these studies one may speculate that increased upwelling will result in increased frequency and intensity of upwelling pulses, causing enhancement of transport. This should lead to more and younger larvae arriving offshore of the nursery grounds but from there they will suffer from two detrimental effects: (1) increased offshore transport; and (2) mortality due to the slow growth rates in cooler waters. Although the successfully recruited larvae will then benefit from a higher enrichment in the nursery grounds, the balance of the triad processes can be negative for recruitment.

Anchovy and, to a greater extent, sardine also have secondary recruitment grounds on the Agulhas Bank on the south coast of South Africa, and in this case spawning and recruitment areas are close to each other. It is possible that, under increased upwelling scenarios, these secondary grounds will become more important. The expected surface heating and increased stratification on the Agulhas Bank due to CC might nonetheless reduce the habitat of young and adult anchovies that are more abundant in cold water (Agenbag *et al.*, 2004). Furthermore, anchovy being more specific than sardine in their preference for various environmental conditions, CC could affected them more than sardine (Twatwa *et al.*, 2005), although the balance between expected positive and negative effects of CC on anchovy spawning habitat is unclear. An increase in temperature and stratification, and a decrease in plankton production on the Agulhas Bank are expected to favor the spawning of this species, but an expected decrease in salinity and mixed layer depth is expected to be unfavorable (Twatwa *et al.*, 2005).

Other examples

The Californian Sardine (*Sardinops sagax*) is one of the few stocks to be managed according to environmental conditions (Conser *et al.*, 2004). The annual quota for USA (California, Oregon, and Washington) is a function of the total stock biomass estimated during the previous year (minus a cut-off value) times an environment-based percentage of biomass above the cut-off value that can be harvested by the fisheries, times the percentage of biomass in USA waters. The environment-based percentage of biomass is a quadratic function of the running average SST at Scripps Pier, La Jolla, California during the 3-year period prior to July 1st of a given year. Due to the parameterization used, only the right (increasing) branch of the quadratic function is active in the model (SST >16.5 °C). Therefore, it is assumed that the productivity of the sardine stock increases when relatively warm-water ocean conditions persist. Nevertheless, it is unlikely that the same assumption could have been used for the earlier decades, and possibly for the next ones. If one assumes that catches and sediment records are reasonable proxies of long-term abundance fluctuations over the last century, as did Norton and Mason (2005), it appears that the sardine abundance in California is not correlated with SST at Scripps Pier, but with cumulative anomalies of this SST series. Although the strong correlations between crude data on sardine landings and several cumulative anomalies of environmental variables found by Norton and Mason (2005) are appealing, they are difficult to interpret. The processes linking the interdecadal "memory" of accumulated environmental data over several decades to interdecadal fluctuations of sardine abundance are unknown. McFarlane *et al.* (2002) also find empirical relationships between the interdecadal changes in the catches of Californian sardine and patterns in climate–ocean conditions, but mention that the processes are still unknown and likely to be driven not only by direct physical forcing, but also by changes in the plankton community. MacCall (in Bakun and Broad, 2002) put forward a hypothesis relating the intensity of the California Current to the recent sardine outburst. Under conditions of a slow, meandering California Current, retention of eggs and larvae spawned in the offshore habitat is greater than under conditions of faster, less meandering flow. Coupling or integration of hydrodynamics, biogeochemical, and ecological models should allow testing of such hypotheses (see also this volume, Chapter 12).

Generalization

Wooster and Bailey (1989) and more recently Myers (1998) indicated that populations located at the limit of a species range are more likely to react to environmental changes than those found at the center of the distribution. Therefore, it is expected that in the northern hemisphere, populations close to the northern limit of the species range will show positive correlations with increasing temperature, while those at the southern limit will show negative correlations. For the southern hemisphere, the situation would be reversed with populations close to their southern limit showing positive correlations.

Changes in circulation are also likely to influence connectivity between populations. Connectivity is pivotal to our understanding of population dynamics, genetic structure, and biogeography (Cowen *et al.*, 2006), but is poorly studied in pelagic fish species, the general belief being that highly mobile species (especially sardine) have frequent opportunities to mix their genes. But connectivity might

be important at smaller time scales for recolonization after stock collapses (see the hypothesis of sardine "offshore habitat corridors" in this volume, Chapter 12). Harley *et al.* (2006) suggest that weakening of alongshore advection due to CC could actually break down certain marine biogeographic barriers that currently prevent range expansions of certain species.

In most of the reviewed SPACC ecosystems, including the Kuroshio and Oyashio Current systems where several interdecadal changes in circulation and fish abundance occurred over the last decades (King, 2005), no consistent long-term patterns coherent with the global warming have been observed since approximately the mid twentieth century. Takasuka *et al.* (2007) presented convincing evidence of dome-shaped relationships between sea surface temperature and recent 3-day growth rates of Japanese sardine and anchovy, but substantiation of the empirical relationships between temperature and stock abundance is still underway. Therefore, a firm generalization of the effect of circulation changes is presently not possible.

Other changes

Changes in riverine input and storm frequency and intensity

In the North Atlantic, some modeling studies suggest that storms and hurricanes are expected to increase moderately in frequency and intensity due to CC (e.g. Knutsona *et al.*, 2001), although there was no evidence for such an increase until recently (Trenberth, 2005). According to the IPCC 2007's WP1 release, the frequency of tropical cyclones globally is projected to decrease. At the same time, increased peak wind intensities and increased mean and peak precipitation intensities are expected, with the possibility of a decrease in the number of relatively weak hurricanes, and increased numbers of intense hurricanes. The IPCC report also predicts a robust pattern of increased subpolar and decreased annual subtropical precipitation over North America and Europe, while subtropical droughts are less obvious over Asia. The precipitation decrease is very likely to occur in European and African regions bordering the Mediterranean, and in winter rainfall in southwestern Australia. These events can be expected to disturb the water column stability resulting in the dispersion of plankton swarms and fish larvae aggregation, an unfavorable situation for the optimal feeding of the latter (Peterman and Bradford, 1987) when storm and critical fish life stages overlap in time and space.

Increases of riverine input are expected under some scenarios of CC, due to increasing precipitation on land, which may or may not be associated with extreme weather events. As a result, the nutrient input should increase near the estuaries of large rivers located on high latitudes such as the Gironde area in the Gulf of Biscay. This increase should favor the nursery areas of some small pelagic fish species such as anchovy. Nonetheless, because the increase is likely to be irregular and mainly linked to heavy storms, it is not obvious that it will match the life cycle history of the pelagic species.

Effect of acidification and UV increase on phyto- and zooplankton community structure

Based on different simulation models and scenarios of CC, Orr *et al.* (2005) suggest that, due to ocean acidification, the Southern Ocean surface waters will begin to become undersaturated in aragonite, a metastable form of calcium carbonate, by the year 2050. The authors estimate that this undersaturation could extend throughout the entire Southern Ocean by 2100. *In vitro* experiments indicate that the aragonite shell of live pteropods exposed to such levels of undersaturation during 2 days show notable dissolution. Because calcite undersaturation is expected to occur later than aragonite undersaturation, coccolithophorid zooplankters and foraminifera might be affected later than pteropods. In SPACC areas, pteropods and coccolithophorids are unlikely to be key species, despite the importance of episodic blooms. The same applies for coralline algae and reef-building scleractinian corals. In contrast, there is more uncertainty regarding the expected detrimental effect of acidification on foraminifera, because the role of these organisms in the trophic web is not fully understood. Because the distribution of aragonite and calcite saturation depth is shallow in the eastern Pacific (Feely *et al.*, 2004), the California and Humboldt ecosystems could be the more affected.

Another concern is the depletion of the ozone layer in response to the increase of CO_2 concentrations (Austin *et al.*, 1992), the resulting increase in ultraviolet radiation at the ocean surface, and the ensuing negative impacts on invertebrate larvae and algae. The response of a given species to UV exposure might depend on the presence of other species (for review see Harley *et al.*, 2006), suggesting the need for more detailed studies.

Increase in diseases due to temperature increases

There are many examples of increase in occurrence or latitudinal shift of diseases in terrestrial and marine ecology, due either to direct response of the pathogenic agent or to the response of its vector. Climate warming can increase pathogen development and survival rates, disease transmission, and host vulnerability, although a subset of pathogens might decrease with warming, releasing hosts from disease (Harvell *et al.*, 2002). As far as we know, less evidence exists in marine ecosystems except for marine mammals, oysters, marine invertebrates and eelgrass (although the mechanisms for pathogenesis are unknown for these last two groups). Furthermore, the growth rates

of marine bacteria and fungi in coral ecosystems could be positively correlated with temperature (Harvell *et al.*, 1999). Some massive mortalities of pelagic fish have been proven to be due to diseases, e.g. in the case of sardine off Australia which was due to a virus (Gaughan, 2002), but are not related to climate change but rather to human introduction of the pathogenic agent. Other massive mortalities, such as the one observed in the Moroccan sardine in 1997, seem more related to abrupt environmental changes (Marek Ostrowski, Institute of Marine Research, Norway, pers.comm.; http://www.imr.no/2004symposium/web/P_Ostrowski.pdf).

Research gaps

Gaps in GCGMs, downscaling and scenarios

Predicting the effect of climate change on marine ecosystems, including those specific to SPACC regions, presents a formidable challenge. There are aspects of climate change that appear likely, in that the predicted trends by independent scientific groups are generally in agreement. For example, it is likely that global temperatures will continue to rise over the next century (e.g. Houghton, 2005; IPCC, 2007) and thus we can conjecture that we will observe changes in the hydrologic cycle, increased stratification of oceanic upper layers, and distributional shifts of marine populations, among others. Other responses are more ambiguous, e.g. the details of species successions, local "sub-gridscale" changes in wind fields. And, finally, there are surprises that lie ahead equivalent to the recently noted ocean acidification.

One of the major problems in trying to forecast the effects of CC is that we are extrapolating outside the range of present observations, and perhaps the past half-million years in terms of types, scales, and rates of change that the Earth System is undergoing (Crutzen and Stoermer, 2000; Steffen *et al.*, 2004). Additionally, ecosystems can amplify a "weak" climate signal (Taylor *et al.*, 2002). Even when considering a single issue related with only one species (e.g. the latitudinal shift of sardine) one is nonetheless dealing simultaneously with a number of processes, direct or indirect, lagged or not, linear or not or, even worse, non-monotonic (Ottersen *et al.*, 2004). Non-monotonic responses are of particular importance for prediction because, if the current range of environmental observation occurs in the monotonic part of the function, the real trend of the response may be in the opposite direction to the predicted one. The problem is aggravated by our limited knowledge of relevant processes, including feedbacks and threshold effects, and by the interaction between any single mono-specific issue and ecosystem interaction (i.e. trophic or behavioural interactions between species). This situation explains why many empirical statistical relationships do not stand the test of time (Myers, 1998).

Specific gaps and shortcomings in our ability to predict future states of SPACC systems include: model resolution (physical, biogeochemical, and ecological), the integration across scales, our ability to assign certainty to our projections, and, of course, the lack of sufficient data to force and validate our models (e.g. Werner *et al.*, 2007).

Gaps in physical–biological models
Resolution and nesting

General Ocean Circulation Models (GCMs) used in making projections of future marine ecosystem states in response to climate are presently run at spatial resolutions of one degree (i.e. grids of 100×100 km; e.g. Sarmiento *et al.*, 2004). However, physical processes determining biogeochemical and biological responses require resolution on the order of kilometers in the open ocean. Model solutions have been shown to yield fundamentally different dynamics depending on the spatial resolution of the model (e.g. McGillicuddy *et al.*, 2003; Komatsu *et al.*, 2006) suggesting that these solutions have not yet converged. The use of regional climate models and methods for downscaling to regional models, e.g. through nesting (Hermann *et al.*, 2002; Snyder *et al.*, 2003; Clark, 2006; Penven *et al.*, 2006; Vikebø *et al.*, 2007) are yielding new insights. However, a point of caution is that increased natural variability at local scales can introduce greater uncertainties in these projections compared to projections at large scales (Houghton, 2005). While in the past decade remarkable advances have been achieved in the realism of basin-scale models (McClean *et al.*, 2006) capable of running at 5–10 km resolution, coastal domains of interest to SPACC species require even higher resolution in order to capture fronts and other topographically controlled flows; such details are presently absent from most advanced basin-scale ocean models. Methodological approaches linking basin-scale models to coastal domains (e.g. Chassignet *et al.*, 2006) and advances in adaptive and unstructured grid refinements appear to be promising (e.g. Pain *et al.*, 2005; Fang *et al.*, 2006). Analogous to nesting physical models, nested approaches for biological processes are described in deYoung *et al.*'s (2004) "rhomboidal" approach to modeling marine ecosystems. In this case, representation of extended food webs in complex marine systems is achieved by concentrating the biological resolution in the main target species of interest, and by making increasing simplifications up and down the trophic scale from the target species. Competitors, prey, and predators would be represented in less detail, leading to a rhomboid-shaped representation of the ecosystem.

Ecological and biogeochemical realism is considerably more difficult to achieve than physical realism. First principles governing the evolution of physical fields are relatively well known, e.g. the constraints imposed by temperature, salinity, and momentum conservation equations. On the other hand, ecosystem models rely much more on empirical relations, relying often on the judgment of the model developers as to what to include and how to parameterize it. Hence, developing a generalized ecosystem model remains elusive. Some of the salient challenges are discussed next.

Lower trophic levels

At geochemical levels, it is essential to include trace elements such as iron (Falkowski *et al.*, 1998; Christian *et al.*, 2002) to model primary production, and there is a need to understand the ocean's acidification and the consequent effect of changes in pH on individual organisms and eventually on ecosystem dynamics (e.g. Orr *et al.*, 2005). Moving up in trophic levels, in order to better represent the biological pump of the carbon cycle, additional phytoplankton, zooplankton, and bacteria groups are needed (e.g. see Le Quéré *et al.*, 2005) but their inclusion is not straightforward (Anderson, 2005; Flynn, 2005). The microbial food web in aquatic ecosystems has been found to have a potentially significant impact on the amount of primary production that is actually available to mesozooplankton, and hence to higher trophic levels (Moloney and Field, 1991) and its parameterization is just now beginning to be quantified (e.g. Smith *et al.*, 2005).

Higher trophic levels (through fish)

The complexity and difficulty of modeling marine ecosystems poses additional challenges when one considers phytoplanktonic, zooplanktonic, and higher trophic level populations. It is important to include not just the total biomass, but also the differences in species' physiology and bioenenergetics in response to changing environmental conditions, as well as their functional roles, ontogeny, behavior, competition, and resulting density-dependent effects, among others. Explicit consideration of zooplankton ontogeny will be required to properly represent the preferred prey of the early life stages of fish, particularly since first feeding larvae are restricted to eating only the small-sized life stages and most fish exhibit a shift towards larger prey as they get older. Promising enhancements in models of marine ecosystems (e.g. Woods, 2005; Ito *et al.*, 2006a,b; Rothstein *et al.*, 2006; Megrey *et al.*, 2007; Rose *et al.*, 2007) are still in an early research stage, but are not yet likely to address most climate change questions quantitatively.

Complex trophic interactions, such as predator–prey competitive interactions, are not commonly included explicitly in marine ecosystem models. In a study of Pacific saury,

Ito *et al.* (2006b) suggested that the inclusion of predation effects on zooplankton resulted in more realistic saury growth. The inclusion of competitive effects when considering co-occurring species in significant abundances cannot be ignored *a priori* and models need to allow for density dependent effects. Another example is that of Megrey *et al.* (2007), who present a model where, in addition to a calculating the bioenergetic component of the target fish, the total numbers of fish are followed thus enabling fish consumption to impose a mortality term on the zooplankton, effectively allowing for density dependent effects on fish growth.

Complexity of spatial and temporal scales
Integration across trophic scales and physical environments is spatially and temporally complex. Not all scales need to be included explicitly when attempting to project future states of marine ecosystems. However, mechanistic understanding needs to be developed before parameterizations can be attempted. For example, at short temporal scales and small spatial scales, observations indicate that feeding by fish larvae on micro-zooplankton occurs most efficiently when micro-zooplankton are concentrated at density gradients such as fronts and thermoclines (Peterman and Bradford, 1987). These processes may be disrupted by turbulence (e.g. Peterman and Bradford, 1987), but such fine scale processes need to be incorporated into models nested inside circulation models (Werner *et al.*, 2001; Yamazaki *et al.*, 2002). More generally, fine scale physical structures are needed to describe the variation in prey distributions, such as patch frequency and distribution, and fish habitat in general (Logerwell *et al.*, 2001). Despite advances in sampling technology and the ability to describe small-scale prey fields, modeling efforts rarely include variability in prey fields at the scales relevant to the individual predator. Behavioral responses to environmental changes are expected, especially in schooling fishes (Fréon and Misund, 1999). A combination of individual-based and population-based models is likely to be required to be able to consider more mechanistic descriptions and to allow fish in the model to dynamically adjust their behavior (e.g. feeding, movement). While advection is critical to understanding the spatial distributions of early life stages of fish, juvenile and adult fish can control their movement on fine to moderate spatial scales.

At the other end of the space–time spectrum, long time scale models of fish populations, e.g. decadal and longer, must consider full life cycle approaches and larger spatial domains. One recent example by Rose *et al.* (2008) presents a modeling approach for following numbers of fish (coupled to a lower trophic model), including how the fish's life cycle must be closed and how each year a new year class needs to be added. With this coupled model, long-term simulations are possible by providing the ability to generate new fish recruits within the simulation model, with spawner–recruit relationships, dependent on the prevailing environmental and climatic conditions, used to estimate the number of new individuals to be added to the population every year from the previous year's spawning biomass.

Uncertainty in marine ecosystem modeling
In the above sections we have outlined some of the challenges that need to be solved in formulating more realistic marine ecosystem models and their responses to climate change. As we attempt to provide projections of future ecosystem state, quantification of model results with levels of uncertainty needs to be implemented routinely. Monte Carlo and ensemble methods exist that can be readily used for this purpose. Quantifying the uncertainty in ecological forecasts will have its own set of challenges different from physical climate models. However, uncertainty estimates should be made part of future ecosystem modeling (Clark *et al.*, 2001).

Quéro (2006) identified three phases of development in modeling: the illusion, the chaos, and the relief. She feels that models have to evolve through these phases before their results can approach the truth with some confidence. To her, ocean carbon models are in the illusion phase (and we can add that this is also true for many other biological models) and terrestrial carbon models are in the chaos phase, whereas climate models could be considered as having reached the relief phase.

Gaps in knowledge of processes and prediction
In addition to modeling gaps listed above, there are many fundamental gaps in our understanding of physical and biological processes that make quantitative predictions difficult. At times, the uncertainties are not just in magnitude of possible changes, but also in their sign. Some of these are discussed next.

Land–ocean exchanges in upwelling systems
The response of climate to land-cover change can vary with atmospheric CO_2 concentration. Proposed future land-cover–upwelling interactions will likely be moderated by changes in CO_2 and the timing of land-cover response to that change, the latter of which will likely vary among eastern boundary current regions (Diffenbaugh *et al.*, 2004). Landscapes in eastern boundary current regions have been, and will continue to be, altered by humans, and anthropogenic land-cover types such as urban and cropland have likely contributed to future land-cover–upwelling interactions. Generally, regional climate models do not explicitly simulate coastal fog. Changes in coastal upwelling can change the coastal fog

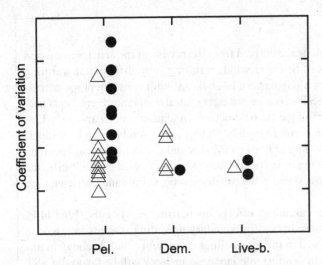

Fig. 14.10. Coefficients of variation of annual larval abundance of exploited and unexploited species off California associated with spawning modes. Filled circles indicate exploited species whereas open triangles indicate unexploited species. Pel., pelagic spawners; Dem., demersal spawners; Live-b., live-bearers. (Reprinted from Hsieh *et al.*, 2006, with permission from Macmillan Publishers Ltd.)

production and create feedbacks that are not presently modeled. For example, increased coastal fog can be a negative feedback through a decrease in short-wave radiation absorbed at the surface, in turn decreasing coastal warming. Complete assessment of the sensitivity of such environmental systems to greenhouse warming requires understanding not only of their sensitivity to forcing but also of their sensitivity to feedbacks.

Behavioral adaptations of marine organisms
Another important unknown for prediction of the effects of CC is the role of behavioral adaptations on the short term and the role of evolutionary processes on a longer time scale but at an unknown rate. Species with short generation times like plankton and, to a lesser extent, small pelagic fishes are expected to have a higher response to rapid CC. However, reductions in population size due to CC, especially if aggravated by overexploitation, can lead to subsequent genetic drift and could restrict a species' potential for adaptation by eliminating heritable traits of ecological importance (for review see in Harley *et al.*, 2006).

Interactions between climate change and other ecosystem stressors
Increase in global connectivity (e.g. ship ballast water; voluntary or involuntary importation of alien species) is likely to increase exposure of marine species to new pathogens and exacerbate the level of mortality due to increased

susceptibility to infection in relation to CC. Similarly, the introduction of alien opportunistic species can benefit from CC for their colonization of new territories (Carlton, 2000; Harley *et al.*, 2006).

Another issue is to understand the intertwining problems of fishing effects and CC. Up until recently, fishing and CC effects were considered as additive, but new analyses suggest that they interact. Hsieh *et al.* (2006) analyzed the California Cooperative Oceanic Fisheries Investigations (CalCOFI) ichthyoplankton data and showed that interannual variability is higher in exploited species than in unexploited ones, including pelagic species (Fig. 14.10). The major process seems to be the decrease in the strength and number of fish older age classes by exploitation, shown for Pacific sardine *Sardinops sagax* and Pacific horse mackerel *Trachurus symmetricus*. Older fish are more fecund than younger ones, produce eggs of higher quality, perform more extended migrations and buffer the interannual variability in recruitment (Beamish *et al.*, 2006). According to Planque *et al.* (in press), the selection of population subunits within meta-populations may also lead to a reduction in the capacity of populations to withstand climate variability and change. Furthermore, Hsieh *et al.* (2006) suggest that, at the ecosystem level, reduced complexity by elimination of species due to overexploitation could lead to reduced resilience to CC perturbations. Differential exploitation of marine resources could also promote increased turnover rates in marine ecosystems, which would exacerbate the effects of environmental changes. Overall, reduction in marine diversity at the individual, population, and ecosystem levels will likely lead to a reduction in the resilience and an increase in the response of populations and ecosystems to future climate variability and change (FAO, 2004; Planque *et al.*, in press). Future studies should not only envisage the sensitivity of species and ecosystems to CC, but also the feedbacks caused by overexploitation, which is decreasing the resilience of natural ecosystems to change (Walther *et al.*, 2002; Ottersen *et al.*, 2006; this volume, Chapter 11).

Finally, upwelling ecosystems are frequently subject to outbursts of species that are naturally present in the ecosystem at low abundance that suddenly increase dramatically in their abundance for a few years before returning to baseline values. This was the case with *Macrorhamphosus* spp. in Morocco (Brethes, 1979) and the trigger fish in West Africa, both species bursting during the 1970s (Caverivière, 1982). But some of these outbursts are suspected to reflect the response of the ecosystem to fishing down (Pauly *et al.*, 1998) or through (Essington *et al.*, 2006) the food web, like the recent invasion of the Namibian shelf by jellyfish and gobies (Lynam *et al.*, 2006). Whether anthropogenic CC will modify the occurrence of these outbursts is still an unanswered question.

Box 14.1. Climate change and small pelagic fish

The following scenarios of climate change (CC) in SPACC regions are derived from the review of the articles discussed in this chapter and the IPCC's 2007 Fourth Assessment Report. Stronger winds resulting from differential warming between land masses and oceans may drive stronger upwelling along eastern boundaries, with consequent lower temperatures close to shore. Coupled with the predicted increased SSTs offshore, stronger spatial temperature gradients may be established on larger scales affecting also the characteristics of the far field ocean circulation's mean and eddy kinetic energy. Latitudinally, warming is expected to be greater in northern high latitudes than in southern high latitudes. Subtropical regions are predicted to become saltier, whereas waters at higher latitudes and the tropics should become fresher and more stratified, therefore reducing the vertical mixing of nutrients. Across ocean basins, in the Pacific the frequency and magnitude of occurrences of El Niño may increase, while in the North Atlantic, storms and hurricanes are expected to increase moderately in frequency and intensity.

These atmospheric and oceanographic changes will have cascading effects on marine ecosystems (Fig. 14.1). An increase in biological productivity could take place near shore in eastern boundary upwelling systems in response to increased winds. Broadly, sardine populations might be favored in these upwelling ecosystems, with anchovy maintaining populations in enhanced upwelling regions near shore. In contrast, the Japanese anchovy will be favored if SST increases, due to an increase in its batch fecundity and its temperature optima for larval growth, which is higher than for Japanese sardine. Offshore and at high latitudes, lower overall biological productivity may result from increased stratification suppressing the mixing of nutrients into the euphotic zone. However, in some cases the increased input of fresh water at high latitudes may stabilize the water column and increase productivity locally in regions where mixing would otherwise be too deep for phytoplankton blooms to be sustained. As such, increases in pelagic fish and cod production may take place in some high latitude systems. A changing climate will influence the delivery of micro-nutrients (e.g. iron) by atmospheric dust deposition to the oceans and, in turn, affect the productivity and phytoplankton composition.

Warming may also change upwelling seasonality, and phenological impacts of CC on plankton and fish are likely (Box 3.1 in this volume, Chapter 3). Warming can also be expected to cause species ranges to shift toward the poles, with species of shorter life cycles and smaller body sizes adapting more quickly. Populations located along the latitudinal boundaries of a species range are more likely to be affected by environmental changes than those found at the center of the distribution. Regardless of the nature of changes in response to CC, not all species will be affected in a similar manner. Exploitation will reduce the resilience of populations and make them more susceptible to future climate variability and change.

Conclusions

An overarching dogma is that, as a result of natural climate change, ecosystems and their environment fluctuate between multiple states but the long-term average does not drift in any one direction. On the other hand, anthropogenically driven climate change is thought to be directional with ecosystems and their environment drifting away from the long-term average unless measures are taken to mitigate or reverse the climate forcing functions. These two "climate change" conditions can also interact so as to result in larger and more frequent "natural" oscillations, but always with a bias driven by the anthropogenic changes. From the compilation of information on the effect of climate change gathered in this chapter and parts of other chapters of this book, one can distinguish between what is predictable, what we hope to be able to predict in a few years, and what we think will be difficult or impossible to predict in the near future as a result of anthropogenically driven climate change. Following the terminology of Jonathan J. Dickau

(http://jond4u.jonathandickau.com/known.htm), this is the distinction between the known, the unknown but knowable, and the unknowable.

1. The known or fairly predictable effects of anthropogenically driven CC on SPACC ecosystems are measurable changes in circulation, stratification, and primary production that will result in changes in the distribution and abundance of many marine organisms through the food web, and in measurable changes in ecosystem services. On average, the oceans' surface temperature will warm by a few degrees, sea ice will retreat in polar regions, especially in the northern hemisphere, stratification will increase, the maximum mixed layer depth (MLD) will become shallower (although possibly modulated by the larger scale wind-stress curl), and the enrichment by upwelling will change substantially in many places, with variations depending on latitude and distance from shore. Many small pelagic species will be disturbed by all these changes. Most of them will

shift their distribution poleward, and there will be more disadvantaged species than advantaged ones. The over-exploited fish species are likely to be more susceptible to anthropogenically driven CC. But, again, this average situation may be modified locally (in time and space). For instance, favored pelagic species will increase their abundance and their geographic distribution, which can result in both a poleward and equatorward expansion.

2. The unknown but potentially predictable effects are the locations where the productivity will increase and those where it will decrease, and the same applies for the areas where trends in temperature, stratification, and depth of the MLD will increase or decrease. Associated latitudinal shifts and changes in abundance are expected to follow, although not necessarily systematically, but these can be detected and known when they occur.

3. The unknowable or unpredictable are the timing and intensity of these changes, the capacity and timing for the individual species to adapt to their new environment through physiological and behavioral adaptations, and the time at which ecosystems will shift from one regime to another. Recent linear positive trends in greenhouse gas concentrations and in global warming observed during the last decades will not always translate into linear trends in the abundance of marine organisms or their latitudinal distribution shifts. Changes could be pseudo-cyclic or sometimes manifest themselves as abrupt interdecadal variability signals (regime shifts) that dominate changes in abundance, distribution and faunistic composition of most vertebrates and invertebrates, shifting from one persistent state to another. Therefore, the succession of changes cannot always expected to be progressive and linear but can also appear as abrupt regime shifts, with several (more than two) possible states.

While we recognize that the above concerns remain, the recent acceleration of efforts by the scientific community to address the issue of the prediction of CC on marine ecosystems has resulted in significant and tangible advances in our understanding of the dynamics of SPACC regions, and we have identified an initial envelope of possible changes that may take place in the coming decades (Table 14.1). We readily admit that, while we can propose and construct different reasonable scenarios, we may even have omitted the most likely ones and, as such, we should be ready for surprises. This point cannot be overemphasized: surprises will likely be the norm rather than the exception and observational programs will never be able to be completely replaced by models. We also recognize that the first-order difficulty of predicting CC itself was made more complicated by the additional intermingling of anthropogenic and natural forcing factors, including those driven by socio-economic aspects.

Indirect effects, feedbacks, threshold effects, non-linear responses, and interactions between different temporal scales (natural and social) are particularly challenging.

Some ecosystem-based management measures have been proposed to mitigate the detrimental effects of the interactions between CC and exploitation, using for instance the framework outlined by King and McFarlane (2006); but early detection of changes or shifts remains a critical issue. One of the encouraging advances is the available technological know-how that enables sustained observations of the relevant physical and biological variables. Observing systems and observatories, e.g. through the Global Ocean Observing System (GOOS; http://www.ioc-goos.org/), promise to provide measurements that will allow us to detect short-term, event-scale changes as they happen, as well as to build up a long-term baseline of the state of our ecosystems' biotic and abiotic components.

It is also clear that fisheries must be considered as part of a complex socioeconomic–ecological system. Therefore, understanding and predicting their future also requires an understanding of the behavior of fish, fishers, and markets and their response to CC (Hannesson *et al.*, 2006a). Elegant strides have been recently taken in developing frameworks that allow for the integration of bio-economic models encompassing results from models of the physics of the ocean and atmosphere (e.g. regional climate models derived from the downscaling of global models) through to the estimated fish abundance to the economic/market responses (e.g. Mullon and Fréon, 2005; Briones *et al.*, 2006). While these models' predictive capabilities (i.e. skills) are presently limited by uncertainties of the physical scenarios and the difficulty to anticipate animal and human behaviors, we believe that such integrated efforts must continue to be nurtured as they will be a key component of future integrated end-to-end observation and modeling approaches to study marine ecosystems, including SPACC systems.

Acknowledgments

We would like to thank Professor John Field (UCT, South Africa) for his comments and contributions, Pierre Lopez (IRD, Sète) for the drawing of some figures and Dr. Carolina Parada (AFSC, USA) for providing us bibliographic references on the CalC system. We are also grateful for the thoughtful comments by two reviewers, Ian Perry and Manuel Barange, who significantly improved our work. This is a contribution of the EUR-OCEANS network of excellence, EC contract no 511106. F. P. Chavez was supported by the David and Lucile Packard Foundation. F. Werner acknowledges support from NSF grant OCE-0535361 and APN project 2004–10-NSY. Pierre Fréon acknowledges the ECO-UP program of IRD, and all coauthors are grateful to GLOBEC for supporting this work.

Table 14.1. *Summary of the main expected outcomes of climate change according to the process, evidence from data and modeling experiments*

Ecosystem level	Process	Evidence from data	Model	Expected outcomes
Physics	Changing wind fields	Upwelling-favorable sea breezes should strengthen, Bakun (1990), Schwing and Mendelssohn (1997), Mendelssohn and Schwing (2002), resulting in SST decrease (e.g. McGregor et al., 2007)	Increased upwelling: Diffenbaugh et al. (2004), Snyder et al. (2003), Auad et al. (2006) Decreased equatorial upwelling: Diffenbaugh (2005) Vecchi et al. (2006)	Stronger upwelling resulting in lower temperature close to shore; SST could increase offshore leading to stronger gradients; Change in upwelling seasonality; Annual mean decrease in the eddy kinetic energy;
	Warmer surface layers, increased stratification due to changes in temperature and salinity	Increased stratification (Roemmich and McGowan, 1995)	Warmer surface ocean in upwelling areas (McGowan et al., 2003; Auad et al., 2006; Behrenfeld et al., 2006; Hashioka and Yamanaka, 2007) or increased freshwater fluxes at high latitudes due to precipitation and melting Arctic regions (Rodhe and Winsor, 2002; ACIA, 2004; Carmack and Wassman, 2006; Doney, 2006; Royer and Grosch, 2006; Schrank, 2007; Stenevik and Sundby, 2007); both latitudes: Sarmiento et al. (2004)	Southern high latitudes warming less than northern high latitudes. Subtropical regions (high evaporation and salinity) become saltier, whereas waters at higher latitudes and the tropics become fresher. Likely increase in the frequency and magnitude of occurrences of El Niño (but see Cane, 2005 who suggests inconclusive evidence). Suppression of the mixing of nutrients into the euphotic zone; decrease in the fertilizing effects of mixing and upwelling.
	Changes in riverine input, storms frequency and intensity	No clear evidence of storm changes up to date (Trenberth, 2005)	Knutsona et al., 2001 IPCC, 2007	In the North Atlantic, storms and hurricanes are expected to increase moderately in frequency and intensity.
	Changes in precipitation and wind may alter dust transport	McTainsh and Strong (2006)	Jickells et al. (2005)	Changes in the delivery of micronutrients to the oceans (e.g. iron), will change the productivity and phytoplankton composition. Different patterns according to the area.
Lower trophic levels (changes in productivity and composition of lower trophic levels)	Bottom-up processes	McGowan et al. (2003)	Behrenfeld et al. (2006); Roemmich and McGowan (1995); Hashioka and Yamanaka (2007)	Decrease in plankton abundance due to increased stratification in offshore upwelling areas (McGowan et al., 2003), or horizontal input of cooler, fresher, nutrient-rich water from the north at higher latitude (Chelton et al., 1982; Freeland et al., 1997; Overland et al., 1999), perhaps increases close to shore due to increased upwelling.
	Phenology and the match–mismatch hypothesis	Beaugrand et al. (2002)	Hashioka and Yamanaka (2007)	Onset of the diatom spring bloom could take place one half-month earlier than in the present-day due to strengthened stratification in NW Pacific. Change in zooplankton composition in the North Sea can lead to different temporal signals of prey occurrence, leading to a mismatch for larval fish feeding.

Table 14.1. *(cont.)*

Ecosystem level	Process	Evidence from data	Model	Expected outcomes
	Food web considerations	Advection of warm and zooplankton-rich Atlantic water from the Norwegian Sea favors fish larvae survival (Sundby, 2000). High temperatures enhance phytoplankton production and trophic transfer from phytoplankton to zooplankton to fish (Pacific hake and Pacific herring: Robinson, 1994; Robinson and Ware, 1994)		Increase in pelagic fish and cod production in high latitude SPACC systems. Sardine regimes might be favored offshore in the Humboldt ecosystem, and possibly in the Californian ecosystem, associated with a lower total pelagic fish production (compared to an anchovy regime), but anchovy favored closer to shore.
	Changes in community structures of plankton (including gelatinous)	Edwards and Richardson (2004), Brodeur *et al.* (1999), Harley *et al.* (2006)	Shifts in phytoplankton species composition (Boyd and Doney, 2002; Sarmiento *et al.*, 2004; Hashioka and Yamanaka, 2007)	Modest changes low-latitude upwelling biomes with increases in the Northern Hemisphere on the order of 1%, and ~2%, decreases in the Southern Hemisphere (Sarmiento *et al.*, 2004). Regional patterns of both higher and lower chlorophyll field stocks in temperate and high latitudes and in coastal upwelling areas, as well as a global reduction in new production between 5–10% (Boyd and Doney, 2002). Change in the dominant phytoplankton group at the end of spring bloom in NW Pacific at least, with a shifts in the dominant phytoplankton group from diatoms to other small phytoplankton (Hashioka and Yamanaka, 2007).
	Distributional changes	Beaugrand *et al.* (2002)		Northward extension of more than 10° in latitude (ongoing in the North Atlantic Ocean) could extend and be observed in other areas.
	Effect of acidification and UV increase on phyto-zooplankton community structure.		Orr *et al.* (2005), Austin *et al.* (1992)	The Southern Ocean surface waters will begin to become undersaturated in aragonite by the year 2050. California and Humboldt ecosystems could be the more affected. Uncertainty regarding the expected detrimental effect of acidification on foraminifera in these upwelling regions. Large uncertainty regarding the depletion of the ozone layer in response to the increase of CO_2 concentrations, the resulting increase in ultraviolet radiation at the ocean surface and possible negative impacts on invertebrate larvae and algae.
Higher trophic levels	Bottom-up process	Skjoldal (2004)		Moderate increase in pelagic fish abundance at high latitude.
	Changes in batch fecundity and larval growth according to temperature	Takasuka *et al.* (2005, 2007)		Increase in SST favors batch fecundity of Japanese anchovy. Higher temperature optima for larval growth of Japanese anchovies compared to sardines. Both processes could favor Japanese anchovy regime if SST increases.

Table 14.1. *(cont.)*

Ecosystem level	Process	Evidence from data	Model	Expected outcomes
	Phenological changes	Stenseth and Mysterud (2002); Beaugrand *et al.* (2003) and Reid *et al.* (2003); Durant *et al.* (2005)		Decrease of match between plankton bloom and fish larvae abundance in high latitude SPACC systems.
	Distributional changes	Southward *et al.* (1995); Quéro *et al.* (1998); Rodríguez-Sánchez *et al.* (2002); Brander *et al.* (2003); Parmesan and Yohe, (2003); Beare *et al.* (2004); Drinkwater (2005); Perry *et al.* (2005)	Conceptual model of Wooster and Bailey (1989) and Myers (1998)	Marine species ranges will shift toward the poles, especially those with faster life cycles and smaller body sizes, and/or at greater depth for some demersal species. But this pattern could vary according to regions. All processes and interactions between the effects of CC on temperature, turbulence, stratification, circulation, light intensity and nutrient concentration are not fully understood. Furthermore these effects can interact with population abundance and exploitation. Populations located at the limit of a species range are more likely to react to environmental changes than those found at the center of the distribution. In the Northern Hemisphere, populations close to the northern limit of the species range should show positive correlations with increasing temperature while those at the southern limit will show negative correlations. For the Southern Hemisphere the situation would be reversed.
	Effects of changes in circulation on recruitment processes (Bakun's triad)	No relevant observations related to CC effects, as far as we know, but large body of literature on present situations, empirical relationships between historical SST (cumulated or not) and proxy fish abundance, e.g. McFarlane *et al.* (2002); Norton and Mason (2005)	No quantitative model addressing specifically the issue of CC, as far as we know, but large body of literature on numerical models of present situations; e.g. empirical model (Conser *et al.*, 2004); IBM models (Mullon *et al.*, 2003), conceptual models such as the meandering flow hypothesis of MacCall (in Bakun and Broad, 2002)	Expected increase of along-shore and offshore transport of fish larvae resulting in higher mortality, not necessarily compensated by enhanced primary production. Sardine should be more favored than anchovy. In the Northern Atlantic, the increase in storm and hurricanes is expected to result in the dispersion of plankton swarm and fish larvae aggregation, an unfavorable situation for the optimal feeding of the latter (Peterman and Bradford, 1987) when storm and critical fish life stages overlap in time and space.
	Increase in diseases due to temperature increases	Harvell *et al.* (2002)		Possible increase in diseases for marine mammals, oysters, marine invertebrates and eelgrass (although the mechanisms for pathogenesis are unknown for these last two groups).

Table 14.1. *(cont.)*

Ecosystem level	Process	Evidence from data	Model	Expected outcomes
General considerations		FAO, 2004; Beamish *et al.* (2006); Hsieh *et al.* (2006); Planque *et al.* (in press)		Exploitation is likely to exacerbate the effects of CC. Reduction in marine diversity at the individual, population and ecosystem levels will likely lead to a reduction in the resilience and an increase in the response of populations and ecosystems to future climate variability and change.

NOTES

1 Intergovernmental Panel on Climate Change.
2 Small Pelagic Fish and Climate Change, a GLOBEC program.
3 Center for Climate System Research/National Institute for Environmental Studies.

REFERENCES

Abraham, C. and Sydeman, W. J. (2004). Ocean climate, euphausiids, and auklet nesting: inter-annual trends and variation in phenology, diet and growth of a planktivorous seabird. *Mar. Ecol. Prog. Ser.* **274**: 235–250.

ACIA. (2004). *Impacts of a Warming Arctic.* Cambridge, UK: Cambridge University Press.

Agenbag, J., Richardson, A. J., Demarcq, H., *et al.* (2004). Estimating environmental preferences of South African pelagic fish species using catch size and remote sensing data. *Progr. Oceanogr.* **59**: 275–300.

Anderson, T. R. (2005). Plankton functional type modelling: running before we can walk? *J. Plankton Res.* **27**: 1073–1081.

Auad, G., Miller, A., and Di Lorenzo, E. (2006). Long-term forecast of oceanic conditions off California and their biological implications, *J. Geophys. Res.* **111** (C09008): doi:10.1029/2005JC003219.

Austin, J., Butchart, N., and Shine, K. P. (1992). Possibility of an Arctic ozone hole in a doubled-CO2 climate. *Nature* **360**: 221–225.

Bakun, A. (1990). Global climate change and intensification of coastal ocean upwelling. *Science* **247**: 198–201.

Bakun, A. (1998). Ocean triads and radical interdecadal stock variability: bane and boon for fisheries management. In *Reinventing Fisheries Management*, Pitcher, T., Hart, P. J. B., and Pauly, D., eds. London: Chapman and Hall, pp. 331–358.

Bakun, A. and Broad, K., eds. (2002). *Climate and Fisheries: Interacting Paradigms, Scales, and Policy Approaches.* New York: International Research Institute for Climate Prediction.

Bakun, A. and Weeks, S. J. (2004). Greenhouse gas buildup, sardines, submarine eruptions and the possibility of abrupt degradation of intense marine upwelling ecosystems. *Ecol. Lett.* **7**: 1015–1023.

Bañón, R., del Rio, J. L., Piñeiro, C., and Casas, M. (2002). Occurrence of tropical affinity fish in Galician waters, north-west Spain. *J. Mar. Biol. Assoc. UK* **82**: 877–880.

Barange, M. (2003). Ecosystem science and the sustainable management of marine resources: from Rio to Johannesburg. *Front. Ecol. Environ.* **1**: 190–196.

Barange, M., Werner, C., Perry, I., *et al.* (2003). The tangled web: global fishing, global climate, and fish stock fluctuations. *IGBP Newsl.* **56**: 24–27.

Barnett, T. P., Pierce, D. W., AchutaRao, K. M. (2005). Ocean science: penetration of human-induced warming into the world's oceans. *Science* **309**: 284–287.

Beamish, R. J., McFarlane G. A., and Benson, A. (2006). Longevity overfishing. *Prog. Oceanogr.* **68**: 289–302.

Beare, D., Burns, F., Jones, E., *et al.* (2004). An increase in the abundance of anchovies and sardines in the north-western North Sea since 1995. *Glob. Change Biol.* **10**: 1209–1213.

Beaugrand, G. and Ibáñez, F. (2004). Monitoring marine plankton ecosystems (2): long-term changes in North Sea calanoid copepods in relation to hydro-meteorological variability. *Mar. Ecol. Prog. Ser.* **284**: 35–47.

Beaugrand, G., Reid, P. C., Ibáñez, F., *et al.* (2002). Reorganization of North Atlantic marine copepod biodiversity and climate. *Science* **296**: 1692–1694.

Beaugrand, G., Brander, K. M., Lindley, J. A., *et al.* (2003). Plankton effect on cod recruitment in the North Sea. *Nature* **426**: 661–664.

Behrenfeld, M. J. and Falkowski, P. G. (1997). A consumer's guide to phytoplankton primary productivity models. *Limnol. Oceanogr.* **42**: 1479–1491.

Behrenfeld, M. J., O'Malley, R. T., Siegel, D. A., *et al.* (2006). Climate-driven trends in contemporary ocean productivity. *Nature* **444**: 752–755.

Blanchard, F. and Vandermeirsch, F. (2005). Warming and exponential abundance increase of the subtropical fish

Capros ape in the Bay of Biscay (1973–2002). C.R. Biolog. Academies des Sciences, **328**: 505–509.

Boyd D. C. and Doney, S. C. (2002). Modelling regional responses by marine pelagic ecosystems to global climate change. *Geophys. Res. Lett.* **29**: 53–61.

Boyer, D. C., Boyer, H. J., Fossen, I., and Kreiner, A. (2001). Changes in abundance of the northern Benguela sardine stock during the decade 1990 to 2000, with comments on the relative importance of fishing and the environment. *S. Afr. J. Mar. Sci.* **23**: 67–84.

Brander, K., Blom, G., Borges, M. F. *et al.* (2003). Changes in fish distribution in the eastern North Atlantic; are we seeing a coherent response to changing temperature? *ICES Mar. Sci. Symp.* **219**: 261–270.

Brethes, J. C. (1979). Contribution à l'étude des populations de *Macrorhamphosus scolopax* (L., 1758) et *Macrorhamphosus gracilis* (Lowe, 1839) des côtes atlantiques marocaines. *Bull. Inst. Pêch. Mar.* **24**: 4–62.

Briones, R., Garces L., and Mahfuzzudin, A. (2006). Climate change and small pelagic fisheries in developing Asia: the economic impact of fish producers and consumers. In *Climate Change and the Economics of the World's Fisheries: Examples of Small Pelagic Stocks*, Hannesson, R., Herrick, Jr. S. and Barange, M., eds., pp. 215–235.

Brodeur, R. D. Mills, C. E. Overland, J. E., *et al.* (1999). Evidence for a substantial increase in gelatinous zooplankton in the Bering Sea, with possible links to climate change. *Fish. Oceanogr.* **8**(4): 296–306.

Bryden, H. L., Longworth H. R., and Cunningham, S. A. (2005). Slowing of the Atlantic meridional overturning circulation at 25°N. *Nature* **438**: 655–657.

Cane, M. A. (2005). The evolution of El Niño, past and future. *Earth Planet. Sci. Lett.* **230**: 227–240.

Carlotti, F. and Radach, G. (1996). Seasonal dynamics of phytoplankton and *Calanus finmarchicus* in the North Sea as revealed by a coupled one-dimensional model. *Limnol. Oceanogr.* **41** (3): 522–539.

Carlotti, F., Giske, J., Werner, F., and Moloney, C. (2000). Modelling zooplankton dynamics. In *Zooplankton Methodology Manual*, Harris, R. P., Wiebe, P., Lenz, J., *et al.*, eds. Academic Press, pp. 571–667.

Carlton, J. T. (2000). Global change and biological invasions in the oceans. In *Invasive Species in a Changing World*, Mooney, H. A., and Hobbs, R. J., Covelo, C. A., eds.: Island Press, pp. 31–53.

Carmack, E. and Wassman, P. (2006). Food webs and physical–biological coupling on pan-Arctic shelves: unifying concepts and comprehensive perspectives. *Progr. Oceanogr.* **71**: 446–477.

Carr, M.-E. (2002) Estimation of potential productivity in Eastern Boundary Currents using remote sensing. *Deep-Sea Res. II* **49**: 59–80.

Caverivière, A. (1982). Le baliste des côtes africaines (*Balistes carolinensis*): biologie, prolifération et possibilités d'exploitation. *Oceanol. Acta* **5**(4): 453–459.

Chassignet, E. P., Hurlburt, H. E., Smedstad, O. M. *et al.* (2006). Generalized vertical coordinates for eddy-resolving global and coastal ocean forecasts. *Oceanogr.* **19**: 20–31.

Chavez, F. P., Pennington, J. T., Castro, C. G. *et al.* (2002). Biological and chemical consequences of the 1997–1998 El Niño in central California waters. *Progr. Oceanogr.* **54**: 205–232.

Chavez, F. P., Ryan, J., Lluch-Cota, S. E., and Ñiquen, M. (2003). From anchovies to sardines and back: multidecadal change in the Pacifc Ocean. *Science* **299**: 217–221.

Chelton, D. B., Bernal, P. A., and McGowan, J. A., (1982). Large-scale interannual physical and biological interaction in the California Current. *J. Mar. Res.* **40**: 1095–1125.

Christian, J. R., Verschell, M. A., Murtugudde, R., *et al.* (2002). Biogeochemical modelling of the tropical Pacific Ocean II. Iron biogeochemistry. *Deep-Sea Res.* II. **49**: 545–565.

Clark, B. M. (2006). Climate change: a looming challenge for fisheries management in southern Africa. *Mar. Policy* **30**: 84–95.

Clark, J. S., Carpenter, S. R., Barber, M. *et al.* (2001). Ecological forecasting: an emerging imperative. *Science* **293**: 657–660.

Clarke, A. J. and Lebedev, A. (1999). Remotely driven decadal and longer changes in the coastal Pacific waters of the Americas. *J. Phys. Oceanogr.* **29**: 828–835.

Conser, R., Hill, K., Crone, P., *et al.* (2004). *Assessment of the Pacific Sardine Stock for US Management in 2005.* Portland, Oregon: Pacific Fishery Management Council.

Costanza, R., D'Arge, R., De Groot, R. *et al.* (1997). The value of the world's ecosystem services and natural capital. *Nature* **387**: 253–260.

Cowen, R. K., Paris, C. B., and Srinivasan, A. (2006). Scaling of connectivity in marine populations. *Science* **311**: 522–527.

Crawford, R. J. M., Underhill, L. G., Shannon, L. V., *et al.* (1991). An empirical investigation of transoceanic linkages between areas of high abundance of sardine. In *Long-term Variability of Pelagic Fish Populations and their Enviroment*, Kawasaky, T., Tanaka, S., Toba, Y., and Tamiguchi, A., eds. New York: Pergamon Press, pp. 319–332.

Crutzen, P. J. and Stoermer, E. F. (2000). The "Anthropocene". *Glob. Change Newsl.* **41**: 12–13.

Cubasch, U., Meehl, G. A., Boer, G. J. *et al.* (2001). Projections of future climate change. In *IPCC: Climate Change (2001): The Scientific Basis.* Cambridge, UK: Cambridge University Press, pp. 525–582.

Cury, P. and Roy, C. (1989). Optimal environmental window and pelagic fish recruitment success in upwelling areas. *Can. J. Fish. Aquat. Sci.* **46**(4): 670–680.

Cushing, D. H. (1969). The regularity of the spawning season of some fishes. *J. Cons. Int. Expl. Mer.* **33**: 81–92.

Cushing, D. H. (1990). Plankton production and year-class strength in fish populations – an update of the match mismatch hypothesis. *Adv. Mar. Biol.* **26**: 249–293.

Davis, A. J., Jenkinson, L. S., Lawton, J. H., *et al.* (1998). Making mistakes when predicting shifts in species range in response to global warming. *Nature*, **391**: 783–786.

de Young, B., Heath, M., Werner, F., *et al.* (2004). Challenges of modeling ocean basin ecosystems. *Science* **304**: 1463–1466.

Diffenbaugh, N.S. (2005). Response of large-scale eastern boundary current forcing in the 21st century. *Geophys. Res. Lett.* **32**: L19718, doi:10.1029/2005GL023905.

Diffenbaugh, N.S., Snyder, M.A., and Sloan, L.C. (2004). Could CO_2-induced land-cover feedbacks alter near-shore upwelling regimes? *Proc. Natl Acad. Sci. USA* **101**: 27–32.

Doney, S.C. (2006). Plankton in a warmer world. *Nature* **444**: 695–696.

Doney, S.C., Lindsay, K., Caldeira, K. *et al.* (2004). Evaluating global ocean carbon models: the importance of realistic physics. *Global Biogeochem. Cycles* **18**: GB3017, doi:10.1029/2003GB002150.

Drinkwater, K.F. (2005). The response of Atlantic cod (*Gadus morhua*) to future climate change. *ICES J. Mar. Sci.* **62**: 1327–1337.

Durant, J.M., Hjermann, D. Ø., Anker-Nilssen, T. *et al.* (2005). Timing and abundance as key mechanisms affecting trophic interactions in variable environments. *Ecol. Lett.* **8**: 952–958.

Edwards, M. and Richardson, A.J. (2004). Impact of climate change on marine pelagic phenology and trophic mismatch. *Nature* **430**: 881.

Enfield, D.B. and Mestas-Nunez, A.M. (2000). Global modes of ENSO and non-ENSO SST variability and their associations with climate. In *El Niño and the Southern Oscillation: Multiscale Variability and Global and Regional Impacts*, Diaz, H.F., and Markgraf, V., eds., Cambridge, UK: Cambridge University Press, pp. 89–112.

Essington, T.E., Beaudreau, A.H. and Wiedenmann, J. (2006). Fishing through marine food webs. *Proc. Nat. Acad. Sci. USA* **103**: 3171–3175.

Falkowski, P.G., Barber, R.T., and Smetacek, V. (1998). Biogeochemical controls and feedbacks on ocean primary production. *Science* **281**: 200–206.

Fang, F., Piggott, M.D., Pain, C.C., Gorman, G.J., and Goddard, A.J.H. (2006). An adaptive mesh adjoint data assimilation method. *Ocean Model.* **15**: 39–55.

FAO. (2004). *The State of World Fisheries and Aquaculture.* Rome: FAO.

Feely, R.A., Sabine, C.L., Lee, K. *et al.* (2004). Impact of anthropogenic CO_2 on the $CaCO_3$ system in the oceans. *Science* **305**: 362–366.

Field, D., Cayan, D., and Chavez, F. (2006). Secular warming in the California Current and North Pacific. *CalCOFI Rep.* **47**: 92–109.

Field, J.G., Hempel, G., and Summerhayes, C.P. (2002). *OCEANS 2020: Science, Trends, and Challenge of Sustainability.* Washington, DC: Island Press.

Fields, P.A., Graham, J.B., Rosenblatt, R.H., and Somero, G.N. (1993). Effects of expected global climate change on marine faunas. *Trends Ecol. Evol.* **8**: 361–367.

Flynn, K.J. (2005). Castles built on sand; dysfunctional plankton models and the failure of the biology-modelling interface. *J. Plank. Res.* **27**: 1205–1210.

Freeland, H., Denman, K., Wong, C.S., *et al.* (1997). Evidence of change in the winter mixed layer in the Northeast Pacific Ocean. *Deep-Sea Res. I.* **44**: 2117–2129.

Fréon, P. and Misund O.A. (1999). *Dynamics of Pelagic Fish Distribution and Behaviour: Effects on Fisheries and Stock Assessment.* Oxford: Blackwell Science.

Fréon, P.J. Alheit, E.D., Barton, S., *et al.* (2006). Modelling, forecasting and scenarios in comparable upwelling ecosystems: California, Canary and Humboldt. In *The Benguela: Predicting a Large Marine Ecosystem*, Shannon, V., Hempel, G., Moloney, C., *et al.*, eds. Amsterdam: Elsevier Series, Large Marine Ecosystems, pp. 185–220.

Gargett, A.E. (1997). The 'optimial stability window': a mechanism underlying decadal fluctuations in North Pacific salmon stocks? *Fish. Oceanogr.* **6**: 109–117.

Gaughan, D.J. (2002). Disease-translocation across geographic boundaries must be recognized as a risk even in the absence of disease identification: the case with Australian *Sardinops*. *Rev. Fish Biol. Fish.* **11**: 113–123.

Gaylord, B. and Gaines, S.D. (2000). Temperature or transport? Range limits in marine species mediated solely by flow. *Am. Nat.*, **155**: 769–789.

Grantham, B.A., Chan, F., Nielsen, K.J. *et al.* (2004). Upwelling-driven nearshore hypoxia signals ecosystem and oceanographic changes in the northeast Pacific. *Nature* **429**: 749–754.

Hannesson, R., Barange M., and Herrick, S.F., Jr. (2006a). *Climate Change and the Economics of the World's Fisheries.* Cheltenham, UK: Elgar Publishing Ltd., New horizons in environmental economics.

Hannesson, R., Herrick, S.F., Jr., and Barange, M. (2006b). On the consequences of climate change in pelagic fish populations: a conclusion. In *Climate Change and the Economics of the World's Fisheries*, Hannesson, R., Barange, M., and Herrick, S.F., Jr., eds. Cheltenham, UK: Elgar Publishing Ltd., New horizons in environmental economics, pp. 296–304.

Hansen, J., Sato, M., Ruedy, R., Lo, K., Lea, D.W., and Medina-Elizade, M. (2006). Global temperature change. *Proc. Nat. Acad. Sci. USA* **103**(39): 14288–14293.

Harley, C.D.G., Hughes, R.A., Hultgren, K.M. *et al.* (2006). The impacts of climate change in coastal marine systems. *Ecol. Lett.* **9**: 228–241.

Harvell, C.D., Kim, K., Burkholder, J.M. *et al.* (1999). Emerging marine diseases: climate links and anthropogenic factors. *Science* **285**: 1505–1510.

Harvell, C.D., Mitchell, C.E., Ward, J.R. *et al.* (2002). Climate warming and disease risks for terrestrial and marine biota. *Science* **296**: 2158–2162.

Hashioka, T. and Yamanaka, Y. (2007). Ecosystem change in the western North Pacific associated with global warming obtained by 3-D NEMURO. *Ecol. Model.* **202**: 95–104.

Hermann, A.J., Haidvogel, D.B., Dobbins, E.L., and Stabeno, P.J. (2002). Coupling global and regional circulation models in the coastal Gulf of Alaska. *Prog. Oceanog.* **53**: 335–367.

Herrick, S.F., Jr., Norton, J., Mason, J., and Bessy, C. (2007). Management application of an empirical model of sardine-climate regime shifts. *Mar. Policy* **31**: 71–80.

Houghton, J. (2005). Global warming. *Rep. Prog. Phys.* **68**: 1343–1403.

Houghton, J. T., Ding, Y., Griggs, D. J., eds. (2001). *Climate Change 2001: The Scientific Basis*. Cambridge, UK: Cambridge University Press.

Hsieh, C. H., Reiss, C. S., Hunter, J. R., *et al.* (2006). Fishing elevates variability in the abundance of exploited species. *Nature* **443**: 859–862.

Huggett, J., Fréon P., Mullon C., and Penven P. (2003). Modelling the transport success of anchovy (*Engraulis encrasicolus*) eggs and larvae in the southern Benguela: the effect of spatio-temporal spawning patterns. *Mar. Ecol. Progr. Ser.* **250**: 247–262.

Hutchings, L., Barange, M., Bloomer, S. F. *et al.* (1998). Multiple factors affecting South African anchovy recruitment in the spawning, transport and nursery areas. *S. Afr. J. Mar. Sci.* **19**: 211–225.

IPCC (2001). *Third Assessment Report, Climate Change 2001*, http://www.ipcc.ch, 4 volumes.

IPCC (2007). *Fourth Assessment Report, Climate Change 2007*, http://www.mnp.nl/ipcc/pages_media/AR4-chapters.html, 4 volumes.

Ito ,S., Kishi, M. J., Megrey, B. A., *et al.* (2006a). Workshop report on sardine and anchovy fluctuations. *PICES Press Newsl.* **14**: 16–17.

Ito, S., Megrey, B. A., Kishi, M. J. (2006b). On the interannual variability of the growth of Pacific saury (*Cololabis saira*): a simple 3-box model using NEMURO.FISH. *Ecol. Model.* **202** (1–2): 174–183.

Jickells, T. D., An, Z. S., Andersen, K. K. *et al.* (2005). Global iron connections between desert dust, ocean biogeochemistry, and climate. *Science* **308**: 67–71.

Kasai, H., Saito, H., Yoshimori, A., and Taguchi, S. (1997). Variability in timing and magnitude of spring bloom in the Oyashio region, the western subarctic Pacific off Hokkaido, Japan. *Fish. Oceanogr.* **6**: 118–129.

King, D. P. F. (1977). Influence of temperature, dissolved oxygen and salinity on incubation and early larval development of the South West African pilchard *Sardinops ocellata*. *Invest. Rep. Sea Fish. Branch S. Afr.* **114**: 1–35.

King, J. R., ed. (2005). Report of the study group on fisheries and ecosystem responses to recent regime shifts. *PICES Sci. Rep.* **28**, 162 pp.

King, J. R. and McFarlane, G. A. (2006). A framework for incorporating climate regime shifts into the management of marine resources, *Fish. Manag. Ecol.* **13**: 93–102.

Knutsona, T. R., Tuleyaa, R. E., Shenb, W., and Ginis, I. (2001). Impact of CO_2-induced warming on hurricane intensities as simulated in a hurricane model with ocean coupling. *J. Climate* **14** (11): 2458–2468.

Komatsu, K., Matsukawa, Y., Nakata, K., *et al.* (2006). Effects of advective processes on planktonic distributions in the Kuroshio region using a 3-D lower trophic model and a data assimilative OGCM. *Ecol. Model.* **202**: 105–119.

Koutsikopoulos, C., Beillois, P., Leroy, C., and Taillefer, F., (1998). Temporal trends and spatial structures of the sea surface temperature in the bay of Biscay. *Oceanol. Acta* **21** (2): 335–344.

Le Cann, B. and Pingree, R. D. (1995). Circulation dans le Golfe de Gascogne: une revue de travaux récents. In *Actas del IV*

Coloquio Internacional sobre Oceanografia del Golfo de Vizcaya, Cendrero, O. and Olaso, I., eds. Santander, Inst. Esp. de Oceanografia, pp. 217–234.

Le Quéré, C. (2006). The unknown and the uncertain in earth system modelling. *Eos* **87**(45): 496–497.

Le Quéré, C., Harrison, S. P., Prentice, I. C. *et al.* (2005). Ecosystem dynamics based on plankton functional types for global ocean biogeochemistry models. *Glob. Change Biol.* **11** (11): 2016–40, doi: 10.1111/j.1365-2486.2005.001004.x

Lehodey, P., Alheit, J., Barange, M. *et al.* (2006). Climate variability, fish and fisheries. *J. Climate* **19**: 5009–5030.

Lett, C., Roy, C., Levasseur, A., *et al.* (2006). Simulation and quantification of enrichment and retention processes in the southern Benguela upwelling ecosystem. *Fish. Oceanogr.* **15** (5): 363–372.

Lett, C., Penven, P., Ayón, P., and Fréon, P. (2007). Enrichment, concentration and retention processes in relation to anchovy (*Engraulis ringens*) eggs and larvae distributions in the northern Humboldt upwelling ecosystem. *J. Mar. Syst.* **64** (1–4): 189–200.

Levitus, S., Antonov, J., and Boyer, T. (2003). Warming of the world ocean, 1955–2003. *Geophys. Res. Lett.* **32**: 1–4.

Logerwell E. A., Lavaniegos, B., and Smith, P.E. (2001). Spatially-explicit bioenergetics of Pacific sardine in the Southern California Bight: are mesoscale eddies areas of exceptional pre-recruit production? *Progr. Oceanogr.* **49**: 391–406.

Lomborg, B. (2001). *The Skeptical Environmentalist: Measuring the Real State of the World*. Cambridge, UK: Cambridge University Press.

Lynam, C., Gibbons, M., Axelsen, B. *et al.* (2006). Jellyfish overtake fish in a heavily fished ecosystem. *Curr. Biol.* **16**(13): 492–493.

McClean, J. L., Maltrud, M. E., and Bryan, O. (2006). Quantitative measures of the fidelity of eddy-resolving ocean models. *Oceanogr.* **19**: 104–117.

McFarlane, G. A. and Beamish, R. J. (2001). The re-occurrence of sardines off British Columbia characterises the dynamic nature of regimes. *Progr. Oceanogr.* **49**: 151–165.

McFarlane, G. A., Smith, P. E., Baumgartner, T. R., and Hunter, J. R. (2002). Climate variability and Pacific sardine populations and fisheries. *Amer. Fish. Soc. Symp.* **32**: 195–214.

McGillicuddy, D. J., Jr., Anderson, L. A., Doney, S. C., and Maltrud, M. E. (2003). Eddy-driven sources and sinks of nutrients in the upper ocean: Results from a 0.1° resolution model of the North Atlantic. *Global Biogeochem. Cycles* **17** (2): 1035, doi:10.1029/2002GB001987.

McGowan, J. A., Bograd, S. J., Lynn, R. J., and Miller, A. J. (2003). The biological response to the 1977 regime shift in the California Current. *Deep-Sea Res.* **50**: 2567–2582.

McGregor, H. V., Dima, M., Fischer, H. W., and Mulitza, S. (2007). Rapid 20th-century increase in coastal upwelling off Northwest Africa. *Science* **315**: 637–639.

McTainsh, G. and Strong C. (2006). The role of Aeolian dust in ecosystems. *Geomorphology*, doi:10.1016/j.geomorph.2006.07.028.

Meehl, G. A., Stocker, T. F., Collins, W. D. *et al.* (2007). Global climate projections. In *Climate Change 2007: The Physical Science Basis. Contribution of Working Group I to the Fourth Assessment Report of the Intergovernmental Panel on Climate Change*, Solomon, S., Qin, D., Manning, M. *et al.*, eds, Cambridge, UK and New York, NY, USA: Cambridge University Press, pp. 747–845.

Megrey, B. A., Rose, K. A., Klumb, R. *et al.* (2007). A bioenergetics-based population dynamics model of Pacific herring (*Clupea harengus pallasii*) coupled to a lower trophic level nutrient-phytoplankton- zooplankton model: description, calibration and sensitivity analysis. *Ecol. Model.* **202** (1–2): 144–164.

Mendelssohn, R. and Schwing, F. B. (2002). Common and uncommon trends in SST and wind stress in the California and Peru–Chile current systems. *Prog. Oceanog.* **53**: 141–162.

Miller, A. J., Cayan, D. R., Barnett, T. P., *et al.* (1994). The 1976–77 climate shift of the Pacific Ocean. *Oceanogr.* **7**: 21–26.

Miller, A. J., Cayan, D. R., and White, W. B. (1998). A westward-intensified decadal change in the North Pacific thermocline and gyre-scale circulation. *J. Climate* **11**: 3112–3127.

Miller, D. C. M., Moloney, C. L., van der Lingen, C. D., *et al.* (2006). Modelling the effects of physical–biological interactions and spatial variability in spawning and nursery areas on transport and retention of sardine *Sardinops sagax* eggs and larvae in the southern Benguela ecosystem. *J. Mar. Syst.* **61**: 212–229.

Moloney, C. L. and Field, J. G. (1991). The size-based dynamics of plankton food webs. I. Description of a simulation model of carbon and nitrogen flows. *J. Plank. Res.* **13**: 1039–1092.

Monteiro, P., van der Plas, A. K., Bailey, G. W. *et al.* (2006). Forecasting low oxygen water (LOW) variability in the Benguela system. In *The Benguela: Predicting a Large Marine Ecosystem*, Shannon, V., Hempel, G., Moloney, C., Rizzoli, P. and Woods, J., eds. Amsterdam: Elsevier Series, Large Marine Ecosystems, pp. 295–308.

Mullon, C. and Fréon, P. (2005). Prototype of an integrated model of the worldwide system of small pelagic fisheries. In *Climate Change and the Economics of the World's Fisheries*, Hannesson, R., Barange, M. and Herrick, S. F., eds. Cheltenham, UK: Edward Elgar Publishing, New horizons in environmental economics, pp. 262–295.

Mullon, C., Fréon, P., Parada, C., *et al.* (2003). From particles to individuals: modelling the early stages of anchovy in the southern Benguela. *Fish. Oceanogr.* **12** (4–5): 396–406.

Myers, R. A. (1998). When do environment–recruit correlations work? *Rev. Fish. Biol. Fisher.* **8**: 285–305.

Norton, J. G. and Mason, J. E. (2005). Environmental influences on California sardine abundance. *CalCOFI Rep.* **46**: 83–92.

Orr, J. C., Fabry, V. J., Aumont, O. *et al.* (2005). Anthropogenic ocean acidification over the twenty-first century and its impact on calcifying organisms. *Nature* **437**: 681–686.

Ottersen, G., Stenseth, N. C., and Hurrell, J. W. (2004). Climatic fluctuations and marine systems: a general introduction to the ecological effects. In *Marine Ecosystems and Climate Variation*, Stenseth, N. C., Ottersen, G., Hurrell, J. W., and Belgrano, A., eds. Oxford: Oxford University Press, pp. 3–14.

Ottersen, G., Hjermann, D., and Stenseth, N. C. (2006). Changes in spawning stock structure strengthen the link between climate and recruitment in a heavily fished cod (*Gadus morhua*) stock. *Fish. Oceanogr.* **15** (3): 230–243.

Overland, J. E., Salo, S., and Adams, J. R. (1999). Salinity signature of the Pacific Decadal Oscillation. *Geophys. Res. Lett.* **26**: 1337–1340.

Pain, C. C., Piggott, M. D., Goddard, A. J. H. *et al.* (2005). Three-dimensional unstructured mesh ocean modelling. *Ocean Model.* **10**: 5–33.

Parada, C., van der Lingen, C. D., Mullon, C., and Penven, P. (2003). Modelling the effect of buoyancy on the transport of anchovy (*Engraulis capensis*) eggs from spawning to nursery grounds in the southern Benguela: an IBM approach. *Fish. Oceanogr.* **12** (3): 170–184.

Parmesan, C. and Yohe, G. (2003). A globally coherent fingerprint of climate change impacts across natural systems. *Nature* **421**: 37–42.

Pauly, D. (2003). Ecosystem impacts of the world's marine fisheries. *IGBP Newsl.* **55**: 21–3.

Pauly, D., Christensen, V., Dalsgaard, J., Froese, R., and Torres, F., Jr. (1998). Fishing down marine food webs. *Science*, **279**: 860–863.

Pennington, J. T., Mahoney, K. L., Kuwahara, V. S., *et al.* (2006). Primary production in the eastern tropical Pacific: A review. *Progr. Oceanogr.* **69** (2–4): 285–317.

Penven, P., Roy, C., Brundrit, G. B. *et al.* (2001). A regional hydrodynamic model of upwelling in the southern Benguela. *S. Afr. J. Mar. Sci.* **97**: 472–475.

Penven, P., Debreu, L., Marchesiello, P., and McWilliams, J. C. (2006). Evaluation and application of the ROMS 1-way embedding procedure to the Central California upwelling system. *Ocean Model.* **12**: 157–187.

Perry, A. L. Low, P. J., Ellis, J. R., and Reynolds, J. D. (2005). Climate change and distribution shifts in marine fishes. *Science* **308**: 1912–1915.

Perry, R. I., Cury, P., Brander, K., *et al.* (in press). Sensitivity of marine systems to climate and fishing: concepts, issues and management responses. *J. Marine Systems*, GLOBEC Special Issue.

Peterman, R. M. and Bradford, M. J. (1987). Wind speed index and mortality rate of a marine fish, the northern anchovy (*Engraulis mordax*). *Science* **235**: 354–356.

PICES (2005). Marine ecosystems of the North Pacific. *PICES Spec. Pub.* **1**: 1–278.

Planque, B. and Frédou, T. (1999). Temperature and the recruitment of Atlantic cod (*Gadus morhua*). *Can. J. Fish. Aquat. Sci.* **56**: 2069–2077.

Planque, B., Fromentin, J. -M., Cury, P., *et al.* (in press). How does fishing alter marine populations and ecosystems sensitivity to climate? *J. Mar. Syst.*

Poulard, J.-C. and Blanchard, F. (2005). The impact of climate change on the fish community structure of the eastern continental shelf of the Bay of Biscay. *ICES J. Mar Sci.* **62**: 1436–1443.

Quéro, J. -C., Du Buit, M. -H., and Vayne, J. -J. (1998). Les observations de poissons tropicaux et le réchauffement des eaux dans l'Atlantique européen. *Oceanol. Acta* **21**: 345–351.

Raven, J.A. (2005). *Ocean Acidification Due to Increasing Atmospheric Carbon Dioxide.* London, UK: Royal Society.

Reid, P.C., Edwards, M., Beaugrand, G., *et al.* (2003). Periodic changes in the zooplankton of the North Sea during the twentieth century linked to oceanic inflow. *Fish. Oceanogr.* **12**: 260–269.

Robinson C.L.K. (1994). The influence of ocean climate on coastal plankton and fish production. *Fish. Oceanogr.* **3**: 159–171.

Robinson, C.L.K. and Ware, D. (1994). Modelling the trophodynamics of pelagic fish and plankton off the west coast of Vancouver Island. *Can. J. Fish. Aquat. Sci.* **51**: 1737–1751.

Rodhe, J. and Winsor, P. (2002). On the influence of freshwater supply on the Baltic Sea mean salinity. *Tellus* **54A**: 175–186.

Rodríguez-Sánchez, R., Lluch-Belda, D., Villalobos, H., and Ortega-García S. (2002). Dynamic geography of small pelagic fish populations in the California Current System on the regime time scale 1931–1997. *Can. J. Fish. Aquat. Sci.* **59**: 1980–1988.

Roemmich, D. and McGowan, J. (1995). Climatic warming and the decline of zooplankton in the California Current. *Science* **267** (5202): 1324–1326.

Rose, K.A., Werner, F.E., Megrey, B.A. *et al.* (2007). Simulated herring growth responses in the Northeastern Pacific to historic temperature and zooplankton conditions generated by the 3-dimensional NEMURO nutrient-phytoplankton-zooplankton model. *Ecol. Model.* **202** (1–2): 184–195.

Rose, K.A., Megrey, B.A., Hay, D.E., *et al.* (2008). Climate regime effects on Pacific herring growth using coupled nutrient–phytoplankton–zooplankton and bioenergetics models. *Trans. Am. Fish. Soc.* **137**: 278–297.

Rothstein, L.M., Cullen, J.J., Abott, M. *et al.* (2006). Modeling ocean ecosystems – the PARADIGM program. *Oceanogr.* **19**: 16–45.

Roy, C., van der Lingen, C.D., Coetzee, J.C., and Lutjeharms J.R.E. (2007) Abrupt environmental shift links with changes in the distribution of Cape anchovy (*Engraulis encrasicolus*) spawners in the southern Benguela. *Afr. J. Mar. Sci.* **29**: 309–319.

Royer, T.C. and Grosch, C.E. (2006). Ocean warming and freshening in the northern Gulf of Alaska. *Geophys. Res. Lett.*: doi:10.1029/2006GL026767.

Ryther, J.H. (1969). Photosynthesis and fish production in the sea. *Science* **166**: 72–76.

Sarmiento, J., Slater, R., Barber, R. *et al.* (2004). Response of ocean ecosystems to climate warming. *Global Biogeochem. Cycles* **18**: GB3003, doi:10.1029/2003GB002134.

Schrank, W.E. (2007). The ACIA, climate change and fisheries. *Mar. Policy* **31**: 5–18.

Schwing, F.B. and Mendelssohn, R. (1997). Increased coastal upwelling in the California Current System. *J. Geophys. Res.* **102**: 3421–3438.

Sissener, E.H. and Bjørndal, T. (2005). Climate change and the migratory pattern for Norwegian spring-spawning herring – implications for management. *Mar. Policy* **29**: 299–309.

Skjoldal, H.R., ed. (2004). *The Norwegian Sea Ecosystem.* Trondheim: Tapir Academic Press.

Smith, S.L., Yamanaka, Y., and Kishi, M.J. (2005). Attempting consistent simulations of Stn. ALOHA with a multi-element ecosystem model. *J. Oceanogr.* **61**: 1–23.

Snyder, M.A., Sloan, L.C., Diffenbaugh, N.S., and Bell, J.L. (2003). Future climate change and upwelling in the California Current. *Geophys. Res. Lett.* **30**: 1823–1826.

Southward A.J., Hawkins S.J., and Burrows, M.T (1995). Seventy years of changes in the distribution and abundance of zooplankton and intertidal organisms in the western English Channel in relation to rising sea temperature. *J. Therm. Biol.* **20**: 127–155.

Steffen, W., Sanderson, A., Tyson, P.D. *et al.* (2004). *Global Change and the Earth System – A Planet Under Pressure.* Global Change – The IGBP Series, New York: Springer-Verlag.

Stenevik, E. and Sundby, S. (2007). Impacts of climate change on commercial fish stocks in Norwegian waters. *Marine Policy* **31**: 19–31.

Stenseth, N.C. and Mysterud, A. (2002). Climate, changing phenology, and other life history traits: Nonlinearity and match-mismatch to the environment. *Proc. Natl. Acad. Sci. USA* **99** (21): 13379–13381.

Sundby, S. (2000). Recruitment of Atlantic cod stocks in relation to temperature and advection of copepod populations. *Sarsia* **85**: 277–298.

Sverdrup, H.U. (1953). On conditions for the vernal blooming of phytoplankton. *J. Cons. Int. Explor. Mer.* **18**: 287–295.

Takasuka, A., Oozeki, Y., Kubota, H., *et al.* (2005). Temperature impacts on reproductive parameters for Japanese anchovy: comparison between inshore and offshore waters. *Fish. Res.* **76**: 475–482.

Takasuka, A., Oozeki, Y., and Aoki, I. (2007). Optimal growth temperature hypothesis: why do anchovy flourish and sardine collapse or vice versa under the same ocean regime? *Can. J. Fish. Aquat. Sci.* **64**(5): 768–776.

Taylor, A.H., Allen, J.I., and Clark, P.A. (2002). Extraction of a weak climatic signal by an ecosystem. *Nature* **416**: 629–632.

Timmermann, A., Ohberhuber, J., Bacher, A., *et al.* (1999). Increased El Niño frequency in a climate model forced by future greenhouse warming. *Nature* **398**: 694–696.

Trenberth, K. (2005). Uncertainty in hurricanes and global warming. *Science* **308**: 1753–1754.

Twatwa, N.M., van der Lingen, C.D., Drapeau, L., *et al.* (2005). Characterising and comparing the spawning habitats of anchovy *Engraulis encrasicolus* and sardine *Sardinops sagax* in the southern Benguela upwelling ecosystem. *Afr. J. Mar. Sci.* **27**(2): 487–499.

Vecchi, G.A., Soden, B.J., Wittenberg, A.T., *et al.* (2006). Weakening of tropical Pacific atmospheric circulation due to anthropogenic forcing. *Nature* **441**: 73–76.

Vikebø, F., Sundby, S., Ådlandsvik, B., and Otterå, O. (2007). Impacts of a reduced thermohaline circulation on transport and growth of larvae and pelagic juveniles of Arcto-Norwegian cod (*Gadus morhua*). *Fish. Oceanogr.* **16**: 216–228.

Visser, M. E. and Both, C. (2005). Shifts in phenology due to global climate change: the need for a yardstick. *Proc. R. Soc. B* **272**: 2561–2569.

Walther, G.-R., Post, E. Convey, P. *et al.* (2002). Ecological responses to recent climate change. *Nature* **416**: 389–395.

Werner, F. E., MacKenzie, B. R., Perry, R. I., *et al.* (2001). Larval trophodynamics, turbulence, and drift on Georges Bank: a sensitivity analysis of cod and haddock. *Sci. Mar.* **65** (Suppl. 1): 99–115.

Werner, F. E., Ito, S., Megrey, B. A., and Kishi, M. J. (2007). Synthesis and future directions of marine ecosystem models. *Ecol. Model.* **202** (1–2): 211–223.

Woods, J. D. (2005). The Lagrangian Ensemble metamodel for simulating plankton ecosystems. *Progr. Oceanogr.* **67**: 84–159.

Wooster, W. S. and Bailey, K. M. (1989). Recruitment of marine fishes revisited. In *Effects of Ocean Variability on Recruitment and an Evaluation of Parameters Used in Stock Assessment Models*, Beamish, R. J. and McFarlane, G. A., eds. *Can. Tech. Rep. Fish. Aquat. Sci.* **108**: 153–159.

Yamazaki, H, Mackas, D., and Denman, K. (2002). Coupling small scale physical processes to biology: towards a Lagrangian approach, In *The Sea: Biological–Physical Interactions in the Ocean*, Vol. 12, Robinson, A. R., McCarthy, J. J. and Rothschild, B. J., eds. New York: John Wiley and Sons, pp. 51–112.

Yasuda, I., Sugisaki, H. Watanabe, Y., *et al.* (1999). Interdecadal variations in Japanese sardine and ocean/climate. *Fish. Oceanogr.* **8**: 18–24.

15 Synthesis and perspective

David M. Checkley, Jr., Andrew Bakun, Manuel Barange, Leonardo R. Castro, Pierre Fréon, Renato Guevara-Carrasco, Samuel F. Herrick, Jr., Alec D. MacCall, Rosemary Ommer, Yoshioki Oozeki, Claude Roy, Lynne Shannon, and Carl D. van der Lingen

CONTENTS

Summary

The Small Pelagic Fish and Climate Change (SPACC) program was created to facilitate research on the dynamics of populations of small pelagic fish, including anchovy and sardine. These populations exhibit large variations in size, extent, and production on the scale of decades. At times, anchovy and sardine alternate in abundance. Collectively, small pelagic fish often occupy a central role in the food web they occur in, often described as a wasp-waist ecosystem. Humans are an integral part of those ecosystems. Variability of populations of small pelagic fish is believed to be due primarily to variations in climate and fishing, but the mechanisms of these relations remain unknown in most cases. It is also uncertain whether these ecosystems alternate between states, e.g. regimes, and whether inherent variability may limit our ability to predict their future states. The fisheries for populations of small pelagic fish are increasingly global in nature. While the global catch of small pelagic fish constitutes approximately one-quarter of the world fish catch and has been relatively constant during the past several decades, the catch of individual taxa and stocks varies much more. The management of these fisheries will be challenged by increasing demand for human consumption and mariculture in light of their finite and variable production, importance within the ecosystem, and unprecedented climate change, and will depend on both science and governance. We recommend continued, global research on climate change effects on small pelagic fish, and its periodic assessment for use by decision makers.

Introduction

This chapter is a partial synthesis of ideas in prior chapters. A general objective of the SPACC program has been to describe and understand the dynamics of populations of small pelagic fish (SPF) in the context of climate variability and change. The ultimate objective has been to contribute to the process of improving the global management of SPF populations, including research into areas identified as gaps in current knowledge and understanding of small pelagic fish dynamics.

Scientific uncertainty and inadequate, or inappropriate, governance, in addition to natural variability of fish stocks, have hampered successful management of SPF stocks, as manifest by collapses of the California sardine in the 1950s, the Peruvian anchoveta and Namibian sardine in the 1970s, and the Japanese sardine in 1990s (this volume, Chapters 3 and 9). Scientific uncertainty includes uncertainty inherent in an ecosystem and our imprecise knowledge and understanding of that system and how it functions, as well as uncertainty of estimating stock levels. Uncertainty also arises because ecosystems are complex and dynamic, and subject to long-term change as well as chaotic and chance events. Limits exist in our ability to predict the future state of SPF populations and ecosystems due, for example, to inadequate knowledge and the inherent variability of future climate states and fish population responses to these, including their ability to adapt to a changing environment by developing new strategies. What are our limits? To what extent can we achieve a mechanistic understanding that allows prediction, with uncertainties and perhaps including a range of scenarios, as opposed to relying on the statistical properties of the past behavior of a population or ecosystem?

While there is consensus among practitioners on many issues, it remains lacking on others. This does not necessarily indicate conflicting views but, rather, that process variability leads to different interpretations in certain situations. Thus, we are still unable to present a conceptual, let alone analytical or quantitative, model of the dynamics of populations of SPF capable of supporting their sustainable management. However, we do have substantial converging views, e.g. that common features are shared by the same taxon, such as anchovy or sardine, in different regions, that these taxa differ significantly from one another across regions, and that important forces on SPF include climate change and fishing. SPACC has used the comparative approach to draw broad lessons and acknowledges the need for, and value of, interdisciplinary work, given the nature of the systems it studies – e.g. climate and ocean physics, chemistry, biology, mathematical and bioeconomic modeling, and social sciences. Temporal and spatial scales, and their interactions, must be considered.

Below, we comment on our scientific understanding of the dynamics of SPF, including their populations and ecosystems, and fisheries, management, and social sciences, including economics. We discuss gaps in our current knowledge and understanding of SPF dynamics and requirements for the future. Finally, we recommend ways to achieve these requirements.

Science

Populations of small, pelagic fish, and their fisheries, fluctuate greatly on time scales of decades. Why does this occur and with what effects?

The combined results of analysis of the paleontological records, formed before intensive fishing, and more recent records, collected during fishing, indicate that populations of small pelagic fish have always fluctuated (this volume, Chapter 4). While year-to-year management does not require full understanding of the cause of these fluctuations, long-term management and investment planning would certainly benefit from a model capable of predicting the long-term dynamics of these populations with and without fishing, or as a combined outcome of the interacting drivers of fishing and the environment. Given that SPF are frequently the dominant forage fish in many, if not all, of the systems in which they occur (Cury *et al.*, 2000), it is plausible that large variations in non-fishing mortality (e.g. the coefficient of natural mortality, M, units y^{-1}) are due to natural top predators or diseases. However, it often seems more likely that the cause of natural fluctuations in population size is from below, either directly by physical forcing (e.g. temperature, Takasuka *et al.*, 2007), indirectly through the food web (e.g. van der Lingen *et al.*, 2006; Rykaczewski and Checkley, 2008), or through a combination of such processes, e.g. "Bakun's triad" (Bakun, 1998). Our inability to predict the dynamics of SPF using traditional, density dependent models is consistent with the importance of extrinsic forcing of their dynamics (Mullon *et al.*, 2009). An alternative to bottom-up and top-down forcing is the interaction between populations of SPF, e.g. competition for food or predation

on early stages by one another (e.g. this volume, Chapter 7). These forces are not mutually exclusive and likely act in concert.

An associated question is whether such forcing and population responses vary continuously or non-linearly between alternative states, or regimes, and whether such changes are reversible? Often time series of population abundance or catch exhibit a hysteresis. Thus, historical populations with fishing often amass biomass gradually until achieving maximal abundance, then decline precipitously (Mullon *et al.*, 2005). Whether this reflects intrinsic population or ecosystem properties or a response to fishing is unknown, due in part to low temporal resolution and imprecision of most paleontological records. A case is the Peruvian anchoveta, *Engraulis ringens* (R. Guevara-Carrasco, unpublished data). The highest recruitment of this species observed in the last two decades was during normal or mildly warm years rather than in cold years. In recent years, anchoveta larvae have been observed in more saline and warmer, subtropical waters. This is contrary to historic records, and consistent with the unexpected appearance of recruits offshore in some years. Thus, the Peruvian system does not seem to have returned to the same state that existed prior to the collapse of the anchoveta fishery in the early 1970s.

Populations of anchovy and sardine (*Sardinops* spp.) are sometimes viewed as alternating in size and/or "replacing" one another (Kawasaki, 1983). Alternation implies direct interaction (e.g. competition or predation), different responses to the same stimulus or co-occurring stimuli, and/or separate responses to simultaneous changes of their respective habitats. Thus, larvae of anchovy and sardine off Japan differ in their temperature for optimal growth (Takasuka *et al.*, 2007). Within the Pacific basin, SPACC systems exhibited these alternations in the last century. However, studies of the paleontological and historical records do not support a strict alternation of populations (this volume, Chapter 4; Fréon *et al.*, 2003). The explanation for the discrepancy between paleontological and recent records might be due to the different lengths of these two types of time series (1000s of years vs. ~ 100 years) and/or the effect of fishing, which has been shown to affect the response of fished populations to climate variability (e.g. Ottersen *et al.*, 2004, 2006; Planque *et al.*, in press). Strong fishing pressure may affect the alternation by increasing the abruptness of population decreases and thus enlarge the duration of the period when one taxon conspicuously dominates the system. Without fishing, this period may be shorter and thus not be noticed in the paleontological record because of its limited resolution (this volume, Chapter 4). Moreover, most SPACC regions have one dominant taxon whose peak abundance, or catch, exceeds that of other species. Anchovy and sardine are similar within their respective taxa but are different from one another (e.g. for the Benguela Current, this volume, Chapter 7). Thus, as the environment varies, it is plausible that sardine and anchovy respond independently and differently, i.e. they "march to the beat of different drummers." This is consistent with their differing properties of population production (Jacobson *et al.*, 2001), with sardine varying more gradually and over a larger time scale than anchovy, and with specific space occupation strategies (Barange *et al.*, 2005). Such major differences at this taxonomic level are also consistent with the shallow genetic histories of these species (Grant and Bowen, 1998; Lecomte *et al.*, 2004). That is, genetic lineages of anchovy and sardine are long (10^7 y) and distinct, while the populations of each have been established only relatively recently (10^4 y) and in similar habitats for each taxon, with sardine in more oceanic regions and anchovy in more coastal, productive regions (this volume, Chapter 3); these patterns vary with population size and exceptions exist (e.g. Barange *et al.*, 1999). The time series of abundance, both paleontological (several thousand years) and historical (to hundreds of years), are consistent with a conceptual model in which populations fluctuate due to extrinsic (climate, fishing, prey, and predation) rather than intrinsic (population size) factors. This, in turn, is consistent with the observed variation of sardine and anchovy being in or out of phase and with similar wavelengths of the "pseudocycles" of the two groups of species. It also allows for apparent global synchrony within or between anchovy and sardine stocks from different oceans, which does not necessarily reflect a global response to a common forcing (Fréon *et al.*, 2003).

Questions exist in regard to climate and its effects on small pelagic fish that limit our ability to predict their dynamics. What is the cause of the physical variability that appears so important to SPF? While ENSO and modes of multidecadal variation have been described, climate forecasts over decades and regions lack necessary accuracy to be useful for decisions. What is the relationship between physical forcing by the climate and the response of SPF? Alheit *et al.* (this volume, Chapter 5) posit that a sequence of physical changes may occur, followed by a relatively rapid response of biological components of the ecosystem. Jacobson *et al.* (2001) show that the specific (per capita) growth rate of SPF populations responds rapidly to change of the environment and precedes a numerical response of the population. Thus, population size lags physical forcing. The differing behaviors of time series of physical (linear or stochastic) and biological (non-linear) variables relevant to climate and fisheries off California indicate that there may be inherent limitations to our ability to predict fish dynamics (Hseih *et al.*, 2005). Once again, we need a mechanistic understanding of how the interdecadal changes observed in physical variables are related to biology to be able to interpret, let alone predict, the dynamics of SPF in a changing environment. Progress

towards a mechanistic understanding has been made in some systems, e.g. the Benguela (Roy *et al.*, 2007; this volume, Chapter 7), California (Rykaczewski and Checkley, 2008; this volume, Chapter 12), Humboldt (this volume, Chapter 5), and Kuroshio (Takasuka *et al.*, 2007) Currents. Some attempts at prediction were made by Fréon *et al.* (this volume, Chapter 14), based on present knowledge, but they remain conjectures.

Wasp-waist (Rice, 1995; Bakun, 1996) is often used to describe the ecosystems with SPACC populations. It connotes a large population or assemblage of SPF, usually dominated at any one time by a single species, which interacts strongly with its prey and predators. Wasp-waist implies that variation in the abundance of SPF is likely to have significant effects on the ecosystem at several levels. This, however, does not preclude bottom-up effects on SPF. Thus, in the Peruvian upwelling ecosystem, parallel changes observed in phytoplankton, zooplankton, and anchoveta abundances are consistent with bottom-up effects, though top-down effects cannot be ruled out (this volume, Chapter 5; Chavez *et al.*, 2008). It may be important to distinguish between individual populations of SPF and the sum of all populations, either regionally or globally. The global catch of all SPF was less variable than that of its component taxa; for 1950–2004, the coefficient of variation was 0.36 for global SPF landings, 0.65 for global anchovy landings, and 0.77 for combined, global sardine, sardinella, and pilchard landings (this volume, Chapter 10). This, in turn, is consistent with the hypothesis that the composition of an ecosystem may change in a relatively subtle manner, e.g. between states dominated by anchovy or sardine, but with little change in system structure and hence the overall flow of matter and energy (Cury and Shannon, 2004; Jarre *et al.*, 2006). This also implies a constant carrying capacity of the ecosystem which, conversely, is believed in some cases to vary over time (e.g. Jacobson *et al.*, 2005). Alternatively, the ecosystem may change in a dramatic fashion, e.g. between states when SPF are dominant in an ecosystem and when they are almost entirely absent, as is currently the situation in the northern Benguela (see this volume, Chapter 8), with drastic changes in system structure and function (van der Lingen *et al.*, 2006b). In the Peruvian upwelling ecosystem, the warm period of mid 1970s to mid 1980s, when sardine was abundant, had similar or higher total biomass of pelagic (sardine, jack mackerel, and mackerel) and demersal (hake and other demersal) taxa to when anchoveta was abundant; the ecosystem may have fundamentally changed (R. Guevera-Carrasco, unpublished data). Another factor to consider in this context is the potential impact of interactions and energy flows between SPF and mesopelagic fish. In the Kuroshio region, myctophids may be a dominant component of the ecosystem in terms of biomass

(Gjøsæter and Kawaguchi, 1980); this is now being investigated using acoustic surveys. The effects of removal by fishing, as opposed to the natural variation of SPF population size, on the ecosystem remain unknown. Shannon *et al.* (this volume, Chapter 8) show that fishing may alter the relative production of pelagic and benthic components of an ecosystem, and that fishing and environmental changes act synergistically to produce the ecosystem dynamics we observe. A certainty is that humans must now be considered an important part of the ecosystem and that over-exploitation decreases resilience of systems to climate change (for review see this volume, Chapter 14). Further, the effects of humans on the ecosystem require consideration of issues such as the non-consumptive value of SPF, e.g. as forage for fish, marine mammals, and birds. SPF play a pivotal and crucial role in the ecosystems in which they are found.

Fisheries

Most SPF stocks are commercially exploited, and many are assessed and managed on the basis of scientific understanding (with some glaring exceptions; this volume, Chapter 9). How can we distinguish between the effects of climate change and fishing on those stocks and their ecosystems, and of what value is this to management?

Fisheries on SPF are local, but the industry is increasingly global. From the physical forcing, with hypothesized teleconnections, to economics, including global trade, a fishery in one region is affected by multiple activities worldwide. This is perhaps most evident in the increasing demand for anchovy and sardine products for mariculture. The early, and large, fisheries for small pelagics were almost entirely for human consumption, especially canning, or agrifeed, e.g. fish meal and oil for poultry and swine. The use of SPF to feed fish, poultry, and swine has grown relative to its use for human consumption and, in particular, an increasing fraction of the world catch of SPF is used for mariculture. It is predicted that, by 2010, 50% of the fish meal and 80% of the fish oil will be used in global mariculture (Delgado *et al.*, 2003). Both are global commodities, the average fish meal ton traveling over 10 000 miles between producer and consumer, and increasingly in demand by mariculture in developing regions such as China (Delgado *et al.*, 2003). Much of the sardine harvested off California is frozen in blocks and shipped to Australia to feed bluefin tuna captured from the wild, "grown out" in pens, harvested, and exported to global markets. In addition, fish meal and oil can be partially substituted for the animal food markets by vegetable proteins (e.g. soybean, corn). The balance between these products is currently under threat by the increased production of biofuels; these displace agrifeed vegetable crops which, in turn, affects the demand for and price of SPF. As the world's population grows, these demands, as well as the direct use by humans, will only increase. SPF stocks will thus be managed under increasing global demand and a limited, variable supply.

This difficult management situation is likely to be aggravated by the issue of bycatches and discards, which are already frequent in SPF due to the implementation of quotas by species and size limitations. SPF tend to shoal by size but different species are frequently found within the same school or in neighboring schools and, as a result, are caught together. Mixed fish schools are more frequent in situations of low abundance of at least one of the two species, for reasons that are still debated (Bakun and Cury, 1999; Fréon and Dagorn, 2000; this volume, Chapter 13). Under a scenario of a decrease in production due to the interaction of heavier exploitation and increased variability in abundance, due to climate change, bycatches and discards of SPF are likely to increase, promoting a negative feedback on the abundance of these species.

Another issue is the effect of climate change on the catchability of SPF. Anchovy off Peru and sardine off California are confined to nearshore waters during El Niño, increasing their availability to fisheries (this volume, Chapter 3, and the entire book). Fishing gears directed to massive capture of SPF take advantage of the highly gregarious behavior of these species, which are found in dense schools, themselves regrouped in clusters. Furthermore, the purse seine (the fishing gear by far the most used to catch SPF) only captures schools located sufficiently close to the surface. Therefore, SPF catchability largely depends on the level of aggregation and distance of fish from the surface. How climate change will affect catchability is difficult to predict, but it is likely that in areas where the depth of the thermocline and/or the oxycline will decrease (increase), fish schools will be distributed in a thinner (wider) upper layer and therefore more (less) vulnerable to purse seiners. Similarly, lower (higher) plankton concentration is expected to decrease (increase) school size.

Ecosystem-based management requires an understanding of the role of SPF in their respective ecosystems. To what degree are fluctuations in abundance natural, and thus presumed to be unavoidable, and to what degree are these due to human activity, including fishing and anthropogenic climate change? The paleontological record shows SPF stocks varied independent of fishing. The hysteresis in stock fluctuations may be due to fishing and, if so, perhaps minimized by decreasing fishing mortality (F, y^{-1}) when a decline is indicated. What are the trade-offs between harvesting small pelagic fish for immediate economic gain and leaving them in the water as forage for higher trophic levels, including predatory fish, squid, marine mammals, and seabirds? The biological, economic, and social consequences of these choices are complex, yet necessary to understand for wise, long-term management. If populations and ecosystems change among states that persist over time (~ regimes), are these changes reversible? For example, whether forced by physics, fishing, or both, is the shift of the Northern Benguela ecosystem from one dominated by small pelagics to one that appears to be dominated by gobies and jellyfish likely to persist, or will anchovy and sardine return to dominate (Boyer and Hampton, 2001)? Does fishing of a stock of SPF alter the resilience of the ecosystem of which it is a part? Decline in biodiversity appears to decrease resilience (Chapin *et al.*, 2000; Planque *et al.*, in press). Does exploitation of a stock have a similar effect? For example, the variation of SPF abundance is believed to have led to decreased abundance of their seabird predators off Peru (Jahncke *et al.*, 2004) and South Africa (Crawford *et al.*, 2006), yet it is unclear if declines in SPF abundance from fishing alter the resilience of the respective ecosystems to climate change and further fishing.

The time scales of change of the interacting elements of small pelagic fish and their fisheries merit consideration. The time scales of investment and capitalization (e.g. in vessels and processing plants) and fish stock fluctuations interact to exacerbate the effects of unfavorable environmental conditions and represent a threat for SPF stocks. This threat is enhanced when overcapacity is high, as in Peru, with the largest monospecific fishery in the world and where overcapacity surpasses 300% (Fréon *et al.*, 2008). On a longer time scale, fishing may alter the genetics of the fished stocks and thus their population dynamics and perhaps resilience to environmental change. Recent studies indicate that the life history characteristics of exploited populations of cod, herring, salmon, plaice, and other taxa have changed over time due to fishing (Jørgensen *et al.*, 2007, and references therein). The "shallow life histories" of SPF stocks worldwide indicate that these stocks may be particularly susceptible to forces such as fishing and climate change (Grant and Bowen, 1998; Lecomte *et al.*, 2004). In fact, small pelagic fish are the subject of the SPACC program due to this susceptibility (e.g. Box 5.1 in this volume, Chapter 5). Anchovy lifespan being shorter than sardine, the former species may be more prone to genetic adaptation in response to climate change and harvest pressure.

If one assumes that SPF stock size varies with the environment, sustainability of the catch of that stock may not be an appropriate goal. Rather, fisheries management must adapt to such variability, e.g. by allowing increased fishing pressure for a particular stock under an improving environment and *vice versa*. The challenge is to know, in real time, when the environment is improving or deteriorating. Fréon *et al.* (2005) proposed a two-level (short- and long-term) management strategy to cope with interannual and interdecadal variations in abundance of pelagic species. This would entail adjusting the quota at the interannual scale,

and adjusting the nominal effort (number of boats) at the interdecadal scale. Other practitioners instead favor adjustments in harvest rates that reflect ecosystem productivity (see this volume, Chapter 9).

Two examples of management illustrate the combined importance of science and governance. The fishery for the Pacific sardine off California may be an example of science and governance resulting in successful management. Longterm variation of sardine productivity is incorporated in the harvest decision rule, allowing for a greater fraction of the assessed stock to be harvested in favorable than unfavorable ocean conditions (Hill *et al.*, 2006). The Magnuson Stevens Act (http://www.nmfs.noaa.gov/msa2007/index.html) created a governance structure that incorporates science in US fishery management. Conversely, the Bay of Biscay anchovy may have suffered from both scientific uncertainty and a lack of adequate governance, the consequence being the recent closure of the fishery. This situation is complicated, however, by the northward shift in distribution of anchovy into the North and Baltic Seas in recent decades, this being detrimental to anchovy in the Bay of Biscay, where the catches are regulated, but favorable to anchovy in the North and Baltic Seas, where they are not regulated (Beare *et al.*, 2004). Because governance decisions involve economic and social considerations in addition to scientific evidence, such considerations should be incorporated into management models of these fisheries. This reinforces our belief that humans are part of the ecosystem and that decisions affecting the ecosystem, and/or its stocks, must consider humans, as acknowledged in the current shift towards an ecosystem approach to fisheries management (cf. this volume, Chapter 11).

Future

We know that the Earth's climate is changing due to human activity and that the human population is growing (IPCC, 2007). How will these two processes affect stocks of SPF in the future? Although SPACC focuses on climate change effects, it is impossible to consider these in isolation from the increasing demand for SPF.

Throughout this book, the effects of climate variability on SPF have been discussed (e.g. see boxes in all chapters). Most of the issues concerning the effects of past climate change on SPF are relevant to our consideration of future climate change. However, future change due to human activity will be unprecedented, at least with regard to the time scales of our present knowledge (e.g. decades to millennia). Thus, the past behavior of SPACC populations may not represent their future behavior, due to changes in both the physical forcing and biological response. Global warming, due to enhanced atmospheric CO_2, will vary regionally and affect hydrology; ocean temperature, stratification, and currents; winds and the magnitude and frequency of events, such as cyclones and perhaps ENSO; and affect the phenology, or timing, of biological events, with potential effects on ecological interactions (e.g. match–mismatch). In addition, approximately half of the CO_2 introduced into the atmosphere will be sequestered in the ocean, altering its chemistry in ways that are only now being imagined and elucidated, and with equally unknown effects on its biota and ecology. Longitudinal shifts in species, e.g. the ongoing poleward shift in distributions of anchovy and sardine into the North Sea (Beare *et al.*, 2004), may affect management as well as have ecological implications. Recent progress in ocean observation techniques and high-resolution models of ocean dynamics (e.g. Guo *et al.*, 2003; Miyazawa *et al.*, 2004; this volume, Chapters 6 and 14) will allow us to test hypothetical mechanisms with the data for SPF and their physical and biological environment. A summary of expected climate change impacts on fish populations is provided in Barange and Perry (in press).

As the human population increases in size and expectations (e.g. quantity and quality of life), the demands on SPF, and other aspects of their ecosystems, will only grow. As discussed above, mariculture will require increasing amounts of SPF for feed, competing with the demand of SPF for human consumption and non-consumptive use (e.g. forage for species at higher trophic levels). The global dimension of these demands indicates the complexity of SPF dynamics and the potential for effects on such issues as food security (cf. Lobell *et al.*, 2008).

Ultimately, scientists must inform decision makers and, ideally, governance will use science to wisely manage resources. Because the future will include unprecedented climate change and demands, it is necessary, now more than ever, to achieve a mechanistic understanding of the dynamics of SPF.

Recommendations

SPACC has benefited from the global comparison of populations of and regions with small pelagic fish. Although each stock and region is unique, common properties exist. Among these are the recognition of the influence of a varying climate on populations, of consistency within and differences between anchovy and sardine, of the pivotal role these stocks occupy in their wasp-waist ecosystems, of the global nature of interactions involving SPF, and of the certainty of future change in climate and demand for SPF. While past studies have focused primarily on the highly productive upwelling regions, future studies will benefit from extending the synthetic approach to multiple systems at a global scale, including, for example, western boundary currents. We feel that these stocks and systems merit continued examination to achieve the best understanding with which to inform decision makers.

To facilitate future research on climate change and small pelagic fish, we recommend an international program like SPACC. This should use the comparative approach and involve scientists from a broad spectrum of disciplines, including climate, fisheries, oceanography and the social sciences. The research focus should include the ecosystem.

International assessments of the state of science in particular areas are effective means by which to inform decision makers. Examples include the Intergovernmental Panel on Climate Change (http://www.ipcc.ch/) and the International Assessment of Agricultural Science and Technology for Development (http://www.agassessment.org/). An assessment is a periodic, critical evaluation of the status of information on a subject for use by decision makers. It is achieved by amassing, evaluating, and synthesizing all the relevant, peer-reviewed literature to arrive at, in an open and transparent manner, a consensus statement relevant to policy. Risk and uncertainty are addressed. While we recognize the need of managers of fisheries for rapid and timely scientific advice, e.g. population assessments, economic evaluations, and near-term forecasts, we also recognize the need for a broad assessment of the science of SPF and climate. Hence, we recommend the periodic, international assessment of climate effects on small pelagic fish.

REFERENCES

Bakun A. (1996). *Patterns in the Ocean: Ocean Processes and Marine Population Dynamics*. San Diego, CA: University of California Sea Grant, in cooperation with Centro de Investigaciones Biológicas de Noroeste, La Paz, Baja California Sur, Mexico.

Bakun, A. (1998). Ocean triads and radical interdecadal stock variability: bane and boon for fisheries management. In *Reinventing Fisheries Management*, Pitcher, T., Hart, P. J. B., and Pauly, D., eds. London: Chapman and Hall, pp. 331–358.

Bakun, A. and Cury, P. (1999). The "school trap": a mechanism promoting large-amplitude out-of-phase population oscillations of small pelagic fish species. *Ecol. Lett.* **2**: 349–351.

Barange, M. and Perry, R. I. (in press). Physical and ecological impacts of climate change relevant to marine and inland capture fisheries and aquaculture. *FAO Fisheries Technical Paper*.

Barange, M., Hampton, I., and Roel, B. A. (1999). Trends in the abundance and distribution of anchovy and sardine on the South African continental shelf in the 1990s, deduced from acoustic surveys. *S. Afr. J. Mar. Sci.* **21**: 367–391.

Barange, M., Coetzee, J., and Twatwa, N. (2005). Strategies of space occupation by anchovy and sardine in the southern Benguela: role of stock size and intra-species competition. *ICES J Mar Sci.* **62**: 645–654.

Beare, D., Burns, F., Jones, E. *et al.* (2004). An increase in the abundance of anchovies and sardines in the north-western North Sea since 1995. *Global Change Biol.* **10**: 1209–1213.

Boyer, D.C. and Hampton, I. (2001). An overview of the living marine resources of Namibia. *S. Afr. J. Mar. Sci.* **23**: 5–35.

Chapin, F.S., Chapin, F.S., Zavaleta, E.S., *et al.* (2000). Consequences of changing biodiversity. *Nature* **405**: 234–242.

Chavez, F.P., Jr., Bertrand, A., Guevera-Carrasco, R., *et al.* (2008). The northern Humboldt Current System: Brief history, present status and a view towards the future. *Prog. Oceanogr.* **79**: 95–105.

Crawford, R.J.M., Dundee, B.L., Dyer, B.M., *et al.* (2006). Trends in numbers of Cape gannets (*Morus capensis*), 1956/1957–2005/2006, with a consideration of the influence of food and other factors. *ICES J. Mar. Sci.* **64**: 169–177.

Cury, P., Bakun, A., Crawford, R.J.M. *et al.* (2000). Small pelagics in upwelling systems: patterns of interaction and structural changes in "wasp-waist" ecosystems. *ICES J. Mar. Sci.* **57**: 603–618.

Cury, P.M. and Shannon, L.J. (2004). Regime shifts in upwelling ecosystems: observed changes and possible mechanisms in the northern and southern Benguela. *Progr. Oceanogr.* **60**: 223–243.

Delgado, C.L., Wada, N., Rosegrant, M.W., *et al.* (2003). *Fish to 2020: Supply and Demand in Changing Global Markets*. Washington, D.C: International Food Policy Research Institute, and Penang, Malaysia: Worldfish Center.

Fréon, P. and Dagorn, L. (2000). Review of fish associative behaviour: toward a generalisation of the meeting point hypothesis. *Rev. Fish Biol. Fisheries* **10**: 183–207.

Fréon, P., Mullon, C., and Voisin, B. (2003). Investigating remote synchronous patterns in fisheries. *Fish. Oceanogr.* **12**: 443–457.

Fréon, P., Cury, P., Shannon, L., and Roy, C. (2005). Sustainable exploitation of small pelagic fish stocks challenged by environmental and ecosystem changes: a review. *Bull. Mar. Sci.* **76**: 385–462.

Fréon, P., Bouchon, M., Mullon, C., *et al.* (2008). Interdecadal variability of anchovy abundance and overcapacity of commercial fleets in Peru. *Progr. Oceanogr.* **79**: 401–412.

Gjøsæter, J. and Kawaguchi, K. (1980). A review of the world resources of mesopelagic fish. *FAO Fisheries Tech. Paper* **193**, 151 pp.

Grant, W.S. and Bowen, B.W. (1998). Shallow population histories in deep evolutionary lineages of marine fishes: Insights from sardines and anchovies and lessons for conservation. *J. Hered.* **89**: 415–426.

Guo, X., Hukuda, H., Miyazawa, Y., and Yamagata, T. (2003). A triply nested ocean model for simulating the Kuroshio – roles of horizontal resolution on JEBAR. *J. Phys. Oceanogr.* **33**: 146–169.

Hill, K.T., Lo, N.C.H., Macewicz, B.J., and Felix-Uraga, R. (2006). Assessment of the Pacific sardine (*Sardinops sagax caerulea*) population for U.S. management in 2006. *NOAA Tech. Mem.*, NOAA-TM-NMFS-SWFSC-386, 85 pp.

Hsieh, C.H., Glaser, S.M., Lucas, A.J., and Sugihara, G. (2005). Distinguishing random environmental fluctuations from ecological catastrophes for the North Pacific Ocean. *Nature* **435**: 336–340.

Hutchings, J.A. (2005). Life history consequences of overexploitation to population recovery in Northwest Atlantic cod (*Gadus morhua*). *Can. J. Fish. Aquat. Sci.* **62**: 824–832.

IPCC (2007). *Fourth Assessment Report, Climate Change 2007*, http://www.mnp.nl/ipcc/pages_media/AR4-chapters. html, 4 volumes.

Jacobson, L.D., De Oliveira, J.A.A., Barange, M. *et al.* (2001). Surplus production, variability, and climate change in the great sardine and anchovy fisheries. *Can. J. Fish. Aquat. Sci.* **58**: 1891–1903.

Jacobson, L.D., Bograd, S.J., Parrish, R.H. *et al.* (2005). An ecosystem-based hypothesis for climatic effects on surplus production in California sardine (*Sardinops sagax*) and environmentally dependent surplus production models. *Can. J. Fish. Aquat. Sci.* **62**: 1782–1796.

Jahncke, J., Checkley, D.M., Jr., and Hunt, G.L. (2004). Trends in carbon flux to seabirds in the Peruvian upwelling system: effects of wind and fisheries on population regulation. *Fish. Oceanogr.* **13**: 208–223.

Jarre, A., Moloney, C.L., Shannon, L.J. *et al.* (2006). Developing a basis for detecting and predicting long-term ecosystem changes. In Shannon, V., Hempel, G., Malanotte-Rizzoli, P., *et al.*, eds., *Benguela: Predicting a Large Marine Ecosystem*, Elsevier Amsterdam: Elsevier, Large Marine Ecosystems Series **14**: 239–272.

Jørgensen C., Enberg, K., Dunlop, E.S. *et al.* (2007). Managing evolving fish stocks. *Science* **318**: 1247–1248.

Kawasaki, T. (1983). Why do some pelagic fishes have wide fluctuations in their numbers? Biological basis of fluctuation from the viewpoint of evolutionary ecology. In *Proceedings of the Expert Consultation to Examine Changes in Abundance and Species Composition of Neritic Fish Resources*, Sharp, G.D. and Csirke, J., eds. *FAO Fish. Rep.* **291**: 1065–1080.

Lecomte, F., Grant, W.S., Dodson, J.J., *et al.* (2004). Living with uncertainty: genetic imprints of climate shifts in East Pacific anchovy (*Engraulis mordax*) and sardine (*Sardinops sagax*). *Mol. Ecol.* **13**: 2169–2182.

Lobell, D.B., Burke, M.B., Tebaldi, C. *et al.* (2008). Prioritizing climate change adaptation needs for food security in 2030. *Science* **319**: 607–610.

Miyazawa, Y., Guo, X., and Yamagata. T. (2004). Roles of mesoscale eddies in the Kuroshio paths. *J. Phys. Oceanogr.* **34**: 2203–2222.

Mullon, C., Fréon P., and Cury, P. (2005). Dynamics of collapse in world fisheries. *Fish Fisher.* **6**: 111–120.

Mullon, C., Fréon, P., Cury, P., *et al.* (2009). A minimal model of the variability of marine ecosystems. *Fish Fisher.* **9**: 1–17.

Ottersen, G., Stenseth, N.C., and Hurrell, J.W. (2004). Climatic fluctuations and marine systems: a general introduction to the ecological effects. In *Marine Ecosystems and Climate*

Variation, Stenseth, N. C., Ottersen, G., Hurrell, J. W., and Belgrano, A., eds. Oxford, UK: Oxford University Press, pp. 3–14.

Ottersen, G., Hjermann, D., and Stenseth, N. C. (2006). Changes in spawning stock structure strengthen the link between climate and recruitment in a heavily fished cod (*Gadus morhua*) stock. *Fish. Oceanogr.* **15** (3): 230–243.

Planque, B., Fromentin, J.-M., Cury, P., *et al.* (in press). How does fishing alter marine populations and ecosystems sensitivity to climate? *J. Mar. Syst.*

Rice, J. (1995). Food web theory, marine food webs, and what climate change may do to northern fish populations. In *Climate Change and Northern Fish Populations*, Beamish, R. J., ed. *Can. Spec. Pub. Fish. Aquat. Sci.* **121**: 561–568.

Roy, C., van der Lingen, C. D., Coetzee, J. C., and Lutjeharms, J. R. E. (2007). Abrupt environmental shift links with changes in the distribution of Cape anchovy (*Engraulis encrasicolus*) spawners in the southern Benguela. *Afr. J. Mar. Sci.* **29**: 309–319.

Rykaczewski, R. R. and Checkley, D. M., Jr. (2008). Influence of ocean winds on the pelagic ecosystem in upwelling regions. *Proc. Nat. Acad. Sci.* USA **105**: 1965–1970.

Takasuka, A., Oozeki, Y., and Aoki, I. (2007). Optimal growth temperature hypothesis: why do anchovy flourish and sardine collapse or vice versa under the same ocean regime? *Can. J. Fish. Aquat. Sci.* **64**: 768–776.

van der Lingen, C. D., Hutchings, L., and Field, J. G. (2006a). Comparative trophodynamics of anchovy *Engraulis encrasicolus* and sardine *Sardinops sagax* in the southern Benguela: Are species alternations between small pelagic fish trophodynamically mediated? *Afr. J. Mar. Sci.* **28** (3/4): 465–477.

van der Lingen, C. D., Shannon, L. J., Cury, P., *et al.* (2006b). Chapter 8. Resource and ecosystem variability, including regime shifts, in the Benguela Current system. In *Benguela: Predicting a Large Marine Ecosystem*, Shannon, L. V., Hempel, G., Malanotte-Rizzoli, P., *et al.* eds. USA, Elsevier, Large Marine Ecosystems Series **14**, pp. 147–184.

Index